SOFT MULTIHADRON DYNAMICS

SOFT MULTIHADRON DYNAMICS

W. Kittel
Radboud University Nijmegen, The Netherlands

E. A. De Wolf
University of Antwerp, Belgium

NEW JERSEY • LONDON • SINGAPORE • BEIJING • SHANGHAI • HONG KONG • TAIPEI • CHENNAI

Published by

World Scientific Publishing Co. Pte. Ltd.
5 Toh Tuck Link, Singapore 596224
USA office: 27 Warren Street, Suite 401-402, Hackensack, NJ 07601
UK office: 57 Shelton Street, Covent Garden, London WC2H 9HE

British Library Cataloguing-in-Publication Data
A catalogue record for this book is available from the British Library.

SOFT MULTIHADRON DYNAMICS

ISBN 981-256-295-8

Printed in Singapore by World Scientific Printers (S) Pte Ltd

Preface

If Maurice Jacob states "La physique est une science expérimentale" [1], this stands as a fact and does not need any proof. If, however, T.D. Lee says [2] "We all know that the basis of physics is experiment", this *does* need a proof. It not only states the fact, but also claims that the fact is known to all of us. This is much more and not always obvious. The proof refers to old Archimed shown in [2] on a sixteenth-century wood engraving (Bibliothèque Nationale, Paris) as sitting in a barrel of water and being described as "erster erfinder scharpffsinniger vergleichung .....". Who can afford not to know about the bath of Archimed, so indeed we all know!

Lee then continues by postulating two laws of *physicists*,

T.D. Lee's first law of physicists:

"Without experimentalists, theorists tend to drift."

T.D. Lee's second law of physicists:

"Without theorists, experimentalists tend to falter".

As an example for the second law, Lee quotes the time development of the famous Michel parameter ρ. This parameter describes the energy distribution for the electron produced in μ decay

$$\mu \rightarrow \mathrm{e} + \nu_\mu + \overline{\nu}_\mathrm{e}$$

and can have values between 0 and 1. First measurements in the late forties and early fifties were compatible with ρ=0. By the middle of the fifties, the value had grown to about 0.5. It reached ρ=3/4, shortly after exactly this value had been predicted from parity non-conservation in 1957, and has stayed there ever since.

The first law is proven with the help of a compilation [2] of the major discoveries from accelerators of the last five decades. With the exception of the anti-nucleon (1955 at LBL) and the intermediate bosons (1983 at CERN), none of the major discoveries was the original reason given (by theorists) for the construction of the corresponding accelerator.

The discovery of the intermediate bosons may be an exception to the first law, but it is by itself one of the most beautiful examples of an interplay between theory and experiment. It is, in fact, an example where a clear prediction by theory through ingenious accelerator development and open-minded science management could lead to experimental verification.

However, in the field of Soft Hadron Production and in particular in Multiparticle Dynamcis, the challenge is just that theory is not a clear guide to an experimentalist. The situation is best described in [3], according to Current Contents one of the 100 most-cited Reviews of Modern Physics articles since 1955:

"It is in our opinion unavoidable that progress in theoretical understanding of these phenomena will require a very close interplay between phenomenological analysis of sufficiently accurate experiments, mainly in terms of various tentative models, and application of the few general theoretical principles which seem to be well established."

As general principles are given solely: Lorentz Invariance, Unitarity and Crossing Symmetry of the S matrix. To this list we should now add the gauge theory of color-SU(3), quantumchromodynamics (QCD).

With some optimism, the situation comes close to that of atomic spectroscopy more than half a century ago. There, careful model independent presentation and analysis of at first very complex data finally triggered the big answers of quantum theory so familiar to us today. We also have a lot of very complex data, but no closed theory, except for QCD, so far only describing the relatively small corner of hard collisions, calculable in perturbative QCD.

While a theory of all strong interactions is the ultimate aim, a differential, but at the same time comprehensive study of experimental data seems the safest way to get there. This is, of course, a way of looking for *questions* rather than *answers*, but in the sense that a proper question is already half answered when at all asked.

The main question we are trying to answer here is, therefore: how can one usefully look at data without being (mis-)led by a model? Decades of work by many physicists has shown that the question is far from simple, but a number of important tools have been developed in a close interplay between experiment and theory.

The purpose of this book is to give a more or less comprehensive account of the development and present status of the field of soft (i.e. non-perturbative) phenomena encountered in the production of (multi-)hadronic final states in the collision of various types of particles at high energies. Rather than being a textbook on a closed field, the book is meant as an introduction and reference for postgraduate students as well as more senior researchers planning to work in the field. Phenomenological models used to describe the data are in general inspired by QCD and we shall repeatedly cross the transition between soft, non-perturbative, and hard QCD. Nevertheless, the emphasis of this book will be on methods and technology applied to experimental data and on the experimental results themselves, rather than on their current phenomenological interpretation. Researchers interested in the concepts of perturbative QCD are referred to the existing literature. Similarly, connections of course exist with the lattice gauge theory of strong interactions in the non-perturbative sector. The main successes of these theories are the bound states of quarks, while the topic of this book will be the dynamics of particle production.

As prerequisites, this book assumes familiarity with the basic notions of elementary particle physics and relativistic kinematics. Multiparticle final states are highly complex systems. Some familiarity with the basis of statistical physics, in particular the concepts of correlation functions, self-similarity and fractality, will be useful for the chapters on particle correlations and density fluctuations. Sufficient references to existing literature will, however, be given for the interested reader.

Good literature already exists on the soft hadronic phenomena of total (hadronic) cross sections as well as elastic scattering and diffraction dissociation. The main issue of this book will, therefore, be soft *multi*particle dynamics. Since these phenomena are very closely related and a lot of new insight is presently being gained from electron-proton collisions at high energies, we cannot, of course, leave these topics untouched and treat those in Chapter 1.

In Chapter 2 we sketch the general scheme of inclusive and exclusive analysis of multiparticle final states and study their overall configuration in momentum space. For demonstration, early results in fully exclusive three-particle final states are analyzed in Chapter 3, while, on the other hand, fully inclusive single-particle spectra are used in Chapter 4. Early and more recently used models for soft hadron production are described in short in Chapters 5 and 6, respectively. In Chapter 7, we summarize the most important statistics tools for the study of multiparticle correlations used in the remaining Chapters on (charged-particle) multiplicity distributions (Chapter 8), two- and three-particle momentum correlations (Chapter 9), self-similarity and fractality in particle production (Chapter 10) and Bose-Einstein correlations (Chapter 11).

An important issue in this is a comparison of experimental results on all types of collisions, from e^+e^-via lepton-hadron and hadron-hadron to heavy-ion collisions.

The authors have profited tremendously from the enthusiasm, the work and the erudition of a large number of teachers, colleagues and students, many of whom, but not all, are quoted in the bibliographies of this volume. The volume would, however, not have materialized without Annelies Oosterhof-Meij and her masterly command of text editing and layout.

January 2005

E.A. De Wolf
W. Kittel

Bibliography

[1] M. Jacob, "Théorie et Expérience", Science et Avenir, Les Secrèts de la Matière, Special Issue N°62 (1987).

[2] T.D. Lee, "Symmetries, Asymmetries, and the World of Particles", Univ. of Washington Press 1988.

[3] L. Van Hove, Rev. Mod. Phys. **36** (1964) 655.

Contents

Chapter 1

Total Cross Sections and Diffraction

1.1 Introduction and synopsis

The total cross section of any type of collision, $A + B \to X$, is the most "inclusive" quantity which characterizes the dynamics of the scattering process. Although by far not the easiest to measure in an experiment, this observable, by its simplicity, lends itself to advanced theoretical analysis, exploiting very general properties of the scattering amplitude such as analyticity, causality and unitarity. Unitarity is of particular importance since it inter-relates, in a non-linear way, the size and energy dependence of total cross sections, as well as that of differential cross sections, to the amplitudes of elastic and numerous inelastic final states that are produced in a high-energy collision. The latter are the main subject of this work.

After nearly two decades outside of the mainstream of high-energy physics, the subject of total cross sections, elastic and inelastic diffraction has made a spectacular come-back stimulated by measurements of the virtual-photon proton cross section in a very wide kinematic range and with unprecedented precision at the ep collider at HERA. The same occurred for diffractive scattering after the observation of "Large Rapidity Gap" events at HERA, and new data on rapidity-gap processes at the highest energy hadron colliders. It has become a field of intense experimental and theoretical research and many detailed aspects have been repeatedly reviewed [1–4]. Comprehensive surveys are given in [5–7] and, in particular, in [8], where also the most recent theoretical developments are treated in great detail.

To develop an understanding of high energy scattering total cross sections and diffraction, a field of research which originated in soft hadron-hadron collisions, it is tempting and traditional to start from a t-channel formalism based on Regge theory. For deep-inelastic scattering (DIS) and hard diffractive processes in hadron-hadron collisions, this leads to the intuitive (and popular) pomeron picture, proposed by Ingelman and Schlein [9] well before the experimental discovery of hard diffraction.

In Regge language, diffraction is associated with pomeron exchange. If the pomeron could be regarded as a color-singlet hadronic component in the target proton, carrying a small fraction ξ of the proton momentum, the virtual photon in diffractive DIS (DDIS) would probe the quark content of the pomeron itself. Just as in the case of the proton, the pomeron could then be characterized by *diffractive* structure functions which describe its quark and gluon content. The pomeron structure functions should be universal if the pomeron were indeed an intrinsic part of the target proton wave function.

However, several striking similarities between inclusive and diffractive data in DIS and in hadron-hadron collisions, have led to the insight [10] that the concept of "the pomeron in the proton" is untenable in Quantum Chromodynamics (QCD) and that the "effective pomeron" is a dynamical effect of the interaction and thus not universal. The data indeed provide strong hints that the underlying short-time, hard scattering sub-processes are identical in inclusive and diffractive DIS, and involve gluons and sea quarks whose Bjorken-x dependence reflects the *inclusive* gluon distribution. The formation of a rapidity gap would then be a soft process happening on a longer time scale. This effectively leads to factorization. For DDIS, factorization can indeed be proven from perturbative QCD (pQCD) [11] but not for soft processes.

Although Regge theory is perfectly valid and beautiful, based on very general properties of the scattering amplitudes, it is plagued by many problems in practical applications which, as happened in the past, also now limit its predictive power. Apart from the fundamental theoretical question how to derive the theory as a strong-coupling limit of QCD, the theory itself provides little insight into the relation between properties of the final states or the dynamical interpretation of its parameters. This becomes particularly important when the formalism is used outside its traditional domain of application, such as in deep-inelastic scattering.

Here, we shall mainly consider an s-channel picture and base the discussion on the Good-Walker approach to diffractive processes. The term *diffraction* as used in high-energy physics is defined in close analogy with classical optics, where coherent phenomena occur when a beam of light meets an obstacle whose dimensions are comparable to its wavelength. To the extent that the propagation and the interaction of extended objects like the hadrons and virtual photons result in the absorption of their wave function due to the many open inelastic channels at high energy, that definition seems most appropriate.

Diffractive scattering is thus explained as a consequence of the differential absorption by the target of the large number of states which coherently build up the initial-state projectile, a hadron or a (virtual) photon, and scatter with different cross sections. This physical picture incorporates from the outset basic quantum mechanics and unitarity, and permits, at least conceptually, a unified treatment of hadron, and real and virtual photon scattering. It is the approach to diffraction as first proposed by Feinberg, Good and Walker [12]. It will allow to appreciate the close inter-relation between the dynamics of high-energy hadronic and deep-inelastic diffraction at very small Bjorken-x and to understand that soft, long-distance physics plays a very important role in both. It is also the view adopted in many presently popular theoretical models based on the color-dipole picture of high-energy interactions [2].

Consider first the collision of two hadrons at high energy, say $\sqrt{s} \geq 10 - 20$ GeV, where $\sqrt{s}$ is the energy in the center-of-mass (cms) of the reaction. Hadrons are composite and extended objects, with typical space dimensions of the order of 1 fm. This is also the effective range of the strong interactions.

Since many types of inelastic interactions can occur, a non-negligible fraction of the wave function describing the two initial hadrons is absorbed as the collision takes place and diffractive phenomena will occur. Both elastic and inelastic diffraction will mainly be characterized by the dimension of the region of space over which the absorption occurs. Inelastic diffraction, however, is the more complicated (and more interesting) of the two processes because hadrons, being composite, have internal degrees of freedom that may be excited.

In an inelastic single-diffractive (SD) collision, one of the colliding objects, say the target, survives the interaction almost unchanged and keeps most of its initial energy. The other, the projectile, diffractively dissociates into a system with the same intrinsic quantum numbers, i.e. the same charge, isotopic spin, baryon number, strangeness, charge conjugation etc. Spin and parity may be changed since some angular momentum may be transferred. Between the projectile fragments and the target a large gap in rapidity is created devoid of particles.

Since the target essentially acts as a passive absorber, SD is mainly an experimental probe of the wave function of the diffracting object. Many of the observable properties of diffraction and of the resulting hadronic final states are encoded in that wave function. The latter can be looked upon as the materialization of the projectile's light-cone wave function [13].

Since a real high-energy photon is known to interact in a hadron-like manner, diffractive dissociation in photoproduction with real photons or photons with very small virtuality, Q^2, as available at HERA in photoproduction, can be expected to resemble closely hadron-hadron diffraction. Increasing the photon virtuality, we enter the domain of deep-inelastic scattering. Here, the $\gamma^\star$ with transverse extension $R_\gamma^\star \sim 1/\sqrt{Q^2} \simeq 2 \times 10^{-14}/Q$ (cm/GeV), starts to probe the internal structure of the target once $R_\gamma^\star$ becomes smaller than the proton radius. Due to the large momentum imparted to a parton in the target, the latter is ejected and gives rise to a "current" quark jet.

In the standard picture of inclusive DIS, a color string-field is formed between the struck parton and the target remnant covering the whole rapidity interval between them. The break-up of the string during the hadronization process fills the rapidity interval with hadrons; the probability of a rapidity gap would (for independent production) be expected to decrease exponentially with increasing gap size. The resulting color separation leads to copious production of hadrons in many different inelastic final states. Consequently, in the standard quark-parton picture of DIS, it is not a priori evident how the target proton can remain unscathed in a sizable fraction (order 10-20%) of events corresponding to diffractive deep inelastic scattering. That fraction shows, as in hadron-hadron collisions, only a very weak dependence on the cms energy (W) of the $\gamma^\star$p collision. Also, at fixed Bjorken-x the DDIS to DIS cross section ratio depends weakly on the virtuality of the photon.

To understand why the mechanism behind DDIS is not completely "point-like" but rather soft, it is necessary to consider the space-time development of the radiation cloud which is formed after the virtual photon splits into an initial $q\bar{q}$. It is thereby convenient to consider the scattering process in the rest frame of the proton.

In the proton rest frame, the $\gamma^\star$ wave function can be viewed as a superposition of a variety of (Fock) states which will be absorbed on the target in differing amounts. Diffraction dissociation is therefore expected to occur due to differential absorption, in a similar way as discussed for the case of hadron-hadron collisions. However, DIS has the major advantage that, at sufficiently large Q^2 ($\geq$ a few GeV^2), the initial small-size $q\bar{q}$ fluctuation will develop a radiation cloud [14] which can partly be described using the calculational tools of *perturbative* QCD (pQCD). For a bound-state system of large size, such as a hadron, such techniques are not (yet) available.

The radiation cloud develops via a cascade, whereby the virtuality of the system is degraded by emission of a set of "perturbative" partons. However, once the cascade enters a regime where the strong coupling is no longer small, it will continue into a non-perturbative region which is not under theoretical control. This corresponds to a regime in which the non-perturbative off-springs have acquired transverse dimensions comparable to the size of the target proton. Following Feynman [15], these non-perturbative partonic states are called "wee partons". These will interact with the target as dressed objects, with a large, hadron-like, cross section. This and the variation with parton number, rapidity and impact parameter of their absorption will induce inelastic diffraction in DIS.

Wee partons, originally discussed by Gribov [16,17], are assumed to be responsible for the soft interactions between hadrons. Combining this with what was said about $\gamma^\star$p collisions, one can expect a strong similarity between DIS and soft hadron-hadron collisions, provided that the radiation cloud created by a virtual photon has had sufficient time to develop its own system of wee partons. This will be the case for a $\gamma^\star$p interaction at large W, $Q^2/W^2 \ll 1$ or at large $1/x$, e.g. in the HERA regime.

The following sections are devoted to a more quantitative presentation of the basic physics picture described above. We shall for a considerable part be concerned with hadron-hadron collisions but it will become clear that, to understand the basic dynamics of $\gamma^\star$p scattering, surprisingly few novel basic dynamical concepts beyond those mentioned will be needed. The emerging picture of the dynamics is simple enough to help develop intuition and provide physical insight which can inspire new avenues of research. Data will be reviewed which illustrate the main points, but we will not attempt to present a complete review of the vast and rapidly evolving subject. This can be found elsewhere [3,8,18].

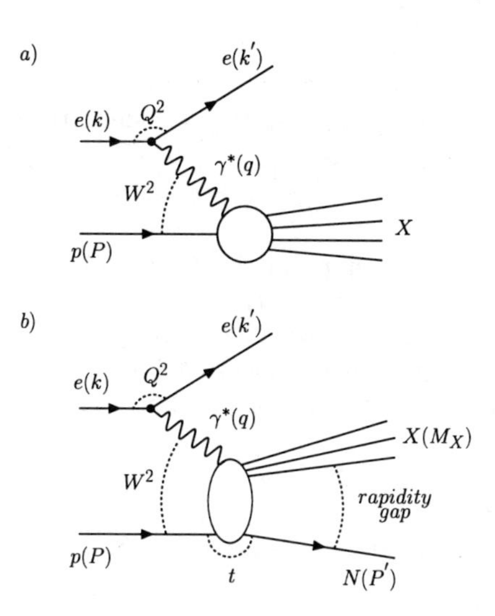

Figure 1.1: Kinematic variables (a) for the reaction e p → e X; (b) for the semi-inclusive reaction e p → e N X with a rapidity gap between the outgoing proton and the hadrons in system X.

This chapter is mainly addressed to experimentalists wishing to broaden their view of the subject. No originality is claimed in the presentation of the material although any errors of interpretation should be attributed solely to the authors.

1.2 Preliminaries

1.2.1 DIS kinematics and cross sections

The standard kinematical variables to describe ep DIS are depicted in Fig. 1.1a. The cms energy squared of the ep system is $s = (P + k)^2$, with P and k the initial-state four-momenta of the proton and electron (or positron), respectively. W, the cms energy of the virtual-photon proton system, is given by $W^2 = (P+q)^2$. The photon virtuality Q^2 and the Bjorken variables x and y are defined as

$$q^2 = -Q^2 = (k - k')^2, \quad x = \frac{Q^2}{2\, P \cdot q} = \frac{Q^2}{W^2 + Q^2 - m_p^2}, \quad y = \frac{P \cdot q}{P \cdot k}. \tag{1.1}$$

Neglecting the proton mass, one has

$$Q^2 = x\ y\ s, \qquad W^2 = Q^2\, \frac{1-x}{x} \simeq \frac{Q^2}{x}, \tag{1.2}$$

the latter expression being valid for $x \ll 1$.

For the "rapidity-gap" process presented in Fig. 1.1b, where a baryon with four-momentum P' is detected in the final state, one defines the additional variables

$$t = (P - P')^2; \quad \xi = \frac{Q^2 + M_X^2 - t}{Q^2 + W^2}; \quad \beta = \frac{Q^2}{Q^2 + M_X^2 - t} = \frac{x}{\xi}. \tag{1.3}$$

The variable ξ (also often called $x_{I\!P}$ in the literature) is the fractional energy-loss suffered by the incident proton. The variable β can naively be thought of as representing the fractional momentum carried by a struck parton in an object — pomeron or reggeon — carrying longitudinal momentum ξ, emitted by the proton and subsequently undergoing a hard scatter. For small $|t|$ one has

$$\beta = \frac{Q^2}{Q^2 + M_X^2}; \quad M_X^2 = \frac{1-\beta}{\beta}\, Q^2; \quad \xi = \frac{(Q^2 + M_X^2)}{W^2}. \tag{1.4}$$

In strict analogy with the total ep cross section

$$\frac{\mathrm{d}^2\sigma}{\mathrm{d}x\mathrm{d}Q^2} = \frac{4\pi\alpha_{\mathrm{em}}^2}{x\, Q^4}\left[1 - y + \frac{y^2}{2(1+R)}\right] F_2(x, Q^2), \tag{1.5}$$

which defines the structure function $F_2(x, Q^2)$ (α_{em} is the QED coupling), the differential cross section for a semi-inclusive (SI) DIS process (Fig. 1.1b) can be written as

$$\frac{\mathrm{d}^3\sigma}{\mathrm{d}x\mathrm{d}Q^2\mathrm{d}\xi} = \frac{4\pi\alpha_{\mathrm{em}}^2}{x\, Q^4}\left[1 - y + \frac{y^2}{2(1+R)}\right] F_2^{\mathrm{SI(3)}}(\xi, x, Q^2). \tag{1.6}$$

Alternatively, in measurements of the diffractive (D(3)) contribution to $F_2(x, Q^2)$, one often uses the definition

$$\frac{\mathrm{d}^3\sigma}{\mathrm{d}\beta\mathrm{d}Q^2\mathrm{d}\xi} = \frac{4\pi\alpha_{\mathrm{em}}^2}{\beta\, Q^4}\left[1 - y + \frac{y^2}{2(1+R)}\right] F_2^{\mathrm{D(3)}}(\xi, \beta, Q^2), \tag{1.7}$$

replacing x by β in Eq. (1.6). $R = \sigma_{\mathrm{L}}/\sigma_{\mathrm{T}}$ is the ratio of the cross sections for longitudinally and transversely polarized virtual photons. Since y is usually small in experiments at HERA, R can be neglected. Equations (1.6) and (1.7) are equivalent since they represent the same experimental data. From an experimental point of view, there is no a priori reason to prefer one over the other and both should be measured. For fixed (x, Q^2) (i.e. W fixed), the ξ dependence of $F_2^{\mathrm{SI,D(3)}}(\xi, x, Q^2)$ reflects that on M_X. Alternatively, for fixed (β, Q^2) (i.e. M_X fixed) the W dependence of $F_2^{\mathrm{SI,D(3)}}(\xi, \beta, Q^2)$ is explored by varying ξ.

The structure function F_2 is related to the absorption cross section of a virtual photon by the proton, $\sigma_{\gamma^*\mathrm{p}}$. For diffractive scattering at high W (low x), we have similarly

$$F_2^{\mathrm{D,SI(3)}}(x, Q^2, \xi) = \frac{Q^2}{4\pi^2\alpha_{\mathrm{em}}} \frac{\mathrm{d}^2\sigma_{\gamma^*\mathrm{p}}^{\mathrm{D,SI(3)}}}{\mathrm{d}\xi\mathrm{d}t}\,. \tag{1.8}$$

1.2.2 Regge formalism

1.2.2.1 Total and elastic cross sections

Since the Regge formalism is so often used in present analyses of diffractive HERA data, we recall here its main ingredients and predictions. A comprehensive introduction to Regge theory can be found in [19]. For small-angle elastic scattering of two hadrons a and b at high s, dominated by pomeron exchange, the Regge scattering amplitude (ignoring the small real part) takes the factorized form

$$\mathcal{A}_{\rm el}^{\rm ab}(s,t) = is\beta_{\rm a}(t)(s/s_0)^{\alpha_{I\!P}(t)-1}\beta_{\rm b}(t). \tag{1.9}$$

Here, s_0 is an arbitrary mass scale, frequently chosen to be of the order of 1 GeV^2. The dependence on the species of the incoming hadron is contained in the form factors or residues, $\beta_{\rm a,b}(t)$, usually parametrized as an exponential $\propto \exp(B_{0;a,b}t)$ and $\alpha_{I\!P}(t)$ is the pomeron trajectory. In its simplest version it is a Regge pole, with intercept $\alpha_{I\!P}(0) = 1+\epsilon$, slightly larger than 1; ϵ controls the large-s or large-W growth of the total and elastic cross sections. The name "pomeron" was first used in [20], but the pomeron concept was introduced by V.N. Gribov [16] and later named in honor of Y. Pomeranchuk [21]. The observed large-s dependence of the cross sections can be accommodated with a linear trajectory $\alpha_{I\!P}(t)$ of the form

$$\alpha_{I\!P}(t) \;=\; \alpha_{I\!P}(0) + \alpha'_{I\!P}t = 1+\epsilon+\alpha'_{I\!P}t;/, \quad \epsilon > 0. \tag{1.10}$$

In Regge theory $\alpha_{I\!P}(t)$, i.e. $\alpha_{I\!P}(0)$ and $\alpha'_{I\!P}$, are universal and independent of the species of the particles colliding. We shall see that their meaning in terms of the particle production dynamics is, at least qualitatively, easy to understand. The energy dependence of the elastic and total cross sections is given by

$$\mathrm{d}\sigma_{\rm el}^{\rm ab}/\mathrm{d}t|_{t=0} \;=\; \frac{1}{16\pi}[\beta_{\rm a}(0)\beta_{\rm b}(0)]^2(s/s_0)^{2\epsilon}, \tag{1.11}$$

$$\sigma_{\rm tot}^{\rm ab} \;=\; \beta_{\rm a}(0)\beta_{\rm b}(0)(s/s_0)^{\epsilon}. \tag{1.12}$$

The meaning of $\alpha'_{I\!P}$ becomes clear if one considers $\mathrm{d}\sigma_{\rm el}^{\rm ab}/\mathrm{d}t$ at small $|t|$, using an exponential approximation

$$\mathrm{d}\sigma_{\rm el}^{\rm ab}/\mathrm{d}t = \frac{1}{16\pi}[\beta_{\rm a}(0)\beta_{\rm b}(0)]^2 e^{B(s)t}(s/s_0)^{2\epsilon} = \frac{\left[\sigma_{\rm tot}^{\rm ab}\right]^2}{16\pi}\, e^{B(s)t}, \tag{1.13}$$

with

$$B(s) = 2\left(B_{0;\rm a} + B_{0;\rm b} + \alpha'_{I\!P}\ln\frac{s}{s_0}\right). \tag{1.14}$$

The energy independent terms $B_{0;\rm a,b}$ originate from the form factors in Eq. (1.9). From pp data $B_{0;p} \approx 2-3$ GeV^{-2}.

Equation (1.14) shows that the forward elastic peak "shrinks" with energy: $B(s)$ increases (here logarithmically) with s. In impact parameter space ($\mathbf{b}$), Eq. (1.9) becomes

$$\mathcal{A}^{\rm ab}(s,\mathbf{b}) = i\,\frac{\beta_{\rm a}(0)\beta_{\rm b}(0)}{8\pi}\frac{(s/s_0)^{\epsilon}}{B(s)}e^{-\mathbf{b}^2/2B(s)}. \tag{1.15}$$

In optical terms, the transverse size of the "interaction region" is Gaussian with $B(s) = \langle \mathbf{b}^2 \rangle$. In the impact-parameter plane, the scattering profile is a disc with a b-dependent opacity, the mean radius of the disc being proportional to $\sqrt{B(s)}$. $B(s)$ contains an s-independent term and a term growing logarithmically with s. The radius expands with s with a rate of growth determined by $\alpha'_{I\!P}$, estimated to be ≈ 0.25 GeV^{-2} [22] (for a similar estimate see [23]). At the same time, the opacity at fixed b is likely to increase, too. In a perturbative QCD picture, this corresponds to an increase of the gluon density in the target or, in the proton rest frame, to an increase of the interaction probability. Since the latter cannot exceed unity, it follows that also the gluon density cannot rise indefinitely.

Concerning the meaning of $\alpha_{I\!P}(0)$, it will become clear in Sect.1.5 that the increase of $\sigma_{\rm tot}$ with s can be attributed to an increase of the number of (wee) partons in projectile and target. This will allow us to relate $\alpha_{I\!P}(0)$ to the rise of the wee-parton density in rapidity or, more generally, to the QCD multiplicity anomalous dimension. Before that, however, it will be necessary to analyze the space-time structure of a fast hadron or γ^* as viewed in the Gribov-Feynman parton model (Sub-Sect. 1.5.5). The emerging picture, suitably modified so as to include (perturbative) QCD effects as well as the (partly) calculable properties of virtual photons will provide a self-consistent framework in which much of the physics studied at HERA, including DDIS, can be understood.

1.2.2.2 Triple-Regge parametrization of the reaction $\rm a + b \to c + X$

Regge theory can be generalized to inclusive reactions, extensively discussed in later chapters. The invariant cross section of an inclusive process, $\rm a + b \to c + X$, is then expressed in terms of a Mueller-Regge expansion, based on Mueller's generalization of the optical theorem [24]. This states that an inclusive reaction $\rm ab \to cX$ is connected to the elastic ("forward") three-body amplitude $A(\rm ab\bar{c} \to ab\bar{c})$ via

$$E\,\frac{{\rm d}^3\sigma}{{\rm d}^3p}({\rm ab} \to {\rm c}X) \sim \frac{1}{s}\,{\rm Disc}_{M^2}\,A({\rm ab\bar{c}} \to {\rm ab\bar{c}}), \tag{1.16}$$

where the discontinuity is taken across the M_X^2 cut of the elastic Reggeized amplitude. Here, E and p are the energy and three-momentum of particle c. For the triple-pomeron diagram, valid in the (diffractive) region of phase space where the momentum fraction of particle c is near one, or $s \gg W^2 \gg (M_X^2, Q^2) \gg (|t|, m_{\rm p}^2)$,

Equation (1.16) has the approximate form

$$E\frac{d^3\sigma}{d^3p} = \frac{1}{\pi}\frac{d^2\sigma}{dt\ d\xi} \simeq \frac{s}{\pi}\frac{d^2\sigma}{dt\ dM_X^2} = f(\xi,t)\cdot\sigma_{I\!Pp}(M_X^2). \tag{1.17}$$

The "flux factor" $f(\xi,t)$ is given by

$$f(\xi,t) = N\ F^2(t)\ \xi^{[1-2\alpha_{I\!P}(t)]}. \tag{1.18}$$

In the above equations, N is a normalization factor, $F(t)$ the form-factor of the pp$I\!P$-vertex; $\sigma_{I\!Pp}$ can be interpreted as the pomeron-proton total cross section. Assuming $\sigma_{I\!Pp} \sim (M_X^2)^\epsilon$, and using Eq. (1.10), Eq. (1.17) and Eq. (1.18) implies that

$$\left.\frac{d^2\sigma}{dt\ d\xi}\right|_{t=0} \sim \frac{s^\epsilon}{\xi^{(1+\epsilon)}} = \frac{(M_X^2)^\epsilon}{\xi^{(1+2\epsilon)}}. \tag{1.19}$$

The two expressions on the right-hand side Eq. (1.19) are equivalent. However, they show that the model predicts different ξ dependences if either s is kept fixed and M_X varied, or if M_X^2 is fixed and s varied. Nevertheless, since ϵ is small, both expressions in Eq. (1.19) show that the triple-Regge $I\!PI\!PI\!P$ contribution in the region $\xi \ll 1$ is of the generic form

$$\left.\frac{d^2\sigma}{dt\ d\xi}\right|_{t=0} \sim \frac{1}{\xi^{(1+\delta)}} \qquad \text{with} \quad \delta \ll 1, \tag{1.20}$$

with a universal $1/\xi$ dependence as long as δ is universal.

Regge theory implies that both σ_{el}/σ_{tot} and σ^{SD}/σ_{tot} increase as s^ϵ, where σ^{SD} is the total single-diffractive cross section. This eventually leads to violation of unitarity since ϵ is found to be positive. Integrated over t and M_X, σ^{SD} grows as $s^{2\epsilon}$.

Although the applicability of Mueller's optical theorem to reactions with (far) off-shell particles has not been proven, it is very frequently used as the starting point in analyses of diffractive phenomena in γ^*p scattering. In that case, s in the above equations has to be replaced by W^2 or by $1/x$ if Q^2 is fixed.

1.2.2.3 Problems

In spite of the elegance of the Regge approach, it has been known for a long time [25] that the theory with a "super-critical pomeron", $\alpha_{I\!P}(0) = 1+\epsilon$ ($\epsilon > 0$), is plagued by unitarity problems as $s \to \infty$ which are especially severe for inelastic diffraction:
i) the power-law dependence, $\sigma_{tot} \propto s^\epsilon$ violates the Froissart-Martin bound [26];
ii) the ratio $\sigma_{el}/\sigma_{tot} \propto s^\epsilon/\ln s$ eventually exceeds the black-disk geometrical bound ($\sigma_{el} \leq \frac{1}{2}\sigma_{tot}$);
iii) the ratio σ^{SD}/σ_{tot} increases as s^ϵ: the diffractive cross section rises faster than the total cross section. This disagrees with experiment not only for hadron collisions [27], but also for deep-inelastic diffraction, where the ratio σ^{SD}/σ_{tot} is found to be essentially independent of W [28,29] at not too small Q^2 (see Sub-Sect. 1.4.3).

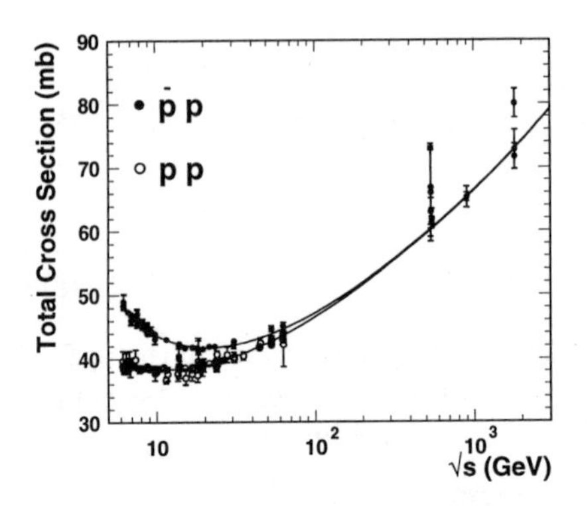

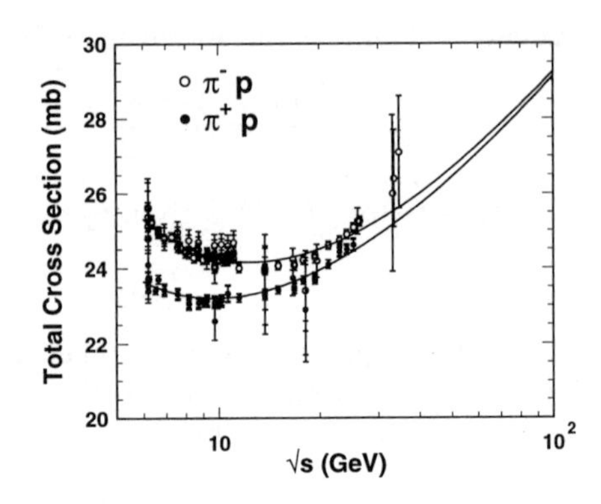

Figure 1.2: Total cross sections measured in hadronic scattering as a function of the cms energy for pp, $\overline{\rm p}$p and $\pi^{\pm}$p scattering. The cross sections show a "universal" rise at high energies of the form $\sigma \sim s^{0.08}$. Data from [30]. Curves are the Donnachie-Landshoff fit, Eq. (1.21) [31].

1.3 Data on total and elastic cross sections

1.3.1 Energy dependence of hadronic total cross sections

The s-dependence of total hadron-hadron cross sections, $\sigma_{\rm tot}$, has been measured for many combinations of hadrons. For a compilation see [30]. Above $\sim$ 20 GeV all hadronic cross sections rise with increasing s. This was first discovered for K^+p collisions in 1970 at the Serpukhov accelerator [32], later confirmed at CERN [33]. The rise of the pp total cross section was first observed at the ISR [34] and later confirmed at Fermilab [35].

For illustration, a compilation of $\overline{\rm p}$p, pp and $\pi^{\pm}$p data is shown in Fig. 1.2. The solid lines are fits which include a component decreasing rapidly with increasing s and a second, rising component which persists at high energies. The important observation by Donnachie and Landshoff (DL) in [31] was that, at high s, all hadron-hadron (and γp [36]) cross sections grow in a similar way. An economical parametrization, shown in the figure, is a sum of two power-law terms in s

$$\sigma_{\rm tot} = X s^{\epsilon} + Y s^{\epsilon'}, \tag{1.21}$$

where the constants X and Y depend on the reaction. This obviously is inspired by Regge theory, the two terms in Eq. (1.21) corresponding to pomeron and "normal" Regge (reggeon) exchanges, respectively. The value of ϵ is not very precisely established. Various recent "global" fits find the data to be compatible with ϵ in the range $0.08 - 0.1$ [31, 37–41]; ϵ' is found to be ~ -0.45 [31]. Global fits to total, elastic and diffractive cross sections performed much earlier yielded similar values for ϵ [42].

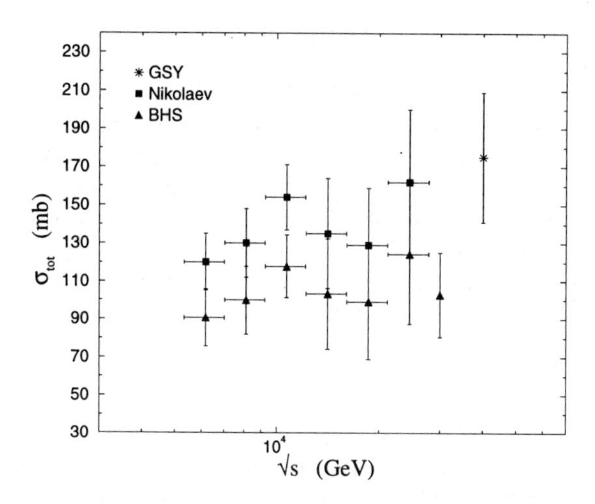

Figure 1.3: Estimates of the pp total cross section derived from cosmic ray experiments by Gaisser, Sukhatme and Yodh (GSY) [44], Nikolaev [45] and Block, Halzen and Stanev (BHS) [46]. Figure from [41].

The DL model was extended [43] to describe virtual-photon proton scattering. It was assumed that the Q^2 dependence would reside in the residue functions, $\beta_a(t, Q^2)$, Eq. (1.9), keeping the Regge intercepts independent of Q^2. This model cannot accomodate the steep rise of F_2 as measured at HERA (cfr. Sub-Sect. 1.3.2 and Fig. 1.8).

In actual fact, the precision of present data does not allow to discriminate between "simple-pole" fits inspired by a Regge-theory, leading to a power-law dependence, and equally valid fits by $\log^2 s$ and $\log s$ (or, for that matter, $e^{\sqrt{\log s}}$) functional dependences [37,47,48]. Recent fits, which also include measurements of the $\gamma\gamma$ total cross section, prefer a "universal" pomeron term of the form $\log^2 s$. They are discussed in [30,39].

The lack of discriminative power of the presently available pp and $\bar{\text{p}}$p data concerning the energy-dependence, and the ensuing uncertainty in predicted cross sections at still larger energies, is due, in part, to conflicting results at the highest $\bar{\text{p}}$p energy [49–52]. Further, in the cosmic-ray energy domain, 6 TeV $< \sqrt{s} \leq$ 40 TeV, shown in Fig. 1.3, differences are due to both experimental and theoretical uncertainties in the determination of $\sigma_{\text{tot}}^{\text{pp}}$ from p-air cross sections. The highest values are quoted by Gaisser, Sukhatme and Yodh [44], together with those by Nikolaev [45]. Substantially lower values are obtained by Block, Halzen and Stanev [46]. For details and references see [41].

Although the significance of Eq. (1.21) has been somewhat over-emphasized in the context of Regge theory, the "universal" high-energy behavior of the total hadronic cross sections, independent of the type of colliding particles, remains an important observation which calls for deeper understanding. It evidently raises the question what type of dynamics is responsible for this universal increase.

Relations between the high energy cross sections of different hadronic processes can be derived from the Additive (constituent) Quark Model (AQM) [53]. This model is based on the following assumptions: i) the constituent quarks (where quark means quark or anti-quark) act as free particles in a hadron, ii) when two hadrons interact, the interaction consists solely of the interaction of a quark from one hadron with a quark from the other hadron, iii) the quark-quark cross section is independent of the nature of the quark, iv) using the impulse approximation, the total amplitude is the sum of all quark-quark amplitudes. For pion-nucleon and the nucleon-nucleon interactions, this leads to the "classical" AQM relation $\sigma_{\text{tot}}^{\pi\text{N}}/\sigma_{\text{tot}}^{\text{NN}} = 2/3$, in agreement with experimental data within a few percent.

Within perturbative QCD, the linear dependence of the total hadron-hadron scattering amplitude on the number of constituent quarks inside the scattered hadrons was first derived in [54,55]. The result was based on the Low-Nussinov model for the pomeron (two-gluon exchange and, therefore, flavor-blind) [56] and explains rather naturally, through "color transparency" [57], the dependence of the total cross section on the size of the bound state or, equivalently, the masses of the constituent quarks, which vary drastically in going from hadrons composed of light quarks to those that contain strange or heavy quarks. The AQM model has recently been modified and extended to photon-induced reactions [58]. For these, data are now available up to quite high energies ($\sqrt{s} \approx 200$ GeV for γp and $\sqrt{s} \approx 100$ GeV for $\gamma\gamma$ inelastic cross sections) [30].

As mentioned in Sub-Sect. 1.2.2, the single-pomeron exchange amplitude violates unitarity, indicating an inconsistency. A way to overcome this problem is to introduce, as required from unitarity, multiple pomeron exchanges (or multiple interactions) in a single scattering process, as shown in Fig. 1.4 [59]. The total amplitude can then be written as the sum of n-pomeron exchange amplitudes $A^{(n)}(s,t)$. For each n-pomeron graph one can define a theoretical "total" cross section applying the optical theorem to the corresponding n-pomeron amplitude

$$\sigma^{(n)} = (-1)^{n+1}\frac{1}{s}\,\text{Im}\left(A^{(n)}\right), \qquad \sigma_{\text{tot}} = \sum_{n=1}^{\infty}(-1)^{n+1}\sigma^{(n)}\,. \tag{1.22}$$

As a simplified model [59] consider only the first two graphs shown in Fig. 1.4, assuming $\sigma^{(n)} \ll 4\sigma^{(2)} < \sigma^{(1)}$ with $n > 2$.

The total cross section then becomes $\sigma_{\text{tot}} = \sigma^{(1)} - \sigma^{(2)}$, where $\sigma^{(1)}$ and $\sigma^{(2)}$ are the cross sections of the one- and two-pomeron exchange graphs, respectively. The energy dependence of the two-pomeron cross section is directly linked to that of $\sigma^{(1)} \sim s^{\epsilon}$ and turns out to be $\sigma^{(2)} \sim s^{2\epsilon}$. The two-pomeron cross section grows faster with energy than the one-pomeron cross section. Since its contribution is negative, this

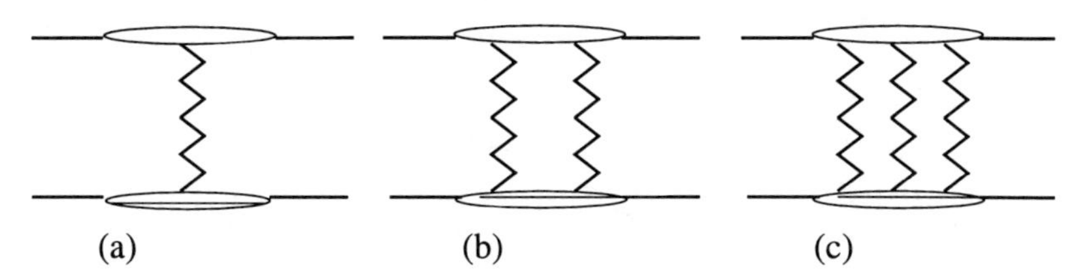

Figure 1.4: Hadron-hadron scattering via pomeron exchange: (a) one-pomeron, (b) two-pomeron, and (c) three-pomeron exchange graphs [59].

leads to a weaker energy dependence of σ_{tot} than in the single-pomeron exchange model and a smaller effective pomeron intercept. It also breaks Regge factorization.

Interestingly, according to the Abramovski, Gribov, Kancheli (AGK) cancellation theorem, the contribution of the two-pomeron graph to the inclusive inelastic single-particle cross section vanishes [60]. Analogously, the factorization violating contributions due to multi-pomeron exchange graphs cancel in all orders. This means that only the one-pomeron graph determines the inclusive particle cross section in the central cms region. Consequently, the energy dependence of the single-particle inclusive spectrum reflects the value of the pomeron intercept in soft hadronic interactions, in a way which is unaffected by multi-pomeron (or screening) effects.

The results of an analysis which addresses this point are shown in Fig. 1.5 [61]. A double-Regge expansion is used, valid at high energies and in the central cms rapidity region ($y = 0$), which predicts the energy dependence

$$\left.\frac{\mathrm{d}\sigma}{\mathrm{d}y}\right|_{y=0} = a_{I\!PI\!P}\, s^{\Delta} + a_{I\!RI\!P}\, s^{(2\Delta-1)/4} + a_{I\!RI\!R}\, s^{-1/2} \,, \tag{1.23}$$

where the a-parameters are reggeon couplings and $1+\Delta$ is the value of the pomeron intercept, unaffected by multi-pomeron absorptive effects. The fit yields $\Delta = 0.170 \pm 0.008$, for negative particle (c^-) production and $\Delta = 0.167 \pm 0.024$ for K^0_S inclusive production. As anticipated above, this is much larger than $\epsilon \simeq 0.08$ deduced from the s-dependence of hadron-hadron total cross sections, which are, however, affected by the re-scattering contributions.

In [62] (see further also [63]) it is argued that the "bare" value of Δ is still larger, since renormalization effects induced by pomeron-pomeron interactions and pion-loops lower its effective value. The correction is estimated to be ~ 0.14. In all, this means that the "bare" pomeron intercept could be as large as 1.3 and thus comparable (see below) to what is measured in deep-inelastic scattering. In DIS, absorption effects due to multi-pomeron exchange are expected to be much smaller than in soft hadronic collisions, due to the short interaction time, and to diminish with increasing Q^2. DIS measurements therefore reflect more directly the properties of the bare pomeron than soft hadron collisions.

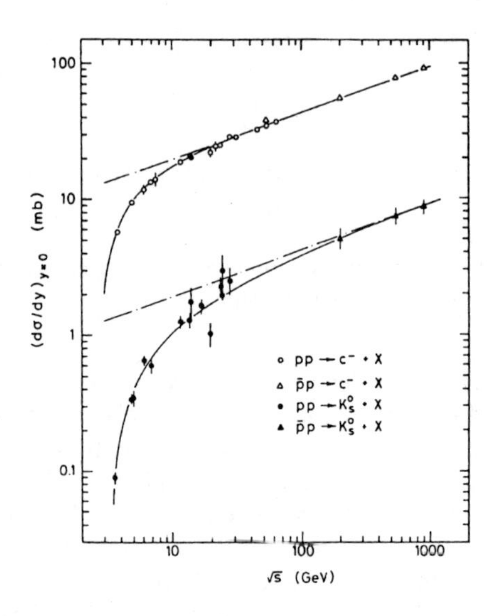

Figure 1.5: Cross sections of negatively charged particles, c^-, and K_S^0 in $pp \to c^- + X$, $p\bar{p} \to c^- + X$, $pp \to K_S^0 + X$ and $p\bar{p} \to K_S^0 + X$ at central rapidity in the cms, $(d\sigma/dy)_{y=0}$. Solid curves are fits by a double-Regge expression, Eq. (1.23), yielding $\Delta \simeq 0.17$ for all reactions. The dashed lines represent the s^Δ term. For refs. to data see [61].

Let us add that the parametrization (1.23) predicts cross sections for negatively charged particles and K_S^0 of 251 ± 26 mb and 25 ± 7 mb, respectively, at LHC energies.

1.3.2 The $\gamma^\star$p total cross section at HERA

1.3.2.1 Energy dependence

The now very precise measurements of the total $\gamma^\star$p cross section or, equivalently, the structure function, $F_2(x, Q^2)$, as a function of Q^2 and W are major achievements of the experiments at HERA. Some results are shown in Fig. 1.6 [64], where they are compared with predictions of the saturation model of Golec-Biernat and Wüsthoff (GBW) [65] and in Figs. 1.10–1.11. This model is discussed further below.

Remembering that $1/\sqrt{Q^2} = R_{\gamma^\star}$ determines the transverse distance which the photon can resolve, we note that for small Q^2 (large $R_{\gamma^\star}$) the cross section has a hadron-like increase with W: the photon acts like a hadron. With increasing Q^2, the photon "shrinks" and becomes more and more point-like: the rise with W becomes stronger.

Parametrizing the x-dependence of $F_2(x, Q^2) \sim (1/x)^{\lambda(Q^2)}$, λ was extracted from

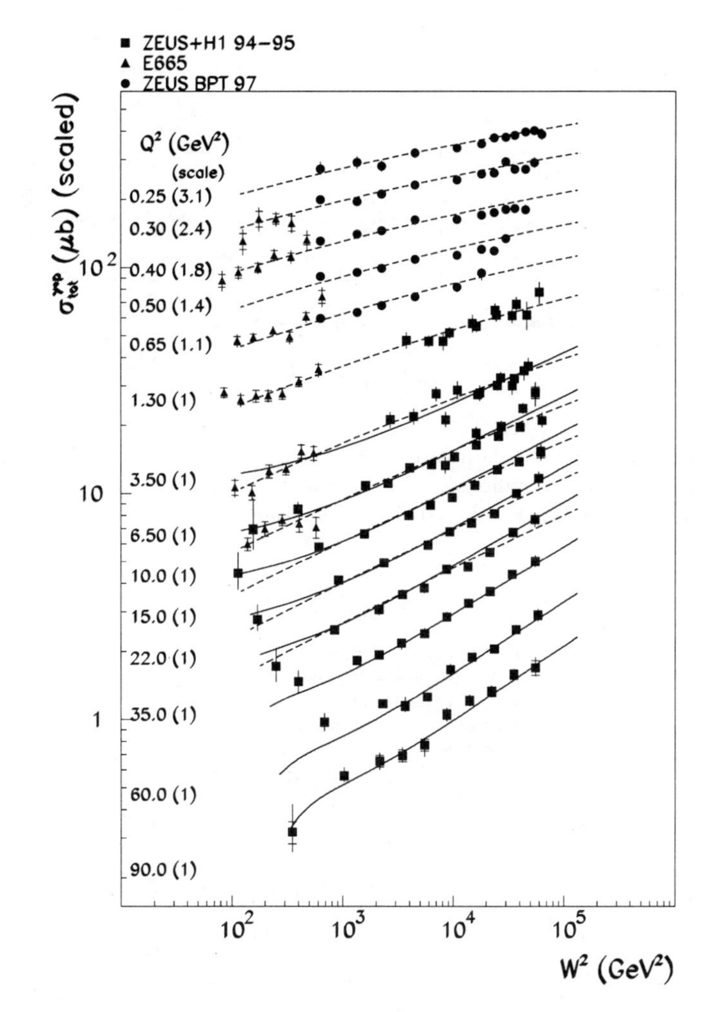

Figure 1.6: $\gamma^{\star}$p cross section as a function of W^2 for fixed intervals of the photon virtuality, Q^2, shown on the left side together with the scale factor applied to the data for better visibility. The full lines show a QCD-fit [66], the dashed lines are a fit by the Golec-Biernat and Wüsthoff saturation model [65]. Figure from [14].

the logarithmic x derivative of F_2 at fixed Q^2 [67, 68],

$$\lambda(x, Q^2) = (\partial \ln F_2 / \partial \ln x)_{Q^2} \, . \tag{1.24}$$

The H1 results at moderate to low Q^2 are shown in Fig. 1.7 [67, 68]. Although not evident a priori, the data here and at larger Q^2 are consistent with no dependence

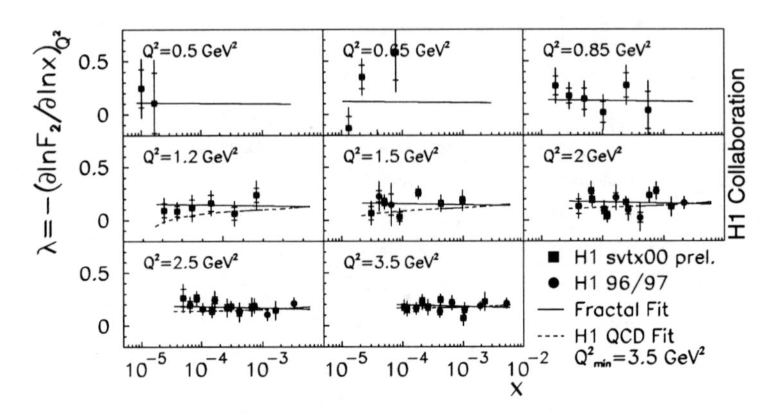

Figure 1.7: The logarithmic x derivative of F_2 in the low Q^2 region [67, 68]. The data are here compared with the predictions of a "fractal" model of the proton structure (solid lines) [69] and extrapolation of an H1 NLO QCD fit [70] (dashed).

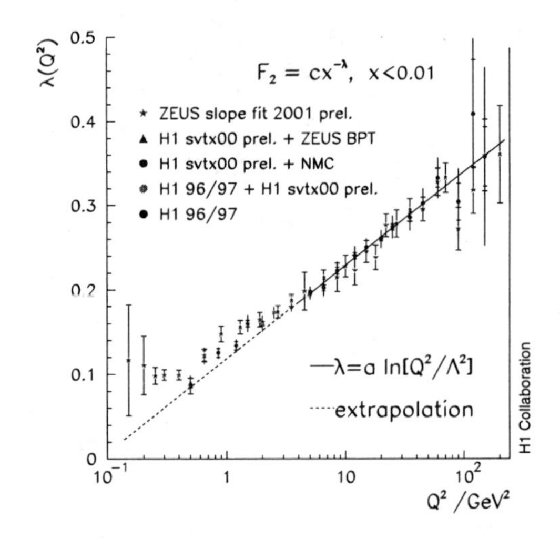

Figure 1.8: The exponent λ_{tot} in $\sigma_{\text{tot}}^{\gamma^* p} \sim (W^2)^{\lambda_{\text{tot}}}$, versus Q^2. The data are compared with a parametrization in which λ grows logarithmically with Q^2 [71].

of λ on x at a fixed Q^2 and $x \lesssim 0.01$, thus indicating a monotonic rise of F_2 with increasing $1/x$ when Q^2 is fixed. There is no evidence for any levelling off of this rise in inclusive electroproduction data from HERA.

Since the logarithmic x derivative is compatible with independence of x, the proton structure function at low x can be parametrized as

$$F_2 = c(Q^2)\, x^{-\lambda(Q^2)} . \tag{1.25}$$

H1 and ZEUS have fitted their data by this form [67,68,72]. The results are shown in Fig. 1.8. Two regions seem to be distinguished. For Q^2 larger than about 3 GeV^2, λ depends logarithmically on Q^2 and c ~ 0.18 is consistent with being independent of Q^2. This behavior is well reproduced by DGLAP-based [73] QCD fits. In contrast, for Q^2 less than about 1 GeV^2, there is evidence for a decrease of $c(Q^2)$ and for deviations of $\lambda(Q^2)$ from the logarithmic dependence on Q^2 as it tends to the value of about 0.1, known to describe hadron-hadron and photoproduction total cross sections, see Sub-Sect. 1.3.1. Around $Q^2 = 1$ GeV^2, a pQCD description based on DGLAP seems to break down. The confinement transition, where partons may no longer be the relevant degrees of freedom, thus seems to take place at a distance scale of around 0.3 fm.

In order to account for the HERA data within a Regge framework, Donnachie and Landshoff have extended their model to a two-pole pomeron model. Besides the soft pomeron, it includes a "hard" pomeron with intercept 1.44 [74]. This allows a description of low-x HERA data up to Q^2 values of a few hundred GeV^2.

If interpreted in terms of Regge exchanges, it is clear that for $\gamma^\star$p collisions "universality" of the trajectory parameters is strongly broken: e.g. $\alpha_{\mathbb{P}}(0)$ depends strongly on Q^2 and a continuous transition is seen between the soft regime and that where a "small-size" $\gamma^\star$ hits a proton. Evidently, for "small-size" virtual photons (large Q^2), the data are well understood in terms of perturbative QCD radiation and DGLAP evolution in Q^2. For $Q^2 \lesssim 1$ GeV^2 this is no longer the case.

1.3.3 Parton densities and the gluon

1.3.3.1 Perturbative QCD fits based on DGLAP evolution

The H1 [70,75] and ZEUS [76] collaborations, as well as various other groups [77,78], have performed QCD fits to extract parton densities using various combinations of HERA and other data. Recent fits are based on the evolution of the parton densities with Q^2 using the DGLAP equations [73] in Next-to-Leading Order (NLO) [79].

With the present HERA neutral (NC) and charged current (CC) data sets, the full set of flavor-separated parton densities can now be extracted using HERA data alone. Figure 1.9 shows the results for the valence densities, the sum of all sea quarks and the gluon density, from H1 NC and CC data only [75], from ZEUS NC data together with other fixed-target DIS experiments [76] and from CTEQ [77]. The latter perform a global fit to many DIS and other data sets. The agreement between the different extractions is reasonable, although differences larger than the quoted error bands are seen between the H1 and ZEUS results.

A most striking aspect of Fig. 1.9 is, of course, the steep rise of the gluon distribution with decreasing Bjorken-x, meaning that the density of gluons in the proton

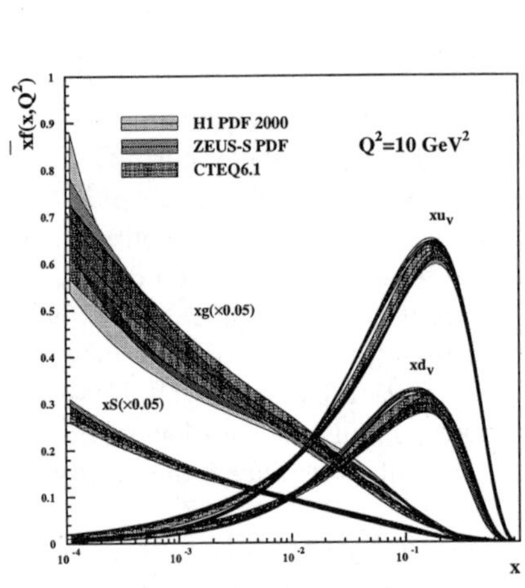

Figure 1.9: Parton densities as obtained by H1 [75], ZEUS [76] and CTEQ [77]. The sea quark (S) and gluon densities are reduced by a factor of 20 [71].

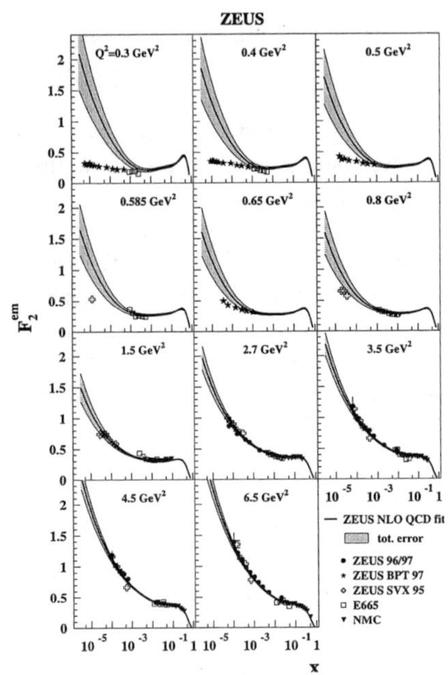

Figure 1.10: ZEUS F_2 data at low Q^2 compared to a ZEUS pQCD fit [76].

becomes very large for $x \to 0$. If the high gluon density is the reason for the strong rise of F_2 as $x \to 0$ or the steep energy dependence of $\sigma(\gamma^*\text{p})$ at large Q^2, then eventually saturation must occur due to multiple interactions (shadowing). At even higher densities, the gluons will recombine via the processes such as gg $\to$ g and the gluon density and hence F_2 will saturate.

These ideas have been first formalized by Gribov, Levin and Ryskin [80] by the addition of a non-linear term, quadratic in the gluon density, to the DGLAP equation for gluon evolution. Saturation has now become a hot topic, not only for HERA physics. The dynamical origin of saturation can be derived from a new form of QCD matter, called Color Gluon Condensate [81].

1.3.3.2 Breakdown of the pQCD DGLAP formalism

The precision of the low-x HERA data on F_2 is now such that the kinematic range of applicability of the DGLAP formalism can be rather precisely delimited. This is illustrated in Fig. 1.10: for Q^2 below $\lesssim 1$ GeV^2, the NLO pQCD fit no longer describes the data. Also, the gluon distribution extracted from the fit turns negative at small x [76]. For Q^2 below about 2 GeV^2, the rise of F_2 towards $x \to 0$ becomes gradually much less pronounced as $Q^2 \to 0$.

1.3.3.3 Self-similar structure of the proton?

Whereas the DGLAP formalism is very successful at high Q^2, pQCD is most likely no longer valid in the low $Q^2 \lesssim 1$ GeV^2 region. There, other models are needed, e.g based on Regge-exchange ideas [82], dipole interactions [65], vector meson dominance (VMD) [83] or other phenomenological parametrizations [84, 85].

An interesting alternative, presented in [69], is based on the idea that the proton structure at small Bjorken-x may be (multi-) fractal (self-similarity and fractals are discussed in Chapter 10). In pQCD the sea-quark densities are driven by gluon emission and gluon splitting. The deeper the proton structure is probed, the more gluon-gluon interactions can be observed. Analogously to fractals, these may show self-similarity, i.e. scaling described by a power law. The notion of self-similarity, applied to *un-integrated* parton densities[1] leads eventually to the parametrization

$$F_2(x, Q^2) = \frac{\mathbf{e}^{\mathcal{D}_0}\, Q_0^2\, x^{-\mathcal{D}_2+1}}{1 + \mathcal{D}_3 - \mathcal{D}_1 \log x} \left(x^{-\mathcal{D}_1 \log(1+\frac{Q^2}{Q_0^2})} (1 + \frac{Q^2}{Q_0^2})^{\mathcal{D}_3+1} - 1 \right), \tag{1.26}$$

with a normalization factor $\mathbf{e}^{\mathcal{D}_0}$. The parameters $\mathcal{D}_i$ (i=1,2,3) are fractal dimensions.

The data, which cover a wide range in Q^2, and now also include the very low-Q^2 transition region to the photoproduction regime, are displayed in Fig. 1.11. The curves show the fit with Eq. (1.26) to H1 [86] and ZEUS [87] data in the range $1.5 \leq Q^2 \leq 120$ GeV^2 (H1), $0.045 \leq Q^2 \leq 0.65$ GeV^2 (ZEUS) and $x < 0.01$. The best-fit parameters are given in Table 1.1

In the limit $Q^2 \to 0$ and fixed W^2 the parametrization (1.26) is linear in Q^2 only for $\mathcal{D}_2 = 1$, so that this parameter is actually fixed from current conservation, reducing the number of parameters to four. The parameter $\mathcal{D}_0$ determines the virtual photon-proton cross section in the photoproduction limit. One finds:

$$\sigma_{\gamma \mathrm{p}} = \left[\frac{4\pi^2 \alpha}{Q^2} F_2(W^2, Q^2) \right]_{Q^2 \to 0} = 189 \pm 3\ \mu\mathrm{b}. \tag{1.27}$$

This is in approximate agreement with the photoproduction total cross sections measured by H1 [88] and ZEUS [72] but which were not included in the fit. To note is that the parameter $\mathcal{D}_3$ turns out to be negative (Table 1.1), thus violating the positivity condition of fractal dimensions. This aspect of the analysis has been criticized in [89] to which we refer for further details.

We end this section by mentioning that the idea of self-similarity, leading to power-law behavior of the cross section, has also been been advocated for inelastic diffraction in deep-inelastic ep scattering [90–92].

---

[1] Conventional, (integrated) quark densities $q_i(x, Q^2)$ are defined as a sum over all contributions with quark virtualities smaller than that of the photon probe, Q^2. The un-integrated parton densities, $f_i(x, Q^2)$ are thus defined via the relation $q_i(x, Q^2) = \int_0^{Q^2} f_i(x, q^2)\, \mathrm{d}q^2$.

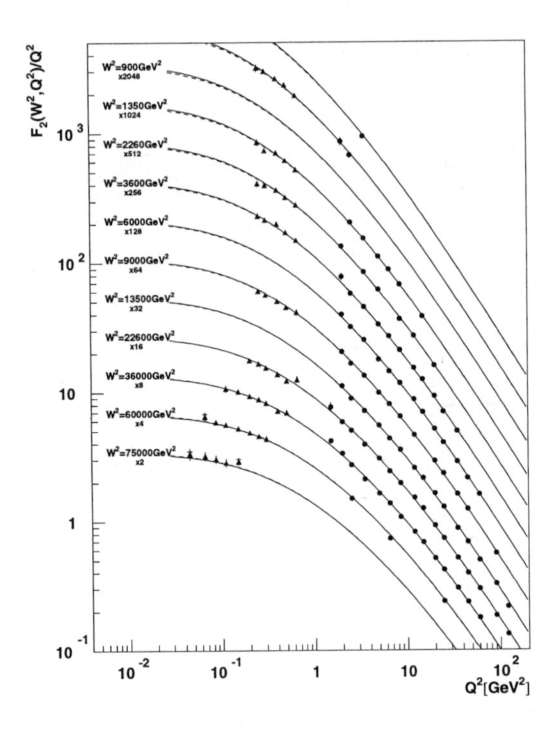

Figure 1.11: Virtual photon-proton cross-section $\sigma_{\gamma^* \mathrm{p}} \propto F_2(W^2, Q^2)/Q^2$ as a function of Q^2 in W^2 bins. H1 (points) and ZEUS (triangles) measurements are shown along with the fit by Eq.(1.26). For details see [69].

1.3.4 Elastic scattering

1.3.4.1 Hadron-hadron interactions

Figure 1.12a shows data on the forward elastic slope (assuming a single-exponential t-dependence) in pp and $\overline{\mathrm{p}}\mathrm{p}$ interactions. The first measurement of the pp2pp experiment at RHIC on elastic scattering of polarized protons at $\sqrt{s} = 200$ GeV, presented in [93], yields a value $B = 16.3 \pm 1.6\,(\mathrm{stat}) \pm 0.9\,(\mathrm{syst})$ GeV/c^2, in the t-range $0.010 \leq |t| \leq 0.019$ (GeV/c)^2, compatible with the world data.

The shrinkage of the diffractive peak with increasing $\sqrt{s}$, expected from Regge theory is clearly seen. Expressed in geometrical or optical terms, the "effective interaction radius" of the proton becomes larger with increasing s.

The values of the slopes are in rough agreement with what is expected for (optical) diffraction on a "black" fully absorbing disk of radius R for which $B = R^2/4$. For a proton with $R \approx 1/m_\pi$ (m_π is the pion mass) this gives $B = 13$ GeV^{-2}, which compares well with the data. However, for scattering on a black disk, $\sigma_{\mathrm{el}}/\sigma_{\mathrm{tot}} = 1/2$,

Table 1.1: The fit of F_2 by Eq. (1.26). First row: all parameters left free; second row: parameter $\mathcal{D}_2$ fixed at 1. The number of F_2 data points is 172.

	$\mathcal{D}_0$	$\mathcal{D}_1$	$\mathcal{D}_2$	$\mathcal{D}_3$	Q_0^2[GeV^2]	χ^2	χ^2/ndf
all fit	0.339	0.073	1.013	-1.287	0.062	136.6	0.82
	±0.145	±0.001	±0.01	±0.01	±0.01		
$\mathcal{D}_2$ fixed	0.523	0.074	*1*	-1.282	0.051	138.4	0.82
	±0.014	±0.001	*const.*	±0.01	±0.002		

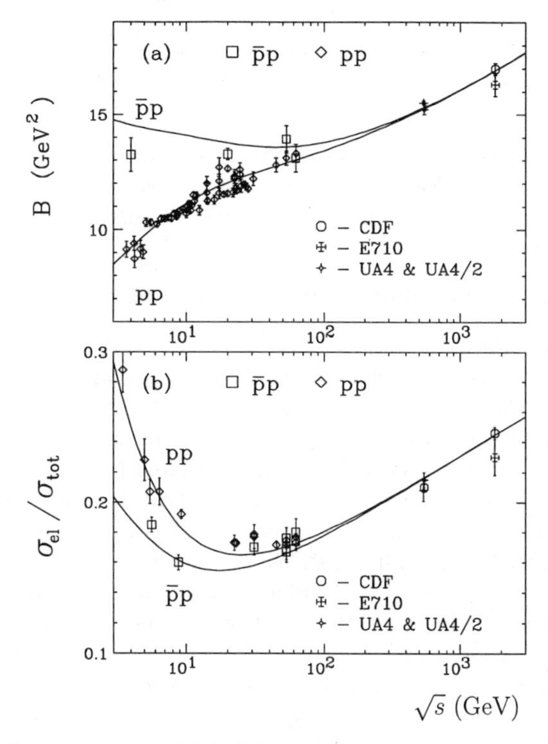

Figure 1.12: (a) Slope parameter $B(s)$, (b) ratio of elastic to total cross section versus $\sqrt{s}$ (in GeV) for $\bar{\mathrm{p}}\mathrm{p}$/pp interactions. The solid lines are Regge fits. For details see [40].

whereas experiment, Fig. 1.12b, shows a value between 1/5 and 1/4 at high s. This means that the proton is not completely black, even at zero impact parameter. This was shown experimentally in [94] (see also the much earlier analysis in [95]). Indeed, since the wave function of the composite hadrons entering the collision is a superposition of states, some of these states will be fully absorbed, while others will pass

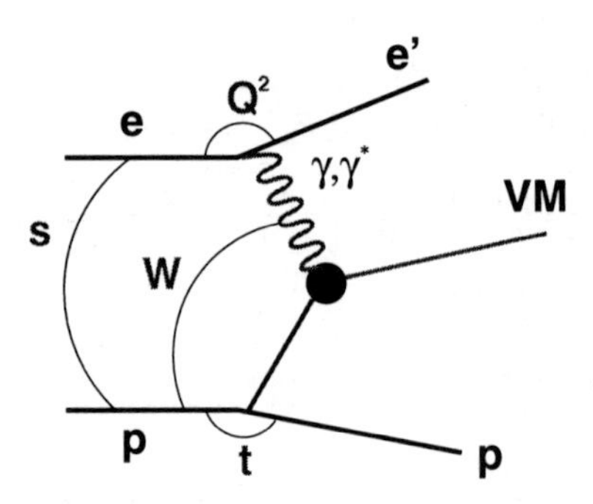

Figure 1.13: Schematic diagram of exclusive vector meson (VM) production in electron proton collisions. The various kinematical quantities are indicated.

through almost unaffected. This agrees with the general idea of color transparency in QCD (see Sect. 1.6.2). Such a mixture of states with very different absorption probabilities will be essential for inelastic diffraction to occur, see Sub-Sect. 1.5.3.

1.3.4.2 Real- and virtual-photon quasi-elastic scattering

The transition from soft to hard processes, seen in the total DIS cross section can, as advocated in [96], be cleanly studied in the diffractive production of vector mesons (VM). The (quasi-)elastic exclusive production of VMs, illustrated in Fig. 1.13 can at high energy be viewed as a two-step process. Consider a Lorentz frame where the proton is at rest. First, a virtual photon is radiated from the incoming charged lepton; second the photon undergoes quantum fluctuations (into vector mesons, or an initial $q\bar{q}$ dipole state, in general) which subsequently scatter elastically off the target. For low virtualities, the fluctuations are mainly into low-mass VMs and one could, therefore, have expected that the dynamics of elastic VM production is similar to that of elastic hadron-hadron scattering. The data, however, show a different and more interesting behavior.

Figure 1.14a shows the evolution of the elastic VM cross sections as a function of W in photoproduction ($Q^2 \approx 0$ GeV^2) [97–99]. The straight lines are power laws of the form $\propto W^\delta$. A clear change in the W dependence is observed. For the light vector mesons ρ, ω and ϕ the slope is $\delta \approx 0.22$, similar to that of the total photoproduction cross section, also shown in Fig. 1.14. For the heavier vector mesons, J/ψ, $\psi(2s)$ and Υ, the measured value of δ is much larger ($\delta \gtrsim 0.8$).

The hierarchy in W-dependence, as well as the absolute value of the cross sections, can easily be understood from the size of the VM bound states, as already hinted to in Sub-Sect. 1.3.1. Whereas the transverse size (or Compton wavelength of a valence quark) of the light VMs is about that of a proton, that of the heavy $q\bar{q}$ bound states is much smaller, even in the photoproduction regime, because of the large mass of

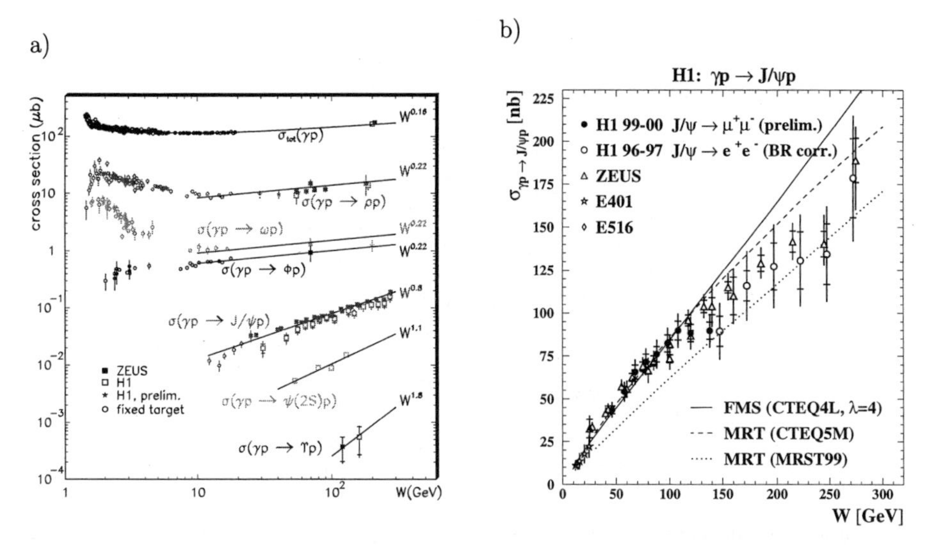

Figure 1.14: a) The W dependence of the elastic cross section of light and heavy vector mesons in photoproduction. The straight lines represent the function W^{δ}, with δ as indicated [99]. b) W dependence of the J/ψ photoproduction cross section [100–104] compared to pQCD models: Frankfurt-McDermott-Strikman (FMS) [105], Martin-Ryskin-Teubner (MRT) [106].

charm and bottom quarks. Due to color screening, the elastic (and total) cross section for such small color-dipoles scattering on the target will, therefore, be much smaller. At the same time, small-size dipoles can resolve the internal structure of the target. The W-dependence of the elastic and total cross section will thus reflect the W or $1/x$ dependence of the quark and gluon distributions in the proton. For elastic scattering, and in models viewing pomeron exchange as the exchange of two gluons, or more generally the exchange of a gluon ladder, in a color-singlet configuration, cross sections will generically depend quadratically on the gluon density, and thus rise steeply with increasing W.

For the heavy vector mesons, the large quark masses provide the hard scale which allows pQCD to be applied. Such models describe well the J/ψ data [100–104] shown in Fig. 1.14b. In particular, the steep rise with increasing energy W is well accounted for [65, 105, 106]. A comprehensive discussion of VM production in the framework of color-dipole models, with further references to the literature can be found in [107].

Whereas in photoproduction, at fixed W, the mass of the VM is the only scale that can be varied experimentally when studying total elastic cross sections, electroproduction, on the other hand, allows to change the photon virtually over a large range, thereby changing the transverse extension, R_{γ^*}, of the photon quantum fluctuations. Besides, both transversely and longitudinally polarized photons participate.

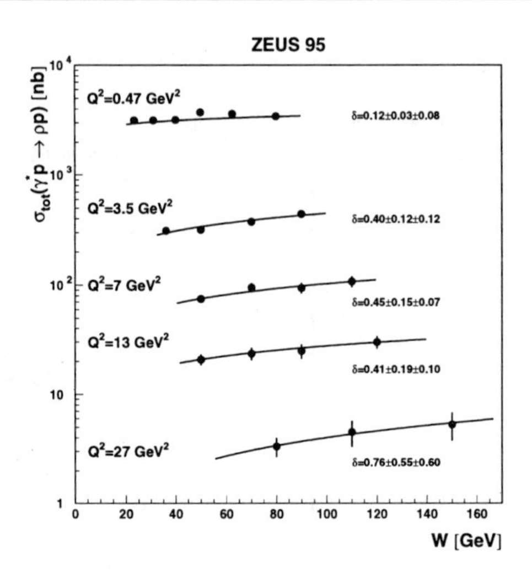

Figure 1.15: W dependence of the cross section $\sigma(\gamma^\star \mathrm{p} \to \rho^0 \mathrm{p})$ for various Q^2 values. The solid lines show a fit with $\sigma_{\gamma^\star \mathrm{p} \to \rho^0 \mathrm{p}} \propto W^\delta$ [111].

From the virtual photon light-cone $\mathrm{q\bar{q}}$ wave function, calculable in QED [108], it can be deduced that two typical $\mathrm{q\bar{q}}$ configurations are of interest: i) a configuration with a large transverse spatial extension and low quark transverse momentum relative to the $\gamma^\star$ momentum direction ("aligned jets" [108]), corresponding to a large-size color dipole, ii) a configuration with small spatial extension and large (anti-) quark transverse momentum, corresponding to a small-size dipole.

For transversely polarized photons, aligned-jet configurations are rare and suppressed as $\sim 1/Q^2$ [109], but they have a large interaction cross section, comparable to a typical hadronic cross section. On the other hand, due to color screening, the more frequent hard (but small-size) fluctuations have a small cross section ($\sim 1/Q^2$). For a longitudinally polarized $\gamma^\star$, the average transverse size of the color-dipole is much smaller than that of a transversely polarized $\gamma^\star$ [110]. It thus becomes clear that elastic electroproduction of VMs are beautiful examples of processes where the interplay between soft and hard dynamics can be studied in detail.

Among the many results now available [112,113] (for a review see eg. [97]), Fig. 1.15 shows DIS measurements of the W-dependence of elastic ρ^0 electro-production as a function of Q^2. For each Q^2 interval, the cross section is assumed to be of the form W^δ. In the same manner as for the $\gamma^\star$p total cross section, the data suggest a marked increase of δ when Q^2 enters a regime where pQCD becomes relevant. However, the errors remain sizable and, in W-regions where the DIS data overlap ($3.5 \leq Q^2 \leq 13.0$ GeV^2), the Q^2-dependence of δ is statistically not yet very significant.

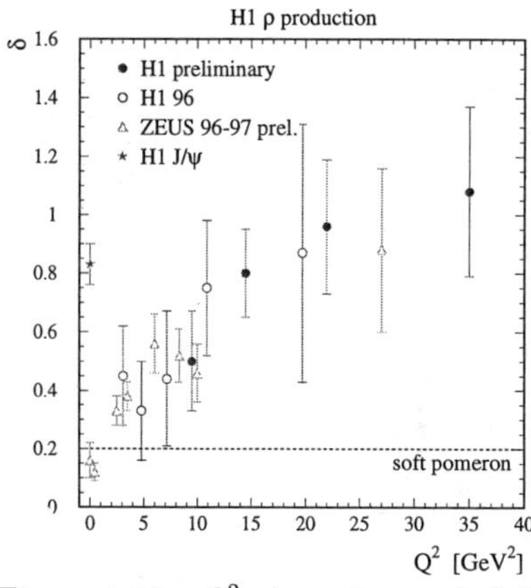

Figure 1.16: Q^2 dependence of δ in $\sigma_{\rm VM} \propto W^\delta$ for ρ^0 [98].

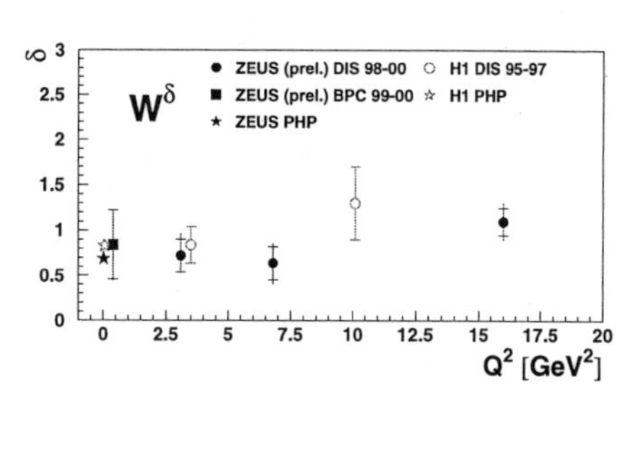

Figure 1.17: As in Fig. 1.16 for J/ψ [98].

The parameter δ is plotted as a function of Q^2 in Fig. 1.16. At low Q^2, δ is consistent with expectations for a soft process; at higher Q^2 the values of δ reach that measured for J/ψ in photoproduction [114], which is also shown in the figure.

In the case of exclusive electroproduction of J/ψ [115,116], Fig. 1.17, no significant Q^2 dependence is seen. The power δ is already large even at $Q^2 = 0$. Due to the large mass of the heavy quark, virtual photon fluctuations interact with the proton in a small-size configuration even at very low Q^2.

A further nice illustration of this point is provided by the Q^2 dependence of the exponential slope of the differential cross section, $d\sigma/dt$, for ρ^0 and J/ψ, Fig. 1.18. At small Q^2, the slope for ρ^0 has a value compatible with that measured in hadron-hadron elastic scattering (see Fig. 1.12a). However, a clear transition is seen in case of the ρ^0, whereas J/ψ production is a hard process already in photoproduction. At $Q^2 \geq 20$ GeV^2, the slopes attain the same value for both mesons. Thus, for ρ^0 the data indicate that the size of the interacting region is changing with Q^2 while there is no apparent change for J/ψ. At the same time, as Q^2 increases, the total elastic cross section itself grows faster with W than in hadron interactions.

Another way to see the soft to hard transition is illustrated in Fig. 1.19 showing a compilation [14] of the exponential slope, at fixed $W = 75$ GeV, as a function of an effective scale, $Q^2_{\rm eff} = Q^2 + m^2_{\rm VM}$, for various vector mesons with mass $m_{\rm VM}$. Plotted versus $Q^2_{\rm eff}$, all data follow more or less a common curve. This is, in fact, expected from theory. The average transverse size of γ^* fluctuations relevant for elastic vector meson production, the so-called scanning radius [110], is estimated to be $\sim C/(m^2_{\rm VM}+Q^2)$, with $C \sim 2$ ($C \sim 6$) for a longitudinally (transversely) polarized γ^* [117]. This follows almost directly from the form of virtual photon Fock state wave function. The elastic vector meson data suggest that the scale $Q_{\rm eff} = Q^2 + m^2_{\rm VM}$ is

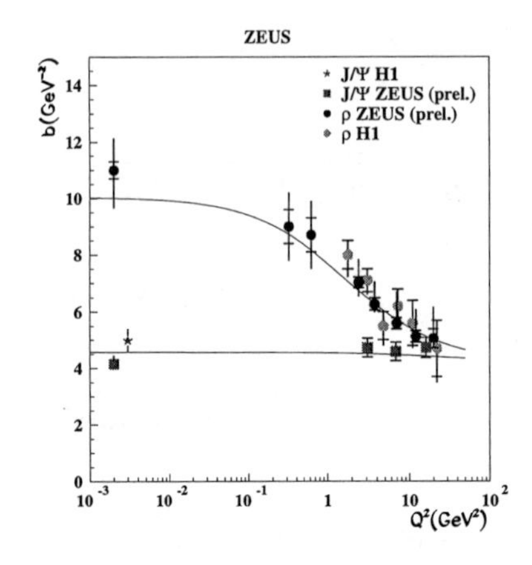

Figure 1.18: The exponential slope, B, of the differential cross section $\mathrm{d}\sigma/\mathrm{d}t$ for ρ^0 and J/ψ [98].

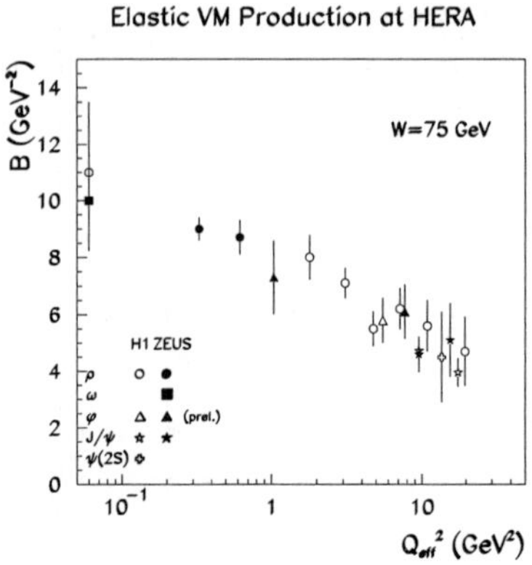

Figure 1.19: The exponential slope, B, as a function of Q^2_{eff}; $Q^2_{\text{eff}} = Q^2$ for ρ and ω, $Q^2_{\text{eff}} = Q^2 + M^2_\phi$ for ϕ, $Q^2_{\text{eff}} = Q^2 + M^2_{\mathrm{J}/\psi}$ for J/ψ [14].

indeed the dynamically relevant variable. For further discussion on this point we refer to [118].

In an optical analog to diffraction, b is related to the radii of the colliding objects, i.e. of the proton and the VM with $b \propto r^2_{\text{VM}} + R^2_{\text{p}}$. The value of $b \approx 4.5\ \text{GeV}^{-2}$, as measured at high Q^2, implies a combined effective interaction radius of the transverse size of the proton [119]. This would imply that the transverse size of the $\mathrm{q}\bar{\mathrm{q}}$ fluctuation producing a J/ψ in photoproduction or a ρ at high Q^2, is already much smaller than that of the proton and no longer contributes significantly to the effective transverse size of the interaction region.

1.3.4.3 Shrinkage

Whereas shrinkage of the elastic diffractive peak is well established in hadron-hadron collisions, the situation for elastic vector meson production in ep interactions is not yet so clear. In the Regge formalism, the differential cross section can be expressed as

$$\mathrm{d}\sigma/\mathrm{d}t \propto W^{4(\alpha_{I\!P}(t)-1)}, \tag{1.28}$$

with a trajectory $\alpha_{I\!P}$ linear in t.

The effective pomeron trajectory can be determined by fitting Eq. (1.28) to the differential cross sections at different t values. An example from ZEUS on electroproduction of J/ψ is shown in Fig. 1.20a-d [120].

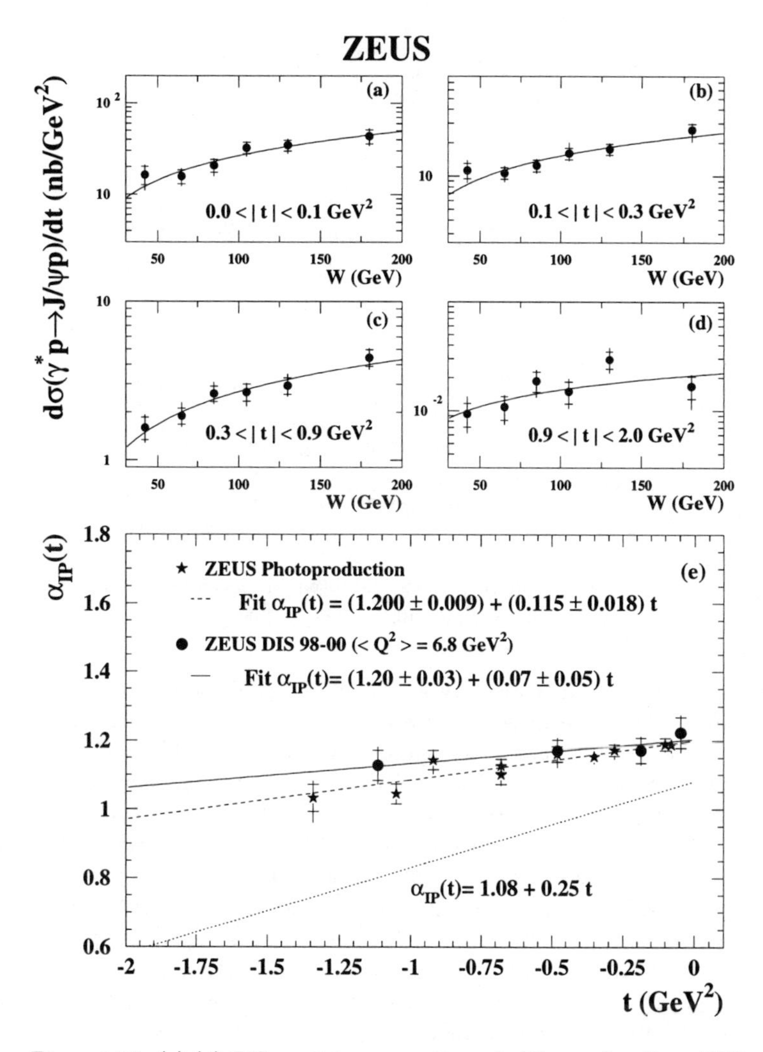

Figure 1.20: (a)-(d) Differential cross sections $d\sigma/dt$ as a function of W for fixed ranges of t [120]; full lines are fits by $W^{4(\alpha_{I\!P}(t)-1)}$. (e) Effective pomeron trajectory: lines are linear fits to the data at $\langle Q^2 \rangle = 6.8\ \text{GeV}^2$ (full) and to the J/ψ photoproduction results (dashed) [101]; the dotted line is the soft pomeron trajectory [22].

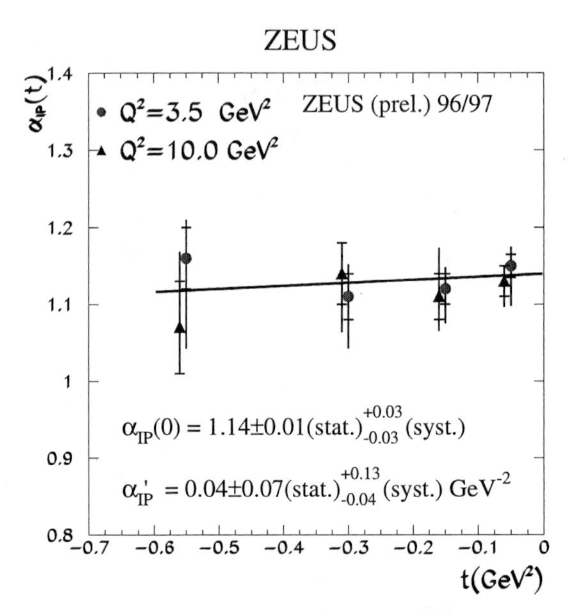

Figure 1.21: Effective pomeron trajectory measured in electroproduction of ρ^0 for two intervals of Q^2. The line is a linear fit to all the data points [124].

The parameters of the trajectory are compared to those measured in photoproduction [101]. They are in good agreement with each other. Similar results are obtained by the H1 collaboration for photoproduced J/ψ's [100]: $\alpha_{I\!P}(0) = 1.20 \pm 0.02$ and $\alpha'_{I\!P} = 0.15 \pm 0.05$ GeV^{-2}. The effective pomeron parameters are in agreement with expectations of pQCD-based models [121, 122], but are not consistent with the trajectory extracted from soft diffractive hadron-hadron processes [22]. Shrinkage is smaller than in soft hadronic collisions but not negligible. This was predicted in [123].

A recent ZEUS measurement [124] of the leading trajectory parameters from exclusive ρ production in DIS, $\gamma^\star\text{p} \to \rho^0\text{p}$ ($1 < Q^2 < 40$ GeV^2), Fig. 1.21, yielded $\alpha_{I\!P}(0) = 1.14 \pm 0.01(\text{stat})^{+0.03}_{-0.03}(\text{syst})$, $\alpha'_{I\!P} = 0.04 \pm 0.07(\text{stat})^{+0.13}_{-0.04}(\text{syst})$ GeV^{-2}. While not yet conclusive, given the errors, this measurement also suggests a value of $\alpha'_{I\!P}$ smaller than that of the "soft" pomeron ($\alpha'_{I\!P} \approx 0.25$ GeV^{-2} [22]).

We return to this point in Sub-Sect. 1.5.2 and find that shrinkage is quite generally expected to be smaller in processes which involve a large transverse momentum scale.

1.3.4.4 Longitudinal and transverse photons

From the decay angular distribution of the vector mesons, and with the additional assumption of s-channel helicity conservation (SCHC), the ratio $R = \sigma_\text{L}/\sigma_\text{T}$, of the

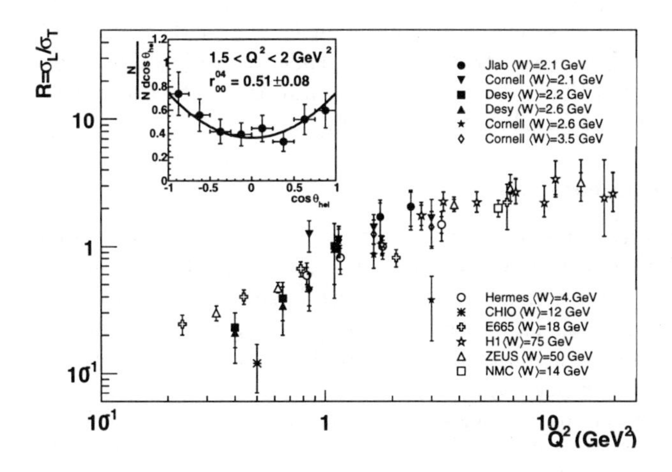

Figure 1.22: The ratio R= $\sigma_{\rm L}/\sigma_{\rm T}$ as a function of Q^2 for ρ^0 electroproduction. Data from [126–135]. The insert shows a $\cos\theta_{\rm hel}$ distribution with a fit to determine r_{00}^{04}. Figure from [135].

cross sections produced by longitudinal and transverse photons can be determined. Precise data on R now exist for the ρ^0.

The decay angular distribution of the π^+ in the ρ^0 helicity rest frame can be written as [125]

$$W(\cos\theta_{\rm hel}) = \frac{3}{4}\left[1 - r_{00}^{04} + (3r_{00}^{04} - 1)\cos^2\theta_{\rm hel}\right], \tag{1.29}$$

where r_{00}^{04} represents the degree of longitudinal polarization of the ρ meson. Assuming SCHC, the ratio of longitudinal to transverse cross sections is

$$R = \frac{\sigma_{\rm L}}{\sigma_{\rm T}} = \frac{1}{\epsilon}\frac{r_{00}^{04}}{1 - r_{00}^{04}}, \tag{1.30}$$

where ϵ is the virtual photon transverse polarization. r_{00}^{04} can be extracted from a fit of the $\cos\theta_{\rm hel}$ distributions by Eq. (1.29), as illustrated in the insert in Fig. 1.22, and can then be used in Eq. (1.30) to determine R.

Figure 1.22 shows a compilation of data on R for ρ^0, as a function of Q^2 [135]. Clearly, the cross section from the longitudinal photon dominates as Q^2 gets larger. This increase is well described by the Martin-Ryskin-Teubner (MRT) model [106] (not shown). The date are well fitted by the parametrization [135]

$$R = 0.75 \pm 0.08 \times (Q^2)^{1.09\pm0.14}, \tag{1.31}$$

motivated by the pQCD prediction that σ_T is power suppressed with respect to σ_L.

Surprising is that R seems to be quite independent of W over a relatively wide range in Q^2 [112]. This means that the W dependence of σ_T is the same as σ_L, leading to the conclusion that the large size configurations of the transversely polarized photon are suppressed for ρ^0 electroproduction.

1.3.4.5 Brief summary

Although more precise measurements are evidently needed, the present data on total and elastic differential cross sections suggest a clear trend. For near-on-shell photons fluctuating into light vector mesons (ρ, ϕ) with large (order 1 fm) transverse extensions, inversely proportional to the Compton wavelength of the light quarks in the meson, the effective pomeron trajectory $\alpha_{I\!P}(t)$ is close to that in soft collisions. For heavier vector mesons (e.g. J/ψ), which are characterized by a smaller transverse size, or in DIS, present data provide some indication for a weaker shrinkage, with $\alpha'_{I\!P}$ smaller than the "soft" value of 0.25 GeV^{-2}. At the same time, the effective intercept $\alpha_{I\!P}(0)$ grows with decreasing size R_{γ^*}. The transition from the soft hadron-like regime to DIS is a smooth one.

1.4 Inelastic diffraction

1.4.1 Experimental signatures

In contrast to forward elastic scattering, which beautifully reflects the wave nature of the particles, the phenomenon of diffraction dissociation, predicted by Good and Walker [12], has no classical analogue. For hadron-hadron scattering, it corresponds to quasi-elastic scattering between the two hadrons, where, in single diffractive dissociation (SD), one of them is excited into a higher mass state retaining its quantum numbers. This *coherent* excitation, illustrated in Fig. 1.23 for SD, requires not only small transverse (ΔP_T) but also small longitudinal (ΔP_L) momentum transfer. This leads to the *coherence condition* (see e.g. [5, 6]):

$$\xi \approx \frac{M_X^2}{s} < \frac{m_\pi}{m_p} \approx 0.1 - 0.2. \tag{1.32}$$

The coherence condition arises from the need to conserve the coherence of the quasi-elastically scattered target and implies that the diffractive mass M_X cannot be too large. For zero-angle production, the minimum four-momentum transfer at which the mass M_X can be produced is $|t_{\text{min}}| = [(M_X^2 - m_p^2)/2p]^2$, with p the incident momentum in the target rest frame. In the transition, the wave number k of the incident hadron varies by an amount $\Delta k \propto \sqrt{|t_{\text{min}}|}$. The condition of coherence follows from the requirement that the wave-number changes little during the passage through the target, so that the waves describing the target before and after the

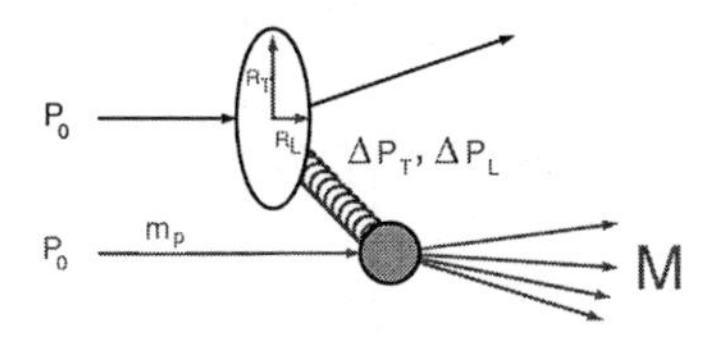

Figure 1.23: Single diffraction dissociation. The invariant mass of the produced hadrons, M, is denoted by M_X in the main text [136].

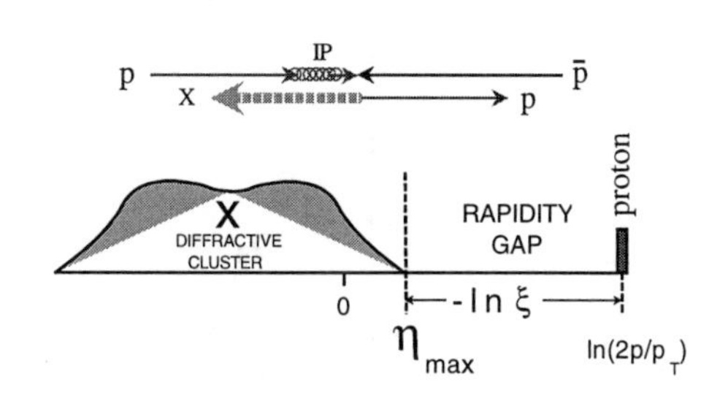

Figure 1.24: Topology for $p\bar{p} \to pX$ [136].

interaction can stay in phase. For DIS kinematics, the minimum value of t required to produce a given M_X from a target with mass m_t is $|t_{\min}| \simeq m_t^2(M_X^2 + Q^2)^2/W^4$. For a typical hadronic radius of 1 fm, $M_X^2 < 0.2\, W^2$.

The generic topology of a single-diffractive (here $\bar{p}p$) event is illustrated in Fig. 1.24. The upper-limit on M_X, Eq. (1.32), implies that the diffractive hadronic final states exhibit a large rapidity gap between the quasi-elastically scattered proton and the dissociation products X of the $\bar{p}$. The width of the gap in (pseudo-)rapidity space measured from the rapidity of the initial-state proton is $\Delta\eta \approx \ln\frac{1}{\xi}$. In collider experiments, diffractive events are thus identified either by detecting directly a "fast" ("leading") proton in a spectrometer, by the presence in the main detector of a large rapidity region devoid of hadrons (a rapidity gap), or by exploiting the characteristic $1/M_X^2$ dependence of diffraction.

Naively, the interaction is often viewed as proceeding via the emission from the proton of a pomeron which subsequently interacts with the $\bar{p}$. In QCD such an object, if it were to exist as a physical entity, must be a color-singlet composed of quarks, antiquarks and gluons. However, such a picture is an unnecessary and often misleading simplification of the dynamics of diffractive dissociation.

1.4.2 Hadron-hadron inelastic diffraction

Evidence for an important diffractive component in the inclusive reaction $p + p \to p + X$, with excitation of large masses, was first established at the ISR by the CHLM collaboration [138]. Figure 1.25 shows single-diffractive pp cross sections from low to high s. The diffractive enhancement becomes less and less prominent as s decreases, in line with the previous discussion about the need to maintain coherence of the target, Eq. (1.32). The M_X-spectrum drops rapidly in the resonance region. Beyond that it levels off and shows an approximate $1/M_X^2$ dependence.

A compilation of measurements [27], now plotted against M_X^2, is shown in Fig. 1.26 for pp and $\bar{p}p$ single diffractive cross sections at $t = -0.05$ GeV^2 (for earlier com-

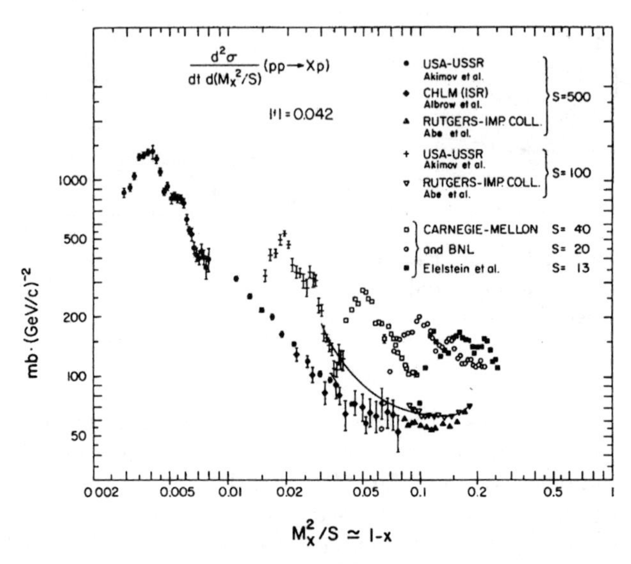

Figure 1.25: Single diffractive pp cross sections [5]. The figure shows how the characteristic $1/M_X^2$ (Regge) behavior of diffraction becomes manifest as s increases [137].

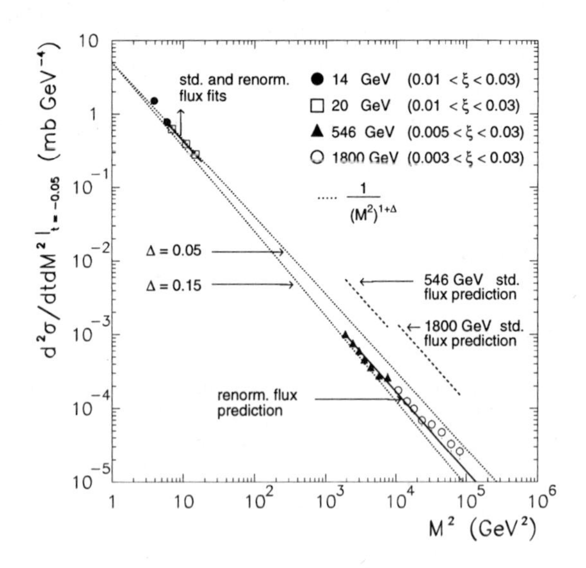

Figure 1.26: Cross sections, $d^2\sigma^{SD}/dM_X^2dt$, for $p + p(\bar{p}) \to p(\bar{p}) + X$ at $t = -0.05$ GeV^2 and $\sqrt{s} = 14$, 20, 546 and 1800 GeV. For a description of the curves see [27].

pilations see [139]). The distribution falls as $1/(M_X^2)^{1+\Delta}$ over the entire M_X region. Quite remarkably, it is independent of s over five orders of magnitude. The data are consistent with the value of $\alpha_{I\!P}(0)-1=\epsilon=0.104$ (denoted Δ in the figure) extracted from the fit in [40] to total and elastic cross sections.

The $1/M_X^2$ scaling shown by the data in Fig. 1.26 implies that, since ϵ is small, the single-diffractive rapidity-gap distribution, $\mathrm{d}^2\sigma^{\mathrm{SD}}/\mathrm{d}t\,\mathrm{d}\Delta\eta$ is nearly independent of s. In Regge models of diffraction, this distribution is related to the "pomeron flux", see Eq. (1.18). Such a weak energy dependence must reflect a fundamental, but not yet fully understood, property of the baryon "re-formation" process in the final state. A strict energy-independence would be consistent with *short-range order* [140], which is a basic property of soft multiparticle production.

The failure of factorization in Regge theory, alluded to in Sub-Sect. 1.2.2 is dramatically illustrated in Fig. 1.27, where the energy dependence of the measured SD cross section is compared with Regge predictions using the standard pomeron flux and a pomeron flux renormalized to unity when integrated over ξ and t [136, 141].

The normalization factor is proportional to $s^{2\epsilon}$ and thus cancels the $s^{2\epsilon}$ dependence of the SD cross section in Regge theory (Sub-Sect. 1.2.2), explaining at the same time the scaling seen in Fig. 1.26. The renormalized-flux prediction is in excellent agreement with the data. Nevertheless, in Sub-Sect. 1.6 we shall see that the data can also be understood in a natural way in the model of Miettinen and Pumplin [142]. This model is based on the Good and Walker interpretation of diffraction [12] and incorporates unitarity from the outset. It neither uses nor needs Regge ideas.

1.4.3 Inclusive diffraction at HERA

In DIS at small x measured at HERA, inelastic diffraction occurs at a rate of $\mathcal{O}(10\%)$ of all events [143]. Although this surprised many in the pQCD community, it had been anticipated even before the advent of QCD [108]. It was also predicted from Regge theory [144]. The occurrence of such diffractive events are indeed difficult to understand in the parton picture on the basis of pQCD alone if one starts from the naive premise that deep inelastic diffraction is necessarily a hard process.

The experimental effort at HERA has concentrated on measurements of the diffractive part, $F_2^{\mathrm{D}(3)}$, of the structure function F_2, Eq. (1.7). The data have been reviewed on many occasions and details can be found in [1, 2, 118, 145]. Here we concentrate on the most salient features of the data.

1.4.3.1 Cross section measurements

Traditionally, the starting point of most of the analyses is based on Eqs. (1.6)-(1.7). The inclusive cross section is written in terms of a "diffractive" structure function F_2^{D} which is a function of four variables, $x_{I\!P}$ (or ξ), t, x and Q^2.

In DIS at sufficiently high Q^2, QCD factorization holds also for DDIS [11]. This

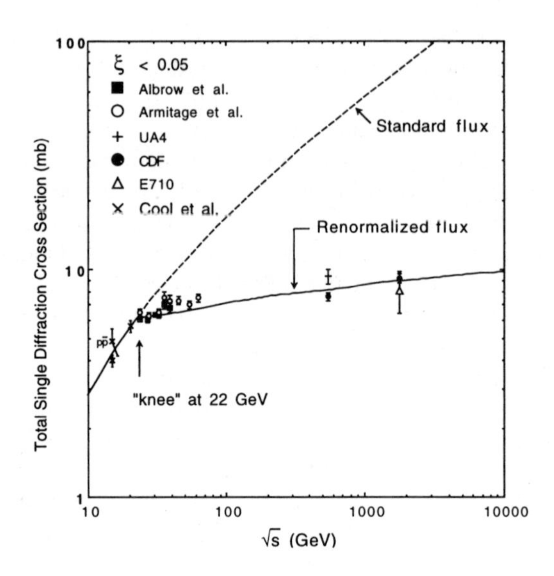

Figure 1.27: Total pp/p̄p single diffraction dissociation cross section data (both p̄ and p sides) for $\xi < 0.05$ compared with predictions based on the standard and a renormalized pomeron [136, 141].

implies that F_2^{D} may be decomposed into "diffractive" parton distributions, which obey the same DGLAP evolution equations that apply also to the inclusive DIS case. Based on experimental evidence, one often assumes Regge factorization [9] whereby F_2^{D} is decomposed into a pomeron flux factor, Eq. (1.18), and the structure function of the pomeron. Since t is often not measured, one usually integrates over the t variable, and this decomposition is written as

$$\frac{\mathrm{d}F_2^{\mathrm{D(3)}}(x, Q^2, x_{I\!P})}{\mathrm{d}x_{I\!P}} = f_{I\!P}(x_{I\!P})\; F_2^{I\!P}(\beta, Q^2), \tag{1.33}$$

where the flux-factor, $f_{I\!P}(x_{I\!P})$, is now taken to be independent of β and Q^2. In Eq. (1.33), it is assumed that only the pomeron contributes. In practice, however, it is often necessary to add a non-leading reggeon contribution in order to fit the data if ξ is larger than $0.1 - 0.2$.

An example of recent measurements by the ZEUS collaboration using their Leading Proton Spectrometer [146] is illustrated in Fig. 1.28, where the $x_{I\!P}$ dependence of $x_{I\!P} F_2^{\mathrm{D(3)}}$ at fixed β and Q^2 is displayed. The curves are the best fit to the data (restricted to $x_{I\!P} < 0.01$) using a flux-factor, as described above, yielding $\alpha_{I\!P}(0) = 1.16 \pm 0.01\,(\mathrm{stat})^{+.04}_{-.01}(\mathrm{syst})$. This value, together with a compilation of other measurements [147], is shown in Fig. 1.29. Also shown in the figure is the Q^2

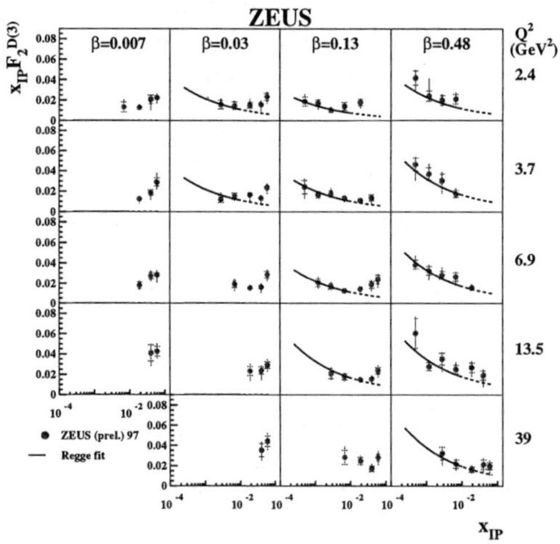

Figure 1.28: ZEUS: $x_{I\!P}$ dependence of $x_{I\!P}F_2^{D(3)}$ at fixed β and Q^2 [98].

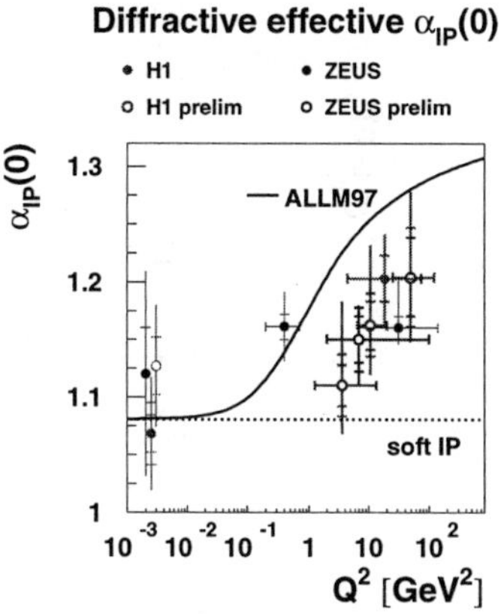

Figure 1.29: Q^2 dependence of $\alpha_{I\!P}(0)$ from DDIS and inclusive DIS. For the curve, see text [98].

dependence of $\alpha_{I\!P}(0)$ derived from the inclusive DIS measurements (cfr. Fig. 1.8) and here represented by the ALLM97 parametrization [148].

Clearly, the DIS data are not compatible with a universal pomeron trajectory. The diffractive measurements indicate some Q^2 dependence, though the errors are still too large for a firm conclusion. Nevertheless, for $Q^2 > 10$ GeV^2, $\alpha_{I\!P}(0)$ from DDIS is significantly higher than that expected from the soft pomeron.

The values of $\alpha_{I\!P}(0)$, extracted from DIS and DDIS using the Regge formalism, are compatible at low Q^2. This means that the W dependence of the diffractive cross section is steeper than that of the total cross section, in agreement with Regge theory expectations. However, there are good indications that this is no longer so at larger Q^2.

For $Q^2 \gtrsim 10$ GeV^2, both the DIS and the DDIS cross sections grow with a similar rate as W increases (see below), at variance with the Regge theory relations $\sigma_{\mathrm{tot}}^{\gamma^\star\mathrm{p}} \sim (W^2)^{\alpha_{I\!P}(0)-1}$ and $(\mathrm{d}\sigma^{\mathrm{D}}_{\gamma^\star\mathrm{p}}/\mathrm{d}M^2\mathrm{d}t) \sim (W^2)^{2(\alpha_{I\!P}(0)-1)}$. Thus, fitting the data by these expressions necessarily leads to a smaller value of $\alpha_{I\!P}(0)$ in DDIS, as seen in Fig. 1.29.

Figure 1.30 shows the ratio of the diffractive cross section and the total $\gamma^\star$p cross section as function of Q^2, for fixed values of β, measured by H1 [29]. It is remarkably flat over a wide kinematic range, indicating that the same pQCD scaling violations occur in both processes.

The ratio is flat also as function of W for fixed M_X values [149], contrary to

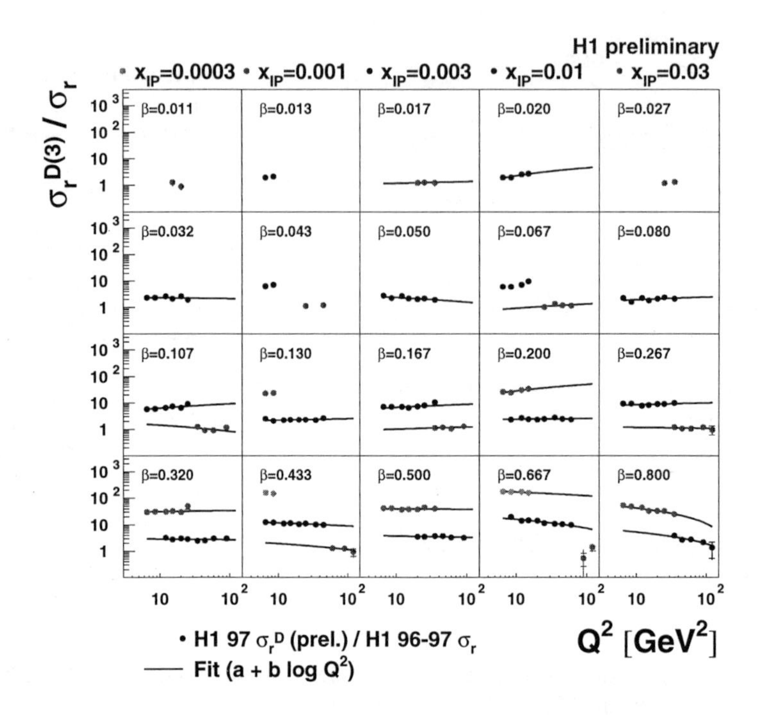

Figure 1.30: H1 preliminary data [29] on the ratio of the diffractive to total cross section as function of Q^2, for fixed values of β.

expectations from Regge phenomenology. This interesting result, first obtained by ZEUS [28], is illustrated in Fig. 1.31, which shows more recent H1 measurements [29]. For Q^2 and M_X (and thus β) fixed, the relative rate of diffractive events is nearly W-independent, except at very small β. Standard triple-Regge theory, without multi-pomeron exchange, predicting an increase as $(W^2)^{2(\alpha_{I\!P}(0)-1)}$, clearly disagrees with data.

As discussed e.g. in [150], pQCD alone cannot explain the constancy of the ratio. The authors conclude that the non-perturbative QCD contribution to diffractive production is essential. On the other hand, a constant ratio is obtained quite naturally in the semi-classical gluon field approach (see Buchmüller in [2, 151] and refs. therein) and in the Soft Color Interaction models (SCI) [152, 153]. It is also correctly predicted in the Golec-Biernat and Wüsthoff model [65]. There, it is a consequence of the crucial assumption that the dipole cross section becomes constant once this system has acquired a large transverse extension. The interaction with the target is then non-perturbative.

The basic physical idea behind the semi-classical and SCI models is that the

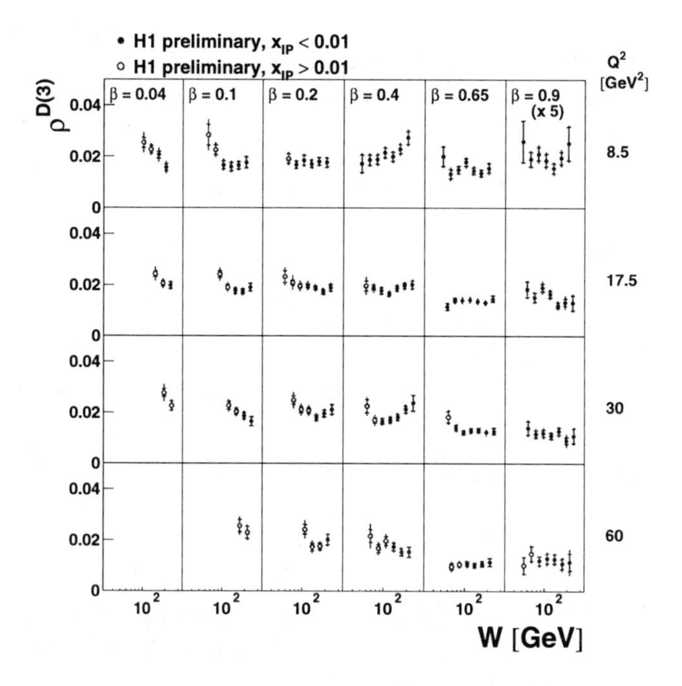

Figure 1.31: H1: measurements of $\rho^{\mathrm{D}(3)}$, the ratio of the diffractive and the inclusive cross section versus W. Data at $\beta = 0.9$ are scaled by a factor of five [29].

hard point-like interaction is the same in DIS and in DDIS ("the $\gamma^\star$ sees the same partons"), but that the difference resides in the difference in time-scales for, on the one hand, the "fast" hard scatter and the long time needed to create a rapidity gap, on the other hand. The latter is a soft process and insensitive to the virtuality of the probe. The factorization theorem for DDIS, proven in [11], makes this idea explicit.

Figure 1.32 shows recent measurements obtained by the H1 Collaboration [29]. The preliminary results are presented as a reduced cross section, $\sigma_r^{\mathrm{D}(3)}$, defined through

$$\frac{\mathrm{d}^3\sigma^{\mathrm{D}}}{\mathrm{d}x_{I\!P}\mathrm{d}x\mathrm{d}Q^2} = \frac{4\pi\alpha^2}{xQ^4}\left(1 - y + \frac{y^2}{2}\right) \times \sigma_{\mathrm{r}}^{\mathrm{D}(3)}(x_{I\!P}, x, Q^2), \tag{1.34}$$

and divided by the "pomeron flux", $f_{I\!P}(x_{I\!P})$. This reduced cross section is equal to the conventional diffractive structure function, $F_2^{\mathrm{D}(3)}$, up to corrections due the longitudinal structure function, $F_{\mathrm{L}}^{\mathrm{D}(3)}$. The Regge-inspired factorized parametrization works well, as can be judged from the overlap of data points at the same β and Q^2 values obtained for different proton momentum losses $x_{I\!P}$. The intercept of the

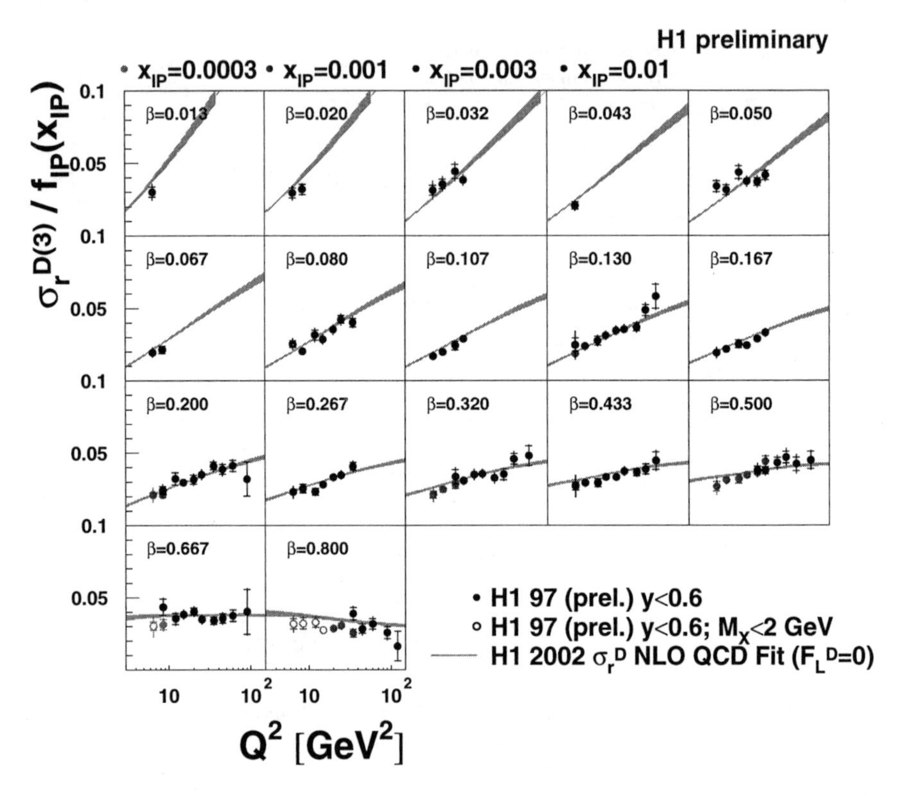

Figure 1.32: H1 data: the diffractive reduced cross section divided by the pomeron flux as a function of Q^2 in intervals of β and for different $x_{I\!P}$ values. The band represents the result of a Next-to-Leading (NLO) QCD fit, assuming $F_L^D = 0$; see Sub-Sect. 1.4.3.

pomeron trajectory extracted from this data is

$$\alpha_{I\!P}(0) = 1.173 \pm 0.018\,(\text{stat}) \pm 0.017\,(\text{syst})^{+0.063}_{-0.035}\,(\text{model}). \tag{1.35}$$

Figure 1.32 shows that large scaling violations are present in most of the phase space up to $\beta \sim 0.6$. Although this is usually thought to reflect the large gluon content of the diffractive exchange, it should be remembered that the regions in β and $x_{I\!P}$ where positive scaling violations are seen correspond to very low Bjorken-x values ($x = \beta x_{I\!P}$) where scaling violations of the *inclusive* cross section are very strong.

1.4.3.2 Next-to-leading order DGLAP fits

Pursuing the (debatable) concept of "the pomeron in the proton", parton distributions of the pomeron have been extracted from a pQCD fit to the measured H1 cross

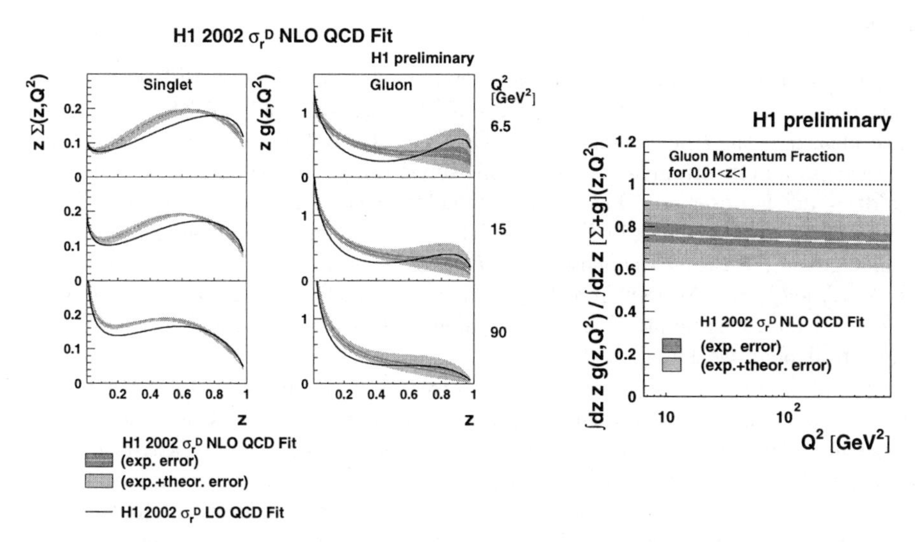

Figure 1.33: H1: Parton density distributions in the pomeron, using a NLO QCD fit (shaded band), and a LO QCD fit (solid line). Quark singlet ($6 * u$ with $u = d = s = \overline{u} = \overline{d} = \overline{s}$) and gluon densities are shown [29].

Figure 1.34: H1: The gluon momentum fraction from a NLO QCD fit [29].

sections by applying Next-to-Leading Order (NLO) DGLAP evolution equations and assuming $F_{\mathrm{L}}^{\mathrm{D}} = 0$ [29]. Using a parametrization based on Chebychev polynomials at a starting scale of $Q_0^2 = 3\ \mathrm{GeV}^2$, quark and gluon densities have been fitted. Sub-leading reggeon exchanges are included assuming them to have a structure function identical to that of the pion. No energy-momentum sum-rule is imposed. The fit, which includes the data shown in Fig. 1.32 together with data at higher Q^2 ($200 < Q^2 < 800\ \mathrm{GeV}^2$, not shown), describes the data well.

The resulting pomeron parton density distributions, extracted with full propagation of statistical, systematic and theoretical errors, are shown in Fig. 1.33. They do not differ much from a Leading-Order (LO) QCD fit. The gluon component is dominant and sizable at large β (denoted by z in the figure). Using these results, QCD factorization [11] in DDIS at HERA energies has been successfully tested in measurements of the β and $x_{I\!P}$ distribution of diffractive jet production [154] and diffractive D* production [155].

As seen from Fig. 1.34, the momentum fraction carried by gluons is estimated to be $75 \pm 15\%$ at $Q^2 = 10\ \mathrm{GeV}^2$, and is almost Q^2 independent. The gluon fraction is similar to, or somewhat larger than that determined from hard diffraction at the Tevatron (see Fig. 1.37 in Sub-Sect. 1.4.4).

1.4.3.3 The forward diffractive peak

A typical feature of diffractive events is the exponential fall-off of the differential cross section with increasing $|t|$ at small values of $|t|$. Both H1 [156] and ZEUS [157,158] have measured the t dependence of the DIS inclusive diffractive cross section using forward proton spectrometers. In the measured range, an exponential dependence $\mathrm{d}\sigma/\mathrm{d}t \sim e^{Bt}$ is observed. The slope parameter B can be determined for different values of $x_{I\!P}$, as illustrated in Fig. 1.35 (left panel). Regge phenomenology predicts shrinkage, an increasing steepness of the t dependence with energy, of the form $B = B_0 + 2\alpha'_{I\!P} \log 1/x_{I\!P}$. As seen from Fig. 1.35 (right panel), the data presently remain inconclusive on this point. The Q^2 dependence of B has been studied in [158] and found to depend weakly, if at all, on Q^2.

It is of interest to note that the slopes measured in inclusive DDIS at HERA are strikingly similar to those measured in hadron single diffraction dissociation. This is consistent with the arguments presented in Sub-Sect. 1.4.3. In contrast, the slopes measured in quasi-elastic electro- and photoproduction of light vector mesons are roughly two times larger at small Q^2, and about the same as in elastic hadron-hadron collisions. However, as seen in Sub-Sect. 1.3.4, they are a decreasing function of Q^2.

1.4.4 Hard diffraction at the Tevatron

Hard diffraction processes are defined as those in which a hard parton-parton scattering occurs together with a diffractive signature, such as a large rapidity gap or

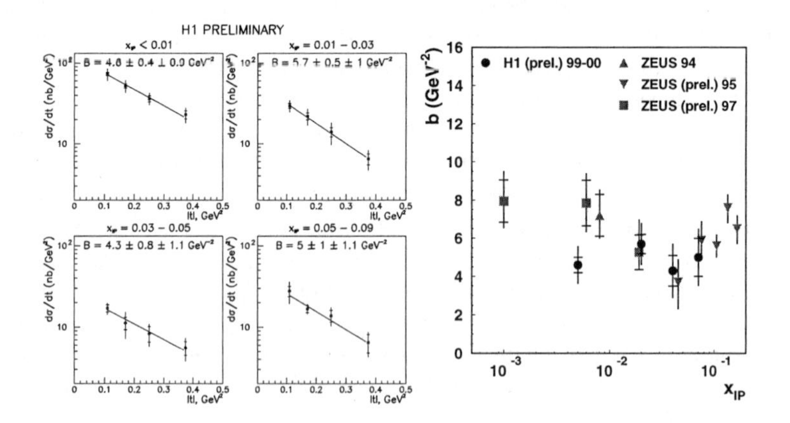

Figure 1.35: Left: The differential cross section $\mathrm{d}\sigma/\mathrm{d}t$ measured with the H1 Forward Proton Spectrometer (FPS) in four different $x_{I\!P}$ bins. The exponential fit by e^{Bt} is also shown. Right: the exponential slope parameter plotted as a function of $x_{I\!P}$. Data from the H1 FPS and the ZEUS Leading Proton Spectrometer (LPS) [156].

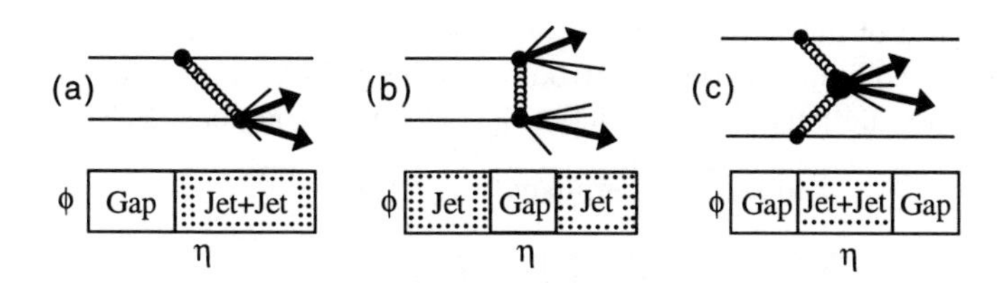

Figure 1.36: Dijet production diagrams and event topologies in pseudorapidity and φ space, for p̄p (a) single diffraction (b) double diffraction and (c) "double pomeron" exchange [176].

a quasi-elastically scattered leading hadron. Events may have forward, central, or multiple rapidity gaps, as shown in Fig. 1.36 for dijet production in p̄p collisions at the Tevatron. In deep-inelastic ep scattering, depicted in Fig. 1.1b, hard diffraction corresponds either to collisions involving a large scale: a highly virtual vector boson ($\gamma^\star$ or weak bosons) or production of large-p_{T} jets.

Hard diffractive processes were first observed at the CERN pp̄ collider by the UA8 experiment [159]: while a diffractively scattered proton was detected in a proton spectrometer, large-p_{T} jets were reconstructed in the central detector. This observation supported the earlier formulated hypothesis of a partonic interpretation of diffraction by Ingelman and Schlein [9]. The same phenomenon occurs in hard hadron collisions producing weak vector bosons or heavy-quark pairs. With the subsequent discovery [143, 160] of events with a large rapidity gap in deep inelastic lepton scattering, a possibility was offered to study hard diffraction using the analysis tools of pQCD.

At the Tevatron collider, hard diffraction is being extensively studied by the D0 [161–163] and CDF [164–174] collaborations, and complements the studies at HERA. A recent overview can be found in [175, 176].

1.4.4.1 Single diffraction, double diffraction and double pomeron exchange

Hard single diffraction processes are studied at the Tevatron through the production of high-p_{T} jets [162, 166] (Fig. 1.36a), and of W bosons [163, 165], J/ψ mesons [170] and b particles [177]. These events are identified either through the detection of the diffractively scattered p̄ in a proton spectrometer or by the presence of a gap in pseudorapidity in the calorimeter and the tracking detector. The production rates, Table 1.2, are at the 1% level compared to the corresponding non-diffractive processes [176].

In the Ingelman-Schlein picture, the pomeron should be universal if it were indeed an intrinsic part of the target proton wave function. However, the fraction of "hard" events produced diffractively is observed to be considerably smaller in hadron-induced events compared to the DDIS/DIS ratio of about 10%. Thus, based on data, one

Table 1.2: Ratio of diffractive to inclusive cross sections for various hard scattering processes in $p\bar{p}$ collisions at the Tevatron, $\sqrt{s} = 1800$ GeV; data from CDF and D0. Last column: predictions of the model of Soft Color Interactions (SCI) [153].

Process	Experiment		Ratio [%]	
			Observed	SCI
W	CDF	[165]	1.15 ± 0.55	1.2
Z	D0	[163]	$1.44^{+0.62}_{-0.54}$	1.0
$b\bar{b}$	CDF	[177]	0.62 ± 0.25	0.7
J/ψ	CDF	[170]	1.45 ± 0.25	1.4
dijets	CDF	[166]	0.75 ± 0.10	0.7
dijets	D0	[162]	0.65 ± 0.04	0.7

must question the Regge-inspired approach of expressing the diffractive cross section as a product of a pomeron flux from the proton times a universal distribution of partons in the pomeron.

Hard double diffractive dissociation (see Fig. 1.36b) is studied through the production of two jets separated by a gap in rapidity attributed to color singlet exchange [161, 162, 174]. The rate of such events has been measured at $\sqrt{s} = 630$ and for $\sqrt{s} = 1800$ GeV. The ratio $R_{630/1800}$ is found to be 2.4 ± 0.7(stat)± 0.6(syst) by CDF and 3.4 ± 1.2 by D0. This decrease with increasing energy is at variance with expectations based on simple BFKL evolution [178], but is predicted by models of soft color recombination [179], the Soft Color Interaction model [153] and others [180–182].

Hard dijet production has also been observed in the central detectors for events containing a scattered $\bar{p}$ identified in the proton spectrometer and a rapidity gap on the other side of the detector (CDF) [166, 171], or a rapidity gap on each side of the detector (D0) [162, 168, 173] (Fig. 1.36c). These events are found to be produced at the 10^{-4} level of the corresponding non-diffractive process, which is consistent with a picture of double pomeron exchange, each pomeron exchange corresponding to a probability at the 1% level.

1.4.4.2 Factorization breaking

Following a procedure similar to the one used by ZEUS [183], the CDF collaboration has determined the partonic content of the pomeron exploiting the different sensitivities of the various processes (dijet, W and b production) to quarks and gluons (see [136] and refs. therein). The production rates were compared to predictions of the model POMPYT [184], which is based on the assumption of a factorizable pomeron flux; a hard parton content of the pomeron was assumed.

A gluon fraction of 0.55 ± 0.15 is found, which is consistent with measurements at HERA (see Fig. 1.37), but the measured rates at the Tevatron are significantly

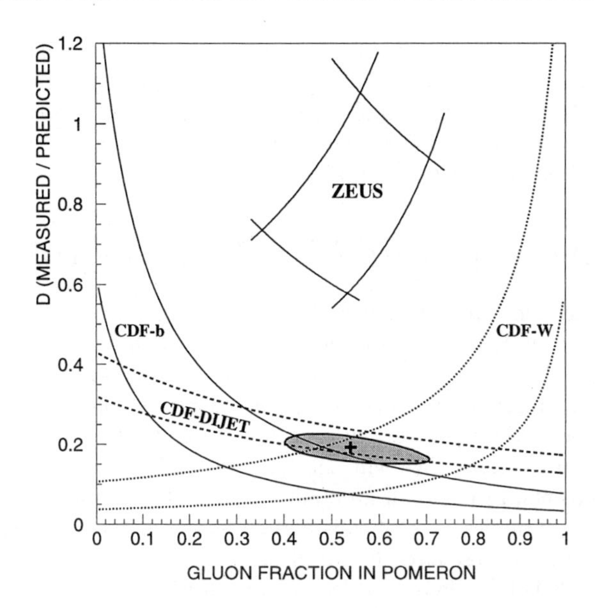

Figure 1.37: Ratio of measured to predicted diffractive rates as a function of the gluon content of the pomeron, for CDF dijet, W and b production, and for a measurement by ZEUS of DDIS and diffractive jet photoproduction. The predictions are from the POMPYT model [184] with a hard pomeron structure. The shaded area is the 1σ contour of a fit to the three CDF results [136, 177].

lower than expected, the reduction factor being 0.18 ± 0.04.

Similarly, predictions for the diffractive production rate of dijets and W bosons [185, 186] and for charm production and double pomeron exchange [187] based on pomeron structure functions extracted from inclusive DDIS, show that QCD factorization, which is valid [11] and verified in DDIS, is strongly broken in hadron-hadron collisions.

The factorization breaking is commonly quantified in terms of a "gap survival probability" [188]. In hadron-hadron scattering, additional interactions between the diffractively scattered particle and remnants of the other beam particle can destroy the rapidity gap, whereas this effect is much weaker in high-Q^2 DDIS. As a result, diffractive rates at the Tevatron are reduced relative to predictions based on HERA DDIS data.

To estimate gap survival probabilities, a variety of approaches has been used based on pomeron flux renormalization [141, 176], multi-pomeron exchange models for screening [180–182, 188], or color neutralization by soft color exchanges [10, 152, 153].

Important progress was made in [189] where it was shown that the breakdown of factorization between production of hard jets observed at HERA and at the Tevatron, can be naturally explained in the Good-Walker picture of diffraction dissociation (see Section 1.5.3). There, the gap survival probability is explicitly obtained in terms of the small-$|t|$ $\bar{p}p$ elastic scattering amplitude. Furthermore, it is shown that no further suppression is expected when going from events with one to events with two rapidity gaps, as is indeed observed experimentally [176].

1.5 A generic picture of high energy collisions

In the foregoing sections, we have reviewed, without aiming for completeness, the vast amount of experimental results on diffractive phenomena accumulated over the past decades. We have devoted special attention to those features of the data which are, in our opinion, the most salient ones. In the following sections we shall attempt to sketch in broad terms the framework which allows to understand much of the physics as it is revealed in the data. We will aim for a general understanding of the basic dynamical ingredients, rather than a detailed technical discussion. The latter is, of course, essential in practical work but is already covered in many excellent reviews and textbooks, such as those cited in Sub-Sect. 1.1.

Perturbative QCD has allowed to clarify many issues, and to produce crisp quantitative predictions in some cases, but the earlier ideas, reaching well beyond pQCD, continue to be of great value. Well before the advent of QCD, and inspired by the ideas of Ioffe, Feinberg, Gribov, Pomeranchuck and others, a basic, although semi-quantitative, understanding of the space-time evolution of a high-energy scattering process was developed [15,17,108]. It testifies to the profoundness of these ideas that, in spite of major developments in the physics of the strong interactions, the physical picture then developed still remains valid to a very large extent. It is therefore useful to review the basic physics ingredients in some more detail. To the experimentalist, they provide a view of the collision dynamics which is simple enough to help develop intuition and physical insight, uncluttered by technical complications.

1.5.1 Unitarity

The importance of unitarity is not always sufficiently appreciated within the community of experimentalists. We devote this section to a description of its main aspects and implications.

The unitarity of the scattering matrix, T, implies close relationships between total cross sections, the elastic scattering amplitude and the amplitudes of inelastic final states. The unitarity relation between states $|i\rangle$ and $|f\rangle$ reads

$$2\,\mathrm{Im}\langle \mathrm{f}|\mathrm{T}|\mathrm{i}\rangle = \sum_{|\mathrm{e}\rangle\langle \mathrm{e}|} \langle \mathrm{f}|\mathrm{T}^{+}|\mathrm{e}\rangle\langle \mathrm{e}|\mathrm{T}|\mathrm{i}\rangle + \sum_{|\mathrm{n}\rangle\langle \mathrm{n}|} \langle \mathrm{f}|\mathrm{T}^{+}|\mathrm{n}\rangle\langle \mathrm{n}|\mathrm{T}|\mathrm{i}\rangle, \tag{1.36}$$

where $\sum_{|e\rangle\langle e|}$ stands for summation and integration over all possible *elastic* intermediate states $|e\rangle$. The second term is the contribution from all possible inelastic states;

$|i\rangle$ is the initial and $|f\rangle$ an *arbitrary* final state.

For forward elastic scattering, $t = 0$ and $|i\rangle \equiv |f\rangle$, Eq. (1.36) immediately leads to the optical theorem. However, the relation has much wider consequences since the state $|f\rangle$ can be any state. It shows that the imaginary part of the amplitude of any particular final state $\langle f|T|i\rangle$ in general receives contributions from all other final states. Such "unitarization effects" will be small only if the "overlap", $\langle f|T|n\rangle\langle n|T|i\rangle$, of the states $|i\rangle, |f\rangle$ with the states $|n\rangle$ happens to be small. This will, therefore, depend crucially on the topology in momentum space of the inelastic states and on the phases of the amplitudes.

The two terms on the right-hand side of Eq. (1.36) are called the elastic and inelastic overlap functions, respectively, and were first introduced by Van Hove [190]. For elastic scattering (and neglecting the real part), we see that the amplitude can, in principle, be calculated from the knowledge of the inelastic final states. This is the so-called s-channel approach to diffractive scattering. It provides an alternative to the t-channel approach in which the diffractive amplitudes are analyzed in terms of their singularities, poles and cuts, in the complex angular momentum plane.

An important result is obtained (valid only at high s) when Eq. (1.36) is written in impact-parameter ($\mathbf{b}$) space[2]. Using angular momentum conservation, one finds

$$2\,\mathrm{Im}\,\mathcal{A}_{\mathrm{el}}(s,b) = |\mathcal{A}_{\mathrm{el}}(s,b)|^2 + \mathcal{G}_{\mathrm{in}}(s,b). \tag{1.37}$$

Here $\mathcal{A}_{\mathrm{el}}(s,b)$ is the elastic amplitude; $\mathcal{G}_{\mathrm{in}}(s,b)$, the inelastic overlap function, is the contribution from all inelastic channels. From Eq. (1.37) follows that $\mathrm{Im}\,\mathcal{A}_{\mathrm{el}}(s,b)$ at impact parameter b is generated by the absorption into the inelastic channels *at the same* impact parameter: "unitarity is diagonal in b-space".

For $\mathrm{Re}\,\mathcal{A}_{\mathrm{el}} = 0$, Eq. (1.37) can be solved easily for $\mathcal{A}_{\mathrm{el}}$ if $\mathcal{G}_{\mathrm{in}}(s,b)$ is known. Alternatively, knowledge of $\mathcal{A}_{\mathrm{el}}(s,b)$ can be used to determine $\mathcal{G}_{\mathrm{in}}(s,b)$ (see e.g. [94]). For DIS, it is presently unknown but of great interest.

Equation (1.37) has the general solution

$$\mathcal{G}_{\mathrm{in}}(s,b) = 1 - e^{-\Omega(s,b)} \tag{1.38}$$

$$\mathcal{A}_{\mathrm{el}}(s,b) = i\,\left\{1 - e^{-\frac{\Omega(s,b)}{2} + i\,\chi(s,b)}\right\}. \tag{1.39}$$

The "opacity" function or eikonal, $\Omega(s,b)$, and the phase $\chi(s,b)$ are arbitrary real functions. The former has a simple meaning: $\exp[-\Omega(s,b)]$ is the probability that *no inelastic* interactions with the target occur; $\mathcal{G}_{\mathrm{in}}(s,b)$ is then the probability that *at least one* inelastic interaction occurs. We further have the general relations

$$\sigma_{\mathrm{el}}(s) = \int \mathrm{d}^2 b\,|\mathcal{A}_{\mathrm{el}}(s,b)|^2, \tag{1.40}$$

[2]For a recent mathematical discussion of the validity of this transformation at finite energies, and for further references, see [191].

$$\sigma_{\rm tot}(s) = 2\int d^2b \, {\rm Im}\mathcal{A}_{\rm el}(s,b), \tag{1.41}$$

$$\sigma_{\rm inel}(s) = \int d^2b \left[2{\rm Im}\mathcal{A}_{\rm el}(s,b) - |\mathcal{A}_{\rm el}|^2(s,b)\right]. \tag{1.42}$$

1.5.2 Elastic diffraction and shrinkage

For scattering on a proton, absorption into inelastic channels will be most important for values of b smaller than the proton radius. From Eq. (1.37) follows that this will generate a large imaginary elastic amplitude at the same impact parameter. The impact parameter profile will be maximum at $b = 0$, where absorption is strongest, and the elastic differential cross section, $d\sigma_{\rm el}/dt$, will be sharply peaked at $t = 0$, its width reflecting the transverse extension of the effective interaction region. The experimental fact that ${\rm Re}\,\mathcal{A}_{\rm el}$ is small at high s implies that elastic scattering can, indeed, be considered as the "shadow" of the inelastic channels.

The physical meaning of the slope, $B(s) = \left[d\log(d\sigma_{\rm el}/dt)/dt\right]_{t=0}$, can also be understood from the shape of $\mathcal{G}_{\rm in}(s,b)$ and Eq. (1.36). Indeed, $\mathcal{G}_{\rm in}(s,b)$ is a measure of the overlap of the amplitude of a given final state with *the same* state but rotated along the incident direction over an angle θ, the elastic scattering angle. For most of the final states $|n\rangle$, the transverse momentum of produced particles, $p_{\rm T}$, relative to the incident direction is sharply cut off, and its average increases slowly with s: the distribution in rapidity-$p_{\rm T}$ space resembles that of a uniformly filled cylinder, sometimes called a "Wilson-Feynman liquid", with short-range correlations only between the hadrons. For such a configuration, it is easily verified that the inelastic overlap function, and thus ${\rm Im}\,\mathcal{A}_{\rm el}$, will fall off as an exponential in t, at small $|t|$, with a slope determined by the mean number of particles produced and by their $\langle p_{\rm T}^2\rangle$. To see qualitatively how shrinkage of the diffraction peak comes about, one may consider a simple model of independent particle production. One finds [190] (see also [192])

$$B(s) \geq {\rm constant} + \langle n\rangle/\langle p_{\rm T}^2\rangle. \tag{1.43}$$

Consequently, in this model $B(s)$ grows with energy like $\langle n\rangle$, the mean multiplicity of produced hadrons if $\langle p_{\rm T}^2\rangle$ is constant, thus implying shrinkage of elastic forward peak. For this estimate, phases of the multiparticle amplitudes are neglected. The phase of the amplitude is related to the position in space-time where the particle is produced [193], and is so far unknown.

Assuming that $\langle n\rangle = \omega_0\,\Delta y = \omega_0\ln(s/s_0)$, we see that $B(s)$ depends on the particle density in rapidity space in *inelastic* final states and on the variance of the transverse momentum distributions. In more rigorous calculations, the second-order transverse momentum transfer correlation function[3] enters in Eq. (1.43) instead of $\langle p_{\rm T}^2\rangle$ [140, 194].

[3] In pQCD "ladder-language", this is the correlation between neighboring propagator transverse momenta.

Although this result is obtained in a very simplistic model, it is quite generic (see e.g. [195]). In processes where $\langle p_{\mathrm{T}}^2 \rangle$ is larger than a soft scale, or large compared to $|t|$, the second term on the right-hand side of Eq. (1.43) will be less important and shrinkage will either be small or absent. This most likely happens in (quasi-)elastic processes where a large scale can be identified, also known as "hard diffraction". Experimental evidence for this from elastic vector meson production was discussed in Sub-Sect. 1.3.4.

Whereas the overlap of the amplitude of two "Wilson-Feynman liquids" will be negligible at large $|t|$, one realizes easily that hard jet emission in the final state will contribute to non-zero values of the overlap function at large $|t|$. This is the basic physical reason for the importance of perturbative QCD in large $|t|$ diffractive scattering.

In a general collision process, and in γ^*p in particular, both ω_0 and $\langle p_{\mathrm{T}}^2 \rangle$ can be expected to be process- and energy-dependent. There is no sound a-priori reason to assume that these quantities, which determine the intercept and slope of the dominant trajectory in Regge theory, are universal.

The arguments given show clearly the connection between general properties of the inelastic final states and Regge trajectory parameters. Since the relevant dynamical quantities, here ω_0 and $\langle p_{\mathrm{T}}^2 \rangle$, are clearly identified, generalization beyond the Regge framework is, at least conceptually, simple to understand.

1.5.3 Inelastic diffraction as a regeneration process

The possibility of inelastic diffraction has been predicted in the seminal papers by Feinberg and by Good and Walker [12]. Consider a projectile (hadron, real or virtual photon, etc.) hitting a target at rest. The projectile, being composite, can be described as a quantum-mechanical superposition of states containing various numbers, types and configurations of constituents. The various states in this superposition are likely to be absorbed in different amounts by the target. As a result, the superposition of states after the scattering is not simply proportional to the incident one. Hence, the process will, besides elastic scattering, also lead to production of inelastic states with the same internal quantum numbers as the projectile. This is the fundamental basis for inelastic diffraction and requires little more than the superposition principle of quantum mechanics, unitarity and the coherence condition, Eq. (1.32).

Assume that the projectile, $|B\rangle$, at a fixed impact parameter, $\mathbf{b}$, from the target is a linear combination of states which are eigenstates of diffraction

$$|B\rangle = \sum_k C_k \, |\Psi_k\rangle \,, \tag{1.44}$$

$$\mathrm{Im}T \, |\Psi_k\rangle = t_k \, |\Psi_k\rangle \,, \tag{1.45}$$

where $\mathrm{Im}T$ is the imaginary part of the scattering operator and the (real) eigenvalue t_k is the probability for the state $|\Psi_k\rangle$ to interact with the target. The eigenvalues or absorption coefficients t_k of course vary with **b**. The states are normalized so that $\langle B|B\rangle = \sum_k |C_k|^2 = 1$. From Eq. (1.44) and Eq. (1.45) one easily derives

$$\frac{\mathrm{d}\sigma_{\mathrm{el}}}{\mathrm{d}^2 b} = \mid \langle B \mid \mathrm{Im}T \mid B\rangle \mid^2 = \left(\sum_k \mid C_k \mid^2 t_k\right)^2 = \langle t_k\rangle^2, \tag{1.46}$$

$$\frac{\mathrm{d}\sigma_{\mathrm{tot}}}{\mathrm{d}^2 b} = 2\ \langle t_k\rangle. \tag{1.47}$$

The cross section for inelastic *single* diffractive production (SD), with elastic scattering removed, is

$$\frac{\mathrm{d}\sigma^{\mathrm{SD}}}{\mathrm{d}^2 b} = \langle t_k^2\rangle - \langle t_k\rangle^2. \tag{1.48}$$

The brackets $\langle\cdots\rangle$ mean that one averages t_k or t_k^2, weighted according to their probability of occurrence, $|C_k|^2$, in $|B\rangle$. We note the important result that inelastic diffraction is proportional to the variance of the cross sections of the diagonal channels. Elastic scattering, on the other hand, is proportional to their mean value squared. Note that if the averages are taken over the components of both of the incoming particles, then Eq. (1.48) is the sum of the cross sections for single and double dissociation.

Equations (1.46)-(1.48), with the condition $0 \leq t_k \leq 1$ lead to the Pumplin bound [196]

$$\frac{\mathrm{d}\sigma^{\mathrm{SD}}(b)}{\mathrm{d}^2 b} + \frac{\mathrm{d}\sigma_{\mathrm{el}}(b)}{\mathrm{d}^2 b} \leq \frac{1}{2}\ \frac{\mathrm{d}\sigma_{\mathrm{tot}}(b)}{\mathrm{d}^2 b}. \tag{1.49}$$

This bound, valid at each impact parameter, is saturated if each Ψ_k is either completely transparent to the target or fully absorbed, so that $\langle t_k\rangle^2 = \langle t_k^2\rangle$.

From Eq. (1.48) follows that, if the variance is zero (e.g. when all states are absorbed with the same strength) there is no inelastic diffraction. Thus if, at very high energies, the amplitudes t_k at small impact parameters are equal to the black disk limit, $t_k = 1$, then diffractive production will be absent in this impact parameter domain and so will only occur in the peripheral b region. This already happens for pp (and p$\bar{\mathrm{p}}$) interactions at Tevatron energies. Hence, the impact parameter structure of inelastic and elastic diffraction is drastically different in the presence of strong absorption. Diffraction will be strongest in regions of b-space where absorption shows the strongest variation, i.e. at the "edges" of the target.

Note that for virtual photon scattering, the purely elastic reaction can be neglected. In this case, the term $\langle t\rangle^2$ in Eq. (1.48) is absent. Comparing virtual-photon proton scattering with hadron-hadron scattering, diffraction in the former corresponds to both the elastic and inelastic diffractive scattering of the projectile fluctuations [197]. For a real or virtual photon interaction, the diffractive cross section on a fully absorbing target reaches 50% of the total cross section [198]. In hadronic collisions, inelastic diffraction would vanish in that case (see Eq. (1.48)), while the elastic cross section would become 50% of the total cross section. Thus, when comparing virtual photon diffraction with hadron diffraction, one should compare to the sum of the elastic and single-diffractive inelastic cross sections in hadron-hadron collisions. This is about 25% of the total cross section in πp and pp collisions at high energy, substantially larger than in DDIS. The smaller cross section in DDIS follows from the fact that although longitudinally polarized as well as transversely polarized $\gamma^{\star}$'s contribute, diffraction is a higher-twist process for the former and strongly suppressed at high Q^2 [197].

For real- and virtual-photon-hadron interactions, very little is known from experiment about the impact parameter profile. It requires a measurement of the t-dependence over a wide range in t. For elastic ρ production it was studied for the first time in [117] (see also [197]), following the method pioneered by Amaldi and Schubert [94].

The overall picture is nicely summarized by A. Mueller [199]: "The increase of σ_{tot} (in DIS) with energy occurs because some regions of impact parameter are changing from grey to black and regions at larger b are going from white (no absorption) to grey. However, the region where absorption shows the strongest variation, and which contributes to diffraction, grows less rapidly than those b-regions giving elastic and highly inelastic scattering. This would explain the observation that the inclusive diffractive cross section grows less rapidly than expected from Regge arguments (cfr. Fig. 1.31). Regge theory is indeed expected to hold for those regions in b where the absorption is weak. Regions of large absorption then correspond to multiple pomeron exchanges."

1.5.4 Ioffe time

It was first observed in QED [200] that photon emission from electrons propagating through a medium occurs over distances which increase with energy. In their seminal paper [201], Gribov, Ioffe and Pomeranchuk demonstrated that at high energies large longitudinal distances, now usually referred to as coherence-lengths, l_c, become important for any kind of projectile, including virtual photons, when considered in the rest frame of the target. The typical time involved is $\mathcal{O}(E/\mu^2)$, with E the energy of the projectile and μ a hadronic mass scale, and basically follows from relativity. For DIS, Ioffe [202] demonstrated that the longitudinal distances involved, measured in the target rest frame and in the Bjorken limit, are growing as

$$l_c \propto \frac{1}{m_N\, x}, \tag{1.50}$$

where m_N is the target mass. It should be noted, however, that scaling violations, which are especially strong at small x, modify Eq. (1.50) and reduce the value of l_c

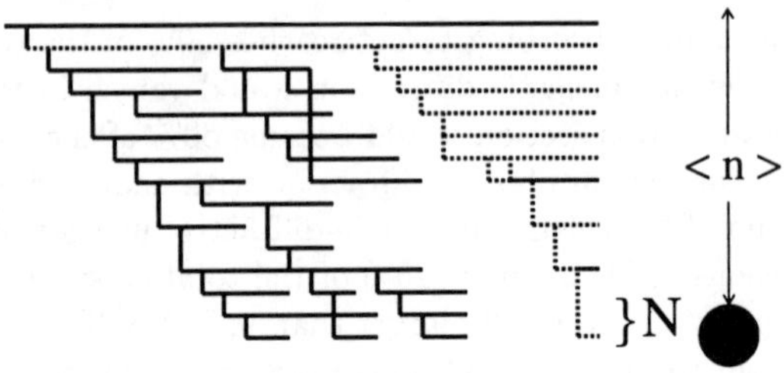

Figure 1.38: A high energy interaction in the parton model [195].

[203]. In addition, l_c depends on the polarization of the virtual photon. Qualitatively, one can understand Eq. (1.50) by considering the energy denominator for the transition of a photon with energy q_0 and virtuality Q^2 to some excited Fock state consisting of quarks and gluons with invariant mass M. The uncertainty principle implies that the lifetime of this state in the nucleon rest frame is the inverse of the energy non-conservation in the $\gamma^* \to$ "M" transition and hence $\sim 2q_0/(M^2 + Q^2)$, which, for M^2 comparable to Q^2, leads to Eq. (1.50) [8].

The value of l_c becomes large for small x or large W. At HERA, for $Q^2 = 10$ GeV^2, the x values range between 10^{-2} and 10^{-4} and l_c corresponds to distances of up to 1000 fm. Pictorially speaking, this means that partonic fluctuations of the virtual photon, the Fock states, are long-lived and travel a substantial distance before interacting. Their state can be considered as frozen during the much shorter time over which the interaction occurs.

1.5.5 The Gribov-Feynman parton model

The parton model views a high-energy interaction of any composite projectile, particle a, with a target, particle b, *at rest* as follows, Fig. 1.38. The fast hadron fluctuates into point-like partons: quarks and gluons. The fluctuations have a lifetime E/μ^2 before interaction with the target occurs. During this time, the partons are in a coherent state which can be described by means of a wave function. Each parton can, in turn, create its own parton cascade, each creating $\langle n \rangle$ partons, resulting eventually in the emission of a total of N soft partons (*"wee" partons* in Feynman's terminology [15]). The latter should not be confused with the partons of pQCD. They are non-perturbative ("dressed") objects due to the long time evolution of the cascade and have acquired a large transverse extension. They interact with a target with a large hadron-like cross section.

For a highly virtual photon, the cascade starts with a $q\bar{q}$ fluctuation (or dipole) of small transverse extension and is followed by an initial evolution stage where the strong coupling remains perturbative and the evolution is calculable in pQCD. However, the non-perturbative end of the cascade is likely to be similar to that originating from a high-energy hadron.

Since gluons rather than quarks will be the dominant component of the cascade, and since gluons carry a larger color charge, it should not have been a surprise to find that the interaction, when viewed in an infinite-momentum frame is, at high energy, driven by the gluon constituents in the proton. The same applies to structure functions of the pomeron and is confirmed by experiment, see Sub-Sect. 1.4.3.

As argued by Gribov [17], in a soft collision, a fast projectile can interact with the target only through its wee components. Indeed, the cross section of interaction of two point-like particles with large relative energy, $\sqrt{s_{ab}}$, is not larger than $\pi\lambda^2 \sim 1/s_{ab} \sim \exp(-\eta_{ab})$ (λ is the wavelength in the cms frame of a and b, while η_{ab} is the relative rapidity). Thus, *only slow partons of the projectile* are able to interact with the target with a non-negligible cross section. If there are n wee partons in total, the interaction cross section is proportional to the probability that *at least one* out of n interacts with the target, i.e. $\propto 1-(1-\delta)^n$, where δ is the interaction probability of a single parton. For small n, the total interaction cross section is proportional to n.

The interaction can also be viewed from the rest-frame of the projectile, or from any collinear rest-frame. The distribution of the wee partons in the rest-frame of particle b is, according to the above arguments, solely determined by particle a and does not depend on the properties of particle b. On the other hand, in the rest frame of particle a the distribution is determined by the properties of particle b. This is possible only if the distribution of partons with rapidities η much smaller than the hadron's rapidity, η_p, *does not* depend on the quantum numbers and the mass of that hadron. It follows that *the distribution of the wee partons with $\eta \ll \eta_p$ should be independent of the projectile and target*, i.e. be universal. Indeed, in the cascade the memory of the initial state is lost after only a few steps, if it resembles a Markov process.

The fact that the wee-parton component of any hadron is independent of the hadron itself, explains semi-qualitatively why hadronic total cross sections show the same energy dependence at large s, as we saw in Sub-Sect. 1.3.1. In addition, if the interaction between wee's is effectively short-range in rapidity (implying that the produced hadrons show short-range rapidity correlations), hadrons produced in regions of rapidity sufficiently far from target and projectile will also show universal properties.

1.5.5.1 Shrinkage

Consider the interaction in the impact-parameter plane. In each step of the cascade the newly emitted parton acquires a certain amount of transverse momentum, k_T. If the emission is purely random in k_T-space, the Nth parton in the cascade will, as the result of a random walk in impact-parameter space, have moved a distance b_N^2 from the origin. On average, and for a completely random process (*which precludes any kind of k_T-ordering of the emissions as e.g. in DGLAP-type evolution*), one has

$$\langle b_N^2 \rangle \propto \frac{1}{\langle k_T^2 \rangle} N = \frac{\omega_0}{\langle k_T^2 \rangle} \ln(s/s_0), \tag{1.51}$$

for $N \propto \ln s$. In this simplified picture, the (transverse) growth of the interaction region with energy is the result of a diffusion process, sometimes referred to as "Gribov diffusion", and $\omega_0/\langle k_\mathrm{T}^2\rangle$ can be identified with $\alpha'_{\mathbb{P}}$ in Eq. (1.14). The argument based on the overlap function, Sub-Sect. 1.5.2, leads to the same result but is more general.

It may be superfluous to mention that wee parton properties, and their interactions, cannot be calculated in pQCD, neither for hadronic collisions nor for deep-inelastic ep scattering, since they are associated with long-wavelength fluctuations in the color fields. For DIS, this ignorance is parametrized in the parton distributions at a small scale. In dipole models, the dipole cross section parametrization takes the same role. In any case, in the small-x region, the wee partons are equally important in both types of scattering processes.

1.5.5.2 Rise of the total cross section and $\alpha_{\mathbb{P}}(0)$

Consider Fig. 1.38. Since each parton in the parton cascade can form its own chain of partons, and so on, this multiplication process will generically (but not in detail) lead to a total mean $N \sim e^{\langle n\rangle}$, if $\langle n\rangle$ is the mean multiplicity in a single chain [195]. With $\langle n\rangle \propto \omega_0 \ln s$ one finds

$$N \propto s^{\omega_0}. \tag{1.52}$$

This can be rewritten in a frame-independent form

$$\sigma_\mathrm{tot} = \sigma_0(\text{projectile}) \times \sigma_0(\text{target}) \times \frac{1}{\mathrm{s}_0} \times \left(\frac{\mathrm{s}}{\mathrm{s}_0}\right)^{\omega_0}. \tag{1.53}$$

The "impact factors", σ_0, are particle-specific but independent of s. The arguments leading to Eq. (1.53) show how the power-law (or Regge) behavior of σ_tot arises. Importantly, ω_0, is independent of any variable related to projectile or target.

For a collision of a small-size (in b-space) virtual photon with a target, Q^2 larger than a few GeV^2, the evolution of ω_0 with Q^2 is calculable in pQCD. This is one of the major theoretical advantages of deep-inelastic scattering over soft hadron-hadron interactions. Evidently, in DIS, the role of s is taken over by W^2 or $1/x$.

From Eq. (1.53) we learn that $\alpha_{\mathbb{P}}(0) - 1$ in Regge theory has to be interpreted as the anomalous multiplicity dimension of the parton cascade. This result is generic. Since the detailed process-specific dynamics of the parton cascade (DGLAP [73], BFKL [204], CCFM [205]) will determine the values and energy dependence of ω_0, we conclude that a "universal" pomeron trajectory with process-independent parameters does not exist.

The power-law form, Eq. (1.53), is a result typical for a self-similar (fractal) branching process with *fixed coupling constant* and ω_0 is related to the fractal dimension. Early pre-QCD examples can be found in [206]. For a running coupling constant, the s-dependence is generally less strong, but faster than any power of $\ln s$.

1.5.5.3 Total cross sections, diffraction and wee-parton multiplicity

Suppose the projectile is a superposition of states with, at a given impact parameter b, n wee partons, each of which can interact with the target with a probability $\delta(b)$. If the structure of the target is ignored, we have (for brevity, we omit the argument b in the following and write σ_{tot} etc. where $d\sigma_{\text{tot}}/d^2b$ is meant)

$$\sigma_{\text{tot}} = \sum \sigma_{\text{tot}}(n) P(n), \tag{1.54}$$

with $P(n)$ the probability that the cascade has produced n such partons. Using conservation of probability (or unitarity), we find

$$\sigma_{\text{tot}}(n) = 2T_{\text{el}}(n); \quad T_{\text{el}}(n) = 1 - \sqrt{1 - \sigma_{\text{inel}}(n)}; \quad \sigma_{\text{inel}}(n) = 1 - (1-\delta)^n. \tag{1.55}$$

The last equation in (1.55) is the probability that *at least one* out of n partons interacts with the target. The previous equations can be compactly expressed in terms of the generating function, $G(z)$, of the multiplicity distribution $P(n)$

$$\begin{aligned} G(z) &= \sum P(n)\,(1+z)^n, & (1.56)\\ \sigma_{\text{tot}} &= 2\sum P(n)[1-(1-\delta)^{n/2}] = 2 - 2\,G(\sqrt{1-\delta}-1), & (1.57)\\ \sigma_{\text{diff+el}} &= \sum P(n)[1-(1-\delta)^{n/2}]^2, & \\ &= 1 - 2\,G\left(\sqrt{1-\delta}-1\right) + G\,(-\delta)\,. & (1.58) \end{aligned}$$

For the ratio of total diffractive (sum of inelastic and elastic) cross section to the total cross section, $R(b)$, at fixed impact parameter, we obtain

$$R(b) = \frac{\sigma_{\text{diff+el}}}{\sigma_{\text{tot}}} = 1 - \frac{1}{2}\,\frac{1 - G(-\delta)}{1 - G(\sqrt{1-\delta}-1)}. \tag{1.59}$$

In the case of total absorption, $\delta \to 1$, the ratio converges towards the black-disk limit of $\frac{1}{2}$, as it should.

Assuming, as an example, $P(n)$ to be Poissonian, we obtain

$$\begin{aligned} \sigma_{\text{tot}} &\approx \frac{1}{2}\delta\,\langle n\rangle, & (1.60)\\ \sigma_{\text{diff+el}} &\approx \frac{\delta^2}{4}\,\langle n^2\rangle = \frac{\delta^2}{4}\left[\langle n\rangle^2 + \langle n\rangle\right], & (1.61) \end{aligned}$$

provided δ or $\langle n\rangle$ or both are small enough. The latter conditions mean that multiple interactions with the target can be neglected, or that the partonic system hitting the target is sufficiently dilute and no saturation takes place.

Equation (1.60) suggests a relation between the total cross section (or F_2 in DIS) and the mean parton multiplicity, which was first tested experimentally in [85] and is

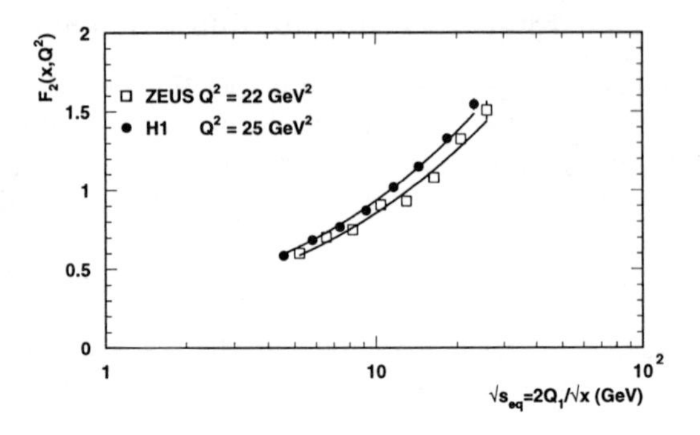

Figure 1.39: tolerance=1000 Comparison of e^+e^- data on average charged-particle multiplicity versus $\sqrt{s}$ and the HERA low-x F_2 data versus $2Q_1/\sqrt{x}$, with $Q_1 = 270$ MeV, for $Q^2 = 22$ GeV^2 (ZEUS) and 25 GeV^2 (H1). The e^+e^- data (solid lines) are represented by phenomenological fit and normalized to F_2 in each Q^2 bin separately [85].

illustrated in Fig. 1.39. There, using a Modified Leading Log (MLLA) pQCD expression for the energy-dependence of $\langle n \rangle$ (Eq. (7.32) in [207]), an excellent description of the x-dependence of $F_2(x, Q^2)$ data at low x was achieved with two free parameters only.

If saturation (parton recombination) effects in the parton cascade in DIS happened to occur, we can expect that the similarity of the energy dependence between mean particle multiplicities in e^+e^- and F_2 will break down for very high multiplicity events. Given the present interest in this topic [65, 208], a dedicated measurement of the W-dependence of semi-inclusive structure functions $F_2^{(n)}(x, Q^2)$, at fixed large final-state multiplicity n, and of its diffractive counterpart, might therefore be of considerable importance.

If $\langle n \rangle$ is assumed to grow as s^ϵ, Eq. (1.60) and Eq. (1.61) predict $\sigma_{\text{tot}} \propto s^\epsilon$ and $\sigma_{\text{diff+el}} \propto s^{2\epsilon}$, the same as in Regge theory, and thus showing the same unitarity violating defects. To obtain a constant ratio, $R(b)$, at each impact parameter, it seems unavoidable to include in the calculation the full multiple-scattering terms and possibly (so far unknown) parton correlations. The approximations made to arrive at Eqs. (1.60) and (1.61), nicely illustrate that the Regge behavior corresponds to the dynamical regime of low parton density; it becomes invalid for high densities.

The role played by correlations among partons can be illustrated using the factorial cumulant expansion of $G(z)$ [209, 210] (see Chapter 7, Eq. (7.46))

$$G(z) = \exp\left\{ \langle n \rangle z + \sum_{2}^{\infty} (z^q/q!)\, f_q \right\}.$$

The cumulants f_q are a measure of the correlations and identically zero for $q > 1$ if the partons are uncorrelated. The inelastic cross section can now be written as

$$\sigma_{\text{inel}} = 1 - G(-\delta) = 1 - \exp\left\{ -\langle n \rangle\, \delta + \sum_{q=2}^{\infty} (-\delta)^q/q!)\, f_q \right\}. \tag{1.62}$$

Comparing Eq. (1.38) with Eq. (1.62), we note that the eikonal function $\Omega(b)$ can be expressed in terms of the cumulant generating function $\ln G(-\delta)$. This shows that not only multiple scattering contributions, but also parton-parton correlations (if $f_q \neq 0$ for $q > 1$) contribute to the total and diffractive cross sections. Such correlations have not been explicitly taken into account, as far as we know, in present pQCD calculations of DDIS, with the exception of [211] using the concept of (Mueller) dipoles in onium-onium scattering.

1.6 Models for diffraction

1.6.1 Diffraction and partons: the Miettinen and Pumplin model

The first detailed calculations of hadronic diffraction in the framework of the parton model were performed by Miettinen and Pumplin (MP) [142]. They are entirely based on unitarity and the Good and Walker mechanism. This work, almost 30 years old and pre-dating QCD, remains of great interest and it is instructive to summarize its main ideas and results.

It is assumed that the diagonal states (Sect. 1.5.3) $|\Psi_k\rangle$ are the states of the parton model, composed of quarks and gluons and a radiation cloud of wee partons. These states are characterized by a definite number N of partons with impact parameters $\mathbf{b}_1, \ldots, \mathbf{b}_N$ and longitudinal momentum fractions, or rapidities, $y_1, \ldots, y_N$.

Since there are parton states which are rich in wee partons, and others with a few or no wees, these states will interact with a target with very different cross sections. Hence, inelastic diffraction will be generated by the mechanism of Good and Walker. The fluctuations in the interaction probabilities t_k, Eq. (1.45), arise from fluctuations in the number of wee partons, fluctuations in y_i and from fluctuations in $\mathbf{b}_i$.

Assuming uncorrelated wee partons, the following expressions are obtained for the total, elastic and single-diffractive cross sections [142, 212]

$$\frac{\mathrm{d}\sigma_{\mathrm{tot}}}{\mathrm{d}^2 b} = 2\left[1 - \exp\left(-G^2 \frac{4}{9} e^{-\frac{1}{3}\cdot\frac{b^2}{\beta}}\right)\right], \tag{1.63}$$

$$\frac{\mathrm{d}\sigma_{\mathrm{el}}}{\mathrm{d}^2 b} = \left[1 - \exp\left(-G^2 \frac{4}{9} e^{-\frac{1}{3}\cdot\frac{b^2}{\beta}}\right)\right]^2, \tag{1.64}$$

$$\frac{\mathrm{d}\sigma^{\mathrm{SD}}}{\mathrm{d}^2 b} = \exp\left[-2G^2 \frac{4}{9} e^{-\frac{1}{3}\cdot\frac{b^2}{\beta}}\right]\left[\exp\left(G^2 \frac{1}{4} e^{-\frac{1}{2}\cdot\frac{b^2}{\beta}}\right) - 1\right]. \tag{1.65}$$

The model depends on two parameters, G^2, the mean number of active partons, and β, which characterizes the absorption strength of the eigenstates on the target. At a given energy, the two parameters can be estimated from experimental data on σ_{tot} and σ_{el} using Eq. (1.63) and Eq. (1.64) and used to predict the inelastic (single) diffractive cross section.

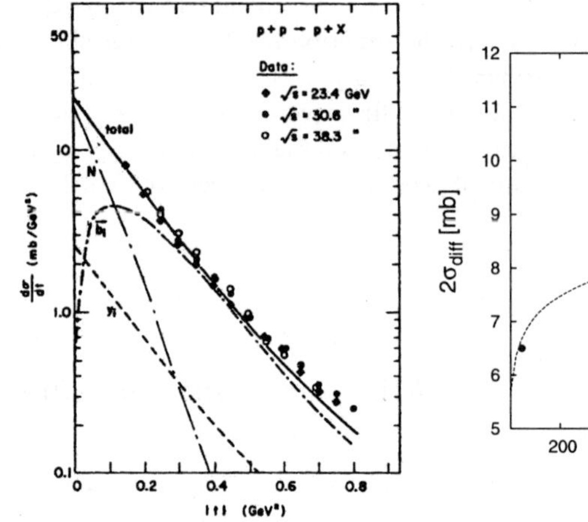

Figure 1.40: Decomposition of $\frac{d\sigma^{SD}}{dt}$ into contributions from fluctuations in number (N), rapidities (y_i) and relative impact parameters (b_i) of wee partons [142].

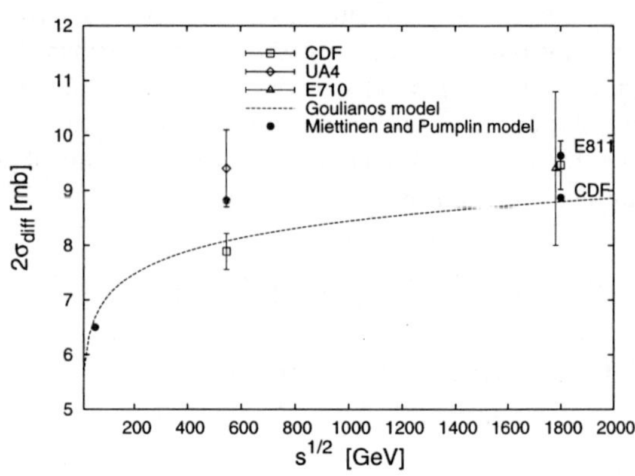

Figure 1.41: Total pp single diffraction cross section data compared with predictions based on the Miettinen and Pumplin model (solid dots) [212] and the flux renormalization model of the Goulianos (dashed curve) [141].

At $\sqrt{s} = 53$ GeV, the calculated inelastic diffractive cross section was found to agree well with data [142]. The y_i fluctuations contribute little (about 10%), whereas fluctuations in b_i and in parton number each account for about 45% of the single diffractive cross section. Also the value at $t = 0$ and the slope of the t distribution are correctly predicted. This is a non-trivial result since the calculated (and measured) slope $B \approx 6.9$ GeV^{-2} is only about half that of elastic scattering, $B \approx 12$ GeV^{-2}. Interestingly, as seen from Fig. 1.40, the small-$|t|$ dissociation is dominated by the large and very steep contribution (slope ≈ 12.2 GeV^{-2}) due to the *fluctuations in the number of partons*. The b_i fluctuations, on the other hand, dominate at large $|t|$. It may be of importance to note that fluctuations of the parton (or dipole) number are not fully included in most modern models of DDIS.

As discussed in Sub-Sect. 1.4.3, the measured t-slopes in DDIS are strikingly similar to those in the pp data, hinting once more to the similarity of the rapidity-gap formation in soft hadron-hadron collisions and in DIS.

The MP-model has recently been used [212] to calculate the inelastic diffractive cross section at $\sqrt{s} = 546$ GeV and $\sqrt{s} = 1800$ GeV using the CDF [49] and E811 [51, 52] measurements of $\sigma_{\rm tot}$ and $\sigma_{\rm el}$. The corresponding values for G^2 and β, presented in Table 1.3, allow to calculate the cross section for single diffractive production.

Table 1.3: Total, elastic and single-diffractive (×2) cross sections, together with the values of G^2 and β.

References	$\sqrt{s}$ (GeV)	σ_{tot} (mb)	σ_{el} (mb)	G^2	β (fm^2)	$2\sigma_{\text{diff}}$ (mb)
[142]	53	43	8.7	2.91	0.235	6.51
CDF [49]	546	61.26±0.93	12.87±0.30	3.12	0.319	8.82
E811 [51,52]	1800	71.71±2.02	15.79±0.87	3.38	0.351	9.63
CDF [49]	1800	80.03±2.24	19.70±0.85	4.20	0.337	8.87

Figure 1.41 shows the results compared with experiments and with the Goulianos flux renormalization model [141]. To take into account the beam and the target dissociation, σ^{SD} is multiplied by a factor of two. The two different values at $\sqrt{s} =$ 1800 GeV correspond to the two different results for σ_{tot} and σ_{el} measured by CDF and E811.

It is seen that the MP-model is consistent with the data over a considerable range of cms energies. Similarly to the model of Goulianos, it predicts a slow rise of the single-diffractive cross section with $\sqrt{s}$, although the values obtained are somewhat higher.

The important result here is that predictions based on a straightforward extrapolation of the model to Tevatron energies agrees very well with the data. No further adjustments are necessary. This contrasts favorably with the Regge description of diffractive dissociation which, at high energies, encounters serious problems related to unitarity, as mentioned in Sub-Sect. 1.2.2. To "save" the Regge picture, Goulianos proposed to renormalize the pomeron flux in a way which restores unitarity and agreement with the data [141] (see Sub-Sect. 1.4.2). Although elegant and effective, this method is not so easily justifiable from theory.

1.6.2 Modern QCD models of diffraction

The model of Miettinen and Pumplin grasps the essential physics which remains valid in the context of DDIS at HERA. In [142] ad-hoc assumptions were needed to build a model of the hadron Fock states. In DIS the light-cone wave functions of the lowest-order $\gamma^\star$ Fock states ($q\bar{q}$, $q\bar{q}$ + gluon) are known [108, 109] and more quantitative results can be obtained. Nevertheless, the interaction of the wee partons needs to be parametrized empirically, as it must be for soft hadron-hadron collisions.

The presently popular models for diffraction in DIS have been reviewed in [2]. They use the same basic concepts discussed in previous sections under various disguises. The most successful of these are merely modernized versions of the Aligned Jet Model [108], which, as is worth to remember, pre-dates QCD. It was the first model to emphasize that even in DIS at large Q^2, non-perturbative, soft physics plays an essential role.

Considered in the target rest frame, the Fock-state wave function of the partonic γ^* fluctuation carries the information on the virtuality of the photon and further depends on its transverse size and the fractional momenta and masses of the partons. In the simplest case of a $\mathrm{q\bar{q}}$ fluctuation, or $\mathrm{q\bar{q}}$ dipole, the wave function is then convoluted with the amplitude for the elastic interaction of the color dipole and the target hadron. At $t = 0$, this amplitude is determined by the cross section for the scattering of the dipole on the target, $\sigma(x, \varrho)$, where ϱ is the relative transverse separation between the q and $\mathrm{\bar{q}}$. It is assumed to be independent of Q^2, in accord with the Gribov-Feynman argument of wee-parton scattering and short-range order in the cascade, but depends on the energy of the collision ($\propto 1/x$ in DIS) and on the impact parameter between dipole and target.

The reason to expect significant contributions from "soft" physics to the total and diffractive DIS cross sections can be easily understood in qualitative terms (see e.g. [8, 65] for a detailed quantitative discussion).

For a highly virtual photon fluctuating into a $\mathrm{q\bar{q}}$ pair, large-ϱ fluctuations are rare; for transversely polarized photons they are suppressed by a factor $\sim 1/Q^2$ [109] as compared to small-size fluctuations. However, such large-size dipoles interact with a large hadron-like cross section. On the other hand, the more probable small-size fluctuations interact with a much smaller cross section ($\sim 1/Q^2$), due to color screening. As a result, in the total DIS cross section, or F_2, both components contribute and show a"leading twist" behavior and Bjorken-scaling. Soft as well as hard interactions contribute at a comparable level.

In inclusive diffractive scattering, the cross section is still proportional to the same probability of finding the appropriate fluctuation within the photon. However, now the *square* of the dipole cross section (see Eq. (1.48)) enters. As a result, small-size dipole configurations are suppressed by a factor $1/Q^4$ and become a higher twist effect in diffraction, whereas the soft fluctuations in DDIS still contribute at leading twist. Consequently, inclusive DDIS is largely non-perturbative.

For the theoretical description of diffraction in DIS, the dipole picture of QCD [109, 213] has proven to be particularly natural and successful. The dipole picture is well-suited because color dipoles, which are two-body color neutral objets characterized by their longitudinal momentum, transverse size and impact parameter, are eigenstates of the interaction at high energy. In addition, the model lends itself well to a theoretical analysis of the phenomenon of parton saturation (see [208] and refs. therein).

1.6.2.1 The dipole saturation model

Consider, as an example, the very successful saturation model of Golec-Biernat and Wüsthoff (GBW) [65], which expands on much earlier work [109]. The physical picture is that in which, in the nucleon rest frame, a photon with virtuality Q^2, emitted by a lepton, dissociates into a $\mathrm{q\bar{q}}$ pair far upstream of the nucleon. This is then followed by the scattering of the color dipole on the nucleon. In this picture, as also assumed in the MP model, the relative transverse separation, ϱ, of the $q\bar{q}$ pair and the longitudinal momentum fraction z of the quark remain unchanged during the collision. The $\gamma^{\star}p$ cross sections, integrated over the impact parameter, take the factorized form (see e.g. [109, 214])

$$\sigma_{T,L}(x, Q^2) = \int d^2\varrho \int_0^1 dz \, |\Psi_{T,L}(z, \varrho, Q^2)|^2 \, \sigma(x, \varrho), \tag{1.66}$$

where $\Psi_{T,L}$ is the wave function of transversely (T) and longitudinally polarized (L) photons.

In Eq. (1.66), all Q^2 dependence is contained in the Fock-state wave function, which further depends on the flavor and mass of the partons. The W- or x-dependence of $\sigma_{T,L}(x, Q^2)$ is therefore solely determined by that of $\sigma(x, \varrho)$. The latter is the principal quantity in the s-channel description of diffractive scattering. Once the dipole cross section is known, Eq. (1.66) enables a parameter-free calculation of the proton structure function and DDIS at small x. In our simple picture, we may interpret it as an effective cross section, the product of the wee-parton flux with the single wee-parton nucleon cross section.

Although the impact-parameter dependence of $\sigma(x, \varrho)$ is not explicitly considered by GBW (only its average enters in Eq. (1.66)), this is clearly of great interest for the t-dependence of diffraction [117]. A dipole saturation model which includes impact parameter dependence is presented in [215].

Turning to diffraction, the differential cross section at $t = 0$ takes the form

$$\left.\frac{d\sigma^{SD}_{T,L}}{dt}\right|_{t=0} = \frac{1}{16\pi} \int d^2\varrho \int_0^1 dz \, |\Psi_{T,L}(z, \varrho)|^2 \, \sigma^2(x, \varrho). \tag{1.67}$$

The form of Eq. (1.67) differs from Eq. (1.66) only by the substitution $\sigma(x, \varrho) \to \sigma^2(x, \varrho)$, in accord with the general formula (1.48)[4].

Comparing to the MP-model, we see that the relative impact parameter and rapidity fluctuations are included here through the photon wave function. However, the important parton-number fluctuations, which largely determine the shape of the forward diffractive peak, and which are also a measure of correlation among the partons, are not explicitly considered.

The energy dependence of $\sigma(x, \varrho)$ follows from the fact that, in low-x DIS, the perturbative evolution of the $q\bar{q}$ dipole results in further "hard" parton multiplication which increases also the wee-parton flux and thus the total cross section.

[4] In the dipole formulation of DIS, and contrary to hadron diffraction, even inelastic DDIS is considered to be purely elastic: the dipole states ($q\bar{q}$) and higher-order Fock-states, $q\bar{q}$ + gluons, are assumed to be orthogonal eigenstates of the diffractive T-matrix, and no regeneration (mixing of the states) occurs. If these states are not orthogonal, they will regenerate and thus add an additional contribution to the diffractive cross section, presently neglected.

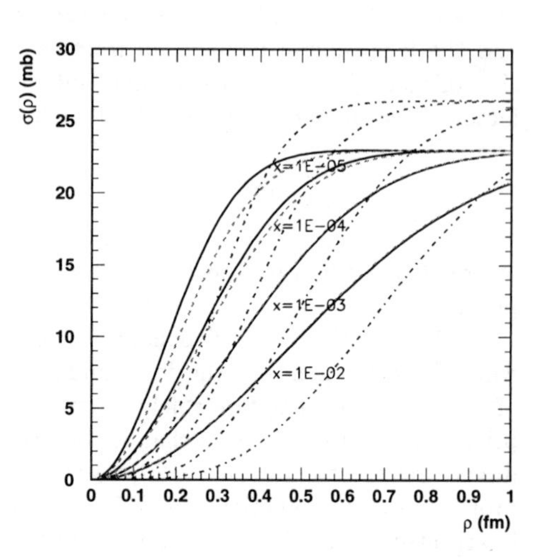

Figure 1.42: The dipole cross section $\sigma(\varrho)$ for various values of Bjorken-x. The GBW parametrization, Eqs. (1.69), (1.70), is shown as the solid curves. Dashed lines correspond to the $1/x$ dependence given by Eq. (1.72) with parameters $n_\mathrm{f} = 3$, $K = 0.288$, $\Lambda = 1.03$ GeV [217]. The dot-dashed lines show $0.05 \times \sigma^2(\varrho)$.

Indeed, due to the bremsstrahlung nature of the soft gluon spectrum $\propto \mathrm{d}z_\mathrm{g}/z_\mathrm{g}$ (z_g is the momentum fraction of the photon carried by a gluon), Fock states with n such gluons give a contribution $\propto \ln(1/x)^n$ to the total photoabsorption cross section, which can be reabsorbed into an energy dependent dipole cross section [109]. For example, in the DGLAP approximation, summing over all n produces the well-known $\exp[2\sqrt{\ln(1/x)\ln(1/\alpha_\mathrm{s}(Q^2))}]$ increase of the $\gamma^\star\mathrm{p}$ cross section and "standard" scaling violations.

The $\mathrm{q\bar{q}}$ dipole-proton cross section $\sigma(x, \varrho)$ has to be modelled although it is known in the perturbative limit of very small dipoles, $\varrho \to 0$, and related there to the inclusive gluon distribution $xg(x, \mu^2)$ of the target [216],

$$\sigma(x,\varrho) = \frac{\pi^2}{3}\alpha_\mathrm{s}[x\,g(x, C/\varrho)]\varrho^2 + \mathcal{O}(\varrho^4), \tag{1.68}$$

C is a scale parameter which cannot be determined in leading-order pQCD. The fact that the cross section vanishes $\propto \varrho^2$ as $\varrho \to 0$ expresses the property of "color transparency" [54, 55, 57], discussed in Sub-Sect. 1.3.1. For a ($\mathrm{q\bar{q}g}$)-dipole system, $\sigma(x, \varrho)$ is roughly a factor 9/4 larger. This largely explains the predominant role of gluons in low-x DIS. Expressed in terms of the "pomeron structure function", it implies that the pomeron is "gluon dominated", see Sub-Sect. 1.4.3.

In the GBW model, the effective dipole cross section is taken to be of the form

$$\sigma(x,\varrho) = \sigma_0\left[1 - \exp\left(-\frac{\varrho^2}{4R_0^2(x)}\right)\right], \tag{1.69}$$

where the x-dependent radius R_0 is parametrized as

$$\frac{1}{R_0^2(x)} = Q_0^2 \left(\frac{x_0}{x}\right)^{\lambda}, \tag{1.70}$$

with $Q_0 = 1$ GeV. The parameters $\sigma_0 = 23$ mb, $x_0 = 3 \,.\, 10^{-4}$ and $\lambda = 0.29$ have been determined by a fit to data on F_2 [65].

As seen in Fig. 1.42, the dipole cross section saturates at a value σ_0 for large-size dipoles where it is entirely non-perturbative. Also, as $x \to 0$, saturation sets in at decreasingly small transverse sizes (thus entering the perturbative regime) and the contribution from large-size dipoles becomes more important.

Since Eq. (1.67) depends on the square of $\sigma(x, \varrho)$, it follows that still larger sizes are involved in diffraction than those dominating the total cross section (dot-dashed lines in Fig. 1.42): non-perturbative soft physics is of even greater importance in DDIS. Saturation effects can, therefore, be expected to play a larger role than in inclusive DIS.

Because of the saturation property of Eq. (1.69), the ratio of the DDIS to DIS cross sections is found to be [65]

$$\frac{\sigma^{\rm SD}}{\sigma_{\rm tot}} \approx \frac{\sigma_0}{8\,\pi\,B}\,\frac{1}{\ln[R_0^2(x)\,Q^2]}, \tag{1.71}$$

thus providing an explanation of the near constancy of that ratio, as shown in Sub-Sect. 1.4.3. In Eq.(1.71), B is the slope of the differential t-distribution, assumed to be a single exponential.

In the GBW model, two essential scales appear: the characteristic transverse size of the $\mathrm{q\bar{q}}$ dipole $\propto 1/Q$, solely determined by the $\gamma^\star$ wave function, and $R_0(x)$.

In a simple picture, $1/R_0^2(x)$ can be interpreted as the mean number of soft partons in the cascade; $R_0(x)$ is their mean relative transverse distance; $Q\,R_0(x) = 1$ defines a critical line. For $1/Q \ll R_0(x)$ the partonic system is dilute, for $1/Q \gg R_0(x)$ the system is densely packed and multiple scattering and parton interactions become important.

It is interesting to note here that the fitted value of λ (≈ 0.29) is quite close to that derived from the cms energy dependence ($\sqrt{s}$) of the mean particle multiplicity in $\mathrm{e^+e^-}$ annihilation, where $\langle n \rangle \sim s^{0.25}$ gives a reasonable fit of the data [218]. Remembering the striking similarity discussed in Sub-Sect. 1.5.5, we have also plotted in Fig. 1.42 the expression (Eq. 1.69) wherein $1/R_0^2(x)$ in Eq. (1.70) is replaced by that of the mean soft-gluon multiplicity, $\mathrm{N_g}$, in a gluon jet with energy squared $\propto 1/x$ [217],

$$\frac{1}{R_0^2} \equiv Q_0^2\,\mathrm{N_g} = K Q_0^2\, y^{-a_1 C^2} \exp\left[2C\sqrt{y} + \delta_{\rm G}(y)\right], \tag{1.72}$$

with K an overall normalization constant, $C = \sqrt{4n_{\rm c}/\beta_0}$, and

$$\delta_{\rm G}(y) = \frac{C}{\sqrt{y}}\left[2a_2C^2 + \frac{\beta_1}{\beta_0^2}[\ln(2y) + 2]\right] + \frac{C^2}{y}\left[a_3C^2 - \frac{a_1\beta_1}{\beta_0^2}[\ln(2y) + 1]\right]. \tag{1.73}$$

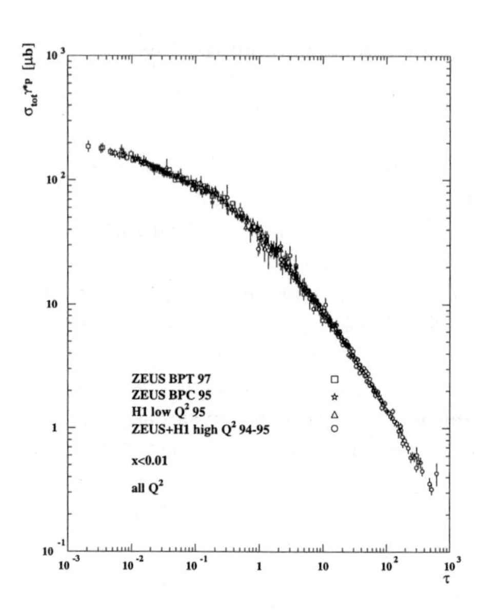

Figure 1.43: "Geometric scaling" for small-x DIS data with $x < 10^{-2}$; $\tau = Q^2 R_0^2(x)$.

Here, $e^y = \sqrt{1/x}\,\frac{1\,(\mathrm{GeV})}{\Lambda}$ and further $\beta_0 = (11n_\mathrm{c} - 2n_\mathrm{f})/3$, $\beta_1 = [17n_\mathrm{c}^2 - n_\mathrm{f}(5n_\mathrm{c} + 3C_\mathrm{F})]/3$, $n_\mathrm{c} = 3$ is the number of colors and $C_\mathrm{F} = 4/3$. The constants a_i are tabulated in [219]. Λ is the QCD scale parameter and n_f the number of active flavors.

The dashed curves in Fig. 1.42 show the dipole cross section as obtained from Eq. (1.72). It is essentially indistinguishable from the GBW parametrization for $x = 10^{-2} - 10^{-3}$, but differences become noticable at smaller x. This follows from the fact that, due to the running of α_s, the multiplicity grows slower than a power in $1/x$ and "saturation" is delayed in comparison with Eq. (1.70), the latter being a result characteristic of a cascade process with a fixed coupling constant.

The results shown in Fig. 1.42 imply that the ansatz in Eq. (1.72) will lead to an equally satisfactory description of F_2 and F_2^D as found in the original GBW analysis [65]. However, contrary to Eq.(1.70), the parametrization Eq. (1.72), involves no free parameters, apart from Q_0 and the normalization constant K which was taken from a fit to $\mathrm{e^+e^-}$ data [217]. In particular, in the former case the important parameter λ follows from theory.

If $1/R_0^2(x)$ is interpreted as the mean number of soft partons confined in the target within a transverse radius R, it is evident that the GBW parameter λ can be identified with the anomalous (multiplicity) dimension, γ_0, of the parton cascade (see [220] for a recent review and further references), which is calculable in pQCD. In pQCD, for a time-like cascade, it is equal to the logarithmic energy derivative of

N_g and given by [217]

$$\gamma_0 = \frac{N'_g}{N_g} \;=\; \sqrt{\bar{\alpha}}\left(1 - 2a_1\sqrt{\bar{\alpha}} - 4a_2\bar{\alpha} + \mathcal{O}(\bar{\alpha}^{3/2})\right). \tag{1.74}$$

The first term, $\sqrt{\bar{\alpha}} \equiv \sqrt{2n_c\alpha_s/\pi}$, is the leading-order term. However, γ_0 is non-linear in $\bar{\alpha}$ and decreases with increasing s due to the running of α_s. Consequently, N_g increases slower than a power of s. A power-law dependence is obtained if α_s is kept constant. Taking, as an example, $\alpha_s = 0.2$ in (1.74) yields $\lambda = 0.30$ at lowest order [221].

The relation between the GBW parameter λ and the multiplicity anomalous dimension has also been derived in the framework of the Balitsky and Kovchegov modified BFKL equation (see [222] for details and references) with the result $\lambda \approx 6\alpha_s/\pi$. For $\alpha_s = 0.2$ this gives $\lambda = 0.38$. Note, however, that (1.74) is a polynomial in $\sqrt{\alpha_s}$, whereas the previous expression is linear in α_s.

A further important result of the GBW-model is that the diffractive structure function F_2^D is found to obey a Regge-like factorization property (except for $\beta \to 1$, where higher-twist contributions from longitudinal photons dominate) with the dependence $F_2^D \propto (1/\xi)^{1+\lambda}$. This corresponds to an effective pomeron intercept $\alpha_{I\!P}(0) \sim 1 + \lambda/2$ [223], in good agreement with data, see Fig. 1.29. This is a highly revealing result, demonstrating on this specific model-example, the generic property that the growth of the cross sections with energy and the proton energy-loss spectrum (or "pomeron flux") are closely connected and determined by the multiplicity anomalous dimension.

Whereas the original GBW-model did not include pQCD Q^2-scaling violations, in subsequent work it was improved by incorporating DGLAP evolution into the dipole cross-section, thereby extending the model applicability to the high Q^2 region (Bartels, Golec-Biernat, Kowalski in [65]).

The above considerations suggest a very simple picture (see also [80] p.8) of the x and Q^2 dependence of F_2 and F_2^D. In the target rest frame at low x, the target is populated with a number of partons proportional to N_g confined within a transverse area πR^2. Since the area "scanned" by the virtual photon $q\bar{q}$ dipole is proportional to $1/Q^2$, the number of partons with which it can interact is proportional to

$$\frac{1}{\tau} \equiv N_g\,\frac{1}{Q^2} \propto \frac{1}{R_0^2}\,\frac{1}{Q^2}. \tag{1.75}$$

One can, therefore, expect that the total cross section will depend only on τ and not on Q^2 and R_0^2 separately, i.e. exhibit "geometrical scaling" [224]. This follows already from dimensional arguments but also agrees with the "universality" hypothesis, advanced in [225], that the physics should depend only on the number of partons per unit of rapidity and per unit of transverse area. This expectation is nicely borne out by the low-x DIS data, as illustrated in Fig. 1.43 [224].

To see the influence of multiple scattering, we return to Eqs. (1.57)-(1.58), which we now apply to the cross section of a dipole of fixed size ϱ interacting with the target at a given impact parameter. We further assume that parton correlations can be neglected. In that case, the generating function is that of a Poisson distribution, $G(z) = \exp(\langle n \rangle z)$. Provided that δ is small enough we obtain

$$\sigma_{\text{tot}}^{\text{dipole}} \simeq 2\left(1 - e^{-\frac{1}{2}\langle n \rangle \delta}\right) \tag{1.76}$$

$$\sigma_{\text{diff+el}}^{\text{dipole}} \simeq 1 - 2\, e^{-\frac{1}{2}\langle n \rangle\, \delta} + e^{-\langle n \rangle \delta}. \tag{1.77}$$

Equation (1.76) is precisely of the GBW (eikonal) form Eq. (1.69). We can identify, following the previous arguments, $\langle n \rangle\, \delta$ with $1/\tau$, the effective number of partons "seen" by the $q\bar{q}$ dipole times the interaction probability.

The above formulae invite further comments on the meaning of the term "saturation". The form of Eq. (1.76) follows from that of the generating function which includes the full Glauber-Mueller multiple scattering series and which implies a levelling off of the dipole cross section. Only for a very dilute parton system, or for a very small dipole can these additional terms be neglected. On the other hand, parton recombination effects, such as gg $\to$ g fusion, when they occur, will induce a weaker $1/x$ dependence of $\langle n \rangle$, compared to that given e.g. by Eq. (1.72). Although the simple semi-classical picture suggests *two distinct origins* of saturation, which might be called "wave function saturation" and "(dipole) cross section saturation", it is not clear if such a distinction is physically justified in more rigorous treatments of the dynamics.

The model results, discussed previously in the target rest frame, can be translated, at least in leading twist, in terms of diffractive parton densities in an infinite momentum frame [223]. The diffractive structure functions are then expressed as the convolution of "diffractive" parton densities *of the proton* with parton cross sections [11]. The evolution with Q^2 at fixed x is the same as that of $F_2(x, Q^2)$. In DDIS, these scaling violations affects the β (or M_X) dependence of the cross section but not the dependence on ξ [226]. However, unlike the case of fully inclusive cross sections, the diffractive structure functions are no longer universal. In particular, they cannot be used directly for hadronic interactions.

An attempt to apply the dipole picture with saturation to soft processes was presented in [227]. There, total and elastic cross sections for hadron-hadron collisions, and the forward elastic slope, were calculated. The saturation mechanism leads to a reasonable description of the data on σ_{tot} and σ_{el} but fails badly for the forward elastic slope, the predicted energy dependence being much to weak.

1.6.2.2 The soft color interaction model

In the foregoing, it has been argued that the pomeron is not an intrinsic part of the proton but a dynamical effect of the interaction. It has also become clear that the formation of a rapidity gap in DIS final states involves long-distance soft physics. It is, therefore, plausible to assume that the underlying short-time, hard scattering sub-processes are identical in inclusive and diffractive DIS. These ideas formed the basis for semi-classical models of diffraction such as the ones proposed in [151] (see also Hebecker in [2]). In these, the formation of a rapidity gap is the result of soft gluon emissions with subsequent color neutralization.

The Soft Color Interaction (SCI) model was originally implemented in the Monte Carlo event generators LEPTO [228] for DIS and PYTHIA [229] for hadron-hadron collisions by Edin, Ingelman and Rathsman [152]. It is, in fact, equivalent to the model of soft re-scattering in DIS proposed by Brodsky et al. [230]. In both, diffraction is the result of soft re-scattering of partons involved in the hard process [10].

The re-scattering interactions are modelled as non-perturbative gluon exchanges with negligible momentum transfer. The exchange of color between the partons participating in a hard interaction and the target remnant, changes the color flow and color string topology in an event. The probability for this to happen is controlled by a free parameter, P, which specifies the probability for each parton to exchange color with the remnant partons.

The SCI model successfully describes the HERA DDIS structure function data, also in normalization [231]. The same model has been applied, without change in the parameter P, to hadron-hadron collisions and, in particular, to hard diffraction at the Tevatron. As illustrated in Table 1.2, it reproduces the data well. For further details we refer to [10] and [153].

1.7 Summary

Over the last decade, the subject of diffraction has become one of the very active fields of experimental and theoretical research in QCD. The revival is, by large, due to the extremely varied experimental program made possible at HERA and at the Tevatron.

In this chapter, we have attempted to describe, mainly in qualitative terms, the close relation between the dynamics of total cross sections and diffraction in hadron-hadron collisions and in deep-inelastic $\gamma^\star$p scattering. This inter-relationship is ultimately a consequence of the fact that at very high energy, the total and, in particular, the diffractive cross section is dominated by the wee components of the target and projectile wave functions, such that long-distance, non-perturbative, physics plays a very important role.

The basic physics can be understood on the basis of a surprisingly small number of dynamical ingredients such as the anomalous multiplicity dimension of parton cascades, which not only determines the rise with energy of the cross sections but also the spectrum of the elastically scattered proton in DDIS, the "pomeron flux".

The discussion of the overlap function illustrates that the small $|t|$ behavior of the quasi-elastic processes is also determined by the anomalous multiplicity dimension and by the transverse-momentum transfer correlation function. These ingredients suffice for a basic understanding of the degree of "shrinkage" of the forward diffractive peaks in soft as well as in hard processes.

The remarkable recent progress in our understanding of diffractive processes, in particular for high-energy ep collisions, is a consequence of the fortunate circumstance that perturbative QCD is able to make reliable predictions for the partonic fluctuations (Fock states) of a virtual photon, and for the subsequent development of these states into a parton shower or radiation cloud, at least in the earliest perturbative phase of the evolution. For a strongly bound system of large size such perturbative techniques are not available.

Bibliography

[1] H. Abramowicz, A. Caldwell, *Rev. Mod. Phys.* **71** (1999) 1275; H. Abramowicz, *Int. J. Mod.Phys.* **A15S1** (2000) 495; eConf C990809 (2000) 495.

[2] M. Wüsthoff and A.D. Martin, *J. Phys.* **G25** (1999) R309 [arXiv:hep-ph/9909362]; A. Hebecker, *Acta Phys. Pol.***B30** (1999) 3777; *Phys. Reports* **331** (2000) 1 [arXiv:hep-ph/9905226]; W. Buchmüller, *Towards the theory of diffractive DIS*, talk presented at "New Trends in HERA Physics", Ringberg Workshop, 1999, [arXiv:hep-ph/9906546].

[3] G. Ingelman, "Diffractive hard scattering", in Proc. *Advanced Study Institute on Techniques and Concepts of High Energy Physics*, edited by T. Ferbel (Kluwer, 1999), p. 597.

[4] M. Wüsthoff and A.D. Martin, *J. Phys.* **G25** (1999) R309.

[5] A. Kaidalov, *Phys. Reports* **50** (1979) 157; G. Alberi and G. Goggi, *Phys. Reports* **74** (1981) 1; K. Goulianos, *Phys. Reports* **101** (1983) 169.

[6] U. Amaldi, M. Jacob and G. Matthiae, *Ann. Rev. Nucl. Sci.* **26** (1976) 385.

[7] M.M. Block and R.N. Cahn, *Rev. Mod. Phys.* **57** (1985) 563.

[8] V. Barone and E. Predazzi, *High-energy Diffraction*, Springer-Verlag, Berlin, 2002.

[9] G. Ingelman and P. Schlein, *Phys. Lett.* **B152** (1985) 256.

[10] S.J. Brodsky, R. Enberg, P. Hoyer and G. Ingelman, *Hard diffraction from parton rescattering in QCD*, [arXiv:hep-ex/0409119].

[11] A. Berera and D.E. Soper, *Phys. Rev.* **D50** (1994) 4328, *Phys. Rev.* **D53** (1996) 6162; M. Grazzini, L. Trentadue and G. Veneziano, *Nucl. Phys.* **B519** (1998) 394; J.C. Collins, *Phys. Rev.* **D57** (1998) 3051.

[12] E.L. Feinberg, *JETP* **29** (1955) 115; A.I. Akieser and A.G. Sitenko, *JETP* **32** (1957) 744; M.L. Good and W.D. Walker, *Phys. Rev.* **D120** (1960) 1857.

[13] S.J. Brodsky, *Diffraction dissociation in QCD and light-cone wavefunctions*, talk presented at the 9th Blois Workshop On Elastic and Diffractive Scattering, Pruhonice, Prague, Czech Republic, [arXiv/hep-ph/0109205].

[14] J. Bartels and H. Kowalski, *Eur. Phys. J.* **C19** (2001) 693.

[15] R.P. Feynman, *Photon-Hadron Interactions*, W.A. Benjamin Inc., 1972.

[16] V.N. Gribov, *JETP Lett.* **41** (1961) 667.

[17] V.N. Gribov, in: *Materials of the 8th Winter School of the Leningrad Institute of Nuclear Physics 1973* (in Russian); [arXiv:hep-ph/0006158] (English translation).

[18] A. Hebecker, *Phys. Reports* **331** (2000) 1, [arXiv:hep-ph/9905226].

[19] P.D.B. Collins, *An Introduction to Regge Theory and High Energy Physics*, Cambridge University Press, 1977.

[20] S.C. Frautschi, M. Gell-Mann and F. Zachariasen, *Phys. Rev.* **D126** (1962) 2204.

[21] I.Ya. Pomeranchuk, *Sov. Phys. JETP* **34** (1958) 499.

[22] A. Donnachie and P.V. Landshoff, *Phys. Lett.* **B348** (1995) 213; G.A. Jaroszkiewicz and P.V. Landshoff, *Phys. Rev.* **D10** (1974) 170.

[23] A. Gaidot et al., *Phys. Lett.* **B57** (1975) 389.

[24] A.H. Mueller, *Phys. Rev.* **D2** (1970) 2963.

[25] R.C. Brower and J.H. Weiss, *Rev. Mod. Phys.* **47** (1975) 605.

[26] M. Froissart, *Phys. Rev.* **D123** (1961) 1053; L. Łukaszuk and A. Martin, *Nuovo Cimento* **A52** (1967) 122.

[27] K. Goulianos and J. Montanha, *Phys. Rev.* **D59** (1999) 114017.

[28] ZEUS Coll., J. Breitweg et al., *Eur. Phys. J.* **C6** (1999) 43.

[29] H1 Coll., *Measurement of diffractive structure function $F_2^{D(3)}$ at HERA*, paper 089 subm. to Int. Europhysics Conf. on High Energy Physics, EPS03, July 2003, Aachen, Germany.

[30] Review of Particle Physics, S. Eidelman et al., *Phys. Lett.* **B592** (2004) 1.

[31] A. Donnachie and P.V. Landshoff, *Phys. Lett.* **B296** (1992) 227.

[32] S.P. Denisov et al., *Phys. Lett.* **B36** (1971) 415.

[33] NA22 Coll., M. Adamus et al., *Phys. Lett.* **B186** (1987) 223.

[34] U. Amaldi et al., *Phys. Lett.* **B44** (1973) 112; S.R. Amendolia et al., *Phys. Lett.* **B44** (1973) 119.

[35] A.S. Carroll et al., *Phys. Lett.* **B61** (1976) 303.

[36] H1 Coll., S. Aid et al., *Z. Phys.* **C69** (1995) 27; ZEUS Coll., S. Chekanov et al., *Nucl. Phys.* **B627** (2002) 3.

[37] J.R. Cudell, K. Kang, S.K. Kim, *Phys. Lett.* **B395** (1997) 311; ibid. *Simple model for total cross-sections*, [`arXiv:hep-ph/9701312`].

[38] J.R. Cudell et al., *Soft pomeron and lower-trajectory intercepts*, [`arXiv:hep-ph/9812429`]; K. Kang et al., *Analytic Amplitude Models for Forward Scattering*, [`arXiv:hep-ph/0111360`].

[39] COMPETE Coll., J.R. Cudell et al., *Phys. Rev.* **D65** (2002) 074024.

[40] R. Covolan, J. Montanha and K. Goulianos, *Phys. Lett.* **B389** (1996) 176.

[41] E.G.S. Luna and M.J. Menon, *Phys. Lett.* **B565** (2003) 123; R.F. Avila, E.G.S. Luna, M.J. Menon, *Phys. Rev.* **D67** (2003) 054020.

[42] M.S. Dubovikov and K.A. Ter-Martirosyan, *Nucl. Phys.* **B214** (1997) 163.

[43] A. Donnachie and P.V. Landshoff, *Z. Phys.* **C61** (1993) 139.

[44] T.K. Gaisser, U.P. Sukhatme and G.B. Yodh, *Phys. Rev.* **D36** (1987) 1350.

[45] N.N. Nikolaev, *Phys. Rev.* **D48** (1993) R1904.

[46] M.M. Block, F. Halzen and T. Stanev, *Phys. Rev. Lett.* **83** (1999) 4926; *Phys. Rev.* **D62** (2000) 077501.

[47] COMPETE Coll., J.R. Cudell et al., *Phys. Rev.* **D61** (2000) 034019; Erratum *Phys. Rev.* **D63** (2001) 059901.

[48] P. Desgrolard, M. Giffon and E. Martynov, *Nuovo Cimento* **A110** (1997) 537.

[49] CDF Coll., F. Abe et al., *Phys. Rev.* **D50** (1993) 5550.

[50] E710 Coll., N.A. Amos et al., *Phys. Rev. Lett.* **68** (1992) 2433.

[51] E811 Coll., C. Avila et al., *Phys. Lett.* **B445** (1999) 419.

[52] E811 Coll., C. Avila et al., *Phys. Lett.* **B537** (2002) 41.

[53] E.M. Levin and L.L. Frankfurt, *Pisma ZhETP* **3** (1965) 105;
H.J. Lipkin and F. Scheck, *Phys. Rev. Lett.* **16** (1966) 71;
J.J.J. Kokkedee and L. Van Hove, *Nuovo Cimento* **A42** (1966) 711;
J.J.J. Kokkedee, The Quark Model, W.A. Benjamin, Inc., New York, 1969;
V.V. Anisovich, Yu.M. Shabelsky and V.M. Shekhter, *Nucl.Phys.* **B133** (1978) 477;
E.S. Martynov, Proc. Workshop "DIQUARKS II", Eds. M. Anselmino and E. Predazzi (World Scientific, Singapore, 1994) p.45.

[54] J.F. Gunion and D.E. Soper, *Phys. Rev.* **D15** (1977) 2617.

[55] E.M. Levin and M.G. Ryskin, *Yad. Fiz. (Sov. Nucl. Phys.)* **34** (1981) 1114.

[56] F. E. Low, *Phys. Rev.* **D12** (1975) 163; S. Nussinov, *Phys. Rev. Lett.* **34** (1975) 1286.

[57] G. Bertsch, S.J. Brodsky, A.S. Goldhaber and J.F. Gunion, *Phys. Rev. Lett.* **47** (1981) 297; A.B. Zamolodchikov, B.Z. Kopeliovich and L.I. Ladidus, *JETP Lett.* **33** (1981) 595.

[58] P. Desgrolard et al., *Eur. Phys. J.* **C9** (1999) 623; *Eur. Phys. J.* **C18** (2000) 359.

[59] R. Engel, *Soft interactions*, Proc. XXXI Int. Symp. on Multiparticle Dynamics, eds. Y. Bai, M. Yu and Y. Wu (World Scientific, Singapore, 2002) p.15.

[60] V.A. Abramovski, V.N. Gribov and O.V. Kancheli, *Sov. J. Nucl. Phys.* **18** (1974) 308.

[61] P.V. Chliapnikov, A.K. Likhoded and V.A. Uvarov, *Phys. Lett.* **B215** (1988) 417.

[62] A.B. Kaidalov, L.A. Ponomarev and K.A. Ter-Martirosyan, *Sov. J. Nucl. Phys.* **44** (1986) 468.

[63] V.A. Khoze, A.D. Martin and M.G. Ryskin, *Eur. Phys. J.* **C18** (2000) 167.

[64] H1 Coll., S. Aid et al., *Nucl. Phys.* **B470** (1996) 4; ZEUS Coll. M. Derrick et al., *Z. Phys.* **C72** (1996) 399; H1 Coll., C. Adloff et al., *Nucl. Phys.* **B497** (1997) 3; ZEUS Coll., J. Breitweg et al., *Phys. Lett.* **B407** (1997) 432; ZEUS Coll., J. Breitweg et al., *Eur. Phys. J.* **C7** (1999) 609; E665 Coll., M.R. Adams et al., *Phys. Rev.* **D54** (1996) 3006; ZEUS Coll., J. Breitweg et al., *Phys. Lett.* **B487** (2000) 53.

[65] K. Golec-Biernat and M. Wüsthoff, *Phys. Rev.* **D59** (1999) 014017; ibid. **D60** (1999) 114023; J. Bartels, K.Golec-Biernat and H. Kowalski, *Phys. Rev.* **D66** (2002) 014001.

[66] A.D. Martin, R.G. Roberts and W.J. Stirling, *Phys. Lett.* **B387** (1996) 419.

[67] H1 Coll., paper 082 subm. to Int. Europhysics Conf. on High Energy Physics, EPS03, July 2003, Aachen, Germany.

[68] H1 Coll., C. Adloff et al., *Phys. Lett.* **B520** (2001) 183.

[69] T. Laštovička, *Eur. Phys. J.* **C24** (2002) 529; *Acta Phys. Pol.* **B33** (2002) 2867.

[70] H1 Coll., C. Adloff et al., *Eur. Phys. J.* **C21** (2001) 33.

[71] P. Newman, *Int. J. Mod. Phys.* **A19** (2004) 1061.

[72] ZEUS Coll., J. Breitweg et al., *Eur. Phys. J.* **C7** (1999) 609.

[73] Yu.L. Dokshitzer, *Sov. Phys. JETP* **46** (1977) 641; V.N. Gribov and L.N. Lipatov, *Sov. J. Nucl. Phys.* **15** (1972) 438; G. Altarelli and G. Parisi, *Nucl. Phys.* **B216** (1977) 298.

[74] A. Donnachie and P.V. Landshoff, *Phys. Lett.* **B518** (2001) 63.

[75] H1 Coll., C. Adloff et al., *Eur. Phys. J.* **C30** (2003) 1.

[76] ZEUS Coll., S. Chekanov et al., *Phys. Rev.* **D67** (2003) 012007.

[77] J. Pumplin *et al.*, *JHEP* **0207** (2002) 01.

[78] A. Martin et al., *Eur. Phys. J.* **C23** (2002) 73.

[79] W. Furmanski and R. Petronzio, *Phys. Lett.* **B97** (1980) 437.

[80] L.V. Gribov, E.M. Levin and M.G. Ryskin, *Phys. Reports* **C100** (1983) 1; *Nucl. Phys.* **B188** (1981) 555.

[81] L. McLerran and R. Venugopolan, *Phys. Rev.* **D49** (1994) 2233; **D49** (1994) 3352; **D50** (1994) 2225. For a recent review see also E. Iancu and R. Venugopolan, `[arXiv:hep-ph/0303204]`.

[82] A. Donnachie and P.V. Landshoff, *Phys. Lett.* **B518** (2001) 63, [arXiv:hep-ph/0105088]; A. Capella, E.G. Ferreiro, A.B. Kaidalov and C.A. Salgado, Proc. 36th Rencontre de Moriond on QCD and Hadronic Interactions, ed. J. Tran Thanh Van (Gioi, Vietnam, 2002) p.239; A. Capella, A. Kaidalov, C. Merino and J. Tran Thanh Van, *Phys. Lett.* **B337** (1994) 358.

[83] J.J. Sakurai, *Currents and Mesons*, University of Chicago Press, Chicago, 1969); H. Fraas and D. Schildknecht, *Nucl. Phys.* **B14** (1969) 543; T.H. Bauer et al., *Rev. Mod. Phys.* **50** (1978) 261, Erratum ibid. **51** (1979) 407; G. Cvetic, D. Schildknecht, B. Surrow and M. Tentyukov, *Eur. Phys. J.* **C20** (2001) 77.

[84] D. Haidt, *Nucl. Phys. Proc. Suppl.* **96** (2001) 166.

[85] A. De Roeck and E.A. De Wolf, *Phys. Lett.* **B388** (1996) 843.

[86] H1 Coll., C. Adloff et al., *Eur. Phys. J.* **C21** (2001) 33.

[87] ZEUS Coll., J. Breitweg et al., *Phys. Lett.* **B487** (2000) 53.

[88] H1 Coll., S. Aid et al., *Z. Phys.* **C69** (1995) 27.

[89] D.K. Choudhury and R. Gogoi, *On fractal dimension of the proton at small x*, [arXiv:hep-ph/0310260].

[90] T. Meng, R. Rittel, Y. Zhang, *Phys. Rev. Lett.* **82** (1999) 2044.

[91] C. Boros, Meng Ta-chung, R. Rittel, K. Tabelow, Y. Zhang, *Phys. Rev.* **D61** (2000) 094010.

[92] R. Rittel, *Nucl. Phys.* (Proc. Suppl.) **99A** (2001) 270.

[93] pp2pp Coll., S.L. Bueltmann et al., *Phys. Lett.* **B579** (2004) 245.

[94] H.G. Miettinen, in Proc. IX th Rencontre de Moriond, Meribel les Allues, Vol. **1**, ed. J. Tran Thanh Van (Editions Frontieres, Orsay, 1974) p.363; U. Amaldi and K.R. Schubert, *Nucl. Phys.* **B166** (1980) 301.

[95] D.S. Ayres et al., *Phys. Rev.* **D14** (1976) 3092.

[96] S.J. Brodsky et al., *Phys. Rev.* **D50** (1994) 3134.

[97] J.A. Crittenden, *Exclusive Production of Neutral Vector Mesons at the Electron-Proton Collider HERA*, Springer Tracts in Modern Physics, Vol. **140**, p.1, Springer, Berlin and Heidelberg, 1997, and references therein.

[98] A. Levy, *Measurements of diffractive processes at HERA*, [arXiv:hep-ex/0301022].

[99] Kai C. Voss, *Vector Meson Production at HERA*, to appear in Proc. 37th Rencontre de Moriond on QCD and Hadronic Interactions, ed. J. Tran Thanh Van (Gioi, Vietnam, 2003).

[100] H1 Coll., C. Adloff et al., *Phys. Lett.* **B483** (2000) 23.

[101] ZEUS Coll., S. Chekanov et al. *Eur. Phys. J.* **C24** (2002) 345.

[102] H1 Coll., *Elastic Photoproduction of* J/ψ *Mesons at HERA*, paper 6-0180 subm. to 32nd Int. Conference on High Energy Physics, ICHEP04, Beijing, 2004.

[103] E401 Coll., M. Binkley et al., *Phys. Rev. Lett.* **48** (1982) 73.

[104] E516 Coll., B.H. Denby et al., *Phys. Rev. Lett.* **52** (1984) 795.

[105] L. Frankfurt, M. McDermott and M. Strikman, *JHEP* **103** (2001) 45.

[106] A.D. Martin, M.G. Ryskin and T. Teubner, *Phys. Lett.* **B454** (1999) 339; *Phys. Rev.* **D45** (2000) 14022 .

[107] J.R. Forshaw, R. Sandapen and G. Shaw, *Phys. Rev.* **D69** (2004) 094013.

[108] J.D. Bjorken, AIP Conf. Proc. No. 6, Eds. M. Bander, G. Shaw and W. Wong (AIP, New York, 1972) p. 151; J.D. Bjorken, J.B. Kogut and D.E. Soper, *Phys. Rev.* **D3** (1971) 1382; J.D. Bjorken and J.B. Kogut, *Phys. Rev.* **D8** (1973) 1341.

[109] N.N. Nikolaev and B.G. Zakharov, *Z. Phys.* **C49** (1991) 607; **64** (1994) 631.

[110] J. Nemchik, N.N. Nikolaev and B.G. Zakharov, *Phys. Lett.* **B341** (1994) 228.

[111] ZEUS Coll., J. Breitweg et al., *Eur. Phys. J.* **C6** (1999) 603; **C2** (1998) 247.

[112] A. Kreisel for the ZEUS Coll., Proc. 9th Int. Workshop on Deep Inelastic Scattering (DIS 2001), eds. G. Bruni, G. Iacobucci, R. Nania (World Scientific, Singapore, 2002) p. 699.

[113] H1 Coll., paper 989 subm. to 31st Int. Conf. on High Energy Physics, ICHEP02, July 2002, Amsterdam, The Netherlands.

[114] H1 Coll., C. Adloff et al., *Phys. Lett.* **B483** (2000) 23.

[115] ZEUS Coll., paper 813 subm. to 31st Int. Conf. on High Energy Physics, ICHEP02, July 2002, Amsterdam, The Netherlands.

[116] H1 Coll., C. Adloff et al., *Eur. Phys. J.* **C10** (1999) 373.

[117] S. Munier, A.M. Stásto and A.H. Mueller, *Nucl. Phys.* **B603** (2001) 427.

[118] G. Iacobucci, *Int. J. Mod. Phys.* **A17** (2002) 3204.

[119] H. Abramowicz, L. Frankfurt and M. Strikman, *Surveys High Energy Phys.* **11** (1997) 51.

[120] S. Chekanov et al., *Nucl. Phys.* **B695** (2004) 3.

[121] S.J. Brodsky et al., *JETP Lett.* **70** (1999) 155.

[122] L. Frankfurt, M. McDermott and M. Strikman, *JHEP* **103** (2001) 45.

[123] N.N. Nikolaev, B.G. Zakharov and V.R. Zoller, *Phys. Lett.* **B366** (1996) 337; J. Nemchik et al., *JETP* **86** (1998) 1054.

[124] ZEUS Coll., *Exclusive Electroproduction of ρ^0 mesons at HERA*, paper 594 subm. to Int. Europhysics Conf., July 2001, Budapest, Hungary.

[125] K. Schilling and G. Wolf, *Nucl. Phys.* **B61** (1973) 381.

[126] P. Joos et al., *Nucl. Phys.* **B113** (1976) 53.

[127] D.G. Cassel et al., *Phys. Rev.* **D24** (1981) 2787.

[128] A.B. Borissov et al., HERMES Report 01-060 (2001).

[129] M. Tytgat, PhD thesis, HERMES Report 01-014 (2001).

[130] W.D. Shambroom et al., *Phys. Rev.* **D26** (1982) 1.

[131] E665 Coll., M.R. Adams et al., *Z. Phys.* **C74** (1997) 237.

[132] H1 Coll., C. Adloff et al., *Eur. Phys. J.* **C13** (2000) 371; S. Aid et al., *Nucl. Phys.* **B468** (1996) 3.

[133] ZEUS Coll., J. Breitweg et al., *Eur. Phys. J.* **C12** (2000) 393.

[134] NMC Coll., N. Arneodo et al., *Nucl. Phys.* **B429** (1994) 503.

[135] CLAS Coll., C. Hadjidakis et al., *Exclusive ρ^0 meson electroproduction from hydrogen at CLAS*, [`arXiv:hep-ex/0408005`].

[136] K. Goulianos, *Nucl. Phys. Proc. Suppl.* **A99** (2001) 9.

[137] K. Goulianos, in Proc. IIIth Rencontre de Moriond, Les Arcs, Vol. **1**, ed. J. Tran Thanh Van (Editions Frontieres, Orsay, 1974) p. 457.

[138] M.G. Albrow et al., *Nucl. Phys.* **B51** (1973) 388; **72** (1974) 376.

[139] A.C. Melissinos and S.C. Olsen, *Phys. Reports* **17** (1975) 77.

[140] See e.g. M. Le Bellac, *Short-range order and local conservation of quantum numbers in multiparticle production*, CERN Yellow Report 76-14; *Acta Phys. Pol.* **B4** (1973) 901.

[141] K. Goulianos, *Phys. Lett.* **B358** (1995) 379; Erratum ibid. **B363** (1995) 268.

[142] H. Miettinen and J. Pumplin, *Phys. Rev.* **D18** (1978) 1696.

[143] ZEUS Coll., M. Derrick et al., *Phys. Lett.* **B315** (1993) 481; *Phys. Lett.* **B346** (1995) 399; H1 Coll., T. Ahmed et al., *Nucl. Phys.* **B429** (1994) 477; *Nucl. Phys.* **B345** (1995) 3.

[144] A. Donnachie and P.V. Landshoff, *Phys. Lett.* **B191** (1987) 309.

[145] H1 and ZEUS Coll., V. Monaco, *Diffractive structure functions*, Proc. 36th Rencontre de Moriond on QCD and Hadronic Interactions, ed. J. Tran Thanh Van (Gioi, Vietnam, 2002) p.247, [`arXiv:hep-ex/0106007`].

[146] ZEUS Coll., paper 823 subm. to 31st Int. Conf. on High Energy Physics, ICHEP02, July 2002, Amsterdam, The Netherlands.

[147] H1 Coll., paper 980 subm. to 31st Int. Conf. on High Energy Physics, ICHEP02, July 2002, Amsterdam, The Netherlands.

[148] H. Abramowicz and A. Levy, report DESY 97-251 (1997).

[149] ZEUS Coll., paper 821 subm. to 31st Int. Conf. on High Energy Physics, ICHEP02, July 2002, Amsterdam, The Netherlands.

[150] E. Gotsman et al., *Energy dependence of σ^{DD}/σ_{tot} in DIS and shadowing corrections*, [arXiv:hep-ph/0007261].

[151] W. Buchmüller, *Phys. Lett.* **B353** (1995) 335; W. Buchmüller and A. Hebecker, *Phys. Lett.* **B355** (1995)573; W. Buchmüller and A. Hebecker, *Nucl. Phys.* **B476** (1996) 203; W. Buchmüller, T. Gehrmann and A. Hebecker, *Nucl. Phys. Proc. Suppl.* **79** (1999) 263; W. Buchmüller, *Towards a theory of Diffractive DIS*, report DESY 99-076, [arXiv:hep-ph/9906546] and references therein.

[152] A. Edin, G. Ingelman and J. Rathsman, *Phys. Lett.* **B366** (1996) 371; *Z. Phys.* **C75** (1997) 57.

[153] R. Enberg, G. Ingelman and N. Tîmneanu, *J. Phys.* **G21** (2000) 712; ibid. *Phys. Rev.* **D64** (2001) 114015.

[154] H1 Coll., paper 987 subm. to 31st Int. Conf. on High Energy Physics, ICHEP02, July 2002, Amsterdam, The Netherlands.

[155] ZEUS Coll., paper 832 subm. to 31st Int. Conf. on High Energy Physics, ICHEP02, July 2002, Amsterdam, The Netherlands.

[156] H1 Coll., *Measurement of semi-inclusive diffractive deep-inelastic scattering with a leading proton at HERA*, paper 984 subm. to 31st Int. Conf. on High Energy Physics, ICHEP02, July 2002, Amsterdam, The Netherlands.

[157] ZEUS Coll., J. Breitweg et al., *Z. Phys.* **C1** (1998) 81.

[158] B. Smalska, *Diffractive results from the ZEUS LPS at HERA*, talk given at DIS2001, Bologna; Proc. 9th Int. Workshop on Deep Inelastic Scattering (DIS 2001), eds. G. Bruni, G. Iacobucci, R. Nania (World Scientific, Singapore, 2002) p. 761; ZEUS Coll., *Diffractive results from the ZEUS Leading Proton Spectrometer at HERA*, paper 823 subm. to 31st Int. Conf. on High Energy Physics, July, 2002, Amsterdam, The Netherlands.

[159] UA8 Coll., A. Brandt et al., *Phys. Lett.* **B297** (1992) 417; ibid. **B421** (1998) 395; R. Bonino et al., *Phys. Lett.* **B211** (1988) 239.

[160] H. Abramowicz, *Proc. of the 19th Int. Symp. on Photon and Lepton Interactions at High Energy LP99*, eds. J.A. Jaros and M.E. Peskin, *Int. J. Mod. Phys.* **A 15S1** (2000) 495.

[161] D0 Coll., B. Abbot et al., *Phys. Lett.* **B440** (1998) 189.

[162] D0 Coll., B. Abbott et al., *Phys. Lett.* **B531** (2002) 52.

[163] D0 Coll., V.M. Abazov et al., *Phys. Lett.* **B574** (2003) 169.

[164] CDF Coll., F. Abe et al., *Phys. Rev.* **D50** (1994) 5535.

[165] CDF Coll., F. Abe et al., *Phys. Rev. Lett.* **78** (1997) 2698.

[166] CDF Coll., F. Abe et al., *Phys. Rev. Lett.* **79** (1997) 2636.

[167] CDF Coll., T. Affolder et al., *Phys. Rev. Lett.* **84** (2000) 5043.

[168] CDF Coll., T. Affolder et al., *Phys. Rev. Lett.* **85** (2000) 4215.

[169] CDF Coll., T. Affolder et al., *Phys. Rev. Lett.* **87** (2001) 141802.

[170] CDF Coll., T. Affolder et al., *Phys. Rev. Lett.* **87** (2001) 241802.

[171] CDF Coll., D. Acosta et al., *Phys. Rev. Lett.* **88** (2002) 151802.

[172] CDF Coll., D. Acosta et al., *Phys. Rev. Lett.* **91** (2003) 011802.

[173] CDF Coll., D. Acosta et al., *Phys. Rev. Lett.* **93** (2004) 141601.

[174] CDF Coll., F. Abe et al., *Phys. Rev. Lett.* **80** (1998) 1156; *Phys. Rev. Lett.* **81** (1998) 5278.

[175] K. Goulianos, *Nucl. Phys.* **B** (Proc. Suppl.) **99** (2001) 9; ibid. p.37.

[176] K. Goulianos, *Hadronic Diffraction, where do we stand?*, Proc. 18th Les Rencontres de Physique de la Vallé d'Aoste, La Thuile, Aosta Valley, Italy, 2004, [ArXiv:hep-ph/0407035].

[177] CDF Coll., T. Affolder et al., *Phys. Rev. Lett.* **84** (2000) 232.

[178] V. Del Duca, W.K. Tang, *Phys. Lett.* **B312** (1993) 225.

[179] O.J.B. Eboli, E.M. Gregores, F. Halzen, *Nucl. Phys. Proc. Suppl.* **71** (1999) 349; *Phys. Rev.* **D61** 034003.

[180] A.B. Kaidalov, V.A. Khoze, A.D. Martin, and M.G. Ryskin, *Eur. Phys. J.* **C21** (2001) 521.

[181] A.B. Kaidalov, V.A. Khoze, A.D. Martin and M.G. Ryskin, *Phys. Lett.* **B559** (2003) 235.

[182] A.B. Kaidalov, V.A. Khoze, A.D. Martin and M.G. Ryskin, *Diffraction of Protons and Nuclei at High Energies*, [arXiv:hep-ph/0303111].

[183] ZEUS Coll., M. Derrick et al., *Phys. Lett.* **B356** (1995) 129.

[184] P. Bruni, G. Ingelman, Proc. Europhysics Conf. on High Energy Physics, eds. J. Carr, M. Perrottet (Editions Frontières, Gif-sur-Yvette, France, 1994) p. 595 and report DESY-93-187.

[185] P. Bruni and G. Ingelman, *Phys. Lett.* **B311** (1993) 317.

[186] L. Alvero et al., *Phys. Rev.* **D59** (1999) 074022.

[187] L. Alvero, J.C. Collins, J.J. Whitmore, arXiv:hep-ph/9806340.

[188] J. Bjorken, *Phys. Rev.* **D47** (1993) 101; J.C. Collins, L. Frankfurt, M. Strikman, *Phys. Lett.* **B307** (1993) 161; E. Gotsman, E. Levin and U. Maor, *Phys. Lett.* **B309** (1993) 199; ibid. *Phys. Rev.* **D60** (1999) 094011.

[189] A. Białas, *Acta Phys. Pol.* **B33** (2002) 2635; A. Białas and R. Peschanski, *Phys. Lett.* **B575** (2003) 30.

[190] L. Van Hove, *Phys. Lett.* **B7** (1963) 69, *Nuovo Cimento* **28** (1963) 798, *Rev. Mod. Phys.* **36** (1964) 655.

[191] V. Kundrat and M. Lokajicek, *High-energy elastic hadron collisions and space structure of hadrons*, [arXiv:hep-ph/0001047].

[192] B. Andersson, *The Lund Model*, Cambridge, Cambridge University Press, 1998.

[193] Z. Koba and M. Namiki, *Nucl. Phys.* **B8** (1968) 413.

[194] A. Krzywicki, *Nucl. Phys.* **B86** (1975) 296; P. Grassberger et al., *Nucl. Phys.* **B89** (1975) 101.

[195] E. Levin, *An Introduction to Pomerons*, [arXiv:hep-ph/9808486]; ibid. *Everything about reggeons*, [arXiv:hep-ph/9710546].

[196] J. Pumplin, *Phys. Rev.* **D8** (1973) 2899.

[197] B. Kopeliovich, B. Povh and E. Predazzi, *Phys. Lett.* **B405** (1997) 361; B. Kopeliovich and B. Povh, *Z. Phys.* **A356** (1997) 467.

[198] N.N. Nikolaev, B.G. Zakharov and V.R. Zoller, *JETP Lett.* **60** (1994) 694.

[199] A.H. Mueller, *Small-x Physics and Parton Saturation in QCD* , [arXiv:hep-ph/9911289].

[200] L. Landau and I.Ya. Pomeranchuk, *Dokl. Akad. Nauk Ser. Fiz.* **92**, (1953) 535.

[201] V.N. Gribov, B.L. Ioffe and I.Ya. Pomeranchuk, *Sov. J. Nucl. Phys.* **2** (1966) 549; *Yad. Fiz.* **2** (1965) 768.

[202] B.L. Ioffe, *JETP Lett.* **9** (1969) 163; *JETP Lett.* **10** (1969) 143; *Phys. Lett.* **B30** (1969) 123.

[203] Y.N. Kovchegov and M. Strikman, *Phys. Lett.* **B516** (2001) 314.

[204] L.N. Lipatov, *Sov. J. Nucl. Phys.* **23** (1976) 338; V.S. Fadin, E.A. Kuraev and L.N. Lipatov, *Phys. Lett.* **B60** (1975) 50; *Sov. Phys. JETP* **44** (1976) 443, **45** (1977) 199; Y.Y. Balitski and L.N. Lipatov, *Sov. J. Nucl. Phys.* **28** (1978) 822.

[205] M. Ciafaloni, *Nucl. Phys.* **B296** (1987) 249; S. Catani, F. Fiorani and G. Marchesini, *Phys. Lett.* **B234** (1990) 339; *Nucl. Phys.* **B336** (1990) 18.

[206] A.M. Polyakov, *Sov. Phys. JETP* **32** (1971) 296, **33** (1971) 850; S.J. Orfanidis and V. Rittenberg, *Phys. Rev.* **D10** (1974) 2892; G. Cohen-Tannoudji and W. Ochs, *Z. Phys.* **C39** (1988) 513; W. Ochs, *Z. Phys.* **C23** (1984) 131.

[207] Yu.L. Dokshitzer, V.A. Khoze, A.H. Mueller and S.I. Troyan, *Basics of Perturbative QCD*, Editions Frontières, 1991.

[208] A.H. Mueller and J.-W Qiu, *Nucl. Phys.* **B268** (1986), 427; A.H. Mueller, *Small-x Physics, High Parton Densities and Parton Saturation in QCD*, Proc. Particle Production Spanning MeV and TeV Energies, eds. W. Kittel, P.J. Mulders, O. Scholten (Kluwer Academic, Dordrecht, 2000) p. 71; ibid. *Parton Saturation: an overview*, [arXiv:hep-ph/0111244].

[209] A.H. Mueller, *Phys. Rev.* **D4** (1971) 150.

[210] E.A. De Wolf, I.M. Dremin and W. Kittel, *Phys. Reports* **270** (1996) 1.

[211] A. Białas and R. Peschanski, *Phys. Lett.* **B378** (1996) 302.

[212] S. Sapeta, *Phys. Lett.* **B597** (2004) 352.

[213] A.H. Mueller, *Nucl. Phys.* **B415** (1994) 373.

[214] J.R. Forshaw and D.A. Ross, *QCD and the Pomeron*, Cambridge University Press, 1996.

[215] H. Kowalski, D. Teaney, *Phys. Rev.* **D68** (2003) 114005.

[216] L. Frankfurt, G.A. Miller and M. Strikman, *Phys. Lett.* **B304** (1993) 1.

[217] I.M. Dremin, J.W. Gary, *Phys. Lett.* **B459** (1999) 341.

[218] OPAL Coll., P.D. Acton et al., *Z. Phys.* **C53** (1992) 539.

[219] A. Capella et al., *Phys. Rev.* **D61** (2000) 074009.

[220] I.M. Dremin, J.W. Gary, *Phys. Reports* **349** (2001) 301.

[221] P. Eden, *Eur. Phys. J.* **C19** (2001) 493.

[222] K. Golec-Biernat, K. Motyka and A.M. Stásto, *Phys. Rev.* **D65** (2002) 074037.

[223] K. Golec-Biernat, M. Wüsthoff, *Eur. Phys. J.* **C20** (2001) 313.

[224] A.M. Staśto, K. Golec-Biernat and J. Kwieciński, *Phys. Rev. Lett.* **86** (2001) 596.

[225] L. McLerran, *AIP Con. Proc.* **490** (1999) 42 [arXiv:hep-ph/9903536].

[226] J. Blümlein and D. Robaschik, *Phys. Lett.* **B517** (2001) 222.

[227] J. Bartels, E. Gotsman, E. Levin, M. Lublinsky and U. Maor, *Phys. Lett.* **B556** (2003) 114.

[228] G. Ingelman, A. Edin and J. Rathsman, *Comput. Phys. Commun.* **101** (1997) 108.

[229] T. Sjöstrand, *Comput. Phys. Commun.* **82** (1994) 74.

[230] S. J. Brodsky, P. Hoyer, N. Marchal, S. Peigné and F. Sannino, *Phys. Rev.* **D65** (2002) 114025.

[231] A. Edin, G. Ingelman and J. Rathsman, in Proc. *Monte Carlo generators for HERA physics*, DESY-PROC-1999-02, p. 280, [arXiv:hep-ph/9912539].

Chapter 2

Inclusive and Exclusive Data Analysis in LPS, Event Shape

2.1 General scheme

The most general reaction of two primary particles A (projectile) and B (target) giving n_f secondary particles C_i reads

$$A + B \rightarrow C_1 + C_2 + \ldots C_n + C_{n+1} + \ldots C_{n_f} \tag{2.1}$$

The right-hand side of 2.1 will be called a *final state* . The multiplicity of final state f, $n_f \geq 2$, is in principle bounded only by available phase space. In practice, except for heavy-ion collisions, its average is of order 5-50 in the reactions we are going to study, much lower than allowed kinematically.

In reaction (2.1), the final state particles $C_1, \ldots, C_n$ are "observed", i.e. their nature is known and their four-momentum is measured. On the other hand, particles $C_{n+1}, \ldots, C_{n_f}$ are "unobserved", either because they are neutral or otherwise not detectable, or because there are just too many of them.

The cross section for final state f can be formally expressed as

$$\sigma_f = \frac{1}{p^*_{\text{in}} s^{1/2}} \int |M_f|^2 \delta\left(\sum_1^{n_f} \mathbf{p}_j - \mathbf{P}\right) \delta\left(\sum_1^{n_f} E_j - E\right) \prod_1^{n_f} \frac{\mathrm{d}^3\mathbf{p}_i}{E_i} \quad , \tag{2.2}$$

where summation over all helicity states is understood and where M_f is the Lorentz-invariant transition matrix element (the big unknown in our game) multiplied by all missing constant factors. The three-vectors $\mathbf{p}_i$ are the momenta and E_i the energies of the n_f outgoing particles (only n of which are actually known), $\mathbf{P}$ and E are the (known) momentum and energy of the total incoming or outgoing system. The product of p^*_{in}, the incoming momentum in the cms and $s^{1/2}$, the total energy in the cms, corresponds to the Møller flux factor.

All possible final states can be considered part of the total cross section

$$\sigma_{\text{tot}} = \frac{1}{p_{\text{in}}^* s^{1/2}} \sum_f \int |M_f|^2 \delta \left(\sum_1^{n_f} \mathbf{p}_j - \mathbf{P} \right) \delta \left(\sum_1^{n_f} E_j - E \right) \prod_1^{n_f} \frac{\mathrm{d}^3 \mathbf{p}_i}{E_i} \quad . \tag{2.3}$$

We have started our considerations with this total cross section (Chapter 1), because it is measurable even when we know nothing (or pretend to know nothing) about the individual final states.

With the energy-momentum four-vectors of n particles known, the Lorentz-invariant n-particle distribution [1, 2] in momentum space reads

$$\begin{aligned} \left(\prod_1^n E_i \right) \frac{\mathrm{d}^{3n} \sigma_n}{\prod\limits_1^n \mathrm{d}^3 \mathbf{p}_i} = \frac{1}{p_{\text{in}}^* s^{1/2}} \Bigg[& |M_f|^2 \delta \left(\sum_1^n \mathbf{p}_j - \mathbf{P} \right) \delta \left(\sum_1^n E_j - E \right) + \\ & + \sum_f{}' \int |M_f|^2 \delta \left(\sum_1^{n_f} \mathbf{p}_j - \mathbf{P} \right) \delta \left(\sum_1^{n_f} E_j - E \right) \prod_{n+1}^{n_f} \frac{\mathrm{d}^3 \mathbf{p}_i}{E_i} \Bigg] \quad , \end{aligned} \tag{2.4}$$

where the sum stands now for summation over final states and permutations in momentum space of particles among $\mathrm{C}_{n+1}, \ldots, \mathrm{C}_{n_f}$ identical to one or more among $\mathrm{C}_1, \ldots, \mathrm{C}_n$.

In the language of Feynman [3], the first term in (2.4) is the *n-particle exclusive* part [2] of the full distribution and depends on the squared cms energy s and $(3n - 4)$ further variables in momentum space. The sum is the *n-particle inclusive* distribution [3, 4] depending on s and $3n$ variables:

$$f_n(s, \mathbf{p}_1, \ldots \mathbf{p}_n) = \left(\prod_1^n E_i \right) \frac{\mathrm{d}^{3n} \sigma_n}{\prod\limits_1^n \mathrm{d}^3 \mathbf{p}_i} \quad , \tag{2.5}$$

although it contains δ-functions related to $n_f = n + 1$, which could still be counted exclusive.

We observe, that going with n from small values to high ones, we project out the exclusive distributions, one after the other, as well as the inclusive distributions of various order. That means that all information is contained in a complete expansion into exclusive final states or, alternatively, in a complete expansion into inclusive distributions of various order. Academically speaking, one can, therefore, fully express an exclusive distribution in terms of inclusive distributions and vice versa. In practice, however, we cannot go high enough in n to do this because of limited statistics and limited imagination. So, we will have to do with projections, projections in momentum space as above and projections in channel space, the latter allowing to study one single reaction at a time. Possible examples of such projections are the total cross section to be understood as inclusive cross section of order zero or the commonly used single-particle distribution, which is of order 1. Obviously, correlations between produced particles can be studied from order two inclusive onwards, or, alternatively, with exclusive distributions of order $n > 2$.

For further connection between the inclusive and exclusive way to study the same thing and for various combinations of these two extreme approaches we refer to [5] and to what follows. Here, we only want to conclude: since the number of particles we can study simultaneously is always small, the inclusive and exclusive approaches should be used as complements if one wants a complete picture. What we need is a 4π detection of all interactions and knowledge about the nature and the four-momentum of the particles, even if they sometimes will be treated as being "unobserved".

2.2 Inclusive LPS analysis and its variables

So far, we have discussed that we should look at data carefully and roughly how we can do that, but still not seen any. Before we can actually start to look, we have to think a little about the immediate tools, the variables to be used to display these data.

2.2.1 Longitudinal and transverse momenta

Figure 2.1 gives so-called Peyrou plots for various particles produced in π^-p interactions at 16 GeV/c beam momentum. In such a plot, the length of the momentum component *normal* to the collision axis (the transverse momentum p_T) is plotted against the component *in* the direction of the incoming pion in the cms (the longitudinal momentum in the cms, $p^*_{\|}$). In Fig. 2.1, one observes that the transverse momentum behaves completely different from the longitudinal one:

- While the longitudinal momentum is *unlimited* (i.e. limited only by phase space), the transverse momentum is *limited* dynamically and its average is about 0.3-0.4 GeV/c;
- while $p^*_{\|}$ *depends* strongly on the nature of the particle, p_T does *not*;
- while $p^*_{\|}$ *depends* strongly on the multiplicity [6], p_T does *not* (Fig. 2.2) (see, however, Sub-Sect. 4.3.1);
- while $p^*_{\|}$ *depends* strongly on the total cms energy, p_T does *not* (see, however, Sub-Sect. 4.3.4).

These experimental facts constitute our fundamental first-order knowledge about soft hadron-hadron collisions and go back to early cosmic ray experiments, i.e. are also valid at energies much higher than those of Figs. 2.1 and 2.2. The knowledge of these facts is extremely useful for the question of data presentation. It allows a natural separation of the three-dimensional momentum vector into an essential scalar longitudinal variable and a much less essential two-dimensional transverse momentum vector $\mathbf{p}_T$ of strongly limited length $p_T = |\mathbf{p}_T|$. This is the basic idea of the longitudinal phase space (LPS) analysis [2] to be discussed below.

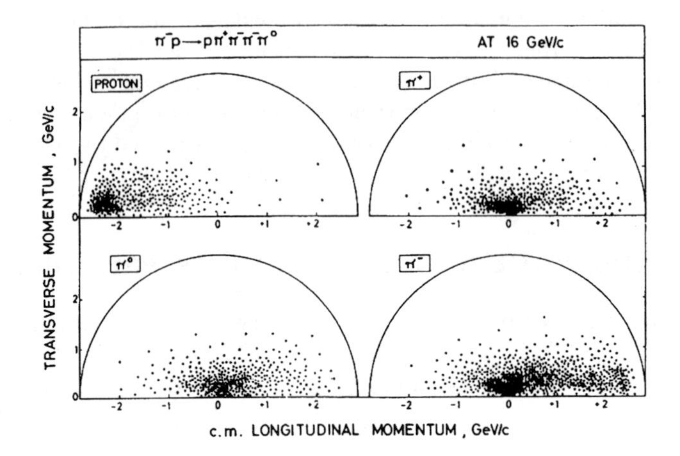

Figure 2.1. Peyrou plots for the particles produced in the reaction $\pi^-\mathrm{p} \rightarrow 2\pi^-\pi^+\pi^0\mathrm{p}$ at 16 GeV/c [6].

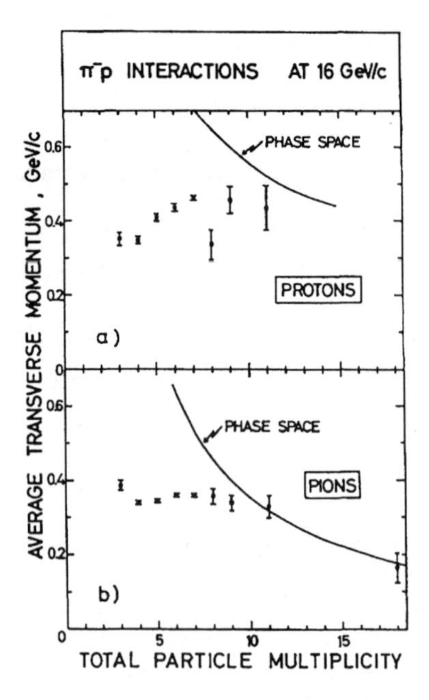

Figure 2.2. Average value of the transverse momentum of protons and pions produced in final states of various total multiplicity. The lines are the phase space limits [6].

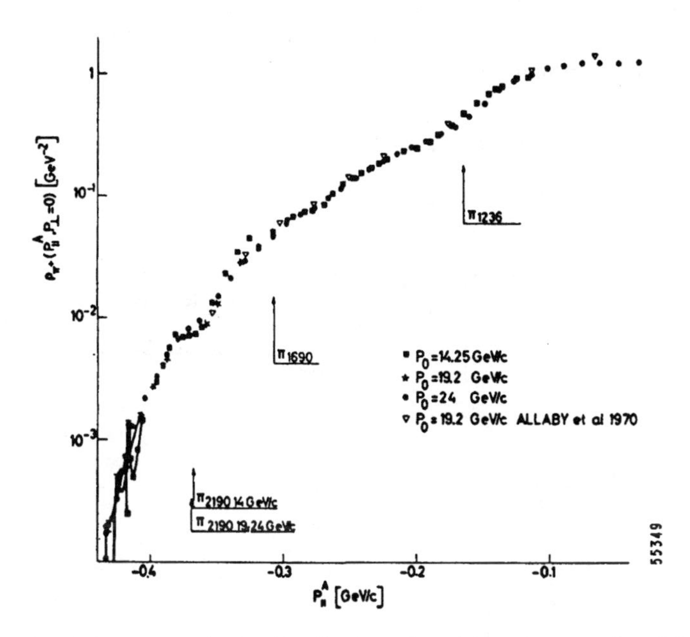

Figure 2.3. Single-particle inclusive distribution of positive pions produced in pp interactions between 14.25 and 24 GeV/c. ρ_{π^+} is defined as $2\sigma_{\mathrm{tot}}^{-1} f_1^{\mathrm{A}}$ in the notation used in the text. The longitudinal momentum is given in the projectile rest frame A. The arrows indicate where structure due to isobar production may be expected [7].

In terms of the longitudinal and transverse momenta, the invariant n-particle distribution of (2.5) can be rewritten as

$$f_n(s, p_{\parallel 1}, \mathbf{p}_{\mathrm{T}_1}, \ldots, p_{\parallel n}, \mathbf{p}_{\mathrm{T}_n}) = \left(\prod_1^n E_i\right) \frac{\mathrm{d}^{3n}\sigma_n}{\prod_1^n \mathrm{d}p_{\parallel_i} \mathrm{d}^2\mathbf{p}_{\mathrm{T}_i}} \quad . \tag{2.6}$$

If the incoming particles are unpolarized, one can integrate over one of the transverse variables. For single-particle distributions, this means that we can use the length p_{T} of the transverse momentum two-vector $\mathbf{p}_{\mathrm{T}}$ as the only transverse variable. For two-particle distributions, e.g. the length of the two transverse momentum vectors and the angle between them would be sufficient. As an example, Fig. 2.3 shows essentially $f_1(s, p_\parallel, p_{\mathrm{T}} = 0)$ for positive pions produced in pp interactions [7], as a function of $p_\parallel$, for various values of s. The longitudinal momentum $p_\parallel$ is measured in the projectile proton rest frame and only very small $p_\parallel$ vales are given. (To normalize the s dependence to that of the total cross section, f_1 is further divided by σ_{tot} in Fig. 2.3, i.e. $\rho_{\pi^+} = 2\sigma_{\mathrm{tot}}^{-1} f_1$).

The great advantage of which the LPS analysis makes use is that, at least for first-order studies, one can integrate over the $2n$ transverse momentum variables without loosing too much information:

$$F_n(s, p_{\|1}, \ldots, p_{\|n}) = \int f_n(s, p_{\|1}, \mathbf{p}_{\mathrm{T}_1}, \ldots, p_{\|n}, \mathbf{p}_{\mathrm{T}_n}) \prod_1^n \mathrm{d}^2\mathbf{p}_{\mathrm{T}_i} \quad . \tag{2.7}$$

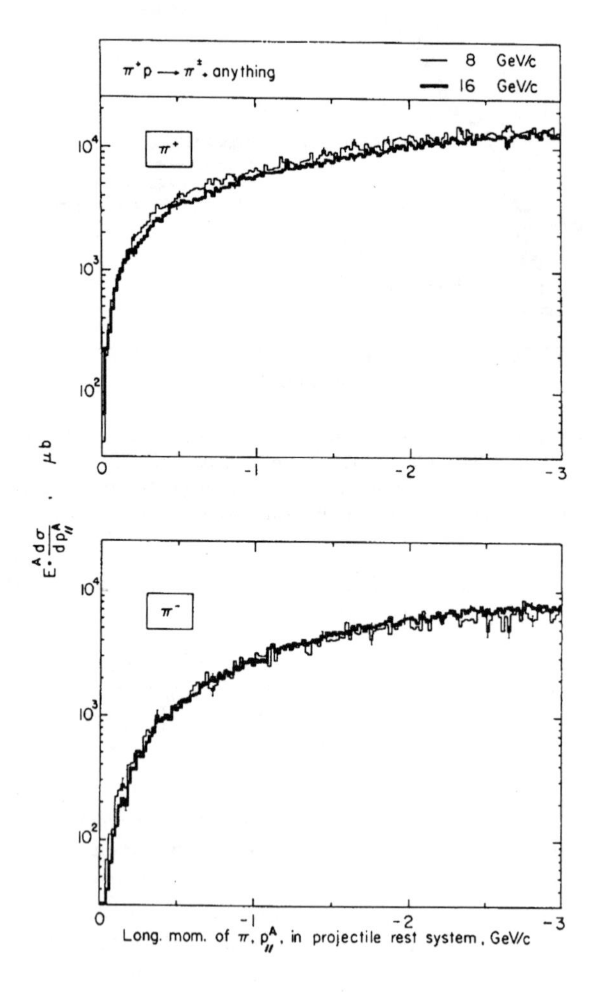

Figure 2.4. Single-particle inclusive distribution of π^+ and π^- produced in π^+p interactions at 8 and 16 GeV/c, integrated over p_{T}. The longitudinal momentum is given in the projectile rest frame [8].

Examples for $F_1(s, p_\parallel)$ are given in Fig. 2.4 for pions produced in π^+p interactions at 8 and 16 GeV/c [8]. Again, $p_\parallel$ is measured in the projectile (here π^+) rest frame and F_1 is given for small $p_\parallel$ values only. We observe that for π^+ production the single-particle distribution in the projectile rest frame [let's call it $F_1^A(s, p_\parallel^A)$] depends on s, while for π^- production it does not. In passing we note that this is as expected by e.g. Chan et al. [9] on the basis of Mueller's generalized optical theorem [10] and called "early limiting" in the case of $AB\bar{C}_1$ forming an exotic system.

Of course, we can plot $F_1(s, p_\parallel)$ as well in other frames. Figure 2.5 shows the same experiment as above, but with $p_\parallel$ in the target rest frame (lab frame) [11]. Again, one can see that $F_1^B(s, p_\parallel^B)$ depends on s for π^+ but not for π^-. A third possibility is $p_\parallel^*$, the longitudinal momentum in the cms. This has the advantage that it covers more easily all momenta and in a more symmetric way.

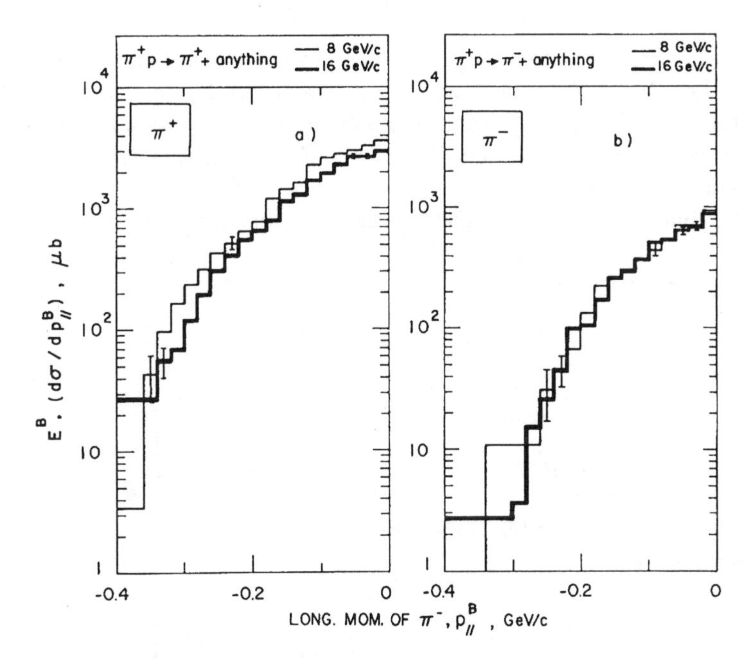

Figure 2.5. Single-particle inclusive distribution of π^+ and π^- produced in π^+p interactions at 8 and 16 GeV/c, integrated over p_T. The longitudinal momentum is given in the target rest frame [11].

2.2.2 The Feynman variable x

In particular, it has turned out practical to use the *reduced* cms longitudinal momentum or Feynman variable [3]

$$x = \frac{p^*_\parallel}{p^*_{\rm in}} \approx \frac{2p^*_\parallel}{s^{1/2}} \tag{2.8}$$

to obtain energy independence of the kinematic range covered. Examples for $F^*_1(s,x)$ are given in Fig. 2.6, again for pion production in π^+p interactions at 8 and 16 GeV/c [11]. Most striking is now the energy dependence for π^- production at $x=0$, to which we will come back later.

The connection between $p^{\rm A}_\parallel, p^{\rm B}_\parallel$ and $p^*_\parallel$ is, of course, through a longitudinal Lorentz transformation of the type

$$\begin{pmatrix} \tilde{p}_\parallel \\ \tilde{p}^x_{\rm T} \\ \tilde{p}^y_{\rm T} \\ \tilde{E} \end{pmatrix} = \begin{pmatrix} \gamma_\parallel & 0 & 0 & \epsilon_\parallel \\ 0 & 1 & 0 & 0 \\ 0 & 0 & 1 & 0 \\ \epsilon_\parallel & 0 & 0 & \gamma_\parallel \end{pmatrix} \begin{pmatrix} p_\parallel \\ p^x_{\rm T} \\ p^y_{\rm T} \\ E \end{pmatrix} \tag{2.9}$$

with $\gamma_\parallel = (1-\beta^2_\parallel)^{-1/2}$ and $\epsilon_\parallel = \gamma_\parallel\beta_\parallel$, where $\beta_\parallel$ is the velocity of one frame as seen from the other. Fig. 2.7 demonstrates for the case of πp interactions at 16 GeV/c, how the three variables $p^{\rm A}_\parallel$, $p^{\rm B}_\parallel$ and x are related to each other in a $p_{\rm T}$- (because E-) dependent way. For small $p_{\rm T}$ values, however, $p^{\rm B}_\parallel$ remains small for $x<0$ and $p^{\rm A}_\parallel$ remains small for $x>0$.

For the limit $s \to \infty$, the connection is best summarized [1] by the following relations:

In the projectile frame one finds

$$\left.\begin{array}{lll} p^{\rm A}_\parallel & \to \dfrac{1}{2}\left[x m_{\rm A} - \dfrac{m^2_{\rm T}}{x m_{\rm A}}\right] & \text{for } x>0 \\ p^{\rm A}_\parallel & \to -\infty & \text{for } x \le 0 \\ \text{but } \; p^{\rm A}_\parallel \sqrt{s} & \to x/2m_{\rm A} & \text{for } x \le 0 \end{array}\right\} , \tag{2.10}$$

where $m_{\rm T} = (m^2 + p^2_{\rm T})^{1/2}$ is the transverse mass.

In the target frame,

$$\left.\begin{array}{lll} p^{\rm B}_\parallel & \to \dfrac{1}{2}\left[x m_{\rm B} - \dfrac{m^2_{\rm T}}{x m_{\rm B}}\right] & \text{for } x<0 \\ p^{\rm B}_\parallel & \to +\infty & \text{for } x \ge 0 \\ \text{but } \; p^{\rm B}_\parallel \sqrt{s} & \to x/2m_{\rm B} & \text{for } x \ge 0 \end{array}\right\} . \tag{2.11}$$

For any intermediate frame defined as having $\beta_\parallel^{\rm A} \to -1$ as seen from the projectile frame and $\beta_\parallel^{\rm B} \to +1$ as seen from target frame

$$\left.\begin{array}{ll} p_\parallel^{\rm I} \to +\infty & \text{for } x > 0 \\ p_\parallel^{\rm I} \to -\infty & \text{for } x < 0 \end{array}\right\} , \tag{2.12}$$

but for the cms

$$p_\parallel^*/s^{1/2} \to x/2 \quad \text{for all } x . \tag{2.13}$$

Experimentally, no frame is a priori superior to any other, and one should plot

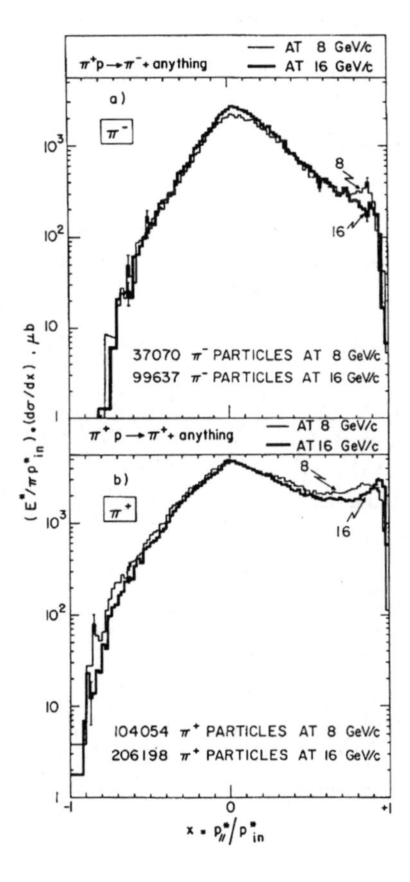

Figure 2.6. Single-particle inclusive distribution of π^+ and π^- produced in π^+p interactions at 8 and 16 GeV/c, integrated over the transverse momentum. The variable used is the reduced longitudinal momentum in the cms [11].

the data in several to gain more insight and information. In testing a certain model, however, one will of course choose frames and variables in which the model predictions come out most sharply, for the particles considered. So, e.g. limiting fragmentation [4] of the projectile (i.e. energy independence at large x, see Sub-Sect. 4.3.1) will be studied in the projectile rest frame, while that of the target will be studied in the target rest frame. For Feynman scaling [3] (i.e. energy independence at small $|x|$), the cms is better suited with x as a variable instead of $p^*_\parallel$, or the rapidity to be discussed below.

Even though at very high energy, relations (2.10) and (2.11) give a direct connection between the two types of distribution for $x \neq 0$,

$$\left.\begin{aligned} f_1^{\mathrm{A}}(s, p_\parallel^{\mathrm{A}}, p_{\mathrm{T}}) &= f_1^*(s, x = (p_\parallel^{\mathrm{A}} + E^{\mathrm{A}})/m_{\mathrm{A}}, p_{\mathrm{T}}) \quad \text{for } x > 0 \\ f_1^{\mathrm{B}}(s, p_\parallel^{\mathrm{B}}, p_{\mathrm{T}}) &= f_1^*(s, x = (p_\parallel^{\mathrm{B}} - E^{\mathrm{B}})/m_{\mathrm{B}}, p_{\mathrm{T}}) \quad \text{for } x < 0 \end{aligned}\right\} \tag{2.14}$$

the region $x = 0$ is lost when plotting in $p_\parallel^{\mathrm{A}}$ or $p_\parallel^{\mathrm{B}}$.

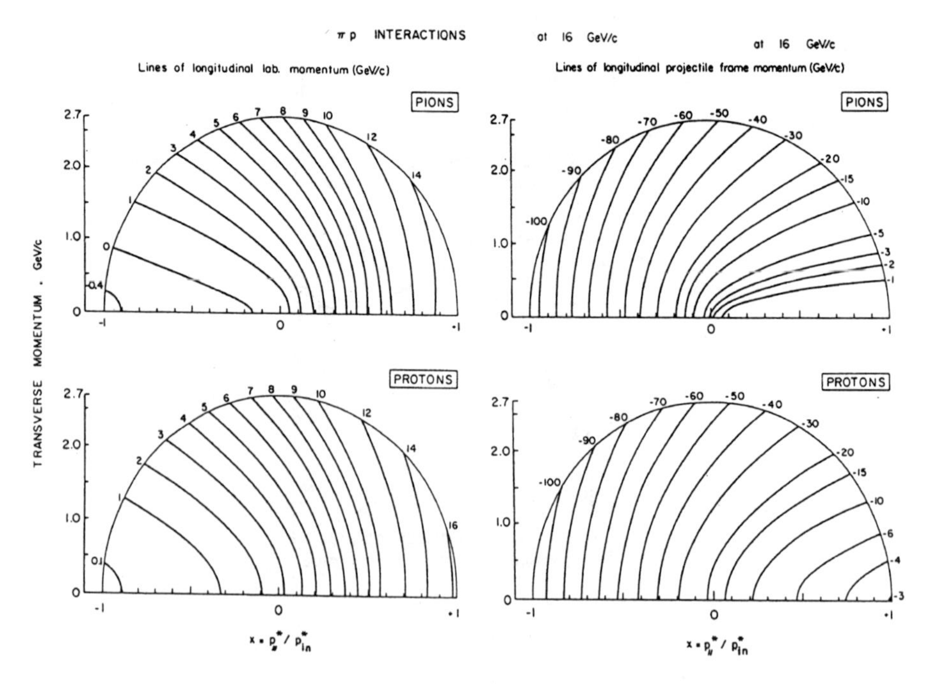

Figure 2.7. Lines of equal longitudinal momentum in the lab. frame (left) and the projectile rest frame (right), as functions of x and p_{T} for pions and protons produced in πp interactions at 16 GeV/c.

2.2.3 Longitudinal rapidities

We have seen that the longitudinal momenta used as variables give a differential description of the particle distribution, but they have disadvantages:

(i) According to (2.4) and (2.5), each event must be weighted with $(\prod_1^n E_i)$ before being plotted.

(ii) According to Fig. 2.6, the distribution has a sharp maximum at $x = 0$ even after weighting. This contains all the events which are intermediate in many ways according to relations (2.12) and (2.13).

(iii) The distribution changes its shape when going from one frame to another.

What we need, therefore, is a variable y which still describes the particle C uniquely in the given frame, but

(i) has a Jacobian

$$\prod_1^n \frac{\mathrm{d}p_{\parallel i}}{\mathrm{d}y_i} = \prod_1^n E_i \tag{2.15}$$

to cancel the factor $(\prod_1^n E_i)$ in (2.5) and (2.6),

(ii) stretches the region around $x = 0$ relative to $x \neq 0$,

(iii) makes differences $(y - y_\mathrm{B})$ invariant under longitudinal Lorentz transformations.

A unique way to describe particle C in momentum space is to give its transverse momentum and a longitudinal Lorentz transformation of the type (2.9) to the longitudinal "rest" frame of particle C:

$$\begin{pmatrix} p_\parallel \\ p_\mathrm{T}^x \\ p_\mathrm{T}^y \\ E \end{pmatrix} = \begin{pmatrix} E/m_\mathrm{T} & 0 & 0 & p_\parallel/m_\mathrm{T} \\ 0 & 1 & 0 & 0 \\ 0 & 0 & 1 & 0 \\ p_\parallel/m_\mathrm{T} & 0 & 0 & E/m_\mathrm{T} \end{pmatrix} \begin{pmatrix} 0 \\ p_\mathrm{T}^x \\ p_\mathrm{T}^y \\ m_\mathrm{T} \end{pmatrix} . \tag{2.16}$$

The trick [3, 12] is to define the new variable as

$$\sinh y = \epsilon_\parallel = \frac{p_\parallel}{m_\mathrm{T}} . \tag{2.17}$$

From Fig. 2.8 it is obvious that the mapping of $p_\parallel$ onto y is such that the region $p_\parallel \approx 0$ is stetched. Further, we get with

$$\cosh y = (1 + \sinh^2 y)^{1/2} \tag{2.18}$$

the second Lorentz variable in (2.16),

$$\cosh y = \gamma_\parallel = \frac{E}{m_\mathrm{T}} \ , \tag{2.19}$$

so that we can rewrite (2.16) as

$$\begin{pmatrix} p_\parallel \\ p_\mathrm{T}^x \\ p_\mathrm{T}^y \\ E \end{pmatrix} = \begin{pmatrix} \cosh y & 0 & 0 & \sinh y \\ 0 & 1 & 0 & 0 \\ 0 & 0 & 1 & 0 \\ \sinh y & 0 & 0 & \cosh y \end{pmatrix} \begin{pmatrix} 0 \\ p_\mathrm{T}^x \\ p_\mathrm{T}^y \\ m_\mathrm{T} \end{pmatrix} \ . \tag{2.20}$$

As demanded, the derivative is

$$\frac{\mathrm{d}p_\parallel}{\mathrm{d}y} = m_\mathrm{T} \frac{\mathrm{d}\sinh y}{\mathrm{d}y} = m_\mathrm{T} \cosh y = E \ . \tag{2.21}$$

The variable y itself is called (longitudinal) *rapidity* and can be expressed in various ways, e.g. via its connection to the longitudinal velocity of particle C,

$$\tanh y = \beta_\parallel = \frac{p_\parallel}{E} \ , \tag{2.22}$$

as

$$y = \frac{1}{2} \ln \left(\frac{1+\beta_\parallel}{1-\beta_\parallel} \right) = \frac{1}{2} \ln \left(\frac{E+p_\parallel}{E-p_\parallel} \right) = \ln \left(\frac{E+p_\parallel}{m_\mathrm{T}} \right) \ . \tag{2.23}$$

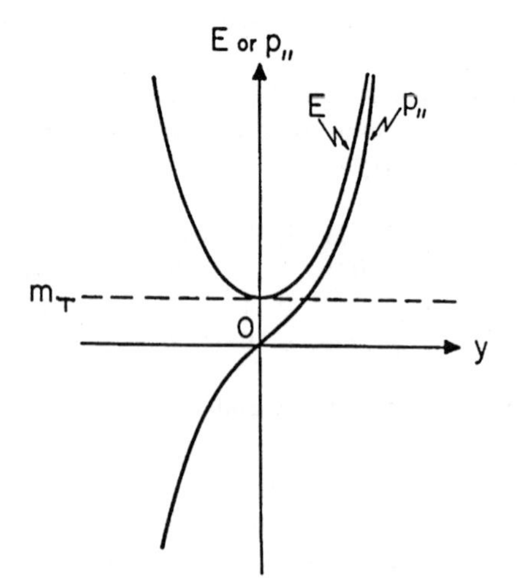

Figure 2.8. Longitudinal momentum $p_\parallel$ and energy E as a function of the longitudinal rapidity y for given transverse mass m_T.

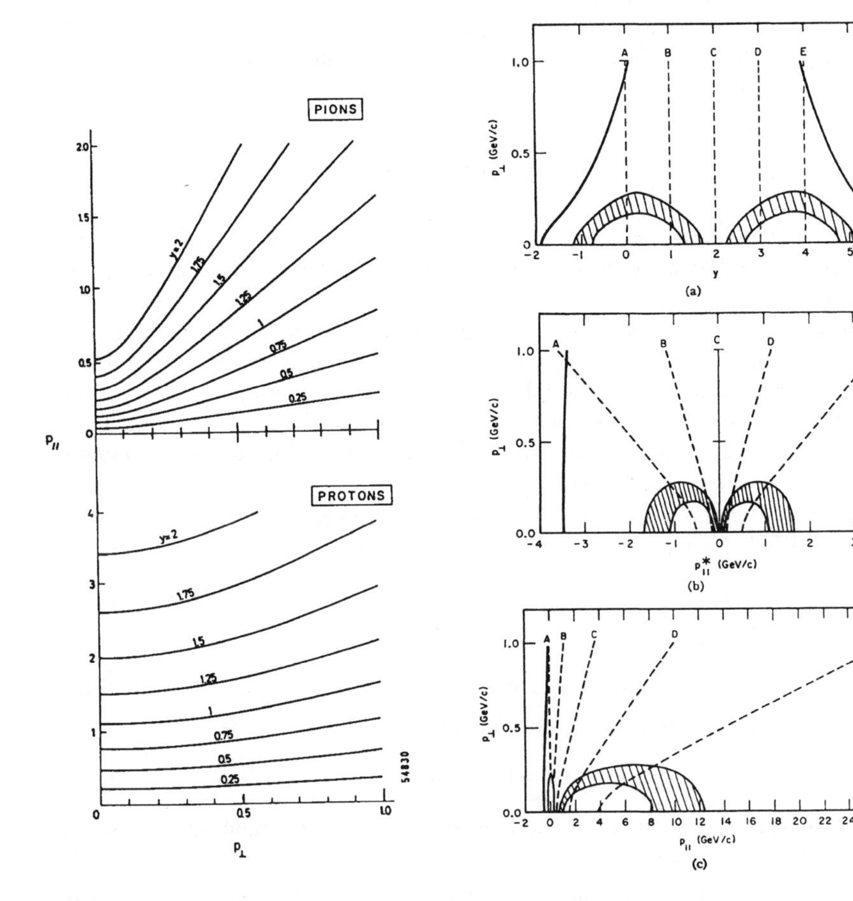

Figure 2.9. Lines of constant longitudinal rapidity y in the $p_{\mathrm{T}}, p_{\|}$ plane for pions and protons, respectively [13].

Figure 2.10. Comparison of the Peyrou plot in the longitudinal rapidity y and the transverse momentum p_{T} (Fig. 2.10a) to the Peyrou plot in the cms (Fig. 2.10b) and target rest (lab.) frame (Fig. 2.10c). The phase space boundary is indicated by a heavy line for the process pp$\rightarrow$ π + anything with beam momentum 25.6 GeV/c. The shaded band denotes the position of pions resulting from the process pp$\rightarrow$ $\Delta(1236)\Delta(1236)$ in the near forward direction. The band corresponds to a mass width of 120 MeV [14].

It can be defined in any frame in which E and $p_{\|}$ can be defined. Figure 2.9 shows y as a function of $p_{\|}$ and p_{T} for pions and protons, respectively [13].

Finally, going from one frame to another by a longitudinal Lorentz boost of velocity $\beta'_{\|}$, we obtain a new rapidity

$$y' = y - \frac{1}{2}\ln\left(\frac{1+\beta'_{\|}}{1-\beta'_{\|}}\right) , \tag{2.24}$$

differing from the old one only by a constant. As a consequence, the shape of the distribution will be the same in all frames connected through longitudinal Lorentz transformations.

Figure 2.10 gives a comparison [14] of generalized Peyrou plots for single-pion distributions in pp interactions at 25.6 GeV/c, in the longitudinal rapidity, the cms longitudinal momentum $p^*_{\|}$ and the lab longitudinal momentum $p^{\mathrm{B}}_{\|}$. Starting from Fig. 2.10b, we see how much the region $p^*_{\|} \approx 0$ is expanded when transformed to y in Fig. 2.10a. We also note how narrow the domain of $p^*_{\|} < 0$ becomes as compared to that of $p^*_{\|} > 0$ when mapped onto $p^{\mathrm{B}}_{\|}$.

One property of the rapidity y is still that its kinematic range increases with increasing s. While for $p^{\mathrm{B}}_{\|}$ and $p^*_{\|}$ this increase goes like

$$p^{\mathrm{B}}_{\|\mathrm{A}} - p^{\mathrm{B}}_{\|\mathrm{B}} = p^{\mathrm{B}}_{\|\mathrm{in}} \simeq s/m_{\mathrm{B}} \tag{2.25}$$

and

$$p^*_{\|\mathrm{A}} - p^*_{\|\mathrm{B}} = 2p^*_{\mathrm{in}} \simeq s^{1/2} , \tag{2.26}$$

respectively, the increase is only logarithmic for y:

$$y_{\mathrm{A}} - y_{\mathrm{B}} = \ln(s/m_{\mathrm{A}}m_{\mathrm{B}}) . \tag{2.27}$$

This logarithmic increase can, in principle, be divided out in the reduced longitudinal rapidity [1] as

$$y_{\mathrm{R}} = \frac{y - y_{\mathrm{B}}}{y_{\mathrm{A}} - y_{\mathrm{B}}} , \tag{2.28}$$

which, in addition, is now the same for any frame connected by a longitudinal Lorentz-transformation. We shall see, however, that particle production is more naturally connected to y than to y_{R}.

In experiments where only production angles Θ ($\tan\Theta = p_{\mathrm{T}}/p_{\|}$), but no masses or momenta can be measured, the so-called pseudo-rapidity η is generally used,

$$\eta = -\ln\tan(\Theta/2) . \tag{2.29}$$

It is approximately equal to the rapidity y for $p \gg m$ and $\Theta \gg 1/\gamma$.

The main conclusion we can draw for the LPS analysis from this chapter is that already at finite energies x is clearly more detailed for target and projectile fragments, while y is more detailed for the central region.

2.2.4 The variable ξ

In recent studies of single-particle inclusive distributions in e^+e^- collisions it has also become customary to use the variable $\xi = -\ln x$ (see Chapter 4).

2.3 Exclusive LPS analysis and its variables

2.3.1 Definitions

The most desirable way to study an exclusive reaction of multiplicity $n \equiv n_f$,

$$A + B \rightarrow C_1 + C_2 + \ldots + C_n \; , \tag{2.30}$$

is to look at the completely differential distribution of the matrix element of (2.2) in full phase space. An analysis in 3n-4 dimensions being necessary ($3n - 5$ if the incoming particles A and B are unpolarized), this method becomes soon impossible as n increases.

In Sub-Sect. 2.2.1 we have recalled that in strong interactions, transverse momenta are limited to small values and are largely independent of the nature of the particle, the multiplicity n and the cms energy $\sqrt{s}$. The longitudinal momenta are unlimited (limited only by phase space) and depend strongly on the nature of the particle, the multiplicity and the energy. Accordingly, in two-body reactions one distinguishes between forward elastic and backward elastic scattering (if C_1 and C_2 are identical to A and B) and various types of exchange processes (if at least one of the produced particles is of a type different from that of the incoming ones). An extension of this two-body classification to many-body reactions is the final-state classification in longitudinal phase space [2,15].

In the longitudinal-phase-space (LPS) analysis of exclusive final states, each individual reaction of type (2.30) is represented by a point with coordinates $(p_{\|1}, \ldots, p_{\|n})$ in an n-dimensional euclidean space S_n. Conservation of the longitudinal momentum in the cms

$$\sum_{i=1}^{n} p^*_{\|i} = 0 \quad , \tag{2.31}$$

defines longitudinal phase space as an $(n-1)$-dimensional hyperplane L_{n-1}. Furthermore, because of conservation of cms energy $s^{1/2}$,

$$\sum_{i=1}^{n} (m_i^2 + \mathbf{p}_{\mathrm{T}i}^2 + p^{*2}_{\|i})^{1/2} = s^{1/2} \quad , \tag{2.32}$$

all points with equal transverse momentum $|\mathbf{p}_{\mathrm{T}i}|$ lie on an $(n-2)$-dimensional hypersurface K_{n-2} defined by (2.32). For the case $m_{\mathrm{T}i} \equiv (m_i^2 + \mathbf{p}_{\mathrm{T}i}^2)^{1/2} = 0$, (2.32) reduces to

$$\sum_{i=1}^{n} |p^*_{\|i}| = s^{1/2} \tag{2.33}$$

and defines a regular polyhedron H_{n-2}. For multiplicity $n = 3$, the polyhedron H_1 is the Van Hove hexagon shown in Fig. 2.11 together with the one-dimensional manifold K_1. For $n = 4$, the polyhedron H_2 is the cuboctahedron in Fig. 2.12.

The phase-space boundary is given by K_{n-2} with all $p_{\mathrm{T}i} = 0$. All reaction points $(p^*_{\|1}, \ldots, p^*_{\|n})$ must lie within this hypersurface, but they stay close to it as long as all transverse momenta are small compared to the values kinematically allowed. An example for a three-body reaction in Fig. 2.13 demonstrates [16] that this is indeed the case, already at 16 GeV/c. At higher energies the points will move even closer to the boundary, so that $n - 2 = 1$ variable is indeed enough to describe the reaction. One possible variable is the so-called Van Hove correlation angle ω, defined in Fig. 2.11.

A generalization of the LPS analysis in the cms to that in an arbitrary frame connected with the cms by a longitudinal Lorentz transformation like (2.9) is possible. One restriction is, unfortunately, that the two constraints of energy and longitudinal momentum conservation degenerate to only one when defined in a frame where all $p_{\|i}$ are of equal sign. Since no data in an arbitrary frame are available, the analysis will always be understood as being performed in the cms.

Another generalization is of course the use of rapidities.

2.3.2 Phase-space effects

We have seen in Sect. 2.2 that an unweighted distribution in the momenta themselves does not yet represent the squared matrix element, because of the particle energy in the denominator of (2.2). Further factors will be introduced when transforming (2.2) to other variables, e.g. the angle ω. Before we start to look at experimental distributions, we therefore have to study the effects of phase space on these distributions. That is, we have to discuss the relativistically invariant volume element $\mathrm{d}V$ in phase space [2].

Because of the different behavior of transverse and longitudinal momenta, we first separate $\mathrm{d}V$ into a transverse element $\mathrm{d}V_\mathrm{T}$ and a longitudinal element $\mathrm{d}V_\|$,

$$\mathrm{d}V = \mathrm{d}V_\mathrm{T}\mathrm{d}V_\| \tag{2.34}$$

with

$$\mathrm{d}V_\mathrm{T} = \delta_2\left(\sum_1^n \mathbf{p}_{\mathrm{T}i}\right) \prod_1^n \mathrm{d}^2\mathbf{p}_{\mathrm{T}i} \tag{2.35}$$

$$\mathrm{d}V_\| = \delta\left(\sum_1^n p^*_{\|i}\right) \delta\left(\sum_1^n (m^2_{\mathrm{T}i} + p^{*2}_{\|i})^{1/2} - s^{1/2}\right) \prod_1^n \frac{\mathrm{d}p^*_{\|i}}{E^*_i} \quad . \tag{2.36}$$

Furthermore, since we have seen that three-body reactions are well described in LPS by an angular variable ω, we generalize this to n-body reactions by going from the n longitudinal momenta $p^*_{\|i}$ to $n-2$ angular variables $\omega_1, \ldots, \omega_{n-2}$, the distance of the event point $(p^*_1, \ldots, p^*_n)$ from the hyperplane L_{n-1},

$$Q = n^{-1/2}\sum_1^n p^*_{\|i} \ , \tag{2.37}$$

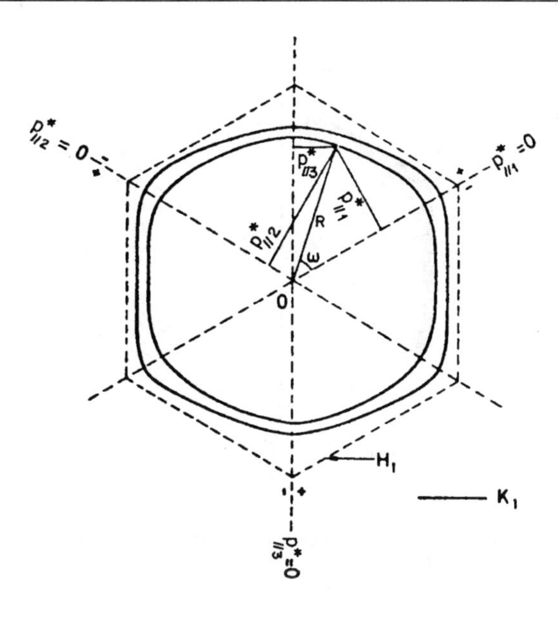

Figure 2.11. Longitudinal phase space plot (Van Hove hexagon) for $\pi\pi$N at c.m. energy $s^{1/2} = 4$ GeV. The inner full curve is K_1 for transverse momenta 0.4, 0.4, 0.5 GeV/c, respectively, the outer one is K_1 for vanishing transverse momenta. The dashed line represents the hexagon H_1 [2].

Figure 2.12. The polyhedron H_2 in the longitudinal phase space of a four-particle reaction [2] (cartoon by R. Sosnowski).

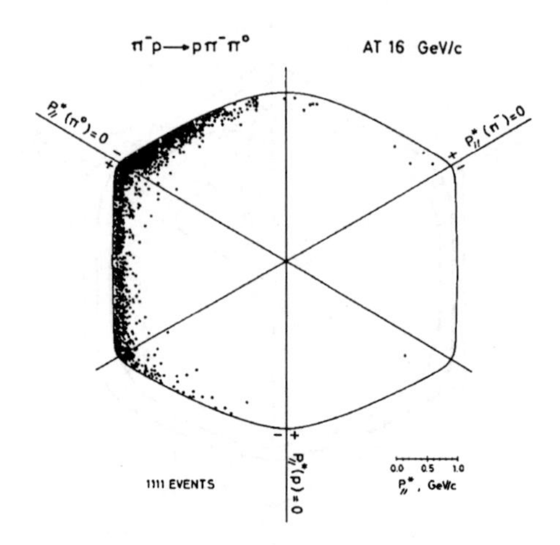

Figure 2.13. Distribution of reaction points in the Van Hove hexagon plot for the reaction $\pi^-\mathrm{p} \to \pi^-\pi^0\mathrm{p}$ at 16 GeV/c [16].

and the radial length in L_{n-1}

$$R = \left[\left(\sum_1^n p_{\parallel i}^{*2}\right) - Q^2\right]^{1/2} \geq 0 \,. \tag{2.38}$$

In these new n variables, the longitudinal volume element can be rewritten as

$$\mathrm{d}V_{\parallel} = n^{-1/2}\delta(Q)\mathrm{d}Q\delta(R - R_0)\mathrm{d}RD_n\mathrm{d}^{n-2}\omega \quad , \tag{2.39}$$

where $\mathrm{d}^{n-2}\omega$ is the element of solid angle $\omega_1, \ldots, \omega_{n-2}$, R_0 is defined by the energy constraint of (2.32),

$$\sum_1^n (m_{\mathrm{T}i}^2 + R_0^2\gamma_i^2)^{1/2} = s^{1/2} \tag{2.40}$$

and the phase space weight function D_n is

$$D_n = \left[\prod_1^n (m_{\mathrm{T}i}^2 + R^2\gamma_i^2)\right]^{-1/2} R^{n-2} \Big/ \frac{\partial}{\partial R}\left[\sum_1^n (m_{\mathrm{T}i}^2 + R^2\gamma_i^2)^{1/2}\right] \quad . \tag{2.41}$$

The γ_i are directional cosines of $(p_{\parallel 1}^*, \ldots, p_{\parallel n}^*)$ in S_n,

$$p_{\parallel i}^* = (R^2 + Q^2)^{1/2}\gamma_i \quad , \quad \sum_1^n \gamma_i^2 = 1 \,. \tag{2.42}$$

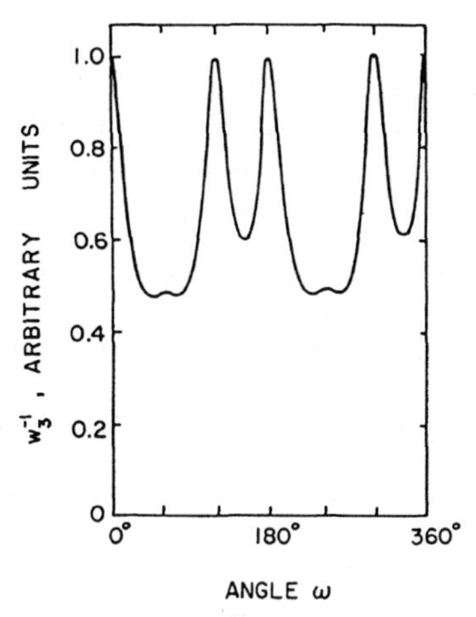

Figure 2.14. Phase-space weight function w_3^{-1} normalized to its maximum value and plotted against the angle ω, for $\pi\pi$N at $s^{1/2} = 4$ GeV with transverse momenta of 0.4, 0.4 and 0.5 GeV/c, respectively [2].

One deduces that the phase space volume varies with the angular variables ω_i (and the transverse momenta) as

$$
\begin{aligned}
\mathrm{w}_n^{-1} &\equiv \frac{\mathrm{d}^{n-2}V_\parallel}{\prod_1^{n-2}\mathrm{d}\omega_i} = n^{-1/2}\int \delta(Q)\mathrm{d}Q\delta(R-R_0)\mathrm{d}RD_n = \\
& n^{-1/2}\left[\sum_1^n(m_{\mathrm{T}i}^2+p_{\parallel i}^{*2})\right]^{-1/2}\left[\sum_1^n p_\parallel^{*2}\right]^{(n-1)/2}\left[\sum_1^n(m_{\mathrm{T}i}^2+p_{\parallel i}^{*2})^{-1/2}p_{\parallel i}^{*2}\right]^{-1} .
\end{aligned}
\tag{2.43}
$$

The weight function w_n^{-1} is plotted in Fig. 2.14 against the angle ω for a $\pi\pi$N final state. The function varies several times by as much as a factor two as ω runs from 0° to 360°. It is largest whenever a pion reaches zero longitudinal momentum. To obtain the distribution of the squared matrix element $|M|^2$ at fixed energy, one, therefore, has to plot each event with weight w_n. By definition, the weighted distribution is flat for cylindrical phase space, so that any structure is due to the matrix element.

For the study of the energy dependence of the weighted LPS distribution it is useful to have a behavior similar to that of the unweighted distribution, e.g. energy independence for pomeron exchange. This is achieved by dividing w_n by its asymptotic energy dependence $\mathrm{w}_n \propto s$. Another factor s^{-1} is contained in the flux factor of (2.2). The quantity we actually will look at, is, therefore, asymptotically equal to $s^{-2}|M|^2$, integrated over transverse momenta.

2.3.3 Kinematics

How are the angular variables ω_i related to the more familiar invariant squared sub-energies s_{ij} of particles C_i and C_j and momentum transfers t_i and t'_i from the projectile and target, respectively? In terms of the four-momenta, s_{ij} can be simply expressed as

$$s_{ij} = (p_i + p_j)^2 = m_i^2 + m_j^2 + 2E_i^* E_j^* - 2p_{\parallel i}^* p_{\parallel j}^* - 2\mathbf{p}_{\mathrm{T}i}\mathbf{p}_{\mathrm{T}j} \ . \tag{2.44}$$

The momentum transfer t_i of particle C_i with respect to the projectile A reads

$$t_i = (p_\mathrm{A} - p_i)^2 = m_\mathrm{A}^2 + m_i^2 - 2E_\mathrm{A}^* E_i^* + 2p_\mathrm{A}^* p_{\parallel i}^* \tag{2.45}$$

and similarly the momentum transfer t'_i with respect to the target B

$$t'_i = (p_\mathrm{B} - p_i)^2 = m_\mathrm{B}^2 + m_i^2 - 2E_\mathrm{B}^* E_i^* + 2p_\mathrm{B}^* p_{\parallel i}^* \ . \tag{2.46}$$

For the three-body reaction $\pi\mathrm{N} \to \pi\pi\mathrm{N}$ at two different energies, the sub-energies s_{12} and s_{23} as well as t_1 and t'_3 are plotted as functions of ω in Fig. 2.15 for fixed values of the transverse momenta [2]. We note the almost linear dependence within large regions of ω when s is large. Furthermore, for given ω (i.e. given $p_{\parallel i}^*$ and $p_{\parallel j}^*$) there is only one solution for s_{ij}, while for given s_{ij} there are four values of ω, depending on the sign of the momenta in the overall cms and in the s_{ij}-cms.

Deviations from the single line representing the function s_{ij} in Fig. 2.15 are due to transverse momenta and are shown to exist in Fig. 2.16. There, the two-pion effective mass $s_{\pi\pi}^{1/2} \equiv M(\pi^+\pi^0)$ is plotted against the correlation angle ω for $\pi^+\mathrm{p} \to \pi^+\pi^0\mathrm{p}$

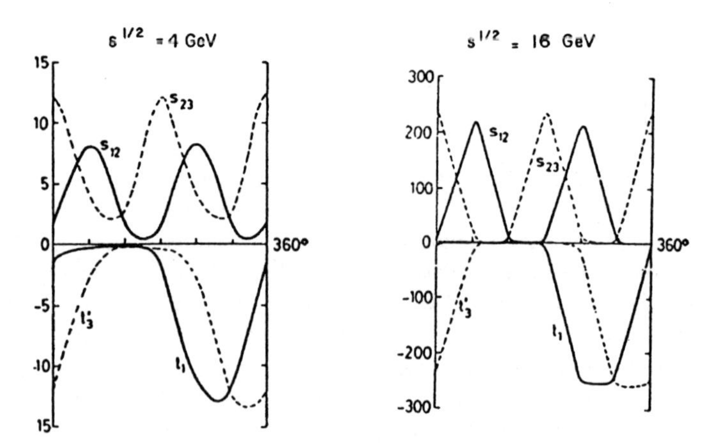

Figure 2.15. The sub-energies s_{12}, s_{23} and the momentum transfer t_1, t'_3 plotted as functions of the angle ω for the reaction $\pi\mathrm{N} \to \pi\pi\mathrm{N}$ at cms energies 4 and 16 GeV, respectively. The transverse momenta are 0.4 GeV/c for the pions and 0.5 GeV/c for the proton [2].

at 16 GeV/c. Only events with a backward proton are considered, corresponding to $60° < \omega < 240°$. As a comparison, the region covered by similar events at 8 GeV/c ($s^{1/2} = 4$ GeV) is given shaded. At 16 GeV/c, the deviation from a single line is considerable only for some events with low $\pi^+\pi^0$ mass. The bulk of the events stay

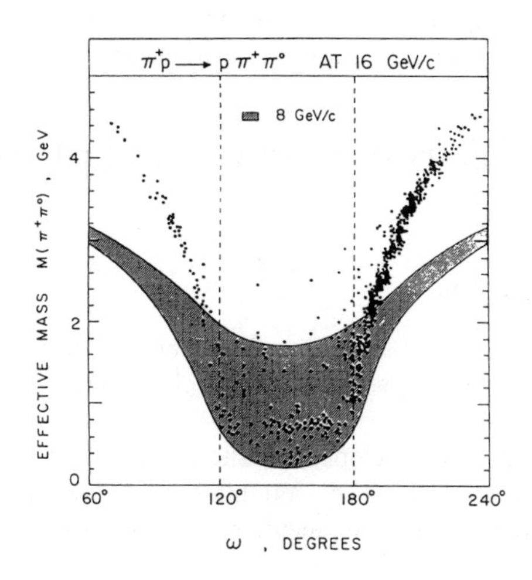

Figure 2.16. Effective mass $M(\pi^+\pi^0)$ as a function of the angle ω for $\pi^+\mathrm{p} \rightarrow \mathrm{p}\pi^+\pi^0$ at 16 GeV/c. The shaded region is the area covered by the experimental points at 8 GeV/c.

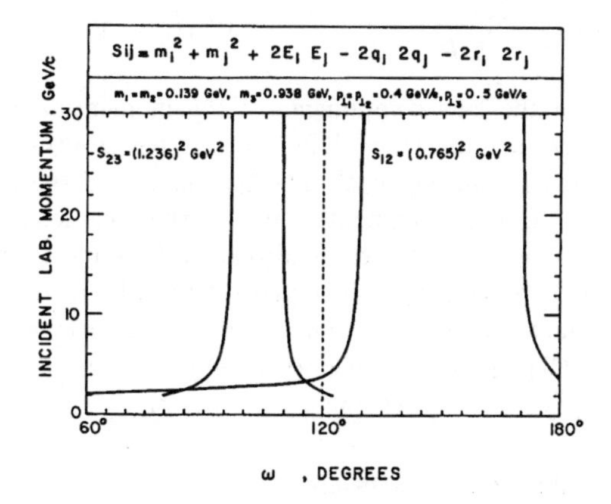

Figure 2.17. The two ω-solutions of Eq. (2.44) for a $(\pi\pi)$ mass in the ρ and $(\mathrm{N}\pi)$ mass in the $\Delta(1236)$ when the proton is backward with respect to the incident pion. The transverse momentum of the pions is fixed to $p_{\mathrm{T}1} = p_{\mathrm{T}2} = 0.4$ GeV/c, that of the proton to $p_{\mathrm{T}3} = 0.5$ GeV/c.

reasonably close together and follow the ideal line s_{ij} of Fig. 2.15.

Finally, in Fig. 2.17, the two ω solutions of (2.44) for backward protons are given for a ρ-decay as well as for a Δ-decay, as functions of the incoming lab. momentum. Mass and transverse momentum are fixed to the values given in the figure. At low incident lab. momentum, the region enclosed by the two corresponding solutions covers a large angular region, but it contracts into the proper sector above 4 GeV/c and becomes independent of the incident lab. momentum above 10 GeV/c.

2.4 Deviations from longitudinal phase space (event shape)

2.4.1 The variables

An alternative way of globally analyzing final states is to define for each event parameters characterizing the degree of alignment of momenta along the collision or the event axis.

This has been originally developed for the investigation of hadron-hadron collisions [17], but later adapted for e^+e^- annihilation [18], where the 'alignment axis', believed to reflect the direction of momenta of the originally produced $q\bar{q}$ pair, varies from event to event.

Generally, the shape of the events in three-dimensional momentum space is defined by the eigenvalues obtained from the diagonalization of the second-rank tensor

$$\boldsymbol{S}_{\alpha\beta} = \frac{\sum p_{i\alpha} p_{i\beta}}{\sum \mathbf{p}_i^2} \ , \qquad (\alpha, \beta = x, y, z) \ , \tag{2.47}$$

where p_{ix}, p_{iy}, p_{iz} are the cms components of the momentum $\mathbf{p}_i$ of particle i in the final state. Let Q_k and $\mathbf{u}_k$ ($k = 1, 2, 3$) be, respectively, the eigenvalues and eigenvectors of the momentum tensor $\boldsymbol{S}$. The Q_k can be ordered in such a way that $0 \leq Q_1 < Q_2 < Q_3$. Due to fragmentation, $\mathbf{u}_3$ (the sphericity axis) is generally aligned along the collision axis in hadron-hadron collisions and along the direction of the original $q\bar{q}$ pair in e^+e^- collisions. The plane defined by $\mathbf{u}_1$ and $\mathbf{u}_2$ is close to the transverse plane of the reaction. One can introduce the variables $S = \frac{3}{2}(Q_1 + Q_2)$ and $Y = \frac{1}{2}\sqrt{3}(Q_2 - Q_1)$. Events with both S and Y small are 'cigar-shaped' and collinear; events with $Y = (1/\sqrt{3})S$ (which implies $Q_1 = 0$) are 'disc-shaped' and coplanar.

The most commonly used variables are:

1. *Sphericity* [19] *and aplanarity* [20]:

$$S = \min\left(\frac{3}{2} \sum \mathbf{p}_{i\mathrm{T}}^2 / \sum \mathbf{p}_i^2\right) = \frac{3}{2}(Q_1 + Q_2), \quad A = \frac{3}{2} Q_1 \tag{2.48}$$

where $\mathbf{p}_i, \mathbf{p}_{iT}$ are the ith particle momentum and its component transverse to that axis for which S is minimal.

2. Parameters C and D [21]

The momentum tensor $\boldsymbol{S}_{\alpha\beta}$ is linearized to become

$$\Theta_{\alpha\beta} = \frac{\sum (p_{i\alpha} p_{i\beta})/|\mathbf{p}_i|}{\sum |\mathbf{p}_i|} \quad . \tag{2.49}$$

The three eigenvalues λ_j of this tensor define

$$C = 3(\lambda_1\lambda_2 + \lambda_2\lambda_3 + \lambda_3\lambda_1) \tag{2.50}$$

$$D = 27\lambda_1\lambda_2\lambda_3 \; . \tag{2.51}$$

3. Spherocity [22]

$$S' = \left(\frac{4}{\pi}\frac{\Sigma|\mathbf{p}_{Ti}|}{\Sigma|\mathbf{p}_i|}\right)^2 \tag{2.52}$$

minimized by the choice of axis analogously to S.

4. Thrust [17], *major and minor, oblateness*

$$T = \max\left(\sum |p_{i\parallel}| / \sum |\mathbf{p}_i|\right) \tag{2.53}$$

where $p_{i\parallel}$ is the component parallel to the axis $\mathbf{n}_T$ for which T is maximal, the thrust axis.

It approaches unity for extreme two-jet (i.e. cigar-like) events and 0.5 for spherical events. Since it is linear in momentum, it is more sensitive to low-momentum particles than sphericity.

A plane through the origin and perpendicular to the thrust axis divides the event into two hemispheres H_1 and H_2. To obtain the thrust major T_{maj}, the maximization of (2.53) is performed within that plane [23]. Thrust minor T_{min} is given by the argument of (2.53) evaluated in the direction perpendicular to both thrust and thrust major axes.

Oblateness is defined [23] by $O = T_{\text{maj}} - T_{\text{min}}$.

For a spherically symmetric distribution $S = 1$, $T = \frac{1}{2}$ and $S' = 1$, whereas for fully aligned events $S = 0$, $T = 1$ and $S' = 0$. The average values $\langle S\rangle, \langle T\rangle, \langle S'\rangle$ and the distribution of S, T, S' at different energies tell us how strongly 'jet-like' the final states actually are and how their shape depends on energy.

5. Heavy-jet mass M_H. The invariant mass of the particles in each hemisphere H_1 and H_2 is calculated and M_H is defined [24] as the larger of the two.

6. Jet broadening variables B_T *and* B_W. For each of the two hemispheres H_k defined above, a quantity

$$B_k = \frac{\sum_{i\epsilon H_k} |\mathbf{p}_i \times \mathbf{n}_T|}{2\sum_i |\mathbf{p}_i|} \tag{2.54}$$

is calculated. From B_k, the two variables are defined [25] as

$$B_{\mathrm{T}} = B_1 + B_2 \quad \text{and} \quad B_{\mathrm{W}} = \max(B_1, B_2) \,, \tag{2.55}$$

with B_{T} being the total and B_{W} the wide-jet broadening.

The latter quantities, 1 - T, M_{H}, B_{T} and B_{W}, are of particular interest for quantitative QCD studies in e^+e^- collisions, because QCD calculations including the resummation of leading and next-to-leading logarithmic terms to all orders (NLLA calculations) exist for these observables [25, 26].

2.4.2 e^+e^- collisions

2.4.2.1 Event shape at fixed energy

The event shape has been studied extensively in e^+e^- collisions to above 200 GeV [27–39]. As an example, Fig. 2.18a shows distributions for thrust T, scaled heavy-jet mass $\rho = M_{\mathrm{H}}^2/s$, total and wide-jet broadening B_{T} and B_{W} and the C-parameter at $\sqrt{s}$ = 189 GeV [27]. The data are compared with the predictions of the QCD based models JETSET PS, HERWIG and ARIADNE to be discussed in Chapter 6. The agreement is satisfactory.

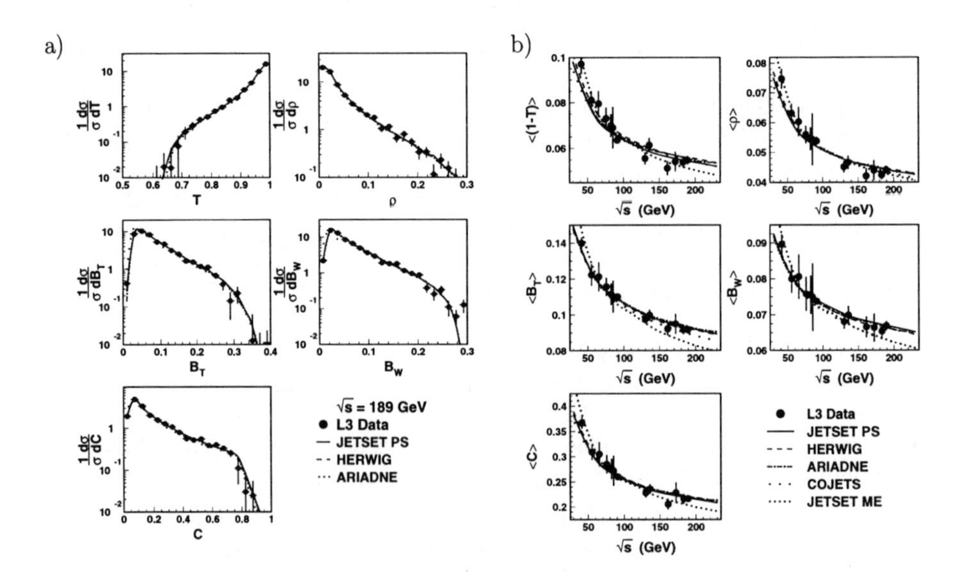

Figure 2.18. a) Distributions for thrust T, scaled heavy-jet mass $\rho = M_{\mathrm{H}}^2/s$, total and wide-jet broadenings, B_{T} and B_{W}, and the C-parameter at $\sqrt{s} = 189$ GeV, in comparison with QCD model predictions. b) The first moments of the same event shape variables as a function of the centre-of-mass energy, compared with several QCD models [27].

In QCD, deviations from longitudinal phase space are directly related to gluon radiation. The running coupling constant α_s can, therefore, be obtained from the event shape variables at each LEP energy by fitting the measured distributions by analytical calculations based on $\mathcal{O}(\alpha_s^2)$ perturbative QCD combined with resummed leading and next-to-leading order terms [40].

In second order, for any event shape variable y, the cumulative distribution

$$R(y', \alpha_s) = \frac{1}{\sigma_{\text{tot}}} \sigma(y < y') \tag{2.56}$$

can be expressed as

$$\ln R(y', \alpha_s) = \alpha_s A(y') + \alpha_s^2 B(y') + \mathcal{O}(\alpha_s^3) \ . \tag{2.57}$$

The functions A and B differ from event shape variable to event shape variable, but can be obtained by integration over the first- and second-order matrix elements.

Alternatively, the cumulative distribution can be expanded in terms of logarithms $L \equiv \ln y'$ as

$$\ln R(y', \alpha_s) = L \cdot f_{\text{LL}}(\alpha_s L) + f_{\text{NLL}}(\alpha_s L) + \mathcal{O}\left(\frac{1}{L}(\alpha_s L)^n\right) , \tag{2.58}$$

where the functions f_{LL} and f_{NLL} only depend on the product of $\alpha_s L$. The first two terms represent the leading and next-to-leading logarithms, where emission of soft gluons from the original quark-antiquark pair is considered up to all orders by resumming large logarithms. The following terms are subleading.

Equation (2.58) can also be expanded as

$$\ln R(y', \alpha_s) = \sum_{n=1}^{\infty} \alpha_s^n \left(a_n L^{n+1} + b_n L^n + \ldots\right) , \tag{2.59}$$

where the terms in $\alpha_s^n L^{n+1}$ represent the leading logarithms, the terms in $\alpha_s^n L^n$ the next-to-leading logarithms. The combination (matching) can thus be obtained by using the exact second-order expression of (2.57) together with the leading and next-to-leading logarithms of (2.58), starting in $\mathcal{O}(\alpha_s^3)$ [26].

The above calculations are performed at parton level and do not include heavy-quark mass effects. To compare the analytical calculations with the experimental distributions, the effects of hadronization and decays have to be incorporated by means of Monte Carlo programs (JETSET, ARIADNE, HERWIG).

The energy dependence of α_s can be obtained by fitting $\ln R(y', \alpha_s)$ at different energies. The result averaged over the event shape variables at each given energy is shown in Fig. 2.19 for L3 [27]. A fit by the QCD evolution equation [41] with $\alpha_s(m_z)$ as a free parameter gives $\alpha_s(m_Z) = 0.1227 \pm 0.0012(\text{exp}) \pm 0.0058(\text{theor})$. This value has to be compared with an all-over average of $\alpha_s(m_Z) = 0.1187 \pm 0.0020$ [41].

These measurements support the energy evolution of the strong coupling constant predicted by QCD.

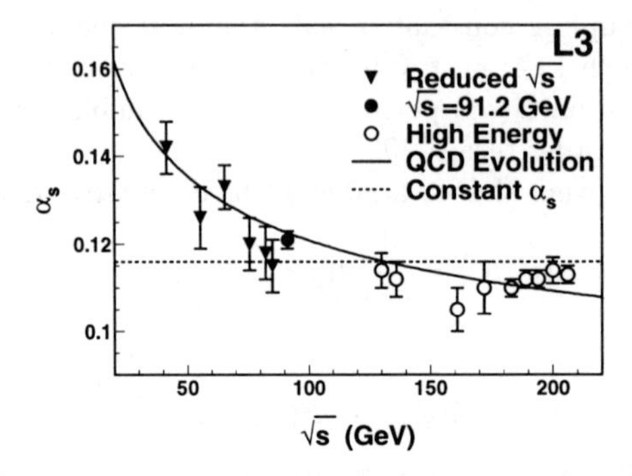

Figure 2.19. α_s measurements from event shape distributions, as a function of the centre-of-mass energy. The solid and dashed lines are fits with the energy dependence of α_s as expected from QCD and with constant α_s, respectively [27].

2.4.2.2 Energy dependence of the event shape

An important test of the models is the energy evolution of the event shape, itself. This shape will be described by the first and second moments of the distributions. The first moments over the event shape variables of Fig. 2.18a are plotted as a function of cms energy in Fig. 2.18b [27]. The parton shower models (JETSET, HERWIG, ARIADNE, COJETS) describe the data, but the $\mathcal{O}(\alpha_s^2)$ matrix element model JETSET ME is too steep. With fixed parameters it cannot be expected to describe the data over a wide energy range.

In terms of analytical calculations, the energy dependence of moments of the event shape variables has also been described [42] as a sum of second-order perturbative contributions and a power-law dependence due to non-perturbative contributions. The first moment of an event shape variable y is written as

$$\langle y \rangle = \langle y_{\text{pert}} \rangle + \langle y_{\text{pow}} \rangle \ , \tag{2.60}$$

where the perturbative contribution $\langle y_{\text{pert}} \rangle$ has been determined to $\mathcal{O}(\alpha_s^2)$ [43]. The power correction term [42], for $1-T$, ρ, and C, is given by

$$\langle y_{\text{pow}} \rangle = c_y \frac{4C_{\text{F}}}{\pi^2} \mathcal{M} \frac{\mu_{\text{I}}}{\sqrt{s}} \left[\alpha_0(\mu_{\text{I}}) - \alpha_s(\sqrt{s}) - \beta_0 \frac{\alpha_s^2(\sqrt{s})}{2\pi} \left(\ln \frac{\sqrt{s}}{\mu_{\text{I}}} + \frac{K}{\beta_0} + 1 \right) \right] \ , \tag{2.61}$$

where only the factor c_y depends on the shape variable y and all the rest is supposed to be a universal power-law form for a renormalization scale fixed at $\sqrt{s}$. The parameter α_0 is related to the values of α_s in the non-perturbative region below an infrared matching scale $\mu_{\text{I}}(= 2\ \text{GeV})$; $\beta_0 = (11n_c - 2n_f)/3$, where n_c is the number of colors and n_f is the number of active flavors; $K = (67/18 - \pi^2/6)C_{\text{A}} - 5n_f/9$ and $C_{\text{F}}, C_{\text{A}}$ are the usual color factors. The so-called Milan factor $\mathcal{M}$ [44] is 1.49 for $n_f = 3$.

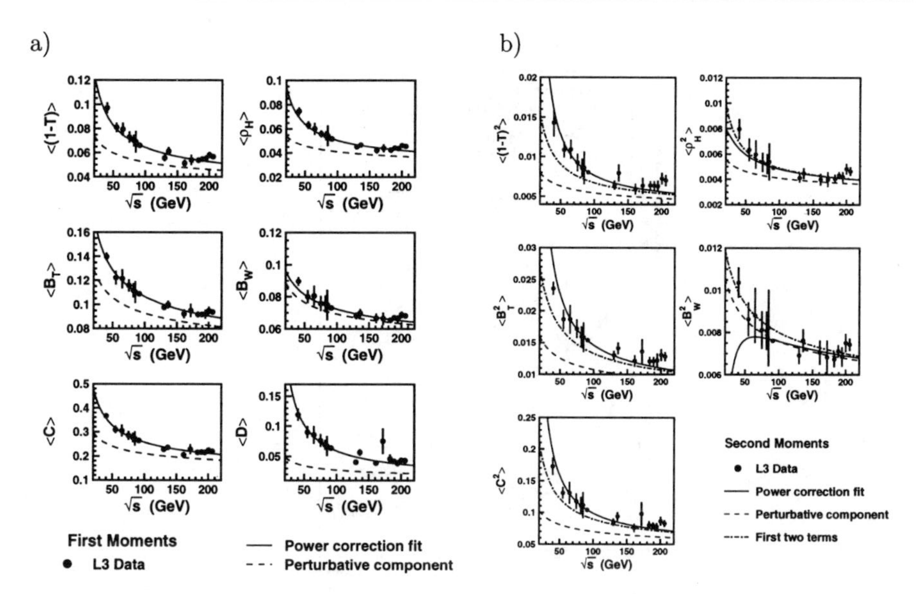

Figure 2.20. a) The first moments of the five event shape variables, $1-T$, ρ, B_T, B_W, C and D compared to the results of a fit including perturbative and power law contributions. b) The second moments of the first five event shape variables, compared to the results of a fit including perturbative and power law contributions. The parameters α_0 and α_s are fixed to the values obtained by the fits to the first moments [27].

For the jet broadening variables an extra multiplicative factor

$$F = \left(\frac{\pi}{2\sqrt{aC_{\mathrm{F}}\alpha_{\mathrm{CMW}}}} + \frac{3}{4} - \frac{\beta_0}{6aC_{\mathrm{F}}} - 0.6137 + \mathcal{O}(\sqrt{\alpha_{\mathrm{s}}}) \right) \tag{2.62}$$

is required with $a = 1$ for B_{T} and 2 for B_{W} and α_{CMW} is related to α_{s} [42]. Furthermore, extra terms are needed in $\langle y_{\mathrm{pert}} \rangle$ and $\langle y_{\mathrm{pow}} \rangle$ for the D-parameter.

Results of fits with $\alpha_{\mathrm{s}}(m_{\mathrm{Z}})$ and α_0 as the only open parameters are given in Fig. 2.20a. While α_0 shows some deviations from its average for D and B_{W}, the six values obtained for α_{s} agree within errors, thus supporting the predicted universality of the power-law behavior. Their average values in L3 are [27]

$$\begin{aligned} \alpha_0 &= 0.478 \pm 0.054 \pm 0.024 \\ \alpha_{\mathrm{s}}(m_{\mathrm{Z}}) &= 0.1126 \pm 0.0045 \pm 0.0039 \;, \end{aligned}$$

where the first errors are experimental, the second are theoretical. Similar results are obtained by the other experiments [35–37, 45]. An example covering the full energy range is given in Fig. 2.21 [36].

It can be seen from Figs. 2.20a and 2.21 that the energy dependence of the mean event shape variables arises from

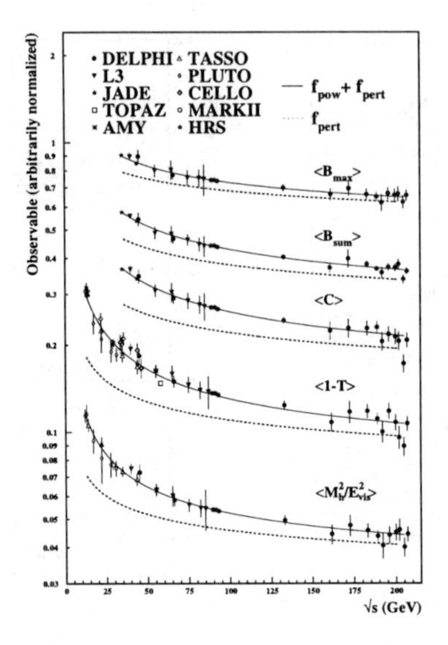

Figure 2.21. Measured values of five mean values indicated, as a function of the centre of mass energy. The solid lines present the results of the fits with (2.60), the dotted lines show the perturbative part only [36].

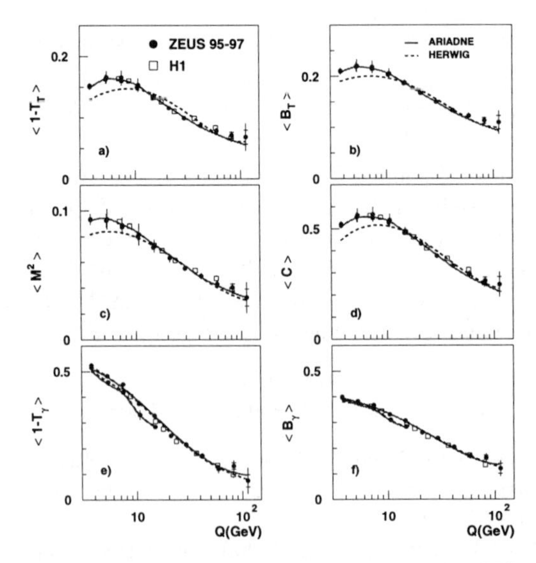

Figure 2.22. Comparison of mean event shape variables in the DIS current-quark region with ARIADNE and HERWIG predictions [46].

1. the logarithmic energy scale dependence of α_s in the perturbative contribution,
2. the power-law behavior of non-perturbative effects.

Furthermore, assuming that the non-perturbative correction to the y distributions causes only a shift, the second moments can be written [47]

$$\langle y^2 \rangle = \langle y_{\rm pert}^2 \rangle + 2\langle y_{\rm pert} \rangle \langle y_{\rm pow} \rangle + A_2/s \ , \qquad (2.63)$$

where A_2/s parametrizes the $\mathcal{O}(1/s)$ contributions. The A_2/s term is expected to be negligible for $1-T$, ρ, C and D. For jet broading the power corrections are more complicated. The fits with α_0 and α_s fixed to the values given above are compared to the data in Fig. 2.20b [27]. The contributions of the power term and the $\mathcal{O}(1/s)$ term are shown separately. Contrary to the expectation, the latter is not small for $1-T$ and C.

Quite striking is the observation of DELPHI [36], that there is no need for power correction terms to describe the energy dependence of any of the inclusive shape means used, when the expansion is done according to so-called renormalization group invariant (RGI) perturbation theory [48–51]. This leads to a considerably improved determination of $\alpha_s(m_Z) = 0.1201 \pm 0.0020$ consistently describing all data and to $\beta_0 = 7.86 \pm 0.32$, the latter corresponding to $n_f = 4.75 \pm 0.44$ active flavors.

2.4.3 Lepton-hadron collisions

Assuming the universality of quark fragmentation, similar event-shape properties are expected for the current-quark region of deep inelastic scattering (DIS) [52]. A natural frame for the comparison of DIS with e^+e^- fragmentation is the Breit frame [53], in which the exchanged virtual photon γ^* is completely space-like [$q = (0, 0, 0, -Q = -2x_B P^{\rm Breit})$], with $P^{\rm Breit}$ being the proton momentum in the Breit frame, and Q and x_B as defined in Chapter 1. In a simple quark-parton model (QPM), this frame gives maximal separation of incoming and outgoing partons. Particles can be assigned to the current region if their z-momentum is negative ($\geq -Q/2$) and to the target region if it is positive ($\leq (1-x_B)Q/2x_B$).

In DIS, the relevant scale is not $\sqrt{s}$ but the square root Q of the virtuality of the exchanged boson. Figure 2.22 shows the mean values for six event shape variables as a function of Q for H1 [54] and ZEUS [46], with T_T, B_T defined with respect to the thrust axis and T_γ, B_y w.r.t. the virtual-photon direction. The data agree with each other and with the predictions of the ARIADNE model and decrease with increasing Q at large Q.

Using Eq. (2.60) with the first term to NLO and the power correction Eq. (2.61) with $\sqrt{s}$ replaced by Q, the α_s value from $(1-T_\gamma)$ and the α_0 value from B_T disagree with the values obtained from the other shape variables [46], but, with the exception of $(1-T_\gamma)$, the values are consistent with those measured in e^+e^- annihilation. The residual inconsistencies in T_γ and B_T suggest the need for higher-order corrections.

2.4.4 Hadron-hadron collisions

For typical hadron-hadron collisions, the axis minimizing S and S' (or maximizing T) is always quite close to the collision axis and the corresponding distributions defined with respect to the collision axis are very similar to those for S, S', T. However, for the purpose of comparing hadron-hadron, lepton-hadron and e^+e^- data, S, S' and T are commonly used directly.

As an example, the decrease in average sphericity $\langle S\rangle$ with increasing energy, generally interpreted as evidence for jet production, is not only a feature of reactions in which single quark effects are expected to dominate, but also a feature of low-p_T hadron-hadron collisions. As is shown in Fig. 2.23, the average shape of the hadronic system at low energies is the same in all three types of collision at given hadronic energy [55], as is its quite dramatic change with energy. Of course, a higher energy hadronic point would be needed to see where flattening-off takes place, there.

The energy dependence of the shape of the sphericity distribution itself [56–59] is shown in Fig. 2.24. As for e^+e^- collisions, one observes a change from a dip at $S = 0$ at low energy to a sharp peak at $S = 0$ at higher energies. Very good agreement is observed between the S distribution in K^-p at 110 GeV/c ($\sqrt{s}$=14.8 GeV) with the PLUTO result at 17 GeV and the K^+p data at 70 GeV/c ($\sqrt{s}$=11.5 GeV) with the TASSO distribution at 13 GeV (for the curves LPS and FF see below).

The normalized 70 GeV/c K^+p rapidity distribution, evaluated with respect to

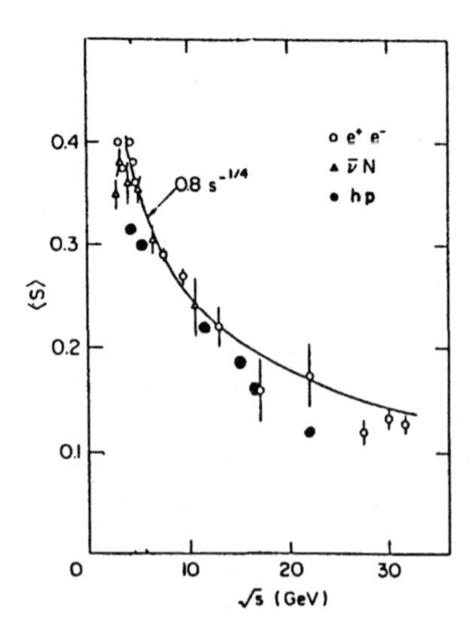

Figure 2.23. Average sphericity as a function of hadronic energy in e^+e^- ($\circ$), $\bar{\nu}N(\triangle)$ and hadron-hadron ($\bullet$) collisions [55].

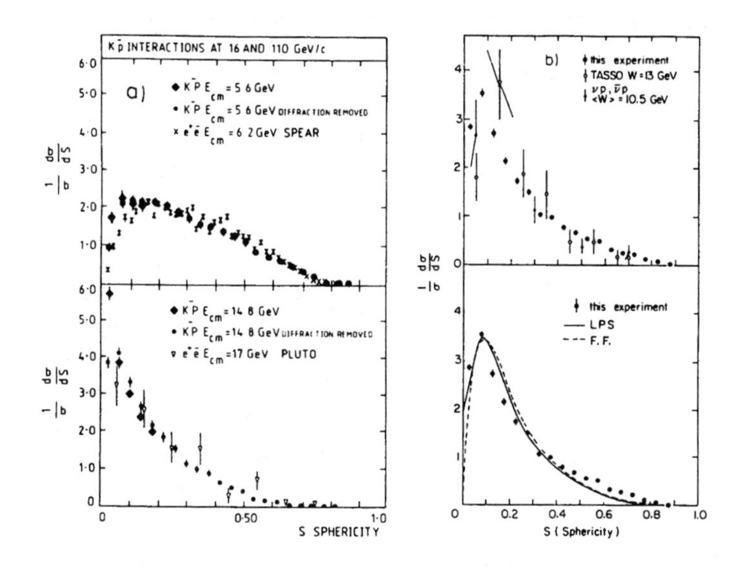

Figure 2.24. a) Sphericity distribution for K^-p (♦, diffraction removed •) at 16 (top) and 110 GeV/c (bottom) compared to e^+e^- at $\sqrt{s}$=6.2 (x) and 17 GeV (▽), respectively [56]. b) Same for K^+p at 70 GeV/c (•), compared to e^+e^- (∘) and $\nu(\bar{\nu})p$ (▲) data at the corresponding energy and to predictions from longitudinal phase space (LPS) (full line) and from Field and Feynman (FF) (dashed) [57].

the thrust axis [57], is compared to e^+e^- results at 13 GeV in Fig. 2.25a. At similar energies, the two rapidity distributions agree remarkably well and it can be expected from the insert (and from earlier results on hadronic rapidity plateaus) that hadron-hadron and e^+e^- data show a similar $\ln\sqrt{s}$ increase of the plateau height at the same energy!

Even the normalized p_T^2 distributions relative to the sphericity axis (Fig. 2.25b) show agreement at 13 GeV/c. There may be a small indication of a larger cross section in the high-p_T tail for the e^+e^- results, but this can be understood from the higher e^+e^- energy. From these energies upwards, the tail of the e^+e^- distribution shows a considerable increase, generally interpreted as evidence for gluon jets. We shall come back to this question in Sect. 4.4.

The conclusion of jet universality up to about 15 GeV is further supported from a comparison of the energy dependence of the average charged-particle multiplicity $\langle n\rangle$, as well as of the average transverse and longitudinal momentum $\langle p_T\rangle$ and $\langle p_\parallel\rangle$ relative to the thrust axis (Fig. 2.25c). Again, at low energies, the hp and e^+e^- data have essentially the same values and the same energy behavior. In particular, the rise in $\langle p_T\rangle$ with $E_{\text{cms}} \equiv \sqrt{s}$, felt to be a characteristic feature of single quark jets, is in fact also a feature of hadronic low-p_T particle production.

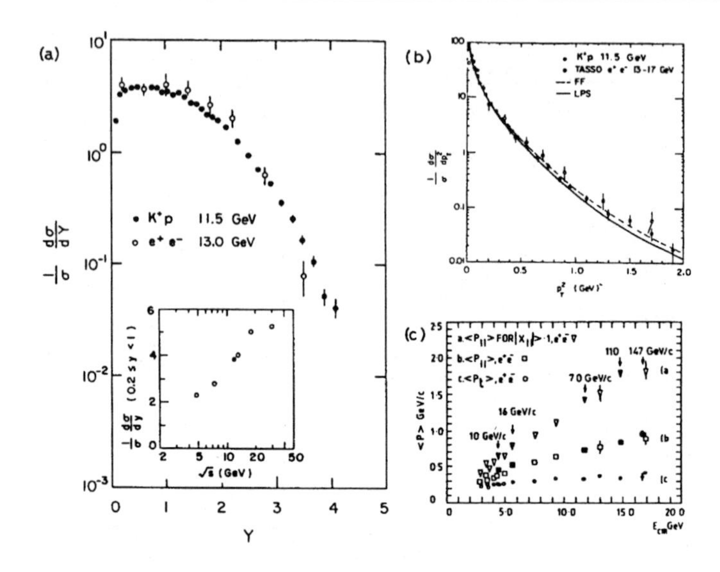

Figure 2.25. a) Rapidity distribution with respect to the thrust axis for non-diffractive K^+p events at 70 GeV/c (•) compared to e^+e^- at $\sqrt{s}$=13 GeV (∘). The insert shows the central region density as a function of cms energy [57]. b) Transverse momentum distribution relative to the sphericity axis for K^+p at 70 GeV/c (•) and e^+e^- at $\sqrt{s}$=13 GeV (∘). The curves correspond to FF (dashed) and LPS (full line) [57]. c) Energy dependence of $\langle p_T \rangle$ and $\langle p_{\|} \rangle$ relative to the thrust axis for hadron-proton (full symbols) and e^+e^- (open symbols) reactions. ∇, $\langle p_{\|} \rangle$ for $|x_F| > 0.1$; $\square$, $\langle p_{\|} \rangle$; ∘, $\langle p_T \rangle$ [56–59].

At this stage, we may wonder whether the agreement between the e^+e^-, lepton-hadron and hadron-hadron data is simply due to kinematics, or has a more fundamental dynamical origin. In [57] the sphericity, thrust, spherocity and other distributions are compared to mere longitudinal phase space (LPS) [60] and to the Field and Feynman (FF) parametrization [61] of quark-parton jets. In all cases, both the FF parametrization as well as LPS more or less describe the data (see e.g. Fig. 2.24b for sphericity). Furthermore, for hadronic reactions one finds that the sphericity, thrust and spherocity axes agree with the beam direction. One may conclude that

(i) FF very closely resembles longitudinal phase space,

(ii) jet universality up to meson-proton cms energies of about 15 GeV turns out to reduce to the equal p_T distributions shown here and to equal average multiplicities (apart from trivial corrections) to be discussed in Chapters 4 and 8 in more detail. The rest follows from independent emission and conservation laws.

The important point is, that this holds equally well for hadronic, deep inelastic and e^+e^- hadron production at these energies.

Now we are ready to start!

Bibliography

[1] L. Van Hove, *Phys. Reports* **1C** (1971) 347.

[2] L. Van Hove, *Phys. Lett.* **28B** (1969) 429 and *Nucl. Phys.* **B9** (1969) 331.

[3] R.P. Feynman, *Phys. Rev. Lett.* **23** (1969) 1415.

[4] J. Benecke, T.T. Chou, C.N. Yang and F. Yen, *Phys. Rev.* **188** (1969) 2159.

[5] Z. Koba, H.B. Nielsen and P. Olesen, *Generating Functionals for Multiple Particle Production, Semi-Inclusive Cross Sections, Momentum Distributions and Correlations* (Niels Bohr Institute) preprint NBI-HE-71-7 (1971).

[6] ABBCCHW Coll., R. Honecker et al., *Nucl. Phys.* **B13** (1969) 571.

[7] CERN-Rome Coll., J.V. Allaby et al., *Comparison of Momentum Spectra of Secondary Particles Produced in 14.25, 19.2 and 24.0 GeV/c* pp *Collisions*, paper presented at the Amsterdam International Conference on Elementary Particles (1971).

[8] Aachen-Berlin-Bonn-CERN-Cracow-Heidelberg-Warsaw Coll., presented at the Amsterdam International Conference on Elementary Particles (1971) and private communication from T. Besliu, R. Stroynowski and H. Wahl.

[9] Chan Hong-Mo, C.S. Hsue, C. Quigg and Jiunn-Ming Wang, *Phys. Rev. Lett.* **26** (1971) 672.

[10] A.H. Mueller, *Phys. Rev.* **D2** (1970) 2963.

[11] Aachen-Berlin-Bonn-CERN-Cracow-Heidelberg-Warsaw Coll., J.V. Beaupré et al., *Phys. Lett.* **37B** (1971) 432.

[12] K.G. Wilson, *Acta Physica Austriaca* **17** (1963) 37.

[13] L. Van Hove, Proc. Colloquium on Multiparticle Dynamics, Helsinki 1971, eds. E. Byckling et al., p. 227.

[14] C.E. De Tar, *Phys. Rev.* **D3** (1971) 128.

[15] A. Białas, A. Eskreys, W. Kittel, S. Pokorski, J.K. Tuominiemi and L. Van Hove, *Nucl. Phys.* **B11** (1969) 479.

[16] ABBCHLV Coll., J. Bartsch et al., *Nucl Phys.* **B19** (1970) 381.

[17] S. Brandt, Ch. Peyrou, R. Sosnowski and A. Wróblewski, *Phys. Lett.* **12** (1964) 57; E. Farhi, *Phys. Rev. Lett.* **39** (1977) 1587.

[18] A.B. Clegg, *Rep. Prog. Phys.* **45** (1982) 887.

[19] J.D. Bjorken and S.J. Brodsky, *Phys. Rev.* **D1** (1970) 1416.

[20] S.L. Wu and G. Zobernik, *Z. Phys.* **C2** (1979) 107.

[21] G. Parisi, *Phys. Lett.* **B74** (1978) 65; J.F. Donoghue, F.E. Low and S.Y. Pi, *Phys. Rev.* **D20** (1979) 2759; R.K. Ellis, D.A. Ross and A.E. Terrano, *Nucl. Phys.* **B178** (1981) 421.

[22] H. Georgi et al., *Phys. Rev. Lett.* **39** (1977) 1237; S. Brandt et al., *Z. Phys.* **C1** (1979) 61.

[23] Mark J Coll., D.P. Barber et al., *Phys. Rev. Lett.* **43** (1979) 830.

[24] T. Chandramohan and L. Clavelli, *Nucl. Phys.* **B184** (1981) 365; L. Clavelli and D. Wyler, *Phys. Lett.* **B103** (1981) 383.

[25] S. Catani, G. Turnok and B.R. Webber, *Phys. Lett.* **B295** (1992) 269; Yu.L. Dokshitzer et al., *JHEP* **01** (1998) 11.

[26] S. Catani, L. Trentadue, G. Turnock and B.R. Webber, *Nucl. Phys.* **B407** (1993) 3.

[27] L3 Coll., B. Adeva et al., *Z. Phys.* **C55** (1992) 39; O. Adriani et al., *Phys. Lett.* **B284** (1992) 471; *Phys. Reports* **236** (1993) 1; M. Acciarri et al., *Phys. Lett.* **B371** (1996) 137; **B404** (1997) 390; **B411** (1997) 339; *Phys. Lett.* **B444** (1998) 569; **B489** (2000) 65; P. Achard et al., *Phys. Lett.* **B536** (2002) 217; CERN-PH-EP/2004-26, *Phys. Reports* **399** (2004) 71.

[28] PLUTO Coll., C. Berger et al., *Z. Phys.* **C12** (1982) 297.

[29] TASSO Coll., M. Althoff et al., *Z. Phys.* **C22** (1984) 307; W. Braunschweig et al., *Phys. Lett.* **B214** (1988) 293; *Z. Phys.* **C41** (1988) 359; **C45** (1989) 11; **C47** (1990) 187.

[30] HSR Coll., D. Bender et al., *Phys. Rev.* **D31** (1985) 1.

[31] JADE Coll., W. Bartel et al., *Z. Phys.* **C33** (1986) 23; P.A. Movilla Fernández et al., *Eur. Phys. J.* **C1** (1998) 461; **C22** (2001) 1; O. Biebel et al., *Phys. Lett.* **B459** (1999) 326; C. Pahl et al., hep-ex/0408123.

[32] AMY Coll., Y.K. Li et al., *Phys. Rev.* **D41** (1990) 2675.

[33] MARKII Coll., A. Peterson et al., *Phys. Rev.* **D37** (1988) 1; G.S. Abrams et al., *Phys. Rev. Lett.* **63** (1989) 1558.

[34] CELLO Coll., H.J. Behrend et al., *Z. Phys.* **C44** (1989) 63.

[35] ALEPH Coll., D. Decamp et al., *Phys. Lett.* **B234** (1990) 209; **B255** (1991) 623; **B257** (1991) 479; **B284** (1992) 163; D. Buskulic et al. *Z. Phys.* **C55** (1992) 209; **C73** (1997) 409; R. Barate et al., *Phys. Reports* **294** (1998) 1; A. Heister et al., *Eur. Phys. J.* **C35** (2004) 457.

[36] DELPHI Coll., D. Aarnio, *Phys. Lett.* **B240** (1990) 271; P. Abreu et al., *Z. Phys.* **C54** (1992) 55; **C59** (1993) 21; *Z. Phys.* **C73** (1996) 11; **C73** (1997) 229; *Phys. Lett.* **B456** (1999) 322; *Eur. Phys. J.* **C14** (2000) 557; J. Abdallah et al., *Eur. Phys. J.* **C29** (2003) 285; **C37** (2004) 1.

[37] OPAL Coll., M.Z. Akrawy et al., *Z. Phys.* **C47** (1990) 505; P.D. Acton et al., R. Akers et al., *Z. Phys.* **C65** (1995) 31; **C68** (1995) 519; G. Alexander et al., *Z. Phys.* **C72** (1996) 191; K. Ackerstaff et al., *Z. Phys.* **C75** (1997) 193; G. Abbiendi et al., *Eur. Phys. J.* **C16** (2000) 185; CERN-PH-EP/2004-044, *Eur. Phys. J. C* (to be publ.).

[38] TOPAZ Coll., I. Adachi et al., *Phys. Lett.* **B227** (1989) 495; K. Nagai et al., *Phys. Lett.* **B278** (1992) 506; Y. Ohnishi et al. *Phys. Lett.* **B313** (1993) 475.

[39] SLD Coll., K. Abe et al., *Phys. Rev.* **D51** (1995) 962.

[40] S. Catani et al., *Phys. Lett.* **B263** (1991) 491; **B272** (1991) 368; **B295** (1992) 269; *Nucl. Phys.* **B407** (1993) 3; S. Bethke et al., *Nucl.Phys.* **B370** (1992) 310; S. Catani and B.R. Webber, *Phys. Lett.* **B427** (1998) 377; Yu.L. Dokshitzer et al., *J. High Energy Phys.* **01** (1998) 011.

[41] Particle Data Group, S. Eidelman et al., *Phys. Lett.* **B592** (2004) 1.

[42] B.R. Webber, *Phys. Lett.* **B339** (1994) 148; Yu.L. Dokshitzer, B.R. Webber, *Phys. Lett.* **B352** (1995) 451; B.R. Webber, hep-ph/9510283; Yu.L. Dokshitzer et al., *Nucl. Phys.* **B511** (1997) 396; Yu.L. Dokshitzer et al., *J. High Energy Phys.* **05** (1998) 003; Yu.L. Dokshitzer et al., *Eur. Phys. J.* **C3** (1999) 1; Yu.L. Dokshitzer, hep-ph/9911299.

[43] S. Catani and M. Seymour, *Phys. Lett.* **B378** (1996) 287.

[44] Yu.L. Dokshitzer, A. Lucenti, G. Marchesini and P. Salam, *Nucl. Phys.* **B511** (1997) 396; *JHEP* **05** (1998) 03.

[45] ALEPH, DELPHI, L3 and OPAL Coll's and the LEP QCD Working Group, paper in preparation.

[46] ZEUS Coll., S. Chekanov et al., *Eur. Phys. J.* **C27** (2003) 531.

[47] Yu.L. Dokshitzer, B.R. Webber, *Phys. Lett.* **B404** (1997) 321; B.R. Webber, *Nucl. Phys.* (Proc. Suppl.) **71** (1999) 66.

[48] A. Dhar, *Phys. Lett.* **B128** (1983) 407; A. Dhar and V. Gupta, *Phys. Rev.* **D29** (1984) 2822; *Pramana* **21** (1983) 207.

[49] J.G. Korner, F. Krajewski and A.A. Pivovarov, *Phys. Rev.* **D63** (2001) 036001.

[50] M. Beneke, *Phys. Lett.* **B307** (1993) 154.

[51] C.J. Maxwell, D.T. Barclay and M.T. Reader, *Phys. Rev.* **D49** (1994) 3480.

[52] M. Dasgupta and B.R. Webber, *Eur. Phys. J.* **C1** (1998) 539.

[53] R.P. Feynman, *Photon-Hadron Interactions*, Benjamin, N.Y. (1972).

[54] H1 Coll., C. Adloff et al., *Phys. Lett.* **B406** (1997) 256; *Eur. Phys. J.* **C14** (2000) 255 and erratum **C18** (2000) 417; **C18** (2000) 293.

[55] K. Fiałkowski and W. Kittel, *Rep. Prog. Phys.* **46** (1983) 1283.

[56] R. Göttgens et al., *Nucl. Phys.* **B178** (1981) 392.

[57] M. Barth et al., *Nucl. Phys.* **B192** (1981) 289.

[58] D. Brick et al., *Z. Phys.* **C15** (1982) 1.

[59] J.M. Laffaille et al., *Nucl. Phys.* **B192** (1981) 18.

[60] L. Van Hove, *Rev. Mod. Phys.* **36** (1964) 655; E.H. de Groot and T.W. Ruijgrok, *Nucl. Phys.* **B101** (1975) 95.

[61] R.D. Field and R.P. Feynman, *Nucl. Phys.* **B136** (1978) 1.

Chapter 3

Three-Particle Exclusive Final States

3.1 Shape and energy dependence

As pointed out in Sect. 2.3, longitudinal phase space (LPS) for a three-particle final state is bounded by a hexagon. A typical distribution of events in three-particle LPS has been given in Fig. 2.13 for the reaction

$$\pi^- \mathrm{p} \to \mathrm{p}\pi^-\pi^0 \tag{3.1}$$

at 16 GeV/c [1]. We have concluded there that the Van Hove correlation angle ω defined in Fig. 2.11 essentially determines the complete longitudinal configuration of one event.

The distribution of reaction (3.1) against the angle ω is shown in Fig. 3.1, before (Fig. 3.1a) and after (Fig. 3.1b) correcting for phase-space effects by means of the weight w_3, defined in (2.43). In our definition, the ω-region considered ($60° < \omega < 240°$) corresponds to the hemisphere of LPS in which the proton goes backward in the cms. Furthermore, $60° < \omega < 120°$ corresponds to a backward π^0 and forward π^-, $120° < \omega < 180°$ to both pions forward, and $180° < \omega < 240°$ to a forward π^0 and backward π^-. The π^0 is longitudinally at "rest" at ω=120°, the π^- at ω=180°.

The main difference between the weighted and unweighted distributions is a relative lowering of the former with respect to the latter in regions where one of the E_i^* is small, i.e. at ω=120° and ω=180°. The figure shows that the weighted distribution, while still having marked structure in ω, is indeed lower than the unweighted one near 120° and 180°. This means that the matrix element is smoother there than longitudinal phase space itself. The three main conclusions to be drawn from Fig. 3.1 and similar ones are:

(i) Peaks in the distributions indicate strong correlations between the particles in the final state.

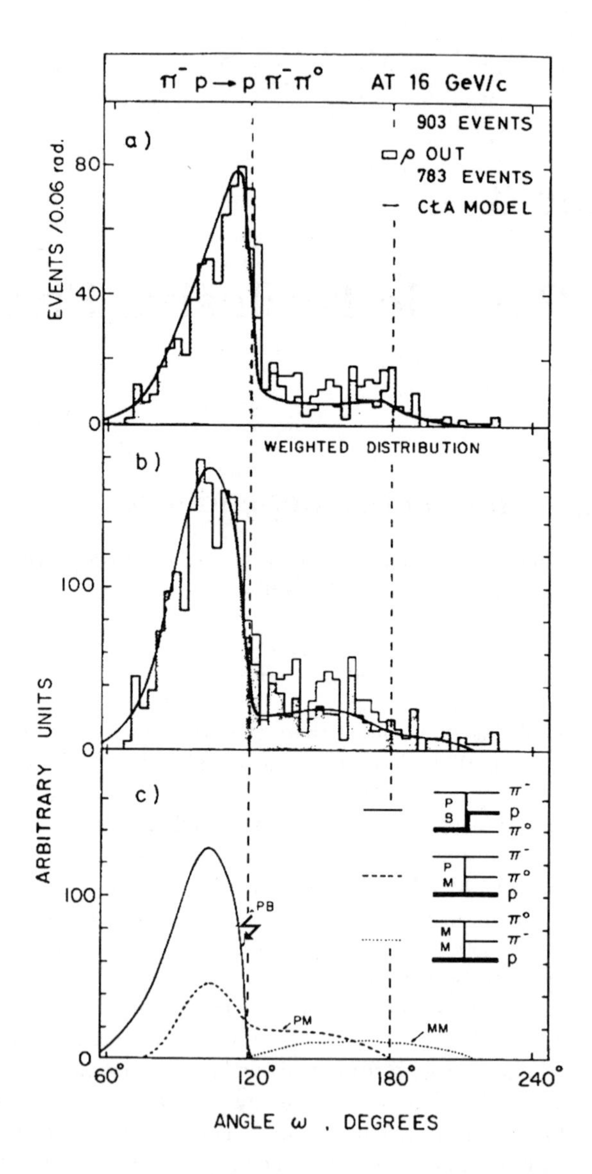

Figure 3.1. Distribution of events (Fig. 3.1a) and of the matrix element averaged over transverse momenta (Fig. 3.1b) against the Van Hove correlation angle ω for the reaction $\pi^-\mathrm{p} \to \pi^-\pi^0\mathrm{p}$ at 16 GeV/c. The solid lines are the distributions calculated in the framework of the CŁA model and are normalized to the distribution after exclusion of the ρ^- band. The individual CŁA exchange graphs considered and their contributions to the total distribution are given in Fig. 3.1c [1].

(ii) The matrix element (weighted distribution) does not have a peak where one of the particles is at rest, i.e. at $60°, 120°, 180°$ or $240°$. This can be taken as evidence against strong double-peripheral configurations, in the sense that two-particle sub-energies $s_{ij}^{1/2}$ and $s_{ik}^{1/2}$ are, in general, not large simultaneously.

(iii) Pronounced peaks occur where vacuum (pomeron) exchange is allowed. For the final state $\pi^-\pi^0$p, this applies for the region $60° < \omega < 120°$.

The above observations are made in a model-independent way by looking at the distribution in the variable describing the complete three-particle final-state. For that reason, there is nothing wrong in now checking predictions of existing models against these experimental distributions.

Before this could be done, an efficient Monte Carlo method had to be developed to generate events only where the matrix element would be non-zero or at least not negligibly small (importance sampling). The method adopted [2] uses as integration variables the momenta of the outgoing particles in the cms, distinguishing between their longitudinal component and their transverse ones. While the longitudinal component is allowed to cover all available phase space, the length of the transverse ones is restricted to a Gaussian p_{T} distribution of energy-independent width comparable to that of the data. The choice of momenta as integration variables makes the method directly applicable to all types of theoretical models.

As an example to demonstrate the methodics, we choose the so-called CŁA model [3], which in spite, or because, of its simplicity was one of the most generally applicable models at these low energies. It is a Reggeized form of a multiperipheral model, in which the amplitude is treated as an incoherent sum of contributions from various multiperipheral graphs. For each graph, the amplitude is

$$|M_n| = \prod_{i=1}^{n-1} \left[\frac{g_i s_i + ca}{s_i + a}\right] \left[\frac{s_i + a}{a}\right]^{\alpha_i(0)} \left[\frac{s_i + b_i}{b_i}\right]^{t_i} , \tag{3.2}$$

where t_i is the squared four-momentum transfer from the beam to particle C_i and s_i is the squared sub-energy of particles C_i and C_j after threshold subtraction; a, b_i, c and g_i are constants and $\alpha_i(0)$ is the intercept of the Regge trajectory exchanged between particles C_i and C_j. In spite of being a multiperipheral model, it is not in contradiction with observation (ii) above, since in the limit of low s_i the corresponding factor in the product of (3.2) approaches smoothly a constant value c. The possible graphs are obtained by considering all permutations of the outgoing particles corresponding to the exchange of non-exotic quantum numbers. They are given for reaction (3.1) in Fig. 3.1c, together with the values of the corresponding squared matrix elements plotted against the correlation angle ω.

After exclusion of the ρ-resonance, which is too sharp an effect to be described by the constant c, the model can fit the shape of the distribution (Figs. 3.1a or 3.1b) surprisingly well. From the contribution of the individual graphs in Fig.3.1c we see that pomeron exchange on the upper vertex essentially determines the shape of the distribution. The amount of baryon exchange needed at the lower vertex is worrying. A closer inspection of the amplitude (3.2) tells us that, within this model, only baryon exchange provides the strong t-cut necessary to reproduce the sharp fall-off of the data when crossing ω=120°.

Using a CLA type model where the constant c is replaced by a four-point Veneziano function [4], the distribution of Fig. 3.1 can be discribed (not shown here) with only pion exchange on the lower vertex, now even with the ρ^- left in. However, the characteristic sharp fall-off at 120° is not reproduced correctly.

We return to the model-independent data analysis and investigate the energy dependence of the distribution and its shape. In particular, we want to see whether the cross section is actually independent of energy in the region $60° < \omega < 120°$, as would be expected from pomeron exchange.

The constancy of the kinematic distribution in ω discussed in Fig. 2.17 makes the study of the energy dependence of the shape of the matrix element dynamically meaningful. The results of a study of a reaction similar to (3.1) is given in Fig. 3.2,

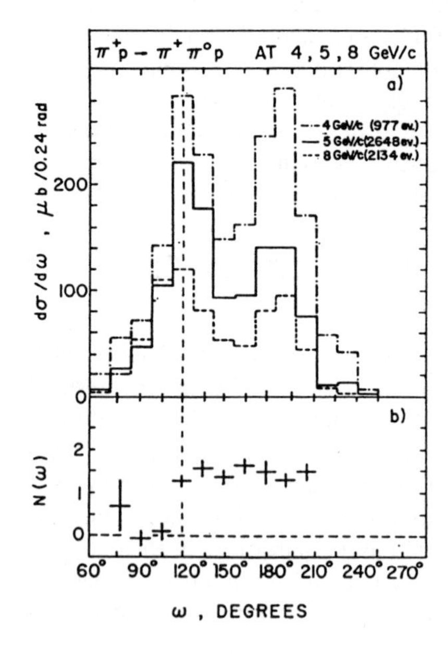

Figure 3.2. (a) Differential cross section $d\sigma/d\omega$ as a function of ω for the reaction $\pi^+p \rightarrow \pi^+\pi^0p$ at 4, 5 and 8 GeV/c. (b) Exponent $N(\omega)$ in the fit with $\sigma \propto p_{\text{lab}}^{-N(\omega)}$, as a function of ω for the same reaction [5].

where the (unweighted) differential cross section $d\sigma/d\omega$ is compared for

$$\pi^+ p \to \pi^+ \pi^0 p \tag{3.3}$$

at 4, 5 and 8 GeV/c [5], and the exponent N in the parametrization

$$\sigma(p_{\text{lab}}) \propto p_{\text{lab}}^{-N} \tag{3.4}$$

is plotted as function of ω. One notes the striking difference in the energy dependence below ($N \approx 0$) and above 120° ($N \approx 1.5$).

Similar results have been obtained with

$$K^+ p \to K^+ \pi^- \Delta^{++}(1232) \tag{3.5}$$

at incident lab. momenta ranging from 5 to 16 GeV/c [6]. In particular, a comparison of this reaction with $K^+ p \to K^0 \pi^0 \Delta^{++}$ suggests the dominance of diffraction dissociation in the former reaction (Fig. 3.3). While one finds $N(\omega) \approx 0$ around $\omega \approx 100°$ for $K^+ \pi^- \Delta^{++}$, corresponding to a diffractively produced low mass ($\pi^- \Delta^{++}$) system, a

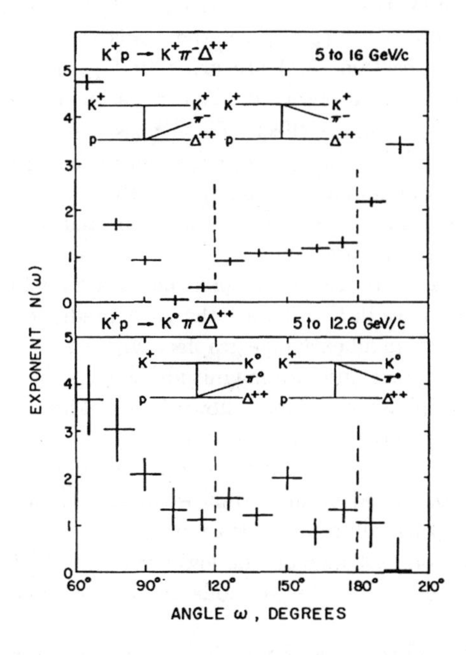

Figure 3.3. Exponent $N(\omega)$ in the fit with $\sigma \propto p_{\text{lab}}^{-N(\omega)}$, as a function of ω for the reactions $K^+ p \to K^+ \pi^- \Delta^{++}(1232)$ and $K^+ p \to K^0 \pi^0 \Delta^{++}(1232)$ between 5 and 16 GeV/c [6].

value of $N \approx 1$ is obtained in the same ω region for $K^0\pi^0\Delta^{++}$ where charge exchange is necessary.

We conclude: The pronounced peak observed in $d\sigma/d\omega$ distributions corresponding to a backward baryon-pion system of the same charge as the incident nucleon is produced predominantly via diffraction dissociation. The question remains, however, what happens to the $\Delta^+_{I=3/2}$ production, which falls into the same ω region as $(N\pi)^+_{I=1/2}$ but proceeds through isospin $T = 1$ exchange in the t channel and, therefore, cannot proceed via diffraction dissociation (see Sect. 3.4).

3.2 Correlation between transverse and longitudinal variables

The use of LPS plots also suggests, in a second approximation, to study transverse momenta in a differential way, e.g. by investigating whether their properties are correlated with the longitudinal momentum configuration of the final state [7]. As discussed above, for a three-particle final state configuration is essentially determined by the correlation angle ω.

For the case of unpolarised incoming particles $2n$-3=3 transverse variables are independent. The average values $\langle p_T \rangle$ of the transverse momentum of the three final state particles of reaction (3.1) are given in Fig. 3.4, after exclusion of ρ^-. Considerable structure can be observed. A correlation between $\langle p_T \rangle$ and $p^*_\parallel$ of a particle is known [8] as "sea-gull" effect. This consists in a $\langle p_T \rangle$ minimum at $p^*_\parallel$=0, which can only partially be explained as phase-space effect. Here, the same effect is expected to lead to a minimum at 120° for $\langle p_T \rangle$ of the π^0 and at 180° for $\langle p_T \rangle$ of the π^-. Transverse-momentum conservation is then expected to induce weaker minima for the other particles at the same angle ω.

As an example for a matrix element, again the CŁA model is used after exclusion of ρ^-. The heavy solid lines in Fig. 3.4 are the CŁA predictions obtained by considering the contributions from all exchange graphs. Without any overall normalization, these curves have the correct absolute magnitude and describe the general structure better than cylindrical phase space (not shown here). The simple model fails, however, just in the most densely populated region of $100° < \omega < 120°$, where the proton curve is too high and the π^0 curve much too low. A more detailed investigation of the $\langle p_T \rangle$ dependence predicted by individual exchange graphs (various thin lines in Fig. 3.4) tells us that this failure is, at least in the case of the π^0, due to just the already suspected baryon exchange on the lower vertex. The behavior is, therefore, better reproduced by the pion-exchange model of [4] (not shown), but also there the situation is not perfect.

The most obvious shortcoming of the CŁA model in this respect is that it does not contain resonance decays. To illustrate the effect of resonances [7] on the ω-dependence of $\langle p_T \rangle$, we give in Fig. 3.5 what is obtained for reaction (3.5), $K^+p \rightarrow$

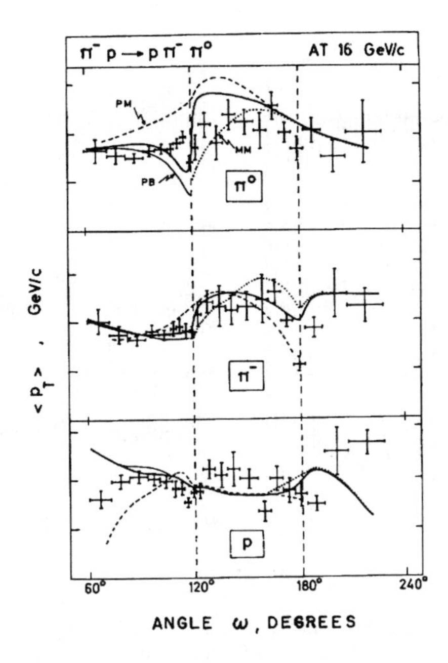

Figure 3.4. Average transverse momentum as a function of the angle ω for $\pi^-\mathrm{p} \to \pi^-\pi^0\mathrm{p}$ at 16 GeV/c after exclusion of ρ^-. The thick solid lines are the absolute CLA model predictions. The other lines give the predictions for the individual exchange graphs indicated in Fig. 3.1c [1].

$\mathrm{K}^+\pi^-\Delta^{++}(1232)$ at 5 GeV/c, when K*(892) and K*(1430) are eliminated (Fig. 3.5a) or left in the data (Fig. 3.5b): the minima at $\omega \approx 120°$ are enhanced and new minima appear at $\omega \approx 165^0$ in the latter case. These two minima correspond to two very significant peaks [6, 7] in the differential cross section $\mathrm{d}\sigma/\mathrm{d}\omega$ corresponding to a longitudinally extremely aligned K* resonance decay. In such a K*-decay the $\langle p_\mathrm{T}\rangle$ values of the decay products are smallest at the ω values under question, especially when the production angle is small, as is generally observed for $\Delta^{++}\mathrm{K}^{*0}$ production.

This spin-dependent argument [7] goes somewhat further than one [9] for the special case of large mass differences and small Q-values: From mere kinematics, one can expect small transverse momenta for small-mass particles produced in a decay of a system of large mass but small Q-value. Since the velocities are the same at zero Q-value, the momenta must be proportional to the masses. Since for given system mass this is a kinematic effect, it should be contained in the constant c of the CLA model. From Fig. 3.4 we can deduce that π^0's are indeed predicted to have the smallest $\langle p_\mathrm{T}\rangle$ values where this effect is expected to be present ($\omega \leq 120°$).

Responsible for the production of such a low Q-value mass system is not kinemat-

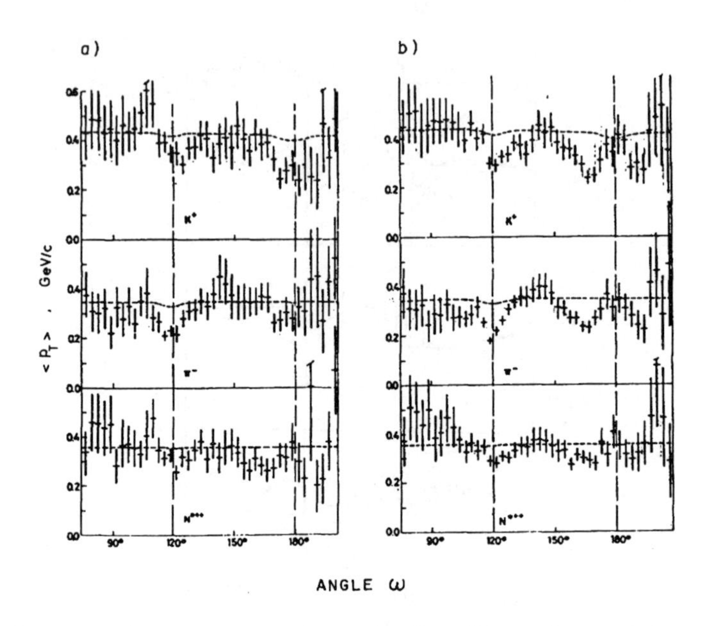

Figure 3.5. Average transverse momenta as a function of the angle ω for $K^+p \to K^+\pi^-\Delta^{++}$ at 5 GeV/c, before (Fig. 3.5 a) and after (Fig. 3.5 b) exclusion of $K_2^{*0}(1430)$ [6].

ics but the matrix element, the effect of which is visible already in the differential cross section $d\sigma/d\omega$. So, we conclude that the transverse momenta display the information contained in $d\sigma/d\omega$ in a complementary way. A model should certainly be able to reproduce both.

3.3 The prism plot

If we want to study a final state in full (longitudinal *and* transverse) phase space simultaneously, we have to go back to $3n-5$ dimensions. In a computer-displayed analysis, ($2n$-2) variables define a generalized equilateral rectangular hyperprism. As (n-2) of these, the authors of [10] use the (n-2) Van Hove angles ω_i. The other n give an n-dimensional energy simplex, a generalization of a Fabri-Dalitz plot. In addition, the remaining (n-3) of the $3n-5$ required variables are defined in a manner adapted to specific problems.

For the case of a three-particle final state, the prism is defined by the total energy of two of the final state particles (Fabri-Dalitz plot) in the xy-plane and the Van Hove angle ω in the z direction. This construction of the prism is shown in Fig. 3.6. As a fourth variable, the radial length R defined in the hexagon can be used in a separate plot, after normalization to its maximum possible value $R_{\max}$ at the phase space boundary.

Fig. 3.7a shows the prism plot populated with Monte-Carlo phase-space events, and Fig. 3.7b populated with 3.9 GeV/c $\pi^+ p \to \pi^+\pi^0 p$ event points. The striking features of the data are a peaking of $R/R_{\max}$ near 0.9 (not shown) and their accumulation into three distinct tubes in the upper half of the prism of Fig. 3.7b, corresponding to backward produced protons. This clustering into tubes is a consequence of the smallness of the transverse momenta and the smallness of the mass of any correlated sub-system in the final state. We can see that, at this low energy, the separation into individual mechanisms, each corresponding to a straight section of the tube, is better

Figure 3.6. The prism plot constructed from pulling a Fabri-Dalitz plot out in the direction of the Van Hove angle ω (cartoon by Suzy Smile).

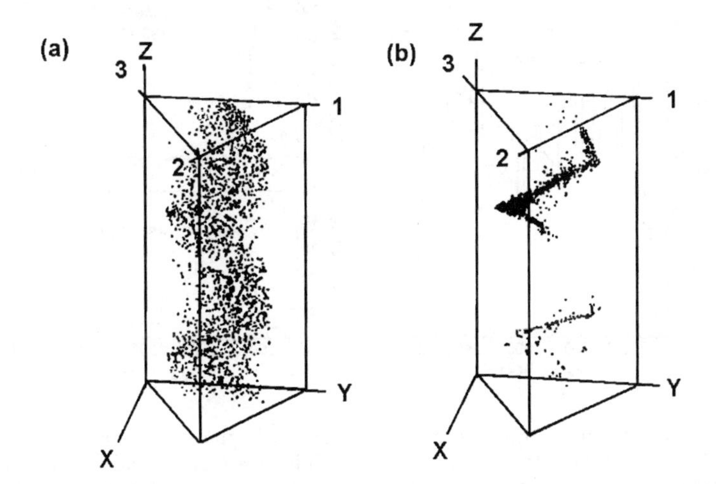

Figure 3.7. Three-particle prism plot for 3.9 GeV/c $\pi^+ p \to \pi^+\pi^0 p$ for a) invariant phase space events and b) experimental data [10].

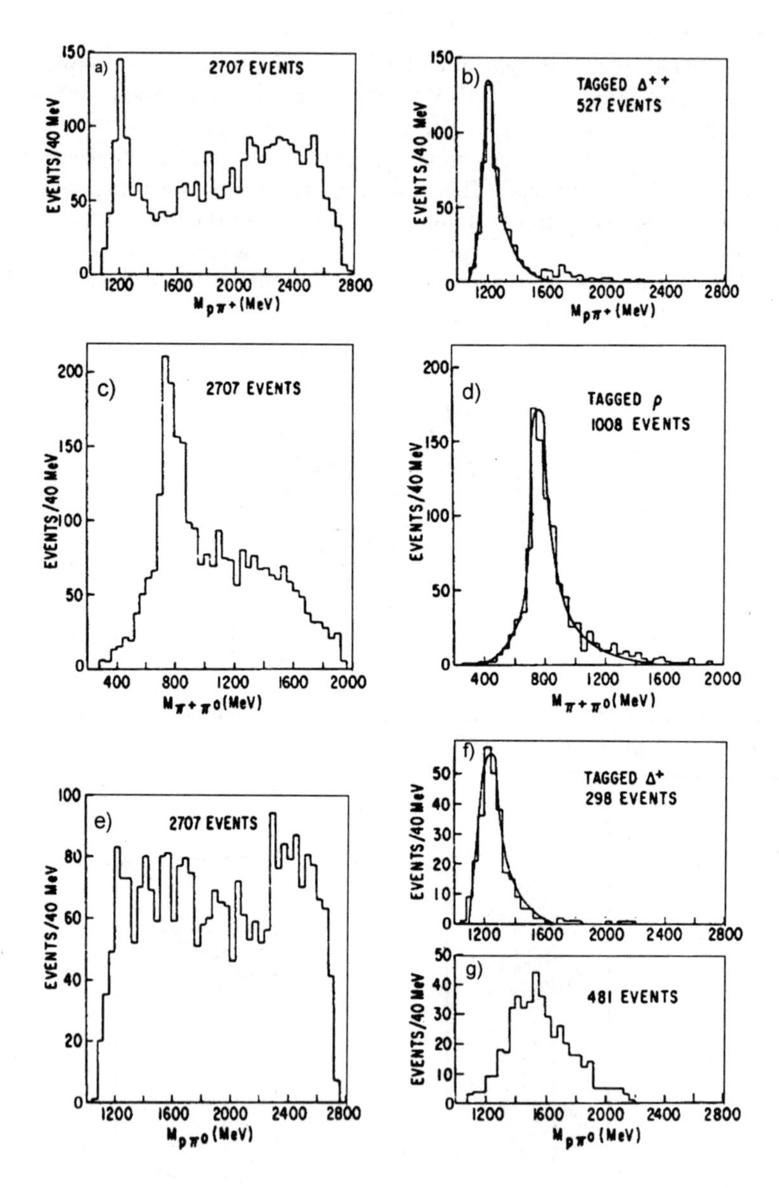

Figure 3.8. Effective mass of the system as indicated in the sub-figures, for the reaction $\pi^+p \to \pi^+\pi^0p$ at 3.9 GeV/c, for all events in sub-figures (a), (c) and (e), and for events in the corresponding sections of the tube in the prism plot in sub-figs. (b), (d), (f) and (g) [10].

in the prism than in a projection of it onto the angle ω or the Fabri-Dalitz plot. In particular, a separation into the quasi two-body final states $\Delta^{++}\pi^0$, $\mathrm{p}\rho^+$, and even a distinction between $\Delta^+\pi^+$ and a diffractively produced $(\mathrm{p}\pi^0)\pi^+$ is possible without any other background. The mass distributions before and after the separation are shown in Fig. 3.8 and speak for themselves.

3.4 Isospin analysis

However, one can even go a few steps further. Resonances overlap in full phase space, within the same sub-system when their Breit-Wigner distributions overlap and in the kinks of the prism plot when they are in different sub-systems. This overlap prevents us from further kinematical separation, but it has the advantage that it allows us to study interferences, if we go back to the amplitude level.

Indeed, a very helpful tool for further decomposition of particular phase-space regions is the separation of different isospin states [11]. It was first used to decompose the $\mathrm{N}\pi$ system in pp collisions [12].

We demonstrate the method on a combined analysis [13] of the $\mathrm{N}\pi$ system in $\pi^+\mathrm{p}$ and $\pi^-\mathrm{p}$ three-body final states at 8 and 16 GeV/c. The study of the energy dependence of the peak in the matrix element (Fig. 3.2) at $\omega \approx 100°$ tells us that this peak is mainly due to diffraction dissociation. However, what about the Δ^+? It also must be contained in the peak and has to be produced by I_E=1 exchange!

The reaction amplitude for production of an $(\mathrm{N}\pi)$ system at the nucleon vertex can be written as a linear combination of three basic exchange amplitudes, each of which corresponds to a definite isospin I_E for the exchanged object and I for the $\pi\mathrm{N}$ system (I_E=2 exchange is neglected and peripheral production is assumed). Exchange diagrams corresponding to the three possible amplitudes $M_I^{I_\mathrm{E}}$ are given in Fig. 3.9.

The integrated cross section for the various charge modes of $(\mathrm{N}\pi)$ production in $\pi^\pm\mathrm{p}$ reactions can be expressed by the phase space integrals [14]:

$$\begin{aligned}
\sigma(1) &\equiv \sigma(\pi^-\mathrm{p}\to\pi^-(\pi^0\mathrm{p})) = \int\left|\frac{1}{\sqrt{3}}M^0_{1/2}+\frac{1}{3\sqrt{2}}M^1_{1/2}+\frac{\sqrt{2}}{3}M^1_{3/2}\right|^2\mathrm{d}V\\
\sigma(2) &\equiv \sigma(\pi^+\mathrm{p}\to\pi^+(\pi^0\mathrm{p})) = \int\left|\frac{1}{\sqrt{3}}M^0_{1/2}-\frac{1}{3\sqrt{2}}M^1_{1/2}-\frac{\sqrt{2}}{3}M^1_{3/2}\right|^2\mathrm{d}V\\
\sigma(3) &\equiv \sigma(\pi^-\mathrm{p}\to\pi^0(\pi^-\mathrm{p})) = \int\left|\frac{\sqrt{2}}{3}M^1_{1/2}+\frac{1}{3\sqrt{2}}M^1_{3/2}\right|^2\mathrm{d}V\\
\sigma(4) &\equiv \sigma(\pi^+\mathrm{p}\to\pi^0(\pi^+\mathrm{p})) = \int\left|\frac{1}{\sqrt{2}}M^1_{3/2}\right|^2\mathrm{d}V
\end{aligned}\tag{3.6}$$

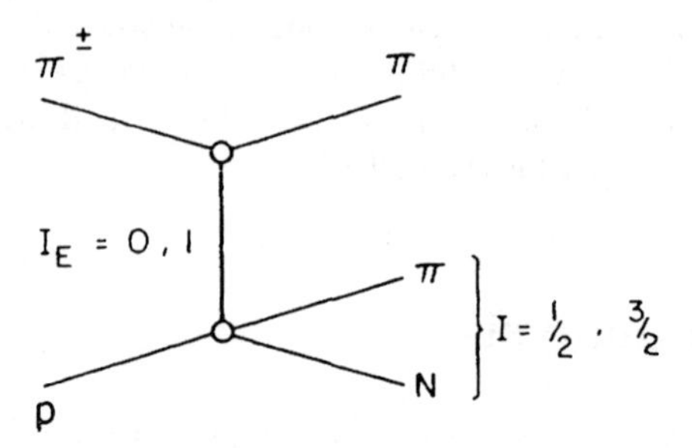

Figure 3.9. The three diagrams corresponding to the amplitudes specified by the exchanged isospin I_{E} and the isospin I of (πN). (Note that the combination $I_{\mathrm{E}} = 0$, $I = 3/2$ is excluded from isospin conservation.)

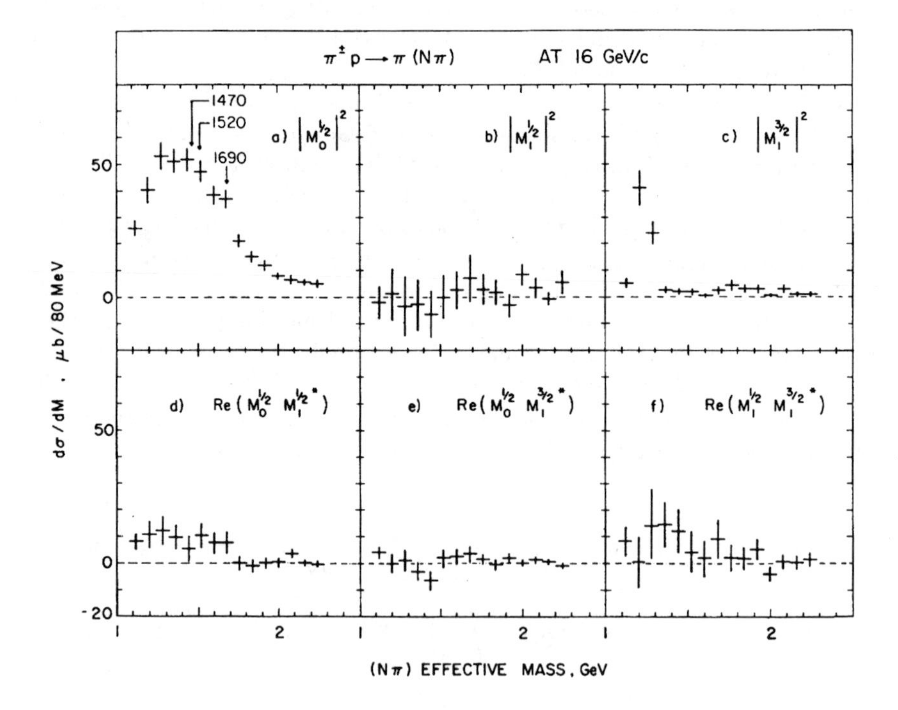

Figure 3.10. Squared amplitudes $|M_I^{I_{\mathrm{E}}}|^2$ and their interference terms as defined in the text, as functions of the (Nπ) mass for $\pi^{\pm}\mathrm{p} \to \pi\pi\mathrm{N}$ at 16 GeV/c [13].

$$\sigma(5) \equiv \sigma(\pi^-\mathrm{p} \to \pi^-(\pi^+\mathrm{n})) = \int \left| \frac{\sqrt{2}}{\sqrt{3}} M^0_{1/2} + \frac{1}{3} M^1_{1/2} - M^1_{3/2} \right|^2 \mathrm{d}V$$

$$\sigma(6) \equiv \sigma(\pi^+\mathrm{p} \to \pi^+(\pi^+\mathrm{n})) = \int \left| \frac{\sqrt{2}}{\sqrt{3}} M^0_{1/2} - \frac{1}{3} M^1_{1/2} + \frac{1}{3} M^1_{3/2} \right|^2 \mathrm{d}V$$

$$\sigma(7) \equiv \sigma(\pi^-\mathrm{p} \to \pi^0(\pi^0\mathrm{n})) = \int \left| \frac{1}{3} M^1_{1/2} - \frac{1}{3} M^1_{3/2} \right|^2 \mathrm{d}V \ .$$

Integration is understood over a proper phase space region $\mathrm{d}V$. The amplitudes $M_I^{I_\mathrm{E}}$ are normalized such that

$$\sum_{i=1}^{7} \sigma(i) = \sum_{I_\mathrm{E},I} \frac{2I+1}{2I_\mathrm{E}+1} \int |M_I^{I_\mathrm{E}}|^2 \,\mathrm{d}V \ . \tag{3.7}$$

Inversion of the system (3.6) fully determines the three integrated squares of the amplitudes and the three integrated interference terms by the first six cross sections, which are measurable. In particular, ρ exchange (I_E=1) leading to an I=1/2 (Nπ) state is expressed by $\int |M^1_{1/2}|^2 \,\mathrm{d}V$ and pomeron or f_2 exchange (I_E=0) by $\int |M^0_{1/2}|^2$ $\mathrm{d}V$.

The three squared amplitudes and the three interference terms are given [13] for $\pi^\pm\mathrm{p} \to \pi^\pm\pi^0\mathrm{p}$ at 16 GeV/c in Fig. 3.10 as a function of the (Nπ) mass. We note that $M^1_{1/2}$ corresponding to $(\mathrm{N}\pi)_{1/2}$ production by e.g. ρ-exchange is compatible with zero. Only its interference with $M^0_{1/2}$ is non-negligible at small masses, giving evidence for possibly stronger ρ-exchange at lower energies.

An analysis [13] of the energy dependence (not shown) indeed tells us that only $|\mathrm{M}_0^{1/2}|^2$ is compatible with diffraction dissociation, while $|\mathrm{M}_1^{3/2}|$ falls fast with increasing energy.

3.5 Partial wave analysis

Besides isospin, we of course also have the angular momentum (spin and orbit) at our disposal. Partial Wave Analysis is a very widely applied method making use of these quantum numbers. In particular, it has been very successfully applied to disentangle 3π and K$\pi\pi$ systems [15] in four-particle final states. Here, we remain with the simpler example of a two-particle sub-system in a three-particle final state, relevant to the next sub-section.

In Fig. 3.11a, a Dalitz plot is shown of the final state of

$$\mathrm{K}^-\mathrm{p} \to \overline{\mathrm{K}}^0\pi^-\mathrm{p} \tag{3.8}$$

at 4.2 GeV/c with about 30 000 events [16]. The effective-mass distribution of the three possible sub-systems is shown in Figs. 3.11b-d, respectively. In this sub-section,

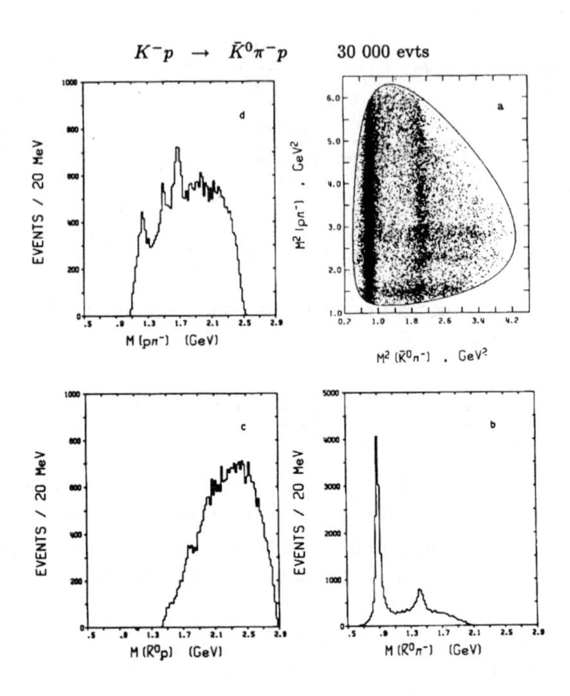

Figure 3.11. (a) Dalitz plot of $M^2(\mathrm{p}\pi^-)$ versus $M^2(\overline{\mathrm{K}}^0\pi^-)$ for the final state $\overline{\mathrm{K}}^0\pi^-\mathrm{p}$. (b) Effective-mass distribution $M(\overline{\mathrm{K}}^0\pi^-)$. (c) Effective-mass distribution $M(\overline{\mathrm{K}}^0\mathrm{p})$. (d) Effective-mass distribution $M(\mathrm{p}\pi^-)$ [16].

we reduce the influence of the $\mathrm{p}\pi^-$ and $\mathrm{p}\overline{\mathrm{K}}^0$ systems by kinematic cuts to less than 4% and concentrate on the $\overline{\mathrm{K}}^0\pi$ system.

For the purpose of the partial wave analysis, the 4 independent phase-space variables describing the three-particle final state are the effective mass $M(\overline{\mathrm{K}}^0\pi)$ of the $\overline{\mathrm{K}}^0\pi$ system, the reduced squared four-momentum transfer $t' = |t - t_{\min}|$ from K^- to the $\overline{\mathrm{K}}^0\pi$ system and the decay angles θ and ϕ of the $\overline{\mathrm{K}}^0\pi$ system.

An element of the $\mathrm{K}\pi$ production density matrix can be written in terms of production amplitudes for angular momentum L, L' and absolute value of the third component M, M' as:

$$\rho^{LL'}_{MM'\eta} = \sum_{\lambda} f^{L}_{M\eta\lambda} f^{L'*}_{M'\eta\lambda} \quad , \tag{3.9}$$

where λ stands for the helicity flip at the nucleon vertex and η corresponds (for $M \neq 0$ only asymptotically) to the exchanged naturality. If no polarization information is available, the $f^L_{M\eta\lambda}$ cannot be determined without assumptions. Assuming spin coherence $f^L_{M\eta 0} = C_\eta f^L_{M\eta 1}$ (with complex proportionality constant C_η) (3.9) reduces to

$$\rho^{LL'}_{MM'\eta} = g^{L}_{M\eta} g^{L'*}_{M'\eta} \quad , \tag{3.10}$$

with

$$g^{L}_{M\eta} = L^{L}_{M\eta} e^{i\varphi^{L}_{M\eta}} \quad (L^{L}_{M\eta} \text{ is real and } \varphi^{L}_{M_\eta} \text{ the phase}) \quad . \tag{3.11}$$

This parametrization automatically satisfies the positivity conditions.

From a maximum-likelihood fit to the decay-angular distribution in the t-channel helicity frame (at fixed Kπ mass and momentum transfer),

$$W(\cos\theta, \phi) = \sum_{\substack{LL' \\ MM'\eta}} L^{L}_{M\eta} L^{L'}_{M'\eta} e^{i(\varphi^{L}_{M\eta} - \varphi^{L'}_{M'\eta})} \mathrm{A}^{L}_{M\eta} \mathrm{A}^{L'*}_{M'\eta} \quad , \tag{3.12}$$

we obtain best estimates for the quantities $L^{L}_{M\eta}$ and $\cos(\varphi^{L}_{M\eta} - \varphi^{L'}_{M'\eta})$. The functions $\mathrm{A}^{L}_{M\eta}$ are linear combinations of spherical harmonics Y^{L}_{M}:

$$\mathrm{A}^{L}_{M\eta}(\cos\theta, \phi) = \lambda_M \{\mathrm{Y}^{L}_{M}(\cos\theta, \phi) - \eta(-1)^M \mathrm{Y}^{L}_{-M}(\cos\theta, \phi)\} \quad , \tag{3.13}$$

where $\lambda_M = \sqrt{1/2}$ if $M \neq 0$, $\lambda_M = 1/2$ if $M = 0$.

From an inspection of the moments $\langle \mathrm{Re}\ \mathrm{Y}^{l}_{m} \rangle$ with l=0,...2L and m=0,...2M (not shown), moments with $l > 4$ or $m > 2$ turn out consistent with zero.

In Fig. 3.12, the results of the partial wave analysis for t' <0.1 GeV^2 are shown as a function of the $\overline{\mathrm{K}}^0\pi$ mass. One can observe that

(i) an S-wave contribution is important over the entire range,
(ii) the P-wave can be identified as the K*(892) (it is dominated by P_+, with P_0 being important, but P_- being small),
(iii) the D-wave can be identified as $\mathrm{K}^*_2(1430)$ and is seen to proceed mainly via D_0 and D_+,
(iv) the interference between S and P_0 can be interpreted in terms of a Breit-Wigner variation,
(v) the phase between P_0 and P_- seems to disagree with phase coherence (which would imply the same phase, modulo 180°),
(vi) the variation of the phase between S and D_0 can be interpreted as the difference of two Breit-Wigner phases.

Similarly, the t' dependence can be determined for the various waves in various mass bins. In order to study K*(892) production, about 9 000 events are selected by the narrow mass cut $0.86 < M_{\mathrm{K}\pi} < 0.92$ GeV. This sample is subdivided into thirteen narrow t' intervals, for $t' < 0.3$ GeV^2. In each of these intervals, the contributions of S, P_0, P_-, P_+ waves and the SP_0 and $\mathrm{P}_0\mathrm{P}_-$ relative phases are determined from a maximum-likelihood fit using a distribution function similar to (3.12), but extended with K*(892) and Kπ S-wave Breit-Wigner amplitudes to describe the mass dependence in the selected mass region. The contributions of P_0, P_-, P_+ waves and the

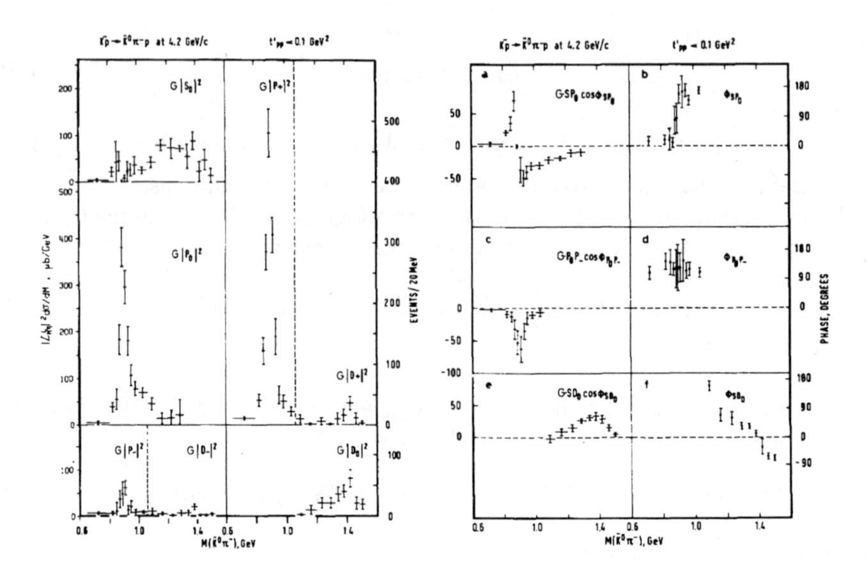

Figure 3.12. Partial-wave contributions, interference and relative phases between partial waves as a function of mass ($t' < 0.1$ GeV^2) [16].

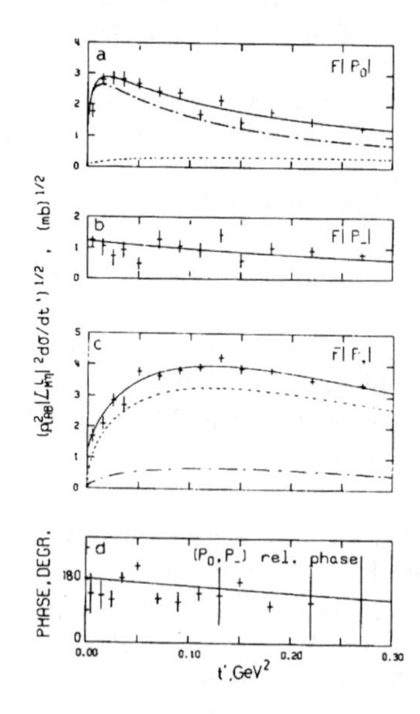

Figure 3.13. P-wave amplitudes and the relative phase of P_0 and P_- as a function of t'. The dashed curve corresponds to an I=0 η-$h_1(\omega$-$f_2)$ exchange contribution needed in addition to I=1 π-b_1 (ρ-a_2) exchange to describe the data [16].

relative P_0P_- phase resulting from this analysis are shown in Fig. 3.13. A significant difference is found for the t' dependence of P_0 and P_+ waves. This difference can be explained within the Reggeon-exchange model and a comparison of these results with results of an analysis [17] of charge-exchange K*(892) production, allows the separation of I_E=0 and I_E=1 exchange amplitudes.

There is one important residual uncertainty in this analysis: interferences are determined from asymmetries in the cosθ distribution, but at least part of these can be due to $(\overline{K}^0\pi^-)$ $(p\pi^-)$ overlap and possibly interference. Since the cross section is much larger for $K^0\pi^-$ than for $p\pi^-$, the influence of this overlap may be small in the $\overline{K}^0\pi$, but will be dominant in the $N\pi^-$ system.

3.6 Analytical multichannel analysis

Mutual overlap and possible interferences, of course, need not be a disadvantage. On the contrary, interference grants a unique possibility to study the relative phase between the overlapping amplitudes. In such a study, all mechanisms contributing to a few-body final state have to be treated simultaneously, however. A method allowing this is the so-called "Analytical Multichannel Analysis" [18].

The various mechanisms contributing to a given final state will be called "channels". Channel i is then described by the distribution function

$$f_i(v) = \sum_{\lambda} |A_{\lambda i}(v)|^2 \quad , \tag{3.14}$$

where the $A_{\lambda i}$ are the channel amplitudes, the summation runs over the nucleon helicities and v represents a complete set of phase-space variables. Interference effects between channel i and channel j are described by the function

$$f_{ij}(v) = 2\sum_{\lambda} \text{Re}\,[A^*_{\lambda i}(v)A_{\lambda i}(v)] \tag{3.15}$$

and the total final state can be written

$$f(v) = \sum_i c_i f_i(v) + \sum_{i<j} c_{ij} f_{ij}(v) \quad . \tag{3.16}$$

If the $f_i(v)$ are normalized, the c_i correspond to the fraction of events in the corresponding channels; c_{ij} contains $\sqrt{c_i c_j}$ and a coherence factor $0 \leq d_{ij} \leq 1$, i.e. $c_{ij} = d_{ij}\sqrt{c_i c_j}$.

Once a guess has been made for the $A_{\lambda i}$ (or the f_i and f_{ij}) and the parameters c_i and d_{ij}, one can calculate for every experimental event α "weights"

$$w_i^\alpha = \frac{c_i f_i^\alpha}{f^\alpha} \text{ and } w_{ij}^\alpha = \frac{c_{ij} f_{ij}^\alpha}{f^\alpha} \quad , \tag{3.17}$$

fulfilling

$$\sum_i w_i^\alpha + \sum_{i<j} w_{ij}^\alpha = 1 \quad . \tag{3.18}$$

The distribution in the kinematical variables obtained by weighting all events of the final state with weight w_i is said to correspond to sample i. The distribution obtained from weight w_{ij} corresponds to the "interference sample". This latter distribution can be negative, as the w_{ij} can be negative, over part or all of the distribution.

Usually, the first guesses will only be rough and a comparison of the weighted with the expected distribution (in all variables) will iteratively lead to an improved parametrization. In practice, it may turn out that not only the size and shape of the f_i has to be adjusted, but also that further functions f_i for channels not yet considered have to be added. Of particular help are experimental events in phase space regions where all f_i vanish. Events corresponding to channels not yet included can then give an entry to a "remainder sample" where all events are weighted with

$$w_R = 1 - \sum_i w_i - \sum_{i<j} w_{ij} \quad . \tag{3.19}$$

The method has first been successfully applied [19] to about 1500 events of the final state in $\mathrm{K^+p \to K^0\pi^+p}$ at 5 GeV/c, where a small cross-section channel $\Delta^{++}(1950)\mathrm{K}^0$ could be isolated. Here, we shall concentrate on the 30 000 event final state (3.8) analysed in [20]. To demonstrate the method, we shall, however, do that in some detail.

As a result of the partial wave analysis of the $(\overline{\mathrm{K}}^0\pi^-)$ system discussed in Sect. 3.5 one can introduce the channels

			to be called
$\mathrm{K^-p \to (\overline{K}^0\pi^-)_S p}$			i
$\mathrm{K^-p \to K^{*-}(892)p}$	K^* in $J^PM\eta$ state	1^-0-	iia
		1^-1-	iib
		1^-1+	iic
$\mathrm{K^-p \to K_2^{*-}(1430)p}$	K_2^* in $J^PM\eta$ state	2^+0-	iiia
		2^+1-	iiib
		2^+1+	iiic

From the Dalitz plot and the invariant-mass distributions of Fig. 3.11 follows the presence of

$$\begin{aligned} \mathrm{K^-p} \to\ & \overline{\mathrm{K}}^0\Delta^0(1232) && \text{iv} \\ & \overline{\mathrm{K}}^0\mathrm{N}^0(1520) && \text{v} \\ & \overline{\mathrm{K}}^0\mathrm{N}^0(1675) && \text{vi} \\ & \pi^-\Sigma^+(1765) \quad . && \text{vii} \end{aligned}$$

The channel distribution functions f_i will be written as a product of a mass-dependent, a t'-dependent and a decay-angle-dependent part. For the $(\overline{\mathrm{K}}^0\pi^-)$ system the functions

$$\begin{aligned} f^{J_1J_2}_{M_1M_2\eta} &= \rho^{J_1J_2}_{M_1M_2\eta}\mathrm{BW}^{J_1}(m)\mathrm{BW}^{J_2\,*}(m)\mathrm{F}^{J_1}_{M_1\eta}(t')\mathrm{F}^{J_2}_{M_2\eta}(t') \\ &\quad\times \mathrm{A}^{J_1}_{M_{1\eta}}(\cos\theta,\phi)\mathrm{A}^{J_2\,*}_{M_{2\eta}}(\cos\theta,\phi) \end{aligned} \tag{3.20}$$

are used, where the functions $|\mathrm{BW}|^2$, F^2 and $|\mathrm{A}|^2$ are normalized and ρ is the $(\mathrm{K}\pi)$ production density matrix, m refers to the $(\mathrm{K}\pi)$ invariant mass, $t' = |t - t_{\mathrm{min}}|$ with t being the four-momentum transfer squared from initial to final proton and θ and ϕ are the decay angles in the $(\mathrm{K}\pi)$ Gottfried-Jackson frame.

BW stands for the relativistic Breit-Wigner function:

$$\mathrm{BW}^J(M) = \sqrt{\frac{M\Gamma(m)}{q}}\;\frac{1}{(m^2 - m_0^2) - im_0\Gamma(m)}\;. \tag{3.21}$$

Here, q is the magnitude of the $\overline{\mathrm{K}}^0$ momentum in the $(\bar{\mathrm{K}}\pi)$ rest frame, m_0 is the resonance mass and Γ is the mass-dependent width approximated by

$$\Gamma(m) = \Gamma_0\left(\frac{q}{q_0}\right)^{2J+1}\frac{m_0}{m}\;, \tag{3.22}$$

where q_0 stands for q calculated at $m = m_0$.

The functions $\mathrm{A}^J_{M\eta}$ are linear combinations of spherical harmonics as used in Sect. 3.5. The parametrization of the functions F depends on the value of M and η of the state they refer to:

$$\begin{aligned} \mathrm{F}^J_{0-} &= \frac{1}{\sqrt{N^J_{0-}}}\sqrt{t}\left\{\frac{\mathrm{e}^{\mathrm{a}t'}}{(t+\mu^2)^2} + \gamma\mathrm{e}^{\mathrm{b}t'}\right\}^{1/2}\;, \\ \mathrm{F}^J_{1-} &= \frac{1}{\sqrt{N^J_{1-}}}\left\{\mathrm{e}^{\mathrm{a}t'} + \gamma\mathrm{e}^{\mathrm{b}t'}\right\}^{1/2}\;, \\ \mathrm{F}^J_{1+} &= \frac{1}{\sqrt{N^J_{1+}}}(t')^{\alpha}\left\{\mathrm{e}^{\mathrm{a}t'} + \gamma\mathrm{e}^{\mathrm{b}t'}\right\}^{1/2}\;. \end{aligned} \tag{3.23}$$

For each channel one has to find the best values for the parameters a, b, α and γ. The symbol μ stands for the pion mass, the $N^J_{M\eta}$ are normalization constants. With such expressions it is possible to describe the t' dependences, as presented in Fig. 3.13, also for higher t' values.

The distribution function for the total $(\overline{K}^0\pi^-)$ system reads

$$f_{K\pi} = \prod_{\substack{J_1J_2\\M_1M_2\eta}} f^{J_1J_2}_{M_1M_2\eta} \quad . \tag{3.24}$$

In the first iteration, only the Breit-Wigner mass dependence is used for the $(p\pi^-)$ and $(p\overline{K}^0)$ systems. In the case of the $(p\pi^-)$ system, this is replaced in the third iteration by

$$\begin{aligned} f_{p\pi} &= \sum_{\substack{J_1J_2\\M_1M_2\eta}} \rho^{J_1J_2}_{M_1M_2\eta} \mathrm{BW}^{J_1}(m)\mathrm{BW}^{J_2 *}(m)\mathrm{F}^{J_1}_{M_1}(t')\mathrm{F}^{J_2}_{M_2}(t') \\ & \times \mathrm{A}^{J_1}_{M_1\lambda\eta}(\theta,\phi)\mathrm{A}^{J_2 *}_{M_2\lambda\eta}(\theta,\phi) \end{aligned} \tag{3.25}$$

with $\theta \equiv \theta_{p\pi}$ and $\phi \equiv \phi_{p\pi}$ the polar and azimuthal angles of the proton in the $(p\pi)$ Gottfried-Jackson frame, m the $(p\pi^-)$ effective mass, and t' the reduced squared four-momentum transfer from K^- to $\overline{K}^0$. The amplitudes $\mathrm{A}^J_{M\lambda\eta}$ are linear combinations of Wigner D functions:

$$\mathrm{A}^J_{M\lambda\eta}(\theta,\phi) = \sqrt{\frac{2J+1}{8\pi}}\left\{\mathrm{D}^{J *}_{M\lambda}(\phi,\theta,0) + i\eta^P(-1)^{J-M}\mathrm{D}^J_{-M\lambda}(\phi,\theta,0)\right\} \quad . \tag{3.26}$$

J, M and P are the spin, the absolute value of its z-component and the parity of the $(p\pi^-)$ system, λ is the helicity of the final-state proton, η is the eigenvalue of the reflection operator with respect to the production plane. At variance with the meson case, it has no obvious physical meaning.

In the case of the $p\overline{K}^0$ system, the Breit-Wigners are complemented by the angular distribution in iteration 6.

As examples for the results after 9 iterations, Fig. 3.14 and 3.15 correspond to the $\overline{K}^0\pi$ S-wave and the P-wave (sub-) channels 1^-0-, 1^-1- and 1^-1+. Except for the S-wave which is not yet flat, the angular distribution corresponds to the particular wave and is as expected. Of particular interest are the differences in the t' distribution for the three waves, typical for pseudo-scalar and vector exchange.

Striking is the difference in reflection into the $p\pi^-$ and $p\overline{K}^0$ systems (lowest row of Fig. 3.15). The Monte Carlo curve superimposed on the $p\pi^-$ mass distribution shows that the two-peak structure reflection from the 1^-1- wave is real and not due to a residual resonance in the $p\pi^-$ system. (Just remember the conventional method of "hand-drawn" background and consider the striking error possibly introduced by that in the earlier literature!) Very similar results are obtained for the three D waves (not shown here).

As an example for a channel necessary to be added during the iterative procedure, the F waves 3^-0- and 3^-1+ are given in Fig. 3.16. Even though the integrated cross section is only 0.2% and 0.5% of the final state cross section, respectively, the angular

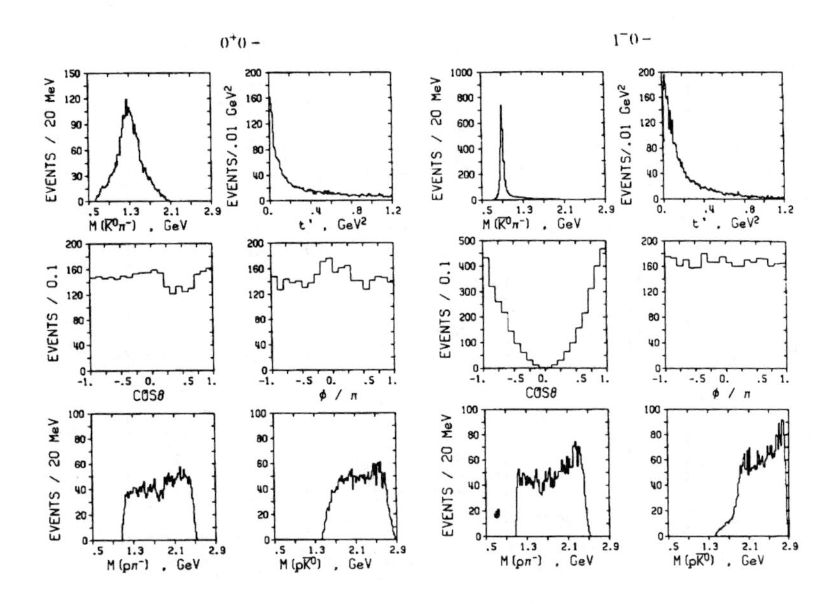

Figure 3.14. $M(\overline{\mathrm{K}}^0\pi^-)$, t' from initial to final proton, $\cos\theta_{\mathrm{K}\pi}$ and $\phi_{\mathrm{K}\pi}$, $M(\mathrm{p}\pi^-)$ and $M(\mathrm{p}\overline{\mathrm{K}}^0)$ for the 0^+0- and 1^-0- $\mathrm{K}\pi$ samples after iteration 9 [20].

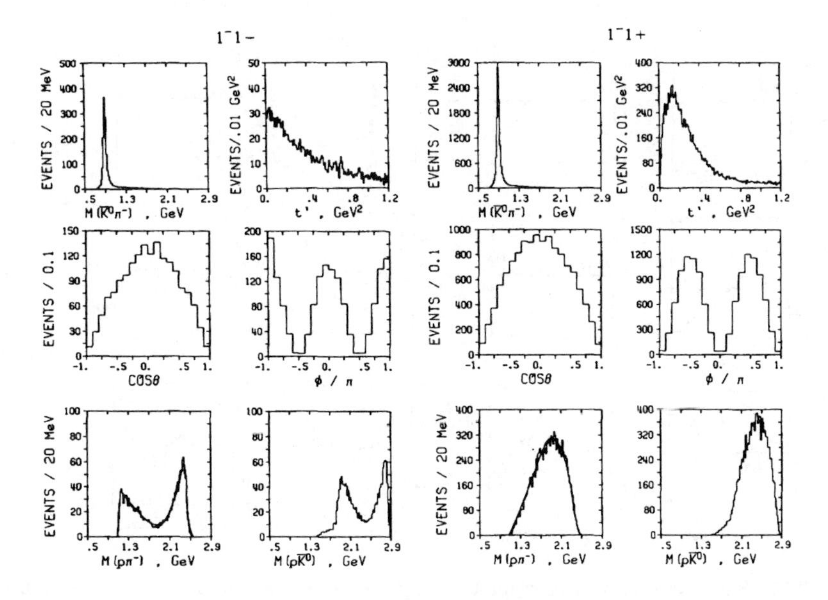

Figure 3.15. Same as Fig. 3.14, but for the 1^-1- and 1^-1+ $\mathrm{K}\pi$ samples [20].

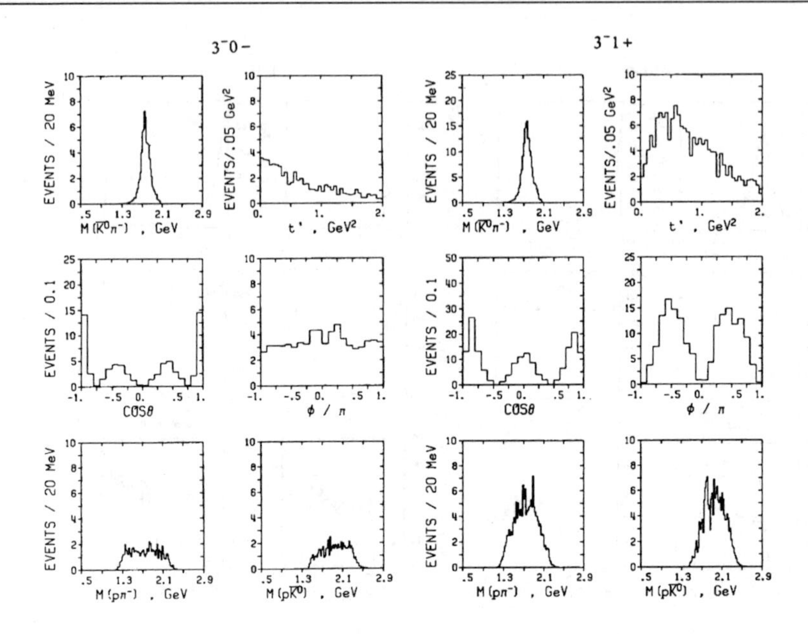

Figure 3.16. Same as Fig. 3.14, but for 3^-0- and 3^-1+ $K\pi$ samples [20].

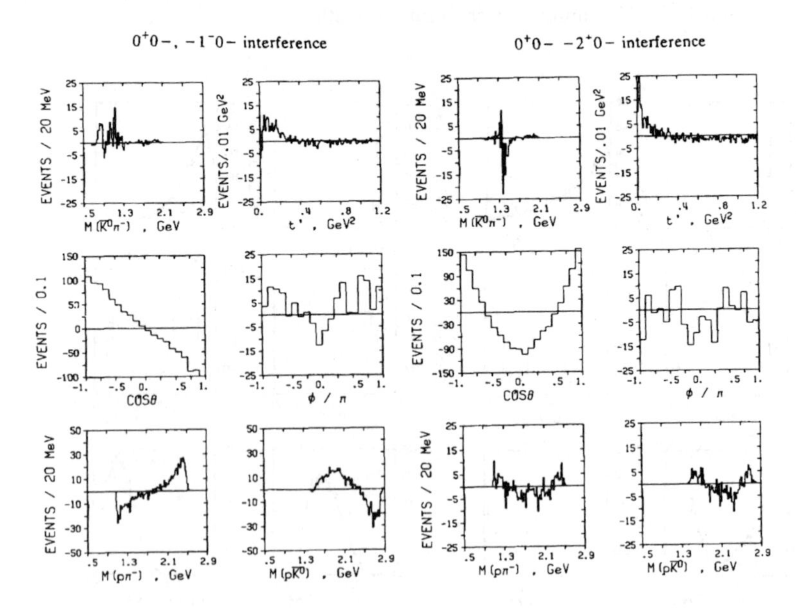

Figure 3.17. Same as Fig. 3.14, but for the interference between 0^+0- and 1^-0-, 0^+0- and 2^+0-, respectively [20].

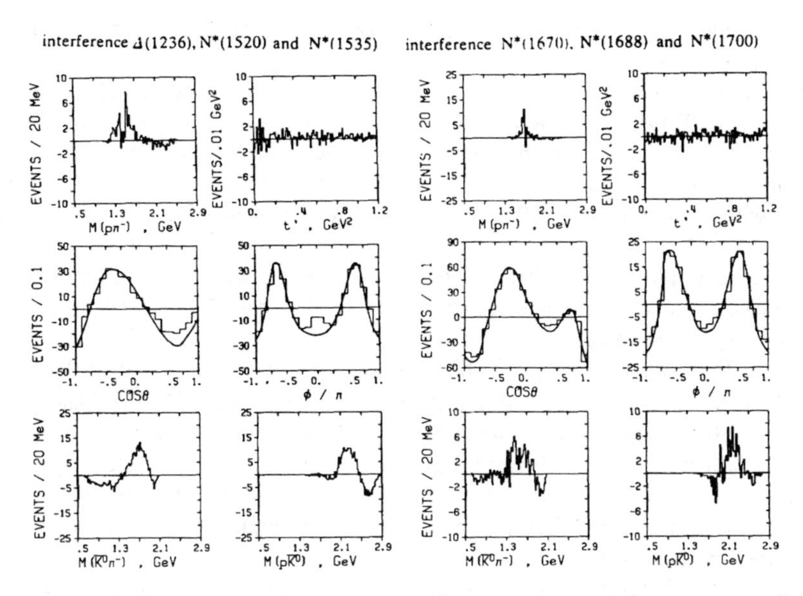

Figure 3.18. $M(\mathrm{p}\pi^-)$, t′ from initial to final kaon, cos$\theta_{\mathrm{p}\pi}$, $M(\overline{\mathrm{K}}^0\pi^-)$ and $M(\overline{\mathrm{K}}^0\mathrm{p})$ for the interference between the nucleon resonances indicated [20].

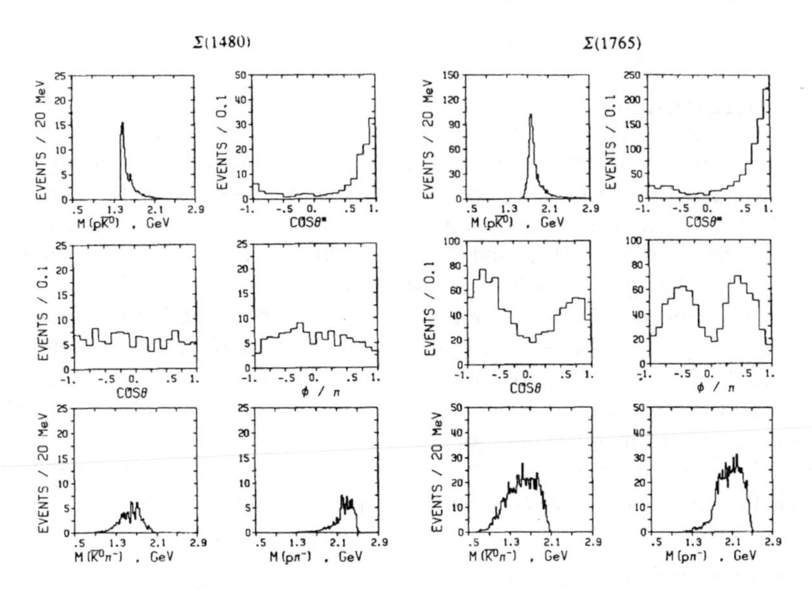

Figure 3.19. $M(\mathrm{p}\overline{\mathrm{K}}^0)$, cos$\Theta^*$ (Θ^* is the c.m. production angle), cosθ_{pK}, $M(\overline{\mathrm{K}}^0\pi^-)$ and $M(\mathrm{p}\pi^-)$ distributions for the $\Sigma(1480)$ and $\Sigma(1765)$ samples after iteration 9 [20].

distribution is clean and the difference in shape of the t' distribution similar to that in the P and D waves.

In Fig. 3.17 we give the distributions for the interference between the (Kπ) partial waves 0^+0- and 1^-0-, 0^+0- and 2^+0-, respectively. For both samples, the t' distributions (which should vanish in each bin) show a number of bins at $t' < 0.3$ GeV^2 with a content comparatively very small, but differing systematically from zero. A possible source of this difficulty may be the use of real functions for the t' dependence.

As further examples for interference effects, we give in Fig. 3.18 the interferences between nucleon resonances. The solid curves on the decay angles correspond to the partial wave expansion (3.25).

Finally, in Fig. 3.19 we give the $\Sigma(1765)$ already introduced in iteration 1 and a three standard deviation spin 1/2 signal near the $\overline{\mathrm{K}}^0$p threshold (0.4% of the final state). Both sub-channels show a forward and backward signal in cosΘ^*, where Θ^* is the cms production angle with respect to the incident K$^-$. The $\Sigma(1765)$ angular distribution is consistent with spin 5/2. A small problem is a loss of events for cos$\theta > 0$. These events are probably lost into the 2^+1+ sample, which indeed shows a rather large and negative error correlation coefficient.

A further refined iteration (e.g. to improve the S-wave of Fig. 3.14) reveals a second Kπ P wave with helicity 0 and strongly interfering with the Kπ S-wave. Its mass is around 1.05 GeV and its width 175 MeV. Its contribution to the final state is 2.3$\pm$0.3%.

3.7 Conclusions

With the help of model independent data analysis, we have moved from an analysis in Longitudinal Phase Space to a complete Multichannel Analysis. While the Longitudinal Phase Space Analysis has demonstrated the correlation of final state particles in the form of diffraction dissociation and resonance production, the increased number of variables of the Prism Plot could show overlap of these mechanisms in full phase space (not just projections of it). These resonances can be separated by means of quantum numbers like isospin, spin and angular momentum and their interferences can be studied when all channels contributing to a final state are treated simultaneously. This Analytical Multichannel Analysis is particularly useful at the permille level of a cross section or branching ratio.

The methods can, of course, be extended to the analysis of four-, five- and even six-particle final states, but a description of that goes beyond the scope of this book [21–27].

What have we learned for the future?

1. The methods are, of course, not limited to the case of strong interactions at low energy. Quite on the contrary, corresponding functions can be expected to

perform even much better in cases where the ambiguity due to the large number of channels is less severe and where amplitudes are better known from existing theories.

2. Experiments have to be *complete* in the sense that acceptance losses should be minimal and the four-vectors of all particles should be known.

3. The analysis has to be done *interactively* and *iteratively*, i.e. has to be supported by computer graphics.

Bibliography

[1] ABBCHLV Coll., J. Bartsch et al., *Nucl Phys.* **B19** (1970) 381.

[2] W. Kittel, L. Van Hove and W. Wojcik, *Comput. Phys. Commun.* **1** (1970) 425.

[3] Chan Hong-Mo, J. Łoskiewicz and W.W.M. Allison, *Nuovo Cim.* **57A**(1968) 93.

[4] S. Humble, *Nucl. Phys.* **B28** (1971) 416; I. Plahte and R.G. Roberts, *Lett. Nuovo Cim.* **1** (1969) 187; G. Veneziano, *Nuovo Cim.* **57A** (1968) 190.

[5] BDNPT Coll., G. Rinaudo et al., *Nucl. Phys.* **B25** (1971) 351.

[6] E. De Wolf, F. Verbeure and O. Czyzewski, *An LPS Analysis of the Reaction* $K^+p \to K^+\pi^-\Delta^{++}(1236)$, paper presented at the Amsterdam International Conference on Elementary Particles (1971).

[7] A. Białas, A. Eskreys, W. Kittel, S. Pokorski, J.K. Tuominiemi and L. Van Hove, *Nucl. Phys.* **B11** (1969) 479.

[8] M. Bardadin-Otwinowska, L. Michejda, S. Otwinowski and R. Sosnowski, *Phys. Lett.* **21** (1966) 351.

[9] E. Yen and E.L. Berger, *Phys. Rev. Lett.* **24** (1970) 618.

[10] J.E. Brau, F.T. Dao, M.F. Hodous, I.A. Pless and R.A. Singer, *Phys. Rev. Lett.* **27** (1971) 1481.

[11] L. Van Hove, R. Marshak and A. Pais, *Phys. Rev.* **89** (1952) 1211.

[12] H. Bøggild et al., *Phys. Lett.* **30B** (1968) 369.

[13] ABBC and ABBCCW Coll., K. Bösebeck et al., *Nucl. Phys.* **B28** (1971) 381 and **B40** (1972) 39; ABBC and NPT Coll., J.V. Beaupré et al., *Nucl. Phys.* **B66** (1973) 93.

[14] Z. Koba, R. Møllerud and L. Veje, *Nucl. Phys.* **B26** (1971) 134.

[15] G. Ascoli et al., *Phys. Rev. Lett.* **25** (1970) 962; ibid. **26** (1971) 929 and *Phys. Rev.* **D7** (1973) 669; J.D. Hansen, G.T. Jones, G. Otter and G. Rudolph, *Nucl. Phys.* **B81** (1974) 403.

[16] ACNO Coll., J.J. Engelen et al., *Nucl. Phys.* **B134** (1978) 14.

[17] P. Estabrooks and A.D. Martin, *Nucl. Phys.* **B102** (1976) 537.

[18] L. Van Hove, *Multidimensional Analysis and Parametric Fitting in Few-Body Hadron Collisions*, in Proc. 4th Int. Symp. on Multiparticle Hadrodynamics, eds. A. Giovannini and S. Ratti, Pavia 1973.

[19] A. Ferrando, S.O. Holmgren, M. Korkea-Aho, L. Montanet and L. Van Hove, *Nucl. Phys.* **B92** (1975) 61.

[20] ACNO Coll., J.J. Engelen et al., *Nucl. Phys.* **B167** (1980) 61.

[21] W. Kittel, S. Ratti and L. Van Hove, *Nucl. Phys.* **B30**(1971) 333.

[22] ABCLV Coll., J. Beaupré et al., *Nucl. Phys.* **B35** (1971) 61, ibid. **B46** (1972) 1, ibid. **B49** (1972) 441; *Phys. Lett.* **41B** (1972) 393; N. Deutschmann et al., *Nucl. Phys.* **B50** (1972) 61, ibid. 80.

[23] R. Arnold et al., *Phys. Lett.* **36B** (1971) 261.

[24] J. Ballam et al., *Phys. Rev.* **D4** (1971) 1946.

[25] E. De Wolf et al., *Nucl. Phys.* **B46** (1972) 333.

[26] N.K. Yamdagni and S. Ljung, *Phys. Lett.* **37B** (1971) 117; World Collaboration of the Reaction pp $\to$ pp$\pi^+\pi^-$, N.K. Yamdagniand M. Gavrilas, *Longitudinal Phase Space Analysis of the Reaction* pp $\to$ pp$\pi^+\pi^-$ *between 4 and 25* GeV/c, paper presented at the Amsterdam International Conference on Elementary Particles (1971).

[27] GHMS Coll., G. Tomasini et al., *Nuovo Cimento* **7A** (1972) 651.

Chapter 4

Single-Particle Inclusive Distributions

Except for the total cross sections shown in Chapter 1, the most simple way to study multiparticle final states according to the expansion (2.4) are the single-particle inclusive distributions. They have been extensively studied in all types of collisions of particles at all energies and we shall have a look at some of these distributions here, before summarizing the main phenomenological models developed to describe multihadron production.

4.1 e^+e^- collisions

We start with e^+e^- collisions as these are the most simple collisions and those best understood in terms of QCD inspired models or even QCD based analytical calculations. Single-particle spectra have been analyzed and compared in the energy range from just a few up to about 200 GeV [1–16] for a large number of types of particles and resonances. Early compilations are given in [17]. A typical example is given in Fig. 4.1 for charged particles produced in e^+e^- collisions at 189 GeV [11]. The variables used are the transverse momentum components in and perpendicular to the event plane, $p_\perp^{\mathrm{in}}$ and $p_\perp^{\mathrm{out}}$, the rapidity y along the event (thrust) axis and the momentum fraction $x_p = 2p/\sqrt{s}$ of the particle, respectively. The distribution in x_p is commonly addressed as fragmentation function.

Models developed to describe multihadron production in e^+e^- collisions will be discussed in Sect. 6.1 below, where also the references are given. We only note already here, that the agreement with the models containing gluon coherence (PYTHIA, HERWIG, ARIADNE) is good over the four orders of magnitude used, except for an underestimate of the high-$p_\perp^{\mathrm{out}}$ cross section also seen in other experiments and other energies. COJETS in particular fails to reproduce the $p_\perp^{\mathrm{in}}$ and y distributions.

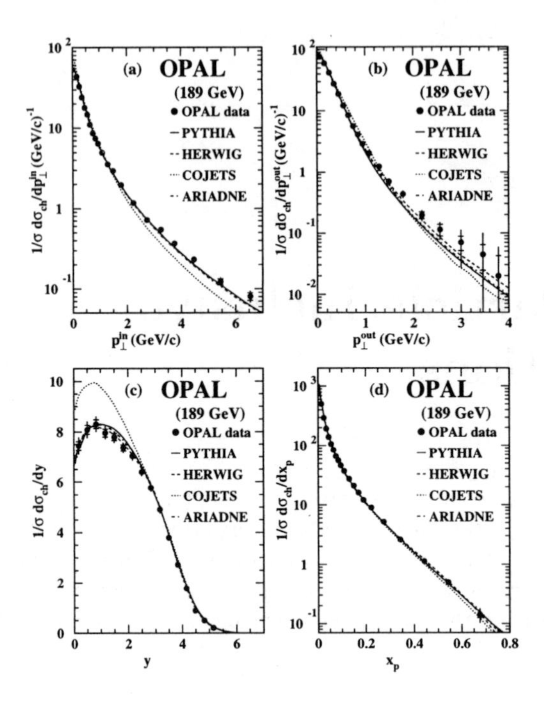

Figure 4.1. Single-particle inclusive distributions in $p_\perp^{\rm in}$, $p_\perp^{\rm out}$, y and x_p for e$^+$e$^-$ collisions at 189 GeV, compared to Monte-Carlo model predictions [11].

4.1.1 Longitudinal, transverse and asymmetry fragmentation functions

For unpolarized e$^\pm$ beams and averaging over h polarizations, the most general form of the differential cross section for h production, e$^+$e$^-$ $\to$ hX, via Z^0 or γ (spin 1) is [18]

$$\frac{\mathrm{d}^2\sigma^{\rm h}}{\mathrm{d}x_p \mathrm{d}\cos\theta} = \frac{3}{8}(1+\cos^2\theta)\frac{\mathrm{d}\sigma^{\rm h}_{\rm T}}{\mathrm{d}x_p} + \frac{3}{4}\sin^2\theta\frac{\mathrm{d}\sigma^{\rm h}_{\rm L}}{\mathrm{d}x_p} + \frac{3}{4}\cos\theta\frac{\mathrm{d}\sigma^{\rm h}_{\rm A}}{\mathrm{d}x_p}\ , \tag{4.1}$$

where θ is the angle between the hadron and the e$^-$ beam. These three contributions are referred to a the transverse, longitudinal and asymmetry fragmentation functions. They can be extracted by analysis of the distribution of the production angle θ for any value of x_p.

Fig. 4.2a gives $F_i(x_E) = (1/\sigma_{\rm tot})\mathrm{d}\sigma_i^{\rm ch}/\mathrm{d}x_E$ $(i = \mathrm{T,L})$ for charged particles produced in Z^0 decay in OPAL [19] as a function of the energy fraction $x_E = 2E/\sqrt{s} \approx x_p$. The longitudinal fragmentation function $F_{\rm L}(x_E)$ falls much more rapidly with increasing x_E than the transverse fragmentation function $F_{\rm T}(x_E)$. The parton shower

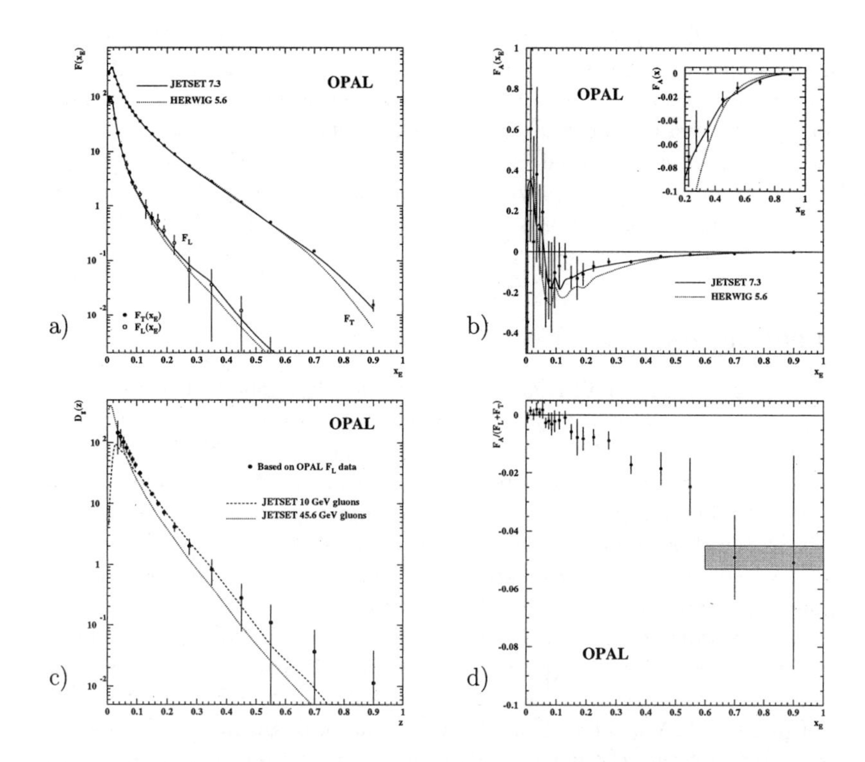

Figure 4.2. a) Transverse and longitudinal fragmentation functions $F_{\mathrm{T}}(x_E)$ and $F_{\mathrm{L}}(x_E)$ for charged particles, as a function of x_E, compared to JETSET and HERWIG results, b) same for the asymmetry fragmentation function $F_{\mathrm{A}}(x_E)$. c) Gluon fragmentation function $D_{\mathrm{g}}(z)$ extracted from F_{L} and F_{T}, compared to predictions of the JETSET model for the fragmentation of gluon jets at 10 and 45.6 GeV. d) The asymmetry fragmentation function as a fraction of the total fragmentation function. The shaded band shows a simple theoretical prediction based on valence dominance [19].

models JETSET and HERWIG are in reasonable agreement with the data over most of the x_E range. The asymmetry fragmentation function $F_{\mathrm{A}} = (1/\sigma_{\mathrm{tot}})\mathrm{d}\sigma_{\mathrm{A}}^{\mathrm{ch}}/\mathrm{d}x_E$ given in Fig. 4.2b differs from zero significantly only for $x_E > 0.15$. Again, the two models and in particular JETSET agree with the data.

It is important to note that F_{L} and F_{T} allow to extract [18] the gluon fragmentation function $D_{\mathrm{g}}(z)$, where z is the fraction of the energy of a gluon carried by a hadron. The OPAL results on this function are shown in Fig. 4.2c and compared to gluon jets produced in JETSET at 45.6 GeV (i.e. half the Z^0 mass) and 10 GeV, respectively.

Furthermore, at large values of x_E the produced hadron h is likely to contain the primary quark from the Z^0 decay [20], and thus preferentially to have the same charge. Allowing for forward-backward asymmetries resulting from parity violation in weak interactions, a negative F_{A} is expected at large x_E for d-, s-, and b-quark events, and a positive one for u- and c-quark events. Since the former are produced more copiously, an overall negative F_{A} is observed at large x_E in Fig. 4.2b.

To show the relative size of the effect, the ratio $R = F_{\mathrm{A}}(x_E)/(F_{\mathrm{T}}(x_E)+F_{\mathrm{L}}(x_E))$, is shown in Fig. 4.2d. Indeed, R grows with increasing x_E, indicating that information about the primary quark charge tends to be most strongly retained by the large-x_E particles.

4.1.2 Leading-particle effect

While Fig. 4.1 corresponds to all charged particles, the fraction of identified pions, kaons and protons is given as a function of momentum in Fig. 4.3 [15]. Pions are seen to dominate at low momentum, but kaons become equally important at large momentum. Protons rise from a few percent at low momentum to about 10% at large momentum. The fractions and their momentum dependence are reproduced qualitatively by the three models shown (see Sect. 6.1), except at large momenta.

This analysis [15] was repeated separately on high-purity light-, c- and b-tagged event samples, corresponding to $Z \to \mathrm{u\bar{u}, d\bar{d}, s\bar{s}}$, Z$\to$ c$\bar{\mathrm{c}}$ and Z$\to$ b$\bar{\mathrm{b}}$, respectively. The charged-hadron fractions in light-flavor events show a behavior similar to that in the flavor-inclusive sample of Fig. 4.3 (including the same differences between data and models). However, as shown in Fig. 4.4, the b:uds and c:uds production ratios show an excess of pions and kaons at $x_p \lesssim 0.1$ for heavy-quark fragmentation followed by a depletion of all three types of particles at larger x_p. This observation is consistent

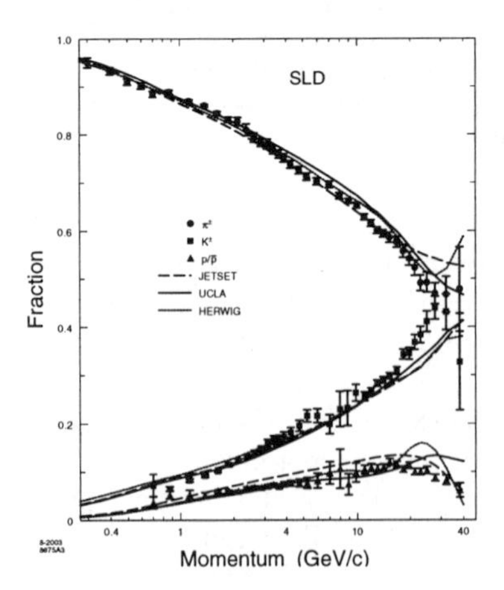

Figure 4.3. Charged-hadron fractions in hadronic Z decay [15].

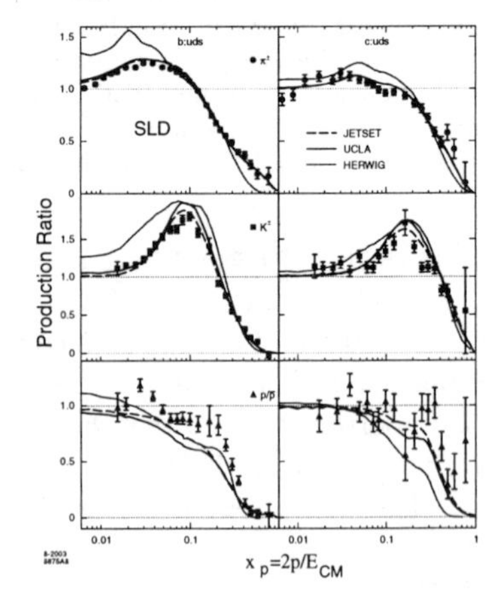

Figure 4.4. Ratios of production rates in b-quark events to those in light-quark events [15].

with the interpretation that a large fraction of the event energy is carried away by a B-hadron, which subsequently decays into a large number of lighter particles (see also last ref. [11]).

A particularly interesting question is, therefore, what happens to the primary quark or antiquark in its fragmentation. Does it end up as a valence constituent of a high-momentum "leading" hadron?

Unfortunately, unlike the heavy bottom and charm quarks, production of additional secondary up, down and strange quarks is dominant in jet development, so that the identification of the hadron containing the primary quark is not trivial. It has been confirmed already by TASSO [22] that identified high-energy pions, kaons, protons or Λ's indeed carry information about the primary quark. Stimulated by that, OPAL [21] applied a highly-dimensional (in quark$\rightarrow$particle space) unfolding method [23] to obtain probabilities $\eta_q^i(x_p)$ for a quark flavor q to develop into a jet in which the particle with the largest x_p above a given minimum is of type i. Fig. 4.5 gives these probabilities under the assumption of hadronization symmetries

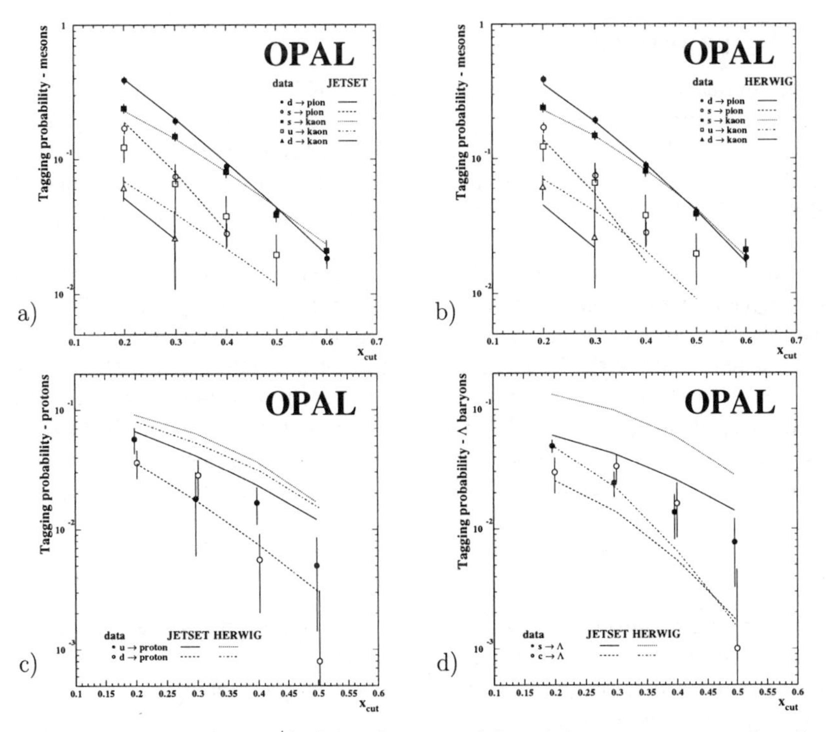

Figure 4.5. Probabilities $\eta_q^i(x_p)$ as a function of the minimum x_p, compared to JETSET and HERWIG predictions [21].

$\eta_{\rm d}^{\pi^\pm} = \eta_{\rm u}^{\pi^\pm}$ and $\eta_{\rm s}^{\rm K^\pm} = \eta_{\rm s}^{\rm K^0}$. In all cases the expected leading-particle pattern is observed:

In particular u- and d-quarks fragment mostly into pions, s-quarks into kaons, leading protons are more frequent in u- than in d-quark jets. The JETSET model reproduces the measurements reasonably well, while HERWIG has deficiencies in baryon production.

SLD [15] can tag quark and antiquark jets using the large forward-backward production asymmetry in collisions of highly polarized electrons with positrons. Fig. 4.6 gives, as a function of x_p, the difference $D_{\rm h}$ between hadron and antihadron differential cross sections in quark jets normalized by their sum. The data are consistent with equal production of hadrons and antihadrons ($D_{\rm h} = 0$) at low x_p, but show a definitive increase towards large x_p and thus a leading-particle effect.

Since baryons contain valence quarks but no valence antiquarks, this is more easily seen in baryon production than in meson production, where dilution effects due to the presence of valence antiquarks and $\rm s\bar{s}$ suppression effects have to be taken into account.

4.1.3 Charge ordering

So, hadronic events produced in $\rm e^+e^-$ annihilation yield two primary quarks ($\rm q\bar{q}$) with opposite quantum numbers. In the string picture, to be discussed in Chapter 6, hadronization results from the break-up of a color flux tube stretching between the q and the $\rm\bar{q}$ as in Fig. 4.7a. Two adjacent hadrons share a $\rm q\bar{q}$ pair created in one of the breaks-up. This is expected to lead to an alternating charge structure along the string.

In the search for such a structure, DELPHI [24] has ordered the charged particles in each event according to their rapidities in the thrust direction as in Fig. 4.7b, with as "tagged" particle (or 3- or 5-particle system) with rank $n_{\rm r} = 1$ that with the largest $|y|$ in its hemisphere and with the rank $n_{\rm r}$ of the other particles given with respect to that. To improve the efficiency for tagging the correct primary-quark charge, the rapidity gap $\Delta y_{\rm tag}$ defined in Fig. 4.7b is required to be larger than a specified value. To increase the sensitivity for original rank order, also the gaps between the other particles are required to exceed a given value (0.5). Fig. 4.7c shows

$$R(n_{\rm r}) = 1 - \frac{N_{\rm s}(n_{\rm r})}{N_{\rm o}(n_{\rm r})} \tag{4.2}$$

as a function of rank n_r in the hemisphere opposite to the tagged particle system for $\Delta y_{\rm tag} > 1$. In Eq. (4.2), $N_{\rm s}(n_{\rm r})$ is the number of particles at rank $n_{\rm r}$ with charge equal, $N_{\rm o}(n_{\rm r})$ that with charge opposite to that of the tagged system.

Two observations can be made:

i) $R(1)$ is large, corresponding to the leading particle in the opposite hemisphere carrying a charge opposite to that of the tagged system. This is consistent with a

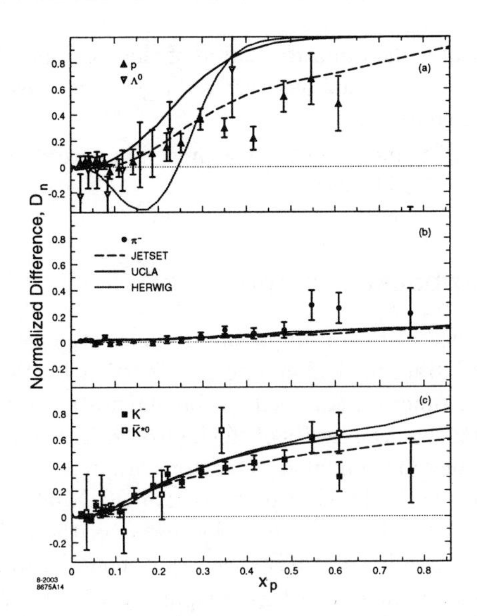

Figure 4.6. Normalized production differences between hadrons and their respective anti-hadrons in light-quark jets [15].

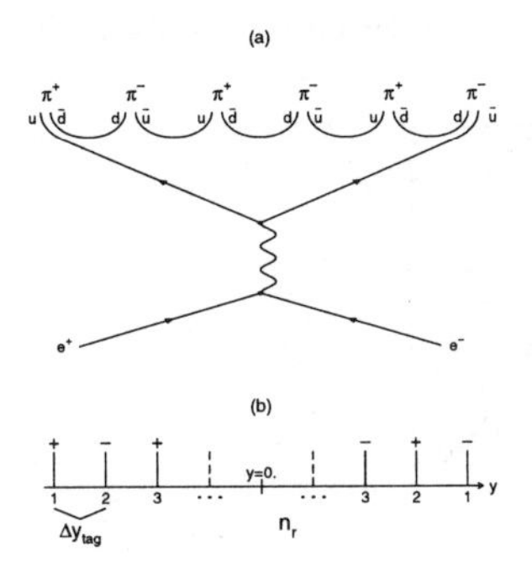

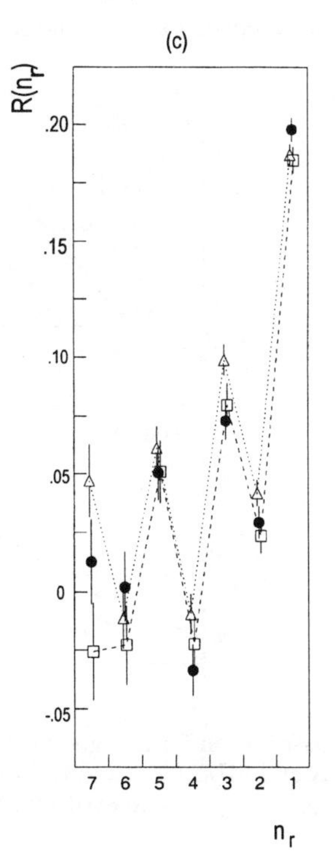

Figure 4.7. a) e^+e^- annihilation event with flavor-ordered string fragmentation, b) charged particles ordered according to their rapidity in the thrust direction, c) $R(n_r)$ as a function of n_r for $\Delta y_{tag} > 1$ (data •, JETSET □, HERWIG △) [24].

leading-particle effect and the opposite charge of the corresponding quark.
ii) The alternating charge structure observed at larger n_r is indeed evidence for charge ordering in quark fragmentation.

Furthermore, the string model JETSET is indeed in agreement with the data, but, somewhat surprisingly, also HERWIG.

4.1.4 The humpbacked shape

Of particular importance in the study of the fragmentation of quarks and gluons is the distribution of the fractional momentum x_p carried by the particular particle, i.e. the fragmentation function (Fig. 4.1d). Fragmentation functions represent the long-distance non-perturbative physics of the hadronization process, in which the observed hadrons are formed from final-state partons originating from a hard scattering process. They cannot be calculated in perturbative QCD, but their scaling violation can be evolved with the hard-process scale by means of the DGLAP evolution [25], from a starting distribution derived from the experimental data.

To avoid the large differences in order of magnitude of the differential cross section and to suppress the phase space factor $1/x_p$, it has become customary to replace x_p by $\xi = -\ln x_p$ as a variable. This leads to the humpbacked distribution shown in Fig. 4.8a, where differences between model predictions are now better resolved. The

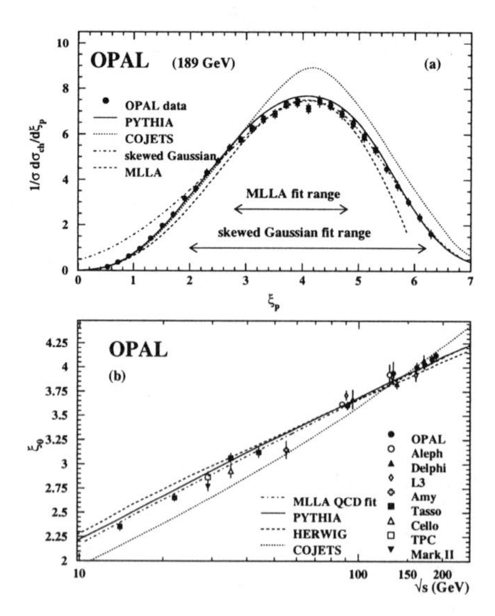

Figure 4.8. a) ξ_p distribution for charged particles produced in e^+e^- collisions at 189 GeV, compared to NLLA and MLLA fits and to Monte-Carlo model predictions. b) Evolution of the peak position ξ_0 with $\sqrt{s}$, compared to a QCD fit and model predictions [11].

shape of the distribution is approximately Gaussian with its maximum position ξ_0 at an intermediate ξ value. This implies that it is not the softest particles, but those with intermediate energies which multiply most often in the fragmentation. In particular, COJETS fails to reproduce large ξ (i.e. small x_p), a region where soft gluon production is reduced due to destructive interference expected from QCD in [26, 27]. Using the most simple concept of local parton hadron duality (LPHD) [28] (see Chapter 6), one expects this low-x_p damping to lead to the Gaussian shape of the ξ distribution. The next to leading-log approximation (NLLA) of QCD, however, leads to a skewed Gaussian [26, 29]. Such a form is fitted to the data in Fig. 4.8a over the range indicated but fails at low ξ (high x_p). The modified leading-log approximation (MLLA) [30] constituting a complete resummation of single and double logarithmic terms and predicting, up to a hadronization correction K^{h}, also the normalization, underestimates the cross section in the tail regions.

The energy evolution of the fragmentation function for charged particles in the range $14 \leq \sqrt{s} \leq 202$ GeV is given in Fig. 4.9 [11]. While there is approximate scaling at the low-ξ_p (high-x_p) side, there is a fast increase of the spectrum at the high-ξ_p side, with the peak position ξ_0 shifting to higher ξ_p values as the energy increases. This trend is followed by the NLLA [29] (Fig. 4.9a) and MLLA [30] (Fig. 4.9b) fits, be it with the limitation already observed in Fig. 4.8a above.

Furthermore, comparing heavy and light quarks at the Z, the peak position ξ_0 is slightly lower for b-quark fragmentation than for light-quark fragmentation (not shown).

4.1.5 The energy evolution of the peak position

The LLA, NLLA and MLLA approximations of QCD predict an almost logarithmic energy evolution of the peak position ξ_0. As shown in Fig. 4.8b [11], MLLA [31] indeed fits the data between 10 and 200 GeV. The PYTHIA and HERWIG predictions are very similar to the MLLA fit, but COJETS (representing a non-coherent parton shower) is too steep.

However, firstly, a lot depends on the careful tuning of COJETS (see [32] for a better agreement). Secondly, as pointed out in [33], both the approximately Gaussian shape of the ξ spectra and the small slope of the logarithmic energy dependence of ξ_0 can be considered a trivial consequence of the transformation of longitudinal (better cylindrical) phase space in rapidity y and transverse mass $m_{\mathrm{T}} = (m^2 + p_{\mathrm{T}}^2)^{1/2}$ into the variables ξ and Θ, where Θ is the production angle with respect to the thrust axis,

$$\frac{\mathrm{d}^2\sigma}{\mathrm{d}\xi \mathrm{d}\cos\Theta} = \frac{2p^3}{E}\frac{\mathrm{d}^2\sigma}{\mathrm{d}y\mathrm{d}m_{\mathrm{T}}^2} \; . \tag{4.3}$$

The fall-off at the high-ξ (low-momentum) side of the humpbacked distribution is an obvious property of the Jacobian $2p^3/E$, which has to be taken into account before exact conclusions about gluon coherence can be deduced from these distributions.

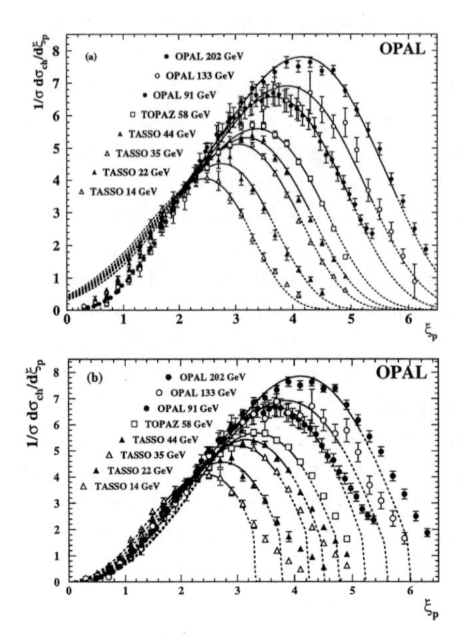

Figure 4.9. Energy evolution of the inclusive ξ_p spectrum [11] in comparison with QCD fits according to a) NLLA [29] and b) MLLA [30].

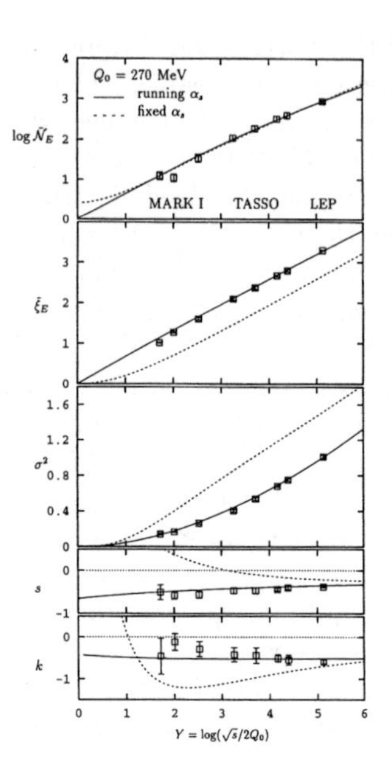

Figure 4.10. The average multiplicity $\bar{N}_E$ and the first four cumulants of the charged-particle energy spectra $E\mathrm{d}n/\mathrm{d}p$ as a function of ξ_E, i.e. the average value $\bar{\xi}_E$, the dispersion σ^2, the skewness s and the kurtosis k, are shown as a function of $Y = \log(\sqrt{s}/2Q_0)$ for $Q_0 = 270$ MeV. Predictions for the limiting spectrum (i.e. $Q_0 = \Lambda$) of MLLA with running α_s, and with fixed α_s, are also shown (for $n_f = 3$) [34].

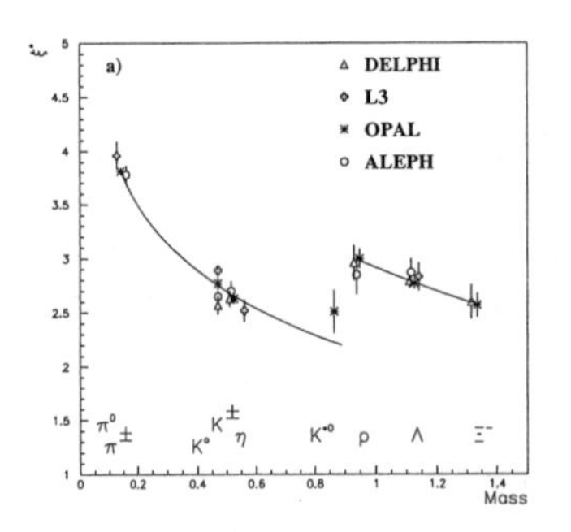

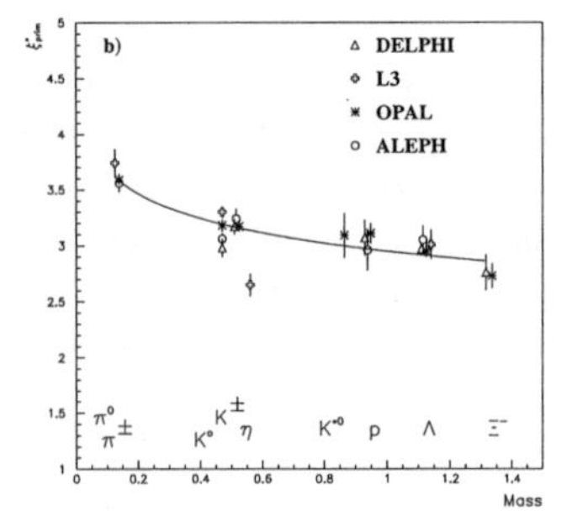

Figure 4.11. a) Dependence of a) the peak position ξ_0 and b) the corrected peak position ξ_0^{prim} on the hadron mass m_h. The full lines correspond to fits by $\xi_0^{(\mathrm{prim})} = \ln(m_\mathrm{o}/m_\mathrm{h})$ [14].

4.1.6 The higher moments

Of course, there is more information in the energy evolution of the fragmentation function than just the shift of the peak position. A compilation of the average charged-particle multiplicity (i.e. the integral) and the first four reduced cumulants of the $\xi_E = -\ln(2E/\sqrt{s})$ distribution are shown as a function of $Y = \log(\sqrt{s}/2Q_0)$ for effective hadron mass $Q_0 = 270$ MeV [34] in Fig. 4.10.

MLLA + LPHD predictions [30] are shown with $\Lambda = Q_0$ for the coupling constant α_s both fixed (dashed line) and running (full line). While there is no hope to be able to describe all moments simultaneously with fixed α_s, running α_s can fit the energy dependence with one single parameter $\Lambda = Q_0$.

4.1.7 The mass dependence of the fragmentation function

Besides for charged particles, the fragmentation function has been studied for a large number of identified hadrons, including resonances, in particular at the high statistics of LEP. The humpbacked shape is observed for all hadrons studied, but the peak position differs from hadron to hadron [11–14]. Indeed, smaller peak values are expected for more massive particles in the framework of NLLA + LPHD [35]. The peak position moves linearly with $-\ln(Q_0/\Lambda)$. In the most simple approach, $Q_0 \approx m_h$ leads to a logarithmic dependence.

Fig. 4.11a [14] shows the peak position ξ_0 for various hadrons as a function of mass m_h at LEP1 [11–14]. The data seem to indicate a different mass dependence of ξ_0 for mesons and baryons, both following a simple $\xi_0 = \ln(m_o/m_h)$, but with a different normalization parameter m_o. However, at LEP the majority of the observed hadrons originate from the decay of heavier hadrons. After correction for the decay effect back to the peak position of the corresponding hadron produced as primary, with the help of JETSET, the corrected $\xi_0^{\text{prim}} = \xi_0 + \Delta\xi_0^{\text{decays}}$ is plotted as a function of the hadron mass in Fig. 4.11b. Except for the η meson, the corrected ξ_0^{prim} now follows one common logarithmic decrease with increasing hadron mass, in good agreement with the MLLA + LPHD expectation. The correction for decay effects is, of course, one step beyond the simple LPHD assumption.

4.1.8 Quark- and gluon-jet differences

In QCD, the color charge of a gluon ($C_A = 3$) is larger than that of a quark ($C_F = 4/3$). In a perturbative cascade, bremsstrahlung will therefore be more intensive in the case of a primary gluon than in that of a primary quark. As a consequence, more but softer particles are expected [36] to be produced in a gluon jet and the gluon jet is expected to be wider than a quark jet of the same energy [37]. Quantitatively, only next-to-MLLA effects turn out to lead to a difference in the shape of the particle spectra in q and g jets [29, 35], which can be expressed as a shift of the limiting distribution.

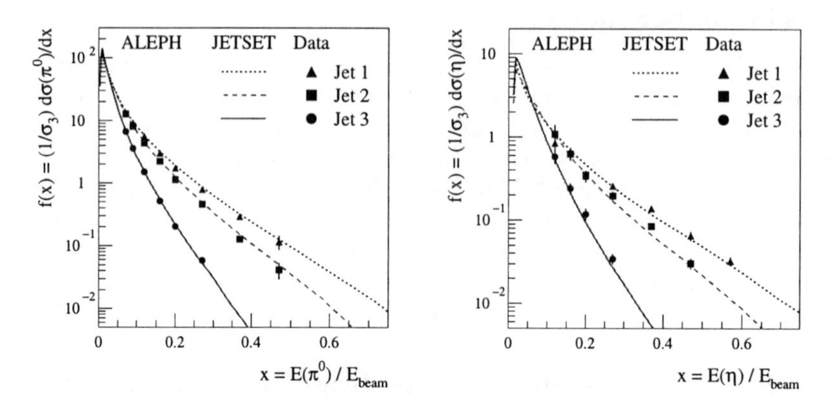

Figure 4.12. The fragmentation function in jets 1-3 for a) π^0 and b) η produced in three-jet events at 91.2 GeV, compared with JETSET7.4 [39].

After a lot of initial confusion due to ambiguities in the gluon-jet isolation, considerable progress has been reported since about 1995 [38–40]. In general, the differences are in qualitative agreement with the analytical expectation and are described quantitatively by the Monte-Carlo models containing gluon interference.

In particular, the following observations have been made:

i) The multiplicity of a gluon jet is higher than that of a light-quark jet (see Chapter 8, where also the gluon jet identification is described), independently of the particle species considered.

ii) The fragmentation function of a gluon jet is considerably softer than that of a quark jet: As two examples, the π^0 and η fragmentation functions are given in Fig. 4.12 for jets 1-3 in three-jet events [39] at 91.2 GeV, where jet 1 is the largest-energy jet (q-enriched) and jet 3 the lowest-energy jet (g-enriched). This can be explained by both, a higher multiplicity of soft gluons radiated in a gluon jet and $\mathrm{g} \to \mathrm{q}\bar{\mathrm{q}}$ splitting or double string formation necessary before hadronization.

iii) Gluon jets are less collimated than quark jets. The mean values of the jet broadening

$$\beta = \frac{\Sigma|\mathbf{p}_i \times \mathbf{r}_{\text{jet}}|}{2\Sigma|\mathbf{p}_i|} , \tag{4.4}$$

where the $\mathbf{p}_i$ are the momenta of the particles belonging to the jet of direction $\mathbf{r}_{\text{jet}}$, are given in Fig. 4.13 as a fuction of the jet scale $\kappa_{\mathrm{H}} = E_{\text{jet}} \cos\Theta/2$, with E_{jet} being the jet energy and Θ the angle to the closest jet. $\langle\beta\rangle$ is independent of κ_{H} for both gluon and quark jets, but at a value ~ 1.5 higher for gluons than for quarks.

The figure also shows lines representing the behavior of $\langle\beta\rangle$ as a function of the scales E_{jet} (dashed line) and $\kappa_{\mathrm{T}} = E_{\text{jet}} \sin\vartheta$, with ϑ being the smaller of the

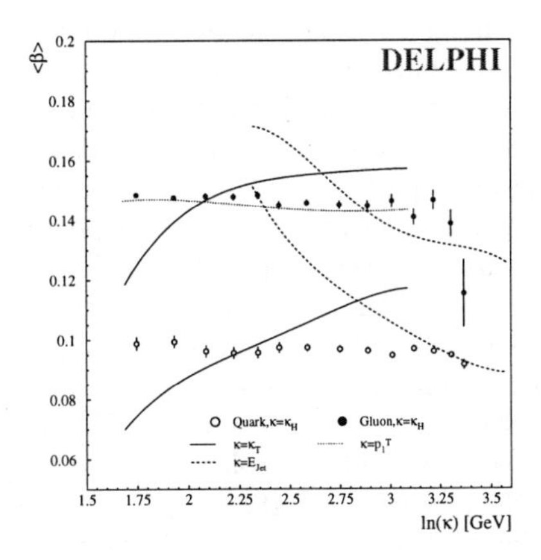

Figure 4.13. Average jet broadening $\langle\beta\rangle$ from charged hadrons in symmetric three-jet events, as a function of various jet scales [40].

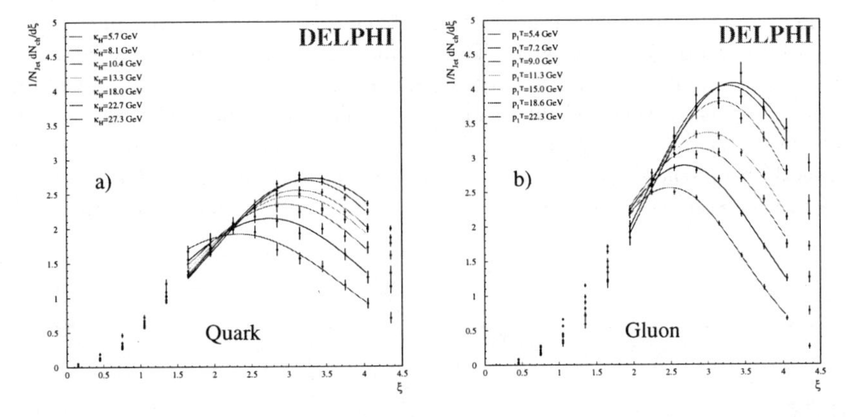

Figure 4.14. ξ distribution for various values of the hardness scales κ_{H} and p_1^{T} for a) quark and b) gluon jets produced in three-jet events at the Z [40]. The lines correspond to Gaussian fits.

two angles with respect to the axis of the most energetic jet (full line). The strong variation of $\langle\beta\rangle$ with these variables suggests that they are not suitable as scaling variables. On the other hand,

$$p_1^{\mathrm{T}} = [(p_{\mathrm{q}} \cdot p_{\mathrm{g}})(p_{\bar{\mathrm{q}}} \cdot p_{\mathrm{g}})/2(p_{\mathrm{q}} \cdot p_{\bar{\mathrm{q}}})]^{1/2} \tag{4.5}$$

coincides with κ_{H} for symmetric three-jet events and can be used as a scale for the gluon jet.

iv) Scaling violations of the gluon fragmentation function are stronger than those of quark jets: This is demonstrated in Figs. 4.14a,b [40] using $\kappa_{\rm H}$ and $p_{\rm 1}^{\rm T} \approx \kappa_{\rm H}$ as a hardness scale, respectively, and is due to the fact that gluon fragmentation is dominated by $C_{\rm A}$ and quark fragmentation by $C_{\rm F}$. In both cases, the scaling violation is observed to be positive for low x_E and negative for high x_E [38]. At fixed x_E, the fragmentation functions show a power-like scaling violation, $D(x_E, \kappa) = a \cdot \kappa^b$, with a power (slope in a log-log plot) depending on x_E. As shown in Fig. 4.15 [40], the slope starts positive at $x_E \lesssim 0.1$ for both quark and gluon jets and then becomes negative for larger x_E values, but the trend is stronger for gluon than for quark jets. The scale dependences are very well reproduced by the DGLAP evolution [25], already used in Chapter 1. From the differences observed, the color factor ratio is measured to be [40] $C_{\rm A}/C_{\rm F} = 2.26 \pm 0.09_{\rm stat} \pm 0.06_{\rm sys} \pm 0.12_{\rm clus,scale}$, in excellent agreement with the SU(3) expectation of 2.25. Furthermore, $\alpha_{\rm s}$ extracted [38] from a DGLAP fit, even though not competitive in precision is well consistent with the world average.

v) A further interesting observation is an extra surplus of baryon production in gluon jets [41–43], as shown in the form of the double ratio $({\rm g/q})_{\rm proton}/({\rm g/q})_{\rm all}$ as a function of ξ_p in Fig. 4.16. The double ratio is near unity at very small ξ_p (highest momenta) and at large ξ_p (small momenta). A strong deviation from unity is, however, visible near $\xi_p = 1$. The deviation is stronger at low energies [42] than at high ones [41] and indicates that baryons are produced directly from colored partons or from strings and that an intermediate state of neutral clusters is disfavoured (see Chapter 6 for the discussion of string and

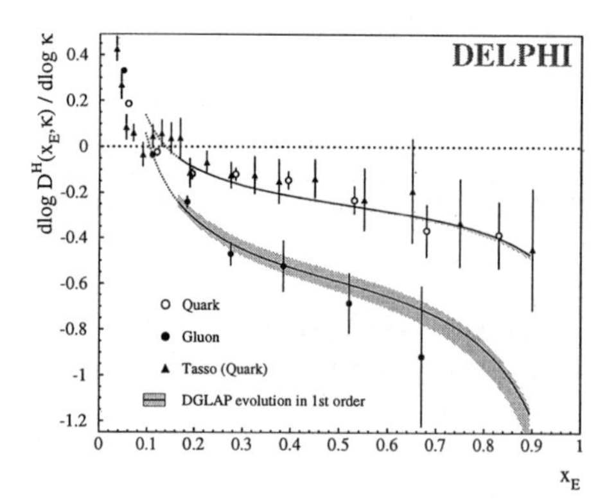

Figure 4.15. Comparison of scaling violation of quark and gluon jets [40].

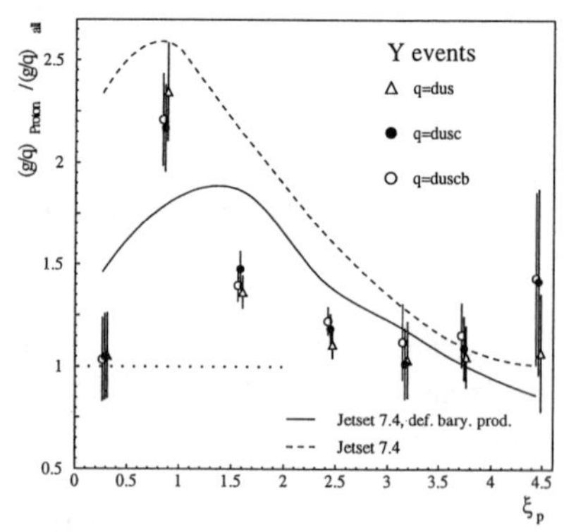

Figure 4.16. Double ratio $({\rm g/q})_{\rm proton}/({\rm g/q})_{\rm all}$ as a function of $\xi_{\rm p}$ for various quark-jet samples [41].

cluster models).

vi) Whereas for gluon jets with a rapidity gap a significant excess of leading systems with total charge zero is found with respect to JETSET and ARIADNE expectations, the corresponding leading systems of quark jets do not exhibit such an excess [44–46]. This may be evidence for so-called octet neutralization [47] (Fig. 4.17a) in addition to triplet neutralization or for color reconnection [48]. No clear understanding has been obtained from Monte Carlo studies, so far, but color reconnection seems disfavored [45, 46, 49].

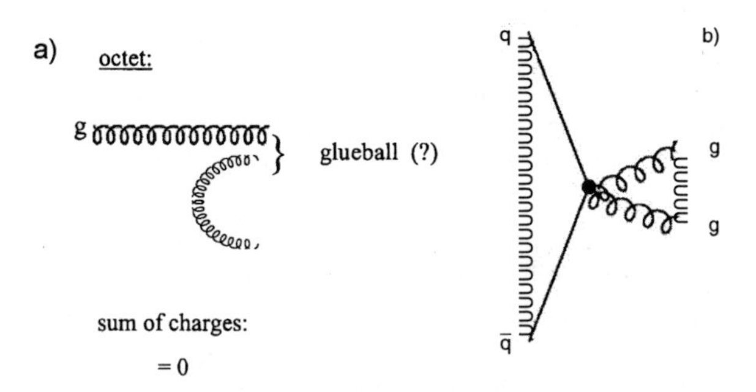

Figure 4.17. a) Octet neutralization in gluon jets, b) reconnection.

4.1.9 Hadronic production rates

Evaluating the integrals over the hadronic cross-section normalized single-particle spectra, one obtains the average multiplicity of the particular hadron per event, i.e., the production rate for that hadron. Systematic studies of the dependence on the particle mass [50–53] have revealed striking exponential or Gaussian regularities with respect to the hadronic mass or the sum of the constituent quark masses.

An example is given in Fig. 4.18 [50], where the rate per spin state $\langle n\rangle/(2J+1)$ of directly produced hadrons is given as a function of the sum of the constituent quark masses relevant for that particular hadron. The direct production rates are obtained from the total measured ones with the help of the string-inspired model of Pei [52]. The lines correspond to fits by

$$\langle n\rangle/(2J+1) = a\exp(-\sum(m_q)_i/T) \qquad (4.6)$$

with different normalization a but the same slope $1/T$. The fit gives $T = 142.4 \pm 1.8$ MeV from which a strange quark suppression factor $\lambda = \exp[-(m_s - \hat{m})/T] = 0.295 \pm 0.006$ can be calculated with a strange to non-strange constituent mass difference of 174 MeV.

It is interesting to note that

i) (apart from the pions, which can be considered the lightest Goldstone particles) all particles within a given SU_3 multiplet follow the same line,

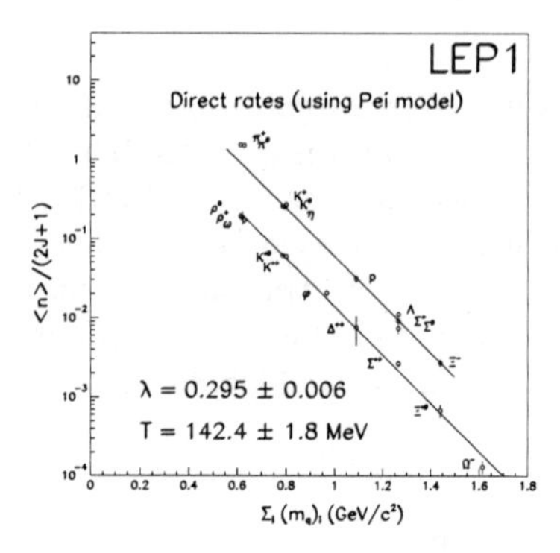

Figure 4.18. The direct hadron production rates as a function of the sum of the corresponding constituent quark masses. The lines correspond to fits according to (4.6) [50].

ii) vector mesons and decuplet baryons (with all quark spins parallel) follow one common line, pseudo-scalar mesons and octet baryons follow the other common line. The former are suppressed by a factor 4-5 ($a_{\mathrm{p,o}}/a_{\mathrm{v,d}} = 0.23 \pm 0.03$) with respect to the latter.

The latter observation suggests an influence of the quark-spin configuration (hyperfine splitting) on the hadron production rate.

While in the string picture of hadron production (see Sect. 6.1) the transverse mass spectrum of the produced quarks follows the Schwinger [54] tunneling formula

$$\frac{\mathrm{d}n_\kappa}{\mathrm{d}^2 p_{\mathrm{T}}^2} \propto \mathrm{e}^{-\pi m_{\mathrm{T}}^2/\kappa} \tag{4.7}$$

with κ being the string tension and $m_{\mathrm{T}} = \sqrt{m^2 + p_{\mathrm{T}}^2}$ being the transverse mass, a thermal equilibrium would lead to the exponential form

$$\frac{\mathrm{d}n}{\mathrm{d}^2 p_{\mathrm{T}}^2} \propto \mathrm{e}^{-m_{\mathrm{T}}/T} \tag{4.8}$$

with T being the temperature of the equilibrium.

It is, of course, difficult to understand how a thermal *equilibrium* can set in during the short time of hadronization in an $\mathrm{e}^+\mathrm{e}^-$ collision. However, as shown by Białas [55], the spectrum of primarily produced partons may already be so close to the thermal one that no secondary collisions are needed to obtain the thermal *distribution*. This is naturally obtained in the string picture, if (4.7) is convoluted with a Gaussian-type fluctuating string tension [55]. The standard value of $\langle\kappa\rangle = 0.9$ GeV/fm would give a value of $T = 170$ MeV. This is only slightly above the value given in Fig. 4.18, thus allowing for some cooling in the expansion before hadronization. The κ fluctuations are readily justifiable in a stochastic picture of the QCD vacuum [56].

It is, however, not possible to extrapolate the exponential dependence to large (transverse) masses. While the hadron yields from the η ($m \cong 0.55$ GeV/c^2) to the

J/ψ ($m \cong 3.1$ GeV/c^2) approximately follow the exponential law [57], this is not the case anymore for the Υ ($m \cong 9.5$ GeV/c^2). This is found [58] to be produced more than 10 orders of magnitude more abundantly.

Similarly, the p_T spectra of Fig. 4.1a,b can be approximated by the exponential Eq. (4.8) at low p_T, but show a clear excess at high p_T. This excess is generally interpreted as due to (semi-)hard contributions and is well reproduced in QCD inspired models, as in Fig. 4.1a,b.

However, a simultaneous study [59] of production rates and m_T spectra of neutral mesons from η to Υ and m_T spectra of π^0's (all in $^{(}\bar{p}^{)}p$ collisions) shows power-law m_T-scaling, with similar values of power and normalization constant for different mesons at fixed collision energy. The appearance of a power law instead of the Boltzmann spectrum was indeed suggested within thermal field theory [60] in the low-temperature and high-mass limit.

4.2 Lepton-hadron collisions

Multihadron production in lepton-nucleon scattering has been studied using electron or muon [61–65] as well as neutrino beams [66–68]. The variables Q^2, x_B and y determining the event kinematics of DIS are given in Eqs. (1.1) and (1.2). The evolution of the fragmentation function $F(x)$ (here called $D(x_F)$) in the forward (current fragmentation) region with the average hadronic energy $\langle W \rangle$ and squared four-momentum transfer $\langle Q^2 \rangle$ is given in Fig. 4.19a [62]. At low Feynman-x, $F(x)$ increases with increasing $\langle W \rangle$. It has been demonstrated [65] that this softening of $F(x)$ is due to hard gluon radiation. At high x, a comparison of E665 and EMC at similar $\langle W \rangle$ but different $\langle Q^2 \rangle$ shows that $F(x)$ decreases there with increasing $\langle Q^2 \rangle$.

From Fig. 4.19b it can be seen that diffraction present in the data of Fig. 4.19a causes a relative hardening of the fragmentation function, but a weak Q^2 dependence remains in the non-diffractive events.

The W and Q^2 dependence of $\langle p_T^2 \rangle$ in the x region 0.2 to 0.4 is given in Fig. 4.19c,d [62]. The dramatic rise of $\langle p_T^2 \rangle$ with increasing W seen in both sub-figures is attributed to hard QCD radiation [65].

The fragmentation of the struck quark in deep inelastic scattering (DIS) can be compared to that of the quarks produced in e^+e^- annihilation, thus allowing to test universality of the fragmentation process. It can, furthermore, be compared to fragmentation of the proton remnant in the target region.

In the Breit frame [71] (see Sub-Sect. 2.4.3), the current region is analogous to a single hemisphere of e^+e^- annihilation at $\sqrt{s} = Q$, so that the scaled-momentum spectra of particles in $x_p = 2p^{\text{Breit}}/Q$ are expected [72–74] to resemble the x_p spectra there. Deviations are, however, expected due to higher-order processes [75] present in DIS but not in e^+e^- annihilation.

Studies of fragmentation in the Breit frame have been reported in [65,69,76]. The

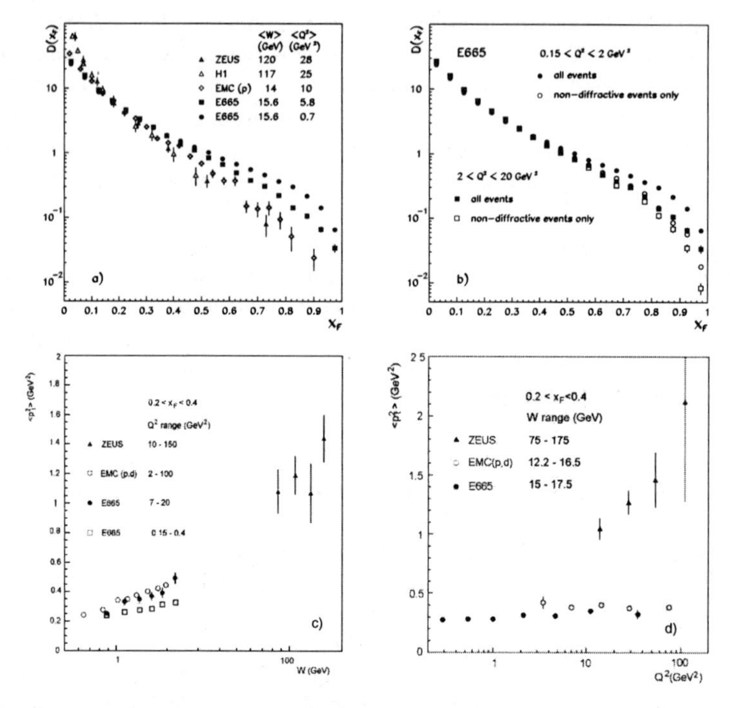

Figure 4.19. The normalized inclusive single-particle Feynman-x distribution (fragmentation functionfor charged hadrons, a) for different experiments b) for all events and non-diffractive events. $\langle p_{\mathrm{T}}^2 \rangle$ for $0.2 < x < 0.4$ as a function of c) W and d) Q^2 [62].

$1/\sigma_{\mathrm{tot}} \mathrm{d}\sigma/\mathrm{d}x_p$ is shown [69] in bins of x_p as a function of Q^2 in Fig. 4.20a. Again, scaling violations are observed here in the form of an increase with increasing Q^2 for small x_p and a decrease for large x_p. The data are compared to $\mathrm{e^+e^-}$ data at $s_{\mathrm{e^+e^-}} = Q^2$. Good agreement is found for the higher Q^2 values, thus supporting the universality of fragmentation. The discrepancy at low x_p and low Q^2 can be attributed to processes like scattering off a sea quark and/or boson-gluon fusion (BGF), not present in $\mathrm{e^+e^-}$ but depopulating the current region of DIS [75].

The universality of quark fragmentation can be extended to jets produced in direct photoproduction. Figs. 4.20b,c show a comparison [70] of the fragmentation function $F(x_E)$ (here called $D(z_E)$) to $\mathrm{e^+e^-}$ [4, 5] and low-x_{B} DIS results [69] at similar jet energy, for charged hadrons $\mathrm{h^\pm}$ and neutral kaons $\mathrm{K^0}$, respectively. At low $z_E \equiv x_E$, the distributions may be affected by differences in color flow and experimental cuts, but above 0.1-0.15 the results are in good agreement with each other. Furthermore, the $\mathrm{h^\pm}$ data even agree with ISR pp data [77] (not shown). The fragmentation function is well described by the PYTHIA/JETSET model. However,

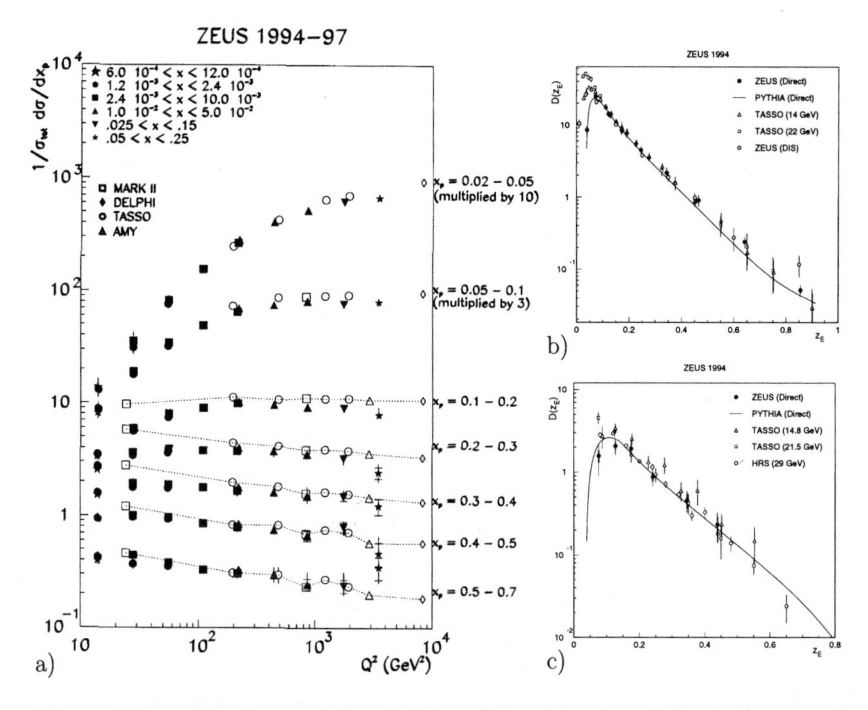

Figure 4.20. The fragmentation function a) for charged particles in the current region of DIS in the Breit frame, compared to e^+e^- data (divided by two) [69], b) to direct photoproduction and e^+e^- [70] and PYTHIA at similar jet energy, c) for K^0 compared to e^+e^- and PYTHIA [70].

for K^0 production, as described in more detail in [70,78], the strangeness suppression factor had to be reduced from the optional value of 0.3 to 0.2.

When plotted as a function of the variable $\xi = \ln(1/x_p)$, the current-region fragmentation function is Gaussian in the neighborhood of the peak [69,76], also here. The evolution with Q is compared to that of e^+e^- annihilation with $E^* \equiv \sqrt{s}$ in Fig. 4.21, in terms of peak position and width (dispersion) of the fitted Gaussian. The solid line is a MLLA/LPHD fit [28–31] to the H1 data [76], in remarkable agreement with the e^+e^- results!

The target region behaves completely different, however. In contrast to that of the current region, its $\xi \equiv \ln(1/x_p)$ distribution does not fall to zero as ξ tends to zero and its integral (the average multiplicity) is about 3-4 times larger.

The general transverse momentum behavior in target and current regions is compared in the form $\langle p_T \rangle$ versus x_p in Fig. 4.22. At large $|x_p|$, $\langle p_T \rangle$ is higher in the

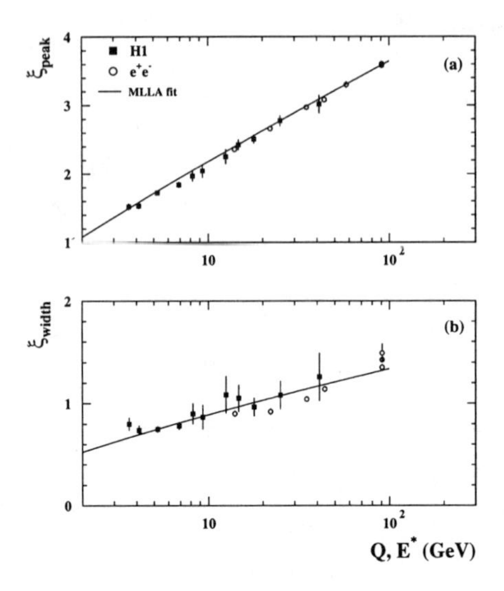

Figure 4.21. Comparison of the evolution in Q and $E^* \equiv \sqrt{s}$, respectively, in DIS and e$^+$e$^-$ collisions, of peak position and width of the fragmentation function. The solid line is a MLLA/LPHD fit to the DIS data [76].

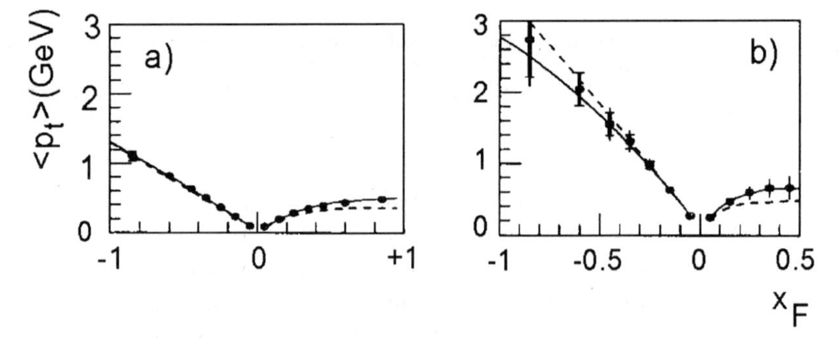

Figure 4.22. $\langle p_{\mathrm{T}} \rangle$ as a function of x_p in the Breit frame of DIS for a) $10 < Q^2 < 20$ GeV^2 and $6 < x < 12 \cdot 10^{-4}$ and b) $160 < Q^2 < 320$ GeV^2 and $2.4 < x < 10 \cdot 10^{-3}$ [69]. The full lines correspond to ARIADNE, the dashed to HERWIG [69].

current region than in the target region and shows a stronger x_p and Q^2 dependence. Thus, the target region has the characteristics of p_{T}-limited (cylindrical) phase space. The ARIADNE MC model gives a good description of the data, while HERWIG shows some discrepancy in the target region due to the lack of primordial transverse momentum of the struck quark.

If isospin invariance holds, the fragmentation function D^{π^0} should be equal to the

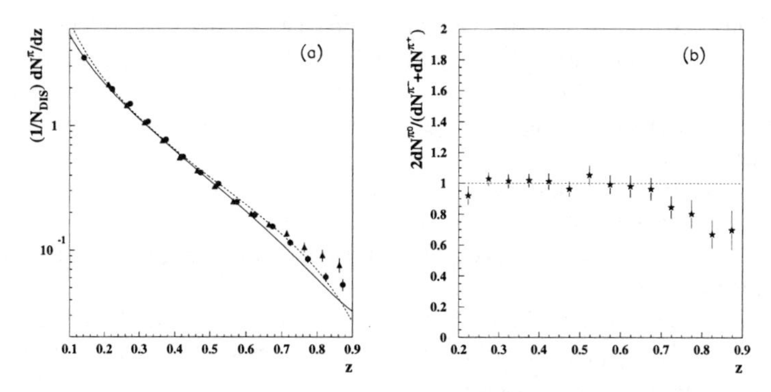

Figure 4.23. a) Single-particle inclusive distributions as a function of $z = \nu/E$ for neutral (circles) and averaged over charged pions (triangles) in e^+p DIS. The full line corresponds to independent fragmentation, the dashed line to parametrization (4.9). b) Ratio of the two distributions [79].

average of D^{π^+} and D^{π^-}, since the quark content of the π^0 is the same as the average of π^+ and π^-. The quark-parton model, therefore, predicts for the single-particle inclusive distribution in $z = \nu/E$ (with $\nu = E - E'$ being the difference of the energy in the target rest frame of the incident and scattered lepton, respectively)

$$\frac{1}{N_{\mathrm{DIS}}}\frac{\mathrm{d}N^{\pi^0}}{\mathrm{d}z} = \frac{1}{2N_{\mathrm{DIS}}}\left[\frac{\mathrm{d}N^{\pi^+}}{\mathrm{d}z} + \frac{\mathrm{d}N^{\pi^-}}{\mathrm{d}z}\right] . \tag{4.9}$$

Fig. 4.23a shows a comparison of the two single-particle spectra by HERMES [79]. An agreement between the data for neutral and charged pions is indeed observed in the ratio of Fig. 4.23b up to $z \sim 0.7$, confirming earlier results from BEBC [80] and EMC [81]. The deviation at larger z-values can be blamed on residual resonance production (as in $\pi^+ + \Delta^0$) affecting different isospin channels differently. Up to $z = 0.7$, the data are well described by independent fragmentation [82] (full line) and by a parametrization of the form

$$\frac{1}{N_{\mathrm{DIS}}}\frac{\mathrm{d}N^{\mathrm{T}}}{\mathrm{d}z} = Nz^{\alpha}(1-z)^{\beta} . \tag{4.10}$$

4.3 (Early) observations in hadron-hadron collisions

The properties of quarks and their interactions are known from hadron spectroscopy as well as from deep inelastic lepton-hadron scattering and from e^+e^- annihilation into jets. From deep inelastic scattering it is known that quarks only carry about half of the nucleon momentum, the remainder being attributed to gluons. In e^+e^- annihilation, planar events with a three-jet structure are interpreted in terms of one of the produced quarks radiating a fast gluon, but also four-jet events have become common at LEP2 energies.

In the context of quarks and gluons (partons), three observations point to their influence on also soft hadronic collisions (see Chapter 5):
(i) Resonance and particle yields in central and fragmentation regions can be understood from quark combinatorics.
(ii) Pion production in the nucleon fragmentation region of soft hadron-hadron collisions reflects the valence quark distribution in the nucleon as observed in moderately deep inelastic lepton-nucleon collisions.
(iii) Soft hadronic particle production resembles moderately low-energy quark fragmentation jets from e^+e^- annihilation and deep inelastic collisions in longitudinal, transverse and multiplicity behavior of the emerging hadrons.

These observations have led to the interpretation that the parton structure of hadrons also governs soft hadron-hadron collisions. To test this unifying concept and to use its far-reaching consequences not only to illuminate the complicated hadron-hadron collisions themselves, but the (soft) hadronization in general, is the basis of a large amount of experimental and theoretical effort in this field. Before discussing the status of this effort in the later chapters, we shall here first review the basic concepts and the three basic experimental observations of above.

4.3.1 Single-particle (and resonance) inclusive spectra

4.3.1.1 Feynman-x and rapidity

Elastic scattering and diffraction dissociation lead to simple final states with relatively few particles. The larger part of the collisions leads to high particle multiplicities with complicated structure in highly-dimensional phase space. The first and simplest approach is then to study an all-inclusive density distribution in one of these dimensions, according to Eq. (2.7).

Fig. 4.24a shows [83] the energy dependence of the invariant distribution

$$F(x) = \int \frac{E^*}{p^*_{\max}} \frac{\mathrm{d}^2\sigma}{\mathrm{d}x\mathrm{d}p_\mathrm{T}^2} \mathrm{d}p_\mathrm{T}^2 \tag{4.11}$$

in the Feynman variable $x = p^*_\parallel / p^*_{\max}$, the component of the particle cms momentum in the beam direction, normalized to its maximum possible value, in $\mathrm{K}^+\mathrm{p}$ collisions. The upper part (mind the change in scale) corresponds to positive particles except for identified protons, the lower part to π^--production. The large-$|x|$ region shows energy scaling and a fall-off of the distribution towards its tails which is steeper for the proton region (large negative x) than for the K^+ region (large positive x). The

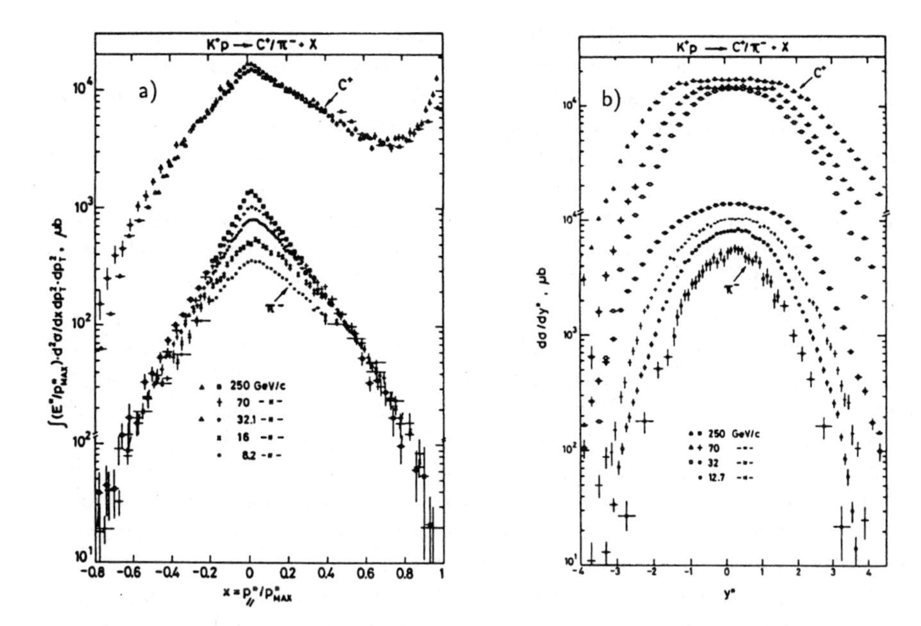

Figure 4.24. a) The invariant Feynman-x distribution for the inclusive reactions $\mathrm{K^+p} \to \mathrm{C^+} + \mathrm{X}$ and $\mathrm{K^+p} \to \pi^- + \mathrm{X}$ between 8.2 and 250 GeV/c; b) the rapidity distribution for the same reactions between 12.7 and 250 GeV/c [83].

large-$|x|$ scaling is in agreement with the early concept of limiting fragmentation [84] stating that at high enough energy, the fragmentation of beam or target is expected to reach an energy independent limit for particles produced with finite momentum in the rest frame of the fragmenting particle. Experimentally, this was convincingly shown to hold at ISR energies for pp collisions between $\sqrt{s} = 31$ and 53 GeV [85].

A scaling violation is, however, observed in the form of an increase of $F(x)$ with increasing beam momentum for the low-$|x|$ (central) region. An alternative variable, expanding the central region, is the rapidity $y = 0.5 \ln[(E + p_{\|})/(E - p_{\|})]$ already defined in Eq. (2.23). The energy dependence for the cms y-distribution for essentially the same data as above is shown in Fig. 4.24b. The distribution widens with increasing energy. In the center, a plateau develops at high energies, reaching a width of about 3 rapidity units at 250 GeV/c beam momentum, and the density increases for all y.

This low-$|y|$ increase with increasing energy is in contradiction to the early hypothesis of so-called Feynman scaling [86], based on the assumption that, asymptotically, interaction between two colliding hadrons occurs only through exchange of partons or parton systems of "wee" longitudinal momentum, i.e. of partons with a non-zero amplitude in both hemispheres.

A lab momentum of 250 GeV/c corresponds to $\sqrt{s}$ = 22 GeV, so not to an asymptotic energy. Therefore, in Fig. 4.26a the central pseudo-rapidity density $\rho(0) = (1/\sigma_{\rm inel})[{\rm d}\sigma/{\rm d}\eta]_{\eta=0}$ is displayed versus $\sqrt{s}$ [83,87,88] up to 1800 GeV, but an upward curvature rather than Feynman scaling is observed. The lines correspond to quark string models to be discussed in Chapter 6. Here, they are only used as examples for a first comparison. The single-string Lund model does not reproduce the rise of the central pseudo-rapidity density in the energy range presented. The two-string model FRITIOF (with hard parton scattering) and a two-string dual parton model (DPM) agree reasonably well with the data up to $\sqrt{s}$=60 GeV, but underestimate the rise for higher energies.

So, Feynman scaling does not hold, but limiting fragmentation does, and it has turned out that this is the case in a much wider range of rapidities than originally proposed, and not only in hadron-hadron [87], but also in hadron-nucleus and nucleus-nucleus collisions [89,90]. Fig. 4.25 shows that in these types of collisions, the particle density in the fragmentation region increases linearly with decreasing pseudorapidity $\eta - Y$ (Y being the beam rapidity) towards the central plateau. The range of this limiting fragmentation region increases with increasing energy, so that the width of the central plateau grows much slower than anticipated.

Abandoning Feynman scaling, Białas and Jezabek [91] show that these features can be understood from a two-step process, where a number of color exchanges take place between two sets of partons (one in each of the colliding hadrons) which are *uniformly distributed in rapidity*, so not just "wee", and the color charges created this

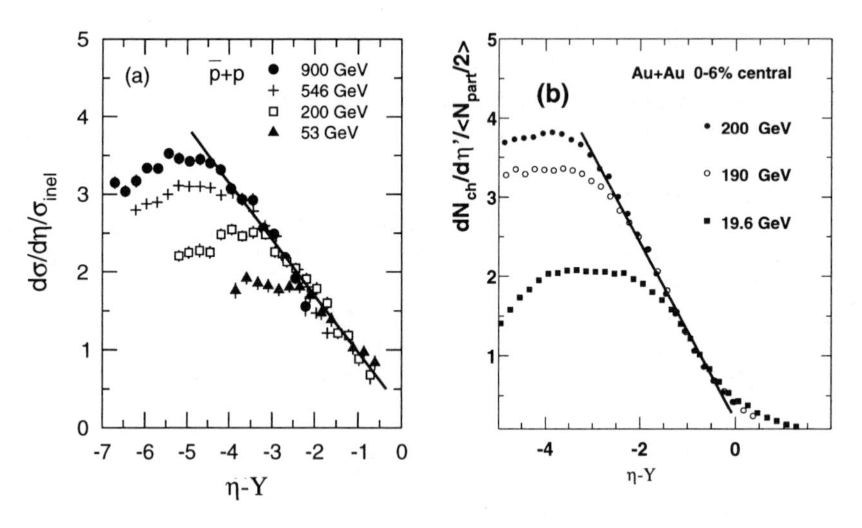

Figure 4.25. Pseudorapidity distribution for a) $p\bar{p}$ collisions between $\sqrt{s}$ = 53 and 900 GeV [87], b) AuAu collisions between 19.6 and 200 GeV [90].

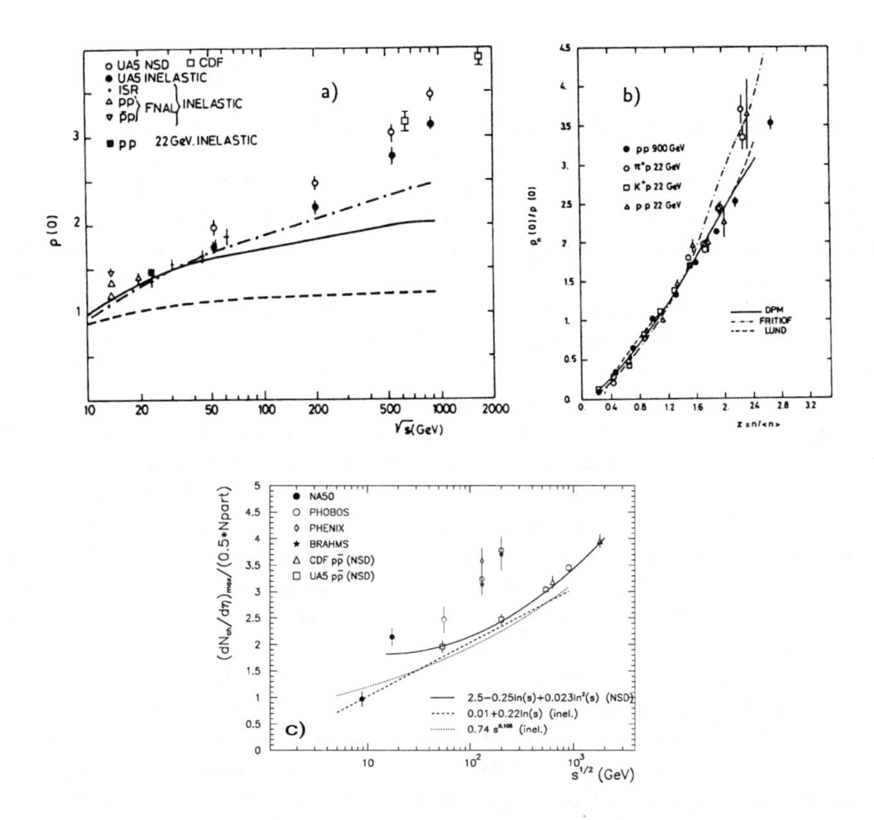

Figure 4.26. a) The central density $\rho(0)$ of the c.m. pseudo-rapidity distribution as a function of $\sqrt{s}$. The solid, dashed and dashed-dotted curves are DPM, Lund and FRITIOF predictions [83]. b) The scaled central pseudo-rapidity density $\rho_n(0)/\rho(0)$ versus $z = n/\langle n\rangle$ for $\pi^+\mathrm{p} \to \mathrm{C}^\pm + \mathrm{X}$, $\mathrm{K}^+\mathrm{p} \to \mathrm{C}^\pm + \mathrm{X}$ and $\mathrm{pp} \to \mathrm{C}^\pm + \mathrm{X}$ at $\sqrt{s} = 22$ GeV, compared with Collider data at $\sqrt{s}$=900 GeV. Model predictions are for $\sqrt{s} = 22$ GeV [83]. c) The central density per participant pair for central heavy ion collisions at SPS and RHIC. The lines are fits to the $\mathrm{p\bar{p}}$ data [92].

way then emit particle clusters by bremsstrahlung or color string decay with a flat distribution of clusters. The saturation in the form of a central plateau then is due to the fact that in the cms only partons can participate with lifetime longer than the time needed for the color exchange.

The UA5 Collaboration [87], furthermore, proposed a form of scaling of the topological charged-particle pseudo-rapidity spectra in the central region of $\bar{\mathrm{p}}\mathrm{p}$ interactions in the range $200 \le \sqrt{s} \le 900$ GeV. UA5 observed that the quantity

$$R_n = \rho_n(0)/\rho(0) \tag{4.12}$$

with

$$\rho_n(0) = (1/\sigma_n)(\mathrm{d}\sigma_n/\mathrm{d}\eta)_{\eta=0} \quad \text{and} \quad \rho(0) = \sum P_n \rho_n(0) \tag{4.13}$$

is nearly energy independent when plotted versus the scaled charged-particle multiplicity $z = n/\langle n \rangle$.

The non-single-diffractive NA22 sample is plotted in Fig. 4.26b, together with the $\mathrm{p\bar{p}}$ data at $\sqrt{s}$ = 900 GeV. Reasonable scaling is indeed seen in the energy range $22 \leq \sqrt{s} \leq 900$ GeV, except at $z > 2$, where the NA22 data are higher. Moreover, the NA22 data show no significant dependence on the type of beam particle. The model predictions are calculated for $\sqrt{s}$ =22 GeV. They are similar to each other for $z < 1.6$ and in agreement with the data. At larger z, and at this energy, the data tend to prefer FRITIOF, although the statistical and systematic uncertainties are large in the highest topology events.

A remarkable difference is observed between the collider $\mathrm{p\bar{p}}$ data [87, 88] and central heavy ion collisions at high energies [92]. In Fig. 4.26c, $\rho(0)$ is given as a function of $\sqrt{s}$ per participating nucleon pair for both types of collisions. While the lower-energy NA50 point is compatible with the $\mathrm{p\bar{p}}$ trend, the higher-energy central heavy ion collisions lead to a $\rho_0(0)$ per participant pair considerably higher than that for $\mathrm{p\bar{p}}$ collisions. Therefore, particle production in the former cannot be explained as a simple superposition of nucleon-nucleon interactions (see e.g. [93] and Sub-Sub-Sect. 8.1.1.2).

Coming back to the humpbacked distribution in $\xi = -\ln x_p$ and the energy evolution of its maximum ξ_0 as shown for $\mathrm{e^+e^-}$ in Fig. 4.8b and compared to ep in Fig. 4.21a, the observed universality has been extended to $\mathrm{p\bar{p}}$ collisions at $\sqrt{s} = 1.8$ TeV [94]. Using $E_{\mathrm{jet}} \sin \Theta_{\mathrm{c}}$ as an evolution variable, with E_{jet} being the jet energy and Θ_{c} its cone opening angle, perfect agreement in ξ_0 values and evolution is observed for all three types of collisions.

4.3.1.2 Transverse momentum distribution

The differential cross section $\mathrm{d}\sigma/\mathrm{d}p_{\mathrm{T}}^2$ for positively charged particles (C^+) and for π^- in hp interactions at 250 GeV/c is plotted in Fig. 4.27a. The data show a significant high-p_{T}^2 tail. Its further increase through ISR, SPS and Tevatron energies [95–97] indicates the onset of a hard-scattering regime, to be studied in more detail in Sub-Sect. 4.3.4. At low p_{T}^2, on the contrary, the exponential slope is largely independent of energy [95–97], and in Fig. 4.27a, one observes no dependence on the type of beam particle. It is, however, smaller for C^+ than for π^-.

The curves in Fig. 4.27a reflect the p_{T} parameters used in the particular versions of Lund, DPM and FRITIOF. All models describe the region $p_{\mathrm{T}}^2 < 1$ $(\mathrm{GeV}/c)^2$ fairly well, but Lund and DPM clearly do not account for the high p_{T}^2 tail of the distributions. Taking into account gluon emission and hard parton scattering processes, FRITIOF describes the inclusive $\mathrm{d}\sigma/\mathrm{d}p_{\mathrm{T}}^2$ distribution better, but still tends to underestimate the data for positive particles in the region $1 < p_{\mathrm{T}}^2 < 2.5$ $(\mathrm{GeV}/c)^2$.

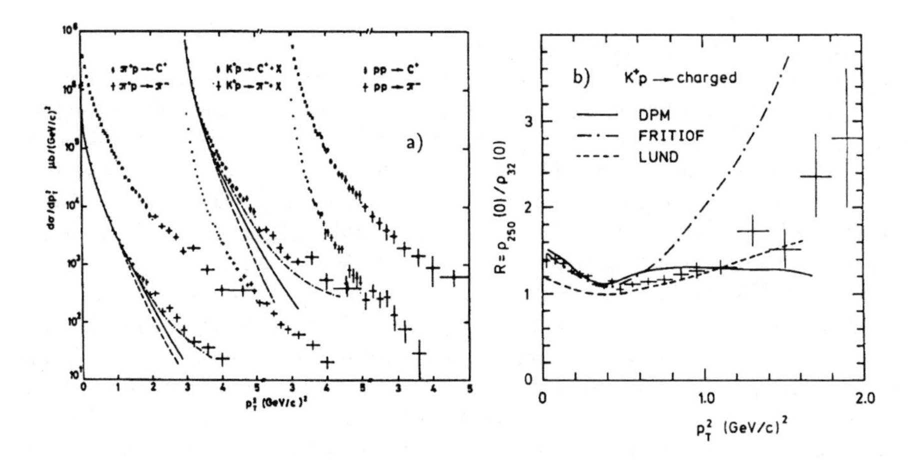

Figure 4.27. a) The $\mathrm{d}\sigma/\mathrm{d}p_\mathrm{T}^2$ distributions for positively charged particles (C^+) and for π^- in hp interactions at 250 GeV/c. The solid dashed and dashed-dotted curves are DPM, Lund and FRITIOF predictions [83]. b) The ratio of the central rapidity density versus p_T^2 for $\mathrm{K}^+\mathrm{p}$ interactions at 250 and 32 GeV/c [83].

In 4.3.1.1 we have analyzed the rise with energy of the central rapidity plateau. Although the effect is well known and also seen in $\mathrm{e}^+\mathrm{e}^-$ annihilation and deep-inelastic lh processes, the dynamical origin of this phenomenon is not fully understood. In [98] it was shown that part of the central plateau rise is of kinematical origin and related to mass effects which are still significant at top ISR energies. In the framework of the DPM, the effect is purely dynamical and due to i) the increasing overlap in rapidity space of the fragmenting valence-quark chains, ii) the contribution from additional chains stretched between quarks and anti-quarks created in the vacuum. In FRITIOF, the rise of the plateau is a consequence of non-scaling behavior in each string separately, due to gluon emission and the presence of hard scatterings. The latter ingredients are thus expected to reflect in the p_T-dependence of the plateau rise.

To investigate this question, we plot in Fig. 4.27b the ratio of $(\mathrm{d}\sigma/\mathrm{d}y)_{y=0}$ for charged particles at 32 and 250 GeV/c, as a function of p_T^2. It is clear that the largest contribution to the total central-plateau increase originates from the *small-*p_T^2 region. The ratio is close to one at p_T^2=0.5 (GeV/c)^2 and increases again for $p_\mathrm{T}^2 > 1$ (GeV/c)^2. A similar energy behavior is observed in ISR data [99]. Furthermore, an additional excess at low p_T and at high p_T is found when comparing heavy ion collisions to hadron-hadron collisions at the same energy per nucleon [100].

Lund gives $R \approx 1$ around p_T^2=0.5 (GeV/c)^2 and $R > 1$ at smaller and larger p_T^2, but on both sides R stays smaller than in the data. At least some of this scaling violation derives from the decay of resonances more abundantly produced at larger energies and from kinematics.

FRITIOF, on the other hand, more successful in describing the overall p_T^2-spectra, fails to account for the energy dependence in the region $p_T^2 > 0.5$ (GeV/c)^2. The onset of hard parton scatters in this model is so strong between 32 and 250 GeV/c that the prediction overshoots the data by a factor of 1.7 at p_T^2=1.0 (GeV/c)^2. DPM describes the rise at small p_T^2 reasonably well, but remains almost constant for p_T^2 >0.75 (GeV/c)^2. However, for DPM the ratio in this p_T^2 region is particularly sensitive to the value of the average primordial quark transverse momentum k_T. If the value of $\langle k_T^2 \rangle$ is lowered from 0.42 (GeV/c)^2 to 0.20 (GeV/c)^2, the rise of the ratio for $p_T^2 > 0.75$ GeV/c^2 is similar as in the Lund prediction.

After a first indication in a cosmic ray experiment [101], the UA1 experiment [102] has established an *increase* of the mean transverse momentum $\langle p_T \rangle$ with increasing charged particle density $\Delta n/\Delta y$ in rapidity. A similar increase has been observed in a second cosmic ray experiment [103], in UA5 [104] and at the Tevatron [96]. Though much weaker at ISR energies, an increase is also seen there [99, 105, 106]. Besides the growth of the effect between ISR and Collider, the correlation between $\langle p_T \rangle$ and $\Delta n/\Delta y$ becomes stronger when low p_T tracks are excluded and when the analysis is restricted to the central region. Explanations have been proposed in terms of possible evidence for a hadronic phase transition in a thermodynamical model [105, 107, 108], small impact parameter scattering in a geometrical model [109] or the production of mini-jets from semi-hard scattering [110–113].

At lower energies, on the other hand, a *decrease* of $\langle p_T \rangle$ with increasing n had been observed. This decrease is mainly visible at the high-n tail of the distribution and is generally interpreted as a phase-space effect.

Comparing in Fig. 4.28a [114] the highest available energy data to intermediate and low energy data, we see that $\langle p_T \rangle$ is surprisingly energy independent for low multiplicities. The slope of $\langle p_T \rangle$ vs. n, on the other hand, is negative for low energies and becomes positive at ISR. This leads to a fast increase of $\langle p_T \rangle$ with increasing energy for high multiplicities. As shown in Fig. 4.28b, this increase depends on the particle type and is faster for heavier particles than for pions [96].

The mini-jet interpretation of the development with increasing density and energy in Fig. 4.28a gets support from the fact that a similar development is seen in e^+e^- collisions [115, 116]. At 91 GeV, part of the e^+e^- collisions lead to a 2-jet topology, part to a three- or more-jet topology. While the first two jets originate from the fragmentation of the original $q\bar{q}$ pair, the third jet corresponds to a gluon radiated off by one of the quarks. Fig. 4.28c shows the average transverse momentum $\langle p_T^{in} \rangle$ in the event plane as a function of the charged-particle multiplicity n for inclusive particle production in e^+e^- collisions at 91 GeV [115], compared to that in 2-jet events and ≥3-jet events. While the latter two still show a decrease of $\langle p_T^{in} \rangle$ with increasing n, the inclusive distribution shows a clear increase. This increase can be interpreted as due to a change from a 2-jet regime at low n to a ≥3-jet regime at large n [117].

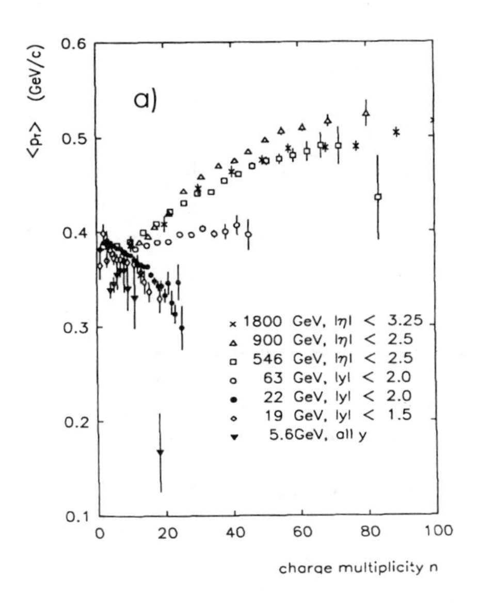

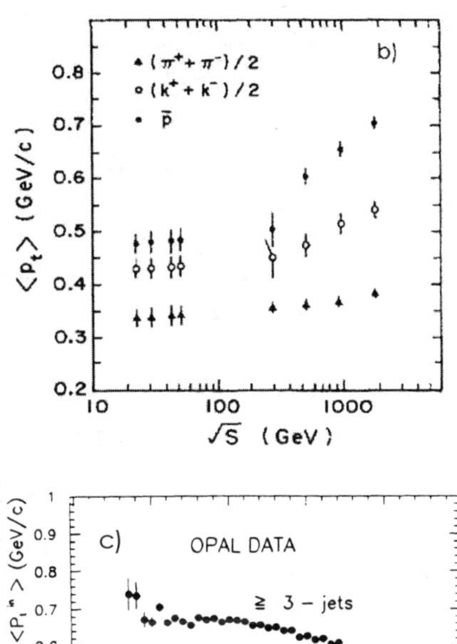

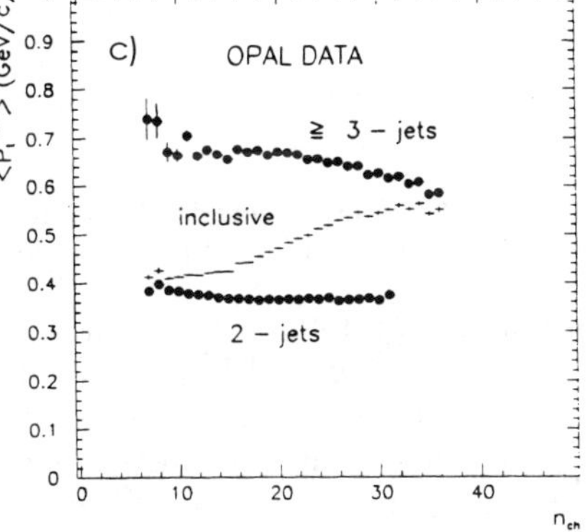

Figure 4.28. The average transverse momentum $\langle p_T \rangle$ as a function of charged-particle multiplicity n a) for hh collisions at $\sqrt{s}$ from 5.6 to 1800 GeV [114], b) as a function of $\sqrt{s}$ for $\bar{p}$, $K^{\pm}$ and $\pi^{\pm}$ [96], c) for inclusive production as well as for 2-jet and $\geq$ 3-jet events in e^+e^- collisions at 91 GeV [115].

4.3.1.3 Differences between quark and gluon jets

The gluon structure function of the proton is considerably softer than the quark ones. UA1 used this difference to perform a statistical separation of quark and gluon jets in two-jet events [118]. In Fig. 4.29a, the fragmentation fuction $D(z)$, with the momentum fraction $z = p_z(\text{track})/p(\text{jet})$ and p_z the momentum component along the jet axis, is shown for both types of jets. The ratio of the two distributions is given in Fig. 4.29b. The softer fragmentation for gluons is indeed also observed in hadronically excited jets, be it with very large errors.

Furthermore, also here gluon jets are observed to be wider than quark jets, and scaling violations are observed to be stronger than gluon/quark jet differences. In Fig. 4.29c, the pure quark and gluon fragmentation functions extrapolated from TASSO [119] are given in bins of z versus the two-jet mass, together with UA1 data. In a detailed comparison of the jet shape in e^+e^-, ep and $p\bar{p}$ collisions [120],

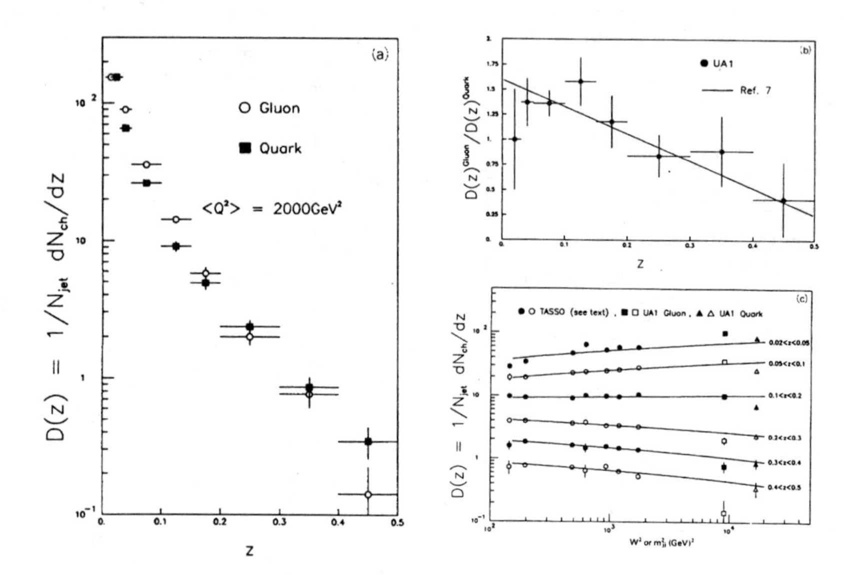

Figure 4.29. a) Fragmentation function $D(z)$ for quark jets and gluon jets, b) their ratio as a function of z. c) W^2 and m_{jj}^2 evolution of the fragmentation function per bin of z [118].

jets are shown to be narrower in the first two (from OPAL and ZEUS) than in the latter (from CDF and D0). This difference can be understood from the abundance of gluon jets in p$\bar{\text{p}}$ collisions at Tevatron energies.

Finally, an analysis of average jet charges demonstrates that gluon jets are neutral, while u($\bar{\text{u}}$)-quark-enriched jet samples show a significant positive (negative) average charge.

4.3.1.4 Resonances

About 50% of the pions shown in Fig. 4.24 come from vector mesons and also tensor mesons and baryon resonances are not negligible as pion sources. So, more direct information on the production mechanism can be expected from the study of resonances.

The dependence of the average ρ^0 multiplicity for $\pi^\pm$p reactions [121] on the squared cms energy s is compared in Fig. 4.30a with that in deep-inelastic μp and ν_μp scattering as a function of W^2, the squared mass of the hadronic system [122]. At given $s = W^2$, the values of $\langle n(\rho^0)\rangle$ found for $\pi^\pm$p and $\mu(\nu_\mu)$p interactions are remarkably similar (mind the change in scale on the vertical axis) and the slopes b of the logarithmic dependence on s and W^2 are the same within errors [123].

In Fig. 4.30b, the Feynman-x distribution is compared for ρ^0 and ω production in deep-inelastic μp collisions [122] and K$^+$p collisions [124] (the μp data are scaled to

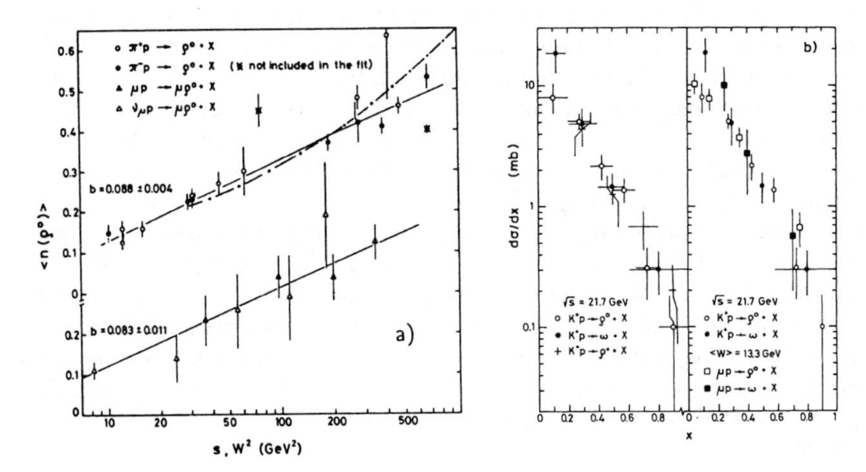

Figure 4.30. a) Average ρ^0 multiplicity in $\pi^{\pm}$p collisions and in deep-inelastic μp and ν_μp reactions as a function of s and W^2, respectively. Straight lines are fits with a logarithmic energy dependence, the dash-dotted line corresponds to the FRITIOF model [121]. b) Comparison of dσ/dx distributions for ρ^0 and ω in μp and K$^+$p interactions [124].

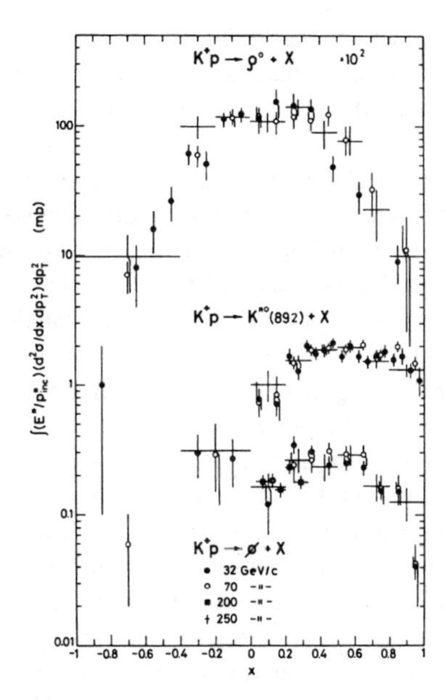

Figure 4.31. Invariant Feynman-x distribution for inclusive resonance production in K$^+$p collisions between 32 and 250 GeV/c [125].

the total inelastic K^+p cross section at $\sqrt{s} = 13.3$ GeV, the average hadronic energy $\langle W \rangle$ in the μp experiment). The μp results are due to u-quark fragmentation and the authors conclude that the differential production rates of ρ^0 and ω are equal within errors. In Kp collisions, ρ production is less affected by diffraction than it may be in πp collisions of Fig. 4.30a. Moreover, the K^+ also carries a u-quark (the $\bar{s}$ would not give a ρ^0 or ω). Indeed, one observes an interesting similarity of ρ^0 and of ω production in K^+p and μp collisions.

The invariant x distribution $F(x)$ for three typical inclusive resonance reactions is shown in Fig. 4.31 for 32 to 250 GeV/c [125]. In all cases, scaling is observed for $F(x)$ for all x. From the K^+ beam, ρ^0 is mainly produced in the central region, K^{*0} and ϕ in the forward region. So, the strange quark of the beam plays a role in forward K^{*0} and ϕ production, but does not contribute to the ρ^0. Since ϕ needs an s-quark in addition to the $\bar{s}$ from the beam, an $s\bar{s}$-pair has to be created in the "sea". The suppression λ of strange to non-strange creation can, therefore, be obtained from the K^{*0}/ϕ ratio.

4.3.2 Particle yields

4.3.2.1 Resonance dominance

An important observation, which has taken place almost simultaneously in e^+e^-, lepton-hadron and hadron-hadron collisions, is that of resonance dominance not only in low multiplicity exclusive final states as discussed in Chapter 3, but also of inclusive reactions.

Experimentally, the estimation of resonance cross sections is a very difficult task, in particular in final states where many particles are involved. Final-state particles are either directly produced in the production process, or they are decay products originating from resonances. In order to find these resonances, effective masses have to be formed for all proper combinations of final-state particles, and the number of possible combinations has to be studied as a function of the effective mass.

Two-particle densities fall exponentially at high masses and an extrapolation of this exponential down to below the low-mass resonances is generally used for a direct estimate of inclusive resonance production. Fig. 4.32a shows the $\pi^+\pi^-$ mass spectrum for $K^-p \to \bar{K}^0\pi^+\pi^- +$ anything and $K^-p \to \Lambda\pi^+\pi^- +$ anything obtained at 10 and 16 GeV/c [126]. In both cases, a small shoulder is visible in the ρ and f regions on a large combinatorial background falling like $\exp(-kM)$ with $k \approx 3$ GeV^{-1}. Multiplication of the spectrum by $\exp(kM)$ grants the ρ and f signals as shown in Fig. 4.32b.

In Fig. 4.30a we have seen that the cross section for ρ production grows logarithmically with available squared cms energy s. The growth is similar to that of $\pi^\pm$ production (not shown), thus leading to a constant $\langle \rho^0 \rangle / \langle \pi^+ \rangle$ or $\langle \rho^0 \rangle / \langle \pi^- \rangle$ ratio of about 15% for meson-proton collisions and 10% for proton-proton collisions. This number is large, if one takes into accound that the denominator *includes* the pions

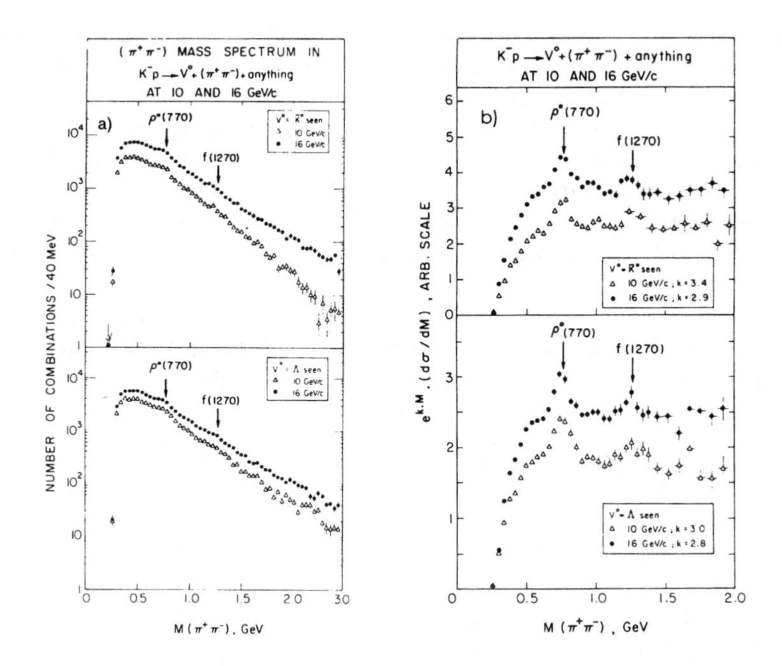

Figure 4.32. a) Effective mass spectra for $\pi^+\pi^-$ production in 10 (Δ) and 16 ($\bullet$) GeV/c K^-p interactions, associated with an $\bar{\mathrm{K}}^0_{\mathrm{seen}}$ and Λ^0, respectively. b) The effective mass spectra $\mathrm{d}\sigma/\mathrm{d}M(\pi^+\pi^-)$ multiplied by a factor $\exp(kM)$ extracted from the slopes in subfigure a) [126].

coming from the decay of the ρ^0 and also that of other resonances.

The percentage of π^- produced via the best known meson resonances are estimated for π^+p interactions at 16 GeV/c in [127]. The authors find that ρ and ω are the dominant mesonic sources of indirect pions and give rise to about 43% of all π. Further sizable contributions come from η, f and K*(890). The ten most important meson resonances are responsible for (51±4)% of inclusive π^- production. Furthermore, proton dissociation into p$\pi^+\pi^-$ could account for 0.6 mb. There are no strong N^{*0} signals observed, but wide and overlapping N* of large masses and multibody decays cannot be excluded and would give further indirect negative pions. In the case of positive pions, a strong Δ^{++} is responsible for a sizable indirect π^+ production.

At $\sqrt{s} = 53$ GeV with the split-field magnet at CERN [128], the background is constructed by combining randomly selected tracks from different events, to define the uncorrelated two-particle mass distribution taking into account detector acceptance and combinatorial problems. The normalization is chosen to give the same number of events in the high-mass region (2-4 GeV/c^2) where correlation effects are expected to be negligible. The result of a fit to the subtracted spectrum with ρ and ω alone is inadequate, particularly in the mass region around 0.6 GeV/c. This is just the region where a $K^{*0}(890) \to K\pi$ decay would contribute with the kaon taken as a pion. A further peak in the subtracted spectrum is caused by $\eta \to \pi^+\pi^-\pi^0$. The fit with $\rho, \omega, K^*(890), \eta$ and f is reasonable and gives the resonance yields shown in Fig. 4.33. The number of charged secondaries originating from the dominating $\rho^{\pm 0}, \omega$ and $K^{*\pm 0}$ contributions is 10.3±1.2. With an average charged-particle multiplicity of pions and kaons per inelastic event of 10.1±1.2, this yields the result that 100±20% (or better, more than 60% at 95% confidence level) of all pions and kaons produced result from vector meson decay.

We can conclude that vector mesons alone account for about half of the produced pions and that tensor mesons are not negligible. It is furthermore to be expected that a number of pions stem from baryonic resonances and diffraction.

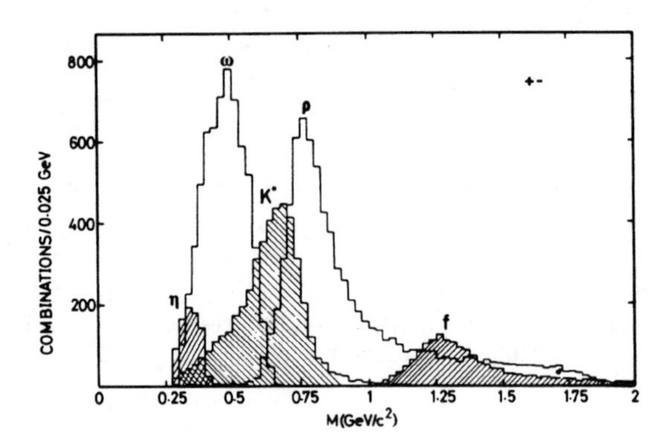

Figure 4.33. Contributions of $\rho^0, \omega, K^*(890), \eta$ and f resonances to the $\pi^+\pi^-$ effective mass spectrum at $\sqrt{s} = 53$ GeV [128].

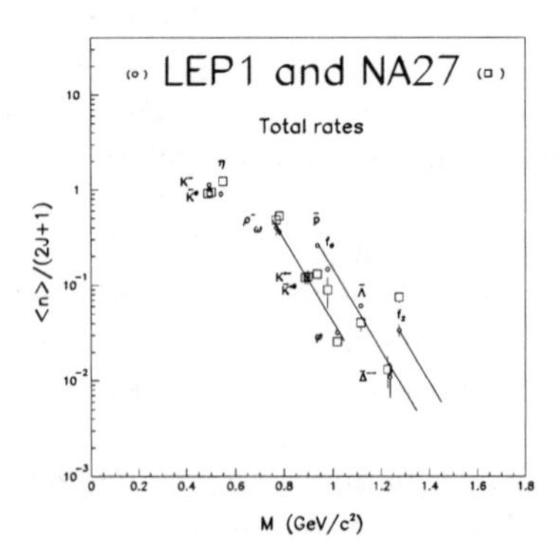

Figure 4.34. Comparison of the total hadron rates, $\langle n \rangle/(2J+1)$, measured at LEP1 (dots) and in pp interactions at a beam momentum of 400 GeV/c (squares) [50].

The observations on the relative particle yield reported for e^+e^- collisions in Sub-Sect. 4.1.9 are also valid for hadronic reactions, at least for particles not resulting from the fragmentation of the incident hadrons [50, 129]. Fig. 4.34 [50] shows production rates measured by NA27 [130] in pp collisions at 400 GeV/c compared to those of the same hadrons at LEP1, after normalization of the K^{*-} rate. In general, the relative rates and their mass dependence measured by NA27 are compatible with those of the LEP1 data.

The small amount of direct pion production is already interesting in itself and will be further discussed in Sect. 5.1. The low direct pion production also has, however, two important consequences. Firstly, it is impossible to study the primary production process with pions alone, resonance effects having to be taken into account. Secondly, resonances can give additional information on the primary production mechanism.

4.3.2.2 Strangeness suppression

Kaons are less copiously produced than pions. Fig. 4.35a shows that the K/π ratio is only about 0.02 at $\sqrt{s} = 5$ GeV, but grows to about 0.09 at 546 GeV [131]. The behavior between 200 and 1800 GeV has been studied in more detail by UA1 [134] in a combination of their data with those of UA5 [135], E735 [96, 136] and CDF [137]. A fit of these data by a function linear in $\ln\sqrt{s}$ still gives a small three-sigma increase, probably due to increasing gluon production. In Fig. 4.35b, we see that the ratio increases with increasing p_T and that the p_T dependence is the same at CERN-Collider and ISR energies. The same even holds down to bubble-chamber results at $11 - 27$ GeV.

A further challenge comes from the comparison of the mid-rapidity K/π ratio in $p^{\pm}p$ and heavy-ion collisions in Fig. 4.35c [132]. For heavy-ion collisions, the K^-/π ratio increases steadily with increasing $\sqrt{s_{NN}}$, but at a factor ~ 1.5 higher than in $p^{\pm}p$ collisions. On the other hand, the K^+/π ratio increases sharply at low energies, goes through a maximum at $\sqrt{s_{NN}} \approx 10$ GeV and then decreases towards the K^-/π ratio at RHIC energies. This behavior is explained by the interplay between a net-baryon density decreasing with increasing $\sqrt{s_{NN}}$ an increasing $K\bar{K}$ pair production rate [138].

Particularly interesting is the ratio of $\sim$1.5 for heavy-ion over $p^{\pm}p$ collisions. Enhanced strangeness production would in fact be expected to be a probe for a deconfined state, e.g. by gluon-gluon fusion into strange quark-antiquark pairs [139]. However, also hadronic mechanisms [140] may enhance strangeness production and, after all, the enhancement is already observed as low as $\sim$3 GeV!

Since pions and kaons are produced via resonances, the K/π ratio does not directly measure the strangeness suppression. A systematic study of particle and resonance yields has been performed [141] with pp interactions at $\sqrt{s} = 52.5$ GeV. As can be seen in Fig. 4.36 the particle yield falls exponentially with particle mass, but separately for strange and non-strange mesons. The line connecting the strange mesons lies about a factor $1/\lambda \approx 3$ lower than that for the non-strange ones. An important exception is the ϕ meson, which is an $s\bar{s}$ quark state. This lies considerably *below* the strange-meson line, in agreement with a double strangeness suppression λ^2.

The strangeness suppression factor has been measured at several energies. The data have been collected [142] as a function of effective energy $\sqrt{\hat{s}_{\text{eff}}} = (s\langle x_1\rangle\langle x_2\rangle)^{1/2}$, with $\langle x_1\rangle$ and $\langle x_2\rangle$ being the average momentum fraction of the beam valence quark and target valence quark, respectively, and are shown in Fig. 4.37. The methods applied to obtain the values differ. In general, the K/π, K^*/ρ or ϕ/ρ ratio was used, but mass effects were not taken into account. Values lie between $\lambda = 0.12$ at the lower energies to $\lambda = 0.35 \pm 0.06$ from the simultaneous fit to Fig. 4.36. There is an indication of an s dependence at small s, but the data are compatible with a constant value of $\lambda = 0.29 \pm 0.02$ at the higher energies within Fig. 4.37. Including SPS [135] and Tevatron [137] results may give evidence for a further slight (one-

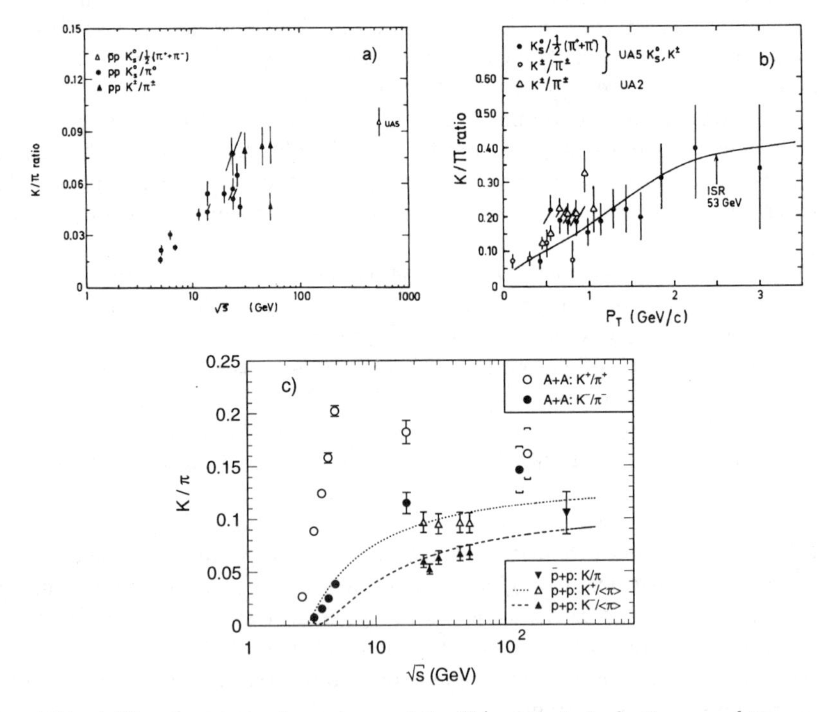

Figure 4.35. a) The cms energy dependence of the K/π ratio in inelastic pp and $\bar{p}p$ events. b) The transverse momentum dependence of the K/π ratio for UA5 and UA2, compared to that at ISR [131]. c) Mid-rapidity K/π ratios versus $\sqrt{s_{\text{NN}}}$ for heavy ion and $p^{\pm}p$ collisions [132]. The curves are parametrizations of pp data [133].

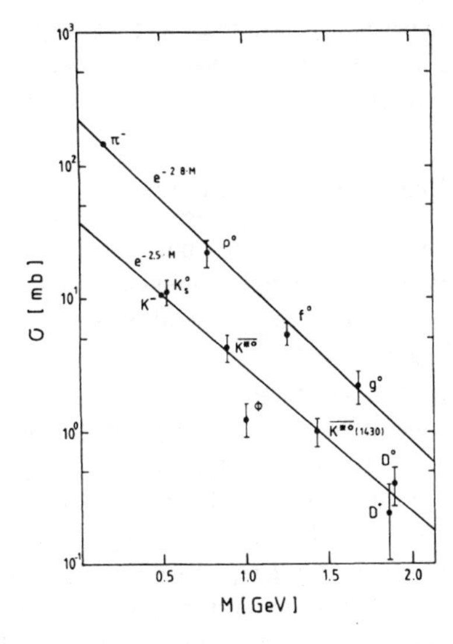

Figure 4.36. Resonance cross section from pp collisions at $\sqrt{s} = 52.5$ GeV as a function of the resonance mass [141].

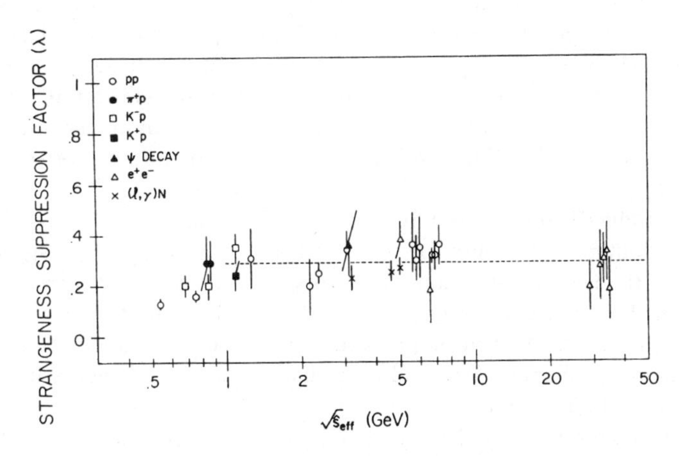

Figure 4.37. Strange quark suppression factor λ as a function of the effective energy [142].

sigma) increase [134, 143] at those energies, but errors are large and the extraction not model independent. Within the large uncertainties, compatibility is observed in λ for hadron-hadron, e^+e^- and lepton-hadron data in Fig. 4.37 and in more recent analysis [50, 78, 144, 145], with $\lambda = 0.295 \pm 0.006$ as the most accurate estimate [50].

The situation is further complicated by the observation [125, 146] that λ depends on the region of phase space studied. For example, as determined by the ratio $\sigma(\phi)/\sigma(K^*_{892})$ in NA22, λ decreases with increasing Feynman-x and drops to $\lambda \approx 0.1$ near $x = 1$ (not shown).

4.3.2.3 Feynman-x and rapidity dependence

Of particular importance in particle or resonance production is their Feynman-x dependence. The yield of particles and resonances differs strongly for different x regions, and the consequent x dependence depends strongly on the quantum numbers of the beam and the produced particle.

A good demonstration for the existence of a quantum number dependence of resonance production is the difference between positive and negative $\Sigma(1385)$ from K^-p collisions studied at 4.2, 10, 14.3 and 16 GeV/c beam momentum [147–149]. The $\Sigma^-(1385)$ is produced symmetrically with respect to $x = 0$, with vanishing cross section for x near unity (see Fig. 4.38 for 10 GeV/c). The $\Sigma^+(1385)$ has a large cross section for all $x < 0$, but is about equal to $\Sigma^-(1385)$ for $x > 0$. In the proton fragmentation region Σ^- production should be small, as it requires double charge exchange, whereas Σ^+ is allowed. The difference shown in Fig. 4.38 for 10 GeV/c then suggests a fragmentation component, the equal part a central component.

The rapidity distribution for $\rho^{\pm 0}$ (Fig. 4.39b) produced from pp at 24 GeV/c [150] is in qualitative agreement with that of $\Sigma^-(1385)$ in K^-p reactions and thus suggestive of being due to largely central production. Here, the total inclusive $\rho^{\pm}$ cross sections had to be estimated by an extrapolation from the cross sections for 1C fit channels. This gives rise to an approximate uncertainty of 20% in the cross sections. It is interesting to note that, within this uncertainty, the distributions for ρ^+, ρ^- and ρ^0 are the same.

For $\rho^{\pm 0}$ produced from π^+p at 16 GeV/c as shown in Fig. 4.39a, equality of the rapidity distribution holds only for $y < 0$. For $y > 0$, only ρ^- is approximately symmetric to the negative y region. The cross section for ρ^+ and ρ^0 stays large for all $y > 0$ and about twice as large for ρ^+ as for ρ^- in the beam fragmentation region. This is again in agreement with suppression of ρ^- in the π^+ fragmentation region due to double charge exchange. We shall come back to these differences in qualitative discussion in Chapters 5 and 6.

In Fig. 4.39c, the rapidity density $(1/\sigma_{\text{inel}})(\text{d}\sigma/\text{d}y)$ is compared for ρ^- produced from pp (circles) and π^+p (crosses). In the whole y region, the distributions are quite similar, in agreement with the expectation of dominant central production in both experiments. As can be deduced from Figs. 4.39a and b, this equality also holds for ρ^+ and ρ^0 production for $y < 0$ where central production is expected to dominate. As expected for a fragmentation component, for $y > 0$ the ρ^0 and ρ^+ production becomes significantly larger in the π^+p than in the pp reactions.

We conclude from this that, in spite of the failure of boost invariance observed in Sub-Sect. 4.3.1.1., there is good evidence for a two-component picture of inclusive particle and resonance production, already at rather low energy. The fragmentation component depends on the produced particle or resonance and on the fragmenting incoming particle. The shape of the central component is universal, i.e., does neither depend on the incoming particles nor on the produced particle or resonance.

Furthermore, it has been noted that only of the order of 10% of the pions are produced directly (the largest pion sources being $\rho^{\pm 0}$ and ω^0) and that strangeness is suppressed by a factor $1/\lambda \approx 3$.

An interesting approach to understand these observations is the extension of the additive quark model of total cross sections to multiparticle production. This approach will be discussed in Sect. 5.1.

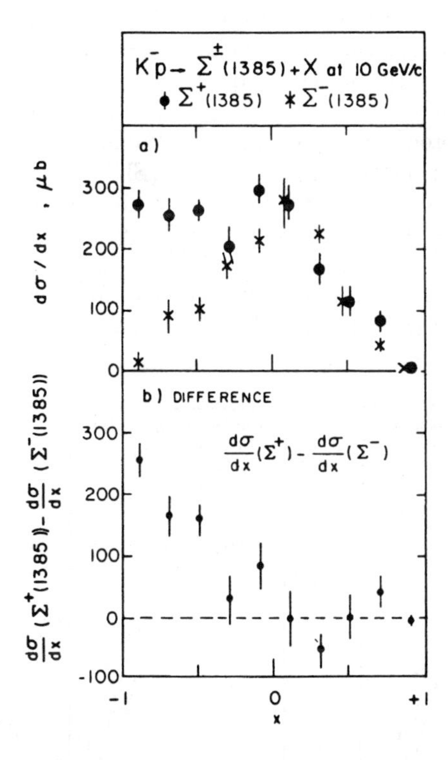

Figure 4.38. a) Differential cross section $d\sigma/dx$ for $\Sigma^+(1385)$ and $\Sigma^-(1385)$ inclusive production at 10 GeV/c. b) Difference between the $d\sigma/dx$ distribution for $\Sigma^+(1395)$ and that for $\Sigma^-(1385)$ [149].

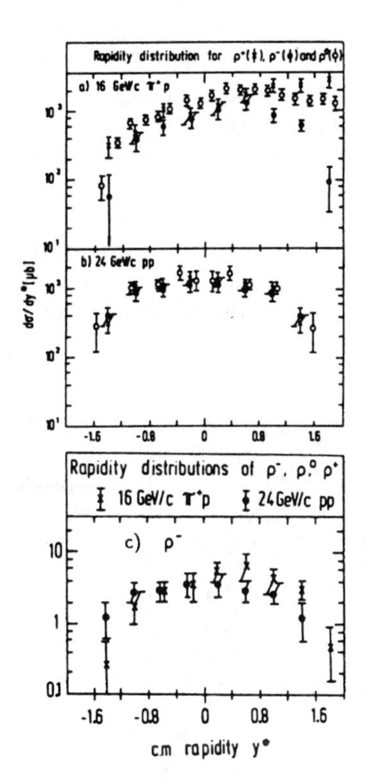

Figure 4.39. Differential cross section $d\sigma/dy$ for inclusive $\rho^{\pm 0}$ production as a function of y in a) π^+p reactions at 16 GeV/c, b) pp reactions at 24 GeV/c. c) Comparison of the inclusive ρ^- density in 16 GeV/c π^+p and 24 GeV/c pp interactions [150].

4.3.3 Reflection of the valence quark distribution

The antiquark distribution in the proton is concentrated at small Bjorken-$x_{\rm B}$ (say, $x_{\rm B} \leq 0.2$, the sea region) and the same is true for gluons which dissociate into a $q\bar{q}$. The presence of an $\bar{q}$ component in the proton structure function implies that the proton, which primarily consists of three quarks, is subject to fluctuations in which extra $q\bar{q}$ pairs are formed. According to the suggestion of Ochs [151], proton fragmentation in the collision with other hadrons may then be viewed as a rearrangement of the pre-existing partons preserving approximately their individual longitudinal momenta.

In the fragmentation region of the proton, the π^+ can be assumed to be composed of a u valence and a $\bar{\rm d}$ sea quark. Since the latter carries very little momentum, we expect to find a $\pi^+ = |{\rm u}\bar{\rm d}\rangle$ with momentum similar to that of the u quark. The same holds for a $\pi^- = |{\rm d}\bar{\rm u}\rangle$ and the d quark. As a consequence, the Feynman-x distribution of a pion in the fragmentation region of an incident proton is expected to be similar to the $x_{\rm B}$ distribution of the valence quark which it shares with the proton. Fig. 4.40a shows [152] that the x distribution of the π^+ produced in pp collisions at ISR is indeed similar to the proton u-quark distribution $u(x_{\rm B}) \equiv F_{\rm u}^{\rm p}(x_{\rm B})$ derived from SLAC data on electron-nucleon deep inelastic scattering. The π^- distribution (Fig. 4.40b) agrees with the proton d-quark distribution $d(x_{\rm B}) \equiv F_{\rm d}^{\rm p}(x_{\rm B})$ up to $x_{\rm B} \approx 0.7$, and is only slightly above $d(x_{\rm B})$ for larger $x_{\rm B}$ values.

Furthermore, the K$^+$ distribution agrees again with $u(x_{\rm B})$, as expected from the fact that it shares a u quark with the proton (Fig. 4.40c). The K$^-$ has no valence

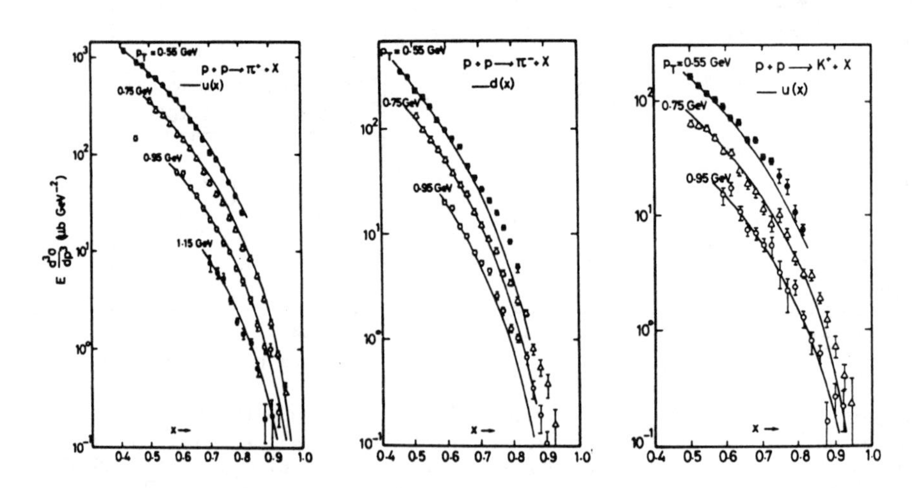

Figure 4.40. Comparison of the invariant π^+, π^- and K$^+$ cross section as a function of Feynman x from pp collisions at $\sqrt{s} = 45$ GeV to the u- and d-quark distribution functions $u(x_{\rm B})$ and $d(x_{\rm B})$, respectively [152].

quark in common with the target proton. Indeed, its x distribution (not shown) falls much more steeply with increasing x than either $u(x_B)$ or $d(x_B)$.

Of course, it is important to know that the observed particle has been produced on its own, and not from a resonance or cluster. Indeed, no short-range correlation is observed between π^- mesons at $|x| \geq 0.6$ and other particles, while K^- mesons at large $|x|$ are produced in close association with other particles [154].

We conclude that the quantum numbers and the momentum distribution of the target-proton valence quarks can be found back in the particles produced in the target fragmentation region.

What is the role of the beam quantum numbers? Fig. 4.41 gives a comparison [153,155] of the ratio of the inclusive invariant cross sections for π^+ and π^- production as a function of x in the target fragmentation region for the reactions $\pi^\pm p \to \pi^\pm X$ at 16 GeV/c to that for pp$\to \pi^\pm X$ (scaling between 19 and 2000 GeV/c). The ratio R is considerably larger for 16 GeV/c π^+p than for pp reactions. On the other hand, R falls below the proton curve for π^-p collisions.

At higher energies, however, the $\pi^\pm$p ratios tend to converge. The approach of R to a high-energy limit is given in Fig. 4.42 for $\pi^\pm$p collisions [156], as a function of $s^{-1/2}$, for several x intervals. The asymptotic limit is consistent with the ratio R for pp reactions. A similar conclusion can be drawn for $K^\pm$p collisions. This means that at $p_{\text{lab}} \geq 200$ GeV/c, the influence of the proton valence quarks alone is observed to govern proton fragmentation in soft hadronic collisions.

A particular significance may be attributed to the value of the ratio R for $x \to 1$.

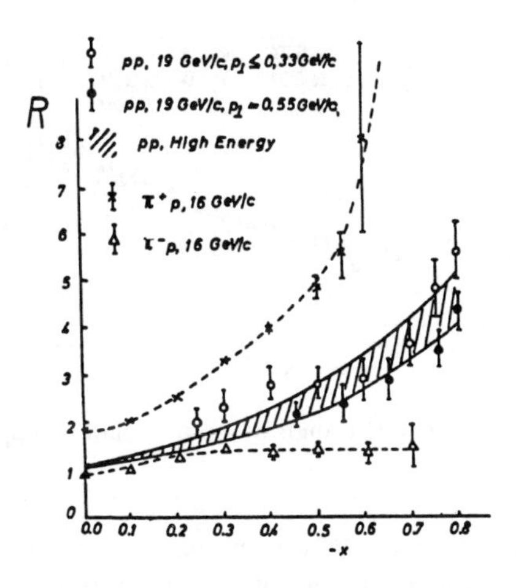

Figure 4.41. π^+/π^- ratio in the proton fragmentation region of 16 GeV/c π^+p collisions (crosses), 16 GeV/c π^-p collisions (triangles) and pp collisions at 10 GeV/c (shaded area and circles) [153].

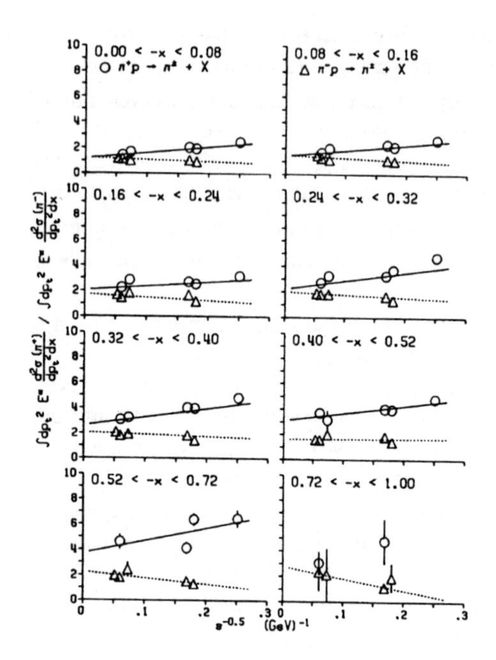

Figure 4.42. The π^+/π^- ratio in the proton fragmentation region for $\pi^\pm$p collisions, as a function of $s^{-1/2}$ for different x intervals [156].

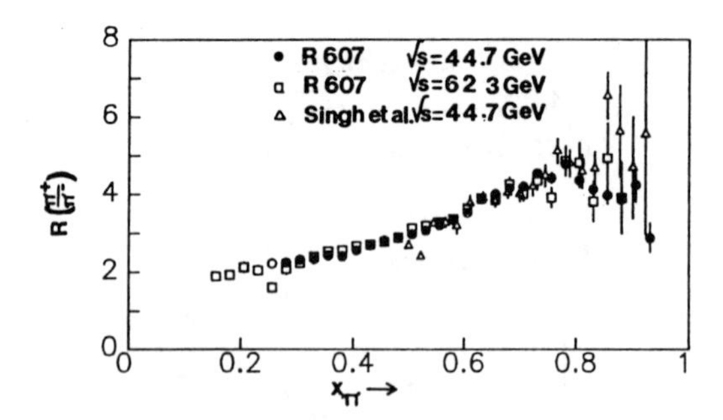

Figure 4.43. π^+/π^- ratio at fixed polar angle as a function of x in pp collisions at $\sqrt{s} = 44.7$ and 62.3 GeV [157].

In a model in which quarks interact by the exchange of vector gluons, the helicity of a fast quark ($x \approx 1$) is the same as the helicity of the proton. Therefore, by examining the wavefunction of the proton

$$|p(\uparrow)\rangle = 2|u(\uparrow)u(\uparrow)d(\downarrow)\rangle - |u(\uparrow)d(\uparrow)u(\downarrow)\rangle + \text{perm} \quad ,$$

one finds $u/d \to 5$ for $x \to 1$. The data [157] show (Fig. 4.43) that $R \approx 5$ is reached from below at $x \approx 0.8$, but that R then suddenly decreases again at highest x values, so that another regime seems to hold there.

The second basic observation of valence-quark reflection in proton fragmentation supports a quark recombination picture of hadron production in the fragmentation region. This mechanism will be discussed in Sect. 5.3.

4.3.4 Jet universality

4.3.4.1 The event shape

A third very important experimental observation is that the criteria usually accepted to define jets in e^+e^- or lepton-hadron collisions are also observed to hold for hadron-hadron collisions, even at low p_T. As an example, the decrease in average sphericity $\langle S \rangle$ with increasing $\sqrt{s}$ has been compared in hadron-proton, $\bar{\nu}$N and e^+e^- collisions in Sect. 2.4. As is shown in Fig. 2.23, at least for $\sqrt{s} \lesssim 20$ GeV the average shape of the hadronic system is the same in all three types of collisions at given hadronic energy, as is its quite dramatic change with energy. Similar conclusions have been drawn in Sect. 2.4 from the shape of the sphericity distribution itself (Fig. 2.24).

4.3.4.2 The p_T development

At low $\sqrt{s}$, even the p_T development of e^+e^- and $\nu(\bar{\nu})$N multiparticle production with energy can be understood from transverse-cut phase space. Figs. 4.44a and b show [158] the $\sqrt{s}$ dependence of $\langle p_T \rangle$ for e^+e^- collisions and the W dependence of $\langle p_T^2 \rangle$ for current fragments of $\nu(\bar{\nu})$N collisions. The curves are from Monte Carlo calculations with a matrix element

$$M \propto \prod_{i=1}^{n} \exp(-Ap_{Ti}) \tag{4.14}$$

with $A = 3.5$ and with a Gaussian multiplicity distribution centred at

$$\langle n \rangle = 3.69 + 0.06 \exp[1.92(\ln 4s)^{1/2}]. \tag{4.15}$$

Up to about 15 GeV there is little need for anything else.

A variable particularly sensitive to deviations from longitudinal phase space is the p_T component $p_{T,\text{in}}$ in the direction corresponding to the second largest eigenvalue of the momentum tensor describing the event shape (see Sect. 2.4). As shown in Fig. 4.45b, the $\langle p_T^2 \rangle_{\text{in}}$ distribution starts to deviate from Field and Feynman [82] and LPS curves in its tail in K^+p collisions at 70 GeV/c [159]. A strong increase of this tail is known from e^+e^- data up to 35 GeV. A comparison of K^-p data from $s = 20.3$ to 205 GeV^2 [160] and μp data [161,162] is shown in Fig. 4.45d. The μ data indeed have a long tail at $100 \leq Q^2 \leq 460$ GeV^2 ($\langle W^2 \rangle = 310$ GeV^2). This tail is reasonably

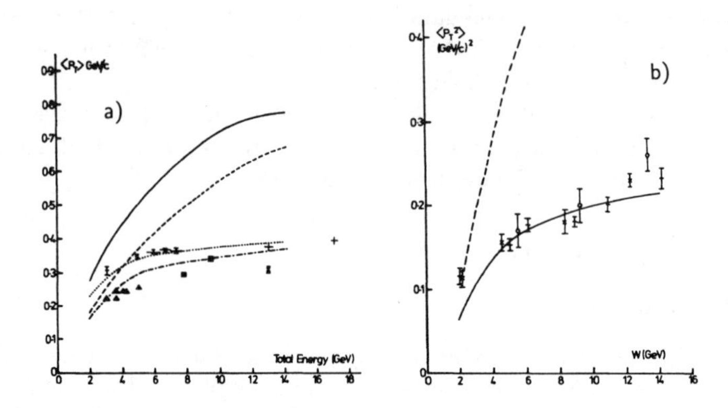

Figure 4.44. a) $\langle p_{\mathrm{T}} \rangle$ with respect to sphericity or thrust axis ($\blacktriangle, \blacksquare, +, \times$), compared to phase space with $A = 0$ (dashed) and $A = 3.5$ (dot-dashed), respectively, corrected to the true axis ($\times$, full and dotted lines), for $\mathrm{e}^+\mathrm{e}^-$ annihilation. b) $\langle p_{\mathrm{T}}^2 \rangle$ in the current fragmentation region of $\nu(\bar{\nu})\mathrm{N}$ collisions compared to phase space, with $A = 0$ (dashed) and $A = 3.5$ (full) [158].

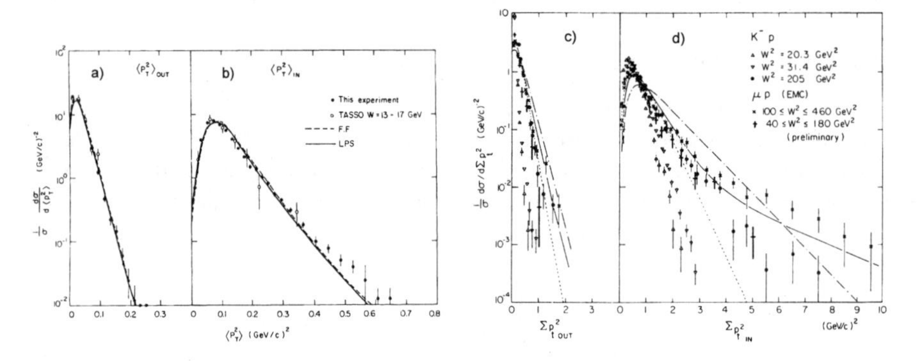

Figure 4.45. a) $\langle p_{\mathrm{T}}^2 \rangle_{\mathrm{out}}$ distribution, b) $\langle p_{\mathrm{T}}^2 \rangle_{\mathrm{in}}$ distribution for $\mathrm{K}^+\mathrm{p}$ collisions at 70 GeV/c ($\sqrt{s} = 11.5$ GeV) ($\bullet$) [159] compared to $\mathrm{e}^+\mathrm{e}^-$ collisions at $\sqrt{s} = 13 - 17$ GeV ($\circ$). The lines correspond to FF (dashed) and LPS (full). c) $\Sigma p_{\mathrm{T,out}}^2$ and d) $\Sigma p_{\mathrm{T,in}}^2$ for $\mathrm{K}^-\mathrm{p}$ collisions and $\mu\mathrm{p}$ data at the W^2 values indicated below [160–162]. The full line is a three-jet Lund fragmentation model, the other lines correspond to a two-jet fragmentation model with two different types of exponential p_{T} fall-off.

well reproduced by a three-jet Lund model (see Chapter 6), but not by the same model without a third (glue) jet. Also the $\mathrm{K}^-\mathrm{p}$ data show a definite increase with rising energy and an indication of a tail at $s = 205$ GeV^2. Within the large errors, the $\mathrm{K}^-\mathrm{p}$ data indeed agree with the $\mu\mathrm{p}$ data [162] at the corresponding hadronic energy. We find it important that this tail can be observed also in hadron-hadron

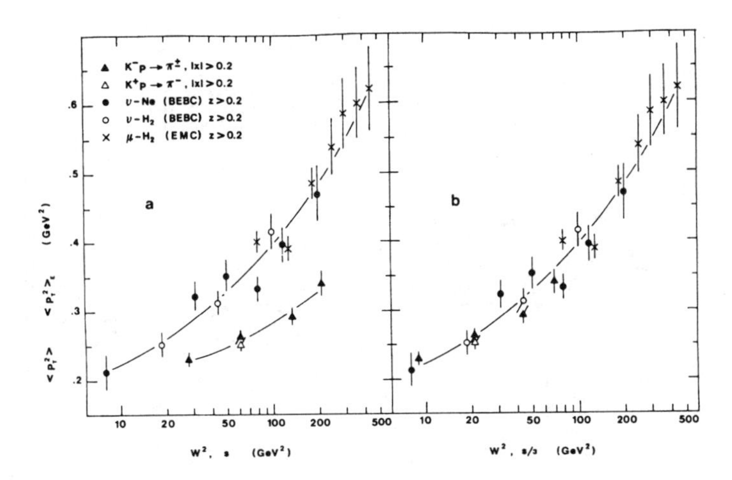

Figure 4.46. $\langle p_{\mathrm{T}}^2 \rangle$ dependence of hadron-hadron collisions compared to that of lepton-hadron collisions in terms of a) W^2 and s and b) W^2 and $s/3$ [163].

collisions, the fact that it may not grow as fast there as in e^+e^- and νN collisions would not be surprising. Since more quarks share the initial momentum in hadronic particle production, according to Fig. 4.44 p_{T} distributions have to be compared at corresponding fractions of energy and not at the total cms energy. Furthermore, the reason for a widening of the $p_{\mathrm{T,in}}$ distribution in hadron-hadron collisions will be due to large-angle quark scattering rather than to gluon emission.

Fig. 4.46 shows $\langle p_{\mathrm{T}}^2 \rangle$ for lepton-hadron and hadron-hadron collisions as a function of $W^2 = s$ and $W^2 = s/3$, respectively. Clearly, the hadron data follow the lepton data when plotted as a function of $W^2 = s/3$. The need for such a shift in energy is not so obvious for $\langle n \rangle$, $\langle p_{\|} \rangle$ or sphericity distributions, since the different sources add in the longitudinal direction.

4.3.4.3 The sea gull

A distribution particularly sensitive to the onset of hard effects in lepton-hadron and e^+e^- collisions has turned out to be the energy dependence of the average transverse momentum of particles produced around Feynman-$|x| = 0.4$.

The dependence of the average transverse momentum on Feynman-x has first been observed in hadron-hadron collisions at lower energies [164]. It has a characteristic shape resembling a sea gull with its head lowered at $x = 0$ and its wings raised around $|x| \approx 0.4$. This "sea-gull effect" is also visible in e^+e^- [165] and lh [166, 167] collisions and qualitative similarities between all three types of collisions (hh, lh and e^+e^-) at comparable energies have been observed [167, 168].

In e^+e^- annihilation, a dramatic rise with cms energy [165] has set in for one

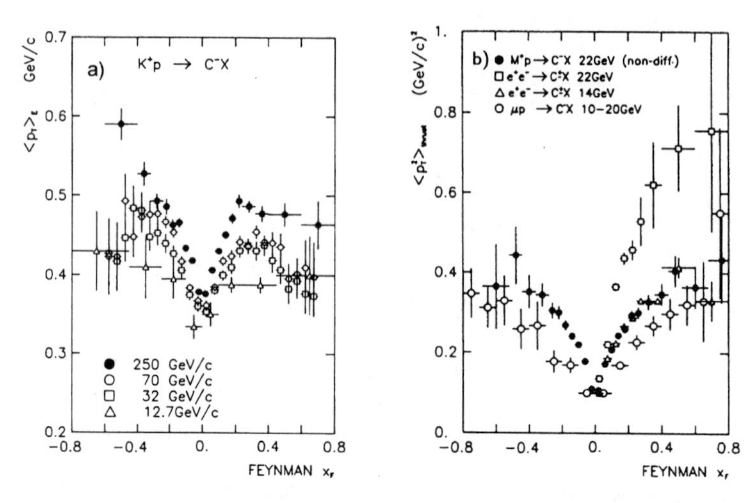

Figure 4.47. a) The energy weighted average transverse momentum $\langle p_{\rm T}\rangle_E$ as a function of Feynman-x for $\rm K^+p \to C^- + X$ between 12.7 and 250 GeV/c incident momentum. b) The average squared transverse momentum $\langle p_{\rm T}^2\rangle_{\rm thrust}$ with respect to the thrust axis for the combined non-single-diffractive $\rm K^+p$ and $\pi^+\rm p$ sample (indicated as $\rm M^+p$) with multiplicity $n \geq 4$ at $\sqrt{s} = 22$ GeV, compared to that for $\rm e^+e^-$ collisions at 14 and 22 GeV, and μp collisions with $10 < W < 20$ GeV [171].

of the wings, as a consequence of the onset of emission of a hard gluon by one of the two leading quarks. This rise is satisfactorily reproduced by a QCD model of independent quark fragmentation [169] and by a string model [170] when hard processes are included. For $\rm e^+e^-$ annihilation, these processes become significant at an energy of about 10 GeV and lead to a rise of $\langle p_{\rm T}^2\rangle$ by a factor of two from 14 to 22 GeV.

Neutrino experiments [166] have shown that already at hadronic masses $W < 10$ GeV, the sea gull is lifting its current-fragmentation wing with increasing W. The EMC collaboration [167] has increased the W range up to 20 GeV and shown that in terms of Lund fragmentation, this effect can be reproduced only if gluon radiation is included.

The point is, that a rise of the sea-gull wings is also observed in hadron-hadron collisions at comparable energy. As in lepton-hadron collisions, the rise may have set in at lower energies [172,173], but is clearly visible in $\rm K^+p$ collisions from 12.7 to 250 GeV/c ($\sqrt{s} \approx 5 - 22$ GeV) in Fig. 4.47a [171].

In Fig. 4.47b, the combined non-single-diffractive $\rm K^+p$ and $\pi^+\rm p$ data are compared to $\rm e^+e^-$ results at $\sqrt{s} = 22$ and 14 GeV and to μp results at $10 < W < 20$ GeV in terms of $\langle p_T^2\rangle_{\rm thrust}$, the average of the square of the particle $p_{\rm T}$ with respect to the thrust axis. In the hh data, the wings of the sea-gull distribution are significantly lower than the (folded) wings from e^+e^- at the same energy, but higher than those from μp collisions with hadronic energy $10 < W < 20$ GeV. The meson fragmentation wing at 22 GeV is consistent with the folded e^+e^- wings at 14 GeV.

From Figs. 4.47a and b we, therefore, conclude that a rise of the sea-gull wings with cms energy is also observed for hh collisions, but the rise is less dramatic than in e^+e^- annihilation. Part of this difference can be explained by heavy quark fragmentation contributing in e^+e^-, but not in hh collisions. Furthermore, the hadronic energy ($\sqrt{s}$ or W) has to be shared by more quarks in lh and hh collisions than in e^+e^- collisions.

4.3.4.4 Charge and energy flow

Being decay products of numerous resonances, a large fraction of the stable hadrons in multiparticle final states are not produced promptly. Therefore, it has been suggested by Ochs and Stodolsky [174] to consider the average flow of energy and of additive quantum numbers (e.g. charge) into a given angular region $\Delta\Omega$ with respect to the direction of the hadronizing parton in a parton jet, or with respect to the incident beam direction in hadron-hadron collisions.

The first data on energy, charge, strangeness and baryon-number flow have been obtained for proton jets in pp collisions at 12 and 24 GeV/c beam momentum [175]. In contrast to the single-particle spectra near cms rapidity $y = 0$, these data show no evidence for scale breaking and thus support the conjecture that cluster-invariant jet measures are of fundamental importance. Moreover, it has been noted that the energy- and charge-flow distributions are nearly equal, implying that the fraction of jet charge equals the fraction of jet energy in any given angular region $\Delta\Omega$. For a single quark jet, Andersson and Gustafson subsequently have proven [176] that the latter property is a natural consequence of the recursive nature of quark cascades. Ochs and Shimada [177] have extended these ideas to the case of meson and proton jets. These authors point out that, particularly for meson jets, the ratio of charge to energy flow in a given angular region is sensitive to the momentum distribution of the valence quarks and could thus help to discriminate between various models.

Results for meson beams have first been obtained in K^+p and π^+p interactions at 32 and 70 GeV/c [178]. These experiments confirm the absence of energy dependence of angular charge and energy flow in the incident momentum range studied. A similar analysis of π^+p, K^+p and pp interactions at 147 GeV/c [179] shows that the charge and energy flows are beam independent and scale between 32 and 147 GeV/c.

High-precision results on energy flow in μp collisions have been presented by the EMC collaboration for the hadronic energy range $4 < W < 20$ GeV [180]. Interestingly, the data exhibit sizeable scale breaking in $dE/d\lambda$, which is another manifestation of the p_T-broadening of the forward (quark) jet. This was shown to be due to hard QCD contributions and multiple soft gluon radiation and is now used with considerable success at HERA [181, 182] to discriminate between QCD models, both in diffractive and non-diffractive DIS.

To describe the flow into an angular region, the authors of [174] define the variable

$$\lambda = \frac{\cot\theta}{E_{\text{jet}}}, \tag{4.16}$$

where θ is the polar angle with respect to the jet axis and E_{jet} the jet energy. For the case of exact Feynman scaling of the single-particle inclusive spectrum for particles of type k

$$E\frac{\mathrm{d}\sigma_k}{\mathrm{d}^3p} = f_k(x, p_{\mathrm{T}}), \tag{4.17}$$

it is shown that the fraction of jet energy $\mathrm{d}\epsilon = \mathrm{d}E/E_{\text{jet}}$ radiated into the angular interval $\mathrm{d}\lambda$ obeys the scaling law

$$\frac{\mathrm{d}\epsilon}{\mathrm{d}\lambda} = \rho(\lambda) \tag{4.18}$$

with

$$\rho(\lambda) = \frac{2\pi}{\sigma_{\text{inel}}}\int \mathrm{d}p_{\mathrm{T}}p_{\mathrm{T}}^2\sum_k f_k(\lambda p_{\mathrm{T}}, p_{\mathrm{T}}). \tag{4.19}$$

In terms of pseudo-rapidity $\eta = -\ln\tan(\theta/2)$, one has

$$\frac{\mathrm{d}\epsilon}{\mathrm{d}\lambda} \approx \frac{\mathrm{d}m_{\mathrm{T}}}{\mathrm{d}\eta} = \sum_k m_{\mathrm{T}}^k(\eta)\frac{\mathrm{d}n_k}{\mathrm{d}\eta}. \tag{4.20}$$

Consequently, $\mathrm{d}\epsilon/\mathrm{d}\lambda$ is equal to the pseudo-rapidity density $\mathrm{d}n_k/\mathrm{d}\eta$ of particles, weighted by their transverse mass $m_{\mathrm{T}} = \sqrt{m^2 + p_{\mathrm{T}}^2}$. From (4.19) and (4.20) one deduces the limiting behavior

$$\lim_{\lambda\to\infty}\frac{\mathrm{d}\epsilon}{\mathrm{d}\lambda} = 0, \qquad \lim_{\lambda\to 0}\frac{\mathrm{d}\epsilon}{d\lambda} = M. \tag{4.21}$$

The mass parameter M is interpreted as the average transverse mass in the central rapidity plateau at $\lambda = 0$ (or rapidity $y = 0$).

Although Feynman scaling is strongly violated at small x, it turns out that the quantity $\mathrm{d}\epsilon/\mathrm{d}\lambda$ has much better scaling properties. Moreover, $\rho(\lambda)$ does not change, in first approximation, if particles are replaced by clusters.

A simple parametrization of $\rho(\lambda)$ was proposed in [175, 183] and deduced from the observation that a fast moving particle appears as a Lorentz-contracted radiation pulse. A generalized Weizsäcker-Williams approach then leads to the suggestion that the energy distribution is a function of one variable only:

$$E\frac{\mathrm{d}\sigma}{\mathrm{d}^3p} = f(Q), \tag{4.22}$$

with $Q^2 = (Mx)^2 + p_\mathrm{T}^2$ and M to be interpreted as an effective hadron mass. For steeply falling $f(Q)$, one finds for $\rho(\lambda)$, the so-called radiation profile [175],

$$\frac{\mathrm{d}\epsilon}{\mathrm{d}\lambda} = \frac{M}{[1 + (M\lambda)^2]^{3/2}} \,. \tag{4.23}$$

The previous arguments can easily be extended to the flow of additive quantum numbers such as charge.

For hadron-hadron collisions, we assign particles with positive Feynman-x to the beam jet, and those with negative Feynman-x to the target jet and use the approximation

$$\lambda = \frac{|x|}{p_\mathrm{T}} \,, \tag{4.24}$$

where the Feynman variable x and the transverse momentum p_T are defined with respect to the beam axis.

In the centre of mass frame of the reaction, the energy and charge flows presented below, are then defined as:

$$\frac{\mathrm{d}Q}{\mathrm{d}\lambda} = N_Q \sum_k \left(\frac{\mathrm{d}n_k^+}{\mathrm{d}\lambda} - \frac{\mathrm{d}n_k^-}{\mathrm{d}\lambda} \right) \,, \tag{4.25}$$

$$\frac{\mathrm{d}E}{\mathrm{d}\lambda} = N_E \sum_k E_k \frac{\mathrm{d}n_k}{\mathrm{d}\lambda} \,, \tag{4.26}$$

where $\mathrm{d}n_k$, $\mathrm{d}n_k^\pm$ are, respectively, the number of charged, positive and negative particles of type k per angular interval $\mathrm{d}\lambda$, and E_k is their energy. N_Q and N_E are normalization factors.

The charge- and energy-flow spectra for π^+p, K$^+$p and pp collisions at 250 GeV/c [184] are displayed in Fig. 4.48a-c and compared to K$^+$p and π^+p data 32 GeV/c [178].

At small λ, the data show a significant increase with $\sqrt{s}$ of $\mathrm{d}Q/\mathrm{d}\lambda$ and $\mathrm{d}E/\mathrm{d}\lambda$. This is compensated by a decrease for $\lambda > 1$. The $\sqrt{s}$-dependence is more clearly apparent in Fig. 4.49b, where the ratios of the 32 and 250 GeV/c K$^+$p data are plotted on a linear scale, both for $\mathrm{d}Q/\mathrm{d}\lambda$ and for $\mathrm{d}E/\mathrm{d}\lambda$. It results in an increase of the parameter M used in a fit by (4.23) between 32 and 250 GeV/c. The scaling violation reflects a gradual p_T-broadening of the meson jets, directly seen in the rise with energy of the sea-gull wing discussed in Sub-Sect. 4.3.4.3 above.

We conclude that the data are consistent with a widening in p_T, at increasing energy, of the beam jet in hadron-hadron collisions; the widening is similar for the energy and the charge flow.

Fig. 4.48a shows a comparison of the energy flow in the current fragmentation region of μp collisions [180] with the π^+p data at 250 GeV/c. The shape of the distributions is quite similar. In the muon-data, however, much larger scaling violations are observed than in hadron-hadron interactions and attributed to hard and soft gluon bremsstrahlung.

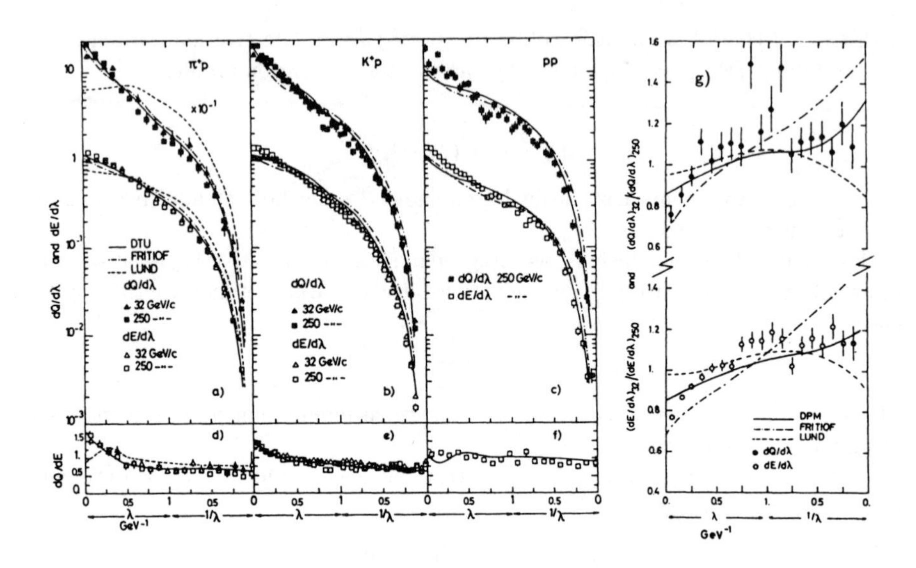

Figure 4.48. a)-f) The charge and energy flow and their ratio for π^+p, K$^+$p and pp interactions at 250 GeV/c, compared to lower energy data. Note the break in the λ-scale. Solid, dashed and dashed-dotted curves are DPM, Lund and FRITIOF predictions. g) The ratios of K$^+$p results at 32 and 250 GeV/c [184].

In Fig. 4.49b the mass parameter M is plotted as a function of the hadronic energy W [180]. The μp data exhibit a strong W-dependence. The curves are μp Lund predictions with and without QCD effects (Λ is the standard QCD mass scale). The M-values derived from the energy-flow distributions in K$^+$p interactions are plotted versus $Q = \sqrt{s}$ as open circles. The weak dependence of M, not inconsistent with the μp "No QCD" curve, might lead to the conclusion that perturbative QCD effects are absent in soft hadron-hadron collisions. However, taking e.g. the DPM or FRITIOF as a guide, it is nevident that the relevant energy scale for comparisons with deep-inelastic processes is not $\sqrt{s}$, but some fraction only of the total cms energy available to each chain.

On more general grounds, the form

$$\sqrt{s} \approx \alpha_s(Q^2)Q^2/\Delta, \tag{4.27}$$

has been suggested [185] as the relation between the cms energy $\sqrt{s}$ in hadron-hadron collisions and Q in hard processes at which similar structure of the final state should be observed. In (4.27), $\alpha_s(Q^2)$ is the running QCD coupling constant and Δ a supposedly energy-independent parameter of the order of 1 GeV. Using this rescaling form with $\Delta = 0.6$ and $\Lambda = 0.4$ GeV as an example, one obtains the data points shown by solid squares in Fig. 4.49b. Although Δ is apriori unknown and the

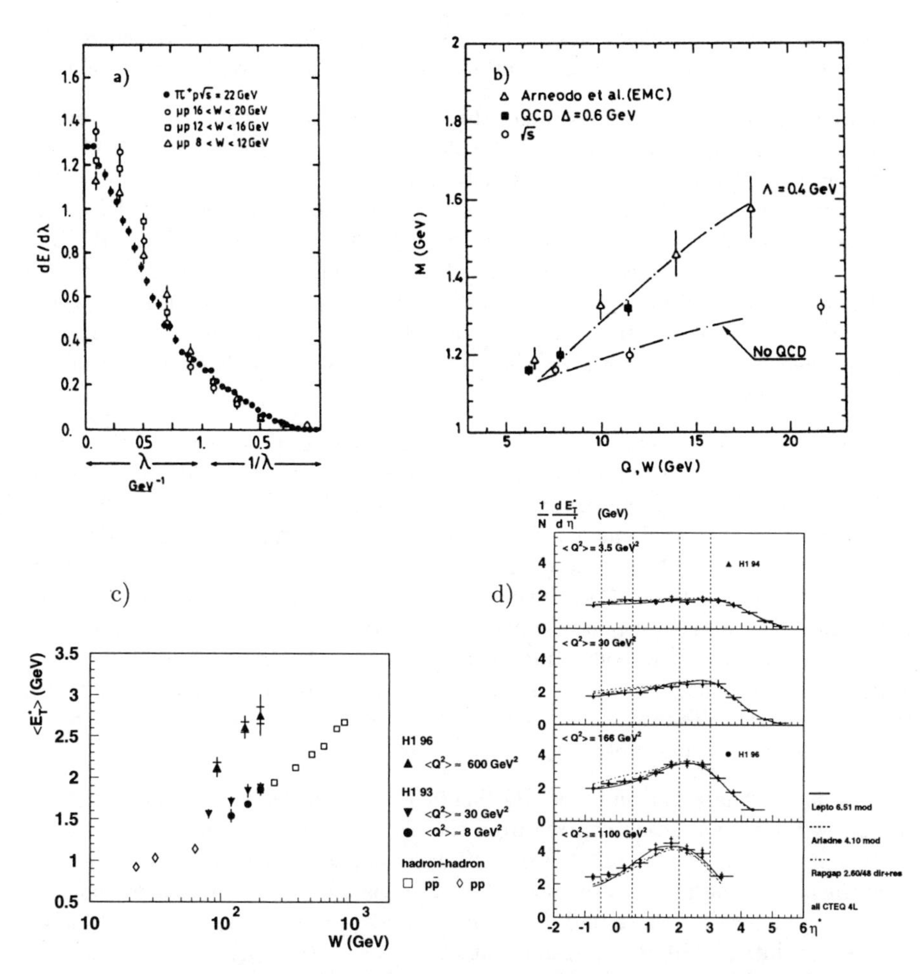

Figure 4.49. a) The energy flow for π^+p interactions at 250 GeV/c [184], compared to μp data [180]. b) Dependence of the mass parameter M on hadronic energy W in μp collisions [180] (triangles); the same for K$^+$p energy flow versus $\sqrt{s}$ (circles) and versus Q (solid squares) [184]. c) Variation of the average E_T^* for $-0.5 < \eta^* < 0.5$ for H1 results at three values of Q^2 (full symbols), compared to hadron-hadron results (open symbols) [181]. d) Transverse energy flow as a function of η^* for fixed $\langle W\rangle \approx 185$ GeV, compared to LEPTO, ARIADNE and RAPGAP [181].

above rescaling procedure therefore rather arbitrary, the example tends to suggest universality of jet phenomena in the sense of [185]. Analoguous ideas along the same line of thought have been advanced by Van Hove and Giovannini [186]. The widening of the meson jet with available fraction of $\sqrt{s}$ is comparable to that of the currect jet in μp collisions, if a reasonable assumption is made on the relevant energy scale.

Extended to much higher W values, Fig. 4.44c compares the average transverse energy $E^*_{i\mathrm{T}} = E^*_i \sin\theta^*_i$ with respect to the photon-proton direction at central-plateau pseudo-rapidity (related to M) as a function of W for H1 [181] at $66 < W < 233$ GeV to that for hadron-hadron data [95,184] at lower and higher values of $W(= \sqrt{s})$. The $\langle E^*_{\mathrm{T}} \rangle$ values increase both with increasing Q^2, and W. At low Q^2, the H1 data interpolate beautifully between the hadron-hadron results. This is consistent with a fluctuation of the virtual photon into a hadronic object interacting with the proton [187] and with the observation [184] that particle production in the central region is largely independent of the nature of the colliding hadrons and only dependent on the total cms energy.

The Q^2 evolution at fixed $W \approx 185$ GeV of the transverse energy flow distribution is shown for H1 in Fig. 4.44d. At low Q^2, i.e. in the photoproduction domain, there is a plateau-like behavior in the current hemisphere of the hadronic system ($\eta^* > 0$). At larger Q^2, i.e. in the DIS domain, a peak develops in the photon fragmentation region, which moves towards $\eta^* = 0$ and obtains a value at maximum about twice that for the proton remnants (at $\eta^* < 0$). The peak position coincides with the origin of the Breit frame [188] and corresponds to partonic structure of the photon and an increase with Q^2 of the transverse momentum of the photon constituent partons [189]. For a diffractive sample (not shown), the shape of the energy flow is found to be similar in the current-region peak, whereas a strongly reduced amount of transverse energy is found in the central and proton regions [181,182].

The simple quark parton model (QPM) in which a quark is scattered out of the proton by the virtual boson emitted from the scattering lepton produces far too low a transverse energy flow for all η^* [181]. QCD allows gluons to be radiated before and after the boson-quark vertex, the boson may fuse via a quark line with a gluon inside the proton or may fluctuate into a hadronic system. At large $x_{\mathrm{B}} \gtrsim 10^{-2}$, standard DGLAP evolution [190] is successful, but at low $x_{\mathrm{B}} \lesssim 10^{-3}$, BFKL effects [191] or their bridging CCFM [192] evolution may be expected from large parton densities. Indeed, early models merely based on DGLAP produced too little transverse energy near the proton remnant for low x_{B}.

At low x_{B} and Q^2, DGLAP and CCFM evolution based approaches implemented within the LEPTO and LDCMC Monte Carlo models (see Chapter 6), respectively, are found [181] not to fully describe the transverse energy flow in the central region, while the DGLAP based RAPGAP including resolved photon processes, and the BFKL related Color Dipole Model as implemented in ARIADNE give a reasonable description. At highest Q^2 the agreement is in general better, except for DGLAP related HERWIG, which has problems with the shape of the distribution for all x_{B} and Q^2.

4.4 Conclusions

In e^+e^- annihilation, hadronic events originate from two primary quarks with opposite quantum numbers. These primary quarks radiate hard as well as (coherent) soft gluons in the form of a parton shower governed by the running coupling α_s. Both hard gluons and primary quarks fragment into hadrons in the form of particle jets, with their quantum numbers preferentially carried by the leading hadron but local quantum number compensation among the particles within the jet. As expected from the SU(3) color charge assignment, gluon jets have higher particle multiplicity than quark jets, a softer fragmentation function with stronger scaling violations, but a harder transverse momentum profile. Gluon jets, furthermore, contain an excess of baryon production and neutral leading systems. Resonances are copiously produced in fragmentation. Surprisingly, production rates of particles and resonances as well as their transverse mass spectra follow a thermal distribution (not necessarily due to a thermal equilibrium!) with a "temperature" not far above the pion mass.

Except for deviations understood from higher-order processes present in DIS but not in e^+e^- collisions, current quark fragmentation in DIS remarkably resembles quark fragmentation in e^+e^- collisions, thus confirming jet universality, while proton remnant fragmentation rather carries the characteristics of transverse-momentum limited (cylindrical) phase space.

For hadron-hadron collisions, meson ($\pi^\pm$,$K^\pm$)-proton collisions have the advantage of being a simple $q_1\bar{q}_2$ system and representing a large flavor variety. Quite surprisingly, this quark flavor is observed to play an essential role in these (soft!) collisions and its fragmentation is found to be similar to the fragmentation of the corresponding quark in e^+e^- collisions and DIS. Furthermore, the quantum numbers and momentum distribution of the target-proton valence quarks can be found back in the particles produced in the target fragmentation region. On the other hand, hadron production in the central region (near zero rapidity) is independent of the quantum numbers of beam or target.

The disadvantage is that meson beams only exist up to 250 GeV lab momentum ($\sqrt{s}$ = 22 GeV), so that they have to be complemented by $p^\pm p$ collisions at higher energies, to study important features, such as scaling violation in the central region or p_T evolution with increasing energy and particle multiplicity.

The fact that jets produced along the beam direction in ordinary hadron-hadron collisions are similar to those produced in e^+e^- annihilation and deep inelastic leptoproduction has led to the assumption of parton fragmentation as a common underlying dynamical mechanism. This fragmentation process will be discussed in Chapter 6.

Bibliography

[1] MARKI Coll., J.L. Siegrist et al., *Phys. Rev.* **D26** (1982) 969.

[2] MARKII Coll., J.F. Patrik et al., *Phys. Rev. Lett.* **49** (1982) 1232; G. Wormser et al., *Phys. Rev. Lett.* **61** (1988) 1057; A. Peterson et al., *Phys. Rev.* **D37** (1988) 1; G.S. Abrams et al., *Phys. Rev. Lett.* **64** (1990) 1334.

[3] JADE Coll., W. Bartel et al., *Z. Phys.* **C20** (1983) 187; **C28** (1985) 343; D.D. Pitzl et al., *Z. Phys.* **C46** (1990) 1.

[4] TASSO Coll., M. Althoff et al., *Z. Phys.* **C17** (1983) 5; *Z. Phys.* **C22** (1984) 307; W. Braunschweig et al., *Z. Phys* **C33** (1986) 13; **C42** (1989) 189; **C46** (1990) 1; **C47** (1990) 167; ibid. 187.

[5] HRS Coll., D. Bender et al., *Phys. Rev.* **D31** (1985) 1; S. Abachi et al., *Phys. Rev.* **D40** (1989) 706.

[6] TPC Coll., H. Aihara et al., *Phys. Rev. Lett.* **52** (1984) 577; **61** (1988) 1263.

[7] AMY Coll., Y.K. Li et al., *Phys. Rev.* **D41** (1990) 2675.

[8] ARGUS Coll., H. Albrecht et al., *Z. Phys.* **C46** (1990) 15.

[9] CRYSTAL BALL Coll., C. Bieler et al., *Z. Phys.* **C49** (1990) 225.

[10] CELLO Coll., H.-J. Behrend et al., *Z. Phys.* **C47** (1990) 1.

[11] OPAL Coll., M.Z. Akrawy et al., *Phys. Lett.* **247** (1990) 617; R. Akers et al., *Z. Phys.* **C63** (1994) 181; **C67** (1995) 389; **C68** (1995) 1; G. Alexander et al., *Phys. Lett.* **B264** (1991) 467; *Z. Phys.* **C72** (1996) 191; **C73** (1997) 569; P.D. Acton et al., *Phys. Lett.* **B291** (1992) 503; **B305** (1992) 407; *Z. Phys.* **C56** (1992) 521; **C63** (1994) 181; K. Ackerstaff et al., *Z. Phys.* **C75** (1997) 193; *Eur. Phys. J.* **C5** (1998) 411; **C7** (1999) 369; G. Abbiendi et al, *Eur. Phys. J.* **C16** (2000) 185; *Eur. Phys. J.* **C27** (2003) 467; *Eur. Phys. J.* **C29** (2003) 463.

[12] L3 Coll., B. Adeva et al., *Phys. Lett.* **B259** (1991) 199; O. Adriani et al., *Phys. Lett.* **B286** (1992) 403; M. Acciarri et al., *Phys. Lett.* **B328** (1994) 223; **B444** (1998) 569; P. Achard et al., *Phys. Reports* **399** (2004) 71.

[13] ALEPH Coll., D. Buskulic et al., *Phys. Lett.* **B292** (1992) 210; *Z. Phys.* **C55** (1992) 209; **C64** (1994) 361; **C66** (1995) 355; **C69** (1996) 379; **C73** (1997) 409; **C74** (1997) 451; A. Heister et al., *Phys. Lett.* **B528** (2002) 19; *Eur. Phys. J.* **C35** (2004) 457.

[14] DELPHI Coll., P. Abreu et al., *Phys. Lett.* **B275** (1992) 231; **B298** (1993) 236; **B311** (1993) 408; **B318** (1993) 249; **B347** (1995) 447; *Z. Phys.* **C65** (1995) 587; *Nucl. Phys.* **B444** (1995) 3; *Z. Phys.* **C73** (1997) 229; *Eur. Phys. J.* **C5** (1998) 585; **C6** (1999) 19; **C18** (2000) 203; W. Adam et al., *Z. Phys.* **C69** (1996) 561.

[15] SLD Coll., K. Abe et al., *Phys. Rev. Lett.* **78** (1997) 3442; *Phys. Rev.* **D59** (1999) 52001; *Phys. Rev.* **D69** (2004) 072003.

[16] TOPAZ Coll., R. Itoh et al., *Phys. Lett.* **B345** (1995) 335.

[17] G.D. Lafferty, P.I. Reeves and M.R. Whalley, *J. Phys.* **G21** (1995) A1; I.G. Knowles and G.D. Lafferty, *J. Phys.* **G23** (1999) 731.

[18] G. Altarelli, *Phys. Reports* **81** (1982) 1; B. Mele and P. Nason, *Nucl. Phys.* **B361** (1991) 626; P. Nason and B.R. Webber, *Nucl. Phys.* **B421** (1994) 473.

[19] OPAL Coll., R. Akers et al., *Z. Phys.* **C68** (1995) 203.

[20] P. Nason and B.R. Webber, *Phys. Lett.* **B332** (1994) 405.

[21] OPAL Coll., G. Abbiendi et al., *Eur. Phys. J.* **C16** (2000) 407.

[22] TASSO Coll., R. Brandelik et al., *Phys. Lett.* **100B** (1981) 357.

[23] J. Letts, P. Mättig, *Z. Phys.*, **C73** (1997) 217.

[24] DELPHI Coll., P. Abreu et al., *Phys. Lett.* **B407** (1997) 174.

[25] Yu.L. Dokshitzer, *JETP* **73** (1977) 1216; V.N. Gribov and L.N. Lipatov, *Sov. J. Nucl. Phys.* **15** (1972) 78; G. Altarelli and G. Parisi, *Nucl. Phys.* **B126** (1997) 298.

[26] Yu.L. Dokshitzer, V.S. Fadin and V.A. Khoze, *Phys. Lett.* **B115** (1982) 242; *Z. Phys.* **C15** (1982) 325.

[27] A.H. Mueller, *Nucl. Phys.* **B213** (1983) 85; **B241** (1984) 141.

[28] Ya.I. Azimov et al., *Z. Phys.* **C27** (1985) 65; **C31** (1986) 213.

[29] C.P. Fong and B.R. Webber, *Nucl. Phys.* **B355** (1991) 54.

[30] Yu.L. Dokshitzer, V.A. Khoze and S.I. Troyan, *Int. J. Mod. Phys.* **A7** (1992) 1875; V.A. Khoze and W. Ochs, *Int. J. of Mod. Phys.* **A12** (1997) 2949.

[31] V.A. Khoze, S. Lupia and W. Ochs, *Phys. Lett.* **B386** (1996) 451.

[32] P. Mazzanti and R. Odorico, *Nucl. Phys.* **B370** (1992) 23.

[33] E.R. Boudinov, P.V. Chliapnikov and V.A. Uvarov, *Phys. Lett.* **B309** (1993) 210; P.V. Chliapnikov and V.A. Uvarov, *Phys. Lett.* **B431** (1998) 430.

[34] S. Lupia and W. Ochs, *Phys. Lett.* **B365** (1996) 339; *Eur. Phys. J.* **C2** (1998) 307.

[35] Yu.L. Dokshitzer, V.A. Khoze and S.I. Troyan, *J. Phys.* **G17** (1991) 1481 and 1602; *Z. Phys.* **C55** (1992) 107.

[36] S.J. Brodsky and J.F. Gunion, *Phys. Rev. Lett.* **37** (1976) 401.

[37] K. Shizuya and S.-H.H. Tye, *Phys. Rev. Lett.* **41** (1978) 787; M.B. Einhorn and B.G. Weeks, *Nucl. Phys.* **146** (1978) 445.

[38] OPAL Coll., G. Alexander et al., *Phys. Lett.* **B265** (1991) 462; P. Acton et al., *Z. Phys.* **C58** (1993) 387; R. Akers et al., *Z. Phys.* **C68** (1995) 179; K. Ackerstaff et al., *Phys. Lett.* **B388** (1996) 659; *Eur. Phys. J.* **C1** (1998) 479; G. Abbiendi et al., **C11** (1999) 217; *Eur. Phys. J.* **C17** (2000) 373; **C23** (2002) 597; **C37** (2004) 25; *Phys. Rev.* **D69** (2004) 032002.

[39] ALEPH Coll., D. Buskulic et al., *Phys. Lett.* **B384** (1996) 353; R. Barate et al., *Eur. Phys. J.* **C16** (2000) 613.

[40] DELPHI Coll., P. Abreu et al., *Z. Phys.* **C70** (1996) 179; *Eur. Phys. J.* **C6** (1999) 19; *Phys. Lett.* **B444** (1998) 491; **B471** (2000) 460; *Eur. Phys. J.* **C13** (2000) 573.

[41] DELPHI Coll., P. Abreu et al., *Eur. Phys. J.* **C17** (2000) 207.

[42] ARGUS Coll., H. Albrecht et al., *Phys. Rev.* **276** (1996) 223.

[43] OPAL Coll., K. Ackerstaff et al., *Eur. Phys. J.* **C8** (1999) 241.

[44] DELPHI Coll., B. Buschbeck and F. Mandl, *Study of the Charge of Leading Hadrons in Gluon and Quark Fragmentation*, DELPHI 2002-0553, CONF587 (June 2002).

[45] ALEPH Coll., G. Rudolph, Rapidity gaps in gluon jets and test of color reconnection models using Z data, ALEPH 2003-008, CONF2 003-005 (July 2003).

[46] OPAL Coll., G. Abbiendi et al., *Eur. Phys. J.* **C35** (2004) 293.

[47] P. Minkowski and W. Ochs, *Phys. Lett.* **B485** (2000) 139.

[48] G. Gustafson, U. Pettersson, P.M. Zerwass, *Phys. Lett.* **B209** (1988) 90; G. Gustafson and J. Häkkinen, *Z. Phys.* **C64** (1994) 659; V.A. Khoze and T. Sjöstrand, *Z. Phys.* **C62** (1994) 281; *Phys. Rev. Lett.* **72** (1994) 28; L. Lönnblad, *Z. Phys.* **C70** (1996) 107; J. Rathsman, *Phys. Lett.* **B452** (1999) 364; G. Corcella et al., *JHEP* **0101** (2001) 010.

[49] L3 Coll., P. Achard et al., *Phys. Lett.* **B581** (2004) 19.

[50] P.V. Chliapnikov and V.A. Uvarov, *Phys. Lett.* **B345** (1995) 313; **B381** (1996) 483; **B423** (1998) 401; P.V. Chlipanikov, *Phys. Lett.* **B462** (1999) 341; **B470** (1999) 263; **B525** (2002) 1; V. Uvarov, *Phys. Lett.* **B482** (2000) 10, **B511** (2001) 136.

[51] F. Becattini, *Z. Phys.* **C69** (1996) 485; F. Becattini and U. Heinz, *Z. Phys.* **C76** (1997) 269.

[52] Yi-Jin Pei, *Z. Phys.* **C72** (1996) 39.

[53] DELPHI Coll., P. Abreu et al., *Phys. Lett.* **B475** (2000) 429.

[54] J. Schwinger, *Phys. Rev.* **82** (1951) 664.

[55] A. Białas, *Phys. Lett.* **B466** (1999) 301.

[56] H.G. Dosch, *Phys. Lett.* **B190** (1987) 177; H.G. Dosch, Yu.A. Simonov, *Phys. Lett.* **B205** (1988) 339; Yu.A. Simonov, *Phys. Lett.* **B307** (1988) 512; O. Nachtmann, in: H. Latal, W. Schweiger (eds.), *Perturbative and Nonperturbative Aspects of Quantum Field Theory* (Springer, Berlin, 1997) p.49.

[57] M. Gaździcki, M.I. Gorenstein, *Phys. Rev. Lett.* **83** (1999) 4009.

[58] E771 Coll., T. Alexopoulos et al., *Phys. Lett.* **B374** (1996) 271.

[59] M. Gaździcki, M.I. Gorenstein, *Phys. Lett.* **B517** (2000) 250.

[60] Sh. Matsumoto, M. Yoshimura, *Phys. Rev.* **D61** (2000) 123508.

[61] EMC Coll., J.J. Aubert et al., *Phys. Lett.* **95B** (1980) 306; **114B** (1982) 373; *Z. Phys.* **C30** (1986) 23; M. Arneodo et al., *Phys. Lett.* **149B** (1984) 415; *Nucl. Phys.* **B258** (1985) 249; *Phys. Lett.* **165B** (1985) 222; *Z. Phys.* **C31** (1986) 1; **C35** (1987) 335; **C35** (1987) 417; **C36** (1987) 527; J. Ashman et al., *Z. Phys.* **C52** (1991) 1; **C52** (1991) 361.

[62] E665 Coll., M.R. Adams et al., *Phys. Lett.* **B272** (1991) 163; *Phys. Rev.* **D48** (1993) 5057; **D50** (1994) 1836; *Z. Phys.* **C61** (1994) 179; **C74** (1997) 237; **C76** (1997) 441 .

[63] NMC Coll., P. Amaudruz et al., *Z. Phys.* **C54** (1992) 239.

[64] H1 Coll., I. Abt et al., *Z. Phys.* **C63** (1994) 377; S. Aid et al., *Z. Phys.* **C72** (1996) 573; C. Adloff et al., *Nucl Phys.* **B485** (1997) 3.

[65] ZEUS Coll., M. Derrick et al., *Z. Phys.* **C67** (1995) 93; **C70** (1996) 1.

[66] WA25 Coll., D. Allassia et al., *Z. Phys.* **C24** (1984) 119.

[67] WA21 Coll., WA47 Coll., G.T. Jones et al., *Z. Phys.* **C25** (1984) 121; **C46** (1990) 25; **C51** (1991) 11; **C54** (1992) 45; **C58** (1993) 375; P.C. Bosetti et al., *Z. Phys.* **C46** (1990) 377.

[68] WA59 Coll., P.J. Fitch et al., *Z. Phys.* **C31** (1986) 51; W. Wittek et al., *Z. Phys.* **C40** (1988) 231; **C44** (1989) 175.

[69] ZEUS Coll., M. Derrick et al., *Phys. Lett.* **B414** (1997) 428; J. Breitweg et al., *Eur. Phys. J.* **C11** (1999) 251; N.H. Brook and I.O. Skillicorn, *Phys. Lett.* **B497** (2001) 55.

[70] ZEUS Coll., J. Breitweg et al., *Eur. Phys. J.* **C2** (1998) 77.

[71] R.P. Feynman, *Photon-Hadron Interactions*, Benjamin, N.Y. (1972).

[72] Yu. Dokshitzer et al., *Rev. Mod. Phys.* **60** (1988) 373.

[73] A.V. Anisovich et al., *Il Nuovo Cimento* **A106** (1993) 547.

[74] K. Charchuła, *J. Phys.* **G19** (1993) 1587.

[75] H.K. Streng, T.F. Walsh and P.M. Zerwas, *Z. Phys.* **C2** (1979) 237.

[76] H1 Coll., S. Aid et al., *Nucl. Phys.* **B445** (1995) 3; C. Adloff et al., *Nucl. Phys.* **B504** (1997) 3.

[77] M. Basile et al., *Nuovo Cimento* **A79** (1984) 1.

[78] ZEUS Coll., M. Derrick et al., *Z. Phys.* **C68** (1995) 29; S. Chekanov et al., *Phys. Lett.* **B553** (2003) 141.

[79] HERMES Coll., A. Airapetian et al., *Eur. Phys. J.* **C21** (2001) 599.

[80] P. Allen et al., *Nucl. Phys.* **B214** (1983) 369; W. Wittek et al., *Z. Phys.* **C40** (1988) 231.

[81] EMC Coll., J.J. Aubert et al., *Z. Phys.* **C18** (1983) 189.

[82] R.D. Field, R.P. Feynman, *Phys. Rev.* **D15** (1977) 2590.

[83] NA22 Coll., M. Adamus et al., *Z. Phys.* **C39** (1988) 311.

[84] J. Benecke, T.T. Chou, C.N. Yang and F. Yen, *Phys. Rev.* **188** (1969) 2159; T.T. Chou and C.N. Yang, *Phys. Rev. Lett.* **25** (1970) 1072; *Phys. Rev.* **D50** (1994) 590.

[85] G. Bellettini et al., *Phys. Lett.* **45B** (1973) 69.

[86] R.P. Feynman, *Phys. Rev. Lett.* **23** (1969) 1415.

[87] UA5 Coll., G. Alner et al., *Z. Phys.* **C33** (1986) 1.

[88] CDF Coll., F. Abe et al., *Phys. Rev.* **D41** (1990) 2330.

[89] R. Holynski, Cracow report INP 1303/PH (1986).

[90] PHOBOS Coll., B.B. Back et al., *Phys. Rev. Lett.* **91** (2003) 052303; nucl-ex/0311009.

[91] A. Białas and M. Jezabek, *Phys. Lett.* **B590** (2004) 233.

[92] WA98 Coll., M.M. Aggarwal et al., *Eur. Phys. J.* **C18** (2001) 651; NA50 Coll., M.C. Abreu et al., *Phys. Lett.* **B530** (2002) 43; PHOBOS Coll., B.B. Back et al., *Phys. Rev. Lett.* **85** (2000) 3100; *Phys. Rev.* **C70** (2004) 021902; PHENIX Coll., K. Adcox et al., *Phys. Rev. Lett.* **86** (2001) 3500; S.S. Adler et al., nucl-ex/0409015; BRAHMS Coll., I.G. Bearden et al., *Phys. Lett.* **B523** (2001) 227 and *Phys. Rev. Lett.* **88** (2002) 202301; STAR Coll., C. Adler et al., *Phys. Rev. Lett.* **87** (2001) 112303.

[93] E.K.G. Sarkisyan, A.S. Sakharov, hep-ph/0410324.

[94] CDF Coll., D. Acosta et al., *Phys. Rev.* **D68** (2003) 012003.

[95] UA1 Coll., C. Albajar et al., *Nucl. Phys.* **B335** (1990) 261.

[96] E735 Coll., T. Alexopoulos et al., *Phys. Rev. Lett.* **60** (1988) 1622, *Phys. Rev.* **D48** (1993) 984; *Phys. Lett.* **B336** (1994) 599.

[97] CDF Coll., F. Abe et al., *Phys. Rev. Lett.* **61** (1989) 1819.

[98] E.A. De Wolf, Proc. XV-th Int. Symp. on Multiparticle Dynamics, Lund, ed. G. Gustafson (World Scientific, Singapore, 1984) p.2.

[99] W. Bell et al., *Z. Phys.* **C27** (1985) 191.

[100] NA35 Coll., H. Ströbele et al., *Z. Phys.* **C38** (1988) 89.

[101] C.M.G. Lattes, Y. Fujimoto and S. Hasegawa, *Phys. Reports* **65** (1980) 151.

[102] UA1 Coll., G. Arnison et al., *Phys. Lett.* **B118** (1982) 167.

[103] JACEE Coll., T.H. Burnett et al., *Phys. Rev. Lett.* **57** (1986) 3249.

[104] UA5 Coll., R.E. Ansorge et al., *Z. Phys.* **C41** (1988) 179.

[105] T. Åkesson et al., *Phys. Lett.* **B119** (1982) 464.

[106] SFM Coll., A. Breakstone et al., *Phys. Lett.* **B132** (1983) 463, *Z. Phys.* **C33** (1987) 333, *Phys. Lett.* **B183** (1987) 227, *Europhys. Lett.* **7** (1988) 131.

[107] L. Van Hove, *Phys. Lett.* **B118** (1982) 138; R. Hagedorn, *Rivista Nuovo Cimento* **6-10** (1983) 1; R. Campanini, *Nuovo Cimento Lett.* **44** (1985) 343.

[108] P. Levai and B. Müller, *Phys. Rev. Lett.* **67** (1991) 1519.

[109] S. Barshay, *Phys. Lett.* **B127** (1983) 129, *Phys. Rev.* **D29** (1984) 1010.

[110] M. Jacob, CERN preprint TH-3515 (1983).

[111] G. Pancheri and C. Rubbia, *Nucl. Phys.* **A418** (1984) 117C; G. Pancheri, Y. Srivastava and M. Pallotta, *Phys. Lett.* **151B** (1985) 453; G. Pancheri and Y. Srivastava, *Phys. Lett.* **B159** (1985) 69.

[112] F.W. Bopp, P. Aurenche and J. Ranft, *Phys. Rev.* **D33** (1986) 1867.

[113] X.N. Wang and R.C. Hwa, *Phys. Rev.* **D39** (1989) 187; X.N. Wang and M. Gyulassy, *Phys. Rev.* **D45** (1992) 844 and *Phys. Lett.* **B282** (1992) 466.

[114] NA22 Coll., V.V. Aivazyan et al., *Phys. Lett.* **B209** (1988) 103.

[115] OPAL Coll., R. Akers et al., *Phys. Lett.* **B320** (1994) 417.

[116] DELPHI Coll., P. Abreu et al., *Phys. Lett.* **B276** (1992) 254.

[117] M. Szczekowski and G. Wilk, *Phys. Rev.* **D44** (1991) R577.

[118] UA1 Coll., G. Arnison et al., *Nucl. Phys.* **B276** (1986) 253.

[119] TASSO Coll., K. Althoff et al., *Z. Phys.* **C22** (1984) 307.

[120] ZEUS Coll., J. Breitweg et al., *Eur. Phys. J.* **C8** (1999) 367.

[121] NA22 Coll., N.M. Agababyan et al., *Z. Phys.* **C46** (1990) 387.

[122] EMC Coll., M. Arneodo et al., *Z. Phys.* **C33** (1986) 167.

[123] A. Wróblewski, Proc. XIV Int. Symp. on Multiparticle Dynamics, eds. P. Yager and J.F. Gunion (World Scientific, Singapore, 1983) p. 573.

[124] NA22 Coll., M. Adamus et al., *Phys. Lett.* **B198** (1987) 292.

[125] NA22 Coll., N.M. Agababyan et al., *Z. Phys.* **C41** (1989) 539.

[126] P. Schmitz et al., *Nucl. Phys.* **B137** (1978) 13.

[127] H. Grässler et al., *Nucl. Phys.* **B132** (1978) 1.

[128] G. Jancso et al., *Nucl. Phys.* **B124** (1977) 1.

[129] M. Szczekowski, *Phys. Lett.* **B357** (1995) 387; M. Szczekowski, G. Wilk, *Phys. Lett.* **B374** (1996) 225.

[130] NA27 Coll., M. Aguilar-Benitez, *Z. Phys.* **C50** (1991) 405.

[131] UA5 Coll., G.J. Alner et al., *Phys. Reports* **154** (1987) 247.

[132] STAR Coll., C. Adler et al., *Phys. Lett.* **B595** (2004) 143.

[133] A.M. Rossi et al., *Nucl. Phys.* **B84** (1975) 269; NA23 Coll., J.L. Bailly et al., *Phys. Lett.* **B195** (1987) 609.

[134] UA1 Coll., G. Bocquet et al., *Phys. Lett.* **B366** (1996) 447.

[135] UA5 Coll., G.J. Alner et al., *Nucl. Phys.* **B258** (1985) 505.

[136] E735 Coll., T. Alexopoulos et al., *Phys. Rev. Lett.* **D64** (1990) 993.

[137] CDF Coll., F. Abe et al., *Phys. Rev.* **D40** (1989) 3791.

[138] F. Wang, *Phys. Lett.* **B489** (2000) 273; F. Wang et al., *Phys. Rev.* **C61** (2000) 064904; F. Wang and N. Xu, *Phys. Rev.* **C61** (2000) 021904; J.C. Dunlop, C.A. Ogilvie, *Phys. Rev.* **C61** (2000) 031901; P. Braun-Munzinger et al., *Nucl. Phys.* **A697** (2002) 902.

[139] J. Rafelski, B. Müller, *Phys. Rev. Lett.* **48** (1982) 1066; **56** (1986) 2334 (erratum); J. Rafelski, *Phys. Reports* **88** (1982) 331; R. Koch, B. Müller, J. Rafelski, *Phys. Reports* **142** (1986) 167.

[140] H. Sorge et al., *Phys. Lett.* **B271** (1991) 37; H. Sorge, *Phys. Rev.* **C52** (1995) 3291.

[141] D. Drijard et al., *Z. Phys.* **C9** (1981) 293.

[142] P. Malhotra and R. Orava, *Z. Phys.* **C17** (1983) 85.

[143] A.K. Wróblewski, Proc. 25th Int. Conf. High Energy Physics, eds. K.K. Phua and Y. Yamaguchi (WSPC, Singapore, 1991) p.125.

[144] H1 Coll., S. Aid et al., *Nucl. Phys.* **B480** (1996) 3.

[145] E665 Coll., M.R. Adams et al., *Z. Phys.* **C61** (1994) 539.

[146] NA22 Coll., M. Adamus et al., *Phys. Lett.* **B198** (1987) 427; I.V. Ajinenko et al., *Z. Phys.* **C44** (1989) 573; N.M. Agababyan et al., *Z. Phys.* **C46** (1990) 525.

[147] M. Bardadin-Otwinowska et al., *Nucl. Phys.* **B98** (1975) 418.

[148] F. Barreiro et al., *Nucl. Phys.* **B126** (1977) 319.

[149] H. Grässler et al., *Nucl. Phys.* **B118** (1977) 189.

[150] K. Böckmann et al., *Nucl. Phys.* **B140** (1978) 235.

[151] W. Ochs, *Nucl. Phys.* **B118** (1977) 397.

[152] J. Singh et al., *Nucl. Phys.* **B140** (1978) 189.

[153] B. Buschbeck, H. Dibon, H.R. Gerhold and W. Kittel, *Z. Phys.* **C3** (1979) 97.

[154] U. Amaldi et al., *Phys. Lett.* **58B** (1975) 206.

[155] B. Buschbeck, H. Dibon and H.R. Gerhold, *Z. Phys.* **C7** (1980) 83.

[156] D. Brick et al., *Z. Phys.* **C13** (1982) 11.

[157] G.J. Bobbink, *Correlations Between High Momentum Particle in Proton-Proton Collisions at High Energy*, Ph.D. Thesis, Utrecht 1981.

[158] A.B. Clegg and A. Donnachie, *Z. Phys.* **C13** (1982) 71.

[159] M. Barth et al., *Nucl. Phys.* **B192** (1981) 289.

[160] R. Göttgens et al., *Nucl. Phys.* **B206** (1982) 349.

[161] J.J. Aubert et al., *Phys. Lett.* **100B** (1981) 433.

[162] P.B. Renton, *Proc. 13th Int. Symp. on Multiparticle Dynamics, Volendam* eds. W. Kittel, W. Metzger and A. Stergiou (World Scientific, Singapore, 1983) p.394.

[163] J.M. Lafaille et al., *Nucl. Phys.* **B192** (1981) 18.

[164] J. Pernegr, V. Šimák and M. Votruba, *Nuovo Cim.* **17** (1960) 129; M. Bardadin et al., *Proc. Sienna Conf. on Elem. Part.* (1963), p.628.

[165] M. Althoff et al., *Z. Phys.* **C22** (1984) 307.

[166] M. Derrick et al., *Phys. Rev.* **D17** (1978) 1; P. Allen et al., *Nucl. Phys.* **B188** (1981) 1; D. Allasia et al., *Z. Phys.* **C27** (1985) 239.

[167] EMC Coll., M. Arneodo et al., *Phys. Lett.* **B149** (1984) 415; *Z. Phys.* **C35** (1987) 417; *Z. Phys.* **C36** (1987) 527.

[168] Ph. Herquet et al., *Properties of Jet-Like Systems Observed in* π^+p, K$^+$p *and pp Interactions*, Proc. XVth Renc. de Moriond 1980, Vol.I, ed. J. Tran Than Van (Editions Frontières 1980) p.215.

[169] P. Hoyer et al., *Nucl. Phys.* **B161** (1979) 34; A. Ali et al., *Z. Phys.* **C2** (1979) 33.

[170] B. Andersson, G. Gustafson, T. Sjöstrand, *Phys. Lett.* **B94** (1980) 211.

[171] NA22 Coll., I.V. Ajinenko et al., *Phys. Lett.* **B197** (1987) 457.

[172] E. Stone et al., *Phys. Rev.* **D5** (1972) 1621; I.V. Ajinenko et al., *Z. Phys.* **C4** (1980) 181; M. Barth et al., *Z. Phys.* **C7** (1981) 187.

[173] A. Borg et al., *Nucl. Phys.* **B106** (1976) 430; J.M. Laffaille et al., *Nucl. Phys.* **B192** (1981) 18.

[174] W. Ochs, L. Stodolsky, *Phys. Lett.* **69B** (1977) 225.

[175] H. Fesefeldt, W. Ochs, L. Stodolsky, *Phys. Lett.* **74B** (1978) 389.

[176] B. Andersson, G. Gustafson, *Phys. Lett.* **84B** (1979) 483.

[177] W. Ochs, T.Shimada, *Z. Phys.* **C4** (1980) 141.

[178] M. Barth et al., *Z. Phys* **C16** (1983) 291.

[179] D.H. Brick et al., *Z. Phys.* **C31** (1986) 59.

[180] EMC Coll., M. Arneodo et al., *Z. Phys.* **C36** (1987) 527.

[181] H1 Coll., T. Ahmed et al., *Phys. Lett.* **B298** (1993) 469; I. Abt et al., *Z. Phys.* **C63** (1994) 377; S. Aid et al., *Phys. Lett.* **B356** (1995) 118; **B358** (1995) 412; *Z. Phys.* **C70** (1996) 609; C. Adloff et al., *Eur. Phys. J.* **C12** (2000) 595.

[182] ZEUS Coll., M. Derrick et al., *Z. Phys.* **C59** (1993) 231; *Phys. Lett.* **B338** (1994) 483.

[183] A. Białas, L. Stodolsky, *Phys. Rev.* **D13** (1976) 199.

[184] NA22 Coll., I.V. Ajinenko et al., *Z. Phys.* **C43** (1989) 37.

[185] J. Kalinowski, M. Krawczyk, S. Pokorski, *Z. Phys.* **C15** (1982) 281.

[186] L. Van Hove, A. Giovannini, *Acta Phys. Pol.* **B19** (1988) 931.

[187] J.D. Bjorken and J.B. Kogut, *Phys. Rev.* **D8** (1973) 1341.

[188] V.A. Khoze and W. Ochs, *Int. J. Mod. Phys.* **A12** (1997) 2949.

[189] Brodsky et al., *Phys. Rev.* **D50** (1994) 3134.

[190] Yu.L. Dokshitzer, *Sov. Phys. JETP* **46** (1977) 641; V.N. Gribov and L.N. Lipatov, *Sov. J. Nucl. Phys.* **15** (1972) 438 and 675; G. Altarelli and G. Parisi, *Nucl. Phys.* **126** (1977) 297.

[191] E.A. Kuraev, L.N. Lipatov and V.S. Fadin, *Sov. Phys. JETP* **45** (1972) 199; Y.Y. Balitsky and L.N. Lipatov, *Sov. J. Nucl. Phys.* **28** (1978) 822.

[192] M. Ciafaloni, *Nucl. Phys.* **B296** (1987) 249; S. Catani, F. Fiorani and G. Marchesini, *Phys. Lett.* **B234** (1990) 339; *Nucl. Phys.* **B336** (1990) 18.

Chapter 5

Early Models

Before entering the discussion of the specific quark-parton models of multiple production at low momentum transfer, let us recall the parton picture of hadrons. It is convenient to describe the hadron in the infinite momentum frame, which idealises the experimental situation of high-energy collisions in the cms frame. The most convenient variable for our purpose is the *Bjorken-*x_{B} defined in this frame as the fraction of the hadron momentum carried by a particular parton. Then, for a proton as seen in deep inelastic scattering (DIS) one makes the following observations:

(i) The valence $\mathrm{u_v}$-quark distribution $x_{\mathrm{B}}u_{\mathrm{v}}(x_{\mathrm{B}})$ is peaked at $x_{\mathrm{B}} \approx 0.2$ and vanishes at $x_{\mathrm{B}} \to 1$ significantly slower than that of the valence $\mathrm{d_v}$-quark $x_{\mathrm{B}}d_{\mathrm{v}}(x_{\mathrm{B}})$. Normalization of $u_{\mathrm{v}}(x_{\mathrm{B}})$ and $d_{\mathrm{v}}(x_{\mathrm{B}})$ is compatible with the usual (uud) picture of the proton.

(ii) Integrating this distribution and summing the contributions from all valence quarks, one finds that valence quarks only carry half of the total momentum of the proton.

(iii) The sea quark or antiquark distribution $x_{\mathrm{B}}q_{\mathrm{s}}(x_{\mathrm{B}})$ is peaked at very low x_{B} values, suggesting a $1/x_{\mathrm{B}}$ singularity in $q_{\mathrm{s}}(x_{\mathrm{B}})$. Above $x_{\mathrm{B}} \approx 0.3$, sea contributions are negligible. Strange sea quarks are suppressed with respect to non-strange ones by the factor $\lambda \approx 0.3$.

(iv) The integral over the sea distribution yields only about 5% as an average fraction of the total proton momentum carried by sea quarks and antiquarks. This leaves nearly half of the momentum to the gluons dominating at low x_{B}.

The above observations are qualitative and correspond to the range of momentum transfer Q^2 in lepton-hadron scattering of the order of a few GeV^2. With increasing Q^2, the distributions decrease at high x_{B} and increase at low x_{B}. This observation is in agreement with an intuitive picture of an increasing number of visible sea constituents with improving resolution. The observed changes are, however, rather slow.

Any model assuming the relevance of parton structure for soft hadronic processes should state what is the fate of quarks and gluons during the collision. In particular, one should indicate if the quarks and gluons act independently, or are combined into three constituent quarks as used, for example, in the early additive quark model. Furthermore, an assumption is needed to specify which and how many of the partons interact during the collision, and which fly through merely as spectators. Finally, a mechanism of hadron formation into the actually observed final state should be specified, indicating how the momenta of the partons are shared among the produced hadrons. Obviously, all those assumptions are in principle subject to experimental tests. As we shall see (see [1] for more details), the interpretation of the experimental results is usually non-unique and one needs quite extensive investigations to select the most likely choices of the possibilities presented above.

In general, however, we would like to convey the message that the hadron structure as seen in DIS is indeed relevant to soft hadronic collisions and the models assuming these connections are surprisingly successful.

5.1 Additive quark model and quark combinatorics

The experimentally observed connections between the valence quark structure of hadrons and their yields in multiple production discussed in more modern interpretation in Sect. 4.3 had, in the early 1970s, led to a revival of interest in an even earlier idea of Satz [2] of applying the additive quark model of total cross sections (Chapter 1) to this type of processes. In 1973, Anisovich and Shekhter [3] proposed a detailed model designed to explain hadron production both in the central and fragmentation regions. We shall present here its basic ideas, some of which were proposed independently by Bjorken and Farrar [4] in their discussion of low and high $p_{\rm T}$ effects.

5.1.1 The central region

In the central region, defined by the *Feynman-x* limit $|x| < 1/3$, multiple production of $\mathrm{q\overline{q}}$ pairs results from the collision of one of the constituent quarks of each of the interacting hadrons (Fig. 5.1). Because of the large number of pairs produced at high enough energy, the final state in the central region is practically independent of the type of colliding hadrons. Following their production, quarks and antiquarks recombine into hadrons, with probabilities statistically equal for all possible combinations.

One can derive a number of predictions already from the simple assumptions given above. Let us denote by α_0, α_1 and α_2 the probabilities of having no, one or two quarks (or antiquarks) free for recombination in some sector of phase space. Statistical equilibrium then demands that the probabilities should not change if we add with equal probabilities one more quark or antiquark. Assuming that each $\mathrm{q\overline{q}}$ system recombines immediately into a meson, and a qqq system into a baryon, we find

$$\alpha_0 = \alpha_1 + \alpha_2 \tag{5.1}$$

$$\alpha_1 = (\alpha_0 + \alpha_2)/2 \tag{5.2}$$

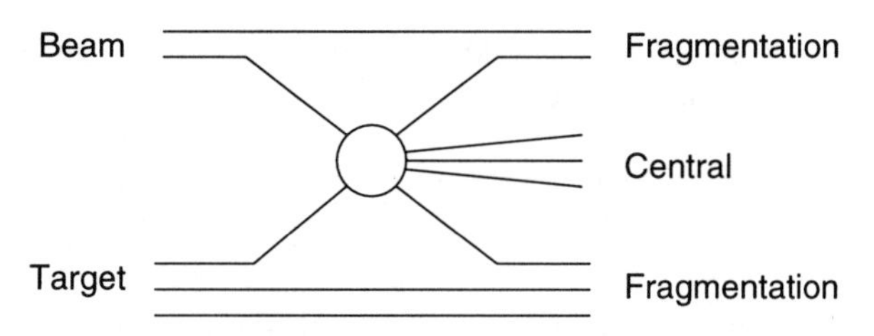

Figure 5.1. Diagram of the additive quark model of multiple production in a meson-proton collision.

$$\alpha_2 \;=\; \alpha_1/2\;, \tag{5.3}$$

since a 'no-quark' state results from adding q to $\overline{\mathrm{q}}$, $\overline{\mathrm{q}}$ to q, q to qq or $\overline{\mathrm{q}}$ to $\overline{\mathrm{qq}}$, a 'q' state comes from adding one quark to a 'no-quark' state or adding $\overline{\mathrm{q}}$ to qq, and finally 'qq' from adding q to q.

Solving these equations, one finds

$$\alpha_0 : \alpha_1 : \alpha_2 = 3 : 2 : 1\;. \tag{5.4}$$

Thus the ratio of baryons to mesons produced should tend to a well-defined value at high energy, where statistical assumptions are reliable. Since a baryon B is formed only by adding one more quark to a 'qq' state, whereas a meson M can appear by adding q to '$\overline{\mathrm{q}}$', or '$\overline{\mathrm{qq}}$' and '$\overline{\mathrm{q}}$' to 'q' and 'qq', respectively, one finds

$$\frac{\mathrm{B}}{\mathrm{M}} = \frac{\alpha_2}{2(\alpha_1+\alpha_2)} = \frac{1}{6}\;. \tag{5.5}$$

To derive well-defined predictions for different flavor and spin states, one adopts two more assumptions. Firstly, to account for the experimentally observed suppression of strange particles, the probability of finding an s-quark is λ times the probability of finding a u(d)-quark. The value of λ may be estimated as in Sub-Sect. 4.3.2 to $\lambda \approx \frac{1}{3}$, but in general it may serve as an adjustable parameter. Secondly, if one considers production of hadrons in the lowest SU_6 multiplets, i.e., 0^- and 1^- mesons and $\frac{1}{2}^+$ and $\frac{3}{2}^+$ baryons, one can assign factors $2J+1$ to count the possible spin states contributing to the recombination probabilities.

Now it is possible to calculate relative probabilities of production for 18 mesons and 18 baryons. The predictions for the ratios of mesons are quoted below, where π, K, ρ and K^* denote any charge state of those particles, since isospin symmetry is built into the model:

π	K	η	η'	ρ	ω	ϕ	K^*
3	3λ	$1+2\lambda^2$	$2+\lambda^2$	9	9	$9\lambda^2$	9λ

Experimentally, it is rather difficult to separate the direct production of pions and kaons from different resonance decay contributions. Therefore, it may be more practical to calculate full yields using known branching ratios of meson and baryon resonances. Corresponding ratios then become

$\pi^\pm$	π^0	$K^\pm$	$K^0, \overline{K}^0$	η	η'	ρ	ω	ϕ	K^*
$36+16\lambda+6\lambda^2$	$39+16\lambda+7\lambda^2$	$12\lambda+4\lambda^2$	$12\lambda+3\lambda^2$	$1+2\lambda^2$	$2+\lambda^2$	9	9	$9\lambda^2$	9λ

To improve the reliability of the model, one should actually also include higher resonances [5,6]. As an example, $J^P = 2^+$ mesons are known to be produced quite frequently in high-energy collisions. This can be done in the SU(6) classification framework by adding the $L = 1$ multiplet containing four SU(3) nonets $J^P = 0^+, 1^+, 1^+$ and 2^+ with statistical weights $2J + 1$. The easiest place to test this idea is the ratio of kaons coming from the $K^*(892)$ and $K_2^*(1430)$ decays compared to the total kaon yield. If the probabilities of the $q\overline{q}$ system to appear in spin $S = 0$ and $S = 1$ combinations obey the simple 1:3 rule, the contribution from K^* decay plus the contribution from K_2^* decay multiplied by $\frac{9}{5}$ (to account for other $L = 1, S = 1$ multiplets) should form about 75% of the total kaon yield [6]. The comparison with data is quite encouraging.

For baryon-antibaryon production we can expect the model to work quantitatively at very high energies only, as the number of produced quark pairs is kinematically limited and the recombination into qqq or $\overline{qqq}$ systems is suppressed. However, one can compare the ratios of various baryon yields, which are less affected by phase-space restrictions than the yields, themselves. Again, this can be done for the two lowest SU(3) multiplets or for the total stable baryon yields using the known branching ratios of resonance decays. We shall discuss the agreement with the data further below. At this point, let us just mention that in a model which is so simple, we can not expect a precise description of the data. Nevertheless, the general agreement of many predictions with the data suggests that the picture of a statistical gas of quarks and antiquarks recombining locally into hadrons is surprisingly successful and may serve as a reasonable first approximation for the central production of hadrons.

5.1.2 The fragmentation region

In the fragmentation region, $|x| > \frac{1}{3}$, the spectra of produced hadrons obviously depend not only on their own quark structure, but also its overlap with that of the initial hadron. One assumes that one of the quarks or antiquarks of the initial hadron is 'lost' in the collision, thus giving rise to the central production. The remaining 'spectator' (anti)quark from a meson (or diquark from a baryon) continues its flight with unchanged momentum and recombines with centrally produced quarks or antiquarks to form a final-state hadron. This idea should, however, be supplemented by some prescription for the recombination probability. Most naively, one would assume that the probabilities to pick up respectively one or two sea quarks or antiquarks are the same as in the central region. This would give a rather unlikely prediction of the baryon/meson ratio of $\frac{1}{2}$ in the meson fragmentation region. Experimentally, we know that baryon production is more strongly suppressed and, moreover, the corresponding ratio depends significantly on x. Thus, quark combinatorics cannot be used directly for quantitative predictions in this case, the general features, however, support the presented idea: The relative yields of various hadrons indeed depend dramatically on their quark structure in the way predicted.

If more of the initial-hadron quarks may be used to build the final one, its yield in the fragmentation region is expected to be bigger than if only one or no quark can be used. Examples for the usefulness of this rule are

$$\begin{gathered}(\pi^+ \to \pi^+) \gg (\pi^+ \to \pi^0) \gg (\pi^+ \to \pi^-) \\ (\mathrm{p} \to \mathrm{p}) \gg (\mathrm{p} \to \Lambda) \gg (\mathrm{p} \to \Xi) \gg (\mathrm{p} \to \Omega^-) \,. \end{gathered} \tag{5.6}$$

To obtain numerical predictions one needs further assumptions. In the original paper of Anisovich and Shekhter [3], arbitrary functions of x and p_{T} were introduced to describe the distribution for each class of processes: $\mathrm{B} \to \mathrm{B}'$ with 2 or 1 common quarks, $\mathrm{B} \to \mathrm{M}$, $\mathrm{M} \to \mathrm{B}$. In each class, various processes were then related by combinatorial coefficients. Fragmentation into hadrons containing no quarks in common with the initial one was assumed to proceed only via the formation and decay of resonances. Due to the scarcity of data, the model was not really tested at that time [7], except for a $(\mathrm{p} \to \pi^+)/(\mathrm{p} \to \pi^-) \approx 2$ prediction which today is known to be quite strongly violated (see Fig. 4.43).

In general, using a constituent quark structure of hadrons in the fragmentation region is not as successful as in the central region. A detailed comparison of quark ideas with data in the fragmentation region was, therefore, not performed until the more specific recombination models were formulated.

A way of using quark combinatorics without involving arbitrary x or p_{T} distributions was presented by Van Hove and collaborators [8]. These authors investigated the overall fate of the valence quarks of an incident nucleon in the fragmentation process in a probabilistic approach. The probabilities that three incident valence qarks will emerge in three distinct hadrons, two hadrons or a single baryon are denoted by A_1, A_2 and A_3, respectively. For each single quark, the probability to recombine into a meson or a baryon is denoted by η and η'. Finally, the probability of picking up a strange quark or antiquark for recombination is suppressed by assigning $\xi, (1-\xi)/2, (1-\xi)/2$ probabilities to s, u, d emergence, respectively. Since $A_1 + A_2 + A_3 = 1$ and $\eta + \eta' = 1$ has to hold, there are only four free parameters to describe the data.

The analysis of data on meson and baryon production resulted in estimates of $\xi \sim 0.1$ (corresponding to $\lambda \approx 0.2$, i.e. slightly stronger strangeness suppression than in the central region), $A_1 = 0.05 - 0.1$, $A_2 = 0.5 - 0.6$, $A_3 = 0.35 - 0.4$ and $\eta' = 0.01 - 0.02$ (which means that formation of more than one baryon is very unlikely).

Later, a meaningful test has become possible from the detailed results on antibaryon production [9], which according to the above picture should proceed via a $p \to B_1B_2\overline{B}_3$ fragmentation process. The results are in disagreement with this, since the predicted yields of antihyperons are systematically too low. The disagreement even increases with increasing strangeness. This suggests that the $B\bar{B}$ pair production cannot be described properly by the fragmentation process alone, but has other (dominant) contributions. Thus, in the later version [10], the model was applied to the $n_B - n_{\bar{B}}$ differences only, focusing on the more precise determination of parameters. The arbitrariness was reduced by neglecting the probabilities for initial valence quarks to emerge in more than one baryon, in agreement with the above results. Thus one is dealing now with the probabilities a_i that the single excess baryon (always present in the proton fragmentation region at high energies) will contain $i = 0, 1, 2, 3$ initial valence quarks. Furthermore, each additional quark needed for the baryon formation may be strange with probability ξ or one of the two non-strange ones (u,d) with probability $\tau = (1-\xi)/2$. In these variables, the $n_B - n_{\bar{B}}$ difference denoted by $n_f(B)$ will read:

$$\begin{aligned}
&n_f(\Omega^-) = \xi^3 a_0 && n_f(\Xi^-) = \xi^2(3\tau a_0 + \tfrac{1}{3}a_1) \\
&n_f(\Sigma^-) = \xi\tau(3\tau a_0 + \tfrac{2}{3}a_1) && n_f(\Sigma^+) = \xi(3\tau^2 a_0 + \tfrac{4}{3}\tau a_1 + \tfrac{1}{3}a_2) \\
&n_f(\Sigma^0) = n_f(\Lambda^0) = \xi(3\tau^2 a_0 + \tau a_1 + \tfrac{1}{3}a_2) && n_f(p) = 4\tau^3 a_0 + \tfrac{62}{27}\tau^2 a_1 + \tfrac{35}{27}\tau a_2 + \tfrac{8}{9}a_3 .
\end{aligned} \tag{5.7}$$

In each case, only the lowest baryon state of the corresponding qqq system is considered (assuming that resonance effects are already included in the probabilistic approach). An exception is the Δ^{++}(uuu), the contribution of which is added to the proton.

Significant constraints already result from the positivity condition for the a_i:

$$n_f(\Sigma^+) + n_f(\Sigma^-) \le n_f(\Lambda^0 + \Sigma^0) \le 2n_f(\Sigma^+) \le n_f(\Lambda^0 + \Sigma^0) + n_f(\Sigma^-) . \tag{5.8}$$

These inequalities agree with the data. More detailed results can be obtained when assuming that each of the valence quarks has an independent fixed probability to emerge in a meson (P) or to stay in a baryon ($1-P$). This is assumed to be valid for non-diffractive processes, to which a probability $\bar{a}$ is assigned. This assumption leads to $a_0 = \bar{a}P^3, a_1 = 3\bar{a}P^2(1-P), a_2 = 3\bar{a}P(1-P)^2$ and only three free parameters are left. Assuming $\xi = 0.1$, the data are satisfactorily described with $\bar{a} = 0.69 \pm 0.30$ and $P = 0.41 \pm 0.14$, or equivalently with $a_0 = 0.05 \pm 0.03$, $a_1 = 0.20 \pm 0.04$, $a_2 = 0.30 \pm 0.16$, $a_3 = 0.45 \pm 0.17$. Thus, in non-diffractive events, the probabilities for the final baryon to have one, two or three initial valence quarks may be quite similar.

The results obtained show that it is possible to understand the observed yields of hadrons in terms of simple combinatorial analysis assuming a standard valence quark structure of particles. Thus, the idea that the quark structure is relevant in soft processes is strongly supported. We show below that a more differential investigation of final states indeed confirms this idea.

5.2 Quark counting rules and perturbative QCD-based approach

Quark counting rules (QCR) are simple rules determining the power n in the $(1-x)^n$ dependence observed in many fragmenation processes $\mathrm{h} \to \mathrm{h}'$ at $x \to 1$:

$$\frac{x}{\sigma_{\text{inel}}} \frac{\mathrm{d}\sigma}{\mathrm{d}x}(\mathrm{h} \to \mathrm{h}') \propto (1-x)^n . \tag{5.9}$$

They originated from the so-called dimensional counting rules for the power dependence of hard processes, discussed by [11] and [12]. According to the ideas of those authors, 'hard' processes (where all the invariant kinematical variables are much larger than hadronic masses) should depend on dimensionless ratios of variables and the only remaining dependence of the amplitude on the cms energy squared s is given by a power determined from dimensional analysis.

5.2.1 Hard processes

For elastic scattering at very high s and momentum transfer squared t, the cross section can be written as

$$\frac{\mathrm{d}\sigma}{\mathrm{d}t}(\mathrm{AB} \to \mathrm{AB}) = (N_{\mathrm{A}} N_{\mathrm{B}})^2 |F_{\mathrm{AB}}(s,t)|^2 , \tag{5.10}$$

where $N_{\mathrm{A}}, N_{\mathrm{B}}$ are structure factors of dimensions $m^{2(n_{\mathrm{A}}-1)}, m^{2(n_{\mathrm{B}}-1)}$ resulting from the dimension m^{-1} of a one-particle state vector in the relativistic normalization, with $n_{\mathrm{A}}, n_{\mathrm{B}}$ denoting the number of elementary constituents (partons) in the colliding hadrons. The amplitude $F_{\mathrm{AB}}(s,t)$ describes the interaction of the parton systems. Thus, the dimension of $|F_{\mathrm{AB}}|^2$ is $m^{-4(n_{\mathrm{A}}+n_{\mathrm{B}}-1)}$ and, since at $s,t \to \infty$ it cannot depend on any dimensional parameters, the resulting dependence of $\mathrm{d}\sigma/\mathrm{d}t$ on s and t should be given by

$$\frac{\mathrm{d}\sigma}{\mathrm{d}t} = s^{2-2(n_{\mathrm{A}}+n_{\mathrm{B}})} f(t/s) . \tag{5.11}$$

This result can be generalized to inelastic processes $\mathrm{AB} \to \mathrm{CD}$, yielding

$$\frac{\mathrm{d}\sigma}{\mathrm{d}t}(\mathrm{AB} \to \mathrm{CD}) = s^{2-(n_{\mathrm{A}}+n_{\mathrm{B}}+n_{\mathrm{C}}+n_{\mathrm{D}})} f(t/s) . \tag{5.12}$$

Similarly, electromagnetic form factors at high t corresponding to scattering of an elementary lepton on a hadron h should behave as [13]

$$F^{\mathrm{h}}(t) \propto t^{1-n_{\mathrm{h}}} = t^{-n_{\mathrm{s}}} , \tag{5.13}$$

where n_{s} is the number of non-interacting (spectator) quarks in a hadron since only one quark interacts point-like with the exchanged photon. As shown in [14] and [15],

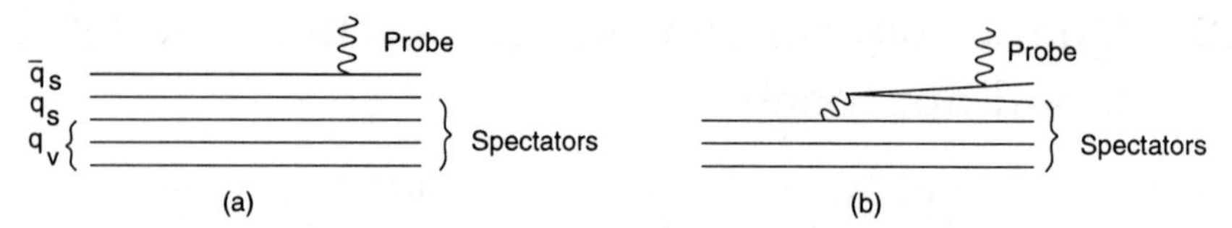

Figure 5.2. Diagrams corresponding to the proton structure function.

the corresponding behavior of inelastic form factors at Bjorken-$x_{\rm B} \to 1$ (related to the x distributon of a quark in a hadron) should be in agreement with the Drell-Yan-West relation between the elastic and inelastic form factors [16, 17] given by

$$F_{\rm q}^{\rm h}(x) \propto (1-x)^{2n_{\rm s}-1} \,. \tag{5.14}$$

Note that if hadrons are built from a fluctuating number of constituents, leading terms in formulae (5.11-5.14) always correspond to the minimal possible number of constituents. With the rule (5.14), the distribution of valence quarks in mesons becomes $F_{\rm v}^{\rm M}(x) \propto (1-x)$ since $n_{\rm s} = 1$, and in baryons $F_{\rm v}^{\rm B}(x) \propto (1-x)^3$ since $n_{\rm s} = 2$.

Obviously, such predictions are oversimplified, since the spin structure of hadrons is neglected. Also the flavor-dependent quark mass effects may be non-negligible. However, this line of research was pursued and found to apply not only to the distribution of a single parton in the parton system (hadron), but also to the probability of any parton (or parton system) to fragment into a parton (or other parton system) carrying x close to unity [18]. Thus, the behavior of the fragmentation functions $D_{\rm h}^{\rm q}(x)$ describing the distribution of the momentum fraction carried by a hadron h originating from quark q can also be estimated by QCR.

Before entering the rules for the case of $\rm h \to h'$ processes, of main interest to us in this context, let us mention that the rule (5.14) is not quite unambiguous when related to sea quarks or gluons. Indeed, the sea quark in question may have belonged to a pair present in the initial hadron state, or may have been created via point-like photon or gluon emission and conversion in the particular fluctuation 'caught' by probing the structure function. Corresponding diagrams, e.g. for the baryon structure function, $F_{\rm q_s}^{\rm B}$, are shown in Fig. 5.2. Although the number of spectators is the same in both cases, it is intuitively plausible that the suppression due to the requirement of pushing all but one parton to $x = 0$ is greater in the first case.

For the electromagnetic point-like interaction, due to the production of such extra pairs, the rule (5.14) had to be replaced by [18]

$$F_{\rm q}^{\rm h}(x) \propto (1-x)^{2n_{\rm H}+n_{\rm PL}-1} \,, \tag{5.15}$$

where $n_{\rm H}$ counts 'passive' spectators present already in a minimal initial state and $n_{\rm PL}$ counts spectators taking part in the point-like creation of additionally necessary partons. Modification of (5.14) to (5.15) is justified by counting the number of 'suppressed' lines of highly virtual partons in corresponding QED diagrams [15].

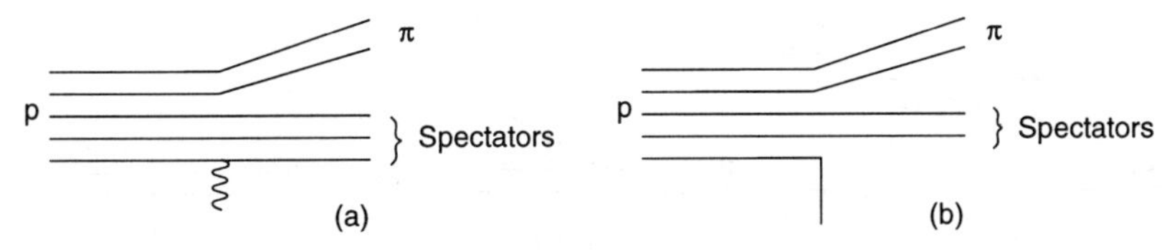

Figure 5.3. a) Gluon and b) quark exchange diagrams for proton fragmentation.

Later, it was argued [19] that the modified rule should indeed also be applied to strongly produced pairs (diagram 5.2b). As we can see, for diagram (b) $n_{\mathrm{PL}} = 2, n_{\mathrm{H}} = 2$ and we find $F^{\mathrm{B}}_{\mathrm{q_s}}(x) \propto (1-x)^5$, whereas diagram (a) yields $F^{\mathrm{B}}_{\mathrm{q_s}}(x) \propto (1-x)^7$. One may expect that at $x \to 1$ diagrams of the type (b) will dominate, but as long as the relative normalization of the two terms is unknown, no clear-cut prediction for a finite range of $x \neq 1$ can be made.

5.2.2 Soft processes

For hadron-hadron interactions at low p_{T}, distributions at high Feynman-x are treated in a similar way. However, the analysis requires specification of the interaction mechanism in the primary hadronic collision followed by the fragmentation process. Two alternative pictures have been proposed: gluon exchange [20,21] or quark exchange [22]. In the first case, valence constituent lines stay in the same fragmentation region before and after the primary collision. Thus, rule (5.14) for the $\mathrm{h} \to \mathrm{q}$ distribution is valid also for the $\mathrm{h} \to \mathrm{h}'$ transition:

$$F^{\mathrm{h}\to\mathrm{h}'}(x) \propto (1-x)^{2n_{\mathrm{s}}-1} , \tag{5.16}$$

where n_{s} is the number of $\mathrm{q}(\overline{\mathrm{q}})$ lines from h absent in the particular h'. In the second case, however, one can get rid of one of the spectators by sending it to the other fragmentation region. The resulting minimal n_{s} is therefore diminished by one unit. Corresponding diagrams, e.g. for $\mathrm{p} \to \pi$, are given in Fig. 5.3.

Further subtleties of respectively valence and sea-quark exchange enter, but the most striking prediction of such a quark-exchange model appears immediately on the level of two-particle correlations. Indeed, if we have gotten rid of a spectator of one of the incoming hadrons, it is bound to appear in the fragmentation of the other one. As a consequence, the fragmentation spectra of the two colliding hadrons do not factorize. A strong negative correlation is expected between the probabilities of finding simultaneously, say, two π^+ mesons in pp collisions at x close to 1 and close to -1, respectively. This idea is in violent disagreement with the ideas of other high-energy models discussed here, which implicitly or explicitly assume 'limiting fragmentation' [23]: inclusive spectra in the high-energy, large-x region (fragmentation region) neither depend on the initial nor final state of the other interacting

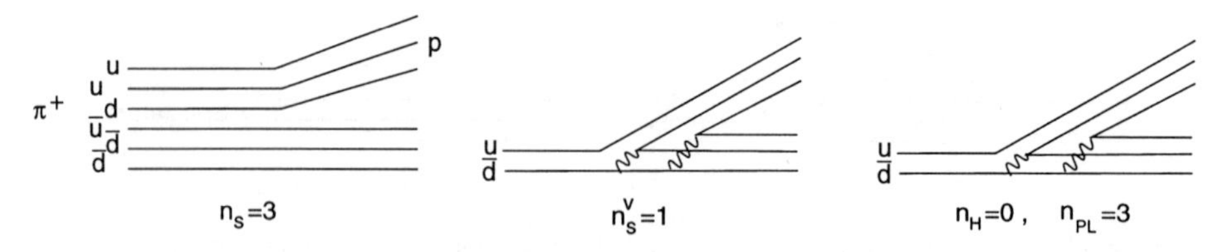

Figure 5.4. Diagrams for the fragmentation process $\pi^+ \to$ p.

hadron. Thus, it was met with some relief when the data were found to contradict the quark-exchange model [24]: none of the predicted correlations exist.

Furthermore, gluon exchange with rule (5.16) was found to predict too steep a fall for hadron spectra, especially for 'unfavored' processes where more sea quarks are needed. This failure has prompted [24] to try neglecting the sea quarks completely and assuming energy equipartition for valence quarks only. Then, instead of (5.16) one expects

$$F^{\mathrm{h}\to\mathrm{h}'}(x) \propto (1-x)^{2n_\mathrm{s}^\mathrm{v}-1} \quad , \tag{5.17}$$

where n_s^v is the number of valence quarks of hadron h which do not appear in hadron h′. Such a rule is more successful, but it underestimates in turn the steepness of baryon spectra in meson fragmentation. Thus, the most promising counting rule seems to be an analogue of rule (5.15), as already justified above [19] for hard processes:

$$F^{\mathrm{h}\to\mathrm{h}'}(x) \propto (1-x)^{2n_\mathrm{H}+n_\mathrm{PL}-1} \, , \tag{5.18}$$

where the symbols are the same as in (5.15). To illustrate the three counting rules we compare in Fig. 5.4 the corresponding diagrams for the $\pi^+ \to$ p fragmentation process. The resulting prediction for $F^{\pi^+\to\mathrm{p}}(x)$ is $(1-x)^5, (1-x)^1$ and $(1-x)^2$, for (5.16), (5.17) and (5.18), respectively.

The counting rules are confronted with data in [19,25,26]. We present in Fig. 5.5 the values of the exponent n in $(1-x)^n$ fits to the data by Gunion supplemented by later data [1]. Solid lines are predictions of counting rule (5.18) and broken lines are values corrected *ad hoc* to account for the suppression of d-quarks as compared to u-quarks in proton structure functions at high x. One can see that the QCR give a reasonable rough estimate of n values, but the data exhibit sytematic differences between various processes of the same class, presumably related to spin and flavor effects. While the difference between u- and d-quarks observed in DIS has already been taken into account by [19], there is clear evidence for $n < 1$ for $\mathrm{K}^+ \to \mathrm{M}$ with M containing an $\bar{\mathrm{s}}$-quark and $1 < n < 2$ for M containing a u-quark. It should also be noted that the x range in which the fit is performed is bound from below (to limit resonance effects), as well as from above (to exclude triple-Regge terms) in a rather arbitrary and not fully consistent way. Since the QCR are derived for the limit $x \to 1$, it is rather difficult to justify why they should work just in the selected x-range.

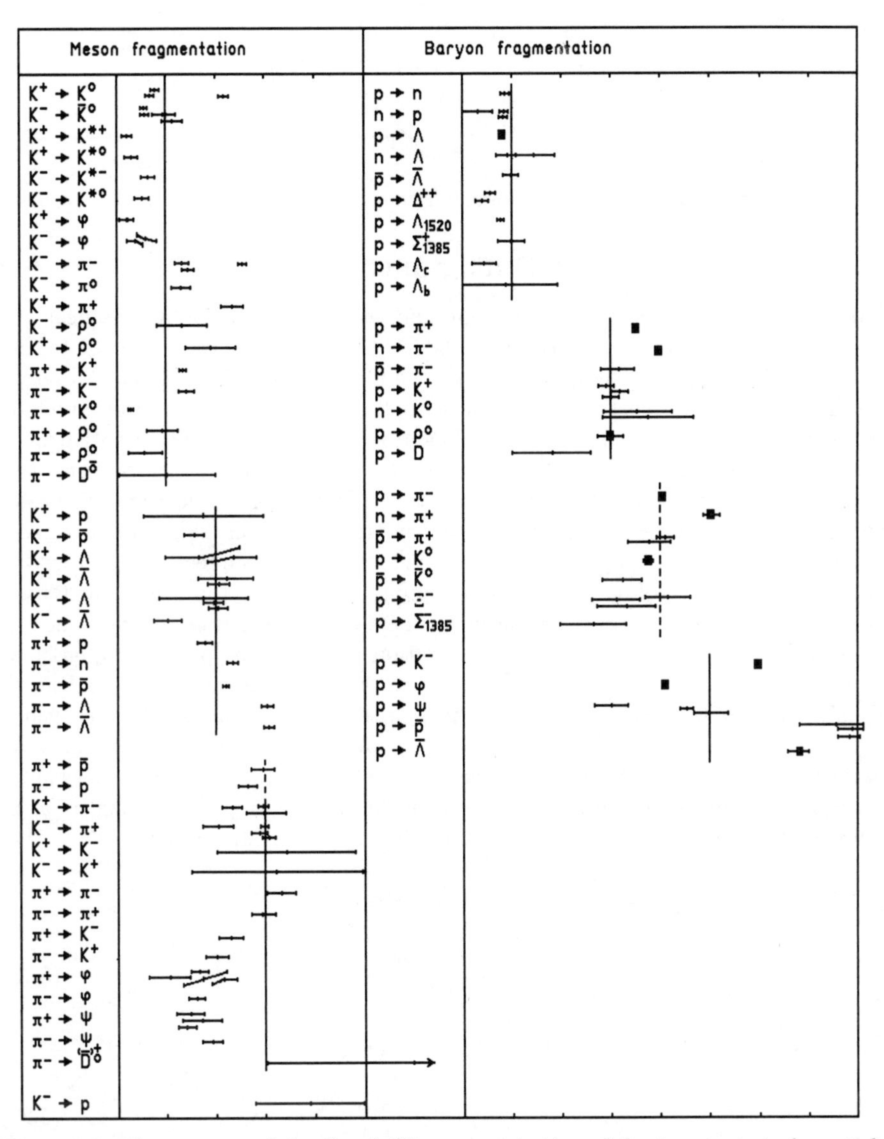

Figure 5.5. The power n of the $(1-|x|)^n$ parametrization of the invariant single-particle inclusive cross section (for $\mathrm{D}, \Lambda_{\mathrm{b}}$ and Λ_c, the non-invariant cross section), compared to the counting rules of [19] (where the broken lines are corrected for the suppression of d-quarks in the proton structure functions). Data are compiled in [1]. The full squares are averages over more than three data points for the same reaction.

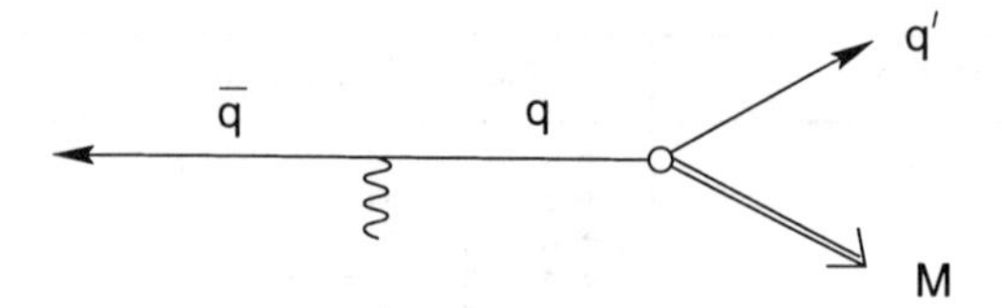

Figure 5.6. Meson production in e^+e^- annihilation.

5.2.3 Perturbative QCD diagrams

More detailed investigations of the origin and the form of the QCR have been initiated in the framework of perturbative QCD diagrams [27]. As already noted repeatedly, the basic problem in applying parton ideas and QCD-based calculations to soft hadronic processes is the apparent absence of large momentum transfer to justify the introduction of point-like constituents and the application of perturbative QCD calculus. It has, however, been suggested that the 'soft' *hadronic final state* may also in fact result from a large Q^2 interaction in the underlying QCD diagram. The basic idea behind this approach is simple and kinematical in origin [28]. To illustrate it, let us consider the probability that e^+e^- annihilation will produce a meson carrying almost all the energy of one of the initial leptons. We assume that, in the first stage of the process, a $q\overline{q}$ pair was produced by the electromagnetic interaction and the final-state meson comes from the fragmentation of one of the produced quarks, as shown in Fig. 5.6. It is most convenient to work in the cms and to use light-cone variables $(p_+ = E + p_\|,\ p_- = E - p_\|,\ \mathbf{p}_\mathrm{T})$, with the 'longitudinal' direction along the momentum of quark q and $\mathbf{p}_\mathrm{T} = (p_\mathrm{T}, 0)$ defining the fragmentation plane. We neglect masses of the 'final-state' quarks $\overline{q}$ and q', as well as that of the meson M, since they are small compared to the total cms energy $W = \sqrt{s}$. Denoting by x the fraction of the p_+ momentum of quark q carried by meson M, we find

$$
\begin{aligned}
p_\mathrm{M} &\approx (x\sqrt{s},\ \frac{p_\mathrm{T}^2}{x\sqrt{s}},\ p_\mathrm{T},\ 0) \\
p_{\mathrm{q}'} &\approx \left(\sqrt{s}(1-x),\ \frac{p_\mathrm{T}^2}{\sqrt{s}(1-x)},\ -p_\mathrm{T},\ 0\right) .
\end{aligned}
\tag{5.19}
$$

Thus, for $p_\mathrm{T} \neq 0$ the requirement of $x \to 1$ is equivalent to $p_\mathrm{q}^2 \to \infty$, since

$$
p_\mathrm{q}^2 = (p_{\mathrm{q}'} + p_\mathrm{M})^2 = \frac{p_\mathrm{T}^2}{x(1-x)} . \tag{5.20}
$$

This means that quark q has to be created far off mass shell to enable the final-state meson to achieve a high x value.

Now, even if we cannot draw and calculate all the QCD diagrams relevant to hadron production, we know what is the suppression of diagrams containing one far-off mass-shell quark line with respect to the 'all soft' diagrams: it is simply given by the quark propagator. Therefore, the simple answer in our case is that we expect the x distribution of the meson to behave as $(1/p_\mathrm{q})^2$ and therefore the fragmentation function $D_\mathrm{h}^\mathrm{A}(x) \propto (1-x)$ when $x \to 1$.

An analysis analogous to that presented for $\mathrm{e}^+\mathrm{e}^-$ annihilation may be performed for any $\mathrm{h} \to \mathrm{h}'$ vertex.

Noting that in the QCD description of $\mathrm{e}^+\mathrm{e}^-$ annihilation one uses diagrams where the first produced $\mathrm{q\bar{q}}$ pair may be very far off mass shell (up to $p_\mathrm{q}^2 \approx s$), the authors of [29, 30] propose that in a typical hadron-hadron collision the interacting quarks may also become highly virtual. This does not contradict the absence of high-p_T particles in those collisions, if a longitudinal momentum transfer non-vanishing at high energies characterizes the underlying quark-quark interactions. This model has not been elaborated to the level of detailed comparison with existing data. One should note, however, an interesting prediction of jet universality with non-linear rescaling of energy between $\mathrm{e}^+\mathrm{e}^-$ and hadron-hadron data [31]. The same central multiplicity is expected in $\mathrm{e}^+\mathrm{e}^-$ at energy $\sqrt{s}$ and in hadron-hadron scattering at energy s/Λ, where Λ is a constant of the order of 1 GeV. This is an alternative to the universality with linear energy rescaling, mentioned in Sub-Sect. 4.3.4.

The spin structure of hadrons, mass effects and the interference between various diagrams makes the direct use of QCD less attractive than originally expected, even when we limit ourselves to diagrams of lowest possible order (as is usually done without really convincing justification). Thus, in spite of the impressive success of QCR to predict reasonably well the x-dependence at $x \to 1$ for structure functions *and* fragmentation functions as well as for inclusive spectra in hadronic collisions, one cannot claim that they may directly justify or contradict models like QRM (or others to be discussed in the following). The remaining freedom in choosing dominant diagrams allows us always to avoid serious discrepancies between QCR and other models, in particular since the applicability range of QCR is not very well defined, except in detailed QCD calculations. The resonance effects are also troublesome for QCR. Nevertheless, QCR are very valuable for rough estimates of expected spectra wherever no data exist and for a systematics of all the existent data on fragmentation processes.

To complete this discussion, we should note that low-order QCD diagrams have also been used to predict the features of particle production in the central region (say $x < 0.1$) and to relate the spectra in hadronic collisions to those of $\mathrm{e}^+\mathrm{e}^-$ annihilation [28]. We shall not discuss this work in detail here, but only mention that calculating gluon bremsstrahlung in the lowest order of perturbative calculus, one finds that for $\mathrm{e}^+\mathrm{e}^-$ the average gluon density at rapidities around zero is growing logarithmically with energy. In contrast, for hadron-hadron interactions mediated by gluon exchange and followed by gluon bremsstrahlung, the leading diagrams cancel. This results in a constant asymptotic limit for the gluon density in the same order of calculations for hadron-hadron collisions.

5.3 The quark recombination model

5.3.1 The idea

The main idea of the quark recombination model (QRM) was formulated by Goldberg [32] and Ochs [33]: A fast final state meson containing one of the valence quarks of the initial hadron has a longitudinal momentum spectrum largely determined by the momentum of this quark before the collision. This explains the success of the phenomenological relation

$$\frac{1}{\sigma_{\text{inel}}}\frac{\mathrm{d}\sigma}{\mathrm{d}x}(\mathrm{A}\to\mathrm{M}) \propto F_{\mathrm{q}}^{\mathrm{A}}(x) , \tag{5.21}$$

where q is the valence quark common to hadron A and meson M and the DIS structure function $F_{\mathrm{q}}^{\mathrm{A}}(x)$ describes the x distribution of the valence quark q in hadron A. It is a straightforward generalization to describe the production of baryons having a valence quark or diquark in common with the initial hadron, although the diquark x distribution is less directly related to DIS data. As we can see, QRM supplements the ideas of quark combinatorics in the fragmentation region by defining the shape of the x distribution for 'favored' fragmentation processes.

The success of relations (5.21), although sometimes quite impressive, should be taken with some caution. Since exact scaling is violated, the structure functions as measured in deep inelastic lepton-hadron scattering actually depend on the squared momentum transfer Q^2. Which value of Q^2 should one choose for comparing the two sides of relation (5.21)? Since one has to do with soft processes, it is customary to use 'moderate' Q^2, in the range of 1-5 GeV^2. There, the scale violating effects are not so crucial for reasonably large x (say $x > 0.2$), but only few attempts have been made to justify this choice. We shall come back to this question later.

A further question arises, whether for the left-hand side of (5.21) cross sections should be used at some fixed value of transverse momentum p_{T} or integrated over the full p_{T} range. Here, however, the approximate factorization of x and p_{T} dependence makes the procedure less ambiguous.

One recognises as the main physical assumption in the interaction mechanism of QRM that the valence quarks just 'fly through' without significant change of their momenta. This is not quite the same picture as used in the original quark combinatorics approach, where one of the valence quarks was actually responsible for the central production and did not appear in the fragmentation region. This is, however, of no practical importance as long as only single-particle distributions in $\mathrm{B}\to\mathrm{M}, \mathrm{M}\to\mathrm{B}$ and $\mathrm{M}\to\mathrm{M}$ processes are investigated. For the $\mathrm{B}\to\mathrm{B}$ processes, as already noted, the original assumptions were modified in the quark combinatorics calculations of Van Hove and collaborators and now correspond to those of QRM.

The idea of non-interacting quarks was actually proposed before the formulation of QRM by Pokorski and Van Hove [34] to account for the large average values of Feynman-x carried by a final-state nucleon (leading-nucleon effect). The fact that this value is approximately $\frac{1}{2}$, apparently the same as the average fraction of proton momentum carried by quarks (as seen in DIS), had led these authors to a simple picture of hadronic collisions. According to this picture, valence quarks fly through undisturbed and consecutively only undergo some final-state interaction to recombine into a hadron, whereas gluons interact strongly and are responsible for multiple production (see also [35–37]).

5.3.2 Detailed modelling

The first detailed model based on the recombination idea was proposed by Das and Hwa [38]. These authors assumed that (5.21) was an approximation of the correct equation describing the recombination of valence quark $\mathrm{q_v}$ and sea antiquark $\overline{\mathrm{q}}_\mathrm{s}$ into the final meson M:

$$\frac{1}{\sigma_{\mathrm{inel}}}\frac{\mathrm{d}\sigma}{\mathrm{d}x}(\mathrm{A}\to\mathrm{M}) = c\iint F^{\mathrm{A}}_{\mathrm{q_v}}(x_1)F^{\mathrm{A}}_{\mathrm{q_s}}(x_2)(1-x_1-x_2)R(x_1,x_2,x)\delta(x-x_1-x_2)\mathrm{d}x_1\mathrm{d}x_2\,. \tag{5.22}$$

Here, instead of one single structure function one uses two, multiplied by a phase-space factor $(1-x_1-x_2)$. The 'recombination function' $R(x_1,x_2,x)$ could be chosen only from plausibility arguments (originally the form x_1x_2/x^2 was used).

We see that for a sea structure function much steeper than the valence one, and for the recombination function $R(x_1,x_2,x)$ relatively flat, one can get a reasonable approximation by using $\int F^{\mathrm{A}}_{\mathrm{q_v}}(x_1)F^{\mathrm{A}}_{\mathrm{q_s}}(x_2)(1-x_1-x_2)\mathrm{d}x_2 \approx F^{\mathrm{A}}_{\mathrm{q_v}}(x_1)$, $R(x,0,x)\approx$ constant, thus recovering relation (5.21). Actually, the errors introduced by these two approximations cancel to a large extent and it is relatively easy to arrange (5.21) to be a good approximation of (5.22) in a wide x range. Again, similar formulae may be constructed for baryon production, although both the three-quark structure function and the recombination function for three quarks are even more arbitrary. We shall comment later on the modifications of the Das-Hwa formula (5.22) introduced by various authors.

Some caution is necessary when one tries to apply the model to the production of particles which have no valence quark in common with the initial hadron. These should be formed from sea quarks and sea antiquarks alone (so-called 'unfavored' processes). Although the general prediction of a steeper x-dependence is confirmed by the data in many cases, the quantitative description is less successful. This may be related to two effects.

Firstly, due to the lack of statistics in the high-x region, it was originally necessary to use data for $|x|$ below 0.5. In this region, the original model as defined by formula (5.22) was known to fail even for 'favored' fragmentation processes (where the initial and final hadron do have a common valence quark). For 'unfavored' processes, such as $p \to K^-, \pi^- \to K^+$, etc, the underestimate of production spectra at low x is even stronger.

The way out of the difficulty is known as 'sea enhancement' [39], where it was argued that the sea-quark structure functions as determined by deep inelastic processes do not really reflect the distribution of partons available for recombination: Also the gluons from the initial hadron may contribute (possibly dissociating first into $q\overline{q}$ pairs) and the hadronic interaction is bound to disturb the original parton distribution. Both effects enhance the density of available slow ('sea') quarks and antiquarks. Thus, the normalization of the function $F_{q_s}^{A}(x_2)$ in (5.22) should be simply adjusted to fit the data for moderate x. Although quite convincing, this proposal limits the predictive power of the recombination model for low Feynman-x values. It may be improved by fixing the 'enhanced sea' normalization from sum rules assigning *all* the proton momentum to valence and 'enhanced sea' quarks and antiquarks, i.e., assuming full conversion of gluons into $q\overline{q}$ pairs [40].

A second effect which may disturb the observed 'unfavored' fragmentation spectra is resonance production. Actually, all the spectra are bound to be affected by the contributions from resonance decay and those contributions may compete with the 'sea enhancement' mechanism described above to explain the observed excess of low-x hadrons [41]. The resonance effects are, however, particularly striking for 'unfavored' fragmentation processes. Here, a resonance may be produced by a 'favored' process and decay to a relatively high-x 'unfavored' product.

Examples for the 'unfavored' processes mentioned above are $p \to \Sigma^{*0} \to K^- p$ and $\pi^- \to K^{*0} \to K^+\pi^-$. A particularly strong effect may be expected for relatively heavy resonances decaying into light particles, as in the process $\pi^- \to \rho^0 \to \pi^+\pi^-$, where it is quite likely for the final π^+ to achieve high x, although the 'direct' $\pi^- \to \pi^+$ fragmentation spectra should fall very steeply in this region. Indeed, Monte Carlo simulations [42] show that in this process at high energy, almost all the events with a positive pion at $x > 0.5$ may be explained by ρ production and its decay.

One must admit that these effects make the recombination model predictions much less clear and unique than the original picture. Only for the high-x region and favored processes may one hope that the resonance effects will not be crucial and, fortunately, those are just the cases where the relations of the type (5.22) and their simplified counterparts (5.21) work best. In any more sophisticated predictions of the model, resonance effects should always be taken into account.

5.3.3 Specific choices of structure and recombination functions

In a more general form, the Das and Hwa formula (5.22) may be written as

$$\frac{1}{\sigma_{\text{inel}}}\frac{d\sigma}{dx}(h \to h') = \int F\{x_i\}R\{x_i\}\delta(\Sigma x_i - x)\prod_i dx_i \ , \tag{5.23}$$

where $i = 1, 2$ if h′ is a meson and $i = 1, 2, 3$ if h′ is a baryon. For the originally considered 'favored' process N → M, where the meson M may contain one of the valence quarks of the initial nucleon N, one usually only considers the recombination of a valence quark q_v with a sea antiquark $\overline{q}_s$.

In Eq. (5.22), the *structure function* was chosen as

$$F(x_1, x_2) = F_v(x_1)F_s(x_2)\rho(x_1, x_2) \ , \tag{5.24}$$

where a phase-space factor $\rho(x_1, x_2)$ behaving as $(1 - x_1 - x_2)$ at $x_1 + x_2 \to 1$ was introduced to assure vanishing of the spectrum (5.23) at $x \to 1$. The factorized single-quark distributions were taken from the Feynman-Field parametrization of deep inelastic lepton-hadron scattering, with the sea distribution enhanced by an arbitrary normalization factor.

The specific choice of $\rho(x_1, x_2)$ is rather arbitrary and the generalization for the three-quark function needed to describe the fragmentation into baryons is not obvious. Therefore, one has proposed [43] to use a generalized Kuti-Weisskopf [44] model for the full nucleon structure function. In this model, the joint x distribution for the set of quarks $\{i\}$ has the form

$$F\{x_i\} = \sum_{v} \left(\prod_{i \in v} h_i^v(x_i) \right) \left[\prod_{s_m \in s} \left(\prod_{i \in s_m} \frac{h_i^s(x_i)}{n_m!} \right) \right] G_v(1 - \Sigma x_i) \ , \tag{5.25}$$

where, v denotes a sub-set of valence quarks in set $\{i\}, s_m$ denote sub-sets of sea quarks of given flavor and n_m is the number of sea quarks in each sub-set. As we can see, sea quarks are indistinguishable, whereas all valence quarks are 'labelled'. The h_i^v and h_i^s are so-called primitive functions, originally fully determined in the model, but in the 'generalized' version allowed to depend on a few free parameters. Finally, $G_v(1 - \Sigma x_i)$ is a correlation function resulting from a sum and integration over all possible configurations of the other partons of the initial hadron (not included in set $\{i\}$), which contains in particular energy-momentum conservation effects. The free parameters of the function $F\{x_i\}$ should be fitted to selected fragmentation data [45] or, as in the earlier version, chosen to agree with the behavior at $x \to 0$ or $x \to 1$ expected from Regge analysis. Obviously, the model allows one to apply QRM to meson fragmentation, for which structure functions are poorly known. It has been noted [43] that, using the Kuti-Weisskopf function, one can get along with a more modest enhancement of the sea than with the *ad hoc* phase-space factor $\rho(x_1, x_2)$.

An obvious drawback of the original Kuti-Weisskopf model is the absence of spin and quark flavor effects, which have to be introduced separately [45] if we want to account, for example, for the difference of π^+ and π^- spectra in proton fragmentation. We should note also that, as seen in (5.25), more than one term should be taken into account, since the recombination of a sea quark with a sea antiquark may also contribute to 'favored' meson spectra. For the recombination into a baryon, e.g. p → Λ, three such terms may contribute, corresponding to the participation of two, one or none of the initial valence quarks. If we use only the term with maximal number of valence quarks, the model should only be applied in the restricted range of high x values, where sea contributions are less important.

The *recombination function* is even more arbitrary. The original choice was

$$R(x_1, x_2) = x^{-2} x_1 x_2 \beta \ , \tag{5.26}$$

assuring the correct kinematical zeros at $(x_1, x_2) \to (0, x)$ or $(x, 0)$. The constant β could, in fact, be a smooth function of x_1/x_2 remaining finite at 0 and ∞. Another choice suggested by the short-range character of correlations observed in the rapidity variable was advocated by [43]. These authors use

$$R(x_1, x_2) \propto \exp\left[-\left(\frac{y_1 - y_2}{\Delta} \right)^2 \right] \ , \tag{5.27}$$

but for the adjustable parameter $\Delta \approx 2$, these two choices give similar results [46].

Efforts were made to reduce the arbitrariness of R for mesons and baryons using processes for which the 'wavefunction', i.e. the distribution of available quarks, is known. Such a situation exists in photoproduction, where the dominant part of the quark and gluon contributions may be calculated. Chang and Hwa [47] considered this process, finding support for the standard choice of $R\{x_i\}$.

Another possibility of determining $R\{x_i\}$ was discussed in [48], using the J/ψ decay believed to proceed via a three-gluon intermediate state. Assuming the conversion of gluons into $q\overline{q}$ pairs, one can recover the recombination function from the hadron spectra. The results are compatible with the proposed forms, although the normalization of the baryon recombination function is found higher than expected.

Finally, the constraints on the choice of the recombination function follow from the so-called jet calculus based on the QCD rules for parton production vertices [49]. This was discussed by various authors. In general, the popular phenomenological choices were found to be acceptable [50], although the presence of more contributions was advocated [51]. In principle, such an analysis should allow us to remove an arbitrariness in the choice of the recombination function, but the results are not fully conclusive and the model becomes unavoidably very complicated when many terms are introduced.

Many competing versions of the QRM, often differing by the specific choice of the structure function or the recombination function, were used to describe more or less restricted classes of reactions. One should mention here the efforts to improve the Kuti-Weisskopf structure function used in the model [52,53], the analysis of differences between fragmentation in annihilation and non-annihilation processes [54].

All those procedures, however, are not accurate enough to determine the exact shape of the recombination function to be recommended for the model. A more general program aimed at the removal of the arbitrariness of wave functions and the recombination function in QRM was advocated by Hwa and collaborators [55] (see Sub-Sect. 5.3.4 below).

The last doubt about the applicability of the recombination model concerns the x range very close to 1 (say $x > 0.9$). Two serious problems appear here. First, the diffractive dissociation processes are known to produce a very strong peak at $x \approx 1$ for a final hadron carrying the same quantum numbers as the initial one. This cannot be described by the QRM, and diffractive events should be removed from inclusive data before testing the model, as is always done with elastic events. This condition is, in fact, common to all the parton models considered here. Elastic and non-elastic diffractive processes are described as shadow effects of non-diffractive processes [56, 57] and should be calculated separately, according to corresponding rules (see Chapter 1). It should be noted that separating high-mass diffractive excitation and double-diffractive processes is not easy experimentally and the resulting 'non-diffractive' spectra should be taken with caution.

Even for the non-diffractive spectra, however, in many cases the distribution seems to fall less steeply in the near vicinity of $x = 1$ and the n powers from fits by the form $C(1-x)^n$ performed for moderate x (say $0.3 < x < 0.9$) are usually too high to describe properly the data at the highest x. This was noted already by Van Hove [36], who proposed to add incoherently a 'triple-Regge' term to QRM spectra. This term is described by the Mueller-Regge model of particle production and may be understood in the parton language as a replacement of one of the initial valence quarks by another one (or $\overline{\mathrm{q}}$ by qq and vice versa in $\mathrm{M} \to \mathrm{B}$ and $\mathrm{B} \to \mathrm{M}$ processes). Such 'quark replacement' may allow the other valence quarks to pass undisturbed (with sea and gluons) from initial to final hadron. Therefore, the x dependence predicted by the Mueller-Regge model is less steep at $x \to 1$ than that predicted by QRM. At lower x, however, such a term becomes negligible, corresponding to a small probability of recombination of many initial partons to a single hadron. The independence of 'triple Regge' and 'QRM' terms and, in particular, a no-interference assumption does not look quite convincing. However, the phenomenological successes of such an approach seem to justify it [58].

5.3.4 The valon model

The development of the recombination model went along various different lines. As noted in the preceding subsection, Van Hove and collaborators [8, 10] as well as the St. Petersburg group [6, 59] concentrated on deriving the predictions of the relative probabilities for the production of different fast hadrons, based on the quark combinatorics approach. Many authors discussed different choices of structure functions and recombination functions, trying to justify them from QCD-based jet calculus. Doing away with arbitrary functions, the 'valon' model of Hwa and collaborators has become particularly popular [47,50,55,60–63]. In this model, the hadrons are composed of constituent valence quarks (called 'valons'), which in turn reveal universal parton structure when probed at sufficiently high momentum transfer. In soft hadronic collisions, one of the valons in each hadron interacts, initiating the fragmentation of all valons into partons, which in turn recombine, becoming valons of the produced fast hadrons. At first glance, the model looks rather complicated, but different functions describing the subsequent dissociation and recombination processes are related to the structure functions as measured in deep inelastic processes and are subject to the QCD rules of scale breaking. Thus, the final predictions for hadron-hadron fragmentation appear to be practically parameter-free.

We shall present here, as an example of model application, the outline of the calculation of proton fragmentation into mesons. First, the x distribution of valons in a proton can be determined from the DIS data on proton structure and its Q^2 dependence. This is because all the scale-breaking Q^2 effects in the proton structure function are assumed to come from changing resolution in probing the parton structure $F^{\mathrm{v}}(x/x', Q^2)$ of a valon, whereas the valon distribution $G_{\mathrm{v}}^{\mathrm{h}}(x')$ itself is Q^2-independent:

$$F^{\mathrm{h}}(x, Q^2) = \sum_{\mathrm{v}} \int_x^1 \mathrm{d}x' G_{\mathrm{v}}^{\mathrm{h}}(x') F^{\mathrm{v}}(x/x', Q^2) \ . \tag{5.28}$$

Assuming the Q^2 evolution at high Q^2 as predicted by QCD, using a moments expansion of (5.28) and remembering correct normalization and average x related to $G_{\mathrm{v}}^{\mathrm{h}}$, one can determine the function $G_{\mathrm{v}}^{\mathrm{h}}(x)$ usually assumed to have a form

$$G_{\mathrm{v}}^{\mathrm{h}}(x) = g_0 x^{j-1}(1-x)^{k-1} \ , \tag{5.29}$$

where normalization conditions yield $j = k/2, g_0 = [B(j,k)]^{-1}$ with B being the Euler β function. A value of $k \approx 2.5 - 3$ is found to fit the data. A more general, although not very strict, argument suggests that as $x \to 1, G_{\mathrm{v}}^{\mathrm{N}}$ should behave as $(1-x)^{k-1}$ if the nucleon structure function behaves as $(1-x)^k$. This would give a similar prediction for k. Note that analogous estimates can be made for mesons and that the multivalon distribution (i.e. the full hadron structure function in valon representation) can be obtained by symmetry principles and sum rules.

Now, the recombination fuction of valons into a hadron is simply the same function as the multivalon distribution in that hadron, since the valon content of a hadron is completely defined. This yields, for example, for pions just the original form (5.26) with $\beta = 1$. Thus, the last unknown elements in the model are the fragmentation functions $F^{\mathrm{v}}(x/x', Q^2)$ describing parton distributions within the valons. They should be estimated separately for cases when parton and valon flavors are the same (favored) and different (unfavored). This is again done by using proton structure functions at relatively low Q^2 and solving moment equations. The evolution of partons into valons of final-state hadrons assumes a sea enhancement as in original QRM, corresponding to gluon conversion enhancing the number of available quarks.

The results of the model agree well with $\mathrm{p} \to \pi^+$ fragmentation data [55, 64]. Flavor effects should be introduced into valence distributions to attempt wider phenomenological application. Indeed, results for p $\rightarrow \pi^{\pm}$ fragmentation with F^{v} different for the u- and the d-valon in a proton look encouraging [65]. However, other versions of the valon model with different choices of relevant functions claim similar successes [66].

In another version, Hwa [67] and other advocates of the valon model do not regard the use of DIS lepton-hadron data, but estimate the valon spectra (and, as a result, the recombination function) from electromagnetic form factors at low Q^2 [68]. Then, the parton distribution inside valons may be determined from one fragmentation process yielding absolute predictions for the spectra of other hadrons [69]. In this approach, one avoids an uncertainty related to the choice of effective Q^2 to which DIS data should be extrapolated to extract structure functions. One should stress that the success of Ochs' relation (5.21) then becomes rather accidental, due to a fortunate choice of Q^2 for DIS data in the original comparison, and to the fact that the Q^2 dependence is rather weak for the relevant range of x.

Recent success could be booked by the model on the questions of flavor asymmetry in the light-quark sea and the internal spin structure of the proton. From the violation of the Gottfried sum rule (NMC, E665) and a Drell-Yan asymmetry (NA51), one deduces that flavor symmetry is strongly violated in the light-quark sea of the proton ($\bar{\mathrm{u}}/\bar{\mathrm{d}} \approx 0.5$). Furthermore, only 1/3 of the proton spin is found to be carried by quarks and antiquarks. Finally, SMC reports an important difference between the proton and neutron spin structure functions at small x.

There obviously is a bridging gap between current and constituent quarks in the proton, reminescent of that having led to the valon approach discussed above. Valons are non-overlapping systems each containing a valence quark and its own sea, internally resolved only from a certain Q^2 threshold onwards. Such a picture has been tried with success [70] to explain the above phenomena.

The recombination picture can be used to determine the valence quark distribution in mesons [26, 64, 71–73], for which no direct information from deep inelastic lepton interactions exists. The results can be given in terms of the power n of the $(1-x)$ distribution of the valence quarks. For a pion, it follows from charge conjugation and isospin invariance that the quark distribution function is the same for both valence quarks. A value of $n = 1.0 \pm 0.1$ has been obtained for the pion structure function. For a kaon the situation is expected to be non-symmetric. The power n is indeed larger than unity for the non-strange valence quark in the kaon, while it is smaller than unity for the strange valence quark. These results are compatible with those extracted via the Drell-Yan model from μ-pair production [74]. One observes that meson valence quarks are harder than those in the nucleon and that strange valence quarks are harder than non-strange ones.

Within the valon model, the full meson structure function can be predicted parameter-free from the universality (i.e. independence from the type of the parent hadron) of F^{v} and the symmetry of u and $\bar{\mathrm{d}}$ valons in the π^+. A recent comparison [75] to ZEUS results on the pion structure function F_2^{π} from leading neutron

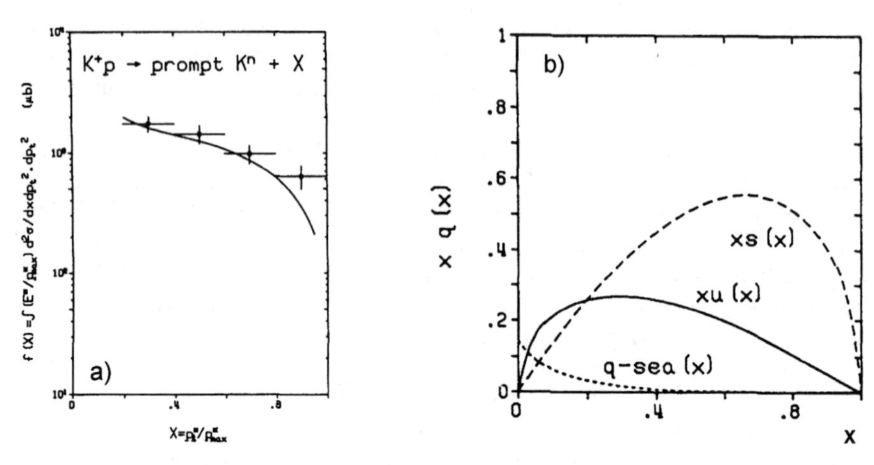

Figure 5.7. a) The Feynman-x dependence of $K^+p \to K^n$ prompt at 70 GeV/c, compared to a fit of the valon model. b) The quark distribution functions as obtained from the fit to the data in a) [77].

production in e^+p collisions [76] not only gives good agreement with the data but also helps to resolve an ambiguity in the normalization of the data.

The real issue, however, is the kaon structure function and the valon model has, indeed, been used to extract it from low-p_T K^n production in 70 GeV/c K^+p collisions [77] and $K^+ \to \pi^\pm$ (and $\pi^+ \to K^\pm$) fragmentation [64] data at 100 GeV/c beam momentum and $p_T = 0.3$ GeV/c [78]. The only free parameter (essentially the power of the strange valence quark distribution) is determined from the x-dependence of K^n production in Fig. 5.7a. The quark distribution functions obtained are given in Fig. 5.7b. The strange quark is observed to be much harder than the u-quark in the K^+. Except for the only free parameter connected with the valon distribution in the kaon, all other parameters specifying the structure of the proton and the pion have been determined independently from data on hard processes. The success of this approach affirms that the hadron structure is important also in soft hadronic processes.

5.3.5 Two-particle distributions

Apart from single-particle distributions, QRM can be used to yield simple predictions for pairs of particles. Let us consider $\pi\pi$-pair production in proton fragmentation. Two valence u-quarks of the incident proton can be used for the production of a $\pi^+\pi^+$ pair, while a u- and a d-quark are at the disposal for a $\pi^+\pi^-$ pair. However, in the case of a $\pi^-\pi^-$ pair only one pion may use the valence d-quark. Consequently, the x-spectrum of the $\pi^-\pi^-$ pair should be suppressed relative to the former cases,

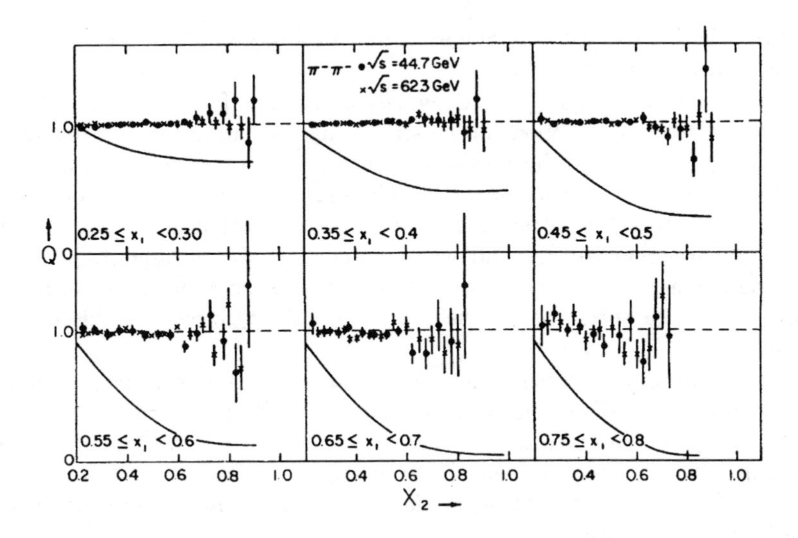

Figure 5.8. The correlation quotient Q for the $\pi^-\pi^-$ combinations in pp-collisions at $\sqrt{s} = 44.7$ GeV (•) and 62.3 GeV (×). The data are shown as a function of the Feynman-x_2 of one pion at several fixed values of x_1 of the other pion. The full curves are predictions from quark exchange counting rules, the broken line from gluon exchange counting rules [82].

in a similar way as the K^- spectrum is suppressed relative to the K^+ spectrum.

To elaborate a more quantitative prediction, one can use Eq. (5.23), in which the hadron structure function integrated over all but four quark variables is needed. Of course, different choices of structure function may lead to different results. Most authors prefer here some version of the Kuti-Weisskopf function. In particular, the authors of [45] have collected a comprehensive set of predictions for double spectra resulting from fits with free parameters to single distributions. Such predictions are in general rather successful (see e.g. [79–81] for early experimental verification).

Of particular interest in connection with the influence of the valence spectators are various types of quantum number correlations between the two different fragmentation regions and within one fragmentation region itself. Long-range correlation of two charged pions, each coming from a *different one* of the two incident protons, has been measured [24, 82] in the form of the two-pion correlation function:

$$Q(x_1, x_2) = \frac{N_{12}(x_1, x_2) \sum_{x_1} \sum_{x_2} N_{12}(x_1, x_2)}{\sum_{x_1} N_{12}(x_1, x_2) \sum_{x_2} N_{12}(x_1, x_2)} , \qquad (5.30)$$

with $N_{12}(x_1, x_2)$ being the content of a two-dimensional (x_1, x_2) bin, for pp→ $\pi\pi$X at $\sqrt{s} = 44.7$ and 62.3 GeV. As an example, the case of $\pi^-\pi^-$ is shown in Fig. 5.8. Over most of the x-range, the $\pi^-\pi^-$ (and also the $\pi^+\pi^-$ and $\pi^+\pi^+$) data are essentially uncorrelated ($Q \approx 1$), in agreement with factorization of the two fragmentation processes. This is expected for gluon exchange (dashed lines), but not for valence or sea-quark exchange (solid lines).

In the search for charge correlations *within the same* fragmentation region, the ratio R of π^+ to π^- production can be studied in association with various triggers. With a π^+ trigger, the spectator system $(\mathrm{u_v d_v d_s})$ should produce equal amounts of π^+ and π^- at fixed $\tilde{x}_\pi = |x_\pi|(1-|x_{\mathrm{tr}}|)^{-1}$. This is confirmed by the data of [83–85] in Figs. 5.9a-c. Whereas the untriggered π^+/π^- ratio rises with increasing $|x_\pi|$, the ratio for a π^+ trigger is close to unity and even slightly falling for the larger $|x_{\mathrm{tr}}|$ values. In comparing this associated ratio to the π^+/π^- ratio measured in charged-current $\bar{\nu}\mathrm{p}$ collisions where the spectator system is the same, the agreement is indeed striking (see the insert in Fig. 5.9a).

As shown in [87], a strongly enriched ud-valence system can also be obtained when requiring a fast $\bar{\mathrm{K}}^0$ in relatively low-energy $\mathrm{K^-p}$ collisions. In this case, one of the proton u-quarks can be absorbed through annihilation by the ū-quark in the K^- beam. Fig. 5.9d shows that, indeed, the associated π^+/π^- ratio is unity when $\bar{\mathrm{K}}^0$ has a large x-value (while the ratio is higher for low x values of $\bar{\mathrm{K}}^0$). A similar conclusion can be drawn from forward Λ.

For a π^- trigger, the spectator system is $(\mathrm{u_v u_v u_s})$ and a strong increase of the ratio R is expected with increasing x. Also this increase is indeed observed in Figs. 5.9a-b, as well as in $\nu\mathrm{p}$ interactions [88] having a similar (uu) spectator system. While these results naturally follow from the quark-recombination picture, calculations [86] shown in Fig. 5.9c indicate that the same trend is expected from the quark-fragmentation view to be discussed in Chapter 6.

In Fig. 5.10, the x-dependence is shown for the production of $\mathrm{K^n}\pi^\pm$ pairs, for three effective mass intervals of the pairs [89]. The solid line is drawn through the $\mathrm{K^n}\pi^-$ data of Fig. 5.10b and repeated in Fig. 5.10a. It coincides with $\mathrm{K^n}\pi^+$ in the backward hemisphere and even describes low-mass $\mathrm{K^n}\pi^+$ pairs up to $x \approx 0.7$. However, pairs containing the $\mathrm{K}^{*+}(890)$ resonance are produced with a much flatter x-dependence than those with lower or higher masses. This observation is confirmed by a resonance-to-background ratio strongly increasing with increasing x even within the K^* mass interval $0.85 < M(\mathrm{K^n}\pi) < 0.93$ GeV (not shown), and by the difference of the invariant cross sections for $\mathrm{K}^{*+}(890)$ and background in the same mass interval in Fig. 5.10c. While a power $n \approx 1.5$ is obtained for $\mathrm{K^n}\pi^\pm$ in the same mass interval, $n = 0.30 \pm 0.06$ for $\mathrm{K}^{*+}(890)$ (after exclusion of quasi-two-body and diffractive channels).

A particularly important lecture is taught by a study of the ABCCILVW Collaboration [90]. The authors show that not only is a $(1-|x|)^n$ dependence with $n = 3.8$ expected for proton fragmentation to pions or pion systems from longitudinal phase space (LPS), but also x_{tr} scaling of the $\tilde{x}$ distribution. So, the dynamics is contained not in these main features, themselves, but only in the deviations from these!

In the K^-p collisions at 110 GeV/c beam momentum, no deviations occur for $\pi^+\pi^+_{\mathrm{tr}}$, $\pi^-\pi^+_{\mathrm{tr}}$ and $\pi^+(\pi^+\pi^+)_{\mathrm{tr}}$, while positive correlation is reported beyond LPS between observed particle and trigger for $\pi^+\pi^-_{\mathrm{tr}}$, $\pi^+(\pi^+\pi^-)_{\mathrm{tr}}$ and $\pi^-(\pi^+\pi^+)_{\mathrm{tr}}$, negative

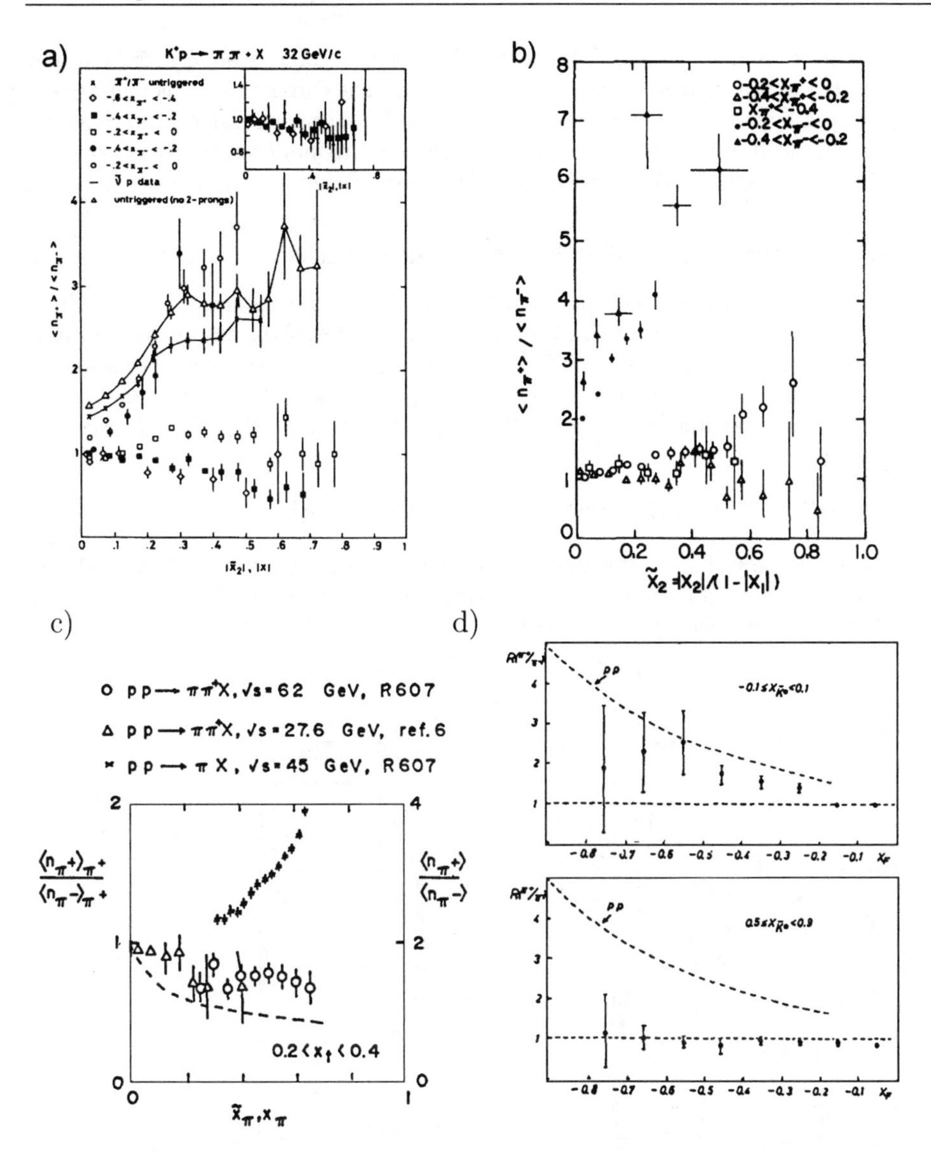

Figure 5.9. a-b) The π^+/π^- ratio as a function of $\tilde{x}_2$, in the proton fragmentation of a) 32 and b) 70 GeV/c K$^+$p collisions, associated with a π^- or π^+ trigger, for several trigger momentum intervals [83,84]. c) The π^+/π^- ratio as a function of $\tilde{x}_\pi$ (respectively x) for pp collisions at the energies indicated, associated with a π^+ trigger (left-hand scale) and no trigger (right-hand scale) [85]. The broken curve comes from QFM [86]. d) The π^+/π^- ratio as a function of x for K$^-$p $\rightarrow$ K$^0\pi^\pm$X at 16 GeV/c, with an $\bar{\text{K}}^0$ in the x-intervals indicated, compared to the π^+/π^- ratio from pp collisions [87].

one for $\pi^-\pi^-_{tr}$ and $\pi^-(\pi^+\pi^-)_{tr}$. Furthermore, violations of x_t scaling in $\tilde{x}$ exist and are related to the charges of the particles involved. The correlations are qualitatively reproduced by the recombination model, but also by the string fragmentation model (to be described in Sect. 6.3). The scaling violations are, however, better reproduced by the string fragmentation model, which includes stronger correlations.

5.3.6 Suppression of valence recombination

The observation of $n(\mathrm{K}^+ \to \mathrm{K}^{*+}) < n(\mathrm{K}^+ \to \mathrm{K}^n\pi^+)$ is in disagreement with the Kuti-Weisskopf [44]) (or longitudinal phase space [91]) model of the valence quarks in a hadron being uncorrelated except for energy-momentum conservation. From uncorrelated valence quarks, powers $n_1 > n_2$ are expected for systems inheriting one or two valence quarks from the incident hadron, respectively [46,92]. For our case this would mean

$$n(\mathrm{K}^+ \to \mathrm{K}^{*+}) \approx n(\mathrm{K}^+ \to \mathrm{K}^0\pi^+) < n(\mathrm{K}^+ \to \mathrm{K}^{*0}) \approx n(\mathrm{K}^+ \to \mathrm{K}^0\pi^-). \tag{5.31}$$

The observation in Fig. 5.10 is in clear contradiction with the first part of (5.31). It can be understood if only one valence quark (the $\bar{\mathrm{s}}$-quark in the K^+) is responsible for K^{*0} or K^{*+} production, the remaining pion being produced from the sea.

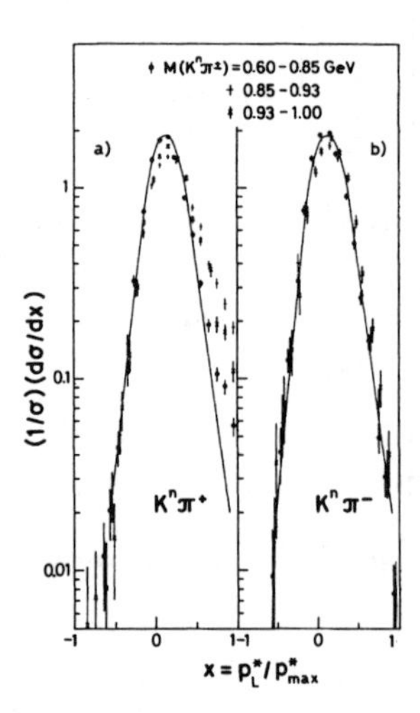

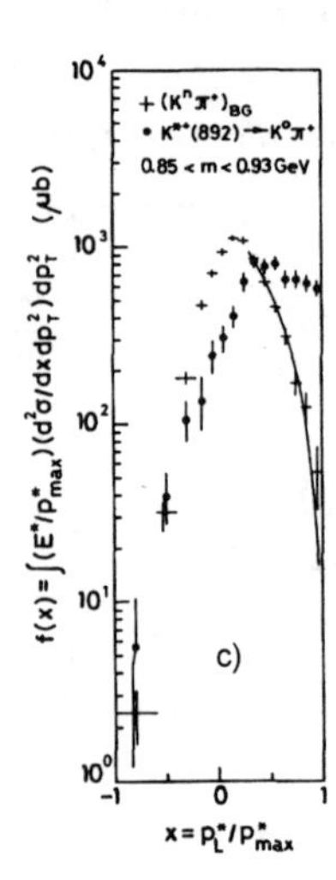

Figure 5.10. a-b) Feynman-x distribution of $\mathrm{K}^n\pi^+$ a) and $\mathrm{K}^n\pi^-$ b) systems from $\mathrm{K}^+\mathrm{p} \to \mathrm{K}^n\pi^\pm\mathrm{X}$ at 70 GeV/c, in intervals of effective mass. The curve corresponds to the data of b). c) Invariant x-distribution for $\mathrm{K}^{*+}(892) \to \mathrm{K}^0\pi^+$ and background in the corresponding mass interval. The curve is a fit by $(1-x)^n$ [89].

This conclusion of a suppression of valence-quark recombination gains support from a direct comparison of $K^{*\pm}$ and K^{*0} (or even $K^\pm$ and K^0) produced in $K^\pm p$ collisions [93–99]. In Fig. 5.11, the Feynman-x spectrum of the two resonances is compared for K^+p at 250 GeV/c [93]. Contrary to the expectation from (5.31), the two spectra look very similar. A similar effect is observed for proton fragmentation [82].

There is, however, one property of the kaon structure function which may make suppression of valence recombination look merely "kinematic" of origin. From Drell-Yan type measurements [74] and from Sub-Sect. 5.3.4, it is known that the momentum fraction carried by the strange valence quark in the $K^\pm$ is approximately twice that carried by the non-strange one. So, the kaon undergoes an interaction in a highly asymmetric valence-quark momentum state. A similar argument holds for proton fragmentation into a quark and a diquark.

On the other hand, the π^+ structure function is symmetric with respect to the u- and $\bar{d}$-valence quarks. A (near) equality of forward ρ^+ and ρ^0 cross sections would, therefore, be a direct consequence of valence-quark recombination. This equality is indeed observed in $\pi^+p \to \rho^{+,0}X$ at 250 GeV/c compared to $K^+p \to K^{*+,0}X$ in Figs. 5.11b and 5.11c [94].

The x-distribution of $\pi^+p \to \omega X$ (Fig. 5.11d) on the other hand, falls faster than $\pi^+p \to \rho^0 X$ at the same energy [94]. Also this difference between ω and ρ^0 cannot be explained by quark recombination. It can be due to additional contributions from i) diffraction dissociation and ii) one-pion exchange (OPE) contributing to the ρ, but not to the ω cross section.

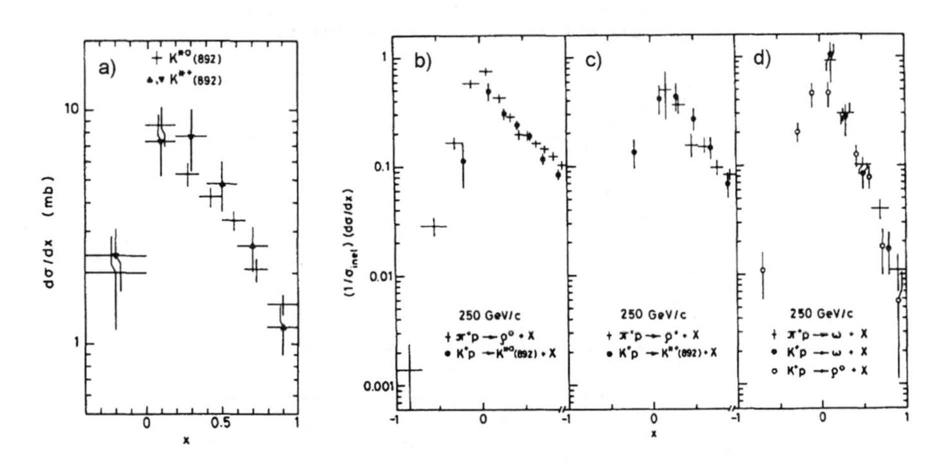

Figure 5.11. a) Feynman-x distribution of $K^{*0}(892)$ and $K^{*+}(892)$ produced in K^+p collisions at 250 GeV/c [93]. b-d) Normalized Feynman-x distributions of vector mesons produced in π^+p and K^+p collisions at 250 GeV/c [94].

Furthermore, the shape of the dσ/dx spectra for, respectively, ρ^0 and K^{*0} (Fig. 5.11b) as well as ρ^+ and K^{*+} (Fig. 5.11c) are surprisingly similar. This runs contrary to the naive expectation, based on quark-recombination ideas, that the K*(892) x-spectrum be flatter than the ρ spectrum since the heavier strange valence quark of the K$^+$, fragmenting into K*(892), carries on average more momentum than a u- or d-valence quark of the π^+ fragmenting into ρ^0, ρ^+ or ω.

The diffractive contribution turns out too small to explain the difference between ρ^0 and ω production at large x. To study the contribution from OPE, the decay angular distribution dσ/dcosΘ in the transversity frame is given in Fig. 5.12 for ρ^0 and K^{*0} production at large x. In this frame, the angle Θ is defined by

$$\cos\Theta = \mathbf{N}\cdot\mathbf{n}(\pi^-) \quad , \tag{5.32}$$

with

$$\mathbf{N} = \frac{\mathbf{n}(\text{target})\times\mathbf{n}(\text{beam})}{|\mathbf{n}(\text{target})\times\mathbf{n}(\text{beam})|} \tag{5.33}$$

being the normal to the reaction plane and $\mathbf{n}(\pi^-)$ a unit vector along the momentum of the π^- from the ρ^0 or K^{*0} decay in the resonance rest frame. For a pure spin J=1 state, the decay angular distribution, integrated over the azimuthal angle, can be written in terms of the resonance density matrix element ρ_{00} as [100]

$$\frac{d\sigma}{d\cos\Theta} = \frac{3}{4}\sigma[(1-\rho_{00}) + (3\rho_{00}-1)\cos^2\Theta] \quad . \tag{5.34}$$

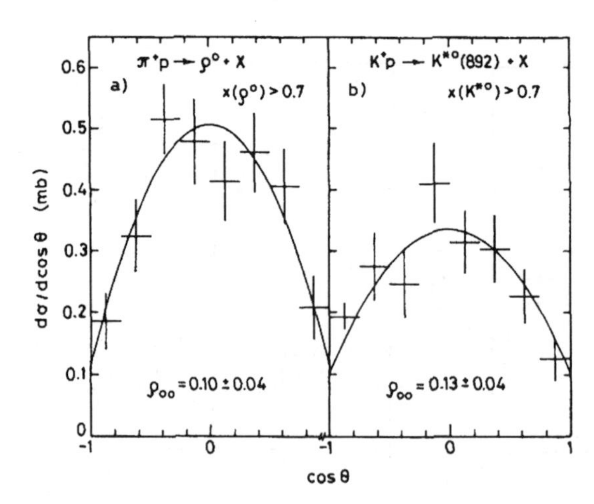

Figure 5.12. dσ/dcosΘ distributions in the tranversity frame for a) ρ^0 in π^+p interactions and b) K^{*0}(892) in K$^+$p interactions at 250 GeV/c for x $\geq$0.7. Solid curves are fits with the form (5.34) [94].

For pure OPE, one has ρ_{00}=0. A fit to the data in Fig. 5.12 gives $\rho_{00}(\rho^0)$=0.10±0.04 and $\rho_{00}(K^{*0})$=0.13± 0.04. The relative contributon of OPE to the $\rho^0(K^{*0}(892))$ cross section at $x \geq 0.7$ thus amounts to 70±12% (61±12%), while in the neighboring region $0.4 \leq x \leq 0.7$ it is ρ_{00}=0.33±0.04, a value suggesting no contribution of pion exchange. From these results one may conclude that a significant part of the cross section difference $\sigma(\rho^0) - \sigma(\omega)$ in $\pi^+ p$ collisions at large x is due to pion exchange.

However, while no spin alignment is observed in e^+e^- collisions at small x_p, the vector mesons tend to develop a preference for helicity zero occupation (i.e. large ρ_{00} in the helicity frame) at large x_p [101–103]. Since small values in ρ_{00} in the transversity frame correspond to large values in the helicity frame, the spin alignment of vector mesons produced in the fragmentation regions of meson-proton and e^+e^- reactions appear to be qualitatively similar. Helicity zero dominance is expected for the latter from a number of models based on perturbative QCD [104].

5.3.7 The fusion model

To complete our presentation of the recombination model, we should discuss the possibility that the final-state hadrons are created as $q\overline{q}$ (or qqq) systems in which the quarks originate from *both* colliding hadrons and not from only one, as assumed above. Such a model, called the quark fusion model, has indeed been constructed [105–108] and has proved quite successful at relatively low energies, in central as well as fragmentation regions. However, at high energies its contribution to the production of low-mass hadrons should be negligible [109]. Although the use of simplified kinematics for quarks which have longitudinal momenta smaller than hadron masses is rather questionable (off-mass-shell effects may become crucial), the general idea of considering quark fusion as a possible low-energy phenomenon seems to be correct. However, the probability of removing one of the valence quarks from the incident hadron by a fusion or annihilation process seems to decrease with energy more slowly than the fusion contributions to the spectra of specific hadrons. It has been observed [110, 111] (see Sub-Sect. 4.3.3) that target fragmentation in hadron-proton collisions, and in particular the π^+/π^- ratio as a function of x, still depends on the quark structure of the beam hadron h and on energy at surprisingly high energies. The systematics of this dependence suggests that for beam hadrons containing a valence $\bar{u}$-quark ($\pi^-, K^-, \overline{p}$), one of the proton valence u-quarks can be annihilated and the π^+/π^- enhancement builds up at much higher energies than for proton fragmentation after interaction with a positive beam. Of course, asymptotically proton fragmentation should not depend on the beam, but the results suggest that 'fusion contamination' becomes negligible only at cms energies above 50 GeV. This should serve as a warning against the use of models assuming the independence of beam and target fragmentation (as all the models described here do) at cms energies below 20-30 GeV.

5.3.8 Hyperon polarization

Parity violation in Λ^0 decay allows for a mixture of s and p waves in the final state, resulting in an asymmetry in its decay distribution,

$$\mathrm{d}W/\mathrm{d}\Theta = (1 + \alpha_\Lambda P_\Lambda \cos\Theta)/2 \ , \tag{5.35}$$

where Θ is the angle between the decay-nucleon momentum and the Λ^0 polarization axis and α_Λ is the Λ^0 decay asymmetry parameter. Since the latter is known from other sources, measurement of the actual asymmetry permits to estimate the Λ^0 polarization P_Λ in inclusive Λ^0 production.

According to parity conservation in strong interactions, the Λ^0 polarization vector has to be normal to the production plane,

$$\mathbf{P} \parallel \mathbf{n} = \frac{\mathbf{p}_\mathrm{b} \times \mathbf{p}_{\Lambda^0}}{|\mathbf{p}_\mathrm{b} \times \mathbf{p}_{\Lambda^0}|} \ , \tag{5.36}$$

where $\mathbf{p}_\mathrm{b}$ is the cms momentum vector of the incident particle and $\mathbf{p}_{\Lambda^0}$ that of the produced Λ^0.

Non-zero Λ polarization in proton fragmentation of proton-proton and proton-nucleus collisions is indeed known to persist up to high energies [112–128]. In Fig. 5.13a the pp points show a negative polarization increasing in absolute value with increasing transverse momentum p_T. Accurate data come from [120] and [121] (see Fig. 5.13b and c). From scattering off deuterium and other nuclei, no difference is found in polarization of Λ produced on proton and neutron targets [129, 130]. Non-zero polarization transverse to the production plane has also been observed in meson-proton collisions [131–143] and in ep scattering [144].

One can conclude that:

(i) Λ^0 produced in proton fragmentation of hadron-proton, hadron-nucleus and electron-proton scattering are polarized transverse to the production plane, along the negative $(\mathbf{p}_\mathrm{p} \times \mathbf{p}_\Lambda)$ axis.

(ii) The polarization increases with increasing p_T and $|x|$ of the Λ^0.

(iii) The polarization is independent of the beam energy (Fig. 5.14) and the projectile type.

(iv) $\bar{\Lambda}$ and protons are not polarized in the proton fragmentation region.

(v) Ξ^0, Ξ^- have the same polarization as the Λ^0.

(vi) $\Sigma^{+,0}$ have a polarization similar to Λ^0 in magnitude, but opposite in sign.

These observations cannot be explained from the Regge model since at high energies RPR-exchange terms dominate and these terms do not give rise to polarization, neither can it be explained from QCD [145].

In the framework of the recombination model the polarization has been explained by [146]. Data for Λ, $\Sigma^{+,0}$, Ξ and $\bar{\Lambda}$ are all consistent with the observation that slow partons preferentially recombine with their spins down in the scattering plane, while

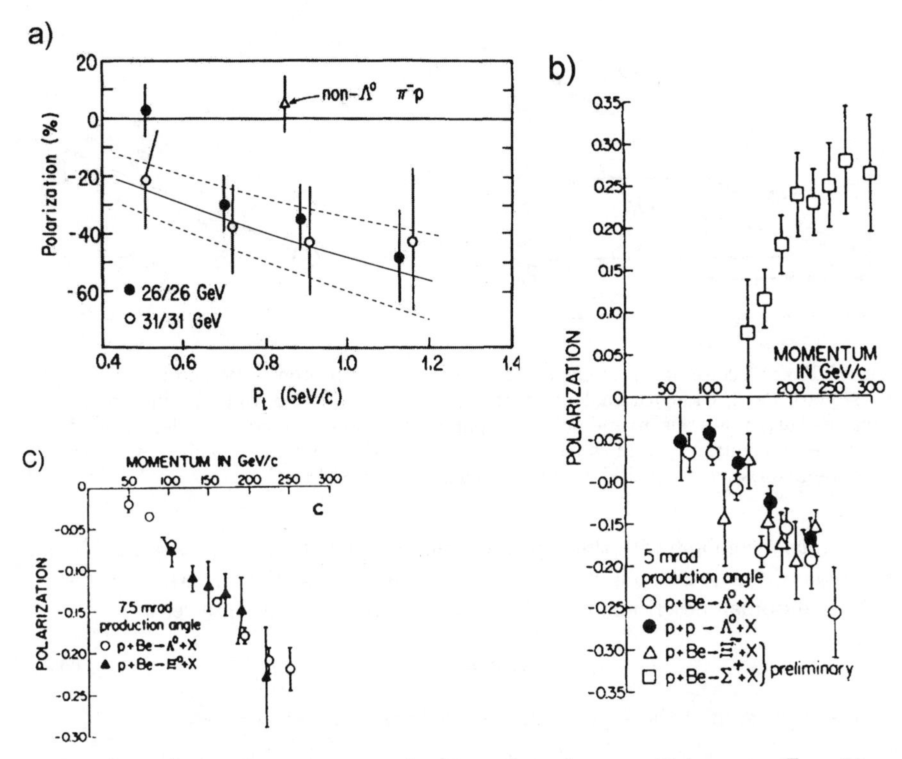

Figure 5.13. a) Λ polarization as a function of p_T for pp collisions at $\sqrt{s} = 53$ and 62 GeV [118]. A polarization measurement for a non-Λ π^-p sample is shown as an open triangle. The lines correspond to the fragmentation model description [147] (the effect of Λ production from Σ^0 or Σ^* decay not included). b), c) Hyperon polarization for pp and pBe collisions at 400 GeV/c, as a function of the hyperon laboratory momentum [120,121].

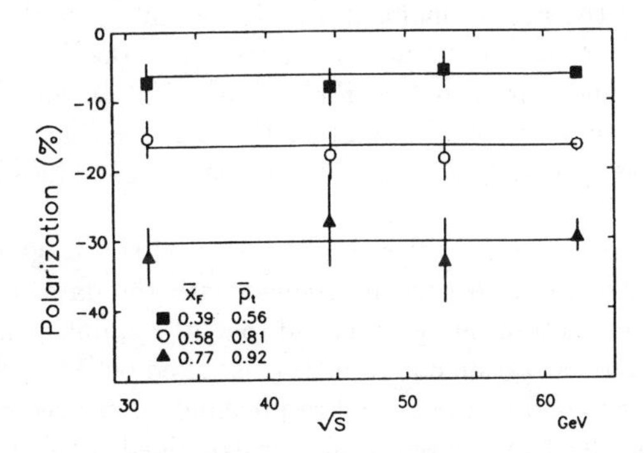

Figure 5.14. Inclusive Λ^0 polarization versus $\sqrt{s}$ for three $x - p_T$ bins. Horizontal lines are drawn at the average P_Λ value for each bin [125].

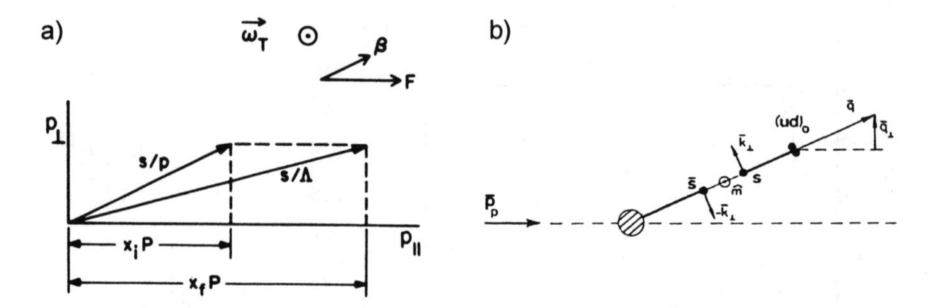

Figure 5.15. a) Vectorial directions of the s-quark in the proton (s/p) and in the Λ (s/Λ). The s-quark carries longitudinal momentum $x_i p$ in the proton and $x_f p$ in the Λ. The direction of $\boldsymbol{\omega}_\mathrm{T} \propto \mathbf{F}\times\boldsymbol{\beta}$ is up and out of the paper. b) A spin-zero ud-quark is scattered with transverse momentum $\mathbf{q}_\perp$; an s$\bar{\text{s}}$-pair is created from the color force field, their transverse momenta give rise to an orbital angular momentum in the direction $\hat{\mathbf{m}} = \mathbf{q}\times\mathbf{k}_\perp/|\mathbf{q}\times\mathbf{k}_\perp|$.

fast partons recombine with their spins up. Such a polarization can arise via Thomas precession of the quark spin in the recombination process (Fig. 5.15a).

A Hamiltonian describing the Λ^0 recombination must contain a term

$$U = \mathbf{s}\cdot\boldsymbol{\omega}_\mathrm{T} \,, \tag{5.37}$$

where $\mathbf{s}$ is the spin of the s-quark and $\boldsymbol{\omega}$ the Thomas frequency

$$\boldsymbol{\omega} = \frac{\gamma^2}{1+\gamma}\frac{\mathbf{F}}{m_s}\times\mathbf{v} \,, \tag{5.38}$$

where $\mathbf{F}$ is a force accelerating the s-quark, $\mathbf{v}$ the s-quark original velocity and m_s its mass. Since the s-quark of the proton is a sea quark, it originally carries a small fraction $x_\mathrm{s} \approx 0.1$ of the proton momentum. To recombine with the (spin-less) ud-diquark into a Λ^0, it becomes a valence quark and, therefore, has to be longitudinally accelerated into the ud direction (see Fig. 5.15a). So, $\boldsymbol{\omega}$ points out upwards the scattering plane. In order to obtain an attractive contribution to the recombination potential, $\mathbf{s}\cdot\boldsymbol{\omega}_\mathrm{T}$ has to be negative, i.e., the s-quark recombines preferentially with spin down.

A quantitative prediction for the Λ polarization at fixed x against p_T and at fixed p_T against x is shown to be in good agreement with the data of [120] and [121] in Fig. 5.16a and b. The picture can be extended to $\mathrm{e^+e^-}$ annihilation and deep inelastic scattering. Hyperon polarization data are also discussed by [148], who relates them to the scalar form of the confining potential responsible for the recombination process.

The polarization effect can, however, also be explained in the quark fragmentation picture to be discussed in Chapter 6. In this model [147], a diquark continues forward as a unit after the collision and a string-shaped color dipole field is stretched between

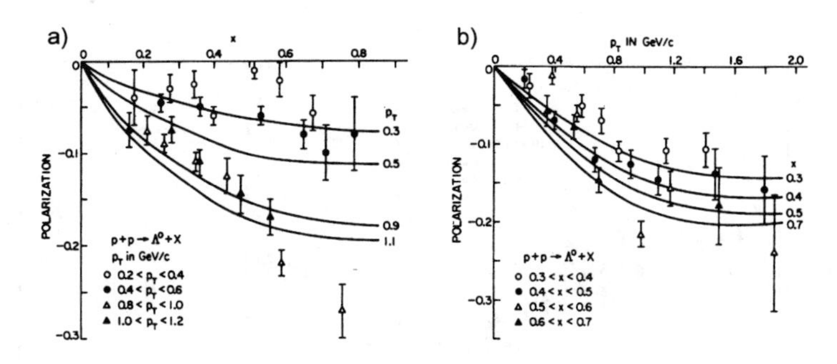

Figure 5.16. Polarization of Λ^0 from pp collisions at 400 GeV/c as a function of x of Λ [120, 121]. (b) Same as (a), but as a function of p_{T} of Λ for the x intervals given. The curves are described in the text [146].

the diquark and the central collision region (see Fig. 5.15b). This field can break up by the production of quark-antiquark pairs (as in $\mathrm{e}^+\mathrm{e}^-$ hadron production). A Λ^0 particle can be formed if an $\mathrm{s}\bar{\mathrm{s}}$ pair is produced in the field of a (ud) diquark (of isospin and spin $I = S = 0$), so that the spin of the Λ^0 is determined by the spin of the s-quark.

The transverse momentum $\mathbf{p}_{\mathrm{T}}$ of the Λ^0 with respect to the beam direction is made up of two contributions, the transverse momentum $\mathbf{q}_{\mathrm{T}}$ of the diquark (giving the direction of the field string) and the (locally conserved) transverse momentum $\mathbf{k}_{\mathrm{T}}$ of the s-quark with respect to the string direction. A pair of massless quark-antiquarks can be produced point-like, but massive quarks have to be produced at a certain distance from each other. Therefore, the pair will have an orbital angular momentum perpendicular to the string; this is assumed to be compensated by the spin of the $\mathrm{s}\bar{\mathrm{s}}$ pair. In a Λ^0 sample of definite p_{T}, we obtain an enhanced number of events with $\mathbf{k}_{\mathrm{T}}$ and $\mathbf{p}_{\mathrm{T}}$ pointing in the same direction. So the observed effect is explained by a sort of trigger bias. The curves in Fig. 5.13a show the model prediction and its upper and lower limits (without inclusion of the effect of Λ^0 production via Σ^0 and Σ^*).

A particularly important test is $\mathrm{K}^- \to \Lambda$ (or $\mathrm{K}^+ \to \bar{\Lambda}$) in the forward direction. In this case, all the asymmetry resides in the leading s- (or $\bar{\mathrm{s}}$)-quark and an asymmetry opposite to that in the target region is expected from the recombination model. As shown in Fig. 5.17, a non-zero polarization is indeed observed in the forward direction for $\mathrm{K}^-\mathrm{p} \to \Lambda$ [136]. A similar observation has been reported for $\mathrm{K}^+\mathrm{p} \to \bar{\Lambda}$ [137]. Considering the sign convention used, the asymmetry is indeed opposite to that of the Λ in target fragmentation. What is surprising is the large absolute value of P_Λ, much larger than expected from a scattering model when using reasonable color factors [149].

In another approach, proposed in [150], the s-quark is polarized by multiple scat-

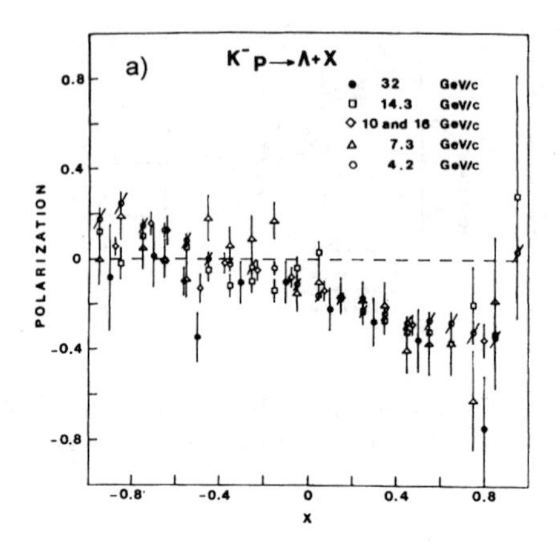

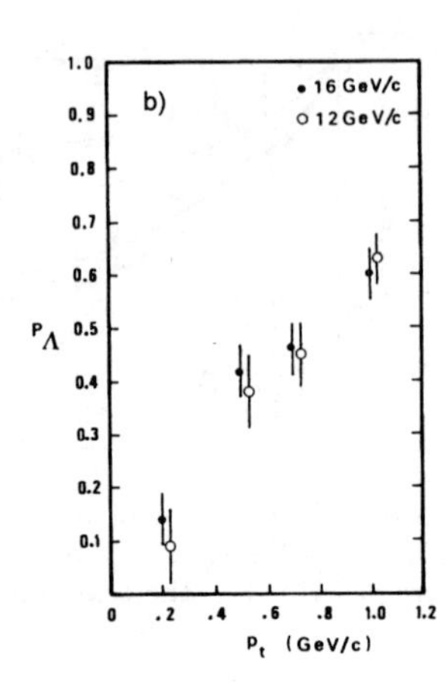

Figure 5.17. Λ polarization for $K^-p\to \Lambda+X$ as a function of a) Feynman-x [136] and b) transverse momentum p_T [141].

tering off the external color field. Since no hadronization is involved, only qualitative predictions can be derived for the Λ^0 polarization in this model.

We conclude that hyperon polarization, which cannot be explained by the triple Regge model, finds an explanation from the valence-quark composition of incident and produced particles. Data on deep inelastic scattering from polarized nucleons [151], however, show that the spin fraction carried by the quarks is much smaller than 1, so that the above quoted models can only hold approximately.

The situation is completely different in $\overset{(-)}{\nu}$p interactions. There, the interaction of the weak current picks out only quarks with negative helicity. Since the target nucleon is initially unpolarized, the remnant diquark must carry compensating polarization [153]. A polarization is then expected (for Λ only as a decay product of other hyperons) *in* the production plane. In the target fragmentation region ($x < 0$) the origin of Λ polarization could, therefore, be transfer from the polarized remnant diquark [153], polarized strange quarks from the target nucleon [154] or both. In the current fragmentation region ($x > 0$) the polarized struck quark transfers its polarization to the Λ [155]. The experimental data are in qualitative agreement with the above parton interpretation [152, 156–158].

Furthermore, polarization normal to the production plane is observed for Λ in [152] but no significant polarization is found for $\bar{\Lambda}$. Its dependence on the transverse momentum of the Λ with respect to the hadronic jet direction is in qualitative agreement with the behavior observed in the hadron-hadron collisions discussed above (see Fig. 5.18).

In high energy e^+e^- collisions, measurement of a longitudinal Λ polarization has

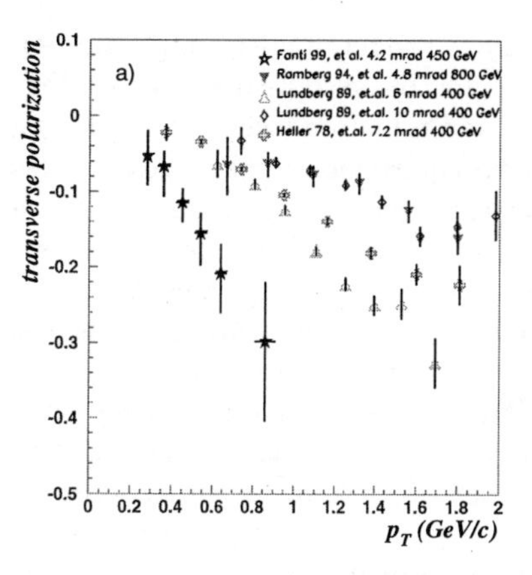

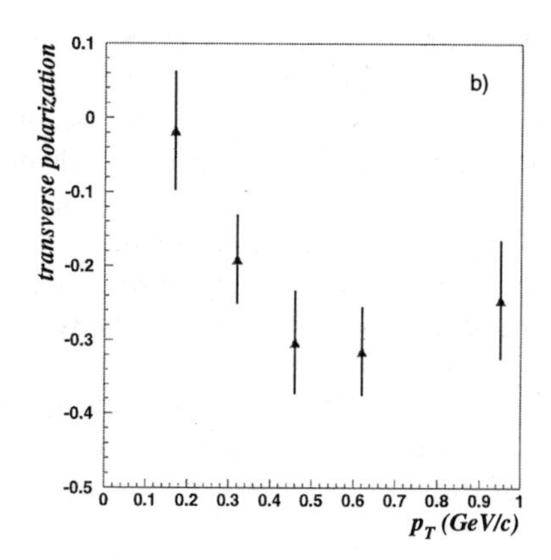

Figure 5.18. Transverse momentum dependence of the transverse Λ polarization in a) hadron-hadron collisions, b) ν CC interactions [152].

been proposed [159] as a method for measuring the polarization of primary quarks resulting from Z^0 exchange and $\gamma - Z^0$ interference. At low energies, the effect should be small and, indeed, the TASSO collaboration [160] finds longitudinal Λ (and $\bar{\Lambda}$) polarization consistent with zero. At the Z^0 energy, however, the longitudinal polarization of Λ^0 is measured to be -0.32 ± 0.07 for $z > 0.3$ by ALEPH [161], and -0.33 ± 0.08 by OPAL [162], in good agreement with the value predicted on the basis of the standard model and the constituent quark model [163].

Transverse polarization is, of course, also possible, but no significant evidence is observed [160–162].

In comparison, Λ and $\bar{\Lambda}$ polarization has also been measured in the Breit frame of polarized muon DIS on (un-polarized) nucleons [164]. Here, a simple naive-quark-model picture could be that the polarized virtual photon is absorbed by a strange quark in the target nucleon sea. This then emerges with its spin aligned in the direction of the photon spin. Hadronizing into a Λ, it is likely to become a valence quark in the Λ and the polarization of the Λ will be that of the strange quark. The results are consistent with an expected [165] trend towards a positive polarization with inceasing Feynman-x, be it with very large errors. A large negative polarization of the Λ is, however, observed at low x, suggesting that target fragmentation spills over into that region.

5.4 Conclusions

It came as a surprise to find out experimentally that the hadron structure as seen in DIS was relevant also in soft hadronic collisions. While a statistical gas of quarks and antiquarks recombining locally into hadrons can serve as a first approximation for central production of hadrons, the spectra of hadrons produced in the fragmentation regions not only depend on their own quark structure, but also its overlap with that of the initial hadron. It appears that also in soft hadron-hadron collisions interacting quarks may become highly virtual. This does not require high-p_T particles in these reactions, since longitudinal momentum transfer can become large for a quark in a particle produced at large Feynman-x. In particular, the so-called valon model uses DIS structure functions, subject to scale breaking, for a set of functions describing subsequent dissociation and recombination processes in soft hadron-hadron collisions, practically without any free parameters. This opens a way to the determination of pion and even kaon structure functions, for which no other direct measurements exist. A particular success of parton based models is the non-zero hyperon polarization observed to persist at high energies, where it cannot be explained from triple-Regge diagrams any more.

An important lesson: In comparing models to experimental data, it is absolutely essential to separate out the effects of mere (longitudinal) phase space, which often by itself can reproduce the "main features" of the data.

Bibliography

[1] K. Fiałkowski and W. Kittel, *Rep. on Prog. Phys.* **46** (1983) 1283.

[2] H. Satz, *Phys. Lett.* **25B** (1967) 220.

[3] V.V. Anisovich and V.M. Shekhter, *Nucl. Phys.* **B55** (1973) 455.

[4] J.D. Bjorken and G.R. Farrar, *Phys. Rev.* **D9** (1974) 1449.

[5] V.M. Shekhter and L.M. Shcheglova, *Yad. Fiz.* **27** (1978) 1070.

[6] V.V. Anisovich, M.N. Kobrinsky and J. Nyiri, *Phys. Lett.* **102B**(1981) 357; *Yad. Fiz.* **34** (1981) 195; ibid. 1576.

[7] V.V. Anisovich, M.N. Kobrinsky, A.K. Likhoded and V.M. Shekhter,*Nucl. Phys.* **B55** (1973) 474.

[8] J. Kalinowski, S. Pokorski and L. Van Hove, *Z. Phys.* **C2** (1979) 85.

[9] M. Bourquin et al., *Z. Phys.* **C5** (1980) 275.

[10] L. Van Hove, *Z. Phys.* **C9** (1981) 145.

[11] V.A. Matveev, R.M. Muradyan and A.N. Tavkhelidze, *Lett. Nuovo Cim.* **7** (1973) 719.

[12] S.J. Brodsky and G.R. Farrar, *Phys. Rev. Lett.* **31** (1973) 1153.

[13] S.J. Brodsky and G.R. Farrar, *Phys. Rev.* **D11** (1975) 1309.

[14] R. Blankenbecler and S.J. Brodsky, *Phys. Rev.* **D10** (1974) 2973.

[15] J.F Gunion, *Phys. Rev.* **D10** (1974) 242.

[16] S.D. Drell and T.M. Yan, *Phys. Rev. Lett.* **24** (1970) 181.

[17] G.B. West, *Phys. Rev. Lett.* **24** (1970) 1206.

[18] R. Blankenbecler, S.J. Brodsky and J.F. Gunion, *Phys. Rev.* **D12** (1975) 3469.

[19] J.F Gunion, *Phys. Lett.* **88B** (1979) 150.

[20] F.E. Low, *Phys. Rev.* **D12** (1975) 163.

[21] S. Nussinov, *Phys. Rev.* **D14** (1976) 246; *Phys. Rev. Lett.* **34** (1976) 1286.

[22] S.J. Brodsky and J.F. Gunion, *Phys. Rev. Lett.* **37** (1976) 402; *Phys. Rev.* **D17** (1978) 848.

[23] J. Benecke, T.T. Chou, C.N. Yang and F. Yen, *Phys. Rev.* **188** (1969) 2159.

[24] G.J. Bobbink et al., *Phys. Rev. Lett.* **44** (1980) 118.

[25] V. Bakken, F.O. Brevik and T. Jacobsen, *Nuovo Cimento* **A66** (1981) 71.

[26] D. Denegri et al., *Phys. Lett.* **B98** (1981) 127.

[27] J.F Gunion, *Proc. Europhysics Study Conf. on Partons in Soft Hadronic Processes*, Erice 1981, ed. R.T. Van de Walle, (Singapore: World Scientific, 1982), p.293.

[28] J.F. Gunion and G. Bertsch, *Phys. Rev.* **D25** (1982) 746.

[29] J. Kalinowski and S. Pokorski, *Acta Phys. Polon.* **D12** (1981) 989.

[30] S. Pokorski and S. Wolfram, *Z. Phys.* **C15** (1982) 111.

[31] J. Kalinowski, M. Krawczyk and S. Pokorski, *Z. Phys.* **C15** (1982) 281.

[32] H. Goldberg, *Nucl. Phys.* **B44** (1972) 149.

[33] W. Ochs, *Nucl. Phys.* **B118** (1977) 397.

[34] S. Pokorski and L. Van Hove, *Acta Phys. Pol.* **B5** (1974) 229.

[35] L. Van Hove and S. Pokorski, *Nucl. Phys.* **B86** (1975) 245.

[36] L. Van Hove, *Acta Phys. Pol.* **B7** (1976) 339; L. Van Hove, *Acta Phys. Austriaca Suppl.* **21** (1979) 621.

[37] R.P. Feynman, *Phys. Rev. Lett.* **23** (1969) 1415; *High Energy Collisions*, ed. C.N. Yang et al. (New York: Gordon and Breach, 1969), p.237.

[38] K.P. Das and R.C. Hwa, *Phys. Lett.* **68B** (1977) 459.

[39] D.W. Duke and F.E. Taylor, *Phys. Rev.* **D17** (1978) 1788.

[40] J. Ranft, *Acta Phys. Pol.* **B10** (1979) 911. .

[41] R.G. Roberts, R.C. Hwa and S. Matsuda, *J. Phys. G: Nucl. Phys.* **5**(1979) 1043.

[42] D. Cutts et al., *Phys. Rev. Lett.* **43** (1979) 319.

[43] E. Takasugi et al., *Phys. Rev.* **D20** (1979) 211.

[44] J. Kuti and V. Weisskopf, *Phys. Rev.* **D4** (1971) 3418.

[45] E. Takasugi and X. Tata, *Phys. Rev.* **D26** (1982) 120.

[46] E. Takasugi and X. Tata, *Phys. Rev.* **D21** (1980) 1838; *Phys. Rev.* **D23** (1981) 2573.

[47] V. Chang and R.C. Hwa, *Phys. Lett.* **85B** (1979) 285.

[48] L.M. Jones, *Phys. Rev.* **D26** (1982) 706.

[49] K. Konishi, A. Ukawa and G. Veneziano, *Phys. Lett.* **78B** (1978) 243 and *Nucl. Phys.* **B157** (1979) 45.

[50] V. Chang and R.C. Hwa, *Phys. Rev.* **D23** (1981) 728.

[51] L.M. Jones, K.E. Lassila, U. Sukhatme and D. Willen, *Phys. Rev.* **D23** (1981) 717.

[52] T. De Grand and H. Miettinen, *Phys. Rev. Lett.* **40** (1978) 612.

[53] H.R. Gerhold, *Nuovo Cimento* **A59** (1980) 373.

[54] M. Markytan et al., *Z. Phys.* **C9** (1981) 87.

[55] R.C. Hwa, *Phys. Rev.* **D22** (1980) 759; ibid. 1593.

[56] E.L. Feinberg and I.Ya. Pomeranchuk, *Suppl. Nuovo Cimento* **3** (1956) 652.

[57] M.L. Good and W.D. Walker, *Phys. Rev.* **120** (1960) 1854.

[58] P.D.B. Collins and A.D. Martin, *Rep. Prog. Phys.* **45** (1982) 335.

[59] V.V. Anisovich, M.N. Kobrinsky and J. Nyiri, *Yad. Fiz.* **35** (1982) 151.

[60] R.C. Hwa and M.S. Zahir, *Phys. Rev.* **D23**(1981) 2539; ibid. **D25** (1982) 2455.

[61] V. Chang and R.C. Hwa, *Phys. Rev. Lett.* **44** (1980) 439.

[62] V. Chang, G. Eilam and R.C. Hwa, *Phys. Rev.* **D24** (1981) 1878.

[63] R.C. Hwa and C.B. Yang, *Phys. Rev.* **C66** (2002) 025204.

[64] R.C. Hwa and C.B. Yang, *Phys. Rev.* **C66** (2002) 025205.

[65] J.J. Dugne, *Lett. Nuovo Cimento* **34** (1982) 279.

[66] F. Amiri and P.K. Williams, *Phys. Rev.* **D24** (1981) 2409.

[67] R.C. Hwa, in *Proc. Europhysics Study Conf. on Partons in Soft Hadronic Processes*, Erice 1981, ed. R.T. Van de Walle (World Scientific, Singapore, 1982) p.137.

[68] R.C. Hwa and C.S. Lam, *Oregon Preprint* OITS-158 (1980).

[69] L. Gatignon et al., *Z. Phys.* **C16** (1983) 229.

[70] F. Arash, *Phys. Rev.* **D52** (1995) 68.

[71] R.C. Hwa and R.G. Roberts, *Z. Phys.* **C1** (1979) 181.

[72] W. Aitkenhead et al., *Phys. Rev. Lett.* **45** (1980) 157.

[73] M. Barth et al., *Nucl. Phys.* **B191** (1981) 39.

[74] J. Badier et al., *Phys. Lett.* **93B** (1980) 354.

[75] F. Arash, *Phys. Lett.* **B557** (2003) 38.

[76] ZEUS Coll., S. Chekanov et al., *Nucl. Phys.* **B637** (2002) 3.

[77] L. Gatignon et al., *Phys. Lett.* **115B** (1982) 329.

[78] A.E. Brenner et al., *Phys. Rev.* **D26** (1982) 1497.

[79] W. Lockman et al., *Phys. Rev. Lett.* **41** (1978) 680.

[80] E. Lehman et al., *Phys. Rev.* **D18** (1978) 3353.

[81] U. Gensch et al., *Z. Phys.* **C17** (1983) 21.

[82] G.J. Bobbink et al., *Nucl. Phys.* **B204** (1982) 173.

[83] E.A. De Wolf et al., *Z. Phys.* **C8** (1981) 189.

[84] M. Barth et al., *Z. Phys.* **C7** (1981) 187.

[85] F.C. Erné, *Proc. 16th Rencontre de Moriond*, Les Arcs, vol II, ed. J. Tran Thanh Van (Editions Frontières) p.565.

[86] B. Andersson, G. Gustafson, L. Holgersson and O. Mansson, *Nucl. Phys.* **B178** (1981) 242.

[87] B. Buschbeck and H. Dibon, *Proc. 12th Int. Symp. on Multiparticle Dynamics*, Notre Dame 1981, ed. W.D. Shephard and V.P. Kenney (World Scientific, Singapore) p.129.

[88] N. Schmitz, *Proc. 12th Int. Symp. on Multiparticle Dynamics*, Notre Dame 1981, eds. W.D. Shephard and V.P. Kenney (World Scientific, Singapore, 1982) p.481.

[89] E.A. De Wolf et al., *Z. Phys.* **C12** (1982) 105.

[90] ABCCILVW Coll., R. Göttgens et al., *Z. Phys.* **C22** (1984) 205.

[91] L. Van Hove, *Rev. Mod. Phys.* **36** (1964) 655.

[92] K. Fiałkowski, *Acta Phys. Polon.* **B11** (1980) 659.

[93] NA22 Coll., N.M. Agababyan et al., *Z. Phys.* **C41** (1989) 539.

[94] NA22 Coll., M. Adamus et al., *Phys. Lett.* **B183** (1987) 425; N.M. Agababyan et al., *Z. Phys.* **C46** (1990) 387.

[95] I.V. Ajinenko et al., *Z. Phys.* **C5** (1980) 177; ibid **C25** (1984)103; P.V. Chliapnikov et al., *Phys. Lett.* **B130** (1983) 432.

[96] M. Barth et al., *Nucl. Phys.* **B223** (1983) 296.

[97] WA27 Coll., E.A. De Wolf et al., *Z. Phys.* **C31** (1986) 13.

[98] Yu.I. Arestov et al., *Z. Phys.* **C6** (1980) 101; Ma Wen-Gan et al.,*Z. Phys.* **C30** (1986) 191.

[99] R. Göttgens et al., *Z. Phys.* **C12** (1982) 323.

[100] J.D. Jackson, *Nuovo Cimento* **34** (1964) 1644.

[101] DELPHI Coll., P. Abreu et al., *Z. Phys.* **C68** (1995) 353; *Phys. Lett.* **B406** (1997) 271.

[102] ALEPH Coll., D. Buskulic et al., *Z. Phys.* **C69** (1995) 393.

[103] OPAL Coll., K. Ackerstaff et al., *Z. Phys.* **C74** (1997) 413 and 437; *Phys. Lett.* **B412** (1997) 210; G. Abbiendi et al., *Eur. Phys. J.* **C16** (2000) 61.

[104] G.R. Farrar and D.R. Jackson, *Phys. Rev. Lett.* **35** (1975) 1416; A.I. Vainstein and V.I. Zakharov, *Phys. Lett.* **B72** (1978) 368; J.F. Donoghue, *Phys. Rev.* **D19** (1979) 2806.

[105] M.B. Green et al., *Nuovo Cimento* **A29** (1975) 128.

[106] A. Donnachie and P.V. Landshoff, *Nucl. Phys.* **B112** (1976) 233.

[107] P.V. Chliapnikov, V.G. Kartvelishvili, V.V. Kniazev and A.K. Likhoded, *Nucl. Phys.* **B148** (1979) 400.

[108] A. Donnachie, *Z. Phys.* **C4** (1980) 161.

[109] M. Teper, *Rutherford Laboratory Preprint* RL-78-022, 1978.

[110] B. Buschbeck, H. Dibon, H.R. Gerhold and W. Kittel, *Z. Phys.* **C3** (1979) 97.

[111] B. Buschbeck, H. Dibon and H.R. Gerhold, *Z. Phys.* **C7** (1980) 83.

[112] A. Lesnik et al., *Phys. Rev. Lett.* **35** (1975) 770.

[113] G. Bunce et al., *Phys. Rev. Lett.* **36** (1976) 1113.

[114] V. Blöbel et al., *Nucl. Phys.* **B122** (1977) 429.

[115] K. Heller et al., *Phys. Lett.* **68B** (1977) 480.

[116] K. Heller et al., *Phys. Rev. Lett.* **41** (1978) 607.

[117] P. Ashlin et al., *Lett. Nuovo Cimento* **21** (1978) 236.

[118] S. Erhan et al., *Phys. Lett.* **B82** (1979) 301.

[119] F. Lomanno et al., *Phys. Rev. Lett.* **43** (1979) 1905.

[120] K. Heller et al., *Proc. 20th Int. Conf. on High Energy Physics*, Madison, ed. L. Durand and L.G. Pondrom, AIP Conf. Proc. **68** (1981) 61.

[121] C. Wilkinson et al., *Phys. Rev. Lett.* **46** (1981) 803.

[122] K. Heller et al., *Phys. Rev. Lett.* **51** (1983) 2025.

[123] NA23 Coll., M. Asai et al., *Z. Phys.* **C27** (1985) 11.

[124] E.C. Dukes et al., *Phys. Lett.* **B193** (1987) 135.

[125] R608 Coll., A.M. Smith et al., *Phys. Lett.* **B185** (1987) 209.

[126] B. Lundberg et al., *Phys. Rev.* **D40** (1989) 3557.

[127] E.J. Ramberg et al., *Phys. Lett.* **B338** (1994) 403.

[128] NA48 Coll., V. Fanti et al., *Eur. Phys. J.* **C6** (1999) 265.

[129] K. Raychaudhuri et al., *Phys. Lett.* **B90** (1980) 319.

[130] S. Dado et al., *Phys. Rev.* **D22** (1980) 2656.

[131] J. Bartsch et al., *Nucl. Phys.* **B40** (1972) 103.

[132] W. Barletta et al., *Nucl. Phys.* **B51** (1973) 479.

[133] A. Borg et al., *Nuovo Cimento* **22A** (1974) 559; H. Abramowicz et al., *Nucl. Phys.* **B105** (1976) 222.

[134] S.U. Chung et al., *Phys. Rev.* **D11** (1975) 1010.

[135] N.N. Biswas et al., *Nucl. Phys.* **D167** (1979) 41.

[136] M.L. Faccini-Turluer et al., *Z. Phys.* **C1** (1979) 19.

[137] M. Barth et al., *Z. Phys.* **C10** (1981) 205, Erratum **C11** (1981)271.

[138] J. Besinger et al., *Phys. Rev. Lett.* **50** (1983) 313.

[139] CERN-USSR and WA27 Coll., I.V. Ajinenko et al., *Phys. Lett.* **121B** (1983) 183.

[140] T. Haupt et al., *Z. Phys.* **C28** (1985) 57.

[141] T.A. Armstrong et al., *Nucl. Phys.* **B262** (1985) 356.

[142] S.A. Gourlay et al., *Phys. Rev. Lett.* **56** (1986) 2244.

[143] S. Barlag et al., *Phys. Lett.* **B325** (1994) 531.

[144] H. Preissner, *Proc. 10th Int. Symp. on Multiparticle Dynamics*, Goa 1979, ed. J.N. Ganguli, P.K. Malhotra and A. Subramanian p.623.

[145] G.L. Kane et al., *Phys Rev. Lett.* **41** (1978) 1689.

[146] T. De Grand and H. Miettinen, *Phys. Rev.* **D23** (1981) 1227; ibid. **D24** (1981) 2419; T. De Grand, *Phys. Rev.* **D38** (1988) 403; T. Fujita and T. Matsuyana, *Phys. Rev.* **D38** (1988) 401; G. Herrera et al., *Phys. Lett.* **B382** (1996) 201.

[147] B. Andersson, G. Gustafson and G. Ingelman, *Phys. Lett.* **B85** (1979)417.

[148] B.V. Struminskii, *Yad. Fiz.* **34** (1981) 1594.

[149] J.M. Gago, R. Vilela Mendes and P. Vaz, *Phys. Lett.* **B183** (1987) 357.

[150] J. Szwed, *Phys. Rev.* **D25** (1982) 735.

[151] SMC Coll., D. Adams et al., *Phys. Lett.* **B357** (1995) 248; *Phys. Rev.* **D56** (1997) 5330.

[152] NOMAD Coll., P. Astier et al., *Nucl. Phys.* **B588** (2000) 3; **B605** (2001) 3.

[153] I.I.Y. Bigi, *Nuovo Cim.* **41A** (1977) 581.

[154] J. Ellis, D. Kharzeev, A. Kotzinian, *Z. Phys.* **C69** (1996) 467; J. Ellis, M. Karliner, D.E. Kharzeev, M.G. Sapozhnikov, *Nucl. Phys.* **A673** (2000) 256.

[155] B. Ma, J. Soffer, *Phys. Rev. Lett.* **82** (1999) 2250; B. Ma, I. Schmidt, J. Soffer, J. Yang, *Eur. Phys. J.* **C16** (2000) 657; *Phys. Rev.* **D62** (2000) 114009.

[156] G.T. Jones et al., *Z. Phys.* **C28** (1985) 23.

[157] WA59 Coll., S. Willocq et al., *Z. Phys.* **C53** (1992) 207.

[158] E632 Coll., D. DeProspo et al., *Phys. Rev.* **B50** (1994) 6691.

[159] A. Bartl, A. Fraas and W. Majerotto, *Z. Phys.* **C6** (1980) 335; J. Ranft and G. Ranft, *Z. Phys.* **C12** (1980) 253.

[160] TASSO Coll., M. Althoff et al., *Z. Phys.* **C27** (1985) 27.

[161] ALEPH Coll., D. Buskulic et al., *Phys. Lett.* **B374** (1996) 319.

[162] OPAL Coll., K. Ackerstaff et al., *Eur. Phys. J.* **C2** (1998) 49.

[163] G. Gustafson and J. Häkkinen, *Phys. Lett.* **B303** (1993) 350.

[164] E665 Coll., M.R. Adams et al., *Eur. Phys. J.* **C17** (2000) 263.

[165] D. Ashery and H.J. Lipkin, *Phys. Lett.* **B469** (1999) 263.

Chapter 6

Fragmentation Models

6.1 Fragmentation models for e^+e^- collisions

At the e^+e^- cms energy $\sqrt{s}$ equal to the Z mass m_Z, about 70% of the collisions lead to multihadronic final states. In general, four different phases are distinguished in the formation of these final states [1,2] (see Fig. 6.1):

i) In a first (electroweak) phase, the e^+e^- pair annihilates into a virtual photon γ^* or a real Z^0, which consequently produces a primary $q\bar{q}$ pair separating with the energy of the γ^*/Z^0. Before annihilation, initial-state QED bremsstrahlung may occur, thus reducing the original e^+e^- energy available for $q\bar{q}$ production.

ii) In a second (perturbative QCD) phase, the initial $q\bar{q}$ may radiate gluons by QCD bremsstrahlung and these gluons may radiate, on their turn. Strong perturbation theory must be used to describe this cascade process. The running coupling constant α_s, however, increases as the cascade proceeds, thus limiting the applicability of perturbative QCD to the first steps.

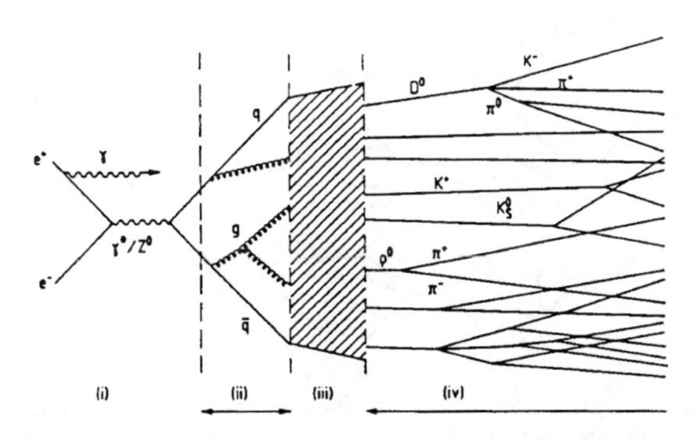

Figure 6.1. Schematic illustration of a hadronic e^+e^- annihilation event [1].

iii) In a third (non-perturbative QCD) phase, a number of colored partons (gluons and quarks) now share the original energy and fragment into a number of colorless hadrons. Since QCD is not well understood at low energy (= soft) scales, the fragmentation has to be described by phenomenological models.

iv) Finally, in a fourth phase, unstable hadrons decay into those hadrons (mainly pions and kaons) which are actually observed in high energy experiments.

Since the electroweak phase i) is well known text-book physics and the hadronic decay phase iv) is described by experimentally determined branching ratios, we shall be concerned here with phases ii) and iii). The final state is a complex system and we shall see in later chapters that complex system methodology is applicable. The intermediate state of a simple $q\bar{q}$ system is the most simple system available, however, and that is why e^+e^- collisions play a major role in the modelling of multihadron production processes.

6.1.1 The perturbative phase

Two basic approaches exist to model the perturbative QCD phase:

6.1.1.1 The matrix-element method

In this approach, Feynman diagrams as those of Fig. 6.2 modifying the Born process $e^+e^- \rightarrow q\bar{q}$ are calculated order by order. Fig. 6.2a corresponds to the first-order gluon radiation $e^+e^- \rightarrow q\bar{q}g$. For massless quarks, the matrix element is given in terms of the scaled parton energy variables $x_1 = 2E_q/\sqrt{s}$, $x_2 = 2E_{\bar{q}}/\sqrt{s}$ and

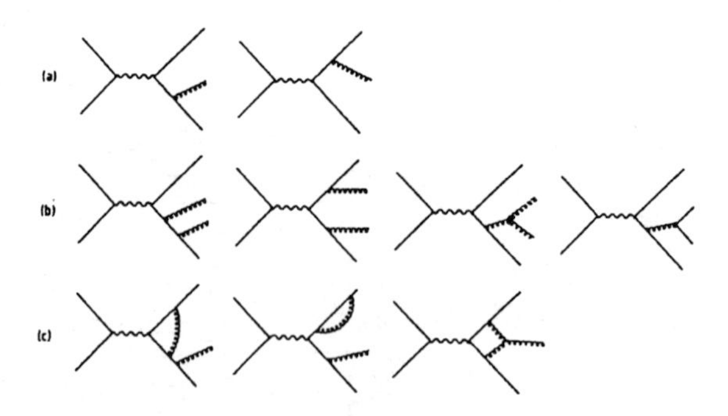

Figure 6.2. Feynman diagrams for three- and four-jet production. a) The two graphs which contribute to three-jet production. b) A few of the graphs which contribute to four-jet production (the ones left out can be obtained by symmetry). c) A few of the loop (vertex and propagator) graphs which contribute to three-jet production in second order [1].

$x_3 = 2E_g/\sqrt{s}$, $(x_1 + x_2 + x_3 = 2)$, as [3]

$$\frac{d\sigma}{dx_1 dx_2} = \sigma_0 \frac{\alpha_s}{2\pi} C_F \frac{x_1^2 + x_2^2}{(1 - x_1)(1 - x_2)} , \tag{6.1}$$

where σ_0 is the lowest-order cross section, α_s is the strong coupling constant and $C_F = 4/3$ the appropriate color factor. Kinematically, x_k of parton k is related to the invariant mass m_{ij} of the other two partons i and j by $y_{ij} \equiv m_{ij}^2/\sqrt{s} = 1 - x_k$.

Experimentally, one does not deal with partons but with jets of hadrons originating from one or several nearby partons defined according to some jet resolution criterion. The most commonly used criterion for a three-parton configuration to be called a two-jet event is $\min(y_{ij}) < y_{cut}$ with y_{cut} of the order 0.01. This automatically takes care of the singularities in 6.1.

In principle, the matrix element method would be the correct approach since it takes into account exact kinematics, full interference and helicity structure. A-priori, it is also justified by the small value $\alpha_s \approx 0.12$ of the running strong coupling constant at the mass of the Z^0, in first order given by

$$\alpha_s(Q^2) = \frac{12\pi}{(33 - 2n_f)\ln(Q^2/\Lambda^2)} , \tag{6.2}$$

where Q is the virtuality scale (e.g. the cms energy, $\sqrt{s}$, in case of e^+e^- annihilation into a $q\bar{q}$ pair), n_f the number of active flavors. The QCD scale parameter Λ, to be determined from experiment, delimits the boundary between perturbative (hard) and non-perturbative (soft) energy regimes.

In practice, the calculations have been carried out in full only up to $\mathcal{O}(\alpha_s^2)$, but become increasingly difficult in higher orders, in particular for loop graphs. These terms can be partially accounted for by choosing a suitable scale μ $(Q^2 = \mu s, \mu \leq 1)$ according to an "experimentally optimized perturbation theory" [4]. To compensate for neglected multiple gluon emission, one must, however, use energy-dependent fragmentation parameters which allow jets to become softer and wider with increasing energy.

6.1.1.2 Parton showering

The second approach is to allow for a *parton shower evolution* of, in principle, all orders, as shown in Fig. 6.3. In the simplest approach, the leading logarithm approximation (LLA), only the leading terms in the perturbative expansion are used. However, various schemes have been developed to take into account various subleading corrections, the most important being MLLA (M for modified), DLLA (D for double) or NLLA (N for next-to-) [5].

The parton shower approach can readily be formulated in terms of a probabilistic picture, suitable for Monte-Carlo event generation, by iterative use of the basic

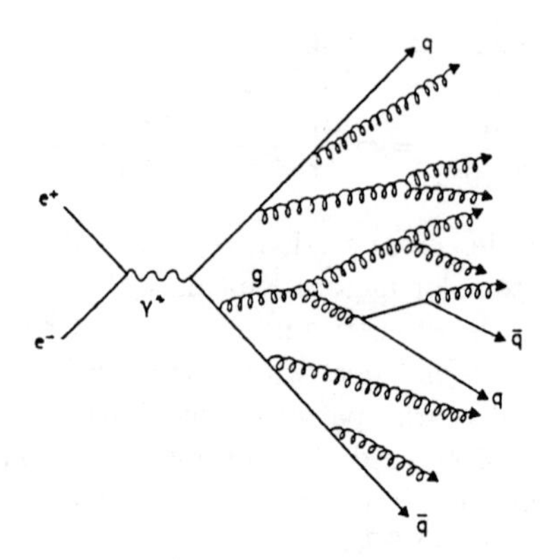

Figure 6.3. Schematic picture of parton shower evolution in e^+e^- events [1].

branchings $q \to qg$, $g \to gg$ and $g \to q\bar{q}$. In that, the probability $\mathcal{P}$ that branching $a \to bc$ takes place during a small change $dt \propto dQ^2_{evol}/Q^2_{evol}$ of the evolution parameter $t = \ln(Q^2_{evol}/\Lambda^2)$ is given by the Altarelli-Parisi equations [6]

$$\frac{d\mathcal{P}_{a\to bc}}{dt} = \int dz \frac{\alpha_s(Q^2)}{2\pi} P_{a\to bc}(z) , \tag{6.3}$$

where z specifies the sharing of four-momentum between the daughters b and c, with b taking the fraction z and c taking $1-z$. The Q^2 scale of α_s is not necessarily the same as the evolution scale Q^2_{evol}. Studies of loop corrections [5] rather suggest the use of $Q^2 = z(z-1)m_a^2 \approx p_T^2$, so that the scale is set by the transverse momentum of a branching rather than by the mass m_a of the branching parton. The functions $P_{a\to bc}(z)$ are the splitting kernels [6, 7]

$$\begin{aligned} P_{q\to qg}(z) &= C_F \frac{1+z^2}{1-z} , \\ P_{g\to gg}(z) &= N_C \frac{(1-z(1-z))^2}{z(1-z)} , \\ P_{g\to q\bar{q}}(z) &= T_R(z^2+(1-z)^2) , \end{aligned} \tag{6.4}$$

with $C_F = 4/3$, $N_C = 3$ and $T_R = n_f/2$.

In the Monte-Carlo application, flavor and four-momentum are conserved at each branching. The singular regions corresponding to the production of very soft gluons are avoided by the (certainly artificial) introduction of an effective gluon mass Q_0. The branching products b and c branch in their turn, and so on, giving a quasi-self-similar tree-like structure until the evolution parameter t reaches t_{min} given by Q_0.

Moving down from a maximum allowed virtuality $t_{\max}$, the probability that no branching of parton a occurs during a small range δt of t values is given by $(1 - \delta t \mathrm{d}\mathcal{P}_{\mathrm{a}\to\mathrm{bc}}/\mathrm{d}t)$. Summing over many small intervals, this becomes

$$\mathcal{P}_{no-em}(t_{\max}, t) = \exp\left(-\int_t^{t_{\max}} \mathrm{d}t' \frac{\mathrm{d}\mathcal{P}_{\mathrm{a}\to\mathrm{bc}}}{\mathrm{d}t'}\right) . \tag{6.5}$$

The probability for a branching at a given t then is the probability for a branching according to (6.3) multiplied by the probability (6.5) that a branching has not already taken place, thus giving an exponential decay law with a t-dependent decay probability.

The probability that a parton starting from a given virtuality t will reach a certain lower limit (e.g. given by Q_0) without branching is

$$S_{\mathrm{a}}(t) = \exp\left(-\int_{t_{\min}}^{t} \mathrm{d}t' \int_{z_{\min}(t')}^{z_{\max}(t')} \mathrm{d}z \frac{\alpha_{\mathrm{s}}(Q^2)}{2\pi} P_{\mathrm{a}\to\mathrm{bc}}(z)\right) . \tag{6.6}$$

This is the so-called Sudakov form factor. It only depends on the single parameter t and can e.g. easily be tabulated for each parton a for later use in a Monte-Carlo program. The no-emission probability (6.5) is simply $S_{\mathrm{a}}(t_{\max})/S_{\mathrm{a}}(t)$. Assigning it a random number R uniformly distributed between 0 and 1, the t value for a branching can be found by solving

$$S_{\mathrm{a}}(t) = S_{\mathrm{a}}(t_{\max})/R . \tag{6.7}$$

The branching is terminated whenever t becomes smaller than $t_{\min}$, i.e. R is chosen smaller than $S_{\mathrm{a}}(t_{\max})$.

The branching probability given by the Altarelli-Parisi equations (6.4) for the first branching of an initial quark or antiquark turns out to have the same singularity structure as the first-order three-jet matrix element (6.1). The parton energy dependence of the nominator $A_{\mathrm{m}}(x_1, x_2) = x_1^2 + x_2^2$ of (6.1) is, however, replaced by the expression

$$A_{\mathrm{s}}(x_1, x_2) = 1 + \frac{1 - x_1}{(1 - x_1) + (1 - x_2)}\left(\frac{x_1}{2 - x_2}\right)^2 + \frac{1 - x_2}{(1 - x_1) + (1 - x_2)}\left(\frac{x_2}{2 - x_1}\right)^2 , \tag{6.8}$$

which is larger than $x_1^2 + x_2^2$ everywhere, reaching a maximum ratio of 20/9 for $x_1 = x_2 = 1/2$ and a minimum ratio of unity in the limit $x_1($ or $x_2) \to 1$. Branchings generated as part of the shower, i.e. according to (6.8), can, therefore, be matched onto the first-order three-jet matrix element by accepting them with a probability $A_{\mathrm{m}}(x_1, x_2)/A_{\mathrm{s}}(x_1, x_2)$.

Particularly interesting corrections beyond leading log are those due to parton coherence effects [8]. These exist in the form of intra-jet and inter-jet coherence:
i) Intrajet coherence leads to a decrease of soft gluon emission inside jets. In so-called coherent-shower Monte-Carlo models, the bulk of these soft-gluon interference effects

is taken into account by an ordering in terms of a decreasing emission angle.
ii) Interjet coherence influences the flow of particles between jets. The interference is constructive or destructive depending on the color configuration.

Angular ordering is built in from the onset in the Marchesini-Webber coherent shower algorithm HERWIG [9]. In this algorithm, $Q^2_{\text{evol}} = E^2\xi$ with $\xi = \frac{p_b p_c}{E_b E_c} \approx 1 - \cos\theta$, E being the energy of the branching parton and θ the opening angle between its branching products with four-momenta p_b, p_c and energies E_b, E_c. At each branching, Q^2_{evol} is selected from the pretabulated Sudakov form factors according to (6.7). Q^2_{evol} defines the allowed range for picking a value of z and the ξ_a value of the branching. The value of z determines the energy sharing, $E_b = zE_a$ and $E_c = (1-z)E_a$. The maximum Q^2_{evol} scale of the two daughters is given by $E_b^2\xi_a$ and $E_c^2\xi_a$, respectively, and $\xi_b < \xi_a, \xi_c < \xi_a$, etc., etc.. So, this algorithm makes use of an evolution variable which explicitely involves polar angles and, therefore, automatically contains polar-angular ordering, while correlations between azimuthal angles are introduced in a later stage.

On the other hand, in the algorithms implemented in PYTHIA/JETSET [10] and CALTECH-II [11], $Q^2_{\text{evol}} = m_a^2$ is used. Angular ordering is not automatically included and must be introduced a-posteriori as an additional constraint on the combination of m^2 and z values allowed for a particular branching, given the values of m^2 and z of the preceding branching.

An interesting alternative to the parton shower algorithms is the dipole picture [12] used in ARIADNE [13]. The string pieces between partons are identified with color dipoles (or color antennas) and the emission of a gluon corresponds to the breaking of a dipole into two. Again, the Sudakov form factor is used to determine the ordering of the emission of a new gluon (or the competing splitting of an existing gluon into a $q\bar{q}$ pair). The natural ordering variable in this case is the transverse momentum of the emitted gluon, defined in an invariant form e.g. as

$$p_T^2 = \frac{s_{12}s_{23}}{s_{123}} \tag{6.9}$$

for a gluon 2 emitted from a dipole between partons 1 and 3, with s_{ij} the squared invariant mass of partons i and j and s_{123} that of the total dipole. That dipole splitting (gluon emission) or that splitting of the chain of dipoles into two independent chains (gluon splitting) is selected which gives the largest p_T. The process is repeated until none of the p_T's are above a given cut-off. The breaking is defined in the rest frame of a dipole and automatically provides angular ordering and non-trivial azimuthal effects from the boost back to the overall cms.

No parton coherence effects are included in the COJET [14] algorithm.

6.1.2 The hadronization or fragmentation phase

The transition from partons to hadrons (hadronization) is, as yet, not understood from QCD and has to be described by a phenomenological model.

Like the approaches used for the parton shower described above, all existing models are of a probabilistic and iterative nature [15]. The whole fragmentation process is described in terms of simple underlying branchings of the type

$$\begin{aligned} \text{jet} &\rightarrow \text{hadron} + \text{remainder-jet} \\ \text{string} &\rightarrow \text{hadron} + \text{remainder-string} \\ \text{cluster} &\rightarrow \text{hadron} + \text{hadron, hadron} + \text{cluster or cluster} + \text{cluster} . \end{aligned}$$

At each branching, the production of new flavors and the sharing of energy and momentum are determined according to probabilistic rules. Several approaches exist and we shall discuss a number of these and compare their success in describing the existing data in the following chapters. The most successful ones can be used to predict properties not yet studied experimentally and the behavior at higher energies. They also tell us which features have to be included into a future theory.

6.1.2.1 Local parton-hadron duality

The most straight-forward approach to hadronization is to assume that the hadronic spectra are proportional to the partonic ones if the cutoff Q_0 of the parton cascade is decreased towards a small value, of the order of the hadron mass itself [16].

This concept of local parton-hadron duality (LPHD) neglects large-distance hadronization effects and applies the perturbative predictions at the partonic level directly to corresponding hadronic distributions. The tremendous advantage is the simplicity of the assumption, in particular where analytical calculations can be performed on the parton level. The concept gets considerable support from its success in reproducing infrared and collinear safe observables, which are insensitive to the addition of a soft particle or to the splitting of one particle into two collinear particles, i.e. are insensitive to the value of Q_0 for small Q_0. Examples are global event shapes like thrust (see Chapter 2) and energy flow or energy-energy correlations. Successes have, however, also been booked with infrared sensitive observables as single-particle inclusive distributions or the energy evolution of multiplicity distributions, after proper rescaling (see Chapters 4 and 8).

6.1.2.2 Independent fragmentation (IF)

In the independent fragmentation approach [17], each parton is fragmented into hadrons in an iterative branching completely independent of the fragmentation of the other partons in the same event. The branching takes place in the overall cms, with each jet fragmentation axis given by the direction of motion of the corresponding parton in that frame. The primary quark q is split into a hadron $q\bar{q}_1$ and an essentially collinear remainder-jet q_1, as shown in the upper three lines of Fig. 6.4. The sharing of energy and momentum is given by a probability distribution $f(z)$, with z being the fraction taken by the hadron and $(1-z)$ that left to the remnant-jet. In

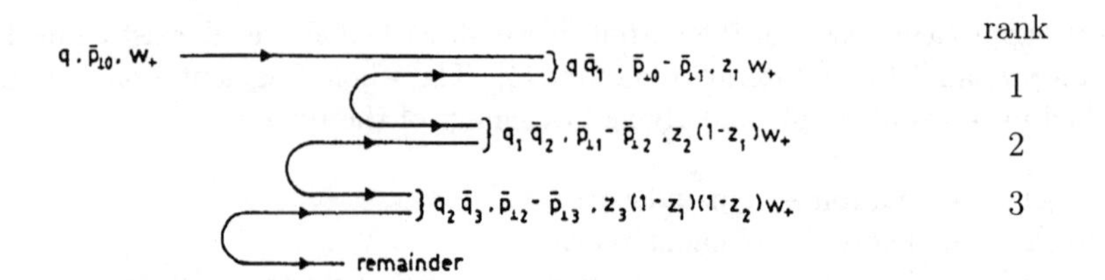

Figure 6.4. The iterative ansatz for flavor q, transverse momentum p_T, and light-cone energy-momentum ($W_+ = E + p_\parallel$) fraction z [1].

an average sense, the latter is assumed to be a scaled-down version of the original jet with the same (energy independent) $f(z)$, so that the process can be iterated as shown in the lower lines of Fig. 6.4. It is, therefore, a self-similar outside-in cascade.

Quark fragmentation functions $D_q^h(z)$, where $D_q^h(z)dz$ is the probability to find a hadron h in a quark jet q carrying an energy-momentum fraction between z and $z + dz$, can be written as

$$
\begin{aligned}
D(z) &= f(z) + \int\int dz' dz'' \delta(z - z'z'') f(1-z') D(z'') = \\
&= f(z) + \int_z^1 \frac{dz'}{z'} f(1-z') D\left(\frac{z}{z'}\right) .
\end{aligned}
\tag{6.10}
$$

The scaling function $f(z)$ gives the probability to find the hadron containing the original quark q_0 at fraction z, $f(1-z')$ the probability that the remnant jet has fraction z' and $D(z'' = z/z')$ gives the probability that this remnant jet produces a particle with energy-momentum fraction $z'' \cdot z' = z$.

Various, parametrizations are used for $f(z)$.

• For hadrons containing light flavors, one generally uses the Field-Feynman [18] parametrization,

$$
f(z) = 1 - a + 3a(1-z)^2 \tag{6.11}
$$

with $a = 0.77$ determined from experiment.

• An alternative to avoid too strong a peaking at $z = 0$ is

$$
f(z) = (1+c)(1-z)^c . \tag{6.12}
$$

• Hadrons containing heavy flavors need a harder fragmentation function, e.g. that of Peterson et al. [19],

$$
f(z) \propto \left[z\left(1 - \frac{1}{z} - \frac{\epsilon_Q}{1-z}\right)^2 \right]^{-1} , \tag{6.13}
$$

with ϵ_Q being a free parameter, but expected to scale with the heavy-quark mass m_Q like $1/m_Q^2$.

As can be seen in Fig. 6.4, each splitting results in the production of a new $q'\bar{q}'$ pair. The flavor of this pair is chosen at random according to the relative probabilities $u\bar{u} : d\bar{d} : s\bar{s} = 1 : 1 : \gamma_s$, with a strange-quark suppression factor γ_s close to 0.3. Normally, only the lowest-lying pseudoscalar and vector (in some cases also tensor) meson multiplets are allowed with relative abundance given by a parameter. Also diquark-antidiquark production is included with corresponding suppression factors.

In addition to the local flavor conservation of the $q'\bar{q}'$ splitting, local transverse momentum conservation is introduced through a vanishing net p_T of the pair. The p_T of the q' is generated according to a Gaussian in the two transverse directions, separately. The transverse momentum of a hadron is the sum of the transverse momenta of its constituents.

A gluon jet may be treated exactly like a quark jet, with the initial quark flavor chosen at random, or the gluon may be split into a pair of parallel q and $\bar{q}$ jets, sharing the energy according to the splitting functions (6.4). Since the gluon is expected to fragment more softly than a quark, the fragmentation function is usually chosen different from that of a quark jet. Furthermore, the fragmentation p_T is allowed to have a mean value different from that in a quark jet.

Independent fragmentation does not a-priori conserve total flavor or energy and momentum. This has to be patched up at the end of the generation and a number of different solutions have been implemented.

6.1.2.3 String fragmentation (SF)

a) simple $q\bar{q}$ *strings*

As the partons move apart, in this picture a color flux tube or vortex line is stretched between a q_0 and an $\bar{q}_0$, as indicated in Fig. 6.5.

The transverse dimensions of the tube are of typical hadronic size, i.e. of order 1 fm. Assuming the tube to be uniform along its length, this leads to a confinement picture with a linearly rising potential. From hadron mass spectroscopy, the amount of energy per unit length is obtained to be $\kappa \approx 1$ GeV/fm. In order to obtain a Lorentz covariant and causal description of the energy flow due to this linear confinement, the dynamics of a relativistic string with massless ends can be used with no transverse degrees of freedom [20, 21].

As the q_0 and $\bar{q}_0$ continue to move apart, the potential energy stored in the string increases and the string will break by the creation of a new pair $q_1\bar{q}_1$, as illustrated in Fig. 6.6. This results in two color-singlet systems $q_0\bar{q}_1$ and $q_1\bar{q}_0$ which will break on their turn if their invariant mass

$$m^2 = \kappa^2(x_2 - x_1)^2 - \kappa^2(t_2 - t_1)^2 \tag{6.14}$$

is large enough. The break-up process proceeds until string pieces of a few GeV mass

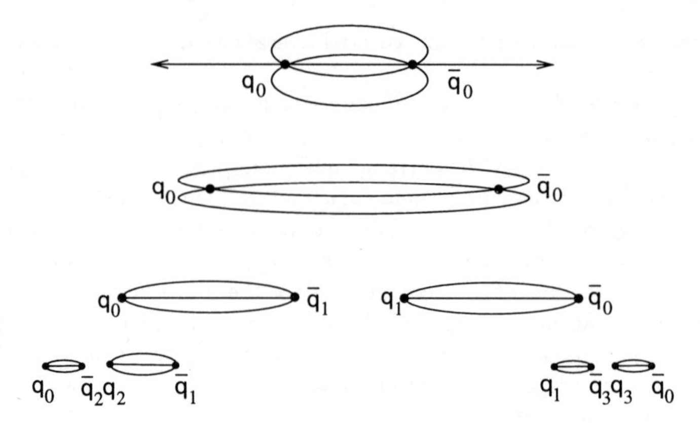

Figure 6.5. The break-up of the color field by $q\bar{q}$ pair creation.

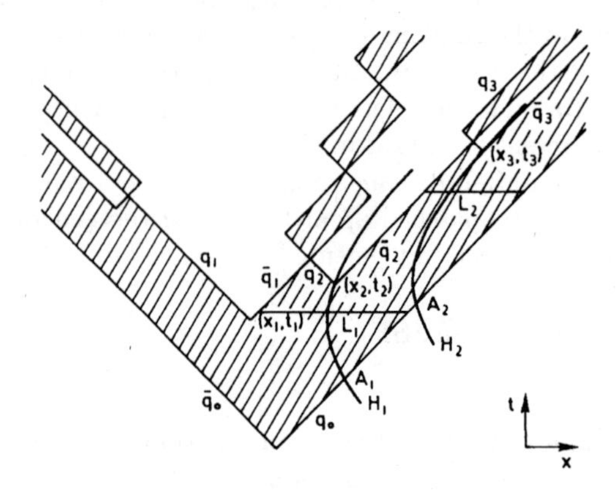

Figure 6.6. The particles q_0 and $\bar{q}_0$ move with large energies in opposite directions. In the field, $q\bar{q}$ pairs are produced at the space-time points (x_1, t_1), $(x_2, t_2), (x_3, t_3)$ etc. Nearby vertices are related by mass constraints. Hadrons are formed by $\bar{q}_1 q_2, \bar{q}_2 q_3$, etc. [20].

(clusters) remain (CALTECH-II) or all remaining string pieces have been identified with on-mass-shell hadrons (Lund) (see Fig. 6.6).

In a relativistic description, the slowest hadrons are produced first in every Lorentz frame, so that the cascade is inside-out. However, Lorentz invariance implies, that the cascade may very well be described by integral equations of an outside-in nature [20]. The different string breaks are causally disconnected, so that the order in which the breaks take place does not matter. The inward ansatz of IF (Fig. 6.4) thereby does get some physical justification. However, the string now connects a fragmenting quark with a fragmenting antiquark and it should not matter on which side the process is started, i.e., the fragmentation function should be symmetric with respect to the two ends.

If the mass of the objects produced in the string break-up is continuous (as in CALTECH-II), a classical constant probability P_0 can be assumed [21] for the string to break up per unit of invariant area A swept out by the string in the $1+1$ dimensional space-time (x,t) plane. This leads to left-right symmetry and to an exponential area decay law [22, 23],

$$\frac{\mathrm{d}\mathcal{P}}{\mathrm{d}A} = P_0 \exp(-P_0 A) , \tag{6.15}$$

describing the probability of a string to break at a point containing the space-time area A within its backward light cone. With $E = \kappa\Delta x$ and $p = \kappa\Delta t$, the string fragment mass is $m^2 = 2\kappa^2 A$ and

$$\frac{\mathrm{d}\mathcal{P}}{\mathrm{d}m^2} = b\exp(-bm^2), \quad b = P_0/2\kappa^2 . \tag{6.16}$$

Since $\langle A\rangle = P_0^{-1}$, on average the string break-up points lie scattered about the hyperbola

$$\tau^2 = t^2 - x^2 = 4/P_0 \sim 2\langle m^2\rangle/\kappa^2 . \tag{6.17}$$

It can be translated into a probability distribution in squared mass m^2 and energy-momentum fraction $z = E + p_\parallel$ of the produced cluster

$$\frac{\mathrm{d}\mathcal{P}}{\mathrm{d}m^2\mathrm{d}z} \propto z^{-1}\exp(-bm^2/z) , \tag{6.18}$$

where b is a free parameter proportional to the probability of break-up per invariant unit of string area.

In the Lund scheme, the left-right symmetric fragmentation function takes the (unique) form [23]

$$f(z) \propto z^{-1} z^{a_\alpha} \left(\frac{1-z}{z}\right)^{a_\beta} \exp(-bm_\mathrm{T}^2/z) \tag{6.19}$$

and describes the probability for a quark of flavor α to combine with an antiquark of flavor β to a meson with transverse mass $m_\mathrm{T} = (m^2 + p_\mathrm{T}^2)^{1/2}$ and energy-momentum fraction z. Experimentally, there is no need to use different values of a for different flavors, at least for the light flavors, so that (6.19) simplifies to

$$f(z) \propto z^{-1}(1-z)^a \exp(-bm_\mathrm{T}^2/z) \tag{6.20}$$

with two (strongly correlated) free parameters a and b. At given m^2 (and $p_\mathrm{T}^2 = 0$, $a = 0$), it corresponds to (6.18). The similarity is striking, since the discrete mass spectrum of Lund implies that string breaks are only allowed along discrete (one-dimensional) hyperbolae (6.14), rather than inside (two-dimensional) areas.

In the CALTECH-II scheme, all transverse momentum is generated during shower evolution and cluster decay and all flavor suppression comes from phase-space effects.

In contrast, in the Lund scheme, quantum mechanical tunneling is assumed to lead to $q'\bar{q}'$ break-ups. In terms of the transverse mass of the q', the probability that the $q'\bar{q}'$ break-up will occur is

$$\exp(-\pi m_{\mathrm{T}}^2/\kappa) = \exp(-\pi m^2/\kappa)\exp(-\pi p_{\mathrm{T}}^2/\kappa). \tag{6.21}$$

A flavor independent Gaussian spectrum is assumed for the p_{T} of q', but as in IF, this p_{T} is locally compensated by the $\bar{q}'$, so that the string has no transverse excitations. A small contribution to the fragmentation p_{T} does, however, come from soft unresolved gluon radiation.

The tunneling also implies heavy-quark suppression

$$\mathrm{u}:\mathrm{d}:\mathrm{s}:\mathrm{c} \approx 1:1:0.3:10^{-11}\ , \tag{6.22}$$

so that charm and heavier quark production is not expected in soft fragmentation. As in IF, an algorithm is necessary to choose between pseudoscalar, vector or tensor mesons. Spin counting would suggest a factor $2J+1$, but heavier states should be disfavored by an unknown amount. In Monte-Carlo implementation, the relative composition is, therefore, left as a free parameter.

An elegant alternative is the UCLA picture [24], in which the relative probability to produce a hadron with given mass m is proportional to the integral over allowed z and p_{T} values

$$\mathcal{P}(m^2) \propto \int_0^1 \mathrm{d}z \int \mathrm{d}p_{\mathrm{T}}^2 z^{-1}(1-z)^a \exp(-bm_{\mathrm{T}}^2/z)\ . \tag{6.23}$$

The only additional weight factors are isospin and spin counting factors. There is no need for quark-level strangeness or diquark suppression, since the exponential term in (6.23) implies a suppression of heavy hadrons and a stiffening of their fragmentation function.

b) $q\bar{q}g$ strings

In a $q\bar{q}g$ event, a string is stretched from the q to the $\bar{q}$ via the g, so that the g causes a kink in the string, carrying energy and momentum (Fig. 6.7a). As a consequence, the gluon has two string pieces attached, so that the ratio of gluon to quark string force is two instead of the usual QCD value of $N_{\mathrm{C}}/N_{\mathrm{F}} = 2/(1-1/N_{\mathrm{C}}^2) \approx 9/4$, i.e. N_{C} is not 3 but infinite.

c) Multi-gluon strings

With several gluons emitted, the full string evolution becomes complicated. The effect of nearby partons is, however, that they drag out a string very much like what would have been dragged out by one single parton with the sum of their momenta.

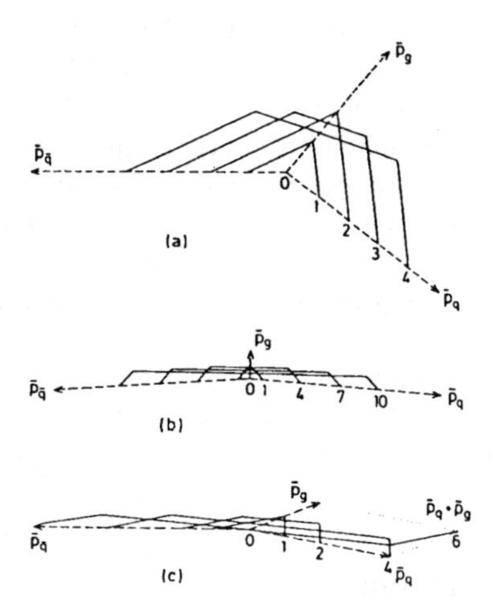

Figure 6.7. The string drawing for (a) an ordinary three-jet event, (b) a three-jet event with a soft gluon, and (c) a three-jet event with a collinear gluon. Dashed lines give the momenta (and hence the trajectories) of the partons. Full lines give the string shape at different times, with numbers representing time in some suitable scale [1].

Furthermore, a soft gluon does not significantly affect the string evolution. So, the string fragmentation scheme is infrared safe with respect to collinear or soft gluon emission (see Figs. 6.7b,c).

In general, however, for multi-gluon states, the string surface in (x, t) space is no longer flat but geometrically bent due to the internal excitation. The solution adopted in JETSET [10] is to project the positions of the break-up points onto the surface of the bent string in a way that their proper times and the masses of the particles produced between them stay the same. Unfortunately, this method does not fulfil the area law on the bent surface, at least not on an event-by-event basis. So, the probability for decay of the string is no longer proportional to the area spanned by the string before it decays.

While this does not cause any observable problems for single-particle inclusive distributions so far, correlations of particles produced across a gluon kink are difficult to handle (see Chapter 11).

An elegant solution which precisely fulfils the area law at every single step in the production process is therefore proposed in [25]. The method is based on the so-called *directrix* [21] defined by the parton energy-momentum vectors laid out in color order. This is given by the perturbative cascade (or, alternatively, by the exact matrix elements), down to some cutoff scale. This cutoff is quite artificial, however, and freezing of the string state at this point, as in JETSET, not fully justified. The directrix corresponds to the orbit of a massless string end quark and, most importantly, describes the whole string surface. Infrared stability is provided in the sense that, due to the least-action nature of the string world surface, small modifications of the directrix only have local influence.

This property can be exploited to continue the partonic states down to the hadronic mass scale by the introduction of additional soft gluons, under the condition that angular ordering is preserved. It is interesting to note that a duality similar to LPHD [16] emerges between the original directrix and the hadronic state, however with hadrons produced in this process always originating from the energy-momenta of two or more partons.

Thanks to the re-formulation of the Lund model, in particular the area law, in terms of transition operators [26], it is now possible to treat string fragmentation and even this multi-gluon fragmentation analytically. The process is implemented into the Monte Carlo program ALFS, which can be run in the framework of PYTHIA/JETSET.

6.1.2.4 Cluster Fragmentation (CF)

Here, the long fragmentation chains of IF and SF described above are replaced by colorless clusters [27] as basic simple, local and universal units from which the hadrons are produced. Ideally, a cluster is characterized by two variables only, its total mass and its total flavor content.

The main difference between cluster fragmentation schemes used in cluster Monte-Carlo models is to what extent string fragmentation is used as an intermediate step.

• In one extreme, matrix elements or parton showers can be used to generate a partonic state, with a string stretched between the partons as described in the SF section above. The string evolution is terminated by a smooth string-mass cut-off procedure and the strings first fragment into clusters, which further decay into the final-state hadrons. This is the CALTECH-II approach [11] illustrated in Fig. 6.8a.

• In the other extreme, gluons remaining at the end of the shower evolution are split non-perturbatively into $u\bar{u}$ and $d\bar{d}$ pairs (the so-called Wolfram ansatz [27]). The quark of one splitting is then combined with the antiquark originating from a nearby splitting with the same color, to form a colorless cluster, which then decays into final-state hadrons. This corresponds to the HERWIG strategy [9] depicted in Fig. 6.8b.

If the shower evolution and/or string breaks are chosen as to give most of the clusters a mass of a few GeV, the cluster mass spectrum can be interpreted as a continuous superposition of wide resonances. Phase space can then be expected to dominate the decay properties, such as the selection of decay channels and the (isotropic) decay kinematics.

To this end, each allowed decay channel of a cluster of mass m is assigned a weight proportional to the density of states, $(2s_1+1)(2s_2+1)2p^*/m$, where s_1, s_2 are the spins of the two hadrons produced and p^* the momentum of the decay products in the rest frame of the cluster. The weight gives the relative probability for the choice to be

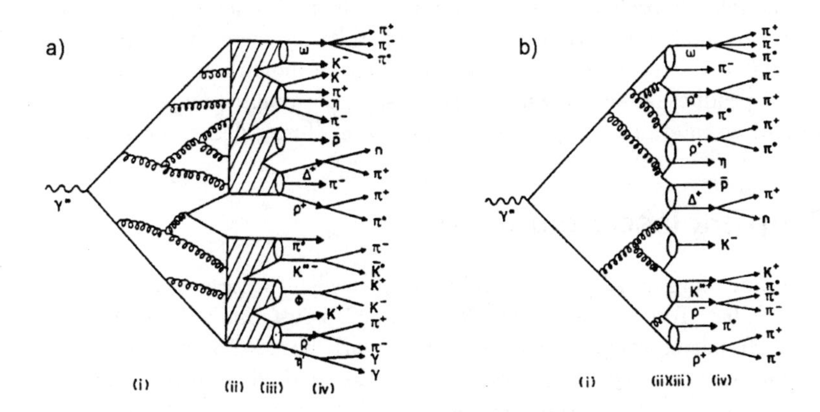

Figure 6.8 Two cluster fragmentation scenarios, both using shower evolution for the pQCD phase (ii), but a) string evolution plus breakup into clusters and b) forced $g \to q\bar{q}$ branchings and cluster formation, for the soft QCD phase (iii) of Fig. 6.1.

retained. In case of rejection, a new flavor is selected and the procedure is repeated. Independently of any quark masses or diquark spins, only the decay-hadron properties can influence the relative production rate.

The transverse momentum is determined by the average energy release in cluster decay and in subsequent resonance decays. The separate longitudinal and transverse fragmentation of IF and SF is replaced by a compact and unified description with very few parameters, where parton showering and the phase space of cluster decay give the full momentum distribution.

Based on the concept of "preconfinement" [28] (i.e. the property of QCD that a parton almost always finds itself nearby a parton carrying opposite color charge, whenever the partons have evolved from high to low virtuality), the parton shower evolution in itself is expected to give a cluster mass spectrum strongly damped at masses above a few GeV. There is a large spread of masses, however, requiring the possibility for cluster decay into more than two particles, e.g. in the form cluster→cluster+cluster or cluster→cluster+hadron. In these cases, the assumption of isotropic decay is dropped and a well-defined longitudinal "string" direction is assumed. The number of these is small (about 2% in default HERWIG), but they contribute with a large hadron multiplicity.

On the low-mass side of the spectrum, Monte-Carlo programs usually assume the cluster to collapse into one single particle (less than 1% in HERWIG). To conserve energy and momentum, surplus or missing four-momentum is redistributed to or from nearby clusters.

Flavors firstly are generated at the string breaks and/or the perturbative $g \to q\bar{q}$

branchings. The relative probabilities are given by explicit parameters in CALTECH-II, while they are implicitely given by the parton mass assignments in HERWIG, but with an arbitrary strength for gluon branching into a diquark-antidiquark pair. Secondly, flavor production occurs when large clusters decay into smaller ones, e.g. by production of an intermediate q$\bar{\text{q}}$ pair and, finally, in cluster decay, itself.

6.2 Deep inelastic collisions

A typical diagram representing parton evolution in deep inelastic scattering (DIS) is given in Fig. 6.9a. To describe it, the Dokshitzer-Gribov-Lipatov-Altarelli-Parisi (DGLAP) [6,7,29] parton evolution has been particularly successful and can be considered a major success of QCD. To lowest order, the method resums the leading logarithmic $(\alpha_s \ln(Q^2/Q_0^2))^n$ contributions, where Q^2 is the squared invariant mass of the exchanged virtual photon and Q_0^2 the cut-off for the perturbative evolution. At small x_B, however, resummation of $(\alpha_s \ln(1/x_B))^n$ has been suggested to lowest order by Balitsky-Fadin-Kuraev-Lipatov (BFKL) [30]. While DGLAP corresponds to a strong ordering of the transverse momenta k_T with respect to the proton direction ($Q_0^2 \ll k_{T1}^2 \ll \dots k_{Ti}^2 \ll \dots Q^2$), BFKL rather leads to a kind of random walk in k_T ($k_{Ti}^2 \approx k_{T(i+1)}^2$), together with strongly decreasing longitudinal momenta. A comparison of the two approaches and their domain of validity in x_B and Q^2 is given in Fig. 6.9b.

A bridge between DGLAP and BFKL is provided by Ciafaloni-Catani-Fiorani-Marchesini (CCFM) [32] by resumming both leading $\ln(1/x_B)$ and $\ln(Q^2/Q_0^2)$ terms.

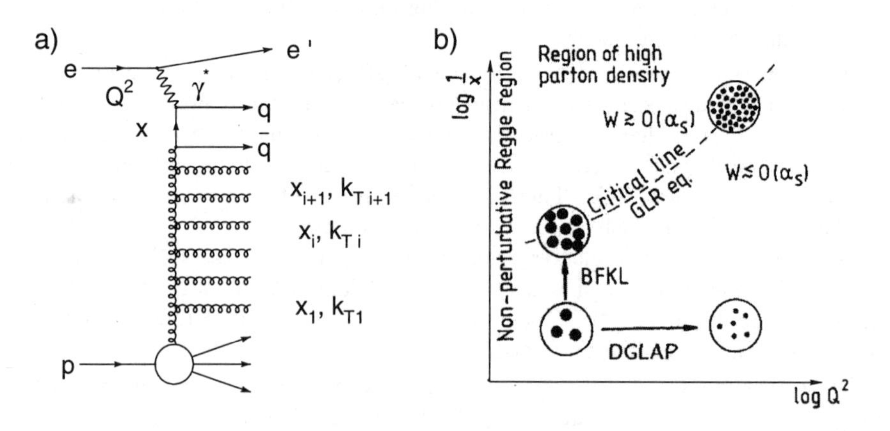

Figure 6.9. a) Generic diagram of parton evolution, b) Schematic evolution of the quark densities in various (x_B, Q^2) regions according to the dominant dynamical effects. The dashed line gives the limit of validity of perturbative QCD (update of [31]).

Monte Carlo models used in DIS are based on one or more of the above approaches.

LEPTO [33] uses first-order QCD matrix elements with additional soft emissions by adding leading-log DGLAP parton showers (MEPS).

RAPGAP [34] also matches exact first-order matrix elements to DGLAP-based leading-log parton showers. However, in addition to the direct photon processes used in LEPTO, it uses resolved photon processes, in which the virtual photon is allowed to have parton structure, itself.

HERWIG [9] is based on leading-log parton showers with matrix element corrections.

ARIADNE [13] is an implementation of the color dipole model [12] in which partons are radiated from color dipoles produced in the hard interaction, with a modified phase space restriction for radiation from an extended source.

LDCMC [35] used the Linked Dipole Chain [36] implementation of CCFM evolution and CDM for higher-order radiation and is part of the ARIADNE package.

CASCADE [37] uses first-order QCD matrix elements with additional parton emissions based on the CCFM equation using backward evolution. An unintegrated gluon density is used as input to the model.

For hadronization, HERWIG uses its cluster hadronization scheme. In all other models, hadronization is performed via string fragmentation as implemented in JETSET.

6.3 Soft hadron-hadron collisions

As pointed out in Chapter 4, it was realized rather early that jets are 'universal', i.e., that the basic characteristics of hadron production along some distinguished momentum axis are the same in all types of collision. Obviously, this could only be an approximate statement, as the different flavor-dependent effects are known to introduce significant corrections (e.g. leading baryon effect, the initial charge retention, heavy quarks etc.). In particular, hadron-induced *proton* fragmentation has to be compared to leptonic *target* fragmentation, while hadronic *meson* fragmentation has to be compared to leptonic *current* fragmentation or to *quark* fragmentation in e^+e^- annihilation. Indeed, the differences between leptonic target and current fragmentation are larger than the differences between different types of collision and the agreement between the latter are even more striking when trivially expected effects are taken into account.

6.3.1 The Lund fragmentation scheme

6.3.1.1 The quark fragmentation model

A quantitative version of the idea relating hadron- and lepton-induced processes has been formulated already in [38]. It follows the observation of a striking similarity between the Feynman-x distribution in meson $\rightarrow$ meson fragmentation processes and the quark $\rightarrow$ meson fragmentation functions, as determined in e^+e^- or lepton-hadron processes. The model, usually referred to as 'quark fragmentation model' (QFM), assumes that the hadron-induced fragmentation should be described as a two-step process. In the first step, one of the hadron's valence quarks or antiquarks, carrying a relatively small fraction of the total momentum, interacts with a quark from the other hadron. In the second step, the 'hadron remnants', carrying most of the initial momentum (and, obviously, the quantum numbers of the remaining valence quark(s)), fragments into final-state hadrons according to the same fragmentation function that describes the fragmentation of a corresponding quark separated from the other in e^+e^- or lepton-hadron collisions.

In general, for a meson M and for a baryon B fragmenting into hadron h, respectively, one writes:

$$\frac{1}{\sigma_{\text{inel}}}\frac{d\sigma}{dx}(\text{M} \rightarrow \text{h}) = \frac{1}{2}[D_{\text{h}}^{\text{q}_1}(x - \Delta x) + D_{\text{h}}^{\overline{\text{q}}_2}(x - \Delta x)] \tag{6.24}$$

$$\frac{1}{\sigma_{\text{inel}}}\frac{d\sigma}{dx}(\text{B} \rightarrow \text{h}) = \frac{1}{3}\sum_{i=1}^{3} D_{\text{h}}^{(\text{qq})_i}(x - \Delta x) \ , \tag{6.25}$$

where Δx denotes the average fraction of initial momentum 'lost' with the interacting quark, and where the valence structure is $(\text{q}_1\overline{\text{q}}_2)$ for M and $(\text{q}_1\text{q}_2\text{q}_3)$ for B. The symbol $(\text{qq})_i$ denotes the valence diquark left after removing quark q_i from the baryon and the D_{h}^{q} are the fragmentation functions of $\text{q}_1, \overline{\text{q}}_2$ or $(\text{qq})_i$ into hadron h. Finally, σ_{inel} stands for the inelastic non-diffractive cross section.

6.3.1.2 The Lund low-p_{T} model

Many theoretical attempts exist to account for the successes of relations (6.24) and (6.25), on the one hand, and to elaborate corresponding models to describe other features of hadronic collisions, on the other. Since analogous single strings are formed in e^+e^- and lepton-hadron collisions (Sections 6.1 and 6.2), the fragmentation functions are the same if the flavor quantum numbers of the fast string ends are chosen correspondingly.

For the simplified case of two mesons colliding, Fig. 6.10 shows an excitation mechanism assumed to be at work in hadronic collisions [39]. Hadrons are represented by color bags in which, due to asymptotic freedom, the color of a quark is not well localized. It is distributed as a blob around the quark and overlaps partially with the color blob of the anti-quark (upper line of Fig. 6.10). As a consequence, the bag is e.g. red in one end, antired in the other and "white" in the center. If in a hadron-hadron collision the colored parts of the bags overlap, the two bags can unite to form one larger bag. This bag is stretched to a string-like color flux tube, the force of which is retarding the meson remnants. As shown in Fig. 6.10 the

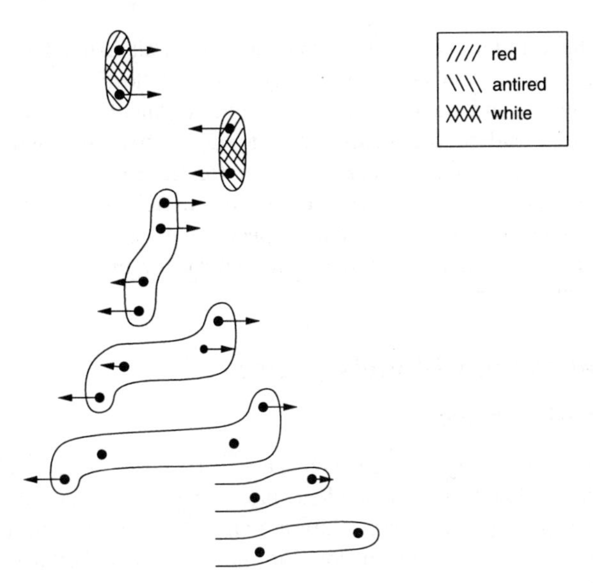

Figure 6.10. A low-p_T hadronic interaction according to the Lund model. The hadrons are presented by color bags [39].

colliding parts will lose their momentum first. When they are stopped, the other (non-interacting) parts will be retarted. The result is that the bag is stretched to the same length as if all the energy was carried by one of the valence quarks of each meson. The interacting quarks are distributed along the string according to the probability distribution $(1 - x)$. Therefore, the valence-quark structure functions in this model are simply:

$$F_I(x) = 2(1 - x), \tag{6.26}$$

$$F_L(x) = \delta(1 - x), \tag{6.27}$$

for the interacting (I) and non-interacting "leading" (L) quarks, respectively.

A similar scheme exists for baryon interactions [39]. An $\bar{q}$ is replaced by a diquark as the non-interacting part of the proton when hit by another hadron (hadron-hadron collision) or a lepton (deep inelastic scattering). A color flux tube is stretched in a stepwise manner, such that one non-interacting valence quark (L) is at the end of the force field followed by its diquark partner (called J for "junction").

The excitation dynamics is, of course, different from that of a $q\bar{q}$ string in e^+e^- annihilation or a q(qq) string in deep inelastic scattering. Nevertheless, when the excitation energy in the color field is divided into final state hadrons, one obtains essentially the same result as for a quark jet or a diquark jet, respectively. The same Lund fragmentation scheme as discussed in Sub-Sect. 6.1.2 is, therefore, used in the default version of this model.

A priori, the Lund model seems, however, less well defined in its application to soft hadron collisions, where the average value of 'lost' x fraction Δx and its distribution may depend on the type of hadron fragmenting. In the model, soft hadronic collisions lead to the formation of a single string or color flux tube, but the positions of initial valence quarks, relevant for the results, are incorporated by rather arbitrary assumptions on their x distribution. Although simple choices of the corresponding functions have led to many spectacular successes of this first version of the model, the results depend on details which can be freely changed if necessary, in particular in the central region.

6.3.2 Dual Parton Models (DPM)

6.3.2.1 The basic scheme

It was necessary to find a model in which the x distribution of the ends of fragmenting strings would follow from independent, more basic considerations. For processes at a small Q^2 scale - such as low-p_{T} interactions - the strong coupling constant α_{s} is too large to be used as an expansion parameter for a perturbative treatment of the process. The underlying idea is that an expansion in powers of $1/N$ should be used instead, where N can either be the number of quark flavors n_{f} or colors n_{c} involved [40]. This is called a topological expansion and has prompted a number of authors to develop a 'dual QFM', motivated by long established connections between the string models and the so-called dual resonance models providing a reliable extrapolation from the Regge limit of high sub-energies to the regions involving resonance production. This developed into the dual topological unitarization scheme (DTU). A comprehensive description of the model and its successes is given in [41].

The DTU approach originated from efforts to understand the pomeron intercept [42], and its initial development [40, 43] was mostly related to the Reggeon calculus framework of hadronic interactions, which will not be discussed here. Veneziano [44] was first to point out that the dual picture of strings, here called 'chains' or 'sheets', in hadron production corresponds to the QCD parton framework. According to this picture, the production spectra derive from the dominant (lowest topology) chains spanned between pairs of valence parton systems [45]. In Fig. 6.11, we show examples of one-chain ($\mathrm{e^+e^-}$, lepton-hadron), two-chain (BB, MB, $\overline{\mathrm{B}}$B non-annihilation) and three-chain ($\mathrm{B\overline{B}}$ annihilation) processes. Ingenious ways of testing this picture by correlation experiments [46] or flavor-dependent effects [47] proved quite successful [41].

6.3.2.2 The structure functions

The important progress is that the dual parton model predicts the distribution of the lost end momentum Δx, or equivalently, the chain energy distribution [48]. For small Δx, the distribution $F_{\mathrm{q}}^{\mathrm{h}}(\Delta x)$ corresponds to the Regge model prediction for

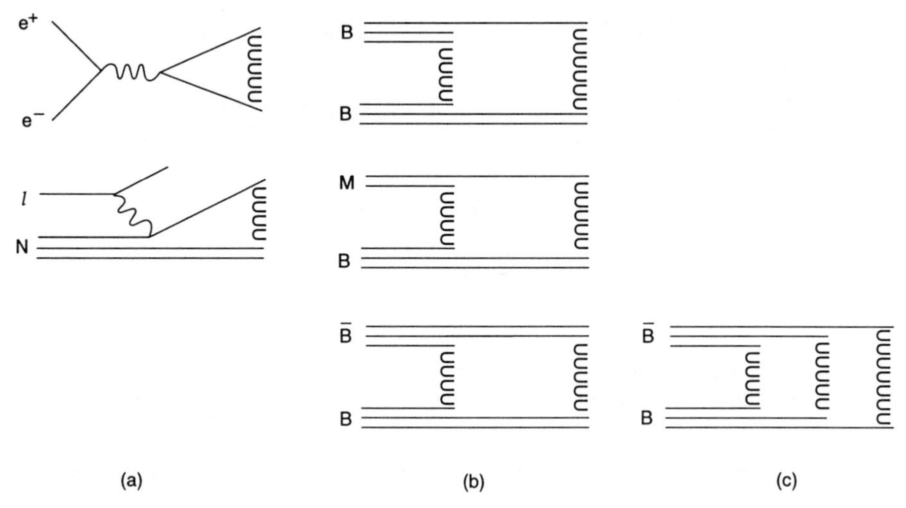

Figure 6.11. DTU diagrams for (a) one-chain, (b) two-chain and (c) three-chain processes.

the valence quark distribution:

$$F_{\mathrm{q}}^{\mathrm{h}}(\Delta x) \propto (\Delta x)^{-\alpha_{\mathrm{q\bar{q}}}(0)} \quad , \tag{6.28}$$

where $\alpha_{\mathrm{q\bar{q}}}(0)$ is the intercept of the leading $\mathrm{q\bar{q}}$ Regge trajectory. For a non-strange valence quark, one finds $F(\Delta x) \propto 1/\sqrt{\Delta x}$. In a similar manner one can find the behavior at Δx close to 1 and the distribution for other quarks and diquarks.

According to the DTU model, the success of relations (6.24) and (6.25) comes from the fact that they approximate well the correct formulae [49]

$$\frac{1}{\sigma_{\mathrm{inel}}}\frac{\mathrm{d}\sigma}{\mathrm{d}x}(\mathrm{M} \to \mathrm{h}) = \frac{1}{2}\left[\int_x^1 F_{\mathrm{q}_1}^{\mathrm{M}}(1-x')D_h^{\bar{\mathrm{q}}_2}\left(\frac{x}{x'}\right)\mathrm{d}x' + F_{\bar{\mathrm{q}}_2}^{\mathrm{M}}(1-x')D_{\mathrm{h}}^{\mathrm{q}_1}\left(\frac{x}{x'}\right)\mathrm{d}x'\right] , \tag{6.29}$$

$$\frac{1}{\sigma_{\mathrm{inel}}}\frac{\mathrm{d}\sigma}{\mathrm{d}x}(\mathrm{B} \to \mathrm{h}) = \frac{1}{3}\sum_{i=1}^{3}\int_x^1 F_{\mathrm{q_i}}^{\mathrm{B}}(1-x')D_{\mathrm{h}}^{(\mathrm{qq})_\mathrm{i}}\left(\frac{x}{x'}\right)\mathrm{d}x' , \tag{6.30}$$

since only one of the chain ends contributes to a high-x spectrum in hadron-hadron collisions, as is the case in $\mathrm{e^+e^-}$ or lepton-hadron collisions. The structure functions $F_{\mathrm{q}}^{\mathrm{h}}$ (always normalized to an integral of one) are not identical to those of valence quarks as seen in DIS, as 'valence q' in the dual model is not the same as 'valence q' in DIS. However, their shape can be predicted from the dual model itself, at least at $x \to 0$ and $x \to 1$. On the other hand, the fragmentation functions $D_{\mathrm{h}}^{\mathrm{q}}$ should be taken from DIS and $\mathrm{e^+e^-}$ data or estimated from the Quark Counting Rules as discussed in Chapter 5. The model, therefore, predicts the effective value of Δx in Eqs. (6.24) and (6.25) and also allows for more elaborate calculations, as, for example,

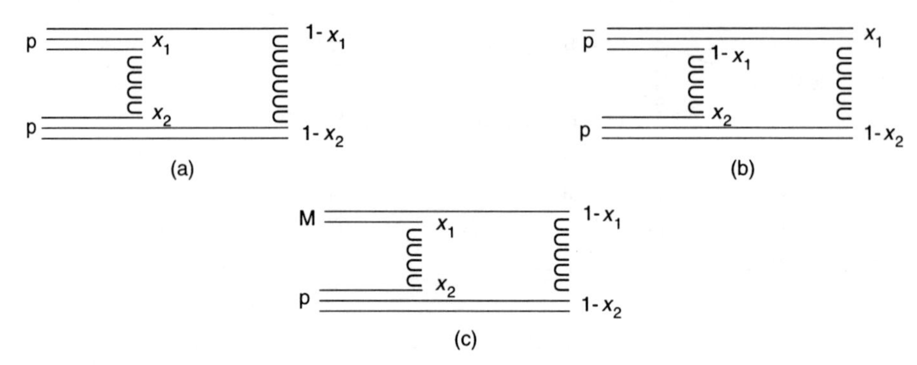

Figure 6.12. Feynman-x variables in DTU diagrams for (a) proton-proton, (b) antiproton-proton and (c) meson-proton collisions.

correlation effects. Moreover, the model now yields predictions for the full range in the x or y variables.

6.3.2.3 The rapidity density

Indeed, let us denote by $(\mathrm{d}n^1/\mathrm{d}y)(y - Y_0, \Delta Y)$ the rapidity density for particles of mass m produced in a single chain of energy $W = m\exp(\Delta Y/2)$ and position Y_0 in the cms. If the chain is spanned between ends at Feynman-x values x_1 and x_2, then W and Y_0 can be calculated as functions of x_1, x_2. For simplified kinematics, neglecting mass effects and transverse momenta, one finds $W = \sqrt{sx_1x_2}$, $Y_0 = \frac{1}{2}\ln(x_2/x_1)$. The rapidity density for baryon-baryon collisions (Fig. 6.12a) in the cms (neglecting flavor differences) can now be written as

$$\begin{aligned}\frac{\mathrm{d}n}{\mathrm{d}y}(\mathrm{A}+\mathrm{B}\to\mathrm{h}) &= \int F_{\mathrm{q}}^{\mathrm{A}}(1-x_1')F_{\mathrm{q}'}^{\mathrm{B}}(x_2')\left[\frac{\mathrm{d}n^1}{\mathrm{d}y}\left(y-\frac{1}{2}\ln\frac{x_2'}{x_1'}\,,\ \ln\frac{sx_1'x_2'}{m^2}\right)+\right.\\ &\left.+\ \frac{\mathrm{d}n^1}{\mathrm{d}y}\left(y-\frac{1}{2}\ln\frac{1-x_2'}{1-x_1'}\,,\ \ln\frac{s(1-x_1')(1-x_2')}{m^2}\right)\right]\mathrm{d}x_1'\mathrm{d}x_2'\,. \qquad (6.31)\end{aligned}$$

For antibaryon-baryon collisions (Fig. 6.12b), $(1-x_1')$ is to be replaced by x_1' and vice versa. For meson-baryon collisions (Fig. 6.12c), the $F_{\mathrm{q}}^{\mathrm{M}}$ are no longer suppressed at $x \to 1$ (in fact, they are symmetric in $x \to 0$ and $x \to 1$), but the formula remains the same. In an analogous way, two-particle densities and correlations can be expressed in terms of the distributions for single chains. Since the single-chain distributions can be measured in $\mathrm{e}^+\mathrm{e}^-$ or lepton-hadron collisions, the model is fully specified.

Equation (6.31) was derived independently by the Saclay group [50], by the Orsay group [49] and in [51]. Although the arguments and some of the details differ, the main results of these papers are very similar. Most of the investigated phenomenological consequences are related to the results of the Orsay group [49], whereas the Saclay group continued a program aiming at a final full understanding of the relationship between QCD and DTU including higher-order corrections [52].

We now will study simple results for global parameters, as average multiplicity and other moments of the multiplicity distribution, which can be deduced without the need for numerical integration of (6.31) and its analogue for double- and multiparticle spectra.

6.3.2.4 Particle multiplicities and multi-chain contributions

By integrating (6.31) over y, one finds for the average multiplicity in a hadron-hadron collision

$$\bar{n}_{\mathrm{hh}} = \bar{n}_1 + \bar{n}_2 \ , \tag{6.32}$$

where $\bar{n}_1, \bar{n}_2$ are average multiplicities for a single chain of type $\mathrm{q} - \mathrm{qq}$ or $\mathrm{q} - \bar{\mathrm{q}}$, further averaged over chain energies W according to the structure functions $F^{\mathrm{A}}, F^{\mathrm{B}}$. If one assumes that the average multiplicity in a single chain grows logarithmically with chain energy (see Chapter 8 below),

$$\bar{n}_i(W^2) = a_i \ln(W^2/s_i) \ , \tag{6.33}$$

one finds by an integration in analogy to (6.31) that

$$\bar{n}_{\mathrm{hh}}(s) = (a_1 + a_2) \ln(s/s_0) \ . \tag{6.34}$$

Thus, asymptotically, one obtaines for pp and νp collisions:

$$\lim_{s\to\infty} \frac{\bar{n}_{\mathrm{pp}}(s)}{\bar{n}_{\nu\mathrm{p}}(s)} = 2 \ . \tag{6.35}$$

Since the parameters a_i should be universal for any single chain, this prediction should actually hold for any non-annihilation (i.e. for any two-chain) hadron-hadron process related to any $\mathrm{e}^+\mathrm{e}^-$ or lepton-hadron process:

$$\lim_{s\to\infty} \frac{\bar{n}_{\mathrm{hh}}(s)}{\bar{n}_{l\mathrm{h}}(s)} = \lim_{s\to\infty} \frac{\bar{n}_{\mathrm{hh}}(s)}{\bar{n}_{\mathrm{e}^+\mathrm{e}^-}(s)} = 2 \ . \tag{6.36}$$

This prediction is somehow embarrassing, at first sight, since the data (see Sect. 8.1) show very similar behavior of average multiplicities as a function of hadronic energy for hadron-hadron and lepton-hadron or $\mathrm{e}^+\mathrm{e}^-$ collisions (in fact, the multiplicities seem to be even higher in the latter class). Obviously, at the available energies we are still far from the limit $s \to \infty$ and actual numerical finite-energy estimates from (6.31) agree with the data quite well [53].

One may, however, find the way out even of the asymptotic prediction (6.36). Within the two-chain model there are even two ways of avoiding this result. Firstly, the average multiplicity for a single chain may grow faster than logarithmically. One can prove that for the dual-based structure functions $F^{\rm h}(x) \propto x^{-1/2}(1-x)^n$ (where n is equal to $-\frac{1}{2}$ for a meson and to $\frac{3}{2}$ for a baryon, both numbers being subject to flavor-dependent upward corrections), the limit (6.36) is valid only if the average multiplicity grows slower than any positive power of s. If, however, one assumes power-like behavior of $\bar{n}_i(s)$:

$$\bar{n}_i(s) = G(s/s_i)^\alpha \ , \tag{6.37}$$

the asymptotic limit (6.36) is replaced by

$$\lim_{s\to\infty} \frac{\bar{n}_{\rm pp}(s)}{\bar{n}_{\nu{\rm p}}(s)} = 2\frac{{\rm B}(\frac{1}{2}+\alpha, \frac{5}{2}){\rm B}(\frac{1}{2}, \frac{5}{2}+\alpha)}{[{\rm B}(\frac{1}{2}, \frac{5}{2})]^2} \qquad \text{with} \quad {\rm B}(x,y) = \frac{\Gamma(x)\Gamma(y)}{\Gamma(x+y)} \ . \tag{6.38}$$

For α values of about $\frac{1}{4}$ (as suggested by some earlier fits) this yields a value close to 1 as seen in the data. Thus, the two-chain model does not necessarily contradict the asymptotic validity of an approximately universal energy dependence of average multiplicities. It should be stressed, however, that the power-like behavior (6.37) does not agree with basic ideas of the dual model and collider data make such fits questionable, phenomenologically (see Sect. 8.1).

A second way to avoid high asymptotic values of the ratio (6.35) and (6.36) in the two-chain model is to assume that the slow chain ends are not the valence quarks, but sea quarks. Then, the relevant x distribution at small x is proportional to $1/x$, as justified by the Regge model or by a bremsstrahlung analogy [54]. To avoid an unphysical singularity, one commonly uses

$$F(x) = C(x^2 + 4\bar{m}_{\rm T}^2/s)^{-1/2} \ , \tag{6.39}$$

where $\bar{m}_{\rm T}^2$ is the average transverse mass squared and the normalization factor C must exhibit energy dependence, decreasing asymptotically as $1/\ln(s/4\bar{m}_{\rm T}^2)$. (Note that previously one could safely neglect the transverse mass at high energies, since the $1/\sqrt{x}$ singularity was integrable). To find approximate analytic results in this case, one can use the simple formula

$$F(x) = C\left\{ \begin{array}{ll} 1/x & \text{for } x > 4\bar{m}_{\rm T}^2/s \\ (s/4\bar{m}_{\rm T}^2)^{1/2} & \text{for } x < 4\bar{m}_{\rm T}^2/s \ , \end{array} \right. \tag{6.40}$$

where

$$C = (1 + \ln s/4\bar{m}_{\rm T}^2)^{-1} \ .$$

For the pp case, such a parametrization, together with the logarithmic energy dependence of $\bar{n}_i(W^2)$, approximately yields

$$\lim_{s\to\infty} \frac{n_{\rm pp}}{n_{\nu{\rm p}}} \approx \frac{3}{2} \ . \tag{6.41}$$

This corresponds to a rapidity length of each chain of $\frac{3}{4}$ of the full available range (since a $1/x$ dependence corresponds to a flat rapidity distribution of the chain end, with average value of $\frac{1}{2}Y_{\text{max}}$). Again, however, the assumption used is not in agreement with dual model ideas. The sea quarks may occur at chain ends only in higher topology diagrams, supposed to be of less importance.

The difficulties of the two-chain model are not really unexpected, since in the dual model the two-chain picture is only the first term of a series and the corrections are quite well defined. We shall show in Chapter 8 that, in fact, a more detailed model study allows us to predict at which energies the modifications of formula (6.31) and corrections to the two-chain picture become necessary [55]. The real asymptotic behavior of the dual model remains, however, rather uncertain.

Analogous difficulties appear with other parameters of the multiplicity distribution, as for example the ratio $D/\bar{n}$ of dispersion D and average multiplicity $\bar{n}$ (see Chapter 8). If the two chains fragment independently, then, from integration of the standard formula analogous to (6.31), we obtain

$$\lim_{s\to\infty} D^2_{\text{pp}}/D^2_{\nu\text{p}} = 2 \tag{6.42}$$

and, as a result [53]

$$\lim_{s\to\infty} \frac{(D/\bar{n})_{\text{pp}}}{(D/\bar{n})_{\nu\text{p}}} = \frac{1}{\sqrt{2}} \,. \tag{6.43}$$

Here, the pp data should be corrected for the diffractive component, which increases the dispersion and decreases the average multiplicity. Even with diffraction removed, however, the ratio (6.43) seems to exceed 1 (see Chapter 8). Moreover, both numerator and denominator are approximately energy-independent, so that an asymptotic convergence of this ratio to its predicted value is very unlikely [56]. Again, at the energies of FNAL and ISR we are far from the asymptotic region, and one can check that the chain energy spread due to $F(x_1)$ and $F(x_2)$ distributions enhances the pp dispersion [53] to the values found experimentally in this range. This effect should decrease at collider energies, since the chain energy spread becomes relatively less important. The collider data, however, support the asymptotic stability of the large $(D/\bar{n})$ ratio in hadronic collisions. A modification of the dual model to explain this and other features of highest energy data are given in [55].

We may note again that more singular structure functions or more steeply rising average chain multiplicities will result in asymptotically non-negligible contributions to the dispersion in hadronic collisions, coming from the chain energy spread. Thus, the two-chain model may explain data on $\bar{n}$ and $D/\bar{n}$ even if the observed trends will continue, but the single-chain description used as input then has to differ significantly from its dual-based characteristics. In fact, the observed constancy of the $D/\bar{n}$ ratio in lepton-hadron and e^+e^- induced hadron production means that also those processes do not look like a single dual chain at high energies. For such a single chain, $D/\bar{n} \to 0$ is expected.

However, these problems, some of which were solved by the natural modifications of the model, should not overshadow the impressive success of its predictions for various hadron spectra. The model describes many of the features of the data, both in the fragmentation and in the central regions.

6.3.2.5 Dual cascade-type models

Another version of a dual fragmentation model, which does not relate the hadronic data to e^+e^- or lepton-hadron results but aims at an understanding of the spectra from a quark-diquark cascade model, has been developed in [57]. In this model, the x distribution of chain ends (quarks, antiquarks and diquarks) is obtained from dual-Regge arguments, as in the versions discussed above. However, the fragmentation spectra are determined by assuming a cascade mechanism in which the original constituents emit mesons or baryons, thus gradually losing their momenta. The free parameters of the momentum-sharing function are fitted to the data. Vector and tensor meson resonances as well as decuplet baryons are taken into account in addition to the stable octet hadrons, and the relative probabilities of their production are fitted. Flavor effects breaking the SU_6 symmetry of initial wave functions are included. Although the global number of parameters is rather large, the model can be regarded as quite successful, since it describes reasonably well very many fragmentation processes in a wide x range ($0.1 < x < 1.0$). The authors regard this as a support for the quark-diquark structure of baryons and as a confirmation of the cascade character of hadon fragmentation processes. As in the other models, diffractive processes are excluded from the analysis. A closely related model has been proposed in [58] and a slightly different picture is presented in [59].

6.3.2.6 The Quark-Gluon-String Model

Also the QGSM [60] is based on dual topological unitarization. It is, in fact, essentially equivalent to a multichain DPM. In addition to 2 strings being formed between valence quark and antiquark and between quark and diquark of the colliding hadrons, respectively, strings are formed between sea quarks and antiquarks of the primordial particles. The distribution functions of valence quarks (diquarks) $F_{V1}(x_1)$ and $F_{V2}(x_n)$ and sea quarks $F_S(x_i)$ in the hadron are of the form:

$$F_{V1}(x_1) = \frac{1}{\sqrt{x_1}} , \quad F_S(x_i) = \frac{1}{\sqrt{x_i}} , \quad \ldots , \quad F_{V2}(x_n) = x_n^\beta , \quad i = 2, \ldots, n-1 , \qquad (6.44)$$

where β=1.5 for a uu- and 2.5 for a ud-diquark in the proton, $\beta = -0.5$ for the u, $\bar{d}$-quark in π^+ and K^+-mesons and $\beta = 0$ for the $\bar{s}$-quark in the K^+-meson. The transverse-momentum distribution of valence and sea quarks in the hadron is of the form $P(p_T^2) \propto \exp(-bp_T^2)$ with $b = 6.25$ (GeV/c)$^{-2}$ for the cylindrical diagram.

The string-breaking algorithm of QGSM is described in [61]. The hadron longitudinal momentum and energy are determined through the variables $z = (E +$

$p_{\parallel})_{\rm h}/(E+p_{\parallel})_{\rm q}$. The quantity z follows the function $f_{\rm q}^{\rm h}(z) = (1+\alpha)(1-z)^{\alpha}$, which at $z \rightarrow 1$ coincides with the fragmentation function $D_{\rm q}^{\rm h}(z)$ of quark or diquark q into hadron h, obtained in [62]. The power α depends on the flavor of the fragmenting constituent, the kind of hadron and the transverse momentum of the hadron relative to the parton direction [62]. At the string break-up, the transverse momentum of the sea quark $\mathbf{p}_{\rm T}$ and antiquark $-\mathbf{p}_{\rm T}$ are assumed to be distributed according to $P(p_{\rm T}^2) = 3b/[\pi(1+b\mathbf{p}_{\rm T}^2)^4]$ with $b = 0.34$ (GeV/c)$^{-2}$. The diquark from the proton compensates the total transverse momentum of the other quarks (antiquarks). Due to an increase of the number of quark-gluon strings with increasing energy, the average $p_{\rm T}$ of quarks and antiquarks, and via the compensation also that of the diquark, increases. The transverse momentum of valence quarks is distributed according to $P(p_{\rm T}^2) = c\exp(-cp_{\rm T}^2)/\pi$ with a slope parameter $c = 10$ (GeV/c)$^{-2}$. The last break of the string is an isotropic two-particle cluster decay. A pseudo-scalar to vector meson ratio PS:V=1:1 is used, without addition of tensor mesons.

The model contains different low-energy interaction mechanisms with cross sections decreasing with increasing energy according to a power law, as well as diffraction dissociation.

6.3.2.7 Monte-Carlo implementations

Several Monte-Carlo programs exist on the basis of dual parton models for hadron-hadron collisions and their extensions to hadron-nucleus and nucleus-nucleus collisions. These are IRIS [63], VENUS [64], QGSM [61], DTUJET [65] and DTUNUC [66]. They differ mainly in the choice of the fragmentation functions. IRIS uses the Lund Monte Carlo, VENUS the fragmentation code AMOR based on the Artru-Mennessier string model [67], QGSM one based on Regge counting rules, while DTUJET and DTUNUC use BAMJET [68].

The Monte-Carlo codes contain soft as well as hard pomeron components and diffractive dissociation. In the nuclear version, VENUS further allows for rescattering of hadrons produced in different strings, but close to each other in space.

In order to use these Monte-Carlo codes to make predictions for pp and nucleus-nucleus collisions at LHC energies, multi-chain effects are essential.

6.3.3 The FRITIOF model

The Lund group proposed yet another interesting dynamical process for soft hadronic interactions [69]. In this model, called FRITIOF, soft hadronic collisions are viewed as an exchange of many soft gluons between the two incident hadrons. As a result of many uncorrelated *longitudinal* energy-momentum exchanges, each of the colliding hadrons becomes excited to a dipole of mass M as if it had been worked upon by a single large momentum transfer. The model further assumes that no net color is exchanged between the incident hadrons. The probability to produce states with mass M_1 and M_2, respectively, is derived from Feynman's expression for the "wee" parton spectrum (i.e. the partons in a hadron at small x) [54] and results in:

$$Prob \sim \frac{dM_1}{M_1^2}\frac{dM_2}{M_2^2} \quad . \tag{6.45}$$

The kinematically allowed region is determined by the conditions $M_1 \geq m_1, M_2 \geq m_2$ and $M_1+M_2 < \sqrt{s}$, where the parameters m_1 and m_2 are cutoff masses, corresponding to the lowest-mass particles with the proper quantum numbers. In principle, these parameters are the masses of the incident particles. However, for a meson beam, this leads to much too high a cross section for interactions with a small excited mass. Therefore, the mass of the ρ and $K^*(892)$ meson is taken as a cutoff value for π^+ and K^+ induced interactions, respectively.

A primordial transverse momentum Q_{2T} is given to the string ends. This is assumed to have a Gaussian distribution with an average $\langle Q_{2T}^2\rangle$ similar to the one adopted in deep-inelastic μp and $\nu(\overline{\nu})$p interactions [70]. In version 7.0, also soft transverse momentum transfer Q_T takes place between the colliding hadrons according to a Gaussian with $\langle Q_T^2\rangle$=0.01 (GeV/c)^2.

In addition, a hard parton-parton elastic scattering (Rutherford parton scattering, RPS) can take place in both versions with a comparatively large transverse momentum. The RPS involves mainly wee partons (gluons) and is usually accompanied by gluon bremsstrahlung. This leads to an increase of the multiplicity of particles produced in the central rapidity region. With a much smaller rate, the RPS also involves the valence quarks and can, therefore, give rise to relatively large transverse momentum of the leading hadrons. In version 7.0 the inclusion of RPS allows to reproduce the high multiplicity tail of the charged-particle distribution measured in the NA22 experiment [71] (see Sub-Sect. 6.3.4 below). At $\sqrt{s}$ = 22 GeV, RPS is responsible for about 12% of the total event sample, while it grows to 28% at 900 GeV.

In the *fragmentation phase* both versions are based on the physical picture, according to which the extended string behaves as a color dipole (antenna) radiating semihard gluons [72] as in e^+e^- annihilation (in the case of an excited meson) or in deep inelastic scattering (in case of an excited baryon).

The gluon radiation is expected to be softer than in e^+e^- annihilation, where charges are point-like and energy concentrations are large. What is used instead is an extension of a soft radiation model for gluon radiation from an extended source successfully used in deep inelastic leptoproduction [73].

In the string c.m., the transverse momentum of radiated gluons is restricted by energy-momentum conservation, $p_T < (M/2)e^{-|y^*|}$ (where M is the dipole mass and y^* is the gluon rapidity in the string c.m.), and by the requirement that the gluon wave length should exceed the transverse size L of the string, $p_T < \sqrt{\frac{\pi M}{L}}e^{-|y^*|/2}$, where L is estimated to be about $L\approx$0.6-0.9fm [69]. Evidently, these restrictions

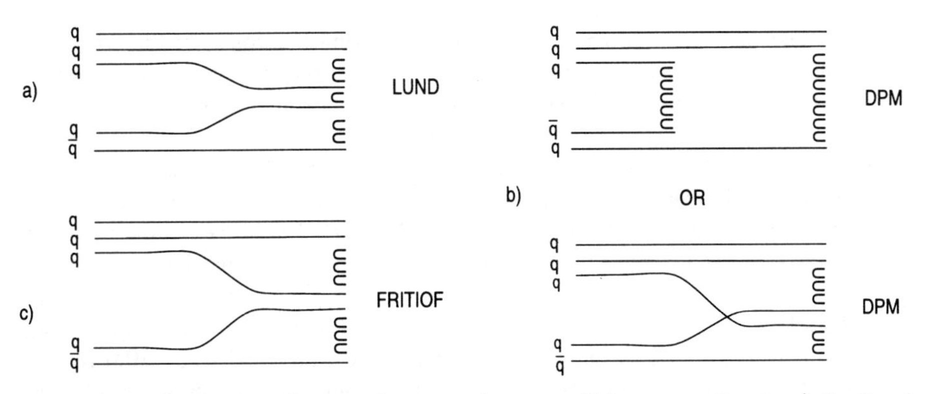

Figure 6.13. Production diagrams for meson-baryon collisions according to a) the Lund model, b) the Dual Parton Model, and c) the FRITIOF model.

limit also the transverse momentum of hadrons produced in string fragmentation, including hadrons with valence quark content, which can acquire a recoil p_T as a consequence of the gluon radiation.

One should stress that the character of the string radiation can be influenced by the RPS included in the FRITIOF7.0 version. The RPS disturbs strongly the color field of the string, which now acts as two dipoles with smaller mass, radiating gluons with smaller p_T as compared with the case of one undisturbed string.

The model does not consider a diquark a point-like entity. Therefore, the energy equivalent for gluon radiation of a diquark string system is additionally reduced by about 50%.

Finally, this model contains also the important process of diffractive-like excitation. Indeed, if the mass of the excited system is close to the cutoff mass mentioned before, it is assumed that the hadron does not become excited at all and retains its original mass. The energy and momentum of this particle are changed accordingly. In this case, one ends up with the situation that one incident hadron is excited, while the second hadron leaves the interaction region unchanged (apart from a loss of energy and momentum).

Like DPM, also FRITIOF has been extended to high energies [69] and to hadron-nucleus and nucleus-nucleus collisions [69,74] and can be used at the LHC. Its Monte-Carlo version, which uses ARIADNE [13] for dipole fragmentation is described in [75].

6.3.4 A first comparison of Lund, FRITIOF and DPM

The final string configuration of the single-string Lund, the two-string DPM and the two-string FRITIOF models are compared in Fig. 6.13. The main differences between the three low-p_T fragmentation models are:

- *Number of strings:* in the Lund model one string is formed; in the DPM the main contribution comes from two-string events, but in principle multi-string events are produced; in FRITIOF always two strings are formed.

- *Valence quarks connected in a single string:* in the Lund model the string connects all valence quarks from the colliding particles; in the DPM strings are formed between color triplets of *different* particles; in FRITIOF strings are formed between color triplets originating from the *same* particle.

- *Possibility for valence quark recombination:* in Lund the valence quarks of an incident hadron can recombine into a single final-state particle; in DPM recombination is excluded since the valence quarks from a hadron end up in different color strings. In FRITIOF, recombination is only possible in diffractive events, since the valence quarks from a hadron are at different string ends.

- FRITIOF includes gluon emission.

- FRITIOF includes hard parton scattering.

- DPM and FRITIOF include primordial transverse momentum.

- Multichain DPM and FRITIOF include diffraction dissociation.

In Fig. 6.14, we show, as a first comparison, the predictions of these three models for the multiplicity distribution of charged particles produced in π^+p collisions at 250 GeV/c [71]. Lund clearly predicts too narrow a multiplicity distribution. The underestimation of the cross section for small multiplicity n can be attributed to the absence of diffraction in this model. The large underestimation of high multiplicities is more fundamental, however, and indicates too limited a degree of "randomness" of particle generation in this model.

The (two-string) DPM also fails for small n, for the same reason as Lund. In the large-n tail, the model prediction starts to deviate from the data for $n > 16$. This is due to the fact that the Monte Carlo includes only interactions where two strings are formed. Multi-chain events are expected to have, on average, a larger number of charged particles than two-string events. Moreover, the events with $n > 16$ represent only $\sim$3% of the total cross section and could, therefore, result mainly from this small cross-section process.

FRITIOF describes the charged-particle multiplicity distribution for $n \leq 18$, but also underestimates the high multiplicity tail. However, the recent version FRITIOF7.0 has been tuned to describe these data [69].

Further comparisons will be shown in the following Chapters.

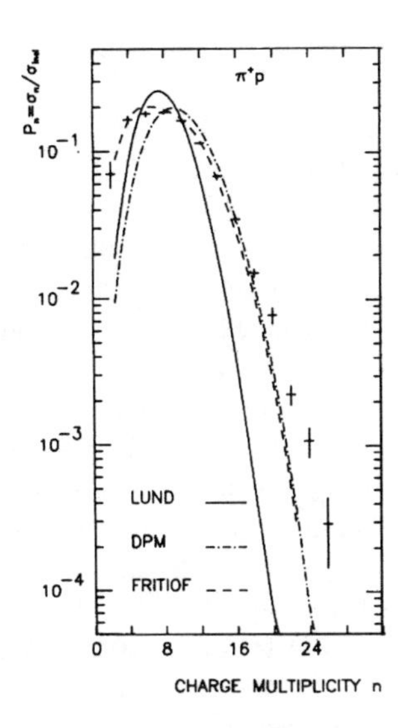

Figure 6.14. The probability distribution $P_n = \sigma_n/\sigma_{\text{in}}$ for π^+p interactions at 250 GeV/c [71]. The curves are predictions of Lund (full line), two-string DPM (dash-dotted) and FRITIOF (dashed).

6.3.5 QFM versus QRM? - Unification Efforts

As may have been noted in the previous sub-sections, the QFM (Sect. 6.3) and QRM (Sect. 5.3) approaches seem to use completely different pictures of hadron fragmentation. In QRM the valence quarks (or antiquarks) of the incident hadron survive the collision without significant loss of momentum and then recombine into final-state hadrons. The final-state hadrons carry, on the average, *more momentum* than the valence quarks themselves, since a small contributon from sea quarks is added. In QFM, one of the valence quarks (or antiquarks) interacts, the remainder flies through with essentially all the momentum and then fragments into hadrons. These hadrons have *less momentum* than the fragmenting remnant. The fragmentation region is populated by fragments of a 'punctured hadron' left after removing one valence (anti)quark. It carries quark or diquark quantum numbers and almost the full initial hadron momentum.

Although these pictures are very different, they yield very similar predictions for single-particle distributions. It was noted in, for example, [76] that for the B $\rightarrow$ M fragmentation, the reference lepton-hadron data used in Eq. 5.21 (single-quark structure function of the proton) and in Eq. 6.25 (diquark fragmentation function into a meson) are in fact quite similar. Thus, both relations are successful and one cannot decide on the basis of a single-particle distribution whether the final meson has originated from recombination of a valence and a sea quark or by fragmentation of a high-momentum diquark. Let us extend this discussion to other fragmentation processes.

For the favored M $\to$ M$'$ process, QRM relates the single-particle spectrum to a distribution $F_{\mathrm{q}}^{\mathrm{M}}(x)$ of quark q in meson M and QFM to the fragmentation function $D_{\mathrm{M}'}^{\mathrm{q}}(x)$ of this quark into meson M$'$. If both mesons are built from only light quarks (as in $\pi^{\pm} \to \rho^0$ or $\pi^0, \pi^{\pm} \to \rho^0$, etc), those two distributions should actually have the same form at $x \to 1$ by the reciprocity relation $D_{\mathrm{M}}^{\mathrm{q}}(x) \propto F_{\mathrm{q}}^{\mathrm{M}}(x)$ (for $x \to 1$) [77]. The two models are really equivalent here. The x distribution of the final-state meson M$'$ is expected to be the same when calculated as fragmentation of a dressed quark or as recombination of the bare valence quark from the initial meson M.

The situation changes, however, for M $\to$ B processes. Here, QFM refers to the q $\to$ B fragmentation function $D_{\mathrm{B}}^{\mathrm{q}}$ and QRM to the same distribution $F_{\mathrm{q}}^{\mathrm{M}}$ as in the previous case. There is no clear reason why these two distributions should be similar. In fact, the baryon spectra are seen to fall faster than the favored meson spectra, contrary to what one expects from naive QRM. One should note, however, that one should not really expect that (5.21) to hold exactly for any final-state hadron, provided only that it contains exactly one quark in common with the initial hadron. The suppression of strange quarks seems to be slightly stronger in the fragmentation region than in the central one. Every such x-dependent flavor effect in the sea distribution is bound to modify (5.21). In the formulation [78] or [79], the recombination function indeed distinguishes between baryons and mesons, and that leads QRM to the successful description of those processes. In [80] a simple way of incorporating both baryon and strangeness suppression effects was proposed as a suppression of processes involving the recombination of a heavy sea quark (or diquark, which is always rather heavy). All those arguments justify the modification of the naive QRM prediction, even if they result in a partial loss of the predictive power.

Finally, in the B $\to$ B$'$ process QRM cannot be applied directly if more than one quark may be in common for B and B$'$, as more contributions appear here. If exactly one quark is common, QFM refers to $D_{\mathrm{B}'}^{\mathrm{q}}$ and QRM to $F_{\mathrm{q}}^{\mathrm{B}}$ and, again, the reciprocity relation allows us to expect equivalence of both approaches.

From the theoretical side, there have been claims of the equivalence of QRM and QFM in the language of topological expansion [81], whereas in terms of perturbative QCD diagrams it seems difficult to understand the similarity [82]. Both QRM and QFM seem to select some of the lowest-order diagrams (different in each case) and the interference terms are absent. Obviously, it is still possible that summing higher-order diagrams one will find results more similar to those of QRM and/or QFM. However, whereas it is quite feasible that 'recombination of bare quarks' and 'fragmentation of a dressed quark system' into single hadrons are two ways of describing the same process, some of the assumptions are definitely contradictory in the two models.

QFM suggests that in a typical collision, one of the valence quarks in a hadron has an x value a few times smaller than the remaining system. So, the hadrons resulting from fragmentation of this slow valence quark will be relatively slow. On the other hand, QRM predicts that all valence quarks should have comparable x values (remaining unchanged during the collision). Their spread as described, for example, by generalized Kuti-Weisskopf functions [83] is not very large. Thus, hadrons containing any of these QRM-type valence quarks will have similar x distributions.

The difference in assumptions, although not easy to present quantitatively without specifying completely the exploited model version, can be tested experimentally. One of the consequences is that in QFM one of the valence quarks does not contribute significantly to the high x region. Therefore, the produced hadron or hadron system that could contain all valence quarks in fact contains one less and should have a similar x spectrum as a system that could contain all but one of those initial valence quarks. On the other hand, in QRM one expects significant differences between these two configurations. As already noted in [84], the distribution of the sum of momenta for all valence quarks of the proton is expected to be flat, but if one of the valence quarks is missing, a decrease proportional to $(1-x)^1$ or similar is expected.

As discussed in Sub-Sect. 5.3.6 above, suppression of valence-quark recombination is indeed observed from the equality of $K^{*\pm}$ and K^{*0} (or $K^{\pm}$ and K^0) production in $K^{\pm}p$ collisions and $\rho^{+,0}$ in π^+p collisions.

Let us mention, however, that the QRM assumption under discussion here is not a crucial and necessary ingredient of the recombination picture. Indeed, the factorized Kuti-Weisskopf form of the initial wave function is just the simplest guess and nothing prohibits the existence of correlations, which could favor the asymmetric configuration of valence quark momenta. This would make the contribution of the slowest valence quark to the fragmentation region unimportant and reconcile the two approaches [85].

Another objection to treating all valence quarks on equal footing is suggested by the successes of the additive quark model. Indeed, as already noted, in the original quark combinatorics approach of [86], one of the constituent quarks (containing obviously a valence quark) was responsible for central production and did not contribute directly to the fragmentation region. It seems likely that in the valon model, the contributions from partons of the interacting valon to the recombination process should also differ from those of spectator valons. Therefore, the results of Sub-Sect. 5.3.6 do not really distinguish between the general QRM and QFM pictures, but rather test one of the assumptions of the particular version of QRM based on the early model of [87] and adopted by most of the later authors.

This short comparison suggests that in all cases either the two models are equivalent, or there are serious reasons to expect corrections to the naive QRM. Since, in addition, structure functions are poorly known in many cases, one can say that single-particle analysis favors the QFM picture. This is better defined, seems to require less corrections and, above all, is a more general picture for hadronization in *all* types of collision. From a practical point of view, it is also better implemented in Monte-Carlo models.

6.3.6 Parton-based Gribov-Regge theory

In DPM, QGSM or VENUS, Gribov-Regge theory [88] has been applied to calculate with some success characteristics of hadron-nucleon, hadron-nucleus and nucleus-nucleus collisions. Is is a multiple scattering theory by construction, with single-pomeron exchange as the elementary interaction (see Chapter 1). Already in hadron-hadron collisions, several pomerons are exchanged and it is important to know how to share the energy between the individual elementary interactions. In general, this is taken into account in treating particle production, but not in calculating cross sections. Even for particle production the subsequent pomerons are often treated differently from the first one.

In the parton models as described earlier in this section, partons of projectile and target are represented by momentum distribution functions and their interactions by elementary parton-parton cross sections, in a factorized form. This factorization hides the complicated multiple scattering structure of the reaction, which has to be reintroduced by eikonalization [89] applicable for hadron-hadron but not uniquely defined for nucleus-nucleus collisions.

Again, no information is provided on how to share the energy between the numerous elementary interactions.

Energy is properly shared between the different interactions taking place in parallel in the parton-based Gribov-Regge approach [90], both when calculating cross sections and particle production. The model includes hard processes in a natural way and can deal with exclusive cross sections without arbitrary assumptions. Essentially a single set of parameters can be used to universally fit e^+e^-, lepton-hadron, hadron-hadron, and the initial stage of nucleus-nucleus collisions.

A problem still to be solved is that of unitarization at very high energies. An expected solution are soft screening corrections due to triple-pomeron interactions.

The approach is realized as the Monte Carlo model NEXUS and has been successfully tested on a number of experimental data.

6.3.7 Geometrical branching and ECCO

A different approach is the eikonal cascade model ECCO [91] based on geometrical branching [92]. The first step of the geometrical branching model is the eikonal formalism in which the geometrical size of the colliding hadrons plays an essential role. The next step is a cut-pomeron expansion, which makes contact with the S-matrix approach to multiparticle production. In both cases, the summation of a series of terms exponentiates the probability of no inelastic collision and imposes unitarity. Each cut pomeron is then treated in the model as a branching process.

A cluster that branches may be regarded as a collection of partons which have no net color and separate into two colorless subcollections. The squared cluster mass is the invariant square of the sum of the 4-momenta of all the partons in the collection. The dynamics of the strong interaction lies in the creation, absorption and rearrangement of all the partons into subsets. If colorless, the latter split from one another without restraining force. The process is too complicated to be tracked in detail, but it can be described by a simple scaling law on the mass distribution,

$$D(m_1, m_2) = \left(\frac{m_1 + m_2}{m}\right)^\gamma , \tag{6.46}$$

which depends only on the mass fractions m_1/m and m_2/m of daughter masses m_1, m_2 and parent mass m, and turns out to be sufficient to specify the dynamics of branching. The power γ is a free parameter to be adjusted from the experimental data. If the number of partons in a collection is large, the probability that they can be arranged into two colorless subcollections is independent of the specific number of partons involved, but depends on the fraction of splitting and on the clustering of the parts.

The basic difference between GBM-ECCO and DPM or FRITIOF is that in the former one assumes the existence of partons in hadrons before their collision, as one does in the parton model. Most of the partons in one hadron pass through those in the other hadron without much interaction except for the ones close in rapidity. In DPM strings are stretched between forward-going diquark and the held-back quark of the other hadron; in FRITIOF hadrons are excited into string-like vortex lines upon collision. In both cases, the stretching and breaking of strings leads to the production of particles. In GMB-ECCO the partons already there rearrange themselves through short-range interaction (in rapidity) into colorless clusters and undergo successive branchings.

The geometrical aspect of hadrons, i.e. the fact that they are extended objects, puts the impact parameter b in a pre-eminent role. The fluctuation in b from event to event leads to fluctuations in p_T. Furthermore, the successive branching is a random process giving rise to fluctuations. So, the model can account for the strong fluctuations, in particular, in high dimensions, to be discussed in Chapter 10. With this taken into account, the model seems to have captured the essence of the dynamical processes involved in soft collisions. Unfortunately, the present version of ECCO is still less refined than the more conventional models. However, besides being the only soft collision model inherently producing fluctuations, it describes the multiplicity distribution and its energy dependence, as well as rapidity and p_T distributions of particles produced in soft collisions.

Two other recent event simulators should be mentioned in comparison, HIJING [93] and the Parton Cascase Model [94]. Although they can both be used for pp collisions, they are basically constructed for heavy-ion collisions. HIJING emphasizes minijet production with the soft part of the production process being adapted from the FRITIOF model. The parton cascade model studies the time evolution of the parton distribution using perturbative QCD, a procedure that is more reliable for hard collisions between partons than for soft interaction. Both models can make significant claims about the simulation of hadron production in high-energy nuclear collisions when hard interactions dominate over soft ones. The aim of GBM-ECCO is to describe soft interactions at energies where hard collisions are unimportant. Thus the two approaches are complementary.

6.4 Conclusions

While the parton excitation phase of the actual particle collision depends on the type of particles colliding, the most successful models are those which converge to a common soft hadronization (or fragmentation) mechanism.

Perturbative shower or dipole evolution can be used for the excitation phase in e^+e^- collisions, DIS and hard hadronic collisions, but a number of phenomenological models had to be developed to describe the (longitudinal) parton excitation in soft hadronic and heavy-ion collisions.

Based on the observations of universality summarized in the previous chapters, a number of phenomenological models were developed to describe, in particular, the common soft hadronization phase. In general, these models either assume a cluster-like or a string-like fragmentation scheme.

In cluster-type models, hadron dynamics is essentially determined by the perturbative phase, with only small corrections in the hadronization phase. As a consequence, coherence is inherent, only few open parameters are needed, strangeness and baryon suppression simply follows from phase space, and, perhaps most importantly, the limited hadron transverse momenta follow naturally. There is, however, a significant dependence on the quite arbitrary shower cutoff parameter Q_0 and string-like treatment turns out to be necessary for massive clusters.

In the string-type models, the emphasis is on the (physically well motivated) string dynamics of hadronization dominating over perturbative effects. As a consequence, there is less dependence on the cutoff Q_0, massive strings are no problem, and strangeness and baryon production is suppressed by tunneling. Disadvantages are the large number of open parameters and that gluon coherence and hadron transverse momenta have to be introduced a-posteriori by hand. Gluon coherence is, however, automatically included in the color dipole radiation approach.

Most of these models are cast into Monte Carlo coding. They differ in their assumptions and number of parameters to be obtained from the data, but in general, are tuned to describe integrated quantities as event shapes, single-particle spectra and (average) multiplicities.

In e^+e^-, lh and high-energy hh collisions (coherent) parton showering is out of question and is needed to explain the large amount of transverse energy and transverse momentum, the large number of produced jets and particles, etc. However, even in hh collisions at moderate energies, at least two strings, but probably more, are needed.

The real issue to be discussed in the remaining chapters is the degree to which these models contain enough stochasticity to reproduce experimentally observed correlations and fluctuations among the final-state particles.

Bibliography

[1] T. Sjöstrand et al., Z Physics at LEP1, Yellow Report CERN 89-08, Vol. 3, p.143.

[2] I.G. Knowles and G.D. Lafferty, *J. Phys.* **G23** (1997) 731.

[3] J. Ellis, M.K. Gaillard, G.G. Ross, *Nucl. Phys.* **B111** (1976) 253.

[4] P.M. Stevenson, *Phys. Rev.* **D23** (1981) 2916.

[5] L.A. Gribov, E.M. Levin, M.G. Ryskin, *Phys. Reports* **100** (1983) 1; A. Bassetto, M. Ciafaloni, G. Marchesini, *Phys. Reports* **100** (1983) 201; B.R. Webber, *Ann. Rev. Nuc. Part. Sci.* **36** (1986) 253.

[6] G. Altarelli, G. Parisi, *Nucl. Phys.* **B126** (1977) 298.

[7] Yu.L. Dokshitzer, *Sov. Phys. JETP* **46** (1977) 641.

[8] A.H. Mueller, *Phys. Lett.* **B104** (1981) 161; B.I. Ermolaev, V.S. Fadin, *JETP Lett.* **33** (1981) 269; Yu.L. Dokshitzer, V.A. Khoze, S.I. Troyan, *Sov. J. Nucl. Phys.* **47** (1988) 1384.

[9] G. Marchesini, B.R. Webber, *Nucl. Phys.* **B310** (1988) 461; I.G. Knowles, *Nucl. Phys.* **B310** (1988) 571 and *Comp. Phys. Commun.* **58** (1990) 271; G. Marchesini et al., *Comp. Phys. Commun.* **67** (1992) 465.

[10] H.-U. Bengtsson, *Comp. Phys. Commun.* **31** (1984) 323; H.-U. Bengtsson and G. Ingelman, *Comp. Phys. Commun.* **34** (1985) 251; H.-U. Bengtsson and T. Sjöstrand, *Comp. Phys. Commun.* **46** (1987) 43; T. Sjöstrand, *Comp. Phys. Commun.* **27** (1982) 243; **28** (1983) 227; **39** (1986) 347; **82** (1994) 74; T. Sjöstrand, M. Bengtsson, *Comp. Phys. Commun.* **43** (1987) 367; T. Sjöstrand et al., *Comp. Phys. Commun.* **135** (2001) 238.

[11] T.D. Gottschalk, *Nucl. Phys.* **B239** (1984) 325; T.D. Gottschalk, *Nucl. Phys.* **B239** (1984) 349; T.D. Gottschalk, D.A. Morris, *Nucl. Phys.* **B288** (1987) 729.

[12] Ya.I. Azimov, Yu.L. Dokshitzer, V.A. Khoze, S.I. Troyan, *Phys. Lett.* **B165** (1985) 147; G. Gustafson, *Phys. Lett.* **B175** (1986) 453; G. Gustafson, U. Pettersson, *Nucl. Phys.* **B306** (1988) 746; B. Andersson, G. Gustafson, L. Lönnblad, *Nucl. Phys.* **B339** (1990) 393; L. Lönnblad, *Z. Phys.* **C65** (1995) 285; **C70** (1996) 107.

[13] L. Lönnblad, *Comp. Phys. Commun.* **71** (1992) 15.

[14] R. Odorico, *Comp. Phys. Commun.* **72** (1992) 238; P. Mazzanti, R. Odorico, *Nucl. Phys.* **B370** (1992) 23.

[15] T. Sjöstrand, *Int. J. Mod. Phys.* **A3** (1988) 751.

[16] Ya.I. Azimov, Yu.L. Dokshitzer, V.A. Khoze, S.I. Troyan, *Z. Phys.* **C27** (1985) 65; **C31** (1986) 213.

[17] A. Krzywicki, B. Petersson, *Phys. Rev.* **D6** (1972) 924; J. Finkelstein, R.D. Peccei, *Phys. Rev.* **D6** (1972) 2606; F. Niedermayer, *Nucl. Phys.* **B79** (1974) 355; A. Casher, J. Kogut, L. Susskind, *Phys. Rev.* **D10** (1974) 732.

[18] R.D. Field, R.P. Feynman, *Nucl. Phys.* **B136** (1978) 1.

[19] C. Peterson, D. Schlatter, I. Schmitt, P. Zerwas, *Phys. Rev.* **D27** (1983) 105.

[20] B. Andersson, G. Gustafson, G. Ingelman and T. Sjöstrand, *Phys. Reports* **97** (1983) 31.

[21] X. Artru, G. Mennessier, *Nucl. Phys.* **B70** (1974) 93; X. Artru, *Phys. Reports* **97** (1983) 147.

[22] K.G. Wilson, *Phys. Rev.* **D10** (1974) 2445.

[23] B. Andersson, G. Gustafson, B. Söderberg, *Z. Phys.* **C20** (1983) 317.

[24] C.D. Buchanan, S.-B. Chun, *Phys. Rev. Lett.* **59** (1987) 1997; *Phys. Lett.* **B308** (1993) 153; S.B. Chun and C.D. Buchanan, *Phys. Rep.* **292** (1998) 239.

[25] B. Andersson, S. Mohanty, F. Söderberg, *Eur. Phys. J.* **C21** (2001) 631; LU-TP-02-13 (2002); B. Andersson, F. Söderberg, S. Mohanty, *Proc. XXX Int. Symp. on Multiparticle Dynamics*, eds. T. Csögő et al. (World Scientific, Singapore, 2001) p.149.

[26] B. Andersson, F. Söderberg, *Eur. Phys. J.* **C16** (2000) 303.

[27] G.C. Fox, S. Wolfram, *Nucl. Phys.* **B168** (1980) 285;
R.D. Field, S. Wolfram, *Nucl. Phys.* **B213** (1983) 65.

[28] D. Amati and G. Veneziano, *Phys. Lett.* **B83** (1979) 87; G. Marchesini, L. Trentadue, G. Veneziano, *Nucl. Phys.* **B181** (1981) 335.

[29] V.N. Gribov and L.N. Lipatov, *Sov. J. Nucl. Phys.* **15** (1972) 438 and 675.

[30] E.A. Kuraev, L.N. Lipatov and V.S. Fadin, *Sov. Phys. JETP* **45** (1972) 199; Y.Y. Balitsky and L.N. Lipatov, *Sov. J. Nucl. Phys.* **28** (1978) 822.

[31] A.D. Martin, *Contemp. Phys.* **36** (1995) 335.

[32] M. Ciafaloni, *Nucl. Phys.* **B296** (1987) 249; S. Catani, F. Fiorani and G. Marchesini, *Phys. Lett.* **B234** (1990) 339; *Nucl. Phys.* **B336** (1990) 18.

[33] G. Ingelman, *Comp. Phys. Commun.* **101** (1997) 108.

[34] H. Jung, *Comp. Phys. Commun.* **86** (1995) 147.

[35] H. Kharraziha and L. Lönnblad, *The Linked Dipole Chain Monte Carlo*, LU-TH97-21, hep-ph/9709424.

[36] B. Andersson et al., *Nucl. Phys.* **B463** (1996) 217; *Z. Phys.* **C71** (1996) 613.

[37] H. Jung, *Comp. Phys. Commun.* **143** (2002) 100.

[38] B. Andersson, G. Gustafson and C. Peterson, *Phys. Lett.* **69B** (1977) 221; ibid. **71B** (1977) 337.

[39] B. Andersson, G. Gustafson, I. Holgersson and O. Månsson, *Nucl. Phys.* **B178** (1981) 242.

[40] G. 't Hooft, *Nucl. Phys.* **B72** (1974) 461; G. Veneziano, *Nucl. Phys.* **B74** (1974) 365; E. Witten, *Nucl. Phys.* **B160** (1979) 57.

[41] A. Capella, U. Sukhatme, C.-I. Tan and J. Tran Thanh Van, *Phys. Reports* **236** (1994) 225.

[42] H. Lee, *Phys. Rev. Lett.* **30** (1973) 719; G. Veneziano, *Phys. Lett.* **43B** (1973) 413; H.M. Chan and J.E. Paton, *Phys. Lett.* **46B** (1973) 228.

[43] G. Veneziano, *Phys. Lett.* **52B** (1974) 220; H.M. Chan et al., *Nucl. Phys.* **B86** (1975) 470; ibid. **B92** (1975) 13; M. Ciafaloni, G. Marchesini and G. Veneziano, *Nucl. Phys.* **B98** (1975) 472; G. Chew and C. Rosenzweig, *Nucl. Phys.* **B104** (1976) 290.

[44] G. Veneziano, *Nucl. Phys.* **B117** (1976) 519.

[45] J. Dias de Deus, *Nucl. Phys.* **B123** (1977) 240; G. Rossi and G. Veneziano, *Nucl. Phys.* **B123** (1977) 507.

[46] A. Giovannini and G. Veneziano, *Nucl. Phys.* **B130** (1977) 61.

[47] J. Dias de Deus and S. Jadach, *Phys. Lett.* **66B** (1977) 81; *Acta Phys. Pol.* **B9** (1978) 249; F. Hayot and S. Jadach, *Phys. Rev.* **D17** (1978) 2308.

[48] C. Chiu and S. Matsuda, *Nucl. Phys.* **B143** (1978) 463; P. Aurenche and L. Gonzalez-Mestres, *Phys. Rev.* **D18** (1978) 2995.

[49] A. Capella, U. Sukhatme, C.I. Tan and J. Tran Thanh Van, *Phys. Lett.* **81B** (1979) 68; *Z. Phys.* **C3** (1980) 329.

[50] G. Cohen-Tannoudji, F. Hayot and R. Peschanski, *Phys. Rev.* **D17** (1978) 2930; G. Cohen-Tannoudji et al., *Phys. Rev.* **D21** (1980) 2699.

[51] H. Minakata, *Phys. Rev.* **D20** (1979) 1656.

[52] M. Bishari, G. Cohen-Tannoudji, H. Navelet and R. Peschanski, *Z. Phys.* **C14** (1982) 7; G. Cohen-Tannoudji et al., *Z. Phys.* **C13** (1982) 219.

[53] J. Dias de Deus, *Phys. Lett.* **100B** (1981) 177; K. Fiałkowski and A. Kotański, *Phys. Lett.* **107B** (1981) 132.

[54] R.P. Feynman, *Phys. Rev. Lett.* **23** (1969) 1415; *High Energy Collisions*, ed. C.N. Yang et al. (Gordon and Breach, New York, 1969) p.237.

[55] A. Capella and J. Tran Thanh Van, *Phys. Lett.* **114B** (1982) 450.

[56] Ch. Berger et al., *Phys. Lett.* **95B** (1980) 313.

[57] K. Kinoshita et al., *Prog. Theor. Phys.* **61** (1979) 165; ibid.**63** (1980) 928; ibid. **63** (1980) 1628;*Z. Phys.* **C8** (1981) 205; *Kagoshima* preprint HE-81-6 (1981), published in Dubna HEP Seminar 1981 p.334.

[58] M. Uehara, *Prog. Theor. Phys.* **66** (1981) 1697.

[59] H. Fukuda and C. Iso, *Prog. Theor. Phys.* **57** (1977) 483; ibid. **57** (1977) 1663; ibid. **58** (1977) 1472.

[60] A. Kaĭdalov, *Phys. Lett.* **B116** (1882) 459; A. Kaĭdalov and K.A. Ter Martirosyan, *Phys. Lett.* **B117** (1982) 247.

[61] N.S. Amelin and L.V. Bravina, *Sov. J. Nucl. Phys.* **51** (1990) 133; N.S. Amelin, L.V. Bravina, L.N. Smirnova, *Sov. J. Nucl. Phys.* **52** (1990) 362.

[62] A.B. Kaĭdalov, *Sov. J. of Nucl. Phys.* **45** (1987) 902.

[63] J.P. Pansart, *Nucl. Phys.* **A461** (1987) 521c.

[64] K. Werner, *Phys. Lett.* **B197** (1987) 225; *Z. Phys.* **C38**(1988) 193; *Nucl. Phys.* **A525** (1991) 501c; K. Werner, *Phys. Reports* **232** (1993) 87.

[65] F.W. Bopp, A. Capella, J. Ranft and J. Tran Thanh Van, *Z. Phys.* **C99** (1991) 51; A. Aurenche, A. Capella, J. Kwiecinski, M. Maire, J. Ranft and J. Tran Thanh Van, *Phys. Rev.* **D45** (1992) 92.

[66] H.J. Möhring and J. Ranft, *UL-HEP-92-2* .

[67] X. Artru and G. Mennessier, *Nucl. Phys.* **B70** (1974) 93.

[68] J. Ranft and S. Ritter, *Acta Phys. Pol.* **B11** (1980) 259; S. Ritter, *Comp. Phys. Commun.* **31** (1984) 393.

[69] B. Andersson, G. Gustafson, B. Nilsson-Almqvist, *Nucl. Phys.* **B281** (1987) 289; LU-TP-87-6, 1987 (unpublished); B. Nilsson-Almqvist and I. Stenlund, *Comp. Phys. Commun.* **43** (1987) 387; B. Andersson, G. Gustafson and Hong Pi, *Z. Phys.* **C57** (1993) 485; Hong Pi, An event generator for interactions between hadrons and nuclei -FRITIOF version 7.0, Lund preprint LU TP 91-28.

[70] M. Arneodo et al., *Phys. Lett.* **B149** (1984) 415; *Z. Phys.* **C36** (1987) 527.

[71] NA22 Coll., M. Adamus et al., *Z. Phys.* **C32** (1986) 475.

[72] G. Gustafson, *Phys. Lett.* **B175** (1986) 453; G. Gustafson, U. Pettersson, *Nucl. Phys.* **B306** (1988) 746.

[73] B. Andersson, G. Gustafson, L. Lönnblad, U. Pettersson, *Z. Phys.* **C43** (1989) 625.

[74] B. Andersson, A. Tai, Ben-Hao Sa, *Z. Phys.* **C70** (1996) 499.

[75] B. Nilsson-Almqvist, E. Stenlund, *Comp. Phys. Commun.* **43** (1987)387; Hong Pi, *Comp. Phys. Commun.* **71** (1992) 173.

[76] F.C. Erné, *Proc. Physics in Collision*, Vol.1, Ed. P. Trower (New York: Plenum, 1982) p.297.

[77] V.N. Gribov and L.N. Lipatov, *Yad. Fiz.* **15** (1972) 1218.

[78] K.P. Das and R.C. Hwa, *Phys. Lett.* **68B** (1977) 459.

[79] E. Takasugi and X. Tata, *Phys. Rev.* **D23** (1981) 2573.

[80] F. Takagi, *Prog. Theor. Phys.* **65** (1980) 1530.

[81] G. Cohen-Tannoudji, A. El Hassouni, J. Kalinowski and R. Peschanski, *Phys. Rev.* **D19** (1979) 3397; A. El Hassouni and O. Napoly, *Phys. Rev.* **D23** (1981) 193.

[82] J.F. Gunion, Proc. Europhys. Conf. on Partons in Soft Hadronic Processes, ed. R.T. Van de Walle (World Scientific, Singapore, 1981) p.293.

[83] J. Kuti and V. Weisskopf, *Phys. Rev.* **D4** (1971) 3418.

[84] L. Van Hove and S. Pokorski, *Nucl. Phys.* **B86** (1975) 245.

[85] L. Van Hove, *Proc. 12th Int. Symp. on Multiparticle Dynamics, Notre Dame*, ed. W.D. Shephard and V.P. Kennedy (World Scientific, Singapore, 1982) p.927.

[86] V.V. Anisovich and V.M. Shekhter, *Nucl. Phys.* **B55** (1973) 455.

[87] S. Pokorski and L. Van Hove, *Acta Phys. Pol.* **B5** (1974) 229.

[88] T. Regge, *Nuovo Cimento* **14** (1959) 951; V.N. Gribov, *Sov. Phys. JETP* **14** (1962) 1395; **30** (1970) 709.

[89] T.K. Gaisser, F. Halzen, *Phys. Rev. Lett.* **54** (1984) 1754; P.L'Heureux, B. Margolis, P. Valin, *Phys. Rev.* **D32** (1985) 1681; X.-N. Wang, *Phys. Reports* **280** (1997) 287.

[90] S. Ostapchenko, T. Thouw, K. Werner, *Nucl. Phys. Proc. Suppl.* **B52** (1987) 3; H. Drescher, M. Hladik, S. Ostapchenko, K. Werner, *Nucl. Phys.* **A661** (1999) 604; *J. Phys.* **G25** (1999) L91; H.J. Drescher, M. Hladik, S. Ostapchenko, T. Pierog, K. Werner, *Phys. Reports* **350** (2001) 93.

[91] R.C. Hwa and J.C. Pan, *Phys. Rev.* **D45** (1992) 106; J.C. Pan and R.C. Hwa, *Phys. Rev.* **D46** (1992) 4890; ibid. **D48** (1993) 168.

[92] W.R. Chen and R.C. Hwa, *Phys. Rev.* **D36** (1987) 760; **39** (1989) 179; R.C. Hwa, ibid. **37** (1988) 1830.

[93] X.N. Wang and M. Gyulassy, *Phys. Rev.* **D44** (1991) 3501.

[94] K. Geiger and B. Müller, *Nucl. Phys.* **B369** (1992) 600.

Chapter 7

Correlations and Fluctuations, the Formalism

7.1 Definitions and notation

In this chapter, we compile and summarize definitions and various relations among the physical quantities used in the sequel of this book. No originality is claimed in the presentation of this material. It merely serves the purpose of fixing the notation and assembling a number of results scattered throughout the literature.

7.1.1 Exclusive and inclusive densities

We start by considering a collision between particles a and b yielding exactly n particles in a sub-volume Ω of the total phase space Ω_{tot}. Let the single symbol y represent the kinematical variables needed to specify the position of each particle in this space (for example, y can be the full energy-momentum four-vector and Ω a sub-volume in four-dimensional energy-momentum space or y can simply be the c.m. rapidity variable of each particle and Ω an interval of length δy). The distribution of points in Ω can be characterized by continuous probability densities $P_n(y_1, \ldots, y_n)$; $n = 1, 2, \ldots$. For simplicity, we assume all final-state particles to be of the same type. In this case, the **exclusive** distributions $P_n(y_1, \ldots, y_n)$ can be taken fully symmetric in $y_1, \ldots, y_n$; they describe the distribution in Ω when the multiplicity is exactly n.

The corresponding **inclusive** distributions are given for $q = 1, 2, \ldots$ by:

$$\begin{aligned}\rho_q(y_1, \ldots, y_q) &= P_q(y_1, \ldots, y_q) \\ &\quad + \sum_{\ell=1}^{\infty} \frac{1}{\ell!} \int_\Omega P_{q+\ell}(y_1, \ldots, y_q, y'_1, \ldots, y'_\ell) \prod_{i=1}^{\ell} \mathrm{d}y'_i \,. \qquad (7.1)\end{aligned}$$

The inverse formula is

$$P_q(y_1,\ldots,y_q) = \rho_q(y_1,\ldots,y_q) + \sum_{\ell=1}^{\infty}(-1)^\ell \frac{1}{\ell!}\int_\Omega \rho_{q+\ell}(y_1,\ldots,y_q,y'_1,\ldots,y'_\ell)\prod_{i=1}^{\ell} \mathrm{d}y'_i \,. \quad (7.2)$$

$\rho_q(y_1,\ldots,y_q)$ is the (symmetrized) number density for q points to be at $y_1,\ldots,y_q$, irrespective of the presence and location of any further points. The probability P_0 of multiplicity zero is given by

$$P_0 = 1 - \sum_{q=1}^{\infty}\frac{1}{q!}\int_\Omega P_q(y_1,\ldots,y_q)\prod_{i=1}^{q}\mathrm{d}y_i \,. \quad (7.3)$$

This suggests to define $\rho_0 = 1$ in (7.1). It is often convenient to summarize the above results with the help of the generating functional[1]

$$\mathcal{G}^{\mathrm{excl}}[z(y)] \equiv P_0 + \sum_{q=1}^{\infty}\frac{1}{q!}\int_\Omega P_q(y_1,\ldots,y_q)\, z(y_1)\ldots z(y_q)\prod_{i=1}^{q}\mathrm{d}y_i \,, \quad (7.4)$$

where $z(y)$ is an arbitrary function of y in Ω. The substitution

$$z(y) = 1 + u(y) \quad (7.5)$$

gives through (7.1) the alternative expansion

$$\mathcal{G}^{\mathrm{incl}}[u(y)] = 1 + \sum_{q=1}^{\infty}\frac{1}{q!}\int_\Omega \rho_q(y_1,\ldots,y_q)\, u(y_1)\ldots u(y_q)\prod_{i=1}^{q}\mathrm{d}y_i \quad (7.6)$$

and the relation

$$\mathcal{G}^{\mathrm{incl}}[z(y)] = \mathcal{G}^{\mathrm{excl}}[z(y)+1] \,. \quad (7.7)$$

From (7.4) and (7.7) one recovers by functional differentiation:

$$P_q(y_1,\ldots,y_q) = \left.\frac{\partial^q \mathcal{G}^{\mathrm{excl}}[z(y)]}{\partial z(y_1)\ldots\partial z(y_q)}\right|_{z=0} \,, \quad (7.8)$$

and

$$\rho_q(y_1,\ldots,y_q) = \left.\frac{\partial^q \mathcal{G}^{\mathrm{incl}}[u(y)]}{\partial u(y_1)\ldots\partial u(y_q)}\right|_{u=0} \,. \quad (7.9)$$

To the set of inclusive number densities ρ_q corresponds a sequence of inclusive differential cross sections:

$$\frac{1}{\sigma_{\mathrm{inel}}}\,\mathrm{d}\sigma = \rho_1(y)\,\mathrm{d}y, \quad (7.10)$$

[1] The technique of generating functions has been known since Euler's time and was used for functionals by N.N. Bogoliubov in statistical mechanics already in 1946 [1]; see also [2]

$$\frac{1}{\sigma_{\text{inel}}}\mathrm{d}^2\sigma = \rho_2(y_1, y_2)\,\mathrm{d}y_1\mathrm{d}y_2\ , \tag{7.11}$$

etc. Integration over an interval Ω in y yields

$$\begin{aligned}
&\int_\Omega \rho_1(y)\mathrm{d}y = \langle n\rangle \\
&\int_\Omega\int_\Omega \rho_2(y_1,y_2)\mathrm{d}y_1\mathrm{d}y_2 = \langle n(n-1)\rangle \\
&\int_\Omega \mathrm{d}y_1 \ldots \int_\Omega \mathrm{d}y_q\rho_q(y_1,\ldots,y_q) = \langle n(n-1)\ldots(n-q+1)\rangle\ ,
\end{aligned} \tag{7.12}$$

where the angular brackets imply the average over the event ensemble and n is the number of particles (multiplicity) in Ω in a given event.

7.1.2 Cumulant correlation functions

The inclusive q-particle densities $\rho_q(y_1,\ldots,y_q)$ in general contain "trivial" contributions from lower-order densities. Under certain conditions, it is, therefore, advantageous to consider a new sequence of functions $C_q(y_1,\ldots,y_q)$ as those statistical quantities which vanish whenever one of their arguments becomes statistically independent of (= uncorrelated with respect to) the others. The quantities with such properties are the correlation functions - also called (factorial) cumulant functions - or, in integrated form, Thiele's semi-invariants [3]. A formal proof of this property was given by Kubo [4] (see also Chang et al. [5]). The cumulant correlation functions are defined as in the cluster expansion familiar from statistical mechanics via the sequence [6–8]:

$$\begin{aligned}
\rho_1(1) &= C_1(1), \\
\rho_2(1,2) &= C_1(1)C_1(2) + C_2(1,2), \\
\rho_3(1,2,3) &= C_1(1)C_1(2)C_1(3) + C_1(1)C_2(2,3) + C_1(2)C_2(1,3) + \\
&\quad + C_1(3)C_2(1,2) + C_3(1,2,3);
\end{aligned} \tag{7.13}$$

and, in general, by

$$\begin{aligned}
\rho_m(1,\ldots,m) &= \sum_{\{l_i\}_m}\sum_{\text{perm.}} \underbrace{[C_1()\cdots C_1()]}_{l_1\,\text{factors}}\underbrace{[C_2(,)\cdots C_2(,)]}_{l_2\,\text{factors}}\cdots \\
&\quad \cdots \underbrace{[C_m(,\ldots,)\cdots C_m(,\ldots,)]}_{l_m\,\text{factors}}.
\end{aligned} \tag{7.14}$$

Here, l_i is either zero or a positive integer and the sets of integers $\{l_i\}_m$ satisfy the condition

$$\sum_{i=1}^{m} i\,l_i = m. \tag{7.15}$$

The arguments in the C_q functions are to be filled by the m possible momenta in any order. In the above relations we have abbreviated $C_q(y_1, \ldots, y_q)$ to $C_q(1, 2, \ldots, q)$; the summations indicate that all possible permutations have to be taken (the number under the summation sign indicates the number of terms). The sum over permutations is a sum over all distinct ways of filling these arguments. For any given factor product there are precisely [7]

$$\frac{m!}{[(1!)^{l_1}(2!)^{l_2} \cdots (m!)^{l_m}]\, l_1! l_2! \cdots l_m!} \tag{7.16}$$

terms. The complete set of relations is contained in the functional identity:

$$\mathcal{G}^{\rm incl}[u(y)] = \exp\{g[u(y)]\} , \tag{7.17}$$

where

$$g[u(y)] = \int \rho_1(y)u(y)\,\mathrm{d}y + \sum_{q=2}^{\infty} \frac{1}{q!} \int_\Omega C_q(y_1, \ldots, y_q)\, u(y_1) \ldots u(y_q) \prod_{i=1}^{q} \mathrm{d}y_i. \tag{7.18}$$

It follows that

$$C_q(y_1, \ldots, y_q) = \left. \frac{\partial^q g[u(y)]}{\partial u(y_1) \ldots \partial u(y_n)} \right|_{u=0} . \tag{7.19}$$

The relations (7.14) may be inverted with the result:

$$\begin{aligned}
C_2(1,2) &= \rho_2(1,2) - \rho_1(1)\rho_1(2) , \\
C_3(1,2,3) &= \rho_3(1,2,3) - \sum_{(3)} \rho_1(1)\rho_2(2,3) + 2\rho_1(1)\rho_1(2)\rho_1(3) , \\
C_4(1,2,3,4) &= \rho_4(1,2,3,4) - \sum_{(4)} \rho_1(1)\rho_3(1,2,3) - \sum_{(3)} \rho_2(1,2)\rho_2(3,4) \\
&\quad + 2\sum_{(6)} \rho_1(1)\rho_1(2)\rho_2(3,4) - 6\rho_1(1)\rho_1(2)\rho_1(3)\rho_1(4).
\end{aligned} \tag{7.20}$$

Expressions for higher orders can be derived from the related formulae given in [9].

From (7.19) follows an important additivity property of the C_q for N_s independent systems fully overlapping in momentum space,

$$C_q^{(N_s)}(y_1, \ldots, y_q) = N_s C_q^{(1)}(y_1, \ldots, y_q) . \tag{7.21}$$

It is often convenient to divide the functions ρ_q and C_q by the product of one-particle densities. This leads to the definition of the normalized inclusive densities and correlations:

$$R_q(y_1, \ldots, y_q) = \rho_q(y_1, \ldots, y_q)/\rho_1(y_1) \ldots \rho_1(y_q), \tag{7.22}$$

$$K_q(y_1, \ldots, y_q) = C_q(y_1, \ldots, y_q)/\rho_1(y_1) \ldots \rho_1(y_q). \tag{7.23}$$

From the additivity (7.21) of the $C_q(y_1,\ldots,y_q)$ follows a dilution property for the normalized correlations

$$K_q^{(N_s)}(y_1,\ldots,y_q) = N_s^{-(q-1)} K_q^{(1)}(y_1,\ldots,y_q)\ . \tag{7.24}$$

From expression (7.17) it can be deduced that, at finite energy, an infinite number of C_q will be non-vanishing: The densities ρ_q vanish for $q > N$, where N is the maximal number of particles in Ω allowed e.g. by energy-momentum conservation. As a consequence, the functional $\mathcal{G}$ is a "polynomial" in $u(y)$. This in turn requires the exponent in (7.17) to be an "infinite series" in $u(y)$. In other words, the higher-order correlation functions must cancel the lower-order ones that contribute to a vanishing density function. Phenomenologically, this implies that it is meaningful to use correlation functions C_q only if the number of correlated particles in the considered phase-space domain Ω is considerably smaller than the average multiplicity in that region [2]. These conditions are not always fulfilled in present-day experiments for very small phase-space cells, with the exception perhaps of heavy-ion collisions.

7.1.3 Correlations for particles of different species

The generating functional technique of Sect. 7.1.1 can be extended to the general situation where several different species of particles are distinguished. This will not be pursued here and we refer to the literature for details [2, 10–12]. Considering two particle species a and b, the two-particle rapidity correlation function is of the form:

$$C_2^{\mathrm{ab}}(y_1,y_2) = \rho_2^{\mathrm{ab}}(y_1,y_2) - f\rho_1^{\mathrm{a}}(y_1)\rho_1^{\mathrm{b}}(y_2), \tag{7.25}$$

with

$$\rho_1^{\mathrm{a}}(y_1) = \frac{1}{\sigma_{\mathrm{inel}}}\frac{\mathrm{d}\sigma^{\mathrm{a}}}{\mathrm{d}y_1};\quad \rho_2^{\mathrm{ab}}(y_1,y_2) = \frac{1}{\sigma_{\mathrm{inel}}}\frac{\mathrm{d}\sigma^{\mathrm{ab}}}{\mathrm{d}y_1\mathrm{d}y_2}. \tag{7.26}$$

Here, y_1 and y_2 are the c.m. rapidities, σ_{inel} the inelastic cross section and a, b represent particle properties, e.g. charge.

The normalization conditions are:

$$\int \rho_1^{\mathrm{a}}(y_1)\mathrm{d}y_1 = \langle n_{\mathrm{a}}\rangle\ ;\ \iint \rho_2^{\mathrm{ab}}(y_1,y_2)\mathrm{d}y_1\mathrm{d}y_2 = \langle n_{\mathrm{a}}(n_{\mathrm{b}} - \delta^{\mathrm{ab}})\rangle\ , \tag{7.27}$$

$$\iint C_2^{\mathrm{ab}}(y_1,y_2)\mathrm{d}y_1\mathrm{d}y_2 = \langle n_{\mathrm{a}}(n_{\mathrm{b}} - \delta^{\mathrm{ab}})\rangle - f\langle n_{\mathrm{a}}\rangle\langle n_{\mathrm{b}}\rangle\ , \tag{7.28}$$

where $\delta^{\mathrm{ab}} = 0$ for the case when a and b are particles of different species and $\delta^{\mathrm{ab}} = 1$ for identical particles, and n_{a} and n_{b} are the corresponding particle multiplicities.

Most experiments use

$$f = 1\ , \tag{7.29}$$

so that the integral over the correlation function vanishes for the case of a Poissonian multiplicity distribution. Other experiments use

$$f = \frac{\langle n_{\rm a}(n_{\rm b} - \delta^{\rm ab})\rangle}{\langle n_{\rm a}\rangle\langle n_{\rm b}\rangle} \tag{7.30}$$

to obtain a vanishing integral also for a non-Poissonian multiplicity distribution.

To be able to compare the various experiments, we use both definitions and denote the correlation function $C_2^{\rm ab}(y_1, y_2)$ when following definition (7.29) and $C_2'^{\rm ab}(y_1, y_2)$ when following definition (7.30). We, furthermore, use a reduced form of definition (7.30),

$$\tilde{C}_2^{\rm ab}(y_1, y_2) = C_2'^{\rm ab}(y_1, y_2)/\langle n_{\rm a}(n_{\rm b} - \delta^{\rm ab})\rangle. \tag{7.31}$$

The corresponding normalized correlation functions

$$K_2^{\rm ab}(y_1, y_2) = \frac{C_2^{\rm ab}(y_1, y_2)}{f\rho_1^{\rm a}(y_1)\rho_1^{\rm b}(y_2)} \tag{7.32}$$

follow the relations

$$K_2' = \frac{1}{f}(K_2 + 1) - 1 \ , \tag{7.33}$$

and $\tilde{K}_2$ is defined as $\tilde{K}_2 = K_2'$. These normalized correlation functions are more appropriate than C_2 when comparisons have to be performed at different average multiplicity and are less sensitive to acceptance problems.

7.1.4 Semi-inclusive correlation functions

The correlation functions defined by expressions (7.20)-(7.33), contain a pseudo-correlation due to the summation of events with different charged-particle multiplicity n and different semi-inclusive single-particle densities $\rho_1^{(n)}$.

The relation between inclusive and semi-inclusive correlation functions has been carefully analyzed in [13]. Let σ_n be the topological cross section and

$$P_n = \sigma_n/\Sigma\sigma_n \ . \tag{7.34}$$

The semi-inclusive rapidity single- and two-particle densities for particles a and b are defined as

$$\rho_1^{(n)}(y) = \frac{1}{\sigma_n}\frac{{\rm d}\sigma_n^a}{{\rm d}y} \quad \text{and} \quad \rho_2^{(n)}(y_1, y_2) = \frac{1}{\sigma_n}\frac{{\rm d}\sigma_n^{ab}}{{\rm d}y_1{\rm d}y_2} \ . \tag{7.35}$$

The inclusive correlation function $C_2(y_1, y_2)$ can then be written as

$$C_2(y_1, y_2) = C_{\rm S}(y_1, y_2) + C_{\rm L}(y_1, y_2) \ , \tag{7.36}$$

where

$$C_{\rm S}(y_1, y_2) = \Sigma P_n C_2^{(n)}(y_1, y_2) \tag{7.37}$$

$$C_L(y_1, y_2) = \Sigma P_n \Delta\rho^{(n)}(y_1)\Delta\rho^{(n)}(y_2) \tag{7.38}$$

with $C_2^{(n)}(y_1, y_2) = \rho_2^{(n)}(y_1, y_2) - \rho_1^{(n)}(y_1)\rho_1^{(n)}(y_2)$ and $\Delta\rho^{(n)}(y) = \rho_1^{(n)}(y) - \rho_1(y)$. In (7.37), C_S is the average of the semi-inclusive correlation functions (often misleadingly denoted as "short-range") and is more sensitive to dynamical correlations. The term C_L (misleadingly called "long-range") arises from mixing different topological single-particle densities.

A normalized form of C_S can be defined as

$$K_S(y_1, y_2) = \frac{C_S(y_1, y_2)}{\sum_n P_n \rho_1^{(n)}(y_1)\rho_1^{(n)}(y_2)} = \frac{\sum_n P_n \rho_2^{(n)}(y_1, y_2)}{\sum_n P_n \rho_1^{(n)}(y_1)\rho_1^{(n)}(y_2)} - 1 \; . \tag{7.39}$$

C'_S and $\tilde{C}_S$ and their normalized forms K'_S and $\tilde{K}_S$ are defined accordingly, with the averages $\langle n \rangle$ and $\langle n_a(n_b - \delta^{ab})\rangle$ replaced by n and $n_a(n_b - \delta^{ab})$, respectively.

Quite a comprehensive account reminding us of possible pitfalls in the interpretation of the various definitions of the correlation function is given in [14], by showing their behavior for more or less "independent" particle production, i.e., with given multiplicity distribution and (n-dependent) semi-inclusive single-particle spectra.

Analogous expressions may be derived for three-particle correlations. They are discussed in Sect. 9.4.

7.1.5 Factorial and cumulant moments

When (e.g. by integration) the parametric function $z(y)$ of Sub-Sect. 7.1.1 is replaced by a constant z, the generating functionals reduce to the generating function for the multiplicity distribution. Indeed, the probability P_n for producing n particles is given by

$$P_n = \sigma_n^{\text{excl}}/\sigma_{\text{inel}} \tag{7.40}$$

and we have

$$G(z) = \sum_{n=0}^{\infty} P_n(1+z)^n = \mathcal{G}^{\text{excl}}[z+1] = \mathcal{G}^{\text{incl}}[z] \tag{7.41}$$

$$= 1 + \sum_{q=1}^{\infty} \frac{z^q}{q!} \int_\Omega \rho_q(y_1, \ldots, y_q)\, \mathrm{d}y_1 \ldots \mathrm{d}y_q \tag{7.42}$$

$$= 1 + \sum_{q=1}^{\infty} \frac{z^q}{q!} \tilde{F}_q \; . \tag{7.43}$$

The $\tilde{F}_q$ are the unnormalized factorial (or binomial) moments

$$\tilde{F}_q \equiv \langle n^{[q]} \rangle \equiv \langle n(n-1)\ldots(n-q+1)\rangle = \int_\Omega \mathrm{d}y_1 \ldots \int_\Omega \mathrm{d}y_q \, \rho_q(y_1, \ldots, y_q)$$

$$= \sum_n P_n\, n(n-1)\ldots(n-q+1). \tag{7.44}$$

This relation can (formally) be inverted. If $P_n = 0$ for $n > N$ then an approximation for P_n is given by:

$$P_n = \frac{1}{n!}\sum_{j=0}^{N-n}(-1)^j \frac{\tilde{F}_{j+n}}{j!} \quad (n = 0, 1, \ldots N), \tag{7.45}$$

and P_n is included between any two successive values obtained by terminating the sum at $j = s$ and $j = s+1$, respectively.

In (7.44), n denotes the multiplicity in Ω and the average is taken over the ensemble of events. All the integrals are taken over the same volume Ω such that $y_i \in \Omega$ $\forall\, i \in \{1, \ldots, q\}$. Using the correlation-function cluster decomposition, one further has

$$\ln G(z) = \langle n\rangle z + \sum_{q=2}^{\infty} \frac{z^q}{q!} f_q. \tag{7.46}$$

The f_q are the unnormalized factorial cumulants, also known as Mueller moments [8]

$$f_q = \int_\Omega \mathrm{d}y_1 \ldots \int_\Omega \mathrm{d}y_q\, C_q(y_1, \ldots, y_q), \tag{7.47}$$

the integrations being performed as in (7.44). The quantities $\tilde{F}_q$ and f_q are easily found if $G(z)$ is known:

$$\tilde{F}_q = \left.\frac{\mathrm{d}^q G(z)}{\mathrm{d}z^q}\right|_{z=0}, \tag{7.48}$$

$$f_q = \left.\frac{\mathrm{d}^q \ln G(z)}{\mathrm{d}z^q}\right|_{z=0} \tag{7.49}$$

and

$$P_q = \left.\frac{1}{q!}\frac{\mathrm{d}^q G(z)}{\mathrm{d}z^q}\right|_{z=-1}. \tag{7.50}$$

Using Cauchy's theorem, this can also be written as

$$P_n = \frac{1}{2\pi i}\oint \frac{G(z)}{(1+z)^{n+1}}\,\mathrm{d}z, \tag{7.51}$$

where the integral is on a circle enclosing $z = -1$. Equation (7.51) is sometimes useful in deriving asymptotic expressions for P_n in terms of factorial moments or cumulants [8, 15].

As a simple example, we consider the Poisson distribution

$$P_n = \mathrm{e}^{-\langle n\rangle}\frac{\langle n\rangle^n}{n!},$$

for which

$$G(z) = \sum_{n=0}^{\infty} P_n \, (1+z)^n = \exp\{\langle n \rangle z\} \quad , \tag{7.52}$$

showing that $f_q \equiv 0$ for $q > 1$. In that case one has:

$$\tilde{F}_q = \langle n(n-1)\ldots(n-q+1)\rangle = \langle n \rangle^q. \tag{7.53}$$

The expressions of density functions in terms of cumulant correlation functions, and the reverse relations, also hold for their integrated counterparts. They follow directly from the equations:

$$1 + \sum_{q=1}^{\infty} \frac{z^q}{q!} \tilde{F}_q = \exp\{\langle n \rangle z + \sum_{q=2}^{\infty} \frac{z^q}{q!} f_q\} \tag{7.54}$$

or

$$\ln\left(1 + \sum_{q=1}^{\infty} \frac{z^q}{q!} \tilde{F}_q\right) = \langle n \rangle z + \sum_{q=2}^{\infty} \frac{z^q}{q!} f_q \tag{7.55}$$

by expanding either the exponential in (7.54) or the logarithm in (7.55) and equating the coefficients of the same power of z. One finds [9]:

$$\begin{aligned}
\tilde{F}_1 &= f_1 \quad , \\
\tilde{F}_2 &= f_2 + f_1^2 \quad , \\
\tilde{F}_3 &= f_3 + 3f_2 f_1 + f_1^3 \quad , \\
\tilde{F}_4 &= f_4 + 4f_3 f_1 + 3f_2^2 + 6f_2 f_1^2 + f_1^4 \quad , \\
\tilde{F}_5 &= f_5 + 5f_4 f_1 + 10 f_3 f_2 + 10 f_3 f_1^2 + 15 f_2^2 f_1 + 10 f_2 f_1^3 + f_1^5 \ ;
\end{aligned} \tag{7.56}$$

and in general:

$$\tilde{F}_q = q! \sum_{\{l_i\}_q} \prod_{j=1}^{q} \left(\frac{f_j}{j!}\right)^{l_j} \frac{1}{l_j!} \ , \tag{7.57}$$

with the summation as in (7.14) and $\sum_{l=1}^{q} i \, l_i = q$.

The latter formula can also be written as:

$$\tilde{F}_q = \sum_{l=0}^{q-1} \binom{q-1}{l} f_{q-l} \tilde{F}_l \ , \tag{7.58}$$

(with $\tilde{F}_0 \equiv 1$, $f_0 \equiv 0$) and is well-suited for computer calculation. An equivalent relation was derived in [8]. The (ordinary) moments:

$$\mu_q = \langle n^q \rangle = \sum_{n=0}^{\infty} n^q P_n \quad , \tag{7.59}$$

may be derived from the moment generating function

$$M(z) = \sum_{n=0}^{\infty} \mathrm{e}^{nz} P_n \ , \tag{7.60}$$

since

$$\mu_q = \left. \frac{\mathrm{d}^q M(z)}{\mathrm{d}z^q} \right|_{z=0} . \tag{7.61}$$

We note the useful relations

$$M(z) = G\left(\mathrm{e}^z - 1\right) \ , \tag{7.62}$$

$$G(z) = M\left(\ln(1+z)\right) \ . \tag{7.63}$$

Moments and factorial moments are related to each other by series expansions. From the identities [16]:

$$n(n-1)\ldots(n-q+1) = \sum_{m=0}^{q} S_q^{(m)}\, n^m \ , \tag{7.64}$$

$$n^q = \sum_{m=0}^{q} \mathcal{S}_q^{(m)}\, n(n-1)\ldots(n-m+1) \ , \tag{7.65}$$

where $S_q^{(m)}$ and $\mathcal{S}_q^{(m)}$ are Stirling numbers of the first and second kind, respectively, follows directly:

$$\tilde{F}_q = \sum_{m=0}^{q} S_q^{(m)}\, \mu_m \ , \tag{7.66}$$

$$\mu_q = \sum_{m=0}^{q} \mathcal{S}_q^{(m)}\, \tilde{F}_m \ . \tag{7.67}$$

Cumulants κ_q can be defined in terms of the moments μ_q in the standard way [9, 17]. They obey relations identical to (7.56). The cumulants are integrals of the type (7.47) of differential quantities known as density moments. These are discussed in [18, 19]. Relations expressing central moments in terms of factorial moments via non-central Stirling numbers are derived in [20].

7.1.6 Combinants

In (7.48) and (7.49), the quantities $\tilde{F}_q$, and f_q are defined as the qth derivative of the generating function $G(z)$ and its logarithm $\ln G(z)$ evaluated at $z = 0$, respectively. In (7.50), the P_q are $1/q!$ times the qth derivative of $G(z)$ evaluated at $z = -1$. Similarly, we can define quantities

$$\tilde{c}_q = \frac{1}{q!} \left. \frac{\mathrm{d}^q \ln G(z)}{\mathrm{d}z^q} \right|_{z=-1} . \tag{7.68}$$

These are the so-called combinants [21, 22] which, in analogy to the factorial cumulants f_q can be understood as "exclusive" correlation integrals.

While the factorial variables are particularly well suited for the study of densely populated phase-space bins (see e.g. (7.44)), combinants are better suited for the study of scarcely populated regions, since

$$\tilde{c}_q = P_q/P_0 - \frac{1}{q}\sum_{n=1}^{q-1} n c_n P_{q-n}/P_0 \ , \tag{7.69}$$

i.e., the calculation of $\tilde{c}_q$ requires only a finite number of probabilities $P_{n\leq q}$. The drawback for full phase space of the requirement of $P_0 > 0$ is more than compensated in small intervals of it. The relation between the factorial cumulants f_q and the combinants $\tilde{c}_q$ can be expressed as

$$f_q = \sum_{n=q}^{\infty} n(n-1)\dots(n-q+1)\tilde{c}_n \ , \tag{7.70}$$

or, in inversed form

$$\tilde{c}_q = \frac{1}{q!}\sum_{n=q}^{\infty}\frac{(-1)^{n-q}}{(n-q)!} f_n \ , \tag{7.71}$$

in analogy to (7.44) and (7.45).

Important features in common with the cumulant moments are:

1. For the Poisson distribution $\tilde{c}_1 = \langle n \rangle$ and $\tilde{c}_{q\geq 2} = 0$;
2. the combinants share the additivity property of the cumulants;
3. for infinitely divisible P_n, both sets of moments are strictly positive.

An advantage over the factorial and cumulant moments is that, according to (7.69), the $\tilde{c}_q$ are *finite* combinations of the probability ratios P_n/P_0 and therefore do not suffer from a bias (empty-bin effect) present at high resolution in the former.

7.1.7 Cell-averaged factorial moments and cumulants; generalized moments

In practical work, with limited statistics, it is almost always necessary to perform averages over more than a single phase-space cell. Let Ω_m be such a cell (e.g. a single rapidity interval of size δy) and divide the phase-space volume into M non-overlapping cells Ω_m of size $\delta\Omega$, independent of m. Let n_m be the number of particles in cell Ω_m. Different cell-averaged moments may be considered, depending on the type of averaging.

Normalized factorial moments [23,24], which have become known as *vertical moments*, are defined as[2]

$$F_q^{\rm V}(\delta y) \equiv \frac{1}{M}\sum_{m=1}^{M}\frac{\langle n_m(n_m-1)\dots(n_m-q+1)\rangle}{\langle n_m\rangle^q} \tag{7.72}$$

$$\equiv \frac{1}{M}\sum_{m=1}^{M}\frac{\int_{\delta y}\rho_q(y_1,\dots,y_q)\prod_{i=1}^q {\rm d}y_i}{\left(\int_{\delta y}\rho(y)dy\right)^q},$$

$$= \frac{1}{M(\delta y)^q}\sum_{m=1}^{M}\int_{\delta y}\frac{\rho_q(y_1,\dots,y_q)\prod_{i=1}^q {\rm d}y_i}{(\bar\rho_m)^q}. \tag{7.73}$$

The full rapidity interval ΔY is divided into M equal bins: $\Delta Y = M\delta y$; each y_i is within the δy-range and $\langle n_m\rangle \equiv \overline{\rho}_m\delta y \equiv \int_{\delta y}\rho_1(q){\rm d}y$.

One may also define normalized *horizontal moments* by

$$F_q^{\rm H}(\delta y) \equiv \frac{1}{M}\sum_{m=1}^{M}\frac{\langle n_m(n_m-1)\dots(n_m-q+1)\rangle}{\langle\overline{n}_m\rangle^q}\ . \tag{7.74}$$

with $\overline{n}_m = \sum_m n_m/M;\quad \langle\overline{n}_m\rangle = \langle n\rangle/M;\quad n = \sum_m n_m$.

Horizontal and vertical moments are equal if $M = 1$. Vertical moments are normalized locally and thus sensitive only to fluctuations within each cell but not to the overall shape of the single-particle density. Horizontal moments are sensitive to the shape of the single-particle density in y and further depend on the correlations between cells. To eliminate the effect of a non-flat rapidity distribution, it was suggested to either introduce correction factors [25] or use "cumulative" variables which transform an arbitrary distribution into a uniform one [26,27].

Likewise, cell-averaged normalized factorial cumulant moments may be defined as

$$K_q(\delta y) = \frac{1}{M(\delta y)^q}\sum_{m=1}^{M}\int_{\delta y}\prod_i {\rm d}y_i\frac{C_q(y_1,\dots,y_q)}{(\bar\rho_m)^q}\ . \tag{7.75}$$

They are related [28] to the factorial moments by[3]

$$\begin{aligned} F_2 &= 1+K_2\ , \\ F_3 &= 1+3K_2+K_3\ , \\ F_4 &= 1+6K_2+3\overline{K_2^2}+4K_3+K_4\,. \end{aligned} \tag{7.76}$$

In F_4 and higher-order moments, "bar averages" appear. They are defined as $\overline{AB} \equiv \sum_m A_m B_m/M$.

[2] Here and in the following we consider rapidity space for definiteness

[3] The higher-order relations can be found in [28]

Besides factorial and cumulant moments, other measures of multiplicity fluctuations have been proposed. In particular, G-moments [29]—known in statistics as frequency moments [9]—were extensively used to investigate whether multiparticle processes possess (multi)fractal properties [30, 31]. G-moments are defined as

$$G_q = \sum_{m=1}^{M}{}' p_m^q, \qquad p_m = n_m/n, \qquad n = \sum_{m=1}^{M} n_m \ . \tag{7.77}$$

Also here, n_m is the number of particles in bin m, the absolute frequency; n is the total multiplicity in an initial interval and M is the number of bins at "resolution" M. Bins with zero content ("empty bins") are excluded in the sum, so that q can cover the whole spectrum of real numbers. For q negative, G_q is sensitive to "holes" in the rapidity distribution of a single event. Note that p_m in (7.77) is not a probability but a relative frequency or "empirical measure" in modern terminology. For small n, G-moments are very sensitive to statistical fluctuations ("noise"), especially for large M. This seriously limits their potential. In attempts to reduce this noise-sensitivity, modified definitions have been proposed in [32].

7.1.8 Multivariate distributions

The univariate factorial moments $\tilde{F}_q$ characterize multiplicity fluctuations in a single phase-space cell and thus reflect only local properties. More information is contained in the correlations between fluctuations (within the same event) in two or more cells. This has led to consider multivariate factorial moments. For non-overlapping cells, the 2-fold factorial moments, also called *correlators*, are defined as:

$$\tilde{F}_{pq} = \langle n_m^{[p]} n_{m'}^{[q]} \rangle, \tag{7.78}$$

where n_m ($n_{m'}$) is the number of particles in cell m (cell m'). A normalized version of the two-fold correlator is discussed in [23] and defined as:

$$F_{pq} = \frac{\langle n_m^{[p]} n_{m'}^{[q]} \rangle}{\tilde{F}_p \tilde{F}_q}. \tag{7.79}$$

For reasons of statistics, these quantities are usually averaged over many pairs of cells, keeping the "distance" D between the cells constant[4]. This averaging procedure requires the same precautions regarding stationarity of single-particle densities as for their single-cell equivalents.

Multi-fold factorial moments are a familiar tool in radio- and radar physics and in quantum optics [33]. There, they relate to simultaneous measurement of photo-electron counts detected in, say M time intervals or M space points, leading to a joint

[4] In one-dimensional rapidity space, D is defined as the distance between the centers of two rapidity intervals; in multidimensional phase space a proper metric must first be defined

probability distribution $P_M(n_1, n_2, \ldots, n_M)$. The importance of multi-fold moments derives from the fact that, e.g. in the simplest case of two cells, $\tilde{F}_{11} = \langle n_m n_{m'} \rangle$ is directly related to the auto-correlation function of the radiation field and obeys, for small cells, the Siegert relation [33], whatever the statistical properties of the field. The higher-order moments are sensitive to higher-order correlations and to the phase of the field.

Factorial moments and factorial correlators are intimately related quantities. In terms of inclusive densities one has:

$$\tilde{F}_{pq} = \int_{\Omega_1} \mathrm{d}y_1 \ldots \mathrm{d}y_p \int_{\Omega_2} \mathrm{d}y_{p+1} \ldots \mathrm{d}y_{p+q}\, \rho_{p+q}(y_1, \ldots, y_p; y_{p+1}, \ldots, y_{p+q}), \tag{7.80}$$

where ρ_{p+q} is the inclusive density of order $p+q$. The integrations are performed over two arbitrary (possibly overlapping) phase-space cells Ω_1 and Ω_2, separated by a distance D.

It should be noted that the definition (7.80) is more general than (7.78). For $\Omega_1 = \Omega_2$ or $D = 0$, (7.80) reduces to the correct definition of $\tilde{F}_2$ whereas (7.78) is, in this case, equal to $\langle n^2 \rangle$ and misses the so-called "shot-noise" term $-\langle n \rangle$.

Factorial moments and factorial correlators of the same order are thus seen to differ only in the choice of the integration domains. Note that for $p \neq q$, definition (7.80) is not symmetric in p and q and a symmetrized version is often used in experimental work:

$$\tilde{F}_{pq}^{(s)} = (\tilde{F}_{pq} + \tilde{F}_{qp})/2\,. \tag{7.81}$$

From (7.80) follows that F_{11} is directly derivable from measured two-particle correlation functions or from appropriate analytical parametrizations. Higher-order correlators involve higher-order density functions which, in general, are unknown.

We now turn to a discussion of multivariate factorial cumulants. For M non-overlapping cells, we introduce the M-variate multiplicity distribution $P_M(n_1, \ldots, n_M)$ and the corresponding moment- and factorial-moment generating functions:

$$M(z_1, \ldots, z_M) = \sum_{n_1=0}^{\infty} \sum_{n_2=0}^{\infty} \cdots \sum_{n_M=0}^{\infty} e^{z_1 n_1 + \cdots + z_M n_M} P_M(n_1, \ldots, n_M)\ , \tag{7.82}$$

$$G(z_1, \ldots, z_M) = \sum_{n_1=0}^{\infty} \sum_{n_2=0}^{\infty} \cdots \sum_{n_M=0}^{\infty} (1+z_1)^{n_1} \cdots (1+z_M)^{n_M} P_M(n_1, \ldots, n_M), \tag{7.83}$$

from which the M-variate moments are easily obtained by differentiation:

$$\mu_{q_1 \ldots q_M} = \langle n_1^{q_1} \ldots n_M^{q_M} \rangle = \left(\frac{\partial}{\partial z_1} \right)^{q_1} \cdots \left(\frac{\partial}{\partial z_M} \right)^{q_M} M(z_1, \ldots, z_M) \Big|_{z_1 = \cdots z_M = 0}\ , \tag{7.84}$$

$$\tilde{F}_{q_1 \ldots q_M} = \langle n_1^{[q_1]} \ldots n_M^{[q_M]} \rangle = \left(\frac{\partial}{\partial z_1} \right)^{q_1} \cdots \left(\frac{\partial}{\partial z_M} \right)^{q_M} G(z_1, \ldots, z_M) \Big|_{z_1 = \cdots z_M = 0}\ . \tag{7.85}$$

The multivariate (ordinary) cumulants $\kappa_{q_1 \dots q_M}$ and multivariate factorial cumulants $f_{q_1 \dots q_M}$ are likewise obtained by replacing $M(.)$ and $G(.)$ in (7.84) and (7.85) by their respective natural logarithms [34]. The same expressions serve to extend the relations between univariate moments and cumulants to their multivariate counterparts.

For $M = 2$ and non-overlapping cells, one has the identity [cfr. (7.54)]:

$$\sum_{l=0}^{\infty} \sum_{m=0}^{\infty} \frac{(z_1)^l (z_2)^m}{l!\,m!} \tilde{F}_{lm} = \exp \left(\sum_{l=0}^{\infty} \sum_{m=0}^{\infty} \frac{(z_1)^l (z_2)^m}{l!\,m!} f_{lm} \right), \tag{7.86}$$

where $\tilde{F}_{00} \equiv 1$ and f_{00} is defined equal to zero. It follows that[5]

$$\begin{aligned}
\tilde{F}_{11} &= f_{11} + f_{01} f_{10}, && (7.87)\\
\tilde{F}_{12} &= f_{12} + f_{01} f_{20} + 2 f_{10} f_{11} + f_{01} f_{10}^2, && (7.88)\\
\tilde{F}_{13} &= f_{13} + f_{01} f_{30} + 3 f_{11} f_{20} + 3 f_{01} f_{10} f_{20} + 3 f_{10} f_{12} + 3 f_{10}^2 f_{11} + f_{01} f_{10}^3; && (7.89)\\
\tilde{F}_{22} &= f_{22} + 2 f_{10} f_{21} + f_{02} f_{20} + f_{01}^2 f_{20} + 2 f_{01} f_{12} + 2 f_{11}^2 + 4 f_{01} f_{10} f_{11} \\
&\quad + f_{02} f_{10}^2 + f_{01}^2 f_{10}^2 \; . && (7.90)
\end{aligned}$$

Similarly, expanding the logarithm in

$$\sum_{l=0}^{\infty} \sum_{m=0}^{\infty} \frac{(-z_1)^l (-z_2)^m}{l!\,m!} f_{lm} = \ln \left(\sum_{l=0}^{\infty} \sum_{m=0}^{\infty} \frac{(-z_1)^l (-z_2)^m}{l!\,m!} \tilde{F}_{lm} \right), \tag{7.91}$$

in powers of s and t and identifying coefficients, the reverse relations follow:

$$\begin{aligned}
f_{11} &= \tilde{F}_{11} - \tilde{F}_{01} \tilde{F}_{10} \; , && (7.92)\\
f_{12} &= \tilde{F}_{12} - \tilde{F}_{01} \tilde{F}_{20} - 2 \tilde{F}_{10} \tilde{F}_{11} + 2 \tilde{F}_{01} \tilde{F}_{10}^2 \; , && (7.93)\\
f_{13} &= \tilde{F}_{13} - \tilde{F}_{01} \tilde{F}_{30} - 3 \tilde{F}_{11} \tilde{F}_{20} + 6 \tilde{F}_{01} \tilde{F}_{10} \tilde{F}_{20} - 3 \tilde{F}_{10} \tilde{F}_{12} + 6 \tilde{F}_{10}^2 \tilde{F}_{11} - 6 \tilde{F}_{01} \tilde{F}_{10}^3, && (7.94)\\
f_{22} &= \tilde{F}_{22} - 2 \tilde{F}_{10} \tilde{F}_{21} - \tilde{F}_{02} \tilde{F}_{20} + 2 \tilde{F}_{01}^2 \tilde{F}_{20} - 2 \tilde{F}_{01} \tilde{F}_{12} - 2 \tilde{F}_{11}^2 \\
&\quad + 8 \tilde{F}_{01} \tilde{F}_{10} \tilde{F}_{11} + 2 \tilde{F}_{02} \tilde{F}_{10}^2 - 6 \tilde{F}_{01}^2 \tilde{F}_{10}^2 \; . && (7.95)
\end{aligned}$$

The quantities $\tilde{F}_{0i}$, $\tilde{F}_{i0}$, f_{0i} and f_{i0} are equal to the single-cell factorial moments and factorial cumulants, respectively. Expressions for $\tilde{F}_{ji}$ (f_{ji}) are obtained from the corresponding expression for $\tilde{F}_{ij}$ (f_{ij}) by permutation of the subscripts. By definition, $f_{01} = \tilde{F}_{01}$ and f_{01} is equal to the average multiplicity in cell 2.

It may be noted that the bivariate relations reduce to the univariate ones (7.56) by simply amalgamating the indices. For example, from

$$\tilde{F}_{12} = f_{12} + f_{01} f_{20} + 2 f_{10} f_{11} + f_{01} f_{10}^2, \tag{7.96}$$

[5] See also [35]

one recovers, by summing the indices

$$\tilde{F}_3 = f_3 + 3f_1f_2 + f_1^3. \tag{7.97}$$

It is shown in [9] (Sect. 13.12) that the above relations, while seemingly complex, have in fact a surprisingly elegant structure, rooted in simple algebraic properties of completely symmetric functions. Further discussion on this point and other useful properties may be found in [34].

Extensions to more than two cells is straightforward, in principle, but involves tedious algebra.

7.2 Poisson-noise suppression

To detect dynamical fluctuations in the density of particles produced in a high-energy collision, a way has to be devised to eliminate, or to reduce as much as possible, the statistical fluctuations (noise) due to the finiteness of the number of particles in the counting cell(s). This requirement can to a large extent be satisfied by studying factorial moments and their multivariate counterparts. It forms the basis of the factorial moment technique, known in optics, but rediscovered in multi-hadron physics in [23,24]. The method rests on the conjecture that the multi-cell multiplicity distribution $P_M(n_1, \ldots, n_M)$ can be written as

$$P_M(n_1, \cdots, n_M) = \int \mathrm{d}\rho_1 \ldots \mathrm{d}\rho_M P_\rho(\rho_1, \ldots, \rho_M) \prod_{m=1}^{M} \frac{(\rho_m \delta)^{n_m}}{n_m!} \exp(-\rho_m \delta). \tag{7.98}$$

The Poisson factors represent uncorrelated fluctuations of n_m around the average $\rho_m \delta = \langle n_m \rangle$ in the m-th interval; δ is here the size of the interval. This can also be written as:

$$P_M(n_1, \cdots, n_M) = \langle \prod_{m=1}^{M} \frac{\langle n_m \rangle^{n_m}}{n_m!} \exp(-\langle n_m \rangle) \rangle_\rho , \tag{7.99}$$

where the outer brackets mean that an average is taken over the probability distribution of the densities ρ_m, which are subject only to dynamical fluctuations. If these are absent, $P_\rho(\rho_1, \ldots, \rho_M)$ is simply a product of δ-functions.

The formulae (7.98-7.99) are formally identical to the expression for the multi-interval photo-electron counting probability distribution in quantum optics and based on the famous Mandel formula [36,37]. The latter relates the probability distribution of *the number of detected photo-electrons* to the statistical distribution of the e.m. field.

In optics, ρ_m has the meaning of a space- or time-integrated field intensity. The ensemble average is calculated from the field density matrix which describes its statistical properties.

Equations (7.98)-(7.99) express $P_M(n_1, \dots, n_M)$ as a linear transformation of $P_\rho(\rho_1, \dots, \rho_M)$ with a "Poisson kernel". This transformation is known as the "Poisson transformation" of P_ρ [38].

The Poisson transform of a single-variable function $f(x)$ is the function $\tilde{f}(n)$ (n integer) defined by the linear transformation

$$\tilde{f}(n) = \int_0^\infty \mathrm{d}x\, f(x) \frac{x^n}{n!} e^{-x}. \tag{7.100}$$

A trivial example is the function $\delta(x - \mu)$ whose transform is the Poisson probability distribution. The Bose-Einstein distribution

$$\tilde{f}(n) = \frac{\mu^n}{(1+\mu)^{n+1}} \qquad (n = 0, 1, \dots), \tag{7.101}$$

is obtained as the Poisson transform of the exponential function $(1/\mu)\exp(-x/\mu)$.

For suitably-behaved functions, the inverse Poisson transform exists. It is closely related to the Laplace-transform of $f(x)$. Several practical methods have been developed to determine the function $f(x)$ from its Poisson transform. Besides methods based on series expansions, the inversion problem may be reduced to an inverse moment problem. This follows from the equality between the factorial moments of $\tilde{f}(n)$ and the ordinary moments of $f(x)$, as further discussed below.

A table of useful transforms for probability distributions and further mathematical properties can be found in [33].

From the basic Poisson transform equation (7.99) it is easily seen that the multifold factorial moment generating function has the simple form

$$G(z_1, \dots, z_M) = \langle \prod_{j=1}^{M} \exp(z_j \rho_j \delta) \rangle_\rho, \tag{7.102}$$

where the statistical average is again taken over the ensemble of densities $\rho_1, \cdots, \rho_M$, as indicated by the subscript.

On the other hand, the (ordinary) moment generating function of the densities is given by:

$$Q(z_1, \dots, z_M) = \int P_\rho(\rho_1, \dots, \rho_M) \exp(\rho_1 z_1 + \dots + \rho_M z_M) \mathrm{d}\rho_1 \dots \mathrm{d}\rho_M \tag{7.103}$$

$$= \left\langle \prod_{j=1}^{M} \exp(z_j \rho_j) \right\rangle_\rho . \tag{7.104}$$

Comparing (7.102) and (7.104), it follows that:

$$G(z_1, \dots, z_M) = Q(\delta z_1, \dots, \delta z_M) . \tag{7.105}$$

This equation implies that the normalized multivariate factorial moments of the multiplicity distribution

$$F_{q_1 \dots q_M} = \frac{\tilde{F}_{q_1 \dots q_M}}{\langle n_1 \rangle^{q_1} \dots \langle n_M \rangle^{q_M}} \tag{7.106}$$

are equal to the normalized multivariate (ordinary) moments of the relative density fluctuation $\rho_m / \langle \rho_m \rangle$. This is the "noise-suppression" theorem [23, 24]. It assumes that the noise is Poissonian (cfr. (7.98)) and that the number of counts in all intervals (the total multiplicity) is unrestricted[6].

The property of Poisson-noise suppression has made measurement of factorial moments a standard technique, e.g. in quantum optics, to study the statistical properties of arbitrary electromagnetic fields from photon-counting distributions. Their utility was first explicitly recognized, for the single time-interval case, in [39] and [5] and later generalized to the multivariate case in [34].

As a complementary (and feasible) approach, it was pointed out [40] that also the probability of finding complete "spike" events with q particles in bin m but no particle outside, $P_M(0, \dots, n_m = q, \dots, 0)$, is equal to the normalized ordinary relative density moment in this bin (see Eq. 7.98) and therefore carries equivalent information on P_ρ.

The authors of [5] further stress the advantages of factorial cumulants compared to factorial moments, since the former measure genuine correlation patterns, whereas the latter contain additional large combinatorial terms which may mask the underlying dynamical correlations (however, see the discussion in Sect. 7.1.2). The generating functional technique and the usefulness of factorial moments have also been discussed in [41] in a different context.

Multivariate factorial cumulants are derived from the (natural) logarithm of the factorial moment generating function. Taking logarithms of both sides of (7.105), one finds that the multivariate normalized factorial cumulants of the counting distribution are equal to the multivariate normalized ordinary cumulants of the densities ($\rho\delta$). This relation, therefore, extends the noise-suppression theorem to cumulants. This property is exploited in many fields from quantum optics [34] to radar-physics and astrophysics (see e.g. [42]).

More recently, the Poisson-noise suppression theorem was generalized [43] from the case of simple particle densities as above, to the case of arbitrary particle variables depending on the cell position m. This allows it to be used in a direct extraction of the dynamics of other types of fluctuations (transverse momentum, erraticity, etc.), in particular also on an event-by-event level.

The method is, however, limited to the suppression of Poissonian noise, thus leaving room for alternative approaches to be described in short in Sects. 7.6 and 7.7 below.

[6] If the sum over all intervals of the number of counts is fixed, a slightly more complicated relation can be obtained if the noise has a Bernoulli (multinomial) distribution [23]

7.3 Sum-rules

In an interesting α-model analysis of factorial correlators [44], scaling relations are derived between single-variate and 2-variate factorial moments which are independent of the dimension of the phase space. The result is stated as follows: If a correlator $F_{11}(D, \delta)$ is effectively independent of δ in a range $\delta < D \leq \delta_0$, then

$$F_{11}(D) = 2F_2(2D) - F_2(D). \tag{7.107}$$

Here, δ is the interval size and D the distance between the intervals.

Similar types of relations—or sum-rules—are well known in optics since the early 1970's. They are exploited in so-called Multi-Cathode and Multiple-Aperture Single-Cathode (MASC) photo-electron counting experiments (see e.g. [45, 46] and refs. therein).

Consider again the multivariate multiplicity distribution $P_M(n_1, \dots, n_M)$ giving the joint probability for the occurrence of n_1 particles in a cell Ω_1, ..., n_M particles in cell Ω_M, with $\Omega_i \cap \Omega_j = 0$, $\forall i, j$ and $i \neq j$. Let n be the number of particles counted in the union of the M cells,

$$n = \sum_{m=1}^{M} n_m \quad . \tag{7.108}$$

The probability distribution of n is given by

$$P(n) = \sum_{n_1=0}^{n} \cdots \sum_{n_M=0}^{n} p_M(n_1, \dots, n_M) \delta_{n, n_1 + \dots + n_M} \cdot \tag{7.109}$$

Define the single-variate factorial moment generating function

$$g(z) = \sum_{n=0}^{\infty} (1+z)^n \, P(n) \, . \tag{7.110}$$

The function $g(z)$ can be expressed in terms of the multivariate generating function (7.83) as:

$$g(z) = G_M(z_1, \dots, z_M)|_{z_1 = z_2 = \dots = z_M = z} \cdot \tag{7.111}$$

Equation (7.111) allows to express factorial moments of n in terms of the multivariate factorial moments of $\{n_1, \dots, n_M\}$. Application of the Leibnitz rule

$$\left(\frac{\mathrm{d}}{\mathrm{d}z}\right)^k f(z) = \sum_{\{a_j\}} \frac{k!}{a_1 ! a_2 ! \dots a_k!} \left(\frac{\mathrm{d}}{\mathrm{d}\, z}\right)^{a_1} f_1(z) \cdots \left(\frac{\mathrm{d}}{\mathrm{d}\, z}\right)^{a_k} f_M(z)$$

to the function

$$f(z) = f_1(z) \cdots f_M(z)$$

leads immediately to the relation

$$\tilde{F}_q = \sum_{\{a_j\}} \tilde{F}^{(M)}_{a_1 \ldots a_M} \frac{q!}{a_1! \ldots a_M!}. \tag{7.112}$$

The summation is over all sets $\{a_j\}$ of non-negative integers such that

$$\sum_{j=1}^{M} a_j = q.$$

Formula (7.112) may be looked upon as a generalization of the usual multinomial theorem for factorial moments[7].

Likewise, taking the natural logarithm of both sides of (7.111), one obtains an identical relation as (7.112) among single-variate and multivariate factorial cumulants.

As an example, for two rapidity bins ($M = 2$) of size δ separated by a distance D, one finds:

$$\begin{aligned} \tilde{F}_2 &= \tilde{F}^{(2)}_{02} + 2\tilde{F}^{(2)}_{11} + \tilde{F}^{(2)}_{20} \,, \\ \tilde{F}_3 &= \tilde{F}^{(2)}_{03} + 3(\tilde{F}^{(2)}_{12} + \tilde{F}^{(2)}_{21}) + \tilde{F}^{(2)}_{30} \,, \\ \tilde{F}_4 &= \tilde{F}^{(2)}_{04} + 4(\tilde{F}^{(2)}_{13} + \tilde{F}^{(2)}_{31}) + 6\tilde{F}^{(2)}_{22} + \tilde{F}^{(2)}_{40} \,. \end{aligned} \tag{7.113}$$

The factorial moments $\tilde{F}_{0i}$ are determined from the single-cell (marginal) counting distribution, whereas the univariate factorial moments $\tilde{F}_q$ are obtained from the sum of the counts in the two cells.

The relations derived in [44] follow immediately from (7.113) by considering two adjacent cells and normalizing properly. Since the derivation of (7.112) is completely general, it obviously holds irrespective of the dimension of phase space.

The relations (7.113) are trivially extended to more than two cells. They allow to measure high-order correlators by varying the distances between the cells. In optics and radar-physics, they are typically used in determining spatial coherence properties of arbitrary e.m. fields.

7.4 Scaling laws

A major part of Chapter 10 is devoted to recent experimental and theoretical research on possible manifestations of scale-invariance in high-energy multiparticle production processes. This work centers around two basic inter-related notions: intermittency and fractality.

[7] See also [12]

If dynamical fluctuations have a typical size δy_0, the factorial moments rise with decreasing bin size as long as $\delta y > \delta y_0$, but saturate for $\delta y < \delta y_0$. If, on the other hand, self-similar fluctuations exist at all scales δy, the factorial moments follow a power law. In particle physics, intermittency is defined, in a strict sense, as the scale-invariance of factorial moments (7.72)-(7.74) with respect to changes in the size of phase-space cells (or bins) say δy, for small enough δy:

$$F_q(\delta y) \propto (\delta y)^{-\phi_q} \qquad (\delta y \to 0). \tag{7.114}$$

The power $\phi_q > 0$ is a constant at any given (positive integer) q and called "intermittency index" or "intermittency slope". The form of (7.114) strictly implies that the inclusive densities ρ_q and the connected correlation functions C_q become singular in the limit of infinitesimal separation ($\delta y \to 0$) in momentum space.

Inspired by the theory of multifractals, scaling behavior of the frequency or G-moments (7.77) has also been looked for in the form

$$G_q(\delta y) \propto (\delta y)^{\tau(q)} \qquad (\delta y \to 0). \tag{7.115}$$

To describe the inter-relation of the two proposals, we briefly discuss the formalism of fractals.

Power-law dependence is typical for fractals [47], i.e. for self-similar objects with a non-integer dimension. These range from purely mathematical ones (the Cantor set, the Koch curve, the Serpinsky gasket etc.) to real objects of nature (coast-lines, clouds, lungs, polymers etc). For reviews see [48–50].

The fractal dimension D_F is defined as the exponent which provides a finite limit

$$0 < \lim_{\epsilon \to 0} N(\epsilon)\epsilon^{D_\mathrm{F}} < \infty \tag{7.116}$$

for the product of ϵ^{D_F} and the minimal number of hypercubes $N(\epsilon)$ of linear size $l = \epsilon$ (Kolmogorov definition) or $l \leq \epsilon$ (Hausdorff definition) covering the object when $\epsilon \to 0$.

To a physicist, the definition becomes more transparent if one considers the relation between the size l of an object and its mass M as a scaling law:

$$M \propto l^{D_\mathrm{F}}. \tag{7.117}$$

For usual objects D_F coincides with the topological dimension (for a line $D_\mathrm{F} = 1$, for a square $D_\mathrm{F} = 2$ and so on). The condition $\epsilon \to 0$ means in practice that such a law should hold in some interval of "rather small" ϵ-values.

The probability $p_i(l)$ to be in a hypercube $N_i(l)$ is proportional to l^{D_F} at small l. Therefore, for a fractal the mean value of the q-th order (ordinary) moment is given by

$$\langle p_i^q(l) \rangle \propto l^{qD_\mathrm{F}} \quad (D_\mathrm{F} = \mathrm{const}) \ . \tag{7.118}$$

Multifractals generalize the notion of fractals, since for these holds

$$\sum_i p_i^q(l) = \langle p_i^{q-1}(l) \rangle \propto l^{\tau(q)} \quad , \tag{7.119}$$

where

$$\tau(q) = (q-1)D_q. \tag{7.120}$$

The D_q are called the Rényi dimensions [29, 51] and depend on q (generally, for multifractals they are decreasing functions of q).

Sometimes it is more convenient to characterize multifractals by spectral properties, rather than by their dimensions.

Let us group all the boxes with a singularity α ($p_i(l) \sim l^\alpha, l \to 0$) into a subset $S(\alpha)$, where α is called the local mass dimension. The number of boxes $\mathrm{d}N_\alpha(l)$ needed to cover $S(\alpha)$ is

$$\mathrm{d}N_\alpha(l) = \mathrm{d}\rho(\alpha) l^{-f(\alpha)} \quad , \tag{7.121}$$

where $f(\alpha)$ is the fractal dimension of the set $S(\alpha)$ related to the Rényi dimension. For the sum of moments one obtains:

$$\sum_{i=1}^{N_i(l)} p_i^q(l) \propto \int \mathrm{d}\rho(\alpha) l^{\alpha q - f(\alpha)} \quad . \tag{7.122}$$

From (7.122), one gets by the saddle-point method:

$$D_q = \frac{1}{q-1} \min_\alpha \left(\alpha q - f(\alpha)\right) = \frac{1}{q-1} \left(\bar{\alpha} q - f(\bar{\alpha})\right) \tag{7.123}$$

with $\bar{\alpha}$ defined as

$$\left.\frac{\mathrm{d}f}{\mathrm{d}\alpha}\right|_{(\alpha=\bar{\alpha})} = q(\bar{\alpha}). \tag{7.124}$$

The notion of Rényi dimensions D_q generalizes the notion of fractal dimension $D_0 = D_\mathrm{F}$, information dimension D_1 and correlation dimension $D_2 = \nu$. A Rényi dimension, therefore, is often called a generalized dimension.

The difference between the usual topological dimension D (i.e. the support dimension) and the Rényi dimension is called the anomalous dimension (or codimension)

$$d_q = D - D_q. \tag{7.125}$$

The multifractal method is a widely used tool in many branches of physics and science in general (cfr. [48, 49, 52]).

A direct relation may be established between the exponents of factorial and generalized moments at comparatively low values of q, much smaller than effective multiplicities contributing to the sum:

$$\phi_q + \tau(q) = (q-1)D. \tag{7.126}$$

Then the exponents are related to Rényi dimension and to codimension as

$$\tau(q) = (q-1)D_q \tag{7.127}$$

and

$$\phi_q = (q-1)d_q. \tag{7.128}$$

According to the general theory [53, 54], there exists "a class of multifractals exhibiting universal properties". They are called universal multifractals and are classified by a Lévy index $0 \le \mu \le 2$ which allows the codimension to be expressed as

$$d_q = \frac{C_1}{\mu - 1} \cdot \frac{q^\mu - q}{q-1} \quad (C_1 = \text{const}). \tag{7.129}$$

The Lévy index μ is also known as the degree of multifractality ($\mu = 0$ for monofractals). Values $\mu < 1$ correspond to so-called "calm" singularities, values $\mu > 1$ correspond to "wild" singularities.

One can proceed further and try to analyse experimental data at two different levels of bin-splitting. For that purpose, it was suggested [55, 56] to study Double Trace Moments (DTM). The procedure is, first, to sum up ν-th-order moments of multiplicity distributions at some bin-splitting level Θ within bins belonging to a single bin of one of the previous steps (having bins of size Δ) and then to calculate their q-th moments at that level

$$\mathrm{Tr}_q^\nu \propto \sum_\Delta (\sum_\Theta n_m^\nu)^q \propto \Delta^{-K(q,\nu)+q-1}. \tag{7.130}$$

It is claimed [55] that "the DTM-technique provides a robust estimate of μ and C_1" for universal multifractals. According to the theory of universal multifractals [53,55], one should observe the following factorizable behavior of "double" exponents $K(q, \nu)$:

$$K(q, \nu) = \nu^\mu K(q, 1) \quad , \tag{7.131}$$

where μ is the Lévy index as in (7.129).

Experimental results on multifractals and generalized multifractals are discussed in Sect. 10.7. A systematic study of the relations between factorial (horizontal and vertical) and associated frequency moments is given in [57]. It has to be kept in mind, however, that the frequency moments do not possess the property of Poisson-noise suppression, essential for the analysis at moderately low multiplicities.

7.5 Bunching-parameter approach

A simple mathematical tool alternative to the normalized factorial moments (7.72-7.74) is the bunching-parameter approach, suggested for high energy applications in [58]. In order to reveal spiky structure of the events, it is only necessary to study the behavior of the probability distribution near the multiplicity $n = q$ by means of the "bunching parameters"

$$\eta_q(\delta y) = \frac{q}{q-1}\frac{P_q(\delta y)P_{q-2}(\delta y)}{P^2_{q-1}(\delta y)} \quad , \quad q > 1 \; . \tag{7.132}$$

As is the case for the normalized factorial moments, the bunching parameters η_q are independent of δy if there are no dynamical fluctuations. For example, $\eta_q = 1$ for all q for the case of a Poissonian probability distribution.

As the $F_q(\delta y)$, the $P_q(\delta y)$ can be averaged over a number M of bins. Assuming approximate proportionality of $\bar{n}_m$ and δy at $\delta y \to 0$ and $P_0(\delta y) \to 1$ for $\delta y \to 0$, one obtains

$$\begin{aligned} \eta_2(\delta y) &\simeq F_2(\delta y) \\ \eta_q(\delta y) &\simeq \frac{F_q(\delta y)F_{q-2}(\delta y)}{[F_{q-1}(\delta y)]^2} \quad , \quad q > 2 \end{aligned} \tag{7.133}$$

or

$$\begin{aligned} \eta_2(\delta y) &\propto (\delta y)^{-\beta_2} \\ \eta_q(\delta y) &\propto (\delta y)^{-\beta_q} \end{aligned} \tag{7.134}$$

with

$$\begin{aligned} \beta_2 &= d_2 \\ \beta_q &= d_q(q-1) + d_{q-2}(q-3) - 2d_{q-1}(q-2) \quad , \quad q > 2 \; . \end{aligned} \tag{7.135}$$

Expressing d_q in terms of the Lévy law approximation (7.129),

$$\beta_q = d_2 \frac{q^\mu + (q-2)^\mu - 2(q-1)^\mu}{2^\mu - 2} \quad . \tag{7.136}$$

In case of monofractal behavior ($\mu = 0$), $\beta_q = 0$ for $q > 2$. In the limit of the log-normal approximation ($\mu = 2$), on the other hand, $\beta_q = d_2$ and all bunching parameters follow the same power law.

The Lévy-law approximation allows a simple description of multifractal properties of random cascade models using only one free parameter μ. In the bunching-parameter approach, one can make an approximation of the high-order bunching parameters to obtain a simple *linear* expression for the anomalous fractal dimensions d_q, still maintaining the number of free parameters at one.

Assuming that high-order bunching parameters can be expressed in terms of the second-order one as

$$\eta_q(\delta y) = [\eta_2(\delta y)]^r \quad , \quad q > 2 \quad , \tag{7.137}$$

the linear expression becomes

$$d_q = d_2(1 - r) + d_2 r \frac{q}{2} \ . \tag{7.138}$$

The use of bunching parameters is interesting, because it gives a general answer to the problem of finding a multiplicity distribution leading to intermittency: according to (7.132), any multiplicity distribution can be expressed as

$$P_q(\delta y) = P_0(\delta y) \frac{[P_1(\delta y)/P_0(\delta y)]^q}{q!} \prod_{\ell=2}^{q} [\eta_\ell(\delta y)]^{q+1-\ell}, \quad q > 1 \ . \tag{7.139}$$

The possible forms of multiplicity distributions with multifractal behavior of d_q (7.125) are discussed in [59].

7.6 The wavelet transform

An increase of factorial and cumulant moments with decreasing bin sizes reflects a widening of a multiplicity distribution, i.e. an increase of multiplicity fluctuations in individual events. This phenomenon can be studied by other methods, as well. In particular, the so-called wavelet transform seems to be suited for that purpose.

The wavelet transform is of particular importance in pattern recognition. This is a more general problem than the fluctuation study itself, since it involves the analysis of individual event shapes, not only the event ensemble, and may become of interest in the analysis of very-high multiplicity events.

It is shown [60] that, for pattern recognition, the wavelet transform is about two orders of magnitude more efficient than ordinary Fourier analysis.

An application of wavelets to multiparticle production processes has been proposed in [61]. The main principle of the wavelet transform is to study the dependence of fluctuations on the phase-space bin size by the so-called difference method. One considers the difference between the histogram of an individual event at a definite resolution to the corresponding histogram at a (e.g., twice) finer resolution. Proceeding step by step, one is able to restore the whole pattern of fluctuations.

Let us explain how this procedure can be applied to an individual event. We consider the one-dimensional projection of the event onto the rapidity interval ΔY. Any n-particle event can be represented by the histograms of particle densities $\rho = \mathrm{d}n/\mathrm{d}y$ at various resolutions. The simplest information is obtained from the value of the average density $\langle \rho \rangle = n/\Delta Y$. To consider the forward-backward correlations, one splits the rapidity interval ΔY into two equal parts and gets the forward and backward average densities $\langle \rho_{\mathrm{f,b}} \rangle = 2n_{\mathrm{f,b}}/\Delta Y$, where $n_{\mathrm{f,b}}$ are the forward (backward) multiplicities with $n_{\mathrm{f}} + n_{\mathrm{b}} = n$. Proceeding further to the J-th step, we approximate the event in terms of the histogram with 2^J bins.

Let us construct now the difference of the two histograms described above. Namely, we subtract the average density from the forward-backward histogram and get another histogram with positive ordinate at one side and negative at the other, demonstrating the forward-backward fluctuations in the event.

Splitting the forward and backward regions further into equal halves, one gets the histogram at $J = 2$. Its difference from the forward-backward histogram at $J = 1$ reveals the fluctuations at finer resolution. Iterating to higher values of J, one studies how fluctuations evolve at ever finer resolution. The set of difference histograms is called the wavelet transform of the event. The above procedure corresponds to the so-called Haar-wavelet transform. Readers interested in mathematical details are referred to [62].

The wavelet transform provides direct information on the evolution of fluctuations at different scales, i.e. on the dynamics of individual high-multiplicity events revealing their clustering (and subclustering) structure. A generalization to factorial (and cumulant) wavelets is possible [61]. The simplest cascade models show such remarkable properties of wavelet transforms [61] as (quasi)diagonalization of their correlation density matrices, scaling exponents etc. It is interesting to note that the equations for the generating functions of wavelet transforms [61] look very similar to the "gain-loss" equations (in particular, to QCD equations). All those features are yet to be studied.

The very first application to experimental data is presented in [63], where wavelet spectra of JACEE events are studied.

7.7 Lévy stable distributions

Another interesting alternative [64], explicitely rejecting the idea of Poisson noise suppression, goes back to the Mandelbrot approach to finance and economics [65] based on the rich class of Lévy stable distributions [66]. This class of distributions generalizes their limiting case of the Gaussian by allowing skewness and heavy tails (and thus large, non-Gaussian fluctuations), without giving up the concept of stability, but by generalizing the central limit theorem.

A distribution P is called *stable* if and only if for each n, there exist constants $c_n > 0$ and γ_n such that

$$S_n \stackrel{d}{=} c_n X + \gamma_n \ , \tag{7.140}$$

where $S_n = X_1 + X_2 + \ldots + X_n$ $(n > 1)$, the sum of n mutually independent random variables X_n with the common distribution P, and X following the same P; the symbol $\stackrel{d}{=}$ stands for equality in distribution, i.e. l.h.s. and r.h.s. follow the same probability law. Furthermore, P is *stable in the strict sense* if and only if $\gamma_n = 0$ for all n.

The normalization constants c_n are given by $c_n = n^{1/\mu}$ with $0 < \mu \leq 2$ being the *characteristic exponent* of P, or the *Lévy index of stability.*

In general, a stable distribution is described by four parameters $\{\mu, \beta, \gamma, \delta\}$, with μ as above, $-1 \leq \beta \leq 1$ the skewness parameter, $\gamma > 0$ the scale parameter, $\delta \epsilon \Re$ the location parameter. Special cases are

$$\begin{array}{lcl} \text{Gaussian} & : & \mu = 2, \beta = 0 \\ \text{Cauchy} & : & \mu = 1, \beta = 0 \\ \text{Lévy} & : & \mu = 1/2, \beta = 1 \; . \end{array}$$

The (fractional absolute) moment

$$E(|X|^p) = \int_{-\infty}^{\infty} \mathrm{d}x |x|^p f(x) \; , \tag{7.141}$$

with $p \epsilon \Re$ and $f(x)$ the density function of a stable distribution with $0 < \mu < 2$, is finite if and only if $0 < p < \mu$.

If the expectation $\bar{X} = \mathrm{E}(X_i)$ and the variance $\sigma^2 = \mathrm{Var}(X_i)$ exist, then S_n has the limiting behavior

$$\lim_{n \to \infty} P \left\{ \frac{S_n - n\bar{X}}{\sigma \sqrt{n}} < x \right\} = \frac{1}{\sqrt{2\pi}} \int_{-\infty}^{x} \mathrm{d}t \exp(-t^2/2) \; . \tag{7.142}$$

This is the *central limit theorem.* It can be seen as a special case of the *generalized* central limit theorem:

There exist constants $a_n = n^{-1/\mu}$, $b_n \epsilon \Re$ and a non-degenerate random variable Z, so that

$$a_n(X_1 + X_2 + \ldots + X_n) - b_n \xrightarrow{d} Z \tag{7.143}$$

if and only if Z is stable, i.e. if and only if $Z \stackrel{d}{=} AY + B$ with $A > 0$ and $B \epsilon \Re$ and Y a random variable with characteristic function

$$\mathrm{E} \exp(iuY) = \exp(-|u|^\mu [1 - i\beta \tan \frac{\pi\mu}{2} \mathrm{sign} u]) \text{ for } \mu \neq 1 \tag{7.144}$$

$$\mathrm{E} \exp(iuY) = \exp(-|u| [1 + i\beta \frac{2}{\pi} \mathrm{sign} u \ln |u|]) \text{ for } \mu = 1 \; . \tag{7.145}$$

The Lévy index μ can be extracted from the experimental data directly, i.e. without any noise suppression. Its deviation from $\mu = 2$ (Gaussian) can then be considered a measure of dynamical fluctuation.

Bibliography

[1] N.N. Bogoliubov, *Problems of Dynamical Theory in Statistical Physics*, Gosizdat, USSR (1946).

[2] L.S. Brown, *Phys. Rev.* **D3** (1972) 748;
K.J. Biebl and J. Wolf, *Nucl. Phys.* **B44** (1972) 301.

[3] T.N. Thiele, *The Theory of Observation*, *Ann. Math. Stat.* **2** (1931) 165.

[4] R. Kubo, *J. Phys. Soc. Japan* **17** (1962) 1100.

[5] R.F. Chang, V. Korenman, C.O. Alley and R.W. Detenbeck, *Phys. Rev.* **178** (1969) 612.

[6] B. Kahn and G.E. Uhlenbeck, *Physica* **5** (1938) 399.

[7] K. Huang, *Statistical Mechanics*, John Wiley and Sons, 1963.

[8] A.H. Mueller, *Phys. Rev.* **D4** (1971) 150.

[9] M.G. Kendall and A. Stuart, *The Advanced Theory of Statistics*, Vol. 1,C. Griffin and Co., London 1969.

[10] Z. Koba, *Acta Phys. Pol.* **B4** (1973) 95.

[11] B.R. Webber, *Nucl. Phys.* **B43** (1972) 541.

[12] H.C. Eggers, *Intermittency, Moments and Correlations in Distributions of Particles Created in High Energy Collisions*, PhD. Thesis, Univ. of Arizona, 1991.

[13] P. Carruthers, *Phys. Rev.* **A43** (1991) 2632.

[14] A.I. Golokhvastov, *Z. Phys.* **C64** (1994) 301.

[15] W.I. Weisberger, *Phys. Rev.* **D8** (1973) 1387.

[16] M. Abramowitz and I. Stegun, *Handbook of Mathematical Functions*, Dover publications, Inc., New York (1964).

[17] H. Cramér, *Mathematical Methods of Statistics*, Princeton N.J. (1946).

[18] D. Weingarten, *Nucl. Phys.* **B70** (1974) 501.

[19] Z. Koba and D. Weingarten, *Nucl. Phys.* **B70** (1974) 534.

[20] M. Koutras, *Discrete Math.* **42** (1982) 73.

[21] S.K. Kauffmann, M. Gyulassy, *J. Phys.* **A11** (1978) 1715.

[22] S. Hegyi, *Phys. Lett.* **B309** (1993) 443; **B318** (1993) 642; **B463** (1999) 126.

[23] A. Białas and R. Peschanski, *Nucl. Phys.* **B273** (1986) 703.

[24] A. Białas and R. Peschanski, *Nucl. Phys.* **B308** (1988) 857.

[25] K. Fiałkowski, B. Wošiek and J. Wošiek, *Acta Phys. Pol.* **B20** (1989) 639.

[26] W. Ochs, *Phys. Lett.* **B247** (1990) 101; *Z. Phys.* **C50** (1991) 339.

[27] A. Białas and M. Gazdzicki, *Phys. Lett.* **B252** (1990) 483.

[28] P. Carruthers, H.C. Eggers and I. Sarcevic, *Phys. Lett.* **B254** (1991) 258.

[29] J. Feder, *Fractals* (Plenum Press, N.Y. and London, 1988).

[30] R.C. Hwa, *Phys. Rev.* **D41** (1990) 1456.

[31] C.B. Chiu and R.C. Hwa, *Phys. Rev.* **D43** (1991) 100.

[32] W. Florkowski and R.C. Hwa, *Phys. Rev.* **D43** (1991) 1548.

[33] B. Saleh, *Photoelectron Statistics* (Springer Verlag, Berlin 1978).

[34] D. Cantrell, *Phys. Rev.* **A1** (1970) 672.

[35] H.C. Eggers, P. Carruthers, P. Lipa and I. Sarcevic, *Phys. Rev.* **D44** (1991) 1975.

[36] L. Mandel, *Proc. Phys. Soc.* **72** (1958) 1037.

[37] L. Mandel, *Proc. Phys. Soc.* (London) **74** (1959) 233.

[38] E. Wolf and C.L. Mehta, *Phys. Rev. Lett.* **13** (1964) 705.

[39] G. Bédard, *Proc. Phys. Soc.* **90** (1967) 131.

[40] F. Takagi and D. Kiang, *Phys. Rev.* **D56** (1997) 5862.

[41] L. Diósi, *Nucl. Instr. Meth.* **138** (1976) 241; 140 (1977) 533.

[42] J.N. Fry, *The Astrophysical Journal* **289** (1985) 10.

[43] Fu Yinghua and Liu Lianshou, *Extracting event dynamics from event-by-event analysis*, hep-ph/0310308, subm. to Phys. Rev. C.

[44] R. Peschanski and J. Seixas, *Scaling Relations between Fluctuations and Correlations in Multiparticle Production*, preprint CERN-TH-5903/90.

[45] C.D. Cantrell and J.R. Fields, *Phys. Rev.* **A7** (1973) 2063.

[46] J. Bures, *Can. J. Phys.* **50** (1972) 706.

[47] B. Mandelbrot, *The Fractal Geometry of Nature*, Freeman, N.Y., 1982.

[48] Ya.B. Zeldovitch, *Sov. Phys. Uspekhi* **152** (1987) 3.

[49] G. Paladin and A. Vulpiani, *Phys. Rep.* **156** (1987) 147.

[50] R. Peschanski, *Int. J. Mod. Phys.* **A6** (1991) 3681.

[51] A. Rényi, *Probability Theory*, North-Holland, 1970.

[52] I.M. Dremin, *Sov. Phys. Uspekhi* **160** (1990) 647.

[53] D. Schertzer and S. Lovejoy, *J. Geo. Res.* **92** (1987) 9693.

[54] J.P. Bouchaud and A. Georges, —em Phys. Rep. **195** (1990) 127;
A.Y. Khintchine and P. Lévy, *C.R. Acad.Sci.* (Paris) **202** (1936) 374.

[55] S.P. Ratti et al., *Z. Phys.* **C61** (1994) 229.

[56] S.P. Ratti et al. (IHSC), *Universal Multifractal Analysis of Multiparticle Production in h−h Collisions at* $\sqrt{s} = 16.7$ *GeV*, Proc. XXI Int. Symp. on Multiparticle Dynamics, Wuhan, China, 1991, eds. Y.F. Wu and L.S. Liu (World Scientific, Singapore, 1992)p.409;
V. Arena et al. (IHSC), *Nuovo Cim.* **108A** (1995) 417.

[57] M. Blažek, *Z. Phys.* **C63** (1994) 263; *Int. J. Mod. Phys.* **12** (1997) 839; Searching for multi-dimensional fractal phenomena in counting experiments, FUSAV/99-03 (1999).

[58] S.V. Chekanov and V.I. Kuvshinov, *Acta Phys. Pol.* **B25** (1994) 1189.

[59] S.V. Chekanov and V.I. Kuvshinov, *J. Phys.* **G22** (1996) 601; S.V. Chekanov, W. Kittel, V.I. Kuvshinov, *Z. Phys.* **C74** (1997) 517.

[60] H. Abarbanel, F. Dyson et al., *Report on Wavelets*, Cornell Mathematics, 1994.

[61] M. Greiner, P. Lipa, and P. Carruthers, *Phys. Rev.* **E51** (1995) 1948;
M. Greiner, J. Giesemann, P. Lipa, and P. Carruthers, Preprint HEPHY-PUB618/95, AZPH-TH/95-03 (1995); P. Lipa, M. Greiner and P. Carruthers, *Acta Phys. Slov.* **46** (1996) 557.

[62] I. Daubechies, *Ten Lectures on Wavelets*, Philadelphia: Society for Industrial and Applied Mathematics (SIAM), 1992.
Y. Meyer, *Wavelets: Algorithms and Operators*, New York; Cambridge University Press, 1992.
G. Kaiser, *A Friendly Guide to Wavelets*, Boston; Birkhauser, 1994.

[63] N. Suzuki et al., *Prog. Theor. Phys.* **94** (1995) 91.

[64] Liu Qin and Meng Ta-chung, *Phys. Rev.* **D69** (2004) 054026.

[65] B.B. Mandelbrot, *The Journal of Business* **36** (1963) 394 and **40** (1967) 393; *Fractals and Scaling in Finance* (Springer, 1997).

[66] W. Feller, *An Introductions to Probability, Theory and Its Applications*, Vol.I and II, John Wiley and Sons (1971).

Chapter 8

Final-State Multiplicity

Although the number of charged particles (the charged-particle multiplicity), n, is only a very global measure of the characteristics of the final state of a high-energy collision, it has proved important in the study of particle production. Independent emission of single particles leads to a Poisson multiplicity distribution. Deviations from this shape, therefore, reveal correlations and dynamics.

8.1 Full phase space

8.1.1 Average multiplicity and its energy dependence

The average number of particles, of all types or of a particular type, is the first moment of the multiplicity distribution and the phase-space integral of the corresponding singe-particle density Eq. (7.12). As such, it does not contain any information on correlations, but is one of the basic observables characterizing hadronic final states and their evolution with increasing energy.

One recent example is the question whether or not color reconnection and/or Bose-Einstein interference between decay products of different W's in $e^+e^- \rightarrow W^+W^-$ (or, more generally, two cross-talking di-jet systems) could affect the charged-particle multiplicity [1–7]. In fact, experimentally [8, 9], no such effect is seen in the full phase-space multiplicity of the W^+W^- system.

8.1.1.1 Early results

An early question was how the average multiplicity and its evolution with energy depend on the type of collision. The average multiplicity $\langle n \rangle$ of charged hadrons produced in $\bar{\nu}$p collisions is plotted in Fig. 8.1a as a function of the squared hadronic energy W^2 for $W^2 < 100$ GeV [10], where it is compared to a fit (solid line) to $\langle n \rangle$ of non-diffractive π^-p interactions [11]. Clearly, the relation $\langle n \rangle_{\bar{\nu}\mathrm{p}} = \langle n \rangle_{\pi^-\mathrm{p}}$ holds well in the considered energy range.

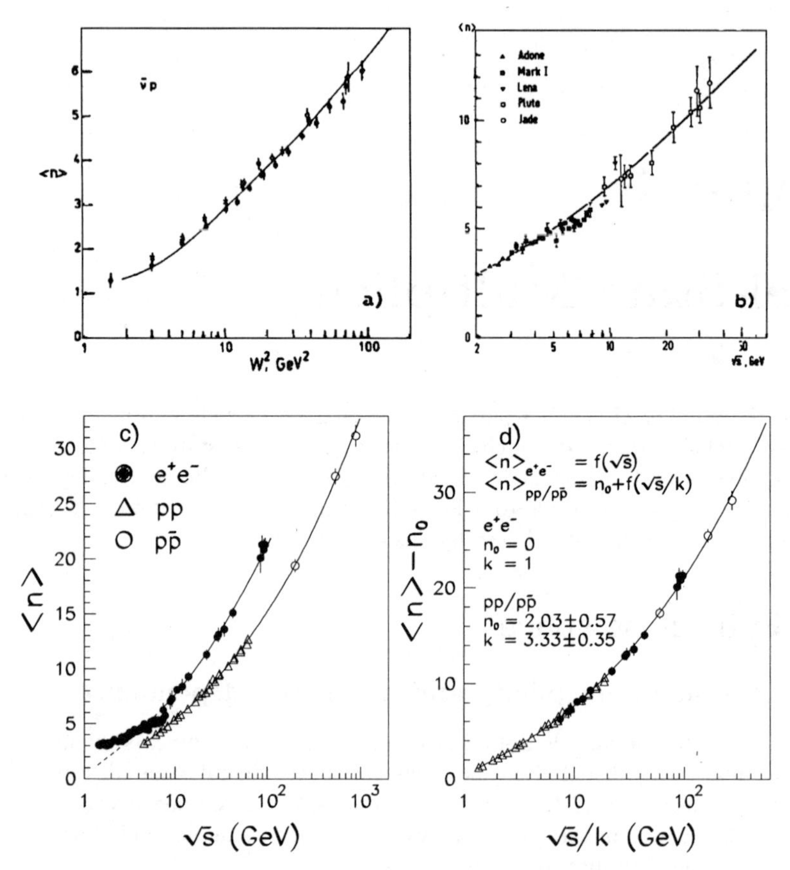

Figure 8.1. Average charged-particle multiplicity of the hadronic system in a) $\bar{\nu}$p collisions [10] (the solid line shows a fit [11] to the non-diffractive component of π^-p collisions), b) e^+e^- annihilation [12] (the solid line shows the prediction from hadronic data according to (8.1)); c) e^+e^- compared to pp$^\pm$ collisions [13]; d) comparison after the transformation $\langle n \rangle \to \langle n \rangle - n_0$, $\sqrt{s} \to \sqrt{s}/k$ for pp$^\pm$ collisions [13].

Similarly, for early e^+e^- annihilation results [12], Fig. 8.1b, one finds [11] that

$$\langle n \rangle_{e^+e^-} = \langle n \rangle_{\pi^+p} + \langle n \rangle_{\pi^-p} - \langle n \rangle_{pp} \quad , \tag{8.1}$$

is well satisfied for $\sqrt{s} < 50$ GeV. Here, the third term on the right-hand side of (8.1) corrects for proton fragmentation, which is obviously absent in e^+e^- annihilation.

8.1.1.2 More recent results

In the meantime we have moved to higher energies, in particular with e^+e^- and $p\bar{p}$ collisions, but not with π^-p collisions. So, other comparisons have been made.

Of course, it is evident that the simple similarities observed above cannot persist. Besides the absence of proton fragmentation in e^+e^- collisions mentioned above, e.g., hard gluon radiation leads to 3- and 4-jet events in e^+e^- collisions, while hard parton scattering leads to 4-jet events in hh collisions. Both mechanisms cause an increase in the number of particles produced, but the relative strength of these two mechanism is different in the two types of collisions.

Nevertheless, Fig. 8.1c gives a comparison [13] of e^+e^- and $pp^\pm$ collisions at higher energies. In both cases, the energy dependence of $\langle n \rangle$ can be well described by [14]

$$\langle n \rangle = a_0 + a_1 \exp(a_2 \sqrt{\ln s}) \tag{8.2a}$$

$$\text{or} \quad \langle n \rangle = c_0 + c_1 \ln s + c_2 (\ln s)^2 \quad , \tag{8.2b}$$

but at given energy, $\langle n \rangle$ is about 25% lower for $pp^\pm$ collisions than for e^+e^- annihilation.

Following an earlier comparison [15], in Fig. 8.1d [13], the average $pp^\pm$ inelasticity and leading system fragmentation have been taken into account by the transformation $\langle n \rangle \to \langle n \rangle - n_0$, $\sqrt{s} \to \sqrt{s}/k$, with n_0 and k as additional fit parameters for the $pp^\pm$ data. Both parametrizations (8.2a) and (8.2b) give excellent combined fits, with similar n_0 and k values. The latter suggest that together the leading $pp^\pm$ systems on average contribute about 2 charged particles, while the energy available to central particle production is about 1/3 of the total energy.

Quite to the contrary and most importantly, this factor 1/3 is not needed when comparing average multiplicity per pair of participants (nucleons) in heavy ion collisions at the nucleon-nucleon cms energy $\sqrt{s_{\rm NN}}$ to that in e^+e^- annihilation at $\sqrt{s_{\rm e^+e^-}}$ [16], for $\sqrt{s_{\rm NN}}$ between 19.6 and 200 GeV. In other words, comparing at given cms energy, heavy ion collisions [17–19], like e^+e^- collisions in Fig. 8.1c, lie higher than $pp^\pm$ collisions, but all three more or less coincide, in particular at an energy ($\sqrt{s_{\rm NN}} = \sqrt{s_{\rm e^+e^-}}$) where n_0 becomes negligible, when $\sqrt{s}/3$ is used for $pp^\pm$ collisions [20].

Such an effect, as well as that already observed in $\rho(0)$ in Fig. 4.26c above, would be expected if the energy of only the interacting quark pair would be responsible for particle production in hadron-hadron collisions, the remaining 2/3 of the energy being carried away by the spectator quarks, while more than one quark per nucleon interacts in a (central) heavy ion collision [20, 21].

Coming back to e^+e^- collisions, in Fig. 8.2, PYTHIA/JETSET, ARIADNE and HERWIG, all based on parton showers including gluon coherence effects, agree with the e^+e^- data, while the second-order matrix element (ME) version of JETSET in Fig. 8.2a does not reproduce the fast rise of $\langle n \rangle$ with increasing cms energy $\sqrt{s}$. COJETS does not contain gluon coherence and $\langle n \rangle$ rises too fast at high energy.

a)

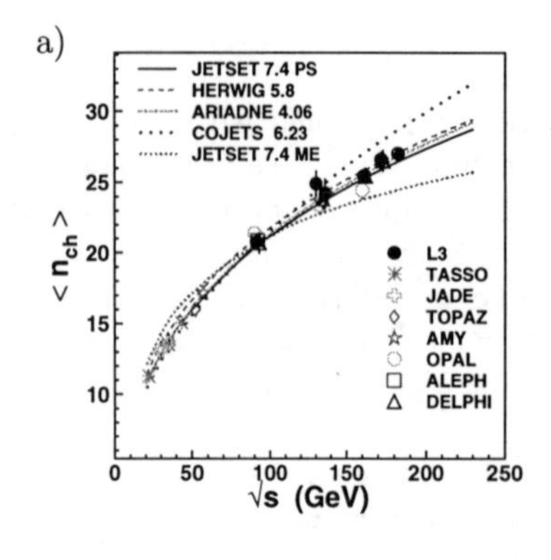

b)

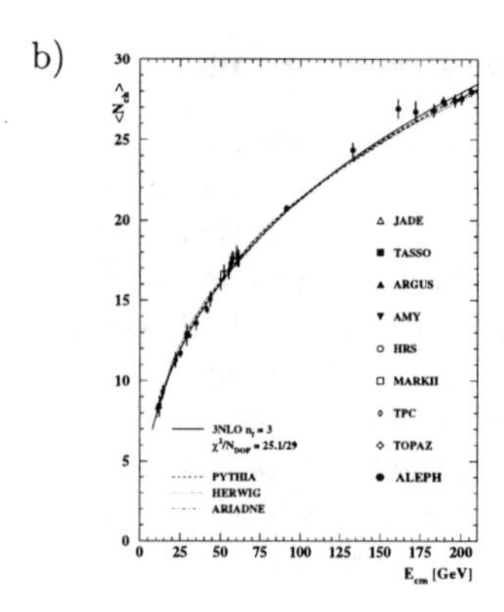

Figure 8.2. Average charged-particle multiplicity as a function of $\sqrt{s} = E_{\mathrm{cm}}$ for e$^+$e$^-$-collisions above $\sqrt{s} = 10$ GeV, compared to a number of models and a 3NLO QCD prediction [22, 23].

As suggested in [24], the energy evolution of the average multiplicity in Fig. 8.2 is closely related to that of the peak position, ξ_0, in Fig. 4.8b, so that energy scaling of $\xi_0/\langle n\rangle$ is expected. This is indeed observed in a comparison of TASSO [25] and DELPHI [26] data.

Analytical QCD predictions exist on the basis of leading (LLA) and next-to-leading (NLLA) corrections [27]

$$\langle n\rangle = a\alpha_{\mathrm{s}}^b \exp(c/\sqrt{\alpha_{\mathrm{s}}})\,[1 + d\sqrt{\alpha_{\mathrm{s}}}] \tag{8.3}$$

and Local Parton Hadron Duality (see Sub-Sect. 6.1.2), where a and d (the latter introduced in [28] to allow for higher-order corrections) are free parameters, $b = 0.49$ and $c = 2.27$ are predicted by QCD and α_{s} is the running strong coupling constant responsible for the energy dependence. This form is in perfect agreement with the e$^+$e$^-$ data up to 206 GeV [13, 29].

However, more recent MLLA [30] and 3NLO [31] predictions exist, the latter taking recoil effects and conservation laws into account. Both contain two fit parameters, the effective QCD scale Λ and a global LPHD normalization constant K. The 3NLO fit with $n_{\mathrm{f}} = 3$ is shown as the full line in Fig. 8.2b. Other fits, including flavor corrections and/or using $n_{\mathrm{f}} = 5$, or using MLLA instead of 3NLO, yield practically the same result [23].

Of particular interest is the multiplicity growth $\gamma_0 = \mathrm{d}\ln\langle n\rangle/\mathrm{d}\ln s = \sqrt{6\alpha_s/\pi}$ in double leading log approximation (DLLA). It is called the anomalous dimension of QCD, Sub-Sub-Sect. 1.6.2.1, and is of order 0.4 at the Z mass. We shall come back to this parameter when discussing fractality of QCD branching in Chapter 10.

An interesting alternative [32] uses the modified leading log approximation (MLLA) plus energy conservation, but avoids LPHD and the sharp low-mass cut-off by the introduction of an inactivation mechanism into the parton cascade. The final hadronization is viewed as a competition between parton fragmentation according to QCD fragmentation kernels and parton inactivation with phenomenological kernels to be determined from the data. The energy dependence of $\langle n \rangle$ is well described by an inactivation rate function, Gaussian in the cluster size k, but the $(\ln s)^2$ dependence of $\langle n \rangle$ in Eq. (8.2) is predicted to be replaced by an $\ln(\ln s)$ behavior at asymptotic energies.

Another way [33] to avoid LPHD is based on the idea that near-mass-shell partons, produced according QCD showering as above, break chiral symmetry spontaneously and independently when converting into hadrons non-perturbatively. In this case, the charged-particle multiplicity increases faster with increasing energy than the parton multiplicity. The difference is small, however, and energies higher than LEP will be needed to see it experimentally.

In Fig. 8.3a the average deep-inelastic (DIS) ep current-jet multiplicity measured at HERA by H1 [34] is compared with the e^+e^- single-hemisphere multiplicity, here

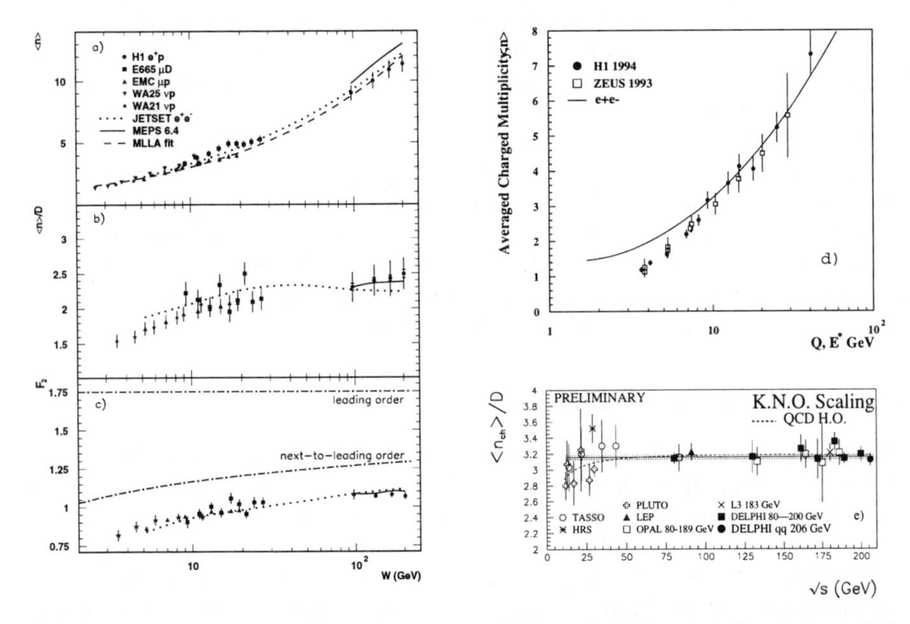

Figure 8.3. a) Hadronic energy W dependence of $\langle n \rangle$ in the full current hemisphere of the γ^*p cms in comparison with fixed-target lh data. b) and c) same as a) for the ratio $\langle n \rangle / D$ and the second normalized factorial moment F_2, respectively [34]. d) Comparison of $\langle n \rangle$ in DIS in the current region of the Breit frame as a function of Q to a parametrization [35] of e^+e^- annihilation as a function of $E^* \equiv \sqrt{s}$ [36]. e) Energy dependence of the $\langle n \rangle / D$ ratio in e^+e^- annihilation; the horizontal lines correspond to a weighted average and its error, the dashed line to QCD with one-loop higher-order corrections [29].

represented by JETSET results for light quark-antiquark pairs (dotted line). The evolution with energy of $\langle n \rangle$ in DIS is indeed the same as that for e^+e^- annihilation and a fit by (8.3) with a and b as free parameters is shown as dashed line. The MEPS 6.4 generator [37] for DIS (solid line), on the other hand, overestimates $\langle n \rangle$.

Alternatively, Fig. 8.3d gives a comparison of $\langle n \rangle$ for e^+e^- as a function of $\sqrt{s}$ and for DIS as a function of Q, studied in the Breit frame [38]. For the simple Quark-Parton Model (QPM), the outgoing struck quark carries $p_z^{\mathrm{Breit}} = -Q/2$ in the negative z-direction ($-Q$ being the momentum of the exchanged virtual boson) and all particles produced with a negative p_z^{Breit} are considered as produced by the fragmentation of the struck quark (current region), while positive p_z^{Breit} corresponds to the fragmentation of the target remnant (target region). The current region is then analogous to a single hemisphere of e^+e^- annihilation. In leading- (and higher-) order QCD processes, the collision is no longer collinear in the Breit frame, but the current and target regions are still well-defined [39].

Fig. 8.3d compares ZEUS [40] and H1 [36] results for $\langle n \rangle$ in the current region as a function of Q to a parametrization [35] of the e^+e^- results as a function of cms energy $E^* \equiv \sqrt{s}$. Indeed, there is very good agreement at large values of Q and E^*, but DIS gives smaller $\langle n \rangle$ at low values. In [36] it is demonstrated that this discrepancy can be explained by leading-order processes (QCD Compton, boson-gluon fusion) which are present in ep but not in e^+e^- and which cause a depletion of the current region of the Breit frame.

In [41] it is demonstrated that the ZEUS data are well reproduced by ARIADNE and HERWIG with default settings, while LEPTO with soft color interaction (SCI) [42] simulates too fast a Q^2 evolution. SCI causes the color string to overlap itself and this causes an increase both of particle number and energy per unit of rapidity. The discrepancy can be largely corrected by removing SCI from the model.

For Figs. 8.3b,c and e, see Sub-Sect. 8.1.2 below.

8.1.1.3 Heavy-quark contribution

Flavor independence of the strong coupling is a fundamental property of QCD. Exploiting the leading particle effect in quark fragmentation to identify the primary quark flavor, the multiplicities per light quark flavor were obtained [43] from a statistical unfolding of the hemisphere opposite to the tagged particle. Consistency with flavor independence was indeed found for u, d and s quark fragmentation, be it within still rather large errors. Flavor symmetry is, however, expected to be broken due to (calculable) mass effects.

In QCD, forward gluons are suppressed in the angular region (dead cone) around the heavy quark [44]. This leads to a lower multiplicity accompanying a heavy quark than a light quark at the same cms energy. However, the particles originating from heavy-quark decay overcompensate this effect in the total multiplicity.

The experimentally observed difference $\delta_{\mathrm{bl}} = \langle n \rangle_{\mathrm{bb}} - \langle n \rangle_{\mathrm{ll}}$ of charged-particle

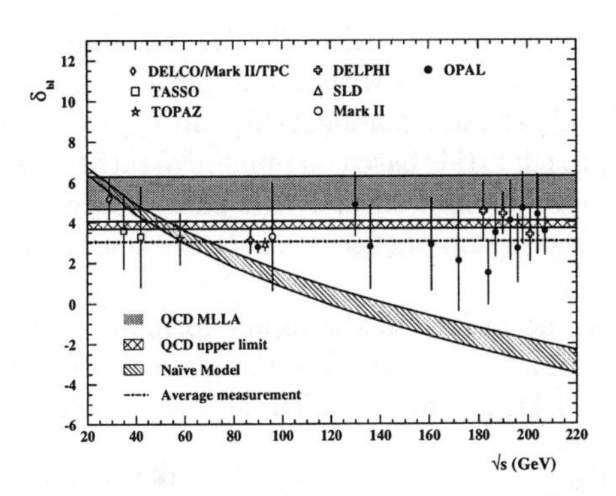

Figure 8.4. The difference in charged-particle multiplicity between b-quark and light-quark events, as a function of the cms energy $E_{\mathrm{cm}} \equiv \sqrt{s}$ [45].

multiplicity in heavy-quark and light-quark events is plotted in Fig. 8.4 as a function of the cms energy [45]. Including mass effects, MLLA QCD calculations + LPHD give an energy independent prediction (first ref. [44]) shown as the upper band, while the cross-hatched band gives an upper bound due to an improved QCD calculation (third ref. [44]; see also second ref. [44] for a similar limit based on phenomenological arguments). The tilted band corresponds to the more naïve assumption [46] that the multiplicity accompanying the heavy hadrons is the same as the multiplicity of a light-quark event at the cms energy left behind after the heavy quarks have fragmented. The high-energy LEP2 data clearly favor the constant behavior of the QCD calculations.

Dispersion D and normalized factorial moment F_2, on the other hand, are observed [47] to be flavor independent.

8.1.1.4 Gluon jets versus quark jets

In jet fragmentation, hadron production can be described as a shower of bremsstrahlung processes, followed by hadronization. Bremsstrahlung is directly proportional to the coupling of the radiated gluon to the radiator (gluon or quark). The probability for a gluon to radiate a gluon is proportional to the color factor $C_{\mathrm{A}} = 3$, that for a quark to radiate a gluon to $C_{\mathrm{F}} = 4/3$. Neglecting gluon splitting into $\mathrm{q\bar{q}}$ pairs, the ratio of the gluon multiplicity radiated from a gluon to that radiated from a quark is, therefore, expected to approach, to leading order and asymptotically, the ratio of the QCD color factors, in case of SU(3) structure of QCD $r = \langle n \rangle_{\mathrm{gg}} / \langle n \rangle_{\mathrm{q\bar{q}}} = C_{\mathrm{A}}/C_{\mathrm{F}} = 9/4$ [48]. Here, $\langle n \rangle_{\mathrm{gg}}$ is the mean number of partons radiated from a virtual color-singlet gg point source and $\langle n \rangle_{\mathrm{q\bar{q}}}$ that radiated from a virtual color-singlet $\mathrm{q\bar{q}}$ point source. Corrections have been calculated analytically to the next-to-next-to-leading order (NNLO) and for energy (but not momentum) conservation [49] and even one order further (3NLO) including energy conservation [50] as $r(y) = r_0(1 - r_1\gamma_0 - r_2\gamma_0^2 - r_3\gamma_0^3)$ with $y = \ln(p\Theta/Q_0)$, p the initial momentum, Θ the jet opening angle, Q_0 the low-mass cut-off, and $r_{1,2,3}$ calculable from QCD.

Another QCD approach [51] is based on numerical rather than analytic techniques. This allows to incorporate more accurately the phase space limits for the emission of soft gluons and the conservation of energy. A comprehensive discussion can be found in [31].

On the other hand, according to the dipole formulation (resumming all leading logarithmic terms) of [52], the evolution of the gluon system multiplicity with scale $L = \ln(s/\Lambda^2)$ is given in MLLA by the differential equation

$$\left.\frac{\mathrm{d}\langle n\rangle_{\mathrm{gg}}(L')}{\mathrm{d}L'}\right|_{L'=L+c_{\mathrm{g}}-c_{\mathrm{q}}} = \frac{C_{\mathrm{A}}}{C_{\mathrm{F}}}\left(1 - \frac{\alpha_0 c_{\mathrm{r}}}{L}\right)\frac{\mathrm{d}\langle n\rangle_{\mathrm{q\bar{q}}}(L)}{\mathrm{d}L} \tag{8.4}$$

with $c_{\mathrm{g}} = 11/6$, $c_{\mathrm{q}} = 3/2$, $c_{\mathrm{r}} = 10\pi^2/27 - 3/2$ and $\alpha_0 = 6/(11 - 2n_{\mathrm{f}}/n_{\mathrm{c}})$, n_c being the number of colors, n_{f} that of flavors. The constant of integration can be obtained at very small scale L_0, where the multiplicity of both the gg and $\mathrm{q\bar{q}}$ systems should be determined by hadronic phase space and should thus be equal: $\langle n\rangle_{\mathrm{gg}}(L_0) \approx \langle n\rangle_{\mathrm{q\bar{q}}}(L_0) = \langle n\rangle(L_0)$.

The average multiplicity in a three-jet event can then be expressed by the two (extreme) alternative formulations [52]

$$\langle n\rangle_{\mathrm{q\bar{q}g}} = \langle n\rangle_{\mathrm{q\bar{q}}}(L_{\mathrm{q\bar{q}}}, \kappa_{\mathrm{Lu}}) + \frac{1}{2}\langle n\rangle_{\mathrm{gg}}(\kappa_{\mathrm{Le}})\ , \tag{8.5a}$$

$$\langle n\rangle_{\mathrm{q\bar{q}g}} = \langle n\rangle_{\mathrm{q\bar{q}}}(L, \kappa_{\mathrm{Lu}}) + \frac{1}{2}\langle n\rangle_{\mathrm{gg}}(\kappa_{\mathrm{Lu}})\ , \tag{8.5b}$$

with $L_{\mathrm{q\bar{q}}} = \ln(s_{\mathrm{q\bar{q}}}/\Lambda^2)$, $\kappa_{\mathrm{Lu}} = \ln(p_{\perp\mathrm{Lu}}^2/\Lambda^2)$, $\kappa_{\mathrm{Le}} = \ln(p_{\perp\mathrm{Le}}^2/\Lambda^2)$ and $p_{\perp\mathrm{Lu}}^2 = s_{\mathrm{qg}}s_{\mathrm{\bar{q}g}}/s$, $p_{\perp\mathrm{Le}}^2 = s_{\mathrm{qg}}s_{\mathrm{\bar{q}g}}/s_{\mathrm{q\bar{q}}}$, $s_{ij} = (p_i + p_j)^2$, the subscripts corresponding to Lu=Lund, Le=Leningrad. The $\langle n\rangle_{\mathrm{q\bar{q}}}(L, \kappa)$ take into account the restriction on the phase space of the quark system imposed by the resolution of a gluon jet at a given transverse momentum. This restricted multiplicity is linked to that of an unrestricted $\mathrm{q\bar{q}}$-system, $\langle n\rangle_{\mathrm{q\bar{q}}}(L)$, via

$$\langle n\rangle_{\mathrm{q\bar{q}}}(L, \kappa_{\mathrm{cut}}) = \langle n\rangle_{\mathrm{q\bar{q}}}(\kappa_{\mathrm{cut}} + c_{\mathrm{q}}) + (L - \kappa_{\mathrm{cut}} - c_{\mathrm{q}})\left.\frac{\mathrm{d}\langle n\rangle_{\mathrm{q\bar{q}}}(L')}{\mathrm{d}L'}\right|_{L'=\kappa_{\mathrm{cut}}+c_{\mathrm{q}}}\ . \tag{8.6}$$

The two formulations (8.5a) and (8.5b) differ in the scale L used for this effect. They further differ in the definition of the evolution scale κ for the gluon contribution. This ambiguity is hoped to be solved by experiment. The sums in Eq. (8.5) seem to imply incoherence, but the shower coherence is taken into account by a proper choice of the evolution scales κ. Nevertheless, just these coherence effects make the strict distinction between contributions from quark and gluon jets look somewhat arbitrary.

Assuming LPHD, a number of experimental attempts have been published to extract the ratio r from the ratio of charged-particle multiplicities in gluon and quark jets [53–65]. Different methods have been applied to distinguish quark and gluon jets, most of which are different from the theoretical definitions, however:

i) CLEO [58] used the hadronic component of $\Upsilon(1S) \to \gamma gg$ decays and radiative $\Upsilon(3S) \to \gamma\chi'_b$ decays. Comparing $(\chi'_b)_{J=0,2} \to gg$ to $(\chi'_b)_{J=1} \to q\bar{q}g$ they observed a larger mean multiplicity for the $J = 0, 2$ decays than for the $J = 1$ decays.

ii) JADE [53] and AMY [56] used e^+e^- three-jet $q\bar{q}g$ events, assuming the lowest-energy jet to be the quark jet and found slightly larger gluon-jet multiplicity.

iii) HRS [55] and DELPHI [59] selected threefold-symmetric $e^+e^- \to q\bar{q}g$ events in which the three jets had similar energy and obtained a ratio consistent with unity (within large errors).

iv) UA2 [54] compared two-jet gg final states from $p\bar{p}$ collisions to e^+e^- quark jets and found an indication for a slightly larger gluon-jet multiplicity.

v) In their early analysis, OPAL [57] identified equal-energy quark and gluon jets in twofold-symmetric $q\bar{q}g$ events by quark-jet tagging and, for the first time, identified a significant difference.

vi) Furthermore, OPAL [57, 61] used rare $q\bar{q}g$ events in which the q and $\bar{q}$ jets appear in the same hemisphere and compared the inclusive g-jet hemisphere multiplicity to half of that of inclusive $q\bar{q}$ events. This method [61] agrees best to the theoretical definition of inclusive gluon jets [48, 49] and comes close to the NNLO and energy-conservation corrected theoretical predictions.

vii) DELPHI [59] used the scale dependence of gluon-jet and quark-jet multiplicities and obtains a C_A/C_F consistent with the QCD expectation.

viii) CDF [62] used the dijet mass dependence of the gluon jet fraction in two-jet events as extracted from HERWIG, to fit MLLA+LPHD calculations to the average charged-particle multiplicities of dijets, to obtain $r = 1.7 \pm 0.3$ and a LPHD conversion constant $K = 0.57 \pm 0.11$.

ix) OPAL [63] and DELPHI [64] also used the evolution of the total charged-particle multiplicity of symmetric three-jet events (so-called Y-events) with their opening angle, to fit a C_A/C_F in full agreement with QCD. The multiplicity of two-gluon color-singlet systems is then obtained by subtracting the quark contribution. This method is generalized to three-jet events of any topology in [64].

x) OPAL [65], furthermore, uses the so-called jet boost algorithm suggested in [52], to specify which particles in a three-jet event to associate with the gluon jet. The algorithm essentially boosts the qg and $g\bar{q}$ dipoles independently to their respective back-to-back frame and then combines them into an event with the color structure of a gg system in a color singlet, i.e. two back-to-back gluon jet hemispheres. This is possible, since the combined $q\bar{q}$ system has the color structure of the gluon. An unbiased gluon jet is defined by the particles in a cone of half angle $\Theta/2$ around the gluon jet axis, where Θ is the qg opening angle in that frame in which it is equal to the $g\bar{q}$ opening angle (symmetric frame).

The theoretical definition of a gluon jet assumes the production of a gluon pair in a color singlet, which is divided into two hemispheres in which the two gluons move in opposite directions. Among the above methods, only i), vi), ix) and x) are consistent with the theoretical definition and are called "unbiased". All other methods depend on jet resolution scales and are called "biased".

Returning to ix), in both experiments, the full event multiplicity $\langle n\rangle_{q\bar{q}g}$ is measured as a function of the opening angle θ_1 between the two low-energy jets determining the full kinematics ($L_{q\bar{q}}, \kappa_{Lu}, \kappa_{Le}$) of the Y shaped events. Equation (8.6) is then used to determine the restricted multiplicities $\langle n\rangle_{q\bar{q}}(L, \kappa_{cut})$ from the unrestricted $\langle n\rangle_{q\bar{q}}(L)$ as measured in inclusive hadronic e^+e^- events (e.g. Fig. 8.2 after correction for b-events). Finally, the $\langle n\rangle_{gg}(\kappa_{Le})$ and $\langle n\rangle_{gg}(\kappa_{Lu})$ are obtained from inversion of Eqs. 8.5a and b, respectively.

Figure 8.5a shows the OPAL results [63] for $\langle n\rangle_{gg}$ as a function of $Q \equiv \kappa_{Le}$ (open circles) and $Q \equiv \kappa_{Lu}$ (asterisks) extracted from Eqs. 8.5a and 8.5b, respectively. Also included in Fig. 8.5a (as triangular symbols) are the direct measurements of CLEO [58] (method i)) and OPAL [61] (method vi)). The HERWIG and JETSET Monte-Carlo generators for the inclusive charged-particle multiplicity of gg events versus $Q \equiv \sqrt{s}$ predict well the direct CLEO and OPAL measurements (triangles) and the OPAL measurements evaluated according to Eq. 8.5b (asterisks) and interpolate well between these measurements. On the other hand, the OPAL measurements evaluated according to Eq. 8.5a (open circles) lie too high with respect to the models and the direct measurements.

The ratio r can then be determined from e.g. 3NLO fits [50] of the $Q \equiv \sqrt{s}$ dependence of $\langle n\rangle_{q\bar{q}}$ [66] of Fig. 8.2 and $\langle n\rangle_{gg}$ of Fig. 8.5a, respectively. The ratio r of these functions is given as "Central results" in Fig. 8.5b, together with the ratios of the first and second derivatives, $r^{(1)}$ and $r^{(2)}$. All three ratios increase with increasing Q towards the theoretically expected limit of $C_A/C_F = 2.25$. The analytic calculations [50–52] (not shown) more or less follow the trend of Fig. 8.5b. In particular, the results agree with an expected hierarchy $r < r^{(1)} < r^{(2)} < 2.25$ for finite energies, suggesting that higher-order corrections are the smaller the higher the order of the derivative.

In Fig. 8.5c, the average gg event multiplicity $\langle n\rangle_{gg}$ obtained from inversion of (8.5a) by DELPHI [64] is compared to $\langle n\rangle_{q\bar{q}}$, where the θ_1 dependence is transformed into an energy dependence. Besides the DELPHI results, also CLEO, TOPAZ and OPAL results are given. The lower line is a fit of the udsc corrected $\langle n\rangle_{q\bar{q}}$ data by (8.3), the upper line an absolute prediction according to (8.4) with $C_A/C_F = 9/4$. It can be seen that the average multiplicity of a gg-system increases about twice as fast with increasing energy as that of a $q\bar{q}$-system.

Fig. 8.5d (upper part) shows the ratio r of gluon- to quark-system multiplicities, ranging from a value of 1.3 at the lower energies to 1.5 at the highest ones. As to be expected from Fig. 8.5c, the full line [52] describes the data. The LO, NLO and 3NLO predictions clearly overestimate the measured ratio, the more the lower the order. The reason is the sensitivity of the value of the average multiplicity to non-perturbative effects.

The lower half of Fig. 8.5d gives the ratio $r^{(1)}$ of the derivatives of the two average multiplicities with respect to the energy. Here, the deviation between the different predictions is less severe, since the multiplicity slope is less sensitive to

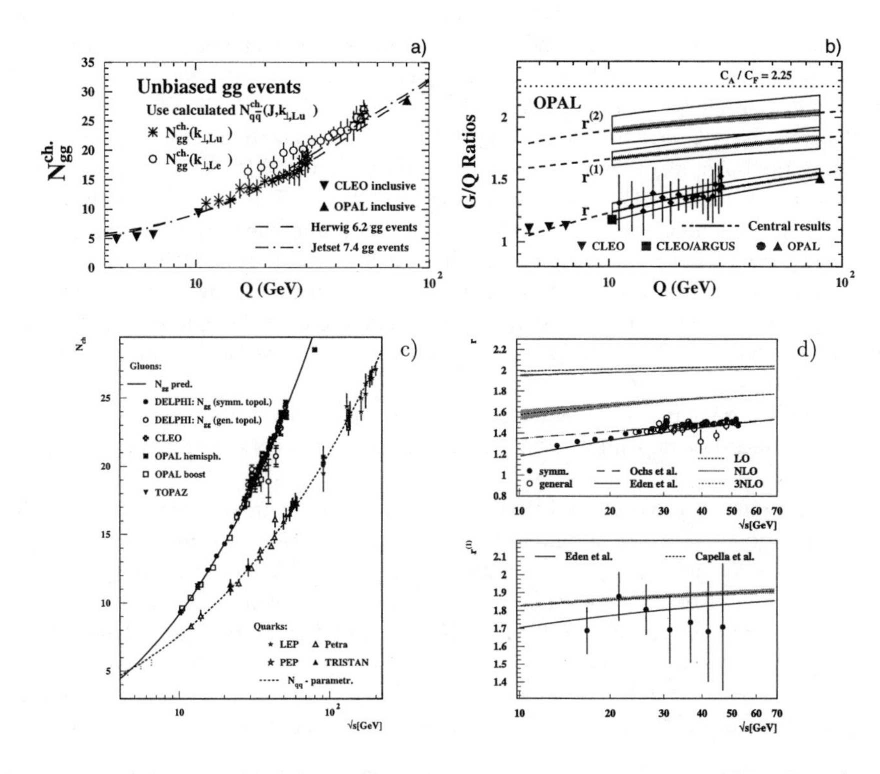

Figure 8.5. a) Average charged-particle multiplicity in gg events extracted from Eqs. (8.5a) (asterisks) and (8.5b) (open symbols), compared to direct CLEO and OPAL measurements and to model predictions [63]. b) The ratios $r, r^{(1)}$ and $r^{(2)}$ extracted from the OPAL results [63], compared to the theoretical limit of $C_A/C_F = 2.25$. c) Average multiplicity in gg events as extracted from Eq. (8.5a) compared to that corresponding to udsc $q\bar{q}$ events [64]. d) The ratio r of the average multiplicities (upper half) and the ratio $r^{(1)}$ of the derivatives, compared to QCD predictions of various order, with and without perturbative effects [64].

non-perturbative effects (see Eq. (8.4)), but data are scarce.

Unfortunately, the two experiments do not consistently resolve the ambiguity between Eqs. (8.5a) and (8.5b). While OPAL [63] favors the second, DELPHI [64] strongly favors the first and even eliminates the second from their analysis of general topology. The main differences are that OPAL anti-tags c- and b-quark events from their sample, while DELPHI applies a hadronization correction and introduces an additional multiplicity off-set to account for b effects. Both experiments, however, observe general agreement with the theoretical expectations and in particular with the expected ratio of the QCD color factors $C_A/C_F = 2.25$.

Extension of the method to three-jet events of general topology [64] gives a more thorough test of the theory. This test is successful, but still cannot definitely resolve the ambiguity between (8.5a) and (8.5b). It does, however allow to increase the available statistics, thus leading to an improved estimate of $\langle n\rangle_{gg}$ for $\sqrt{s}$ between 16 and 52 GeV, as shown in Fig. 8.6, and of the ratio $C_A/C_F = 2.277 \pm 0.02 \pm 0.05$ and $C_A/C_F = 2.093 \pm 0.05 \pm 0.08$ according to Eqs. (8.5a) and (8.5b), respectively.

Using the boost algorithm (method x) sketched above), OPAL [65] obtains results consistent with their earlier ones denoted $N_{gg}^{ch}(k_{\perp,Lu})$ in Fig. 8.5a, but with largely improved precision. Combined with the CLEO data and the OPAL inclusive result at large Q, the ratio r and its evolution can be well reproduced by HERWIG, both on the parton and hadron level, and by the numerical solution [51], while the 3NLO [50] calculation and an exact fixed-α_s [69] calculation lie too high. So, energy conservation and phase space limits, more properly treated in the numerical solution, play an important role.

The QCD prediction of an enhanced particle multiplicity in gluon jets relative to quark jets does not depend on the particle species. This is confirmed by the LEP data [70–72].

If one prefers to avoid the LPHD assumption, one can, of course, try (at the cost of a number of other assumptions) to extract the ratio r directly from subjet multiplicities N_g and N_q in gluon and quark jets [73,74]. In e^+e^- collisions, this ratio was indeed found to be larger than unity [59, 60]. In $p\bar{p}$ collisions, identification of gluon and quark jets is more difficult. D0 [75] uses the energy dependence of the fractions of gluon and quark jets [76] from $\sqrt{s} = 630$ to 1800 GeV to statistically extract $r \equiv (\langle N_g\rangle - 1)/(\langle N_q\rangle - 1) = 1.84 \pm 0.15^{+0.22}_{-0.18}$. This ratio is in agreement with expectations from HERWIG and with a resummation calculation [74].

8.1.1.5 Coherence of gluon radiation

Soft gluon coherence was necessary to describe e.g. the exact shape of the hump-backed distribution in Sub-Sect. 4.1.4 or the energy dependence of the average multiplicity in Fig. 8.2 above.

The cross section for soft gluon radiation in three-jet events varies strongly with the topology of these events [77]. A particular advantage is that soft radiation out

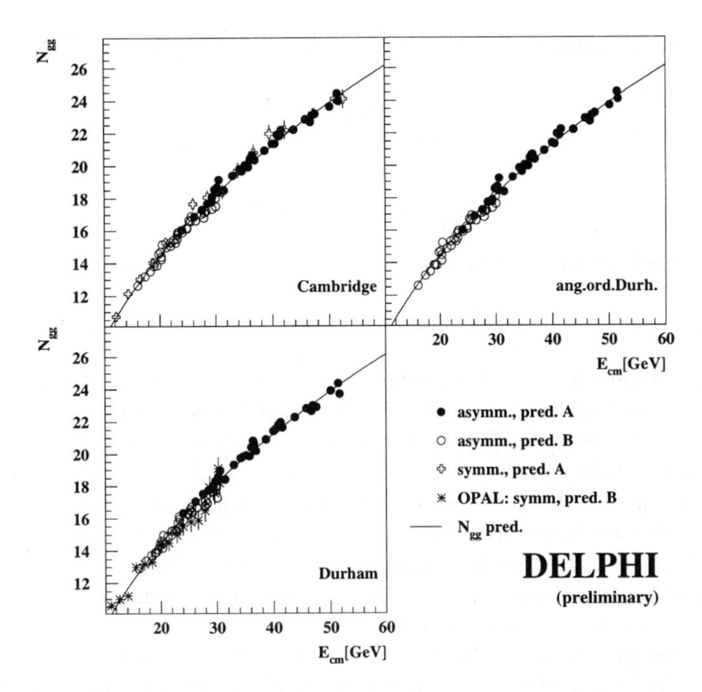

Figure 8.6. The average charged-particle multiplicity $\langle n \rangle_{\text{gg}}$ of a two-gluon color-singlet system as a function of an effective cms energy $\sqrt{s} \equiv E_{\text{cm}}$ as obtained from three-jet event multiplicities, according to three different jet-finding algorithms [64].

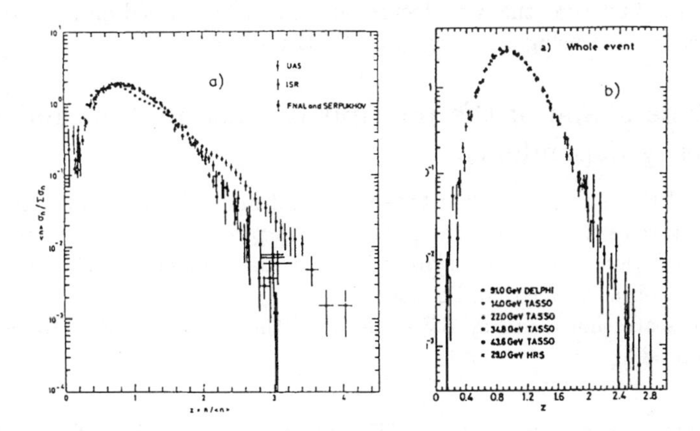

Figure 8.7. The shape of the multiplicity distribution in KNO form for a) $\text{pp}^{\pm}$ collisions [67], b) e^+e^- collisions [68].

of the $q\bar{q}g$ plane is dominated by the simple leading-order QCD contribution [78]

$$d\sigma_3 = d\sigma_2 \frac{C_A}{C_F} r_t , \tag{8.7}$$

where

$$r_t = \frac{1}{4}\{\widehat{qg} + \widehat{\bar{q}g} - \frac{1}{n_c^2}\widehat{q\bar{q}}\} \tag{8.8}$$

with the so-called antenna functions

$$\widehat{ij} = 1 - \cos\Theta_{ij} = 2\sin^2\frac{\Theta_{ij}}{2} .$$

Here, $d\sigma_2$ is the cross section for emission perpendicular to the $q\bar{q}$ axis in a two-jet event, n_c is the number of colors and Θ_{ij} is the angle between partons i and j. The third term in r_t is due to destructive interference. The hard gluon radiated parallel to the quark or antiquark gives $d\sigma_3 = d\sigma_2$, since soft particles cannot resolve the quark and the gluon. A gluon recoiling against a parallel ($q\bar{q}$) pair resembles an initial two-gluon state in a color singlet and gives $d\sigma_3 = C_A/C_F \cdot d\sigma_2$.

In a comparison of the average charged-hadron multiplicities in cones perpendicular to the event plane of three-jet events to those in cones perpendicular to the event axis in two-jet events, DELPHI [79] finds

1. excellent agreement of the data with Eq. (8.7) in absolute value and Θ_{ij} dependence when using the interference term in (8.8), but too large a prediction without,
2. a fitted value of $C_A/C_F = 2.211 \pm 0.014 \pm 0.053$,
3. good agreement with LPHD.

This is the first time the color factor ratio could be verified on a hadron multiplicity measurement using a leading-order QCD prediction.

8.1.2 The shape of the multiplicity distribution and its energy dependence

The shape of the multiplicity distribution P_n, and in particular its deviation from a Poissonian, reveals the amount of correlation in the production of final-state particles. Positive correlations lead to a distribution wider than Poisson, negative correlations to a distribution narrower than Poisson.

Examples are shown in Fig. 8.7 in terms of the so-called KNO (Koba-Nielsen-Olesen) form [80],

$$\psi(z) \equiv \langle n \rangle P_n \tag{8.9}$$

as a function of $z = n/\langle n \rangle$, for $pp^\pm$ [67] and for e^+e^- collisions [68]. While $pp^\pm$ collisions lead to a wide distribution, widening with increasing energy, e^+e^- collisions lead to a relatively narrow distribution with only little energy dependence (see also Figs. 8.3b,c,e).

To arrive at more quantitative statements, the shape can be fitted by an analytical form parametrizing the multiplicity distribution in terms of two or more free parameters or, alternatively, it can be studied in terms of its moments of rank $q \geq 2$.

8.1.2.1 The negative-binomial parametrization

One of the most striking phenomena emerging from studies of multiplicity distributions is the wide occurrence of the negative-binomial distribution

$$P_n(\bar{n}, k) = \frac{1}{n!}\frac{\Gamma(k+n)}{\Gamma(k)}\left(\frac{\bar{n}}{k}\right)^n\left(1+\frac{\bar{n}}{k}\right)^{-n-k} \quad . \tag{8.10}$$

For the two independent parameters, one usually chooses the average multiplicity[1] $\bar{n}$ and a parameter k describing the shape of the distribution. The dispersion D is given by

$$(D/\bar{n})^2 = C_2 - 1 = 1/\bar{n} + 1/k. \tag{8.11}$$

From (8.10), the negative binomial is wider than Poisson as long as k is positive and finite. In the limit $k \to \infty$ the negative binomial reduces to the Poisson distribution

$$P_n = e^{-\bar{n}}\bar{n}^n/n! \quad . \tag{8.12}$$

If k is a negative integer, the negative binomial becomes a (positive) binomial distribution, which is narrower than Poisson.

The usefulness of the negative-binomial distribution in describing full-phase-space multiplicity distributions was already shown in the early seventies [81]. However, the interest was revived by the observation of the UA5 collaboration [82] that the charged-particle multiplicity of non-diffractive pp and $\mathrm{p\bar{p}}$ collisions is well described by the even component of a negative-binomial distribution, from a centre of mass energy $\sqrt{s} = 10$ to $\sqrt{s} = 546$ GeV.

Moreover, the same collaboration found for non-diffractive $\mathrm{p\bar{p}}$ collisions at $\sqrt{s} =$ 546 GeV, that not only the full-phase-space multiplicity distribution appears to be of this type, but also the distribution within central pseudo-rapidity intervals [83] (see Sect. 8.2 below). Since then, negative binomials have been successfully fitted to multiplicity distributions in full and in limited phase space for hh, hA and AA collisions at other energies [13,84–94], as well as for lh [34,95,96] and $\mathrm{e^+e^-}$ collisions [35,97–100]. An example is given in Fig. 8.8a.

Based on these findings, a large number of possible physical interpretations have been given for negative-binomial or negative-binomial-like distributions. Summaries can be found in [101]. In general, the interpretations can be classified [102] as being of (partial) stimulated emission or of cascading type. A number of critical comments

[1]We denote by $\bar{n}$ the average over the distribution as distinct from $\langle n \rangle$, the average over the experimental sample.

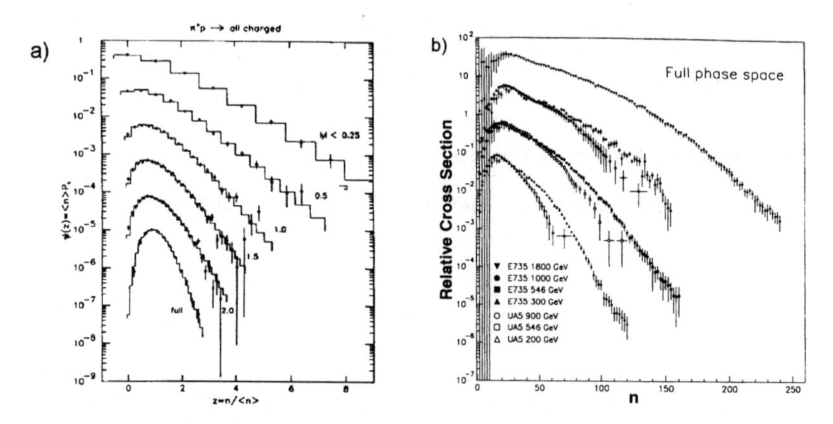

Figure 8.8. a) Charged-particle multiplicity distribution for non-single diffractive π^+p data $\sqrt{s} = 22$ GeV in different rapidity intervals $|y| < y_{\text{cut}}$ and full phase space, plotted in KNO form, together with the best-fit negative binomials [85]; b) Charged-particle multiplicity distribution for non-single-diffractive p$\bar{\text{p}}$ data as measured by UA5 and E735 at various collider energies. Data from the two experiments at the corresponding energy are normalized to each other over a range of n just past the peak of the distribution [91].

on the applicability of negative binomials in full phase space, mainly based on the influence of the conservation laws, can be found in [103].

At the highest energies of $\sqrt{s} = 900$ GeV [90] and 1800 GeV [91,93], a shoulder is building up at high n (Fig. 8.8b), so that the distribution can no longer be fitted by a single negative binomial. This is interpreted in terms of the presence of two components, one corresponding to conventional soft physics, the other to QCD semi-hard mini-jets [104], one to a pure birth, the other to a Poisson process [105] or, alternatively, to multiparton collisions and multichain production [91,106].

Another approach [107] is to understand particle production as a two-cascade process, where the first cascade is responsible for the partons or strings, the second for their fragmentation into hadrons. The composition of two Poisson distributions, each describing one of these two Markov type branching processes, can lead to oscillations in P_n at the upper SPS and Tevatron energies.

A similar structure, though less pronounced, is becoming visible in e^+e^- collisions at the Z^0 [99], but is not observed in DIS [34] so far (see Sect. 8.2 below).

8.1.2.2 Energy dependence (KNO scaling, NO scaling or log-KNO scaling?)

Koba, Nielson and Olesen [80] have shown that, if Feynman scaling [108] holds, the function $\psi(z)$ of (8.9) becomes asymptotically ($n \to \infty$, $\langle n \rangle \to \infty$, z fixed) independent of $\sqrt{s}$. Note that (8.9) corresponds to rescaling the P_n curves corresponding to the collision energy by stretching the vertical axis and shrinking the horizontal axis both by $\langle n(s)\rangle$, thus maintaining normalization. If $\psi(z)$ is indepent of $\sqrt{s}$, then also its normalized moments $C_q = \langle n^q\rangle/\langle n\rangle^q = \int_0^\infty z^q\psi(z)\mathrm{d}z$.

Even though the original derivation from Feynman scaling turned out to be wrong [109], Feynman scaling is known to be violated (see Sub-Sect. 4.3.1), and the increase of $\langle n\rangle$ with s faster than logarithmic in Fig. 8.1, KNO scaling is often claimed for full-phase-space and single-hemisphere multiplicity distributions in e^+e^-, lepton-hadron and medium energy (ISR) hadron-hadron collisions [103, 110]. It was demonstrated [111] even earlier, however, that KNO scaling should hold approximately more generally in any model based on a scale-invariant stochastic branching process with an energy-independent coupling constant.

With high-energy (SPS and Tevatron) $p\bar{p}$ collisions, UA5 [67, 84, 90, 112] and E735 [91] and to some extent also UA1 [113] could show that the scaling at ISR energies was accidental and that KNO scaling is in fact violated in hh collisions up to at least 2 TeV (see Fig. 8.9a for the energy dependence of the C_q moments up to 0.9 TeV). In e^+e^- collisions, we may still be in the domain of accidental scaling near 100 GeV and Fig. 8.9b indeed shows rather constant C_q moments [28, 29]. However, a wide minimum at $\sqrt{s} \approx 30-50$ GeV and an onset of an increase above that cannot be excluded even in e^+e^- collisions.

In Fig. 8.3 b and c, the ratio $\langle n\rangle/D$ [114] and the normalized factorial moment F_2 for the current hemisphere in DIS are given [34] as a function of the hadronic energy

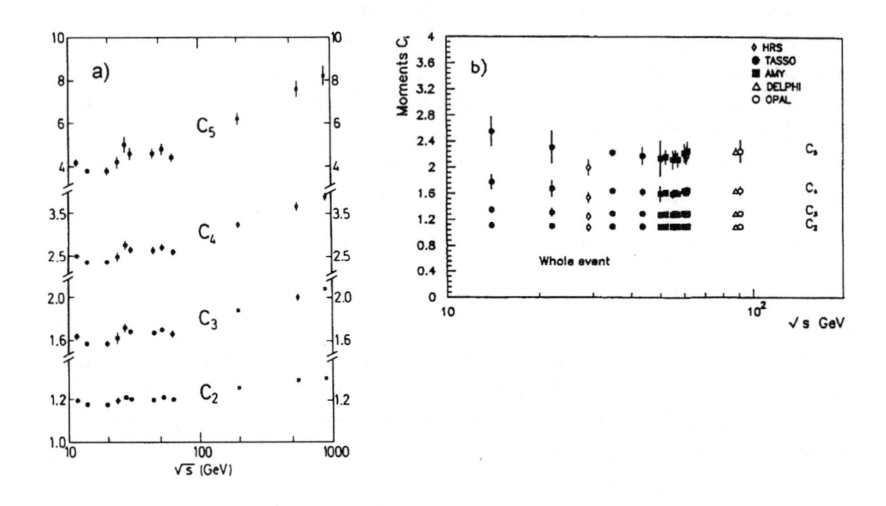

Figure 8.9. The energy variation of the C-moments of the charged-particle multiplicity distribution for a) non-single diffractive $pp^\pm$ collisions [90], b) e^+e^- collisions [29].

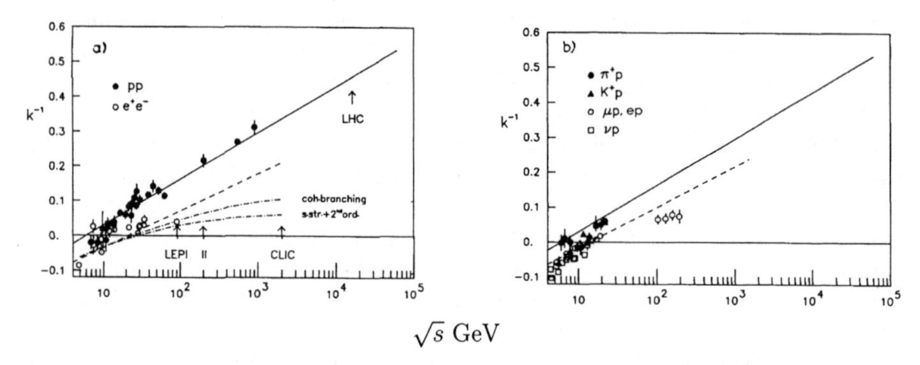

Figure 8.10. a) Negative-binomial parameter 1/k for non-single diffractive $\mathrm{pp}^{\pm}$ collisions and for $\mathrm{e}^+\mathrm{e}^-$ annihilation as a function of $\ln\sqrt{s}$. The full line is a linear fit to the $\mathrm{pp}^{\pm}$ data, the dashed line to the $\mathrm{e}^+\mathrm{e}^-$ data below 50 GeV. The dash-dot lines correspond to the predictions from coherent gluon branching and single-string plus second order corrections in JETSET, as indicated. b) Negative-binomial parameter $1/k$ for non-diffractive $\pi^+\mathrm{p}$ collisions and for lh collisions as a function of $\ln\sqrt{s}$. The full line is the linear fit to the $\mathrm{pp}^{\pm}$ data of sub-figure (a). The dashed line is a fit to the μp data (below 50 GeV) [117].

W, in Fig. 8.3e $\langle n\rangle/D$ for $\mathrm{e}^+\mathrm{e}^-$ annihilation as a function of $\sqrt{s}$ [29]. Both moments should be constant if KNO scaling holds. The $\langle n\rangle/D$ data are indeed constant above ~25 GeV, within errors, but a steady rise is observed for F_2. Both agree with the JETSET expectations for $\mathrm{e}^+\mathrm{e}^-$ single-hemisphere multiplicities. It is interesting to note that also next-to-leading order of QCD violates KNO scaling at HERA energy.

P_n is a discrete distribution, but at high energies, it can be approximated by a continuous probability density $f(x)$ either as $P_n \simeq f(x=n)$ or as $P_n \approx \int_n^{n+1} f(x)\mathrm{d}x$. A formulation for discrete distributions, valid also at finite energies, was proposed by Golokhvastov [115] (KNO-G scaling).

In connection with the negative binomial it is important to note that, according to Feynman scaling, it should be the factorial moments

$$F_q \equiv \langle n(n-1)...(n-q+1)\rangle/\langle n\rangle^q = k(k+1)...(k+q-1)/k^q \qquad (8.13)$$

which are expected to be constant [80, 116], and the reduced moments C_q only in the approximation $\langle n\rangle \approx \bar{n} \gg q$. This has been an infortunate misunderstanding for a long time, which considerably held up development in this field. In fact, from Eq. (8.11), a constant $\langle n\rangle/D$ at non-zero $1/\langle n\rangle$ would require an *increasing* $1/k$, in particular up to the LEP energy range! Furthermore, contrary to the C_q, the F_q and k tend to finite limits as $\bar{n}\to 0$ and therefore provide a better measure of the shape of a multiplicity distribution at small $\bar{n}$.

In Fig. 8.10, a compilation [117] of the parameter $1/k$ is given as a function of $\ln\sqrt{s}$ for hh, lh and $\mathrm{e}^+\mathrm{e}^-$ collisions. At given $\sqrt{s}$, the value of $1/k$ is lower for $\mathrm{e}^+\mathrm{e}^-$ and lh than for hh collisions. However, in all cases, $1/k$ increases with increasing energy from a negative value at low energies to a positive one above a certain Poisson-like ($1/k = 0$) transition point. The transition point is at $\sqrt{s} \approx 5$ GeV for hh and $\sqrt{s} \approx 20 - 30$ GeV for e^+e^- collisions, but the increase is quite similar up to about 50 GeV. At LEP energies, $1/k$ tends to flatten for e^+e^- collisions [35, 99, 100]. So, contrary to hh collisions, an approach to KNO scaling may be observed for e^+e^- collisions. It is important to note, however, that the flattening takes place at a positive $1/k$ value. So, the distribution is wider than Poisson, even for e^+e^- collisions.

The increase of $1/k$ for e^+e^- collisions up to 45 GeV and flattening above that is reproduced by both the JETSET (Fig. 8.10) and HERWIG models. At higher energies more than 2nd order corrections are needed but coherent branching predicts negative-binomial-like multiplicity distributions up to the highest energies ($\sqrt{s} = 2$ TeV) [118]. Above $\sqrt{s} \approx 25$ GeV these are wider than Poisson ($1/k > 0$).

The KNO form (8.9) can be generalized [119] as

$$P(n, s) = \frac{1}{\lambda(s)} \psi\left(\frac{n + c(s)}{\lambda(s)}\right) , \tag{8.14}$$

with an energy-dependent scale parameter $\lambda(s)$ corresponding to $\langle n \rangle$ and an energy-dependent location parameter $c(s)$, associated with leading particle effects. Even though there is no experimental evidence for an energy-dependent shift at very high energies (i.e. $c(s) = 0$), this form has led [120] to the alternative ansatz

$$P(\ln n, s) = \frac{1}{\lambda(s)} \varphi\left(\frac{\ln n + c(s)}{\lambda(s)}\right) . \tag{8.15}$$

This ansatz is based on Polyakov's original self-similar, scale-invariant branching model interpretation [111] leading to the negative-binomial type scaling function (gamma distribution in z^μ, i.e. the rescaled multiplicity to the power μ)

$$\psi(z) \propto a(z) \exp(-z^\mu) , \quad \mu > 1 , \tag{8.16}$$

with $a(z)$ being a monomial in z. In the language of QCD, taking into account higher-order effects responsible for energy-momentum conservation in parton jets [121], this reads

$$\psi(z) = \frac{\mu D^{\mu k}}{\Gamma(k)} z^{\mu k - 1} \exp(-[Dz]^\mu) , \tag{8.17}$$

with $k = 3/2$, $\mu = (1 - \gamma_0)^{-1}$ and D being a γ_0 dependent scale parameter.

Obviously, this distribution shows KNO scaling (s-invariance) for fixed coupling. On the contrary, violation of KNO scaling is expected from the running of $\alpha_s(s)$ in the form of a tail of $\psi(z)$ widening with increasing s (Fig. 8.11a and b).

Rewriting the Polyakov-Dokshitzer form (8.17) as

$$\psi(x) = \frac{\mu}{\Gamma(k)} \exp(k\mu x - e^{\mu x}) , \quad x = \ln(Dz) , \tag{8.18}$$

the multiplicity scaling violated by QCD effects is recovered by plotting $\mu^{-1}\psi(\mu x)$ as a function of μx, i.e. by a location and scale change of $P(\ln n, s)$ governed by the QCD anomalous dimension (Fig. 8.11c). This property is referred to as log-KNO scaling [120].

It comes as a surprise that even the non-diffractive $\mathrm{p\bar{p}}$ collisions [90] fall onto the $\mathrm{e^+e^-}$ curve (see Fig. 8.11d).

Restricting oneself to n values above the shoulder observed in the E735 data in Fig. 8.8b, log-KNO scaling is observed also there, but the scaling function looks different ($k = 1/2$, μ decreasing from 2.2 to 1.7 between 300 and 1800 GeV). Note

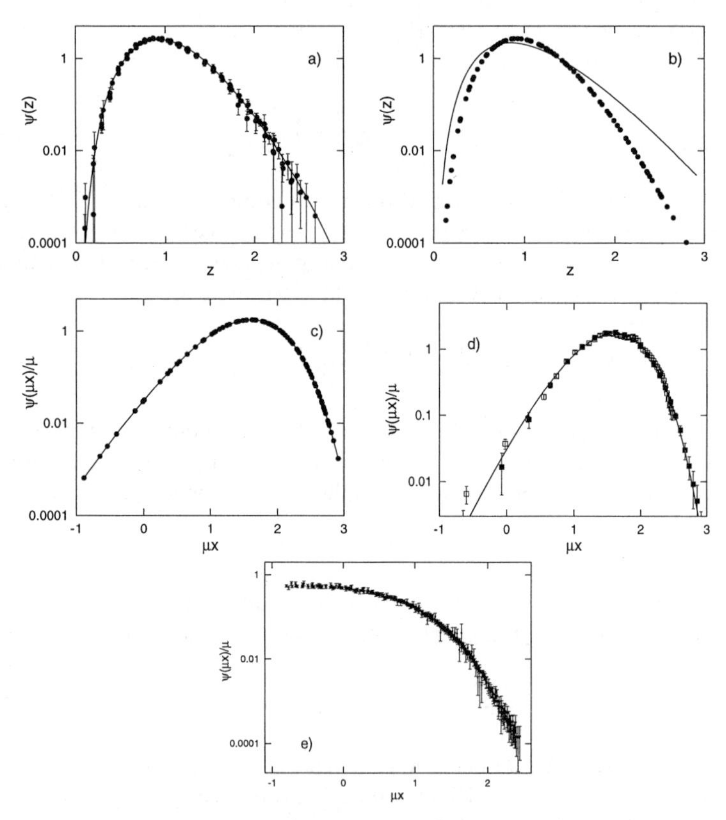

Figure 8.11. The LEP1 data [35, 99, 100, 122] for the full-phase-space multiplicity distributions (dots) compared to Eq. (8.17) with $k = 5$ and $\mu = 5/3$ (solid line). b) $\psi(z)$ with $\mu = 5/3$ (dots) evolves to $\psi(z)$ with $\mu = 1$ (solid line) at $s \to \infty$. c) The scaling behavior is recovered in $\psi(\mu x)/\mu$ versus μx. d) Comparison of the UA5 non-diffractive data at $\sqrt{s} = 546$ GeV [90] and the OPAL data at $\sqrt{s} = 91.2$ GeV [35]. The solid curve is (8.18) with $k = 5$. e) Log-KNO scaling of the E735 data [91, 120].

that scaling is observed within the two $p\bar{p}$ experiments, but not between the two. This difference, already visible in Fig. 8.8b, is a severe experimental problem, which will have to be solved by future experiments.

8.1.3 Higher moments

Alternatively, the shape of a distribution can be analyzed in terms of its moments (see Chapter 7). Such an approach is even preferred, both theoretically (generating function) and experimentally (stability).

The normalized factorial moment F_q of rank q of the multiplicity distribution P_n expresses the normalized phase-space integral over the q-particle density function. If particles are produced independently, the multipicity distribution is a Poissonian (for which $F_q \equiv 1$ for all $q > 1$). If the particles are positively correlated, the distribution is broader than Poisson and the normalized factorial moments are larger than one.

The normalized cumulant moment K_q expresses the normalized phase-space integral over the q-particle correlation function, i.e., it describes the genuine correlations between q particles.

Since $|K_q|$ and F_q increase rapidly with increasing q, it is useful to define [123] the ratio $H_q = K_q/F_q$ reflecting the genuine q-particle correlation relative to the q-particle density.

The H_q calculated for the gluon multiplicity distribution at different orders of perturbative QCD [123] are given in Fig. 8.12a. For DLLA, H_q decreases to 0 as q^{-2}. For MLLA, H_q decreases to a negative minimum at $q = 5$, and then rises to approach 0 asymptotically. For NLLA, H_q decreases to a positive minimum at $q = 5$ and goes to a positive constant value for large q. For NNLLA, finally, H_q decreases to a negative first minimum for $q = 5$ and shows quasi-oscillation around 0 for $q > 5$. Assuming the validity of LPHD, such a behavior is also expected for the charged-particle multiplicity distribution.

Indeed, a negative minimum for $q = 5$ followed by quasi-oscillations about $H_q = 0$ for higher q is observed by SLD [124] (Fig. 8.12b). Earlier, the H_q moments were calculated [125,126] from charged-particle multiplicity distributions of e^+e^- experiments between 22 and 91 GeV. Similar behavior with a negative first minimum followed by quasi-oscillation about 0 was obtained. Since the behavior of the H_q agrees qualitatively with that predicted in NNLLA, it seems attractive to interpret this agreement as a confirmation of the NNLLA prediction, and SLD [124] and others [31] indeed adopted this interpretation. However, the same behavior was observed in hadron collisions between 20 and 900 GeV [125,127] and even in hA and AA collisions [128], so that the QCD interpretation has to be taken with a grain of salt.

In Fig. 8.13 it is shown [47] that also H_q calculated from JETSET 7.4 PS agree well with the data in spite of the parton-shower modelling not being at the NNLLA order of pQCD and, therefore, not expected to oscillate! Also the H_q calculated from HERWIG 5.9 show an oscillatory behavior, but with amplitudes much larger than in

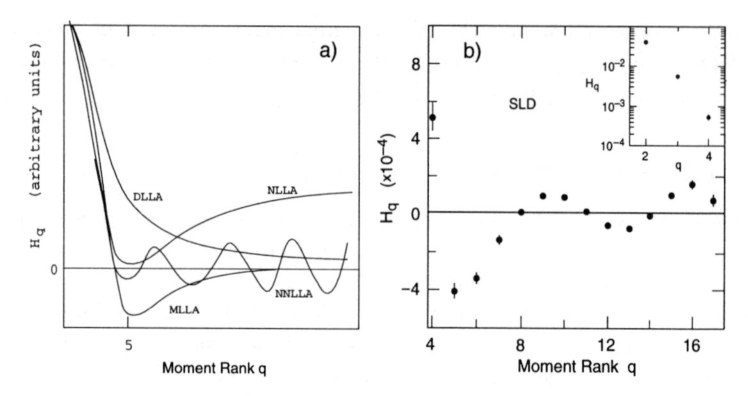

Figure 8.12. a) Predictions at different order of perturbative QCD for the H_q as a function of the rank q. [123] b) H_q versus q measured by the SLD collaboration. [124]

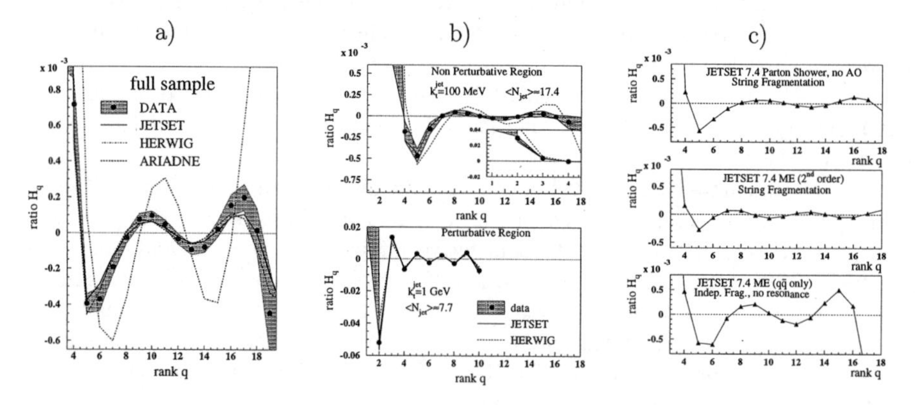

Figure 8.13. a) H_q obtained from the charged-particle multiplicity distribution. ; b) H_q for sub-jet multiplicities, for $k_t^{\rm jet} = 0.1$ and 1 GeV (top to bottom). c) H_q for the charged-particle multiplicities obtained from different Monte-Carlo generators [47].

the data.

One can assume that the sub-jet multiplicity distribution is related to the parton multiplicity distribution at the energy scale corresponding to the jet resolution. By choosing a scale where perturbative QCD should be applicable ($\gtrsim 1$ GeV), one should be able to test the pQCD predictions for the behavior of the H_q without the LPHD assumption. It has been shown [30, 129] that sub-jet mean multiplicities agree well with the NLLA prediction in the perturbative region, i.e., for values of transverse momentum cut $k_t^{\rm jet} \gtrsim 1$ GeV.

The sub-jet H_q distributions are shown in Fig. 8.13b for two typical values of $k_t^{\rm jet}$. A first minimum at $q = 4$ followed by quasi-oscillation about 0, is seen for $k_t^{\rm jet} = 100$

MeV. This behavior is qualitatively similar to that for the charged-particle multiplicity distribution. For $k_t^{jet} = 1$ GeV these oscillations have disappeared, however! In the perturbative region, H_q alternates between positive and negative values for each q (mind the scale!), but does not show the characteristic minimum at $q = 5$.

Although good agreement with NLLA is found for the production rates of soft gluons, none of the predictions made at different orders of pQCD for the H_q is observed at LEP1 energy. Even though pQCD describes collective processes (as the average sub-jet multiplicity), it seems to be unable to describe the shape of these distributions.

Given the success of JETSET, one can attempt to find that aspect of the Monte Carlo generators which is responsible for the agreement, e.g. by varying several options in JETSET and studying their influence on the behavior of the H_q [47]. Indeed, in all cases, H_q has a negative first minimum near $q = 5$ and quasi-oscillations for higher q. Some examples are shown in Fig. 8.13c. This Monte-Carlo study shows that one can reproduce the dip at $H_q \approx 5$ without the need for NNLLA. The behavior may be related to energy-momentum conservation included in the MC models, but properly approximated within the analytic approach first only in NNLLA.

The parton showering of ARIADNE is very similar to that of DLLA, but with the additional advantage that exact energy-momentum conservation is applied at each branching [130]. In [131,132], ARIADNE is, therefore, allowed to shower down to the very low scale $Q_0 \approx \Lambda_{eff}$ of a few hundred MeV. In this approach, the hadronization model is replaced by a more extensive parton shower and strict LPHD. Qualitative agreement with the data of Figs. 8.13a and b suggests that energy-momentum conservation is indeed responsible for a large part of the effect.

In addition, the reason for the transition from the narrow odd-even oscillation in the perturbative region (lower Fig. 8.13b) to the wide quasi-oscillation pattern observed in the non-perturbative region (upper Fig. 8.13b) is beautifully demonstrated in [132] be means of the above parton-level ARIADNE simulation, as well as by analytical MLLA calculations including energy (but not momentum) conservation. On the hadron level, the Poissonian transition with the H_q becoming positive takes place at $\sqrt{s} \approx 30$ GeV (see Fig. 8.10). For jets produced at $\sqrt{s} = 91$ GeV, this takes place at $k_t^{jet} \approx 0.3$ GeV [132]. For larger k_t^{jet} values, energy-momentum conservation forces the multiplicity distribution to be narrower than Poisson, thus leading to odd-even oscillations.

For lower k_t^{jet} values, a transition from jets to individual hadrons takes place. The strong variations of all moments in that region is well reproduced by the parton-level ARIADNE, with low cut-off $k_t^{jet} \geq Q_0 \gtrsim \Lambda$. Because of the large α_s in this region, extension of the analytical calculations into that region is not justified a-priori. However, the comparison of the analytical DLLA and MLLA results with ARIADNE show that convergence is not in danger there, so that main characteristics of a unified description of the correlations between hadrons and the correlations between jets can indeed be obtained from the MLLA evolution equations, as well.

Still, it would be important to see how much is really due to QCD and how much to mere energy-momentum conservation. If indeed demonstrated by QCD showering, the question remains, why the oscillation is also visible in other reactions.

In e^+e^- collisions at LEP1, the oscillation of H_q can also be reproduced quantitatively by a weighted superposition of negative binomials, associated with two- and more-jet production [47, 126] or by the simple parametrization of the multiplicity distribution in terms of a truncated modified negative-binomial distribution [133] (see Sub-Sect. 8.2.4 below for a definition) and it has been shown [47] that the tail truncation of the multiplicity distribution is responsible for most of the oscillation and part, but not all, of the first minimum.

More generally, it has been shown [134] that H_q oscillations appear if the multiplicity distribution at large $n/\langle n\rangle$ decreases faster than exponential (see also Sect. 8.2.3 below). This is not generally the case in non-truncated single-negative-binomial type distributions. A decrease faster than exponential can be due to (natural, i.e., limited statistics) truncation as above, but also dynamical reasons exist. In heavy-ion collisions [135], with ν independent elementary collisions of average multiplicity $\bar{n}$ and dispersion d, average multiplicity and reduced dispersion are

$$\begin{aligned} \langle n\rangle &= \langle\nu\rangle\bar{n} \\ \text{and}\quad D^2/\langle n\rangle^2 &= \frac{\langle\nu^2\rangle - \langle\nu\rangle^2}{\langle\nu\rangle^2} + \frac{1}{\langle\nu\rangle}\frac{d^2}{\bar{n}^2}\,. \end{aligned} \tag{8.19}$$

In general, the first term of (8.19) (fluctuation in the number of elementary collisions) is large, so that the second term (contribution from within the individual collisions) can be neglected. If, however, the analysis is restricted to a fixed impact parameter (e.g. very central collisions), the distribution becomes particularly narrow, due to the suppression by $1/\langle\nu\rangle$ of the second term of (8.19), with a tail dropping faster than exponential, thus leading to oscillations.

8.2 Limited phase-space domains

Multiplicity distributions can be studied in full phase space as well as in restricted domains. Firstly, if different basic sub-processes contribute in different regions of phase space, a study in various limited parts is appropriate. Secondly, while energy-momentum and charge conservation influence the multiplicity distribution for full phase space, the distribution in the central region is largely free from these constraints and hence can give a more direct measure of the production mechanism.

8.2.1 Shape and energy dependence

In Fig. 8.14a-d, the UA5 charged-particle multiplicity distribution at 546 GeV [83] is given for several central pseudo-rapidity intervals $|\eta| < \eta_{\rm cut}$. The distributions in

Figs. 8.14a,c and d are in KNO form, those in Fig. 8.14b in P_n vs. n. One observes a strong dependence of the shape of the distribution on the η_{cut} value. In KNO form, the distribution widens as η_{cut} is reduced. A similar behavior is seen for e$^+$e$^-$ collisions at $\sqrt{s} = 29$ GeV [97] for rapidity intervals $|y| < y_{\mathrm{cut}}$ in Fig. 8.14e.

In Fig. 8.14a-d, furthermore, four models are compared to the UA5 data, PYTHIA, FRITIOF, the Dual Parton Model and a Three-Fireball Model [136, 137]. At first sight (on logarithmic scale), the models are all reasonably good and even follow the change of the shape with decreasing η_{cut}. At second inspection, PYTHIA, FRITIOF and TFM tend to be too wide, in particular at larger η_{cut}, while DPM tends to be too narrow. For e$^+$e$^-$, both JETSET with 2nd order corrections and HERWIG have

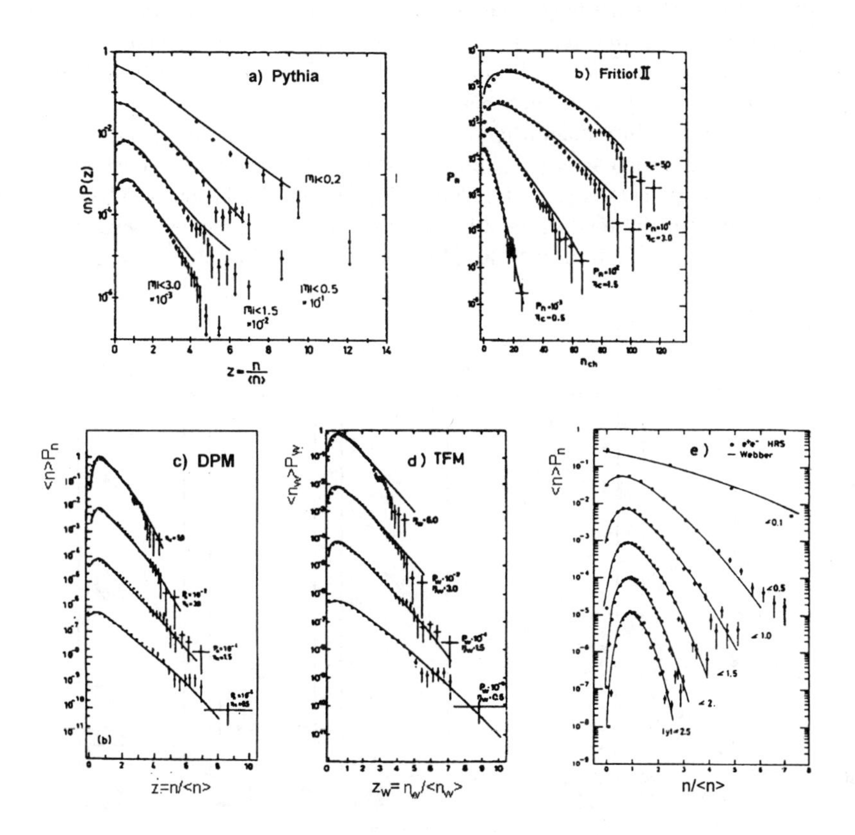

Figure 8.14. Charged-particle multiplicity distribution for a)-d) non-single diffractive p$\bar{\mathrm{p}}$ collisions in central pseudo-rapidity intervals $|\eta| \leq \eta_{\mathrm{cut}}$ [83], compared to PYTHIA, FRITIOF, DPM and TFM, e) e$^+$e$^-$ collisions at $\sqrt{s} = 29$ GeV in central rapidity intervals $|y| \leq y_{\mathrm{cut}}$ [97], compared to HERWIG.

been compared [118] and, as shown in Fig. 8.14e for the case of HERWIG, describe the distributions at 29 GeV.

On the other hand, Fiałkowski [138] shows that the dependence on the rapidity cut can be understood already from a "minimal model" of independent cluster emission, when the full phase space multiplicity distribution is taken from experiment.

The multiplicity distribution in the central region of rapidity was first suggested to obey KNO scaling between ISR and SPS collider energies. A comparison of the NA22 data at 22 GeV [85] with UA5 Collider data [90] then allowed for a systematic study over a large energy range. In Fig. 8.15a, the energy variation of the C_2 to C_4 moments of non-diffractive charged-particle multiplicity distributions is shown for $0.25 \leq \eta_{\rm cut} \leq 2.5$. The moments for the two bigger intervals are seen to increase approximately linearly with $\ln \sqrt{s}$, while a decrease (possibly to a minimum at SPS energies) is seen in the two smaller intervals. This decrease is a trivial consequence of all $\langle n^q \rangle$ becoming equal for $\langle n \rangle < 1$ and the C_q being dominated by a term behaving as $\langle n \rangle^{-q+1}$ [139] (see further in Sect. 8.4 below).

Except for the absolute magnitude of the moments being lower than in hh collisions at the same hadronic energy, a similar result is obtained for the current region of deep-inelastic e^+p scattering by H1 [34] in Fig. 8.15b. Violation of KNO scaling is seen in the intermediate-size intervals. This result is in agreement with those of earlier analyses in the Breit frame [40, 140]. So, KNO scaling does not hold, even here.

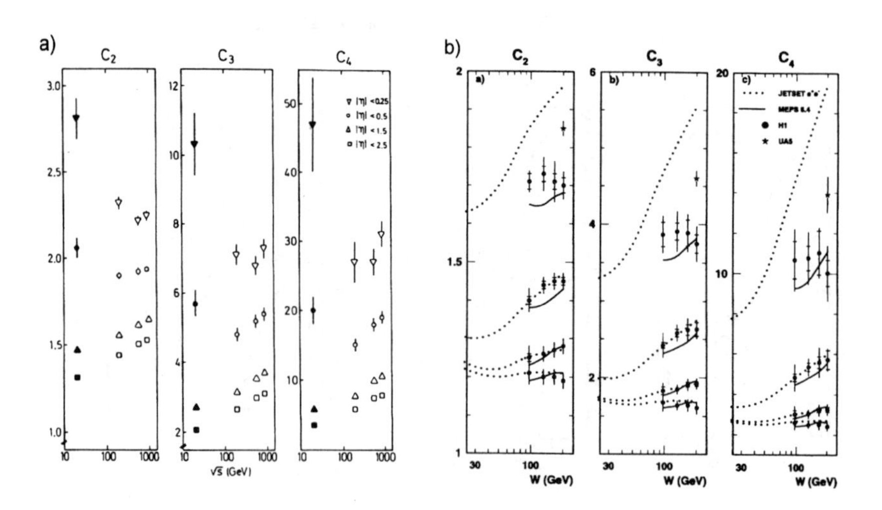

Figure 8.15. a) The energy variation of the moments C_2 to C_4 of the $pp^{\pm}$ charged-particle multiplicity distribution, in the symmetric pseudo-rapidity intervals indicated [85, 90]; b) Same for DIS current fragmentation in domains (from top to bottom) $1 < \eta^* < 2$, $1 < \eta^* < 3$, $1 < \eta^* < 5$ [34]. The $\sqrt{s} = 200$ GeV $p\bar{p}$ point is shown for $1 < |\eta^*| < 2$.

A widening of the multiplicity distribution with $\sqrt{s}$ and/or with decreasing rapidity interval as observed in Figs. 8.7a and 8.14 has been obtained from DPM. [141] The violation of KNO scaling as shown in Fig. 8.7a is now also an ingredient of the FRITIOF Monte Carlo model. It can, furthermore, be more or less described by a statistical bootstrap model [142], within cluster models [138, 143], a three-fireball model [136], the bremsstrahlung model of [144], geometrical models [127, 145, 146], two-component models [147–152], branching models [153–157], a quantum-statistical approach [158], or chiral symmetry breaking [33]. Future results from LHC or a linear collider may be able to distinguish. An older cluster model [159] predicts a decrease of the moments with increasing energy and can, therefore, be excluded.

A test more stringent than the violation of KNO scaling for full phase space in Fig. 8.7a is the way KNO scaling is violated in central rapidity intervals in Fig. 8.15. Apart from DPM, which predicted Fig. 8.15 to look just as it looks, not many models stick out their head. The authors of [144] blame KNO scale breaking on finite energy corrections due to energy conservation and, therefore, would expect less energy dependence in the central region, but the data show an increase also there. In the "minimal" cluster model of [138], on the other hand, the increase of the moments should be similar in the central region as in full phase space (except in the very center). Some hope to be able to describe the trend of Fig. 8.15 exists within the Three Fireball Model [136], even though absolute values are far off, in particular in the central region.

What is said about the violation of KNO scaling, of course, only holds for hh collisions. However, there is no doubt that scaling of the fragmentation function is broken also in lh and e^+e^- collisions, in this case due to QCD corrections. In Fig. 8.15b, the dotted line corresponds to the JETSET representation for e^+e^- collisions, the full line to the HEPS 6.4 generator [37] for DIS.

KNO scaling has kept many of us busy for a long time and still is keeping some of us, but we believe that it has to be replaced by NO scaling. Of course, it is difficult to imagine that the distributions will keep widening for ever. So, scaling may still set in at some future energy [153, 160]. For a generalization to a logarithmic scaling in case of multiplicative cascade processes see [120] in Sub-Sect. 8.1.2.

8.2.2 Negative-binomial fits

In Fig. 8.16, the parameter $1/k$ is shown as a function of $\ln\sqrt{s}$ for non-single diffractive $pp^\pm$ collisions [83, 85, 90] in central pseudo-rapidity η intervals (and full phase). The energy dependence is flatter for $|\eta| < 0.5$ and full phase space than for intermediate window sizes, but never flat or decreasing as were the C_q moments in Fig. 8.15. Furthermore, $1/k$ is shown in the corresponding rapidity intervals for e^+e^- [97, 100] and μp [95] data. The following observations can be made:

1. At similar energy and interval size, e^+e^- collisions lead to a narrower multiplicity distribution (lower $1/k$) than hh collisions.

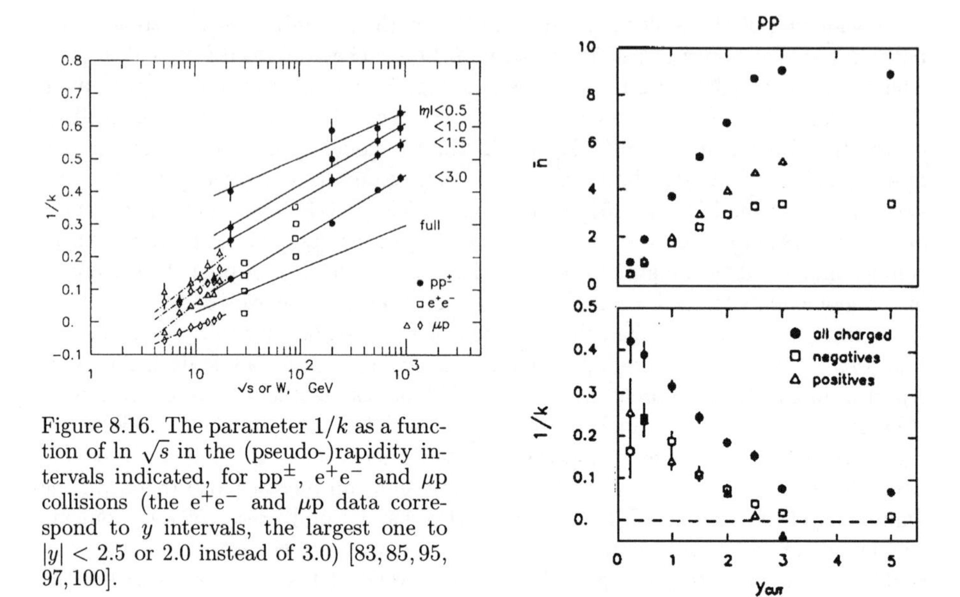

Figure 8.16. The parameter $1/k$ as a function of $\ln \sqrt{s}$ in the (pseudo-)rapidity intervals indicated, for $pp^{\pm}$, e^+e^- and μp collisions (the e^+e^- and μp data correspond to y intervals, the largest one to $|y| < 2.5$ or 2.0 instead of 3.0) [83, 85, 95, 97, 100].

Figure 8.17. The negative-binomial parameters determined for the multiplicity distribution of all charged, negative and positive particles in symmetric rapidity intervals $|y| < y_{\text{cut}}$, as a function of y_{cut}, for pp collisions at $\sqrt{s} = 22$ GeV. Full phase space corresponds to $y_{\text{cut}} = 5$ [85].

2. The multiplicity distribution is significantly wider than Poisson also for e^+e^- collisions, in particular for small central rapidity intervals.
3. μp collisions lead to $1/k$ values between those for hh and e^+e^- collisions.
4. The increase of $1/k$ with $\ln \sqrt{s}$ or $\ln W$ is similar for all types of collisions.

The results from the negative-binomial fits obtained for all charged particles as well as for negatives and positives taken separately are displayed in Fig. 8.17, for non-single diffractive pp data at $\sqrt{s} = 22$ GeV [85]. The values for the parameters $\bar{n}$, $1/k$ are shown as a function of y_{cut}.

As expected from the general shape of the rapidity distribution, the average multiplicity $\bar{n}$ first increases linearly with increasing y_{cut} and saturates when approaching the maximum rapidity range.

The parameter $1/k$ decreases for increasing y_{cut}. This (together with a decreasing $1/\bar{n}$) corresponds to the narrowing of the KNO distribution with the increase of the rapidity range as seen for the UA5 and HRS data in Fig. 8.14. From the charges taken separately, it can be seen that the decrease is slightly faster for positives (triangles) than for negatives (squares). In the region $y_{\text{cut}} < 2$, the $1/k$ values for positives and negatives are compatible with each other, but a factor two *smaller* than those for all charged particles combined.

A further important observation (not shown) [85,161] is that $1/k$ does not considerably depend on the size $\Delta\varphi$ of an azimuthal angular interval in the plane transverse to the collision axis.

8.2.3 Interpretation

8.2.3.1 Geometrical models

A difference in multiplicity distribution for e^+e^- and hh collisions is most naturally described by geometrical models [145, 146], where the widening in hh collisions is interpreted as due to the superposition of different impact parameters. It is an important challenge for these models, however, that even the e^+e^- distributions are wider than Poisson!

8.2.3.2 (Partial) stimulated emission

The last two observations in Sub-Sect. 8.2.2 above ($1/k$ smaller for positives and negatives taken separately and independent of $\Delta\varphi$) disfavor the class of (partial) stimulated emission models [101, 102] as an interpretation of the success of the negative binomial. In these models, $1/k$ is the fraction of particles already present that is responsible for the production of a further one. This fraction would be expected to *increase* if the sample is reduced to identical particles or to a smaller angular region.

Of course, the two observations are not sufficient to fully exclude stimulated emission, since this could still operate at the level of (neutral) bosons decaying into the observed charged particles. This decay, however, represents a cascading mechanism, so that then both mechanisms would be needed.

8.2.3.3 Pure cascading

Ekspong [162], as well as Giovannini and Van Hove [102], independently introduced the concept of a set of particles of common ancestry ('clan') in the cascade process. In this picture, the negative-binomial distribution follows from a composition of a Poisson and a logarithmic distribution, in which clans produced according to a Poisson distribution fragment into the final hadrons on the average according to a logarithmic distribution. The average number of particles per clan $\bar{n}_c$ and the average number of clans $\bar{N}$ are directly derived from the negative-binomial parameters as

$$\bar{n}_c = \frac{\bar{n}}{k} / \ln\left(1 + \frac{\bar{n}}{k}\right) \quad \text{and} \quad \bar{N} = k \ \ln\left(1 + \frac{\bar{n}}{k}\right) \quad . \tag{8.20}$$

In this approach, $1/k$ is a measure for the aggregation of particles, i.e., the probability for two particles to stem from one clan compared to the probability for them to come from two separate clans. It, however, has to be noted that the cascading picture is more general than the negative binomial and that the clan-model parameters can be determined in a model independent manner from the rapidity-gap probability P_0 [163] as $\bar{n}_c = \bar{n}/(-\ln P_0)$ and $\bar{N} = -\ln P_0$.

In Fig. 8.18a,b, the variation of $\bar{n}_c$ and $\bar{N}$ with growing rapidity interval $|y| < y_{\rm cut}$ is shown [117] for hh data between $\sqrt{s} = 5.6$ and 900 GeV and e$^+$e$^-$ collisions at 29 GeV. At 22 GeV, most clans are truncated by the rapidity cut $\eta_{\rm cut} < 1.5$. For increasing $\eta_{\rm cut}$, the truncation effect gets weaker and $\bar{n}_c$ gets larger. For $\eta_{\rm cut} > 1.5$, the average clan is no longer truncated appreciably and $\bar{n}_c$ no longer increases. One notices even a decrease of $\bar{n}_c$ for $\eta_{\rm cut} > 2$, suggesting that the clans located in the outer rapidity ranges are smaller in $\bar{n}_c$ than the central ones.

At 546 GeV, the maximum is reached at $\eta_{\rm cut} \approx 4$ and is considerably larger ($\bar{n}_c \approx 3.8$ instead of 1.6). However, the variation of the average number of clans $\bar{N}$ with $\eta_{\rm cut}$ is described by approximately the same curve for all energies. An increase in energy is affecting the size $\bar{n}_c$ of clans, but not their average number $\bar{N}$.

The e$^+$e$^-$ data between 14 and 91 GeV are compared to each other and to the JETSET PS model in Fig. 8.18 c) and d) [99]. Also here, the average number of clans is approximately independent in fixed rapidity intervals, while the average number of particles per clan increases strongly with increasing energy.

These results imply that in both types of collisions the multiplicity increase for fixed rapidity is due not to an increased clan density, but to the average clan getting larger in the number of particles.

Comparison of the data at $\sqrt{s} = 22$ GeV [85] with the HRS e$^+$e$^-$ data at $\sqrt{s} = 29$ GeV [97] in Fig. 8.18 a) and b) shows that for e$^+$e$^-$ annihilation, clans (whatever their dynamical origin) are substantially smaller and more numerous than in hh collisions at comparable energy.

QCD jets are described in LLA in terms of Markov branching processes of negative-binomial type and dominated by gluon self interaction. The non-linear QCD vertex inherently leads to a self-similar fractal structure of jets. Are clan ancestors identical to these bremsstrahlung gluon jets produced independently? At parton level, this seems to be the case and quark jets dominate e$^+$e$^-$ and lepton-hadron collisions [118, 164] while gluon jets dominate high energy hadron-hadron collisions [118], but primary hadrons are identified with these ancestors at the hadron level [165]. However, if the primary hadrons play such an important role, why are clans so much larger and fewer in number in hh than in e$^+$e$^-$ collisions? This needs further clarification.

8.2.3.4 Correlation of positives and negatives

The NA22 data on symmetric rapidity intervals [85] have been analyzed [166] in terms of independent cluster emission and strongly correlated production of negative and positive particles from these clusters. It is argued that a constant ratio of $k_{\rm ch}/k_-$ is expected throughout a wide interval of central rapidity. As shown in Fig. 8.19, this

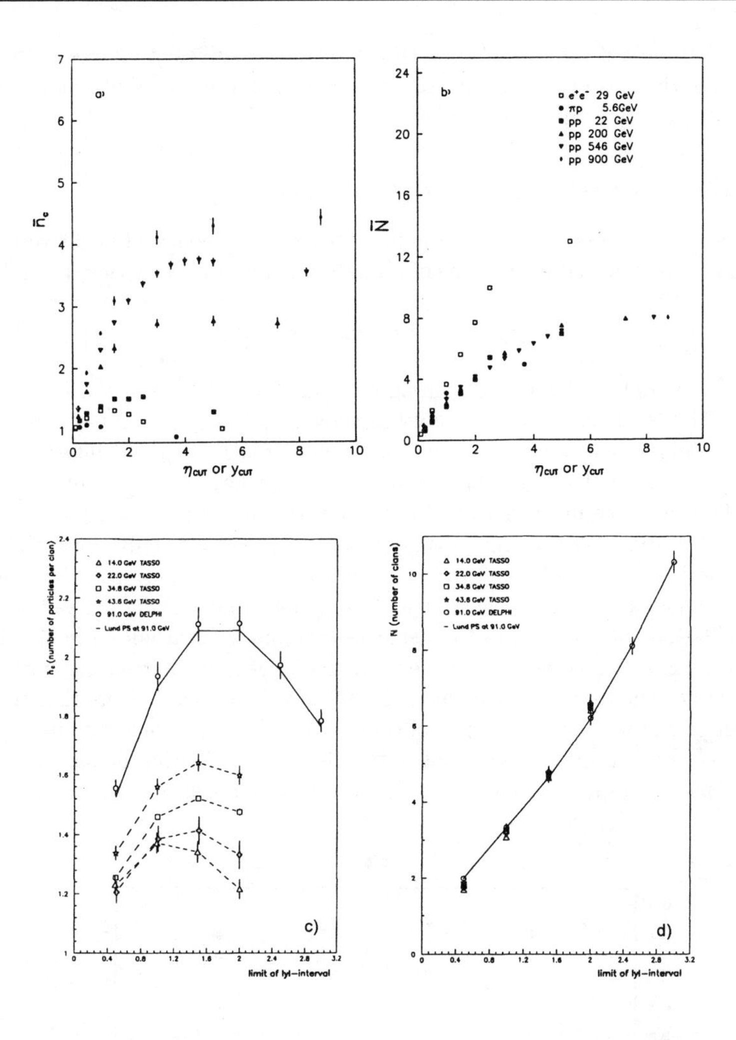

Figure 8.18. a) The average number $\bar{n}_c$ of particles in a clan and b) average number $\bar{N}$ of clans, for different (pseudo-)rapidity intervals $|\eta| < \eta_{\rm cut}$, for hh data between 5.6 and 900 GeV compared to e$^+$e$^-$ annihilation at 29 GeV (in the case of the e$^+$e$^-$ data intervals are in rapidity y) [117]; c)d) same for e$^+$e$^-$ data between 14 and 91 GeV [99].

is indeed observed in the NA22 data.

Furthermore, the approximate relation

$$\frac{k_{\rm ch}}{k_-} = 1 - \frac{k_{ch}}{\bar{n}_{ch}} \tag{8.21}$$

is expected to hold if the rapidity interval $|y| < y_{\text{cut}}$ is sufficiently large compared to the spread of the cluster in rapidity. In agreement with the spread of a clan described above, this equality is approximated for $y_{\text{cut}} \geq 1.5$ only.

8.2.3.5 How good is it, the negative binomial?

To answer the question whether the multiplicity distribution is of negative-binomial type in every (connected or disconnected) sub-domain of phase space, the relation

$$\bar{n}_{D_0}/k_{D_0} = \bar{n}_{D_1}/k_{D_1} + \bar{n}_{D_2}/k_{D_2} \tag{8.22}$$

is expected to hold [167] when combining intervals D_1 and D_2 into D_0, D_0 is equal to $D_1 \cup D_2$. The NA22 collaboration [85] has shown that at $\sqrt{s} = 22$ GeV the relation is indeed fulfilled for transverse momentum and azimuthal angular intervals and for rapidity intervals in the region $|y| < 0.5$, but *not* for larger rapidity intervals. This was the first evidence for deviations from a negative binomial.

Later, deviations also became visible in the multiplicity distributions, themselves (see Sub-Sect. 8.1.2. and Figs. 8.7a and 8.8b above for $\mathrm{p\bar{p}}$ collisions). The DELPHI $\mathrm{e^+e^-}$ results [99] are given in Fig. 8.20a. A shoulder at intermediate z-values leading to an oscillation with respect to the negative binomial is clearly seen in the data for $|y| < 1.5$ and 2.0. It becomes even more prominent when, in addition, the particles are restricted to originate from a given hemisphere (not shown here). Note that this effect is different from that observed in $\mathrm{p\bar{p}}$ collisions [90]. There, it is most prominent in full phase space, but disappears for small intervals. The oscillating deviation from the negative binomial (and from HERWIG) is beautifully demonstrated in Fig. 8.20b [100].

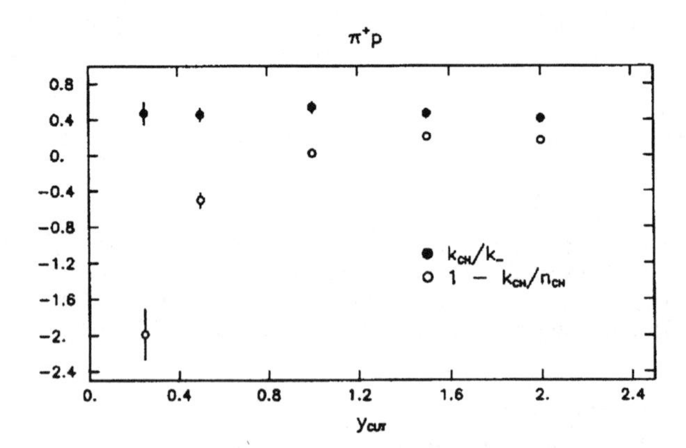

Figure 8.19. The ratio k_{ch}/k_{-} of the negative-binomial parameter for all charged and for negative particles from $\pi^+\mathrm{p}$ data at 22 GeV in rapidity intervals $|y| < y_{\text{cut}}$. This ratio is compared to $1-k_{\text{ch}}/\bar{n}_{\text{ch}}$ [85, 166].

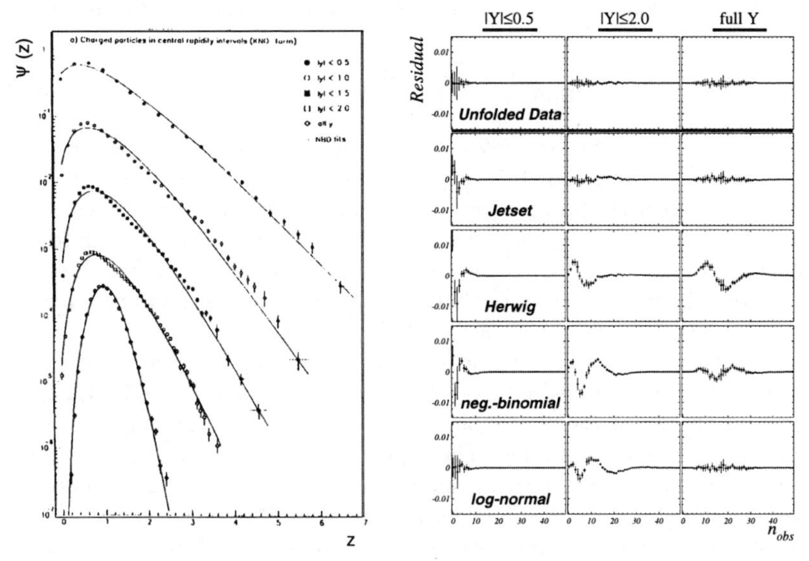

Figure 8.20. a) Charged-particle multiplicity distribution in central rapidity intervals fitted by negative binomials [99]. b) Comparison with various model predictions [100].

The shoulder in the e^+e^- data can be explained by the superposition of multiplicity distributions for 2-, 3- (and 4-) jet events. Furthermore, the distribution for each of these classes of events is more or less reproduced by a single negative binomial. The shoulder can thus be associated with the radiation of hard gluons leading to one or more extra jets in the hadronic final state. Except for a slight underestimation of the highest multiplicities, the full structure can be reproduced by the Lund Parton Shower model.

8.2.4 Beyond the negative binomial

A number of fits to multiplicity distributions also exist with functions related to, but more general than the negative binomial and, in general, with more parameters.

8.2.4.1 The "modified" negative binomial

The generating function of the negative binomial reads

$$G_{\mathrm{NB}}(z) = \frac{1}{[1 - r(z-1)]^k} \tag{8.23}$$

with $r = \bar{n}/k$. If one convolutes a binomial, and a negative binomial, the generating function becomes [168]

$$G_{\mathrm{MNB}}(z) = \left[\frac{1-\Delta(z-1)}{1-r(z-1)}\right]^k \tag{8.24}$$

with $r = \Delta + \bar{n}/k$. This can be derived from a pure birth process with the initial condition given by the binomial [155, 169]. It is assumed that k is integer and corresponds to the number of sources of particle production at some initial stage of the interaction. These sources (clans etc.) branch independently of each other into neutral clusters. The parameter $-\Delta$ corresponds to the cluster decay probability into a pair of charged hadrons, ($1+\Delta$ to that into a pair of neutral hadrons) [168] or, alternatively, to the production rate of the initial sources [169].

The parametrization describes hh, e^+e^- and lh charged-particle multiplicities in full phase space and in restricted intervals [155, 168] in general better than the negative binomial itself, but at the cost of one more parameter, even though this can be fixed at an energy-independent value. The dependence of the parameters on the size of the rapidity interval is explained by simple assumptions on the hit probabilities for charged decay products. The shoulder structure for e^+e^- at 91 GeV, however, requires a superposition of two modified negative binomials, with parameters k and $k+1$.

It is interesting to note that the full-phase-space distributions for e^+e^- and lh collisions are described with energy independent Δ and a k value approaching 7 for both. The significance of this value is not (yet) clear.

8.2.4.2 The (shifted) log-normal distribution

A scale-invariant bi-variate multiplicative branching mechanism leads, via the central limit theorem, to the log-normal form $f(\tilde{n})$ [103]. In this approach, P_n is related to the continuous probability density $f(\tilde{n})$ as

$$P_n = \int_n^{n+1} f(\tilde{n})\mathrm{d}\tilde{n} \tag{8.25}$$

with the mean continuous multiplicity $\langle\tilde{n}\rangle$ approximately given by $\langle\tilde{n}\rangle = \langle n\rangle + 0.5$:

$$P_n(\langle\tilde{n}\rangle; N, \sigma, \mu, c) = \int_{n/\langle\tilde{n}\rangle}^{(n+1)/\langle\tilde{n}\rangle} \frac{N}{\sqrt{2\pi}\sigma}\frac{1}{\tilde{z}+c}\exp\left(-\frac{[\ln(\tilde{z}+c)-\mu]^2}{2\sigma^2}\right)\mathrm{d}\tilde{z} \tag{8.26}$$

with $\tilde{z} = \tilde{n}/\langle\tilde{n}\rangle$ being the scaled continuous multiplicity and N, σ, μ and c being parameters of which, due to normalization conditions, only two are independent. The distribution explicitly satisfies KNO-G scaling, with $\langle\tilde{n}\rangle$ growing as a power of the energy.

Comparison of the log-normal distribution with e^+e^-, lp and $\mathrm{p\bar{p}}$ data can be found in [34, 35, 100, 103, 170]. Again, the fits are, in general, better than those of the negative bionomial, but not in intermediate-size intervals and at the cost of an increase in number of parameters.

8.2.4.3 The "generalized" negative binomial or HNBD

At pre-asymptotic energies, the (discrete) multiplicity distribution P_n can be expressed as the Poisson transform of the asymptotic scaling function $\psi(z)$,

$$P_n = \int_0^\infty \psi(z) \frac{(\bar{n}z)^n}{n!} e^{-\bar{n}z} dz \ . \tag{8.27}$$

Guided by the incorporation of the pQCD-based characteristics,

i) H_q ratios displaying non-trivial sign-changing oscillations,
ii) high-n tail being squeezed (i.e. faster than exponential) with respect to the double-logarithmic approximation (DLLA),
iii) recoil effects leading to factorial moments F_q being dominated by the Γ-function of rescaled rank q/μ with $\mu = (1-\gamma_0)^{-1}$ and $\gamma_0 \propto \sqrt{\alpha_s}$ the QCD anomalous dimension,

Hegyi [134] introduced a generalized negative binomial by extending the validity of the asymptotic KNO scaling form of the negative binomial (the gamma distribution) to $\mu \neq 1$ powers of the scaling variable $z = n/\bar{n}$:

$$\psi(x) = \frac{|\mu|}{\Gamma(k)} \lambda^{\mu k} z^{\mu k - 1} \exp[-(\lambda z)^\mu] \ , \tag{8.28}$$

which is the generalized gamma distribution with shape parameter $k > 0$ and scale parameter $\lambda > 0$.

A Poisson transformation of $\psi(x)$ gives the generalization

$$P_n = \begin{cases} \frac{1}{n!\Gamma(k)} H^{2,0}_{0,2} \left[\frac{\bar{n}\Gamma(k)}{\Gamma(k-1/\mu)} \middle| \begin{matrix} -- \\ (n,1),(k,1/\mu) \end{matrix} \right] & \text{for } \mu < 0 \\ \frac{1}{n!\Gamma(k)} H^{1,1}_{1,1} \left[\frac{\Gamma(k+1/\mu)}{\bar{n}\Gamma(k)} \middle| \begin{matrix} (1-n,1) \\ (k,1/\mu) \end{matrix} \right] & 0 < \mu < 1 \\ \text{negative binomial} & \mu = 1 \\ \frac{1}{n!\Gamma(k)} H^{1,1}_{1,1} \left[\frac{\bar{n}\Gamma(k)}{\Gamma(k+1/\mu)} \middle| \begin{matrix} (1-k,1/\mu) \\ (n,1) \end{matrix} \right] & \mu > 1 \end{cases} , \tag{8.29}$$

where $H(\cdot)$ is called the Fox generalized hypergeometric function [171], while the negative binomial is recovered for $\mu = 1$. The possible utility of the Poisson transform of the generalized gamma distribution (8.26) was first mentioned in [172]. For $|\mu| > 1$, P_n is not infinitely divisible and the factorial cumulants exhibit non-trivial sign-changing oscillations. The additional parameter μ, therefore, measures the degree of violation of infinite divisibility for $|\mu| > 1$. Since the factorial moments are dominated by the Γ-function of rank q/μ for large q, the function indeed can reproduce possible enhancement ($\mu < 1$) and suppresion ($\mu > 1$) of large-n multiplicity fluctuations with respect to the negative binomial.

The strength of this (somewhat complicated) parametrization is that it involves as special and limiting cases the Poisson transform of many asymptotic probability density functions, as e.g. gamma ($\mu = 1$), chi ($\mu = 2$), Pearson type V ($\mu = -1$), Weibull ($k = 1, \mu > 0$), Frèchet ($k = 1, \mu < 0$), log-normal ($k \to \infty, \mu \to 0$), Pareto ($k \to 0, \mu \to \infty$).

Fitting the HNBD to various types of experimental data (in particular those where the simple negative binomial has problems) in general leads to satisfactory results [134]. Two types of departure from $\mu = 1$ emerge:

i) For inelastic hh [134] and deep inelastic e^+p scattering [34], the dominant source of discrepancy is at small multiplicities (in the case of hh and perhaps also in e^+p caused by the diffractive component). This can be reproduced successfully by the Weibull case ($k = 1$) with $\mu > 1$ decreasing with decreasing bin size (e^+p) and increasing energy (hh).

ii) Non-diffractive $p\bar{p}$ collisions at 900 GeV [90] and high-energy e^+e^- results [35, 99, 100, 122, 173, 174] can be reproduced by the Poisson transformed log-normal law ($k \to \infty, \mu \to 0$). For the latter, the sign-changing oscillations of the H_q (in principle caused by $\mu > 1$) can be fully reproduced after taking into account the truncation effects.

In addition, e^+e^- results in the PETRA range [98] give negative μ values (of order $\mu = -10$) and $k = 1$ at the lower energies (22 - 25 GeV), but $k \to \infty$, $\mu \to 0$, at LEP. So, there is a change of regime in this energy range.

The shoulder structure of the high-energy $p\bar{p}$ and e^+e^- distributions is of course not resolved by this unimodal parametrization, so that, also here, a weighted sum of two or more functions would be needed.

The ubiquity of the $\mu \to 0$ limit for the high-energy e^+e^- and $p\bar{p}$ data finds a natural explanation based on renormalization group arguments for asymptotic KNO scaling [175].

8.2.4.4 The generalized Glauber-Lachs, generalized Laguerre or partially coherent laser distribution

$$P_n(N,S,k) = \frac{(N/k)^n}{(1+N/k)^{n+k}} \exp\left(-\frac{S}{1+N/k}\right) \mathrm{L}_n^{k-1}\left(-\frac{kS/N}{1+N/k}\right) \quad \text{with } N+S = \bar{n} \ , \tag{8.30}$$

has been suggested on the basis of quantum statistics [176]. In this picture, particles are emitted from k identical cells, which have a chaotic part N and a coherent part S and L_n^{k-1} is the nth Laguerre polynomial of order k. The reduced dispersion is given by

$$(D/\bar{n})^2 = 1/\bar{n} + \frac{N}{\bar{n}}(1 + \frac{S}{\bar{n}})/k \; . \tag{8.31}$$

The distribution becomes Poisson if $N/S \to 0$ and negative binomial if $S/N \to 0$. The narrow KNO form for e^+e^- can be explained [176] with a very small N/S ratio and k chosen at a small integer value.

Fits to the UA5 data with all parameters left free give results very similar to a negative binomial, except for the most narrow pseudo-rapidity window $|\eta| < 0.2$. When k is fixed to an integer, it is not possible to get good fits in all regions analyzed. However, $k = 1$ is unacceptable only for $|\eta| < 3.0$. The chaoticity decreases with increasing η_{cut} when k is fixed, but increases if k is left free.

Further work would be needed, in particular with respect to energy dependence and the distinction of negative and positive charges, but in fact the model is already excluded from the NA22 and UA5 results discussed at the end of Sub-Sect. 8.2.2, above.

8.2.4.5 Parton branching

In parton branching [153–155] the number of gluons can, in some special cases, follow a negative binomial. However, the parameter k is related to the initial number of quarks and it is difficult to understand why this should decrease with increasing energy.

On the other hand, it can be deduced from coherent branching [153,154] that the multiplicity distribution follows the Furry-Yule form

$$P_n^{\text{Br}}(\bar{n}, k') = \binom{n-1}{n-k'} \left[\frac{\bar{n}/k' - 1}{\bar{n}/k'}\right]^n \left(\frac{\bar{n}}{k'} - 1\right)^{-k'} = P_{n-k'}^{\text{NB}}(\bar{n} - k', k') \tag{8.32}$$

with $1/k' = 1/k + 2/\bar{n}$ and k' interpreted as the average number of initial clusters, and with

$$(D/\bar{n})^2 = 1/k' - 1/\bar{n} \; . \tag{8.33}$$

As indicated in (8.32), the branching distribution P_n^{Br} follows the negative binomial after a transformation of variables and the difference between these two distributions is apparent only at low energy. Like the negative binomial, (8.32) approaches the gamma distribution when n and $\bar{n}$ grow much larger than k'. At low energies, the Furry-Yule distribution provides a less precisely fitting distribution than the negative binomial, but k' remains constant at $k' \approx 3.3$ from $\sqrt{s} = 20$ to 1000 GeV, at least when "mini-jets" are removed.

The e^+e^- data cannot be fitted with this distribution, but it has been shown [118] that the coherent branching of HERWIG comes close to a negative binomial up to at least 2 TeV.

8.2.4.6 The generalized gamma or Krasznovszky-Wagner approximation

The negative binomial only describes non-diffractive multiplicity distributions. A distribution flexible enough to successfully describe non-diffractive and non-elastic hh distributions, as well as e^+e^- and nuclear data is [177]

$$P_n(m,\bar{n},A) = \frac{2m}{\bar{n}\Gamma(A)} F^A(A) \left(\frac{n}{\bar{n}}\right)^{mA-1} \exp[-\mathrm{F}(A)\left(\frac{n}{\bar{n}}\right)^m] \tag{8.34}$$

with

$$F(A) = \frac{\Gamma^m(A+1/m)}{\Gamma^m(A)} \quad . \tag{8.35}$$

It was obtained from a generalized geometrical model in the impact parameter representation. The energy dependent parameters A and $\bar{n}$ are related to the reduced dispersion via

$$(D/\bar{n})^2 = \frac{\Gamma(A+2/m)}{\Gamma^2(A+1/m)}\Gamma(A) - 1 \tag{8.36}$$

and to the normalized moments via

$$C_q = \frac{\Gamma(A+q/m)}{\Gamma^q(A+1/m)}\Gamma^{q-1}(A) \tag{8.37}$$

and m is a (real) positive number independent of energy, but depending on the type of collision. In a stochastic number evolution picture [101], m measures the degree of nonlinearity. The distribution has KNO scaling behavior if A becomes constant in the limit $s \to \infty$. Thus, A is the scaling violation parameter.

This distribution contains as special cases a number of classical distributions such as Weibull ($A = 1$), gamma ($m = 1$), exponential ($m = A = 1$). Furthermore, its Poisson transform (with $m > 1$) was shown to give a generalization of the negative binomial that is asymptotically consistent with QCD (contrary to the log-normal and negative binomial, themselves) [178].

8.2.4.7 Neutral clusters

In [179], the distribution in the number n^* of neutral objects is assumed to follow a gamma distribution

$$\bar{n}^* P_{n^*} = \frac{\mu^\mu}{\Gamma(\mu)} z^{\mu-1} e^{-\mu z} \tag{8.38}$$

with $z = n^*/\bar{n}^*$, $\mu^{-1} = (D^*/\bar{n}^*)^2$, and the neutral objects to decay into a charged hadron pair or to become a neutral hadron with probability q and $1-q$, respectively. If $q \approx \frac{1}{2}$, the decay is binomial and the gamma distribution becomes an apparent negative binomial. The author achieves good results for diffractive and non-diffractive hh as well as for two-jet e^+e^- collisions. Sub-Sub-Section 8.2.3.4 and Fig. 8.19, however, tell us that either not all charged particles come from neutral clusters or that the rapidity range of these clusters is large. It would be important to have more information on separate charges at higher energies.

8.2.4.8 Comparison

While the various distributions are usually claimed by the authors to fit existing data better than the negative binomial, systematic comparison of all distributions on the same set of data, for full phase space as well as for limited regions, under the same conditions, is practically non-existent. Partial exceptions are hadron-hadron data at $\sqrt{s} = 22$ and 900 GeV.

The NA22 $\sqrt{s} = 22$ GeV data were used [180] to compare the negative binomial, generalized Laguerre, Krasznovszky-Wagner, and branching distributions. The generalized Laguerre distribution was found to follow the negative binomial in all its aspects, providing a good fit whenever the negative binomial does and vice versa. Because of the strong correlation of the parameters, k cannot be determined from the fits. However, for all k, S is considerably larger than N. The Krasznovszky-Wagner distribution gives the best fit for full phase space, but is slightly worse than negative binomial or generalized Laguerre in intermediately sized intervals. The branching distribution, on the other hand, does not fit at these low energies.

At 900 GeV [181], as many as eight distributions and models were tried (negative binomial, statistical bootstrap, three-fireball, Laguerre, branching, convolution of Poisson and two $1/n^2$ distributions, convolution of two negative biomials, sum of two negative binomials). A convolution is applicable whenever two different and independent production mechanisms contribute within all events of a sample. However, if each individual event belongs to one of two possible classes, the resulting distribution is a weighted sum of the two classes. In case each class follows a negative binomial, the distribution is

$$P_n = (1-\omega)P_{\rm NB}(\bar{n}_1, k_1) + \omega P_{\rm NB}(\bar{n}_2, k_2) \ . \tag{8.39}$$

Because of the shoulder observed in the full-phase-space distribution at 900 GeV, only the weighted sum of two negative binomials (with as many as 5 open parameters!) gives an acceptable fit, there.

So, the UA5 data are compatible with a two-class picture, such as one-chain plus two-chain, jet plus no-jet, hard plus soft. It is interesting to note that the weight ω and the class multiplicities $\bar{n}_1, \bar{n}_2$, fitted to grow from $\omega \approx 0$ at ISR to $\omega \approx 1/3$ with $\bar{n}_2 \approx 2\bar{n}_1$ at 900 GeV, are compatible with the fraction of "mini"-jet events and their average multiplicity found at 900 GeV [182]. Furthermore, the first component is found compatible with KNO scaling, while only the second one becomes wider with increasing energy.

8.3 Information-entropy scaling

As an alternative quantity characterizing the final-state multiplicity distribution P_n, a (momentum-integrated) information entropy can be defined [183, 184] as

$$S = -\sum_n P_n \ln P_n \quad . \tag{8.40}$$

It is a measure of the uncertainty associated with a multiplicity distribution. A wide distribution gives more uncertainty and a larger value of S than a sharply peaked one. Important properties of S are:

i) This variable describes the general pattern of particle emission. The total entropy produced from ν statistically independent sources (e.g. clans or super-clusters) is just the sum of entropies of the individual sources:

$$S = \sum_{i=1}^{\nu} S_i \; . \tag{8.41}$$

ii) Distortion of the multiplicity scale leaves S invariant, so does insertion of zeros or mutual permutation. In particular, in full phase space, the entropy is the same when calculated from all charged particles or negatives (i.e. charged pairs) only.

iii) From the identity

$$S - \ln\langle n\rangle = -\frac{1}{\langle n\rangle}\sum \langle n\rangle P_n \ln(\langle n\rangle P_n) \tag{8.42}$$

follows for large $\langle n\rangle$

$$S - \ln\langle n\rangle = \frac{1}{c}\int_0^\infty \psi(z)\ln\psi(z)\mathrm{d}z \tag{8.43}$$

with $\psi(z)$ normalized as

$$\int \psi(z)\mathrm{d}z = \int z\psi(z)\mathrm{d}z = c \; , \tag{8.44}$$

where $c = 2$ for all charged particles and $c = 1$ for negatives (or pairs).

iv) For the geometric distribution $\psi(z) = \exp(-z)$, an upper bound is

$$S - \ln(\langle n\rangle/c) \leq 1 \; , \tag{8.45}$$

so that at high enough $\sqrt{s}$ [184]

$$\langle n\rangle \simeq (\sqrt{s})^\kappa \; .$$

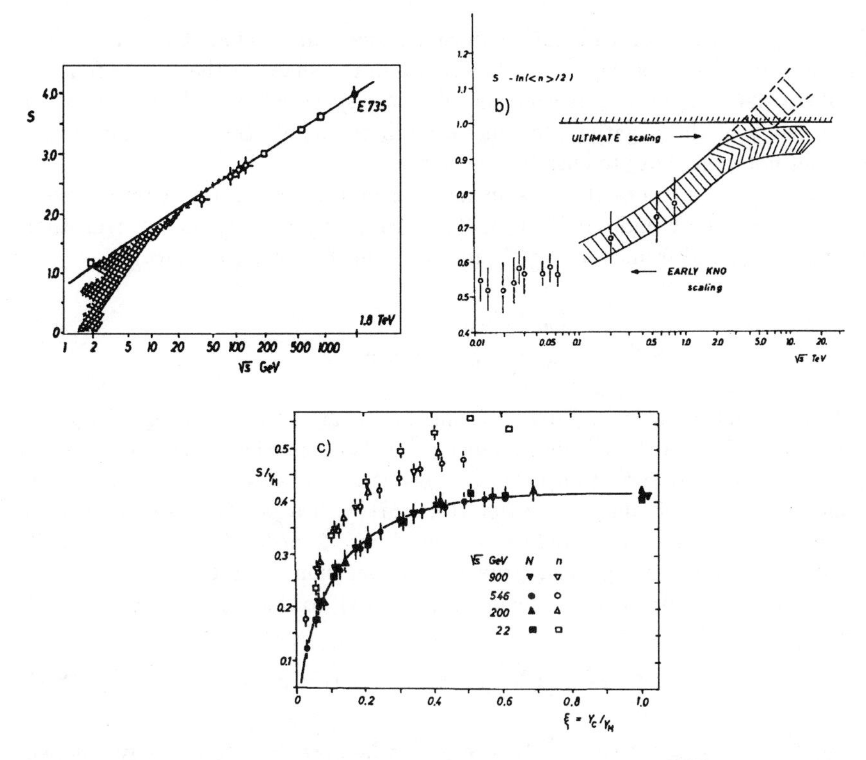

Figure 8.21. a) Entropy of the charged-particle multiplicity distribution as a function of $\sqrt{s}$, b) entropy of $\psi(z)$ from (8.45) as a function of $\sqrt{s}$, c) entropy per unit rapidity for rapidity windows $\xi_{\rm cut}$ (denoted ξ in the figure) [184].

As is shown in Fig. 8.21a, at high energies ($\sqrt{s} \gtrsim 20$ GeV), the value of S increases linearly with $\ln \sqrt{s}$ [184]. Extrapolating back, the intercept is near $\ln m_\pi$. Since the maximum rapidity is $y_{\rm max} = \ln(\sqrt{s}/m_\pi)$, it follows that the entropy per rapidity unit $\kappa \equiv S/y_{\rm max}$ is constant. This constancy is indeed observed up to Tevatron energies [90, 93, 185] with $\kappa = 0.437 \pm 0.004$.

From Fig. 8.21b it is clear that the limit (8.45) is not reached at present collider energies. However, at LHC the multiplicity distribution is expected to be governed by (8.45). So, either the entropy increase of Fig. 8.21a must slow down, or $\langle n \rangle$ must grow faster than presently indicated.

Approximate scaling is observed for the function $\kappa(\xi)$ between NA22 at $\sqrt{s} = 22$ GeV and CDF at 1800 GeV [93, 184], when n is restricted to negatives (or the number of oppositely charged pairs), with $\xi = y_{\rm cut}/y_{\rm max}$ and $y_{\rm cut}$ being half the size of a central rapidity window (Fig. 8.21c).

From the constancy of κ for full phase space, one can conclude that the entropy per rapidity unit does not depend on energy. From the shape of the scaling function $\kappa(\xi)$ follows that the entropy reaches its full-phase-space value κ already at ξ=0.5, where $\bar{n}$ and k of the negative binomial are still changing. So, the fragmentation region does not contribute to entropy production.

It is interesting to note that the increase in multiplicity is the main source of entropy of high-energy hadronic matter. The entropy from fluctuation of transverse momentum, S_{p_T}, and from rapidity, S_y, increases much slower than $\ln\langle n\rangle$:

$$\begin{aligned} S_{p_T} &\approx \ln\langle p_T\rangle \\ S_y &\approx \ln y_{\max} \approx \ln\ln\sqrt{s}\ . \end{aligned} \tag{8.46}$$

In hadron-nucleus collisions, an extra contribution ΔS_A to the entropy of negative particles comes from a fluctuating number of nucleons participating [184]. Indeed $\Delta S_{Ne} = 0.3 \pm 0.05$ is observed for hNe collisions, independent of energy. Also heavy-ion collisions show a similar S-behavior [186, 187] with a fast increase of S in the central rapidity region and a rapid saturation of $S/y_{\max}$ above $\xi = 0.4$. The results suggest that, with increasing energy, the entropy per pion saturates.

The information entropy S can be generalized [184] to the Rényi order-q information entropy

$$I_q = \frac{1}{1-q} \ln \sum_n (P_n)^q \quad , \quad \text{with } I_1 = S \ , \tag{8.47}$$

and κ to $D_q = I_q/y_{\max}$. Also the D_q turn out to be approximately energy independent. Comparing hadron-hadron to e^+e^- data, one observes a small but significant difference ($D_q(\text{hh}) > D_q(e^+e^-)$), increasing with increasing order q.

There are hints that this behavior can be understood from cascading processes [188]. The question remains, how heavy-ion collisions behave w.r.t. the additivity of entropy as observed in hA collisions, and whether entropy differences can be used as a signature for a quark-gluon plasma.

Entropy scaling is an interesting concept, but, contrary to KNO or log-KNO scaling where a whole function is considered, it only concerns one single number.

8.4 Rapidity gap probability

Interesting information on higher-order correlations is contained already in the $n = 0$ bin of the multiplicity distribution in a given phase space (e.g. rapidity) bin. Using (7.41) with $\langle n\rangle$ as the average number of particles in bin Δy, the probability of detecting no particles in Δy is related to the generating function through

$$P_0(\Delta y) = G(z = -1) \tag{8.48}$$

and can be used [189] as a generating function for $P_n(\Delta y)$:

$$P_n(\Delta y) = \frac{(-\langle n\rangle)^n}{n!}\left(\frac{\partial}{\partial\langle n\rangle}\right)^n P_0(\Delta y)\ , \tag{8.49}$$

where the differentiation is carried out with the correlation functions of (7.47) fixed. Its dependence on the (higher-order) cumulants is, according to (7.46),

$$\ln P_0(\Delta y) = \sum_{q=1}^{\infty}\frac{(-1)^q}{q!}f_q = \sum_{q=1}^{\infty}\frac{(-\langle n\rangle)^q}{q!}K_q\ , \tag{8.50}$$

thus involving cumulants of all orders. Applying the so-called linked-pair ansatz to the normalized cumulant moments K_q [190] gives

$$K_q = A_q K_2^{q-1}\ . \tag{8.51}$$

If the linking coefficients A_q are independent of $\sqrt{s}$ and Δy, as confirmed by the analysis of UA1 and UA5 data up to $q = 5$ [190], then the quantity

$$\chi = -\ln P_0(\Delta y)/\langle n\rangle = \tag{8.52}$$

$$= \sum_{q=1}^{\infty}\frac{A_q}{q!}(-\langle n\rangle K_2)^{q-1} = \chi(\langle n\rangle K_2) \tag{8.53}$$

only depends on the moment product $\langle n\rangle K_2$ [191]. Note that $\langle n\rangle = -\ln P_0(\Delta y)$ for the Poisson distribution, so that χ measures the amount of deviation from independent emission, involving correlations of all orders. The scaling feature was in fact already derived in [192] for the study of void probability in galaxy clustering, where this scaling is found to hold and χ is found to follow $A_q = (q-1)!$, i.e. is equal to the linking parameters of the NBD, $\chi = \ln(1+\langle n\rangle K_2)/(\langle n\rangle K_2)$ [193].

In high energy collisions, the scaling was shown to hold [191] in the NA22 hydrogen- and nuclear-target data for $\Delta y < 1$ and χ agrees with the NBD expectation up to $\langle n\rangle K_2 \approx 1$, but falls below for larger values.

A systematic study of P_0 values of UA5 in various central and non-central rapidity bins [90] was done in [194] (Fig. 8.22a). Contrary to the galaxy data of [193], most of the points fall somewhat below the NBD expectation (dashed), in agreement with the increase of the linking parameters being somewhat weaker than expected by the NBD [190]. In general, they, however, stay above the full line, representing the simple case of $A_q = 1$ for all q ("minimal model"). The strongest deviation from the NBD scaling curve appears for the most non-central rapidity bins indicating violation of translation invariance of the correlation.

The UA5 data are scarce and have large errors for $\langle n\rangle K_2 \gtrsim 3$. This region has been extended to 30 in [195] from heavy-ion collisions at 60 A GeV (Fig. 8.22b) and 200 A GeV. Again, the results lie between the scaling curves expected from NBD

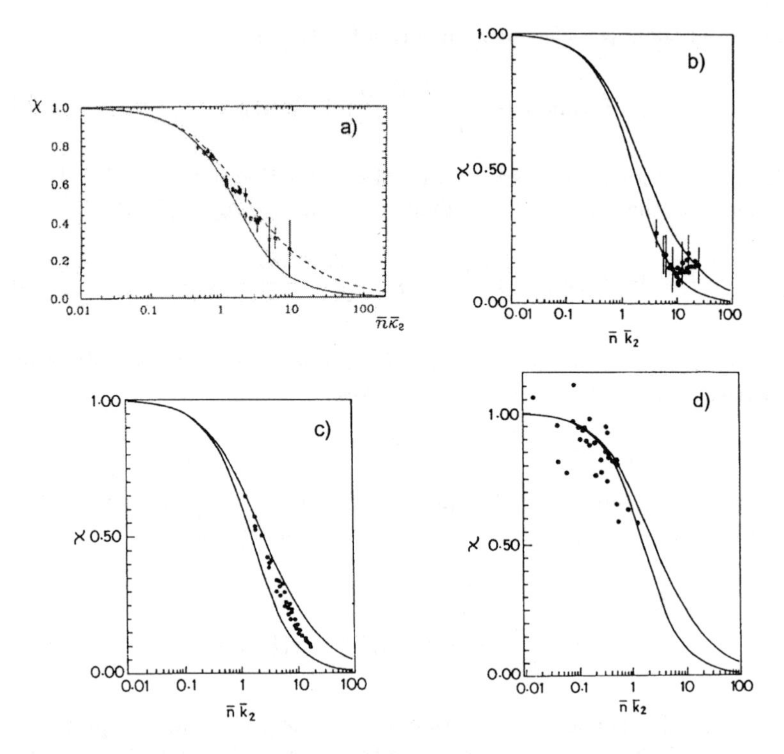

Figure 8.22. Scaled rapidity gap probability χ as a function of $\langle n\rangle K_2$ for a) UA5 data [194], b) ^{16}O-AgBr, c) FRITIOF and d) independent emission at 60 A GeV [195]. The upper curve corresponds to the NBD, the lower one to the minimal model.

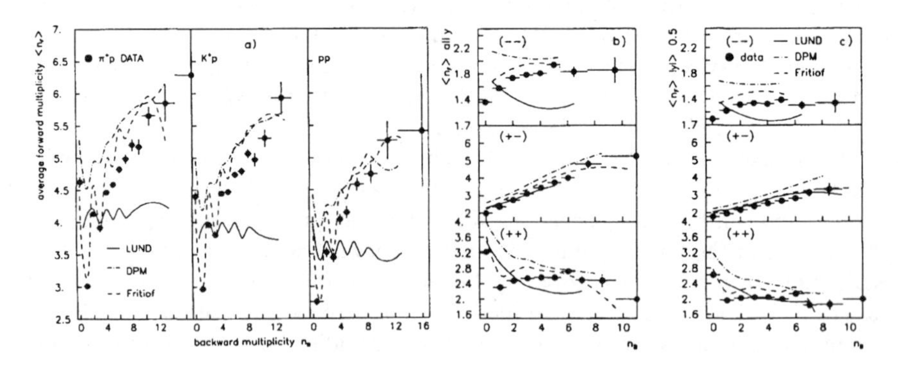

Figure 8.23. a) The average number of charged forward particles versus the number of charged backward particles for π^+p, K$^+$p and pp collisions at $\sqrt{s}$=22 GeV, with MC predictions as indicated. b) The same for different charge combinations in the combined data sample. c) Same for particles with $|y| > 0.5$ [196].

and from the minimal model. The authors also compared to the expectations from FRITIOF (Fig. 8.22c) and from random production of particles in η-space (Fig. 8.22d). While FRITIOF lies within the region bounded by the NBD and the minimal model, be it distributed over a wider $\langle n\rangle K_2$ range than the data, random particle production is limited to $\langle n\rangle K_2 < 1$ and scattered over a wide range in χ, showing no scaling.

A simple connection exists [194] to the clan model parameters $\bar{N}$ (the average number of clans) and $\bar{n}_c$ (the average number of particles in a clan) in an interval Δy, as defined in (8.20). Clans are assumed to be produced independently (according to a Poissonian) and must contain at least one particle. So, $P_0(\Delta y)$ is the Poisson value for the probability of having no clans in Δy, i.e.

$$\bar{N} = -\ln P_0(\Delta y) , \tag{8.54}$$

while

$$\bar{n}_c = \frac{\langle n\rangle}{\bar{N}} = \chi^{-1} . \tag{8.55}$$

8.5 Forward-backward correlations

The correlation between the charged-particle multiplicity in one hemisphere with that in the other was studied in a wide range of $\sqrt{s}$, from lowest-energy bubble chamber experiments to the Tevatron, and for all types of collisions. Examples are NA22 [196], NA27 [197], the ISR [198], UA5 [199], E735 [200], ν [201,202], EMC [203], TASSO [98,204], HRS [97,205], DELPHI [99], OPAL [206], ZEUS [207].

The average charged-particle multiplicity $\langle n_{\mathrm{F}}\rangle$ in the forward hemisphere is given as a function of the charged-particle multiplicity n_{B} in the backward hemisphere for the NA22 experiment [196] in Fig. 8.23. A comparison to three low p_{T} models shows that the single-chain Lund model, which reproduces the $\mathrm{e^+e^-}$ data at comparable energies [204,205] does not reproduce the hh data at all. The two-chain FRITIOF and a two-chain version of the Dual Parton Model slightly overestimate the correlation. In all three models oscillations are visible between odd and even n_{B}.

In Fig. 8.23b, the same distributions are shown, but now for $(--)$, $(+-)$ and $(++)$ charge combinations, separately. The correlation is dominated by unlike-charged particles (note the difference in scale). This observation is extended to $\mathrm{e^+e^-}$ collisions by TASSO [98] and DELPHI [99] (not shown here). All three models are able to reproduce the $(+-)$ hp data, while FRITIOF does quite well also for $(--)$ and $(++)$. For the latter, an anti-correlation is expected from Lund. The $\mathrm{e^+e^-}$ results can be reproduced by the JETSET PS model. From Fig. 8.23c, one can see that the correlation is not completely gone when the influence from short-range order is suppressed by eliminating the central region.

The actual range of the correlation in hh collisions is investigated by the UA5 collaboration [199], who give the slope b defined from

$$\langle n_{\mathrm{B}}\rangle = a + b n_{\mathrm{F}} \tag{8.56}$$

for two windows of one unit in pseudo-rapidity as a function of the size of the gap separating them. From Fig. 8.24a it is clear that a correlation persists up to a gap size of $\Delta\eta = 6$. This (long-range) correlation effect can be well reproduced by the UA5 Cluster Monte Carlo [199] of Poisson like clusters and a negative binomial total charged-particle multiplicity. It can also be reproduced by the upgraded version of FRITIOF and reasonably well by DPM, but is overestimated in PYTHIA. The energy dependence of this effect between 22 and 900 GeV is given in Fig. 8.24b [196].

Among the other models mentioned in Sect. 8.2, forward-backward correlations, and in particular differences found for hh and e^+e^- collisions, are expected and discussed in the geometrical models [145, 146, 154].

In Fig. 8.25, a compilation is given for the correlation strength b in hh, lh and e^+e^- collisions, when no central region is excluded. For hh collisions (Fig. 8.25a) an approximately linear rise of b is found with increasing $\ln s$ with no saturation, so far. As for $1/k$ in Fig. 8.10, the slope b is lower for lh collisions in Fig. 8.25b than for hh collisions, but the energy dependence is the same, at least up to the highest values of W available. Furthermore, b is lower for e^+e^- than for hh and lh collisions and the energy dependence is flatter.

This is not completely unexpected. If particle production follows a negative binomial with no further correlations from conservation laws or dynamics, $1/k$ and b are related [116, 208] by

$$b = \frac{\langle n_{\rm B}\rangle/k}{1 + \langle n_{\rm F}\rangle/k} \quad . \tag{8.57}$$

The curves derived from the fits in Fig. 8.10 are drawn in Fig. 8.25. Relative to the overall negative binomial, there is an *anti-correlation* building up at high energy in hh collisions. This deviation can be expected if particles are produced in clusters [116, 208] or pairs [136, 209]

From the increase of $1/k$ in Fig. 8.10, a positive slope b is also expected for higher energy lh and e^+e^- collisions. This is shown by the dashed lines in Fig. 8.25. Indeed, the TASSO [98], DELPHI [99] and OPAL [206] points are well above $b = 0$, thus establishing positive forward-backward correlations in e^+e^- collisions, as well.

Fig. 8.26a gives the OPAL data in comparison to JETSET, HERWIG and COJETS. As observed in hh collisions (Fig. 8.24), the slope b is reduced when excluding a central rapidity gap but does not vanish (not shown). In Fig. 8.26b, the data are plotted separately for 2-, 3- and 4-jet events. From this figure, the positive correlation observed in the inclusive sample can be understood as a consequence of hard gluon radiation. However, 2-jet events still show a positive correlation. Fig. 8.26c, therefore, gives a Monte-Carlo comparison of $\langle n_{\rm F}\rangle$ versus $n_{\rm B}$ for 2-jet $b\bar{b}$ events, 2-jet light-flavor events and all 2-jet events. Again, a residual positive correlation of the 2-jet events appears, due to a superposition of event classes with different mean multiplicity, here those originating from light flavors and those from b quarks.

Nevertheless, although less pronounced, a positive correlation still remains in both the light-quark and b-quark 2-jet samples. Fig. 8.26d, therefore, compares

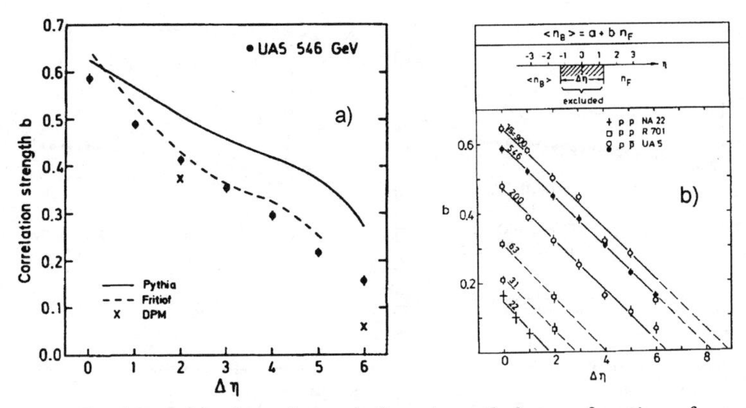

Figure 8.24. The forward-backward correlation strength b as a function of an excluded central gap $\Delta\eta$ a) for UA5 [199] b) for the energies indicated [196].

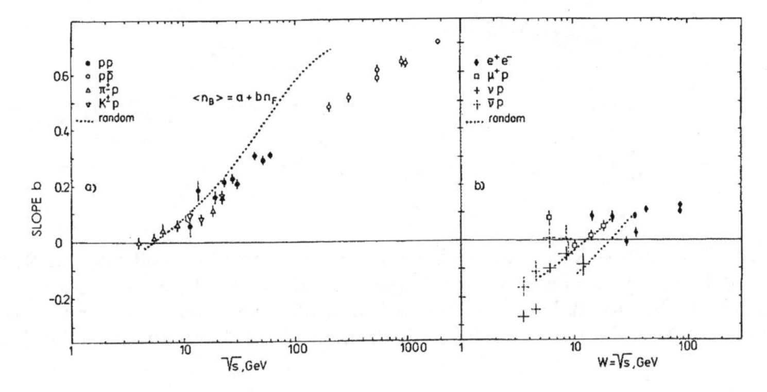

Figure 8.25. Compilation of the values of the correlation strength b for a) hh, b) lh and e^+e^- collisions, as a function of $\sqrt{s}$ and W, respectively [196]. The lines are obtained from those in Fig. 8.10 via Eq. (8.57).

the JETSET correlation in b-quark 2-jet events in the central region ($|y| \leq 1.0$) for partons, primary hadrons and final charged particles. The correlation is dominated by the hadronic decay processes, probably due to a spread of the decay of centrally produced resonances into both hemispheres.

In ZEUS [207], the correlations are studied in the Breit frame. Negative long-range correlations have been predicted [210] due to the kinematics of the first-order QCD effects [210]. ZEUS uses as a normalized correlation measure [202]

$$\kappa = \sigma_c^{-1}\sigma_t^{-1}(\langle n_c n_t\rangle - \langle n_c\rangle\langle n_t\rangle) \tag{8.58}$$

where n_c and n_t are the multiplicities in the current and target regions, and σ_c, σ_t

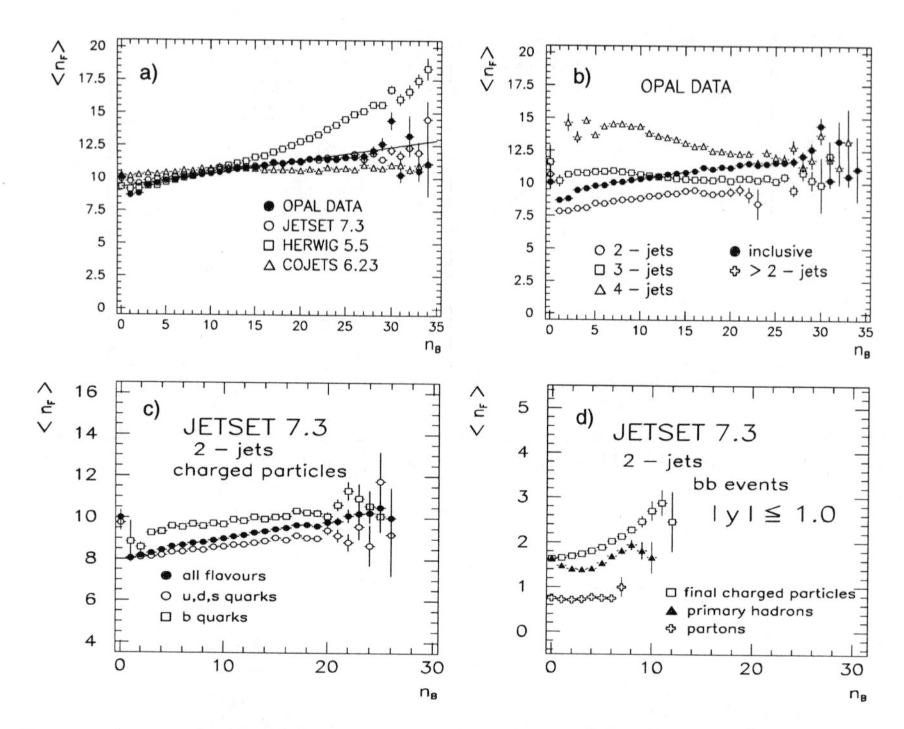

Figure 8.26. Forward-backward multiplicity correlations in e^+e^- collisions at 91.2 GeV, a) compared to model predictions, b) compared to those for 2-, 3- and 4-jet events, c) predicted by JETSET for light-quark, b-quark and all 2-jet events, d) for b-quark 2-jet events at $|y| \leq 1.0$, at the parton, primary hadron and final charged-particle level [206].

the standard deviations of the corresponding multiplicity distributions. Its advantage is the simplicity of the boundary condition ($-1 \leq \kappa \leq 1$) allowing a quantitative estimate of the correlation strength, its easier accessibility for analytical QCD calculations, and its low sensitivity to the experimental losses of particles from the proton remnants. It is related to the slope b via

$$\kappa = \frac{\sigma_c}{\sigma_t} b \ . \tag{8.59}$$

Fig. 8.27 gives κ as a function of Q^2 and Bjorken-x_B. Anti-correlations are indeed observed for all values of Q^2 and x_B, as predicted in [210]. Even though the absolute size also decreases with increasing Q^2 and x_B, the correlation does not become positive in the Q^2 and x range measured. To compare to Fig. 8.24, note that $W \approx Q\sqrt{\frac{1}{x_B} - 1}$ reaches values above 100 GeV at large Q^2, while the current jet has energy $Q/2 \approx 20 - 30$ GeV, where e^+e^- forward-backward correlations are positive, already.

In the UA5 results [199], the variance $\langle z^2 \rangle$ of the asymmetry parameter $z = n_{\rm F} - n_{\rm B}$ is observed [211] to be a linear function of n. This led the authors to propose the binomial

$$P(n,z) = \psi(n/\langle n \rangle) C^{n/2}_{n_{\rm F}/2} [B(n)]^{-1} \tag{8.60}$$

with $C^{n/2}_{n_{\rm F}/2}$ the binomial term and $[B(n)]^{-1}$ a normalization factor. This expression leads to the simple relation $\langle z^2 \rangle_n = 2n$ observed in UA5. The fact that the correlation is larger in hh than in e$^+$e$^-$ collisions at the same energy is then solely due to the range of the impact parameter in the former.

Lim et al. [212] inserted the negative binomial for ψ and replaced the binomial factor by the generalization $C^{n/r}_{n_{\rm B}/r}$, with r being a free parameter to be interpreted as the average size of the cluster prior to the production of the final-state particles and fitted to the slope b of (8.56). This cluster size was found to vary from 1.14 at 24 GeV to 2.62 at 1800 GeV for p$\bar{\rm p}$ collisions with an indication for saturation at the highest energies [200], and from 0.831 at 14 GeV to 1.124 at 43.6 GeV for e$^+$e$^-$ collisions.

A generalization allowing for non-linear dependence of $\langle n_{\rm B} \rangle$ on $n_{\rm F}$ and obtaining r from the single-hemisphere data directly was used by Chen et al. [213] to extract a cluster size of $r = 1.45$ from the 91 GeV DELPHI data [99] and predicting 1.60 for 133 GeV e$^+$e$^-$ collisions.

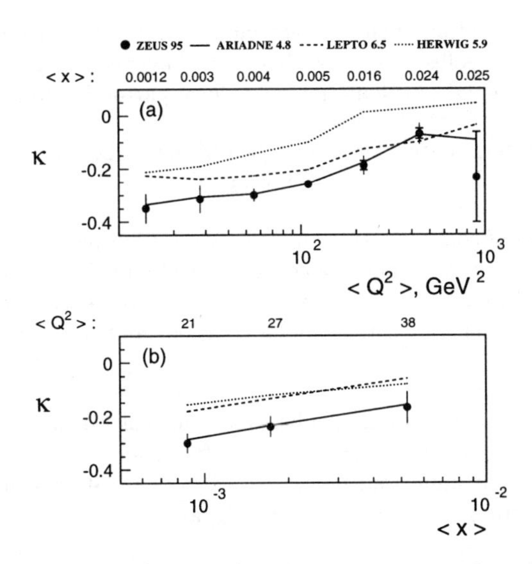

Figure 8.27. a) The evolution of the coefficient of correlations κ with predominant variation in Q^2 for data and MC predictions. b) The same quantity where predominantly $x_{\rm B}$ varies [207].

In the geometrical model as discussed in [211,212] the contributions of stochastic and non-stochastic components are fundamental. An extension [214] is based on the weighted superposition of negative-binomial distributions, each describing a different event class, as already applied in Sub-Sects. 8.1.2 and 8.2.3 [104]. Independent (Poissonian) clan production can be considered a stochastic process, while the logarithmic multiplicity distribution inside clans can be considered a non-stochastic Markovian process dominated by gluon self-interaction. The relative importance of these two mechanisms may be different in different event classes. An interesting difference is that these two classes are interpreted to correspond to soft and semi-hard (mini-jet) events in $p\bar{p}$ collisions, but to two- and three-jet events in e^+e^- collisions. In the latter, clans are more or less confined to their own hemisphere, thus leading to $b \approx 0$ for each of the classes separately (see Fig. 8.26b), so that the positive values of b observed in the full sample are mainly due to the superposition of the two event classes. To the contrary, in the former, clans do leak into the other hemisphere and contribute to the relatively larger value of b. As could be the shoulder in Sub-Sect. 8.1.2, and the dip in Sub-Sect. 8.1.3, the linear increase observed in Fig. 8.25a, can be readily reproduced by this superposition. If still found at LHC, a third component would, however, be required.

8.6 Conclusions

1. The energy evolution of the average multiplicity is similar for all types of collisions. For e^+e^- collisions it is well described by *coherent* parton shower models and by analytic QCD calculations plus local parton hadron duality.

2. The difference between light- and heavy-quark jets and between gluon and quark jets can be understood well from QCD.

3. For all types of collisions, the multiplicity distribution gets wider than Poisson (i.e. correlations exist) above a certain transition energy depending on type of collision and rapidity interval considered. At given energy, multiplicity distributions are wider for hh than for lh collisions and wider for lh than for e^+e^- collisions.

4. For all types of collisions, in full phase space as well as in limited regions of it, the negative binomial is a surprisingly successful parametrization of the multiplicity distribution, but important deviations exist. The latter have stimulated work on a large variety of extensions and alternatives.

5. Up to the highest energy reached so far, KNO scaling neither holds for full-phase-space multiplicity distributions, nor for multiplicity distributions in limited intervals. For all types of collisions, a log-scaling law may be an interesting

candidate for a replacement. Also entropy scaling deserves further investigation.

6. Higher-rank cumulant moments show a characteristic dip plus an oscillation with increasing rank, also expected in NNLLA of QCD. It will, however, be important to distinguish how much of the effect is due to mere energy-momentum conservation and/or other trivial effects, and how much to dynamics.

7. Positive forward-backward correlations exist for all types of collisions at energies above the Poissonian transition point. At given energy, they are stronger in hh than in lh collisions and stronger in lh than in e^+e^- collisions. They grow with increasing energy, but less fast than would be expected from the widening of the negative binomial. For hh collisions at Collider energies, they are positive over a gap of at least 6 units in rapidity.

8. Important information on the higher-order moments is already contained in the rapidity-gap (or empty-bin) probability. Scaled to its Poisson expectation, it forms a measure of deviation from independent emission. In the existing data, it deviates clearly from independent emission, but also from the negative binomial.

Bibliography

[1] T. Sjöstrand and V.A. Khoze, *Z. Phys.* **C62** (1994) 281.

[2] B.R. Webber, Physics at LEP2, eds. G. Altarelli, T. Sjöstrand and F. Zwirner, CERN 96-01, Vol.2 (1996) 161.

[3] L. Lönnblad and T. Sjöstrand, *Phys. Lett.* **B351** (1995) 293 and *Eur. Phys. J.* **C2** (1998) 165.

[4] V. Kartvelishvili, K. Kvatadze and R. Møller, *Phys. Lett.* **B408** (1997) 331.

[5] K. Fiałkowski, R. Wit, *Acta Phys. Pol.* **B28** (1997) 2039.

[6] Š Todorova-Nová and J. Rames, hep-ph/9710280 (1997).

[7] DELPHI Coll., P. Abreu et al., *Phys. Lett.* **B416** (1998) 233.

[8] OPAL Coll., K. Ackerstaff et al., *Eur. Phys. J.* **C1** (1998) 395; G. Abbiendi et al., *Phys. Lett.* **B453** (1999) 153.

[9] DELPHI Coll., P. Abreu et al., *Eur. Phys. J.* **C18** (2000) 203.

[10] S. Barlag et al., *Z. Phys.* **C11** (1982) 283; M. Derrick et al., *Phys. Rev.* **D25** (1982) 624; H. Grässler et al., *Nucl. Phys.* **B223** (1983) 269.

[11] A. Wróblewski, Proc. XIV Int. Symp. on Multiparticle Dynamics, Lake Tahoe, ed. P. Yager and J.F. Gunion (World Scientific, Singapore, 1984) p.573.

[12] C. Bacci et al., *Phys. Lett.* **B86** (1979) 234; J.L. Siegrist et al., *Phys. Rev.* **D26** (1982) 969; B. Niczyporuk et al., *Z. Phys.* **C9** (1981) 1 and M.S. Alam et al., *Phys. Rev. Lett.* **49** (1982) 357; Ch. Berger et al., *Phys. Lett.* **B95** (1980) 313; W. Bartel et al., *Z. Phys.* **C20** (1983) 187.

[13] P.V. Chliapnikov and V.A. Uvarov, *Phys. Lett.* **B251** (1990) 192.

[14] A. Bassetto, M. Ciafaloni and G. Marchesini, *Phys. Lett.* **B83** (1979) 207 and *Nucl. Phys.* **B163** (1980) 477; W. Furmanski, R. Petronzio and S. Pokorski, *Nucl. Phys.* **B155** (1979) 253.

[15] M. Bardadin-Otwinowska, M. Szczekowski, A.K. Wróblewski, *Z. Phys.* **C13** (1982) 83.

[16] PHOBOS Coll., B.B. Back et al., nucl-ex/0301017 and 0410022.

[17] PHOBOS Coll., B.B. Back et al., *Phys. Rev. Lett.* **85** (2000) 3100; **87** (2001) 102303; **88** (2002) 022302; *Phys. Rev.* **C65** (2002) 061901.

[18] PHENIX Coll., K. Adcox et al., *Phys. Rev. Lett.* **86** (2001) 3500; S.S. Adler et al., nucl-ex/0409015 and 0410003.

[19] BRAHMS Coll., I. Arsene et al., nucl-ex/0410020.

[20] E.K.G. Sarkisyan, A.S. Sakharov, hep-ph/0410324.

[21] S. Eremin, S. Voloshin, *Phys. Rev.* **C67** (2003) 064905.

[22] L3 Coll., M. Acciarri et al., *Phys. Lett.* **B444** (1998) 569; P. Achard et al., *Phys. Reports* **399** (2004) 71.

[23] ALEPH Coll., A. Heister et al., *Eur. Phys. J.* **C35** (2004) 457.

[24] P.V. Chliapnikov and V.A. Uvarov, *Phys. Lett.* **B431** (1998) 430.

[25] TASSO Coll., W. Braunschweig et al., *Z. Phys.* **C47** (1990) 187.

[26] DELPHI Coll., P. Abreu et al., *Phys. Lett.* **B459** (1999) 397.

[27] B.R. Webber, *Phys. Lett.* **B143** (1984) 501.

[28] DELPHI Coll., P. Abreu et al., *Phys. Lett.* **B372** (1996) 172.

[29] OPAL Coll., G. Abbiendi et al., *Eur. Phys. J.* **C16** (2000) 185; DELPHI Coll., P. Abreu et al., *Eur. Phys. J.* **C25** (2002) 493; P. Abreu, N. Anjos, *Charged Particle Multiplicity in* e^+e^- *Annihilations into* $q\bar{q}$ *at* $\sqrt{s} = 206$ *GeV*, DELPHI 2002-051, CONF585 (June 2002).

[30] V.A. Khoze and W. Ochs, *Int. J. Mod. Phys.* **A12** (1997) 2949.

[31] I.M. Dremin, J.W. Gary, *Phys. Reports* **349** (2001) 301.

[32] R. Botet, M. Płoszajczak, *Z. Phys.* **C76** (1997) 257.

[33] J. Ellis, M. Karliner and H. Kowalski, *Phys. Lett.* **B235** (1990) 341.

[34] H1 Coll., S. Aid et al., *Z. Phys.* **C72** (1996) 573.

[35] OPAL Coll., P.D. Acton et al., *Z. Phys.* **C53** (1992) 539.

[36] H1 Coll., C. Adloff et al., *Nucl. Phys.* **B504** (1997) 3.

[37] G. Ingelman, Proc. 1991 Workshop on Physics at HERA, DESY Vol. 3 (1992) 1366.

[38] R.P. Feynman, *Photon-Hadron Interactions*, Benjamin, New York, 1972.

[39] K.H. Streng, T.F. Walsh and P.M. Zerwas, *Z. Phys.* **C2** (1979) 237.

[40] ZEUS Coll., M. Derrick et al., *Z. Phys.* **C67** (1995) 93.

[41] ZEUS Coll., J. Breitweg et al., *Eur. Phys. J.* **C11** (1999) 251.

[42] A. Edin, G. Ingelman and J. Rathsman, *Phys. Lett.* **B366** (1996) 371.

[43] OPAL Coll., G. Abbiendi et al., *Eur. Phys. J.* **C19** (2001) 257.

[44] B.A. Schumm, Y.L. Dokshitzer, V.A. Khoze and D.S. Koetke, *Phys. Rev. Lett.* **69** (1992) 3025; J. Dias de Deus, *Phys. Lett.* **B338** (1994) 80 and **B355** (1995) 539; V.A. Petrov and A.V. Kisselev, *Z. Phys.* **C66** (1995) 453.

[45] OPAL Coll., R. Akers et al., *Z. Phys.* **C61** (1994) 209 and *Phys. Lett.* **B352** (1995) 176; G. Abbiendi et al., *Phys. Lett.* **B550** (2002) 33; DELPHI Coll., P. Abreu et al., *Phys. Lett.* **B347** (1995) 447; **B479** (2000) 118 and **B492** (2000) 398 ; P. Abreu and A. De Angelis, *Hadronization properties of b quarks compared to light quarks in* e^+e^- *annihilation into* $q\bar{q}$ *at* $\sqrt{s} = 206$ *GeV*, DELPHI 2002-052 CONF 586; SLD Coll., K. Abe et al., *Phys. Rev. Lett.* **72** (1994) 3145 and *Phys. Lett.* **B386** (1996) 475; DELCO Coll., M. Sakuda et al., *Phys. Lett.* **B152** (1985) 399; TPC Coll., H.Aihara et al., *Phys. Lett.* **B184** (1997) 299; TASSO Coll., W. Braunschweig et al., *Z. Phys.* **C42** (1989) 17; TOPAZ Coll., K. Nagai et al., *Phys. Lett.* **B278** (1992) 506; MARKII Coll., P.C. Rowson et al., *Phys. Rev. Lett.* **54** (1985) 2580.

[46] A. Kisselev, V. Petrov, O. Yushchenko, *Z. Phys.* **C41** (1988) 521.

[47] L3 Coll., P. Achard et al., *Phys. Lett.* **B577** (2003) 109; D.J. Mangeol, Ph.D. Thesis, Univ. of Nijmegen (2002).

[48] J. Brodsky and J.F. Gunion, *Phys. Rev. Lett.* **37** (1976) 402; K. Konishi, A. Ukawa, G. Veneziano, *Phys. Lett.* **B78** (1978) 243.

[49] A.H. Mueller, *Nucl. Phys.* **B241** (1984) 141; J.B. Gaffney, A.H. Mueller, *Nucl. Phys.* **B250** (1985) 109; E.D. Malaza, B.R. Webber, *Phys. Lett.* **B149** (1984) 501; E.D. Malaza, *Z. Phys.* **C31** (1986) 143; I.M. Dremin and R. Hwa, *Phys. Lett.* **B324** (1994) 477; I.M. Dremin and V.A. Nechitailo, *Mod. Phys. Lett.* **A9** (1994) 1471.

[50] A. Capella et al., *Phys. Rev.* **D61** (2000) 074009.

[51] S. Lupia and W. Ochs, *Phys. Lett.* **B418** (1998) 214, S. Lupia, *Phys. Lett.* **B439** (1998) 150.

[52] P. Edén and G. Gustafson, *JHEP* **09** (1998) 015; P. Edén, G. Gustafson and V. Khoze, *Eur. Phys. J.* **C11** (1999) 345; P. Edén, *Eur. Phys. Phys.J.* **C19** (2001) 493.

[53] JADE Coll., W. Bartel et al., *Z. Phys.* **C21** (1983) 37.

[54] UA2 Coll., P. Bagnaia et al., *Phys. Lett.* **144B** (1984) 291.

[55] HRS Coll., M. Derrick et al., *Phys. Lett.* **165B** (1985) 449.

[56] AMY Coll., M. Ye, Proc. 25th Int. Conf. on High Energy Physics, eds. K.K. Phua and Y. Yamaguchi (World Scientific, Singapore, 1991) p.889.

[57] OPAL Coll., G. Alexander et al., *Phys. Lett.* **B265** (1991) 462; P. Acton et al., *Z. Phys.* **C58** (1993) 387; R. Akers et al., *Z. Phys.* **C68** (1995) 179; G. Alexander et al., *Z. Phys.* **C69** (1996) 543; *Phys. Lett.* **B388** (1996) 659; K. Ackerstaff et al., *Eur. Phys. J.* **C1** (1998) 479; G. Abbiendi et al., *Euro Phys. J.* **C11** (1999) 217.

[58] CLEO Coll., M.S. Alam et al., *Phys. Rev.* **D46** (1992) 4822; **D56** (1997) 17.

[59] DELPHI Coll., P. Abreu et al., *Z. Phys.* **C56** (1992) 63; *Z. Phys.* **C70** (1996) 179; *Eur. Phys. J.* **C4** (1998) 1; *Phys. Lett.* **B449** (1999) 383.

[60] ALEPH Coll., D. Buskulic et al., *Phys. Lett.* **B346** (1995) 389, **B384** (1996) 353; R. Barate et al., *Z. Phys.* **C76** (1997) 191.

[61] J.W. Gary, *Phys. Rev.* **D49** (1994) 4503.

[62] CDF Coll., T. Affolder et al., *Phys. Rev. Lett.* **87** (2001) 211804.

[63] OPAL Coll., G. Abbiendi et al., *Eur. Phys. J.* **C23** (2002) 597.

[64] DELPHI Coll., K. Hamacher et al., Proc. 30th Int. Conf. on High Energy Physics, Osaka, Japan, eds. C. Lim and T. Yamanaka (World Scientific, Singapore, 2001) p.388; M. Siebel and K. Hamacher, ABS 247 at ICHEP02, DELPHI 2002-060, CONF 594 (July 2002); M. Siebel, K. Hamacher, J. Drees, DELPHI 2004-030, CONF 705 (July 2004).

[65] OPAL Coll., G. Abbiendi et al., *Phys. Rev.* **D69** (2004) 032002.

[66] I.M. Dremin and J.W. Gary, *Phys. Lett.* **B459** (1999) 341.

[67] UA5 Coll., G.J. Alner et al., *Phys. Lett.* **B138** (1984) 304.

[68] H. Müller, Proc. 25th Int. Conf. on High Energy Physics, eds. K.K. Phua and Y. Yamaguchi (South East Asia Theor. Phys. Assoc., Singapore, 1991) p.872.

[69] I.M. Dremin and R.C. Hwa, *Phys. Lett.* **B324** (1994) 477; *Phys. Rev.* **D49** (1994) 5805.

[70] DELPHI Coll., P. Abreu et al., *Phys. Lett.* **B401** (1997) 118; *Eur. Phys. J.* **C17** (2000) 207.

[71] L3 Coll., M. Acciari et al., *Phys. Lett.* **B407** (1997) 389.

[72] OPAL Coll., K. Ackerstaff et al., *Eur. Phys. J.* **C8** (1999) 241; G. Abbiendi et al., *Eur. Phys. J.* **C17** (2000) 373.

[73] S. Catani, Yu.L. Dokshitzer, F. Fiorani, B.R. Webber, *Nucl. Phys.* **B383** (1992) 419.

[74] M.H. Seymour, *Nucl. Phys.* **B421** (1994) 545; *Phys. Lett.* **B378** (1996) 279; J.R. Forshaw and M.H. Seymour, *J. High Energy Phys.* **09** (1999) 009.

[75] D0 Coll., V.M. Abazov et al., *Phys. Rev.* **D65** (2002) 052008.

[76] A.D. Martin, R.G. Roberts, W.J. Stirling, R.S. Thorne, *Eur. Phys. J.* **C4** (1998) 463.

[77] Y.I. Azimov et al., *Phys. Lett.* **B165** (1985) 147.

[78] V.A. Khoze, S. Lupia and W. Ochs, *Phys. Lett.* **B394** (1997) 179; *Eur. Phys. J.* **C5** (1998) 77.

[79] DELPHI Coll., J. Abdallah et al., CERN-PH-EP-2004-18, Phys. Lett. B to be published.

[80] Z. Koba, H.B. Nielsen and P. Olesen, *Nucl. Phys.* **B40** (1972) 317.

[81] M. Garetto and A. Giovannini, *Lett. Nuovo Cimento* **7** (1973) 35;
A. Giovannini et al., *Nuovo Cimento* **24A** (1974) 421;
N. Suzuki, *Prog. Theor. Phys.* **51** (1974) 1629;
W.J. Knox, *Phys. Rev.* **D10** (1974) 65.

[82] UA5 Coll., G.J. Alner et al., *Phys. Lett.* **B160** (1985) 199.

[83] UA5 Coll., G.J. Alner et al., *Phys. Lett.* **B160** (1985) 193.

[84] UA5 Coll., G.J. Alner et al., *Phys. Lett.* **B167** (1986) 476.

[85] NA22 Coll., M. Adamus et al., *Phys. Lett.* **B177** (1986) 239; *Z. Phys.* **C32** (1986) 475 and *Z. Phys.* **C37** (1988) 215.

[86] LEBC-MPS Coll., R. Ammar et al., *Phys. Lett.* **B178** (1986) 124.

[87] NA5 Coll., C. De Marzo et al., *Phys. Rev.* **D26** (1982) 1019; F. Dengler et al., *Z. Phys.* **C33** (1986) 187.

[88] NA23 Coll., J.L. Bailly et al., *Z. Phys.* **C40** (1988) 215.

[89] SFM Coll., A. Breakstone et al., *Nuovo Cim.* **102A** (1989) 1199.

[90] UA5 Coll., R.E. Ansorge et al., *Z. Phys.* **C43** (1989) 357.

[91] E735 Coll., A.O. Bouzas et al., *Z. Phys.* **C56** (1992) 107; T. Alexopoulos et al., *Phys. Lett.* **B435** (1998) 453.

[92] NA35 Coll., J. Bächler et al., *Z. Phys.* **C57** (1993) 541.

[93] CDF Coll., F. Rimondi, Proc. XXIII Int. Symp. on Multiparticle Dynamics, Aspen 1993, eds. M. Block and A. White (World Scientific, Singapore, 1994) p.400; F. Abe et al., *Phys. Rev.* **D50** (1994) 5550.

[94] E802 Coll., H.J. Tannenbaum, *Mod. Phys. Lett.* **A9** (1994) 89; T. Abbott et al., *Phys. Rev.* **C52** (1995) 2663.

[95] EMC Coll., M. Arneodo et al., *Z. Phys.* **C35** (1987) 335.

[96] WA29 Coll., D. Allasia et al., DFUB 88/9 (1988).

[97] HRS Coll., M. Derrick et al., *Phys. Lett.* **B168** (1986) 299; *Phys. Rev.* **D34** (1986) 3304.

[98] TASSO Coll., W. Braunschweig et al., *Z. Phys.* **C45** (1989) 193.

[99] DELPHI Coll., P. Abreu et al., *Z. Phys.* **C50** (1991) 185; ibid. **C52** (1991) 271; ibid. **C56** (1992) 63.

[100] ALEPH Coll., D. Decamp et al., *Phys. Lett.* **B273** (1991) 181; D. Buskulic et al., *Z. Phys.* **C69** (1995) 15.

[101] P. Carruthers and C.C. Shih, *Int. J. Mod. Phys.* **A2** (1987) 1447;
A. Giovannini, Proc. XXVI Int. Symp. on Multiparticle Dynamics, Faro 1996, eds. J. Dias de Deus et al. (World Scientific, Singapore, 1997) p.232.

[102] A. Giovannini and L. Van Hove, *Z. Phys.* **C30** (1986) 391;
L. Van Hove and A. Giovannini, Proc. of XVII Int. Symp. on Multiparticle Dynamics, Seewinkel (1986), eds. M. Markytan et al. (World Scientific, Singapore, 1987) p. 561.

[103] R. Szwed and G. Wrochna, *Z. Phys.* **C29** (1985) 255;
R. Szwed, G. Wrochna and A.K. Wróblewski, *Acta Phys. Pol.* **B19** (1988) 763; *Mod. Phys. Lett.* **A5** (1990) 1851 and **A6** (1991) 245; S. Carius and G. Ingelman, *Phys. Lett.* **B252** (1990) 647.

[104] A. Giovannini, R. Ugoccioni, *Phys. Rev.* **D59** (1999) 094020; J. Dias de Deus, R. Ugoccioni, *Phys. Lett.* **B469** (1999) 243.

[105] T. Mizoguchi, M. Biyajima, G. Wilk, *Phys. Lett.* **B301** (1993) 131; M. Biyajima et al., *Phys. Lett.* **B515** (2001) 470.

[106] S.G. Matinyan and W.D. Walker, *Phys. Rev.* **D59** (1999) 034022.

[107] V.D. Rusov et al., *Phys. Lett.* **B504** (2001) 213.

[108] R.P. Feynman, *Phys. Rev. Lett.* **23** (1969) 1415.

[109] J. Huskins, *Nucl. Phys.* **B46** (1972) 547, P. Olesen, *Nucl. Phys.* **B47** (1972) 157.

[110] W. Ochs, Proc. XX Int. Symp. On Multiparticle Dynamics, Gut Holmecke 1990, Eds. R. Baier and D. Wegener (World Scientific, Singapore, 1991) p.434.

[111] A.M. Polyakov, *Sov. Phys. - JETP* **32** (1971) 296 and **33** (1971) 850;
S.J. Orfanidis and V. Rittenberg, *Phys. Rev.* **D10** (1974) 2892;
W. Ochs, *Z. Phys.* **C23** (1984) 131;
G. Cohen-Tannoudji and W. Ochs, *Z. Phys.* **C39** (1988) 513.

[112] UA5 Coll., K. Alpgård et al., *Phys. Lett.* **B121** (1983) 209.

[113] UA1 Coll., C. Albajar et al., *Nucl. Phys.* **B335** (1990) 261.

[114] A. Wróblewski, *Acta Phys. Pol.* **B4** (1973) 857.

[115] A.I. Golokhvastov, *Sov. Jour. Nucl. Phys.* **27** (1978) 430 and **30** (1979) 128.

[116] W.A. Zajc, *Phys. Lett.* **B175** (1986) 219.

[117] W. Kittel, Proc. Workshop on Physics with Future Accelerators, La Thuile 1987, ed. J. Mulvey, CERN 87-7, Vol. II., p.424.

[118] A. Giovannini, L. Van Hove, *Acta Phys. Pol.* **B19** (1988) 495; L. Van Hove and A. Giovannini, *Acta Phys. Pol.* **B19** (1988) 917 and 931.

[119] O. Czyżewski and K. Rybicki, *Nucl. Phys.* **B47** (1972) 633; M. Blažek, *Z. Phys.* **C32** (1986) 309.

[120] S. Hegyi, *Phys. Lett.* **B466** (1999) 380; **B467** (1999) 126; *Nucl. Phys.* **B** (Proc. Supp.) **92** (2001) 122; Proc. XXXth Symp. on Multiparticle Dynamics, Tihany, eds. T. Csörgő et al. (World Scientific, Singapore, 2001) p.423.

[121] Yu.L. Dokshitzer, *Phys. Lett.* **B305** (1993) 295; LU-TP/93-3 (1993).

[122] L3 Coll., B. Adeva et al., *Z. Phys.* **C55** (1992) 39.

[123] I.M. Dremin, *Phys. Lett.* **B313** (1993) 209; I.M. Dremin and V.A. Nechitaĭlo, *Sov. Phys. JETP Lett.* **58** (1993) 881; *Mod. Phys. Lett.* **A9** (1994) 1471; I.M. Dremin, *Physics-Uspekhi* **37** (1994) 715; I.M. Dremin and R.C. Hwa, *Phys. Rev.* **D49** (1994) 5805.

[124] SLD Coll., K. Abe et al., *Phys. Lett.* **B371** (1996) 149.

[125] I.M. Dremin et al., *Phys. Lett.* **B336** (1994) 119.

[126] A. Giovannini, S. Lupia, R. Ugoccioni, *Phys. Lett.* **B342** (1995) 387; **B374** (1996) 231; **B388** (1996) 639.

[127] N. Nakajima, M. Biyajima and N. Suzuki, *Phys. Rev.* **D54** (1996) 4333; Wang Shaoshun et al., *Phys. Rev.* **D56** (1997) 1668.

[128] A. Capella et al., *Z. Phys.* **C75** (1997) 89; I.M. Dremin et al., *Phys. Lett.* **B403** (1997) 149.

[129] OPAL Coll., R. Akers et al., *Z. Phys.* **C63** (1994) 363.

[130] S. Lupia, W. Ochs, J. Wosiek, *Nucl. Phys.* **B540** (1999) 405.

[131] W.J. Metzger, Proc. XXIX Int. Symp. on Multiparticle Dynamics, eds. I. Chung and I. Sarcevic (World Scientific, Singapore, 2000) p.238.

[132] M.A. Buican, C. Förster, W. Ochs, *Eur. Phys. J.* **C31** (2003) 57.

[133] N. Suzuki, M. Biyajima and N. Nakajima, *Phys. Rev.* **D53** (1996) 3582; ibid. **D54** (1996) 3653; ibid. **D54** (1996) 4333.

[134] S. Hegyi, *Phys. Lett.* **B387** (1996) 642; **B388** (1996) 837; **B414** (1997) 210; **B417** (1998) 186.

[135] J. Dias de Deus, C. Pajares, C.A. Salgado, *Phys. Lett.* **B407** (1997) 335.

[136] Cai Xu, Liu Lian-Shou, *Nuovo Cimento Lett.* **37** (1983) 495; X. Cai, W. Chao, T. Meng and C. Huang, *Phys. Rev.* **D33** (1986) 1287;
W. Chao, T. Meng and J. Pan, *Phys. Lett.* **B176** (1986) 211; *Phys. Rev.* **D35** (1987) 152 and *Phys. Rev. Lett.* **58** (1987) 1399.

[137] K. Kudo and E.R. Nakamura, *Phys. Lett.* **B191** (1987) 195.

[138] K. Fiałkowski, *Phys. Lett.* **B169** (1986) 436; **B173** (1986) 197.

[139] Wu Yuanfang and Liu Lianshou, *Phys. Rev.* **D41** (1990) 845.

[140] H1 Coll., S. Aid et al., *Nucl. Phys.* **B445** (1995) 3.

[141] A. Capella, A. Staar, J. Tran Thanh Van, *Phys. Rev.* **D32** (1985) 2933.

[142] R. Hagedorn, *Suppl. Nuovo Cimento* **3** (1965) 147; G.J.H. Burgers, Ch. Fuglesang, R. Hagedorn, V. Kuvshinov, *Z. Phys.* **C46** (1990) 465.

[143] E.L. Berger and G.C. Fox, *Phys. Lett.* **B47** (1973) 162.

[144] A. Białas, I. Derado and L. Stodolsky, *Phys. Lett.* **B156** (1985) 421.

[145] S. Barshay, *Nucl. Phys.* **238** (1984) 277 and *Z. Phys.* **C32** (1986) 513; *Mod. Phys. Lett.* **A2** (1987) 693 and **A6** (1991) 55; S. Barshay and E. Eich, *Phys. Lett.* **B178** (1986) 431 and *Mod. Phys. Lett.* **A1** (1986) 459.

[146] T.T. Chou, C.N. Yang and E. Yen, *Phys. Rev. Lett.* **54** (1985) 510; T.T. Chou and C.N. Yang, *Phys. Rev.* **D32** (1985) 1692.

[147] Y.K. Lim and K.K. Phua, *Phys. Rev.* **D26** (1982) 1785; C.H. Kam, Y.K. Lim and K.K. Phua, *Z. Phys.* **C26** (1984) 381.

[148] G. Pancheri, Y. Srivastava and M. Pallota, *Phys. Lett.* **B151** (1985) 453; G. Pancheri and Y.N. Srivastava, *Phys. Lett.* **B159** (1985) 69 and **B182** (1986) 199.

[149] S. Rudaz and P. Valin, *Phys. Rev.* **D34** (1986) 2025; P. Valin, *Z. Phys.* **C34** (1987) 313.

[150] T.K. Gaisser et al., *Phys. Lett.* **B166** (1986) 219; A.D. Martin and C.J. Maxwell, *Z. Phys.* **C34** (1987) 71.

[151] V. Gupta and N. Sarma, *Z. Phys.* **C41** (1988) 413.

[152] A. Ballestrero, B. Carazza, *Z. Phys.* **C50** (1991) 61.

[153] B. Durand and I. Sarcevic, *Phys. Lett.* **B172** (1986) 104; *Phys. Rev.* **D36** (1987) 2693; I. Sarcevic, *Phys. Rev. Lett.* **59** (1987) 403.

[154] A. Giovannini, *Nucl. Phys.* **B161** (1979) 429; C.S. Lam and M.A. Walton, *Phys. Lett.* **B140** (1984) 246; D.C. Hinz and C.S. Lam, *Phys. Rev.* **D33** (1986) 3256; R.C. Hwa and C.S. Lam, *Phys. Lett.* **B173** (1986) 346 and C.S. Lam, Proc. 21st Rencontre de Moriond (1986), ed. J. Tran Thanh Van (Editions Frontières) Vol.II, p.241; C.C. Shih, *Phys. Rev.* **D33** (1986) 3391 and **D34** (1986) 2710; C.K. Chew, D. Kiang and H. Zhou, *Phys. Lett.* **B186** (1987) 411; R.C. Hwa, *Phys. Rev.* **D37** (1988) 1830; A.V. Batunin, *Phys. Lett.* **B212** (1988) 495; A.H. Chan and C.K. Chew, *Nuovo Cimento* **101A** (1989) 409.

[155] M. Biyajima, T. Kawabe and N. Suzuki, *Phys. Lett.* **B189** (1987) 466; M. Biyajima, K. Shirane and N. Suzuki, *Phys. Rev.* **D37** (1988) 1824.

[156] P.V. Chliapnikov and O.G. Tchikilev, *Phys. Lett.* **B222** (1989) 152; **B223** (1989) 119; **B235** (1990) 347.

[157] Y.P. Chan, K. Young, D. Kiang, T. Ochiai, *Phys. Rev.* **D42** (1990) 3037.

[158] M. Biyajima et al., *Phys. Rev.* **D43** (1991) 1541; G.N. Fowler et al., *Phys. Rev.* **D37** (1988) 3127; *J. Phys.* **G16** (1990) 1439; E.M. Friedlander, F.W. Pottag and R.M. Weiner, *J. Phys.* **G15** (1989) 431.

[159] F. Hayot and H. Navelet, *Phys. Rev.* **D30** (1984) 2322.

[160] P. Carruthers and I. Sarcevic, *Phys. Lett.* **B189** (1987) 442.

[161] B. Åsman, Ph. D. Thesis, Univ. of Stockholm (1985).

[162] G. Ekspong, Proc. XVI Int. Symp. on Multiparticle Dynamics, Kiryat Anavim (1985), ed. J. Grunhaus (Editions Frontières, World Scientific, 1986) p. 309.

[163] S. Hegyi, *Phys. Lett.* **B274** (1992) 214; S. Hegyi and T. Csörgő, *Phys. Lett.* **B296** (1992) 256; A. Giovannini, S. Lupia, R. Ugoccioni, *Z. Phys.* **C59** (1993) 427.

[164] A. Giovannini, S. Lupia, R. Ugoccioni, *Z. Phys.* **C70** (1996) 291.

[165] F. Becattini, A. Giovannini and S. Lupia, *Z. Phys.* **C72** (1996) 491.

[166] J. Dias de Deus, *Phys. Lett.* **B178** (1986) 301 and **B185** (1987) 189.

[167] L. Van Hove, *Physica* **147A** (1987) 19.

[168] P.V. Chliapnikov and O.G. Tchikilev, *Phys. Lett.* **B242** (1990) 275; ibid. **B282** (1992) 471; P.V. Chliapnikov, O.G. Tchikilev and V.A. Uvarov, *Phys. Lett.* **B352** (1995) 461; O.G. Tchikilev, *Phys. Lett.* **B382** (1996) 296; ibid. **B388**(1996) 848; ibid. **B393** (1997) 198; ibid. **B471** (2000) 400.

[169] N. Suzuki, M. Biyajima and G. Wilk, *Phys. Lett.* **B268** (1991) 447; N. Suzuki, M. Biyajima, N. Nakajima, *Phys. Rev.* **D53** (1996) 3582; ibid. **D54** (1996) 3653; ibid. 4333.

[170] WA21 Coll., C.J. Jones et al. (WA21), *Z. Phys.* **C54** (1992) 45.

[171] C. Fox, *Trans Amer. Math. Soc.* **98** (1961) 395; A.M. Mathaiand, R.K. Saxena, *The H-Function with Applications in Statistics and Other Disciplines* (Wiley Eastern, 1978).

[172] P. Carruthers, Proc. LESIP I Workshop, eds. D. Scott and R. Weiner (World Scientific, Singapore, 1985) p.390.

[173] ALEPH Coll., R. Barate et al., *Phys. Reports* **294** (1998) 1.

[174] OPAL Coll., K. Ackerstaff et al., *Z. Phys.* **C75** (1997) 193.

[175] S. Hegyi, *Phys. Lett.* **B411** (1997) 321; *Asymptotic multiplicity scaling: a renormalization group perspective* (1997), hep-ph/9709326.

[176] M. Biyajima, *Prog. Theor. Phys.* **69** (1983) 966; M. Biyajima and N. Suzuki, *Prog. Theor. Phys.* **73** (1985) 918; P. Carruthers and C. Shih, *Phys. Lett.* **B137** (1984) 425; M. Weiner, Proc. 2nd Int. Workshop on Local Equilibrium in Strong Interaction Physics, Santa Fe, 1986 (World Scientific, Singapore) p. 106; M. Blazek, *Czech. J. Phys.* **B38** (1988) 705.

[177] S. Krasznovszky and I. Wagner, *Nuovo Cimento* **A76** (1983) 539; *Can. J. Phys.* **62** (1984) 330; *Europhys. Lett.* **5** (1988) 395; *Phys. Lett.* **B213** (1988) 103; **B228** (1989) 159; **B241** (1991) 605; **B295** (1992) 320; **B306** (1993) 403; S. Hegyi and S. Krasznovszky, *Phys. Lett.* **B235** (1990) 203; **B241** (1990) 605 and 611.

[178] S. Krasznovszky, *Canad. J. Phys.* **76** (1998) 4553.

[179] K. Goulianos, *Phys. Lett.* **B193** (1987) 151.

[180] P. Geraedts, T. Haupt, W. Kittel, *Z. Phys.* **C44** (1989) 331.

[181] Ch. Fuglesang, *Multiparticle Dynamics*, Festschrift L. Van Hove, edts. A. Giovannini and W. Kittel (World Scientific, Singapore, 1990) p. 193.

[182] UA1 Coll., C. Albajar et al., *Nucl. Phys.* **B309** (1988) 405.

[183] A. Wehrl, *Rev. Mod. Phys.* **50** (1978) 221.

[184] V. Šimák, M. Šumbera and I. Zborovský, *Phys. Lett.* **B206** (1988) 159; M. Pachr, V. Šimák, M. Šumbera and I. Zborovský, *Mod. Phys. Lett.* **A7** (1992) 2333; Proc. XIXth Int. Symp. on Multiparticle Dynamics 1988, eds. D. Schiff and J. Tran Than Van (Editions Frontières, Gif-sur-Yvette, 1988) p.399; Proc. XXVth Int. Symp. on Multiparticle Dynamics 1995, eds. D. Bruncko et al. (World Scientific, Singapore, 1996) p.617.

[185] E735 Coll., T. Alexopoulos et al., *Multiplicity distributions of charged particles from proton-antiproton collisions at the Tevatron collider at* $\sqrt{s} = 1.8$ *TeV*, The Vancouver Meeting, Particles and Fields 1991, eds. D. Axen and D. Bryman (World Scientific, Singapore) p.748.

[186] D. Ghosh et al., *Canad. J. Phys.* **70** (1992) 667.

[187] A. Mukhopadhyay et al., *Phys. Rev.* **C47** (1993) 410.

[188] P. Brogueira, J. Dias de Deus and I.P. da Silva, *Phys. Rev.* **D53** (1996) 5283.

[189] P. Carruthers, *Recent trends in strong interaction physics phenomenology*, Arizona preprint AZPH-TH/90-98 (1990); P. Carruthers, *Galaxy correlations, counts and moments*, Arizona preprint AZPH-TH/91-08 (1991).

[190] P. Carruthers, H.C. Eggers, Q. Gao and I. Sarcevic, *Int. J. Mod. Phys.* **A6** (1991) 3031; P. Carruthers, H.C. Eggers and I. Sarcevic, *Phys. Lett.* **B254** (1991) 258.

[191] E.A. De Wolf, Proc. Ringberg Workshop on Multiparticle Production, eds. R.C. Hwa et al. (World Scientific, Singapore, 1992) p.222.

[192] S.D.M. White, *Mon. Not. R. Astron. Soc.* **186** (1979) 145.

[193] J. Fry, R. Giovanelli, M.P. Haynes, A. Molott and R.J. Scherrer, *Astrophys. J.* **340** (1989) 11.

[194] S. Hegyi, *Phys. Lett.* **B274** (1992) 214.

[195] D. Ghosh et al., *Astroparticle Phys.* **15** (2001) 329 and *Multiparticle production in ^{16}O-AgBr interaction at 60 A GeV - evidence of void probability scaling*, Jadavpur preprint 2000.

[196] NA22 Coll., V.V. Aivazyan et al., *Z. Phys.* **C42** (1989) 533.

[197] Yuanxiu Ye et al., *Phys. Lett.* **B382** (1996) 196.

[198] S. Uhlig et al., *Nucl. Phys.* **B132** (1978) 15.

[199] UA5 Coll., K. Alpgård et al., *Phys. Lett.* **B123** (1983) 361;
G.J. Alner et al., *Nucl. Phys.* **B291** (1987) 445;
R.E. Ansorge et al., *Z. Phys.* **C37** (1988) 191.

[200] E735 Coll., T. Alexopoulos et al., *Phys. Lett.* **B353** (1995) 155.

[201] ABBCLMO Coll., H. Grässler et al., *Nucl. Phys.* **B223** (1983) 269.

[202] D. Zieminska, *Phys. Rev.* **D27** (1983) 502.

[203] EMC Coll., M. Arneodo et al., *Nucl. Phys.* **B258** (1985) 249.

[204] TASSO Coll., M. Althoff et al., *Z. Phys.* **C29** (1985) 347.

[205] HRS Coll., M. Derrick et al., *Z. Phys.* **C35** (1987) 323.

[206] OPAL Coll., R. Akers et al., *Phys. Lett.* **B320** (1994) 417.

[207] ZEUS Coll., J. Breitweg et al., *Euro. Phys. J.* **C12** (2000) 53.

[208] P. Carruthers and C.C. Shih, *Phys. Lett.* **B165** (1985) 209;
M. Braun, C. Pajares, V.V. Vechernin, *Phys. Lett.* **B493** (2000) 54.

[209] M. Biyajima et al., *Z. Phys.* **C44** (1989) 199.

[210] S.V. Chekanov, *J. Phys.* **G25** (1999) 59.

[211] T.T. Chou and C.N. Yang, *Phys. Lett.* **B135** (1984) 175; **B167** (1986) 453 and **B171** (1986) 486 (E); **B193** (1987) 531; *Int. J. Mod. Phys.* **A2** (1987) 1727.

[212] S.L. Lim, C.H. Oh and K.K. Phua, *Z. Phys.* **C43** (1989) 621; ibid. **C54** (1992) 107.

[213] L.K. Chen, D. Kiang and C.K. Chew, *Phys. lett.* **B408** (1997) 422.

[214] A. Giovannini and R. Ugoccioni, *Phys. Rev.* **D66** (2002) 034001; *Phys. Lett.* **B558** (2003) 59.

Chapter 9

Experimental Results on Correlations

In this chapter, we review experimental results on "classical" correlations, a subject with a long history in particle physics. It was instrumental in establishing fundamental concepts of hadrodynamics, such as short-range order, which are essential ingredients of all popular Monte-Carlo models of hadronization. With the exception of Bose-Einstein interferometry, the field lay dormant for several years, but was revived with the introduction of generalized concepts. The data cover a variety of multiparticle production processes ranging from e^+e^- annihilation to nucleus-nucleus collisions.

Interest in correlation functions received a vigorous boost when their intimate connection with factorial moments was realized (see Chapters 7 and 10). Both are now explored with novel techniques. These offer promising perspectives towards a long overdue unified approach to correlation phenomena, including Bose-Einstein correlations to be discussed in Chapter 11.

Multiplicity distributions inspired many early ideas on scale-invariance and phase-transition analogies in multiparticle production, such as Koba-Nielsen-Olesen scaling [1] and the Wilson-Feynman liquid picture with clustering fluctuations [2]. However, the major part of the data relate either to full phase space or to sizable portions of it. It remains an interesting task to explain the *large-scale* properties of multiplicity distributions in terms of correlation function behavior at *small momentum distances*, the main subject of these last three chapters. Of course, the factorial moments to be discussed in Chapter 10 are just another representation of multiplicity distributions and their increase with decreasing bin size reveals the evolution of the multiplicity distribution.

9.1 Rapidity correlations

The study of correlation effects in particle production processes provides information on hadronic production dynamics beyond that obtained from single-particle inclusive spectra. Correlations in rapidity y, as defined in Sect. 7.1, have first been studied in the early '70ies on fixed-target $\pi^{\pm}$p and pp data in the $p_{\text{LAB}} = 10 - 30$ GeV/c range [3–7] and shortly later at Serpukhov, FNAL and ISR energies [8–15] (for early reviews see [16, 17]). Strong y-correlations have been observed in all experiments in one form or another, depending on the specific form of the correlation function, type of interaction, kind of particles, the kinematic region under consideration, etc. The main conclusions were:

1. Two-particle correlations are strong at small interparticle rapidity-distances $|y_1 - y_2|$ (see Fig. 9.1), i.e. are of short range, but most of them are pseudo-correlations due to the difference of semi-inclusive single-particle distributions.
2. The "correlation length" in y does not depend on the collision energy, but
3. the ridge along $y_1 = y_2$ gets elongated as the energy increases.
4. The correlations depend on the two-particle charge combination and on transverse momentum.
5. There also exists a small positive long-range correlation for large $|y_1 - y_2|$, associated with the diffractive component.

These early observations have stimulated their interpretation in terms of cluster models (e.g. [18–20]), with a Gaussian correlation function of half width about one

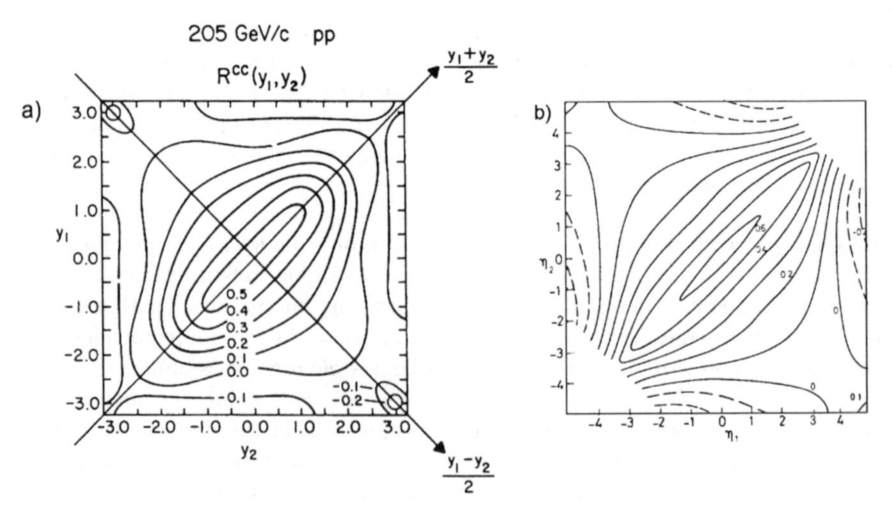

Figure 9.1. Contours of the two-particle correlation function, $R^{\text{cc}}(y_1, y_2)$, from pp interactions at a) $p_{\text{LAB}} = 205$ GeV/c [16], b) $\sqrt{s} = 63$ GeV [11].

rapidity unit. We shall see in the next chapter, however, that this conclusion is mainly due to the relatively large bin sizes chosen in early analysis. Even though already Foà [17] had realized a triangular (rather than Gaussian) shape of the correlation function, this misconception has stayed with us for more than 30 years and is one of the most difficult prejudices to be eliminated.

9.1.1 Correlations in hadron-hadron collisions

In Fig. 9.2a the pseudo-rapidity correlation function $C_2(\eta_1, \eta_2)$ as defined in (7.25) is given for $\eta_1 = 0$, as a function of $\eta_2 = \eta$, for the energy range between 63 and 900 GeV [21]. Whereas $C_2(0, \eta)$ depends on energy, the correlation $C_{\rm S}$ defined in (7.37) does not strongly depend on energy and has a full width of about 2 units in pseudo-rapidity (Fig. 9.2c). The function $C_{\rm L}$ defined in (7.38) is not a two-particle rapidity correlation proper, but derives from the multiplicity distribution and the difference in the single-particle distribution function for different multiplicities. As can be seen in Fig. 9.2b, the dominant pseudo-correlation $C_{\rm L}$ is considerably wider than $C_{\rm S}$ and increases with energy (the 63 GeV data are from [11]). It does not correspond to any dynamical effect within individual events, however.

In Fig. 9.3, the semi-inclusive correlation $C_2^{(n)}(\eta_1, \eta_2)$ for $\rm p\bar{p}$ collisions at 900 GeV [22] is compared to the UA5 Cluster Monte Carlo (MC) GENCL [23], as well as to the FRITIOF 2 and PYTHIA Monte Carlos, for charged-particle multiplicity $34 \leq n \leq$

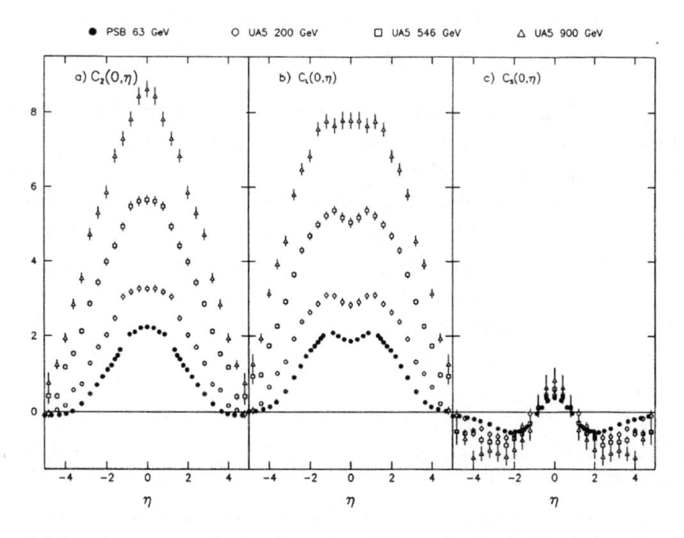

Figure 9.2. a) The charge correlation function $C_2(\eta_1, \eta_2)$ plotted for $\rm p\bar{p}$ collisions at fixed $\eta_1 = 0$ versus η_2 at 63, 200, 546 and 900 GeV, b) the contribution $C_{\rm L}$ and c) the contribution $C_{\rm S}$ [21].

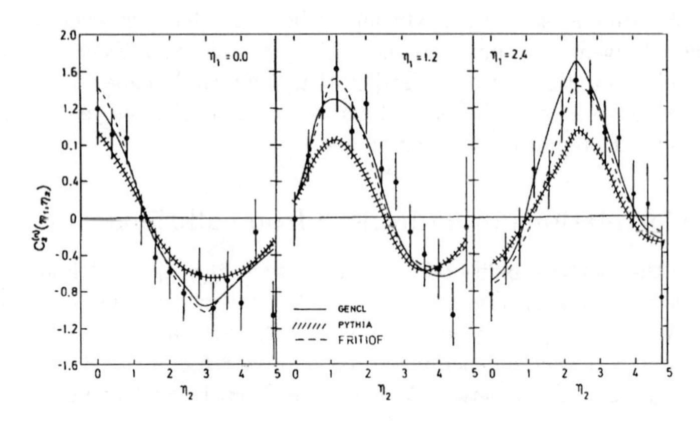

Figure 9.3. The semi-inclusive correlation function $C_2^{(n)}(\eta_1, \eta_2)$ for $34 \leq n \leq 38$ p$\bar{\text{p}}$ collisions at 900 GeV, compared to the UA5 Cluster MC, PYTHIA and FRITIOF 2 [22].

38. The Cluster MC is designed to fit just these short-range correlations, but also FRITIOF 2 is doing surprisingly well (see however Sub-Sect. 10.4.4 below).

At lower energy, the NA23 Collaboration [24] studied the correlation of different combinations of charged particles in pp collisions of $\sqrt{s} = 26$ GeV in terms of $K_2(y_1, y_2)$ defined in (7.32). Only events with charged-particle multiplicity $n > 6$ are used. The positive correlations for $y_1 \approx y_2$ are in agreement with those found earlier at $\sqrt{s} = 53$ GeV [25] and interpreted in terms of the production of clusters of particles.

The NA23 data are compared to single-string Lund and to a two-chain Dual-Parton Model (DPM) in Fig. 9.4. The one-string model (without gluon radiation) does not at all describe the correlation in the data. The two-chain model does better, but remains unsatisfactory. Somewhat better but still insufficient agreement is obtained by renormalizing the MC events to the experimental multiplicity distribution (not shown). The effect of Bose-Einstein correlations in the (++) and (– –) data is found to be insignificant, as may be expected for data integrated over transverse momentum p_{T} and azimuthal angle φ. Obviously, more chains, possibly with higher p_{T}, are needed to explain short-range order with fragmentation models, even below $\sqrt{s} \approx 30$ GeV.

NA22 results for $C_2(0, y_2)$ and $\tilde{C}_2(0, y_2)$ (Eqs. 7.25, 7.31) for π^+p and K$^+$p collisions at $\sqrt{s}$=22 GeV [26] are compared with FRITIOF 2, a 2-string DPM and QGSM predictions in Fig. 9.5a,b. FRITIOF and 2-string DPM largely underestimate the correlation. QGSM reproduces $C_2^{--}(0, y_2)$ very well and even overestimates $C_2^{++}(0, y_2)$ and $C_2^{+-}(0, y_2)$. It has been verified that the differences between QGSM and FRITIOF or DPM are not due to the different treatment of tensor mesons (only included in the latter two).

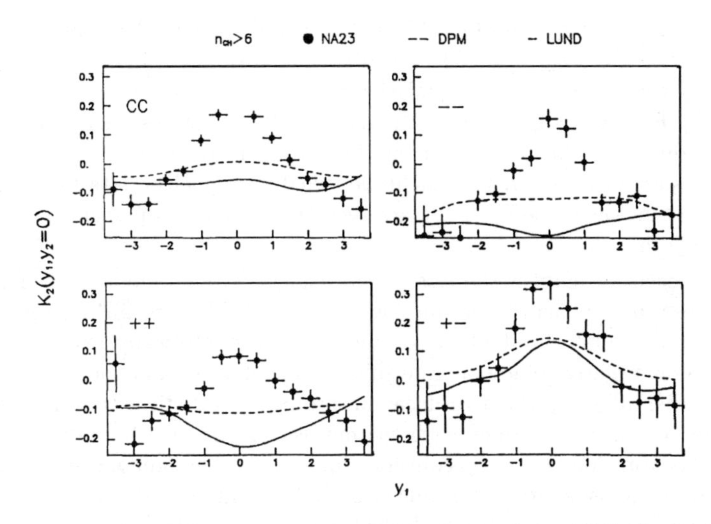

Figure 9.4. Normalized correlation function $K_2(y_1, y_2 = 0)$ for (CC), (– –), (+ +) and (+ –) combinations in $n > 6$ pp collisions at 360 GeV/c, compared to predictions from single chain Lund and a two-chain DPM [24].

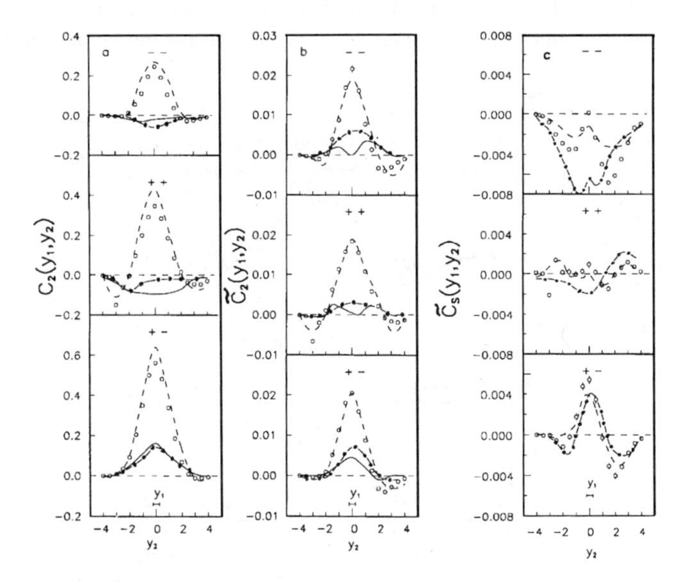

Figure 9.5. a,b) Correlation functions $C_2(0, y)$ and $\tilde{C}_2(0, y)$ for M$^+$p reactions as compared with calculations in FRITIOF (– · – · –), DPM (——) and QGSM (– – –) (non-single-diffractive sample). c) Correlation function $\tilde{C}_S(0, y_2)$ for M$^+$p reactions as compared with FRITIOF (– · – · –) and QGSM (– – –) [26].

In Fig. 9.5c, FRITIOF and QGSM are compared to the NA22 data in terms of the contribution $\tilde{C}_S(0, y_2)$. The $(+-)$ short-range correlation is reproduced reasonably well by these models. For equal charges, however, the strong anti-correlation predicted by FRITIOF is not seen in the data. QGSM contains a small like-charged correlation due to a cluster component, but still underestimates its size. Similar discrepancies are observed in semi-inclusive (fixed multiplicity) data for each charge combination (not shown here). They are even larger than in the inclusive data, also in the QGSM model.

A more recent attempt is the Rossendorf collision (ROC) model [27]. This is designed to describe soft hadron production from threshold up to ISR energies by first producing intermediate fire-balls according to an asymptotically exponential mass distribution, a negative-binomial multiplicity distribution and longitudinal phase space, and then allowing them to decay according to statistical considerations and phase space. In Fig. 9.6, the model (histogram) is compared to C_2 for different charge combinations from 200 GeV/c pp collisions (dots) [16]. The pronounced peaks in the data are indeed satisfactorily reproduced by ROC, while PYTHIA (solid line) and a model version without correlations (dashed line) are not sufficient.

From this brief survey, we conclude that in hadron-hadron collisions two-particle correlations are dominated by multiplicity effects. They are badly reproduced and generally underestimated in commonly used models. Improvement can be obtained either via a large number of chains (as in QGSM) or intermediate clusters or fire-balls (as in ROC).

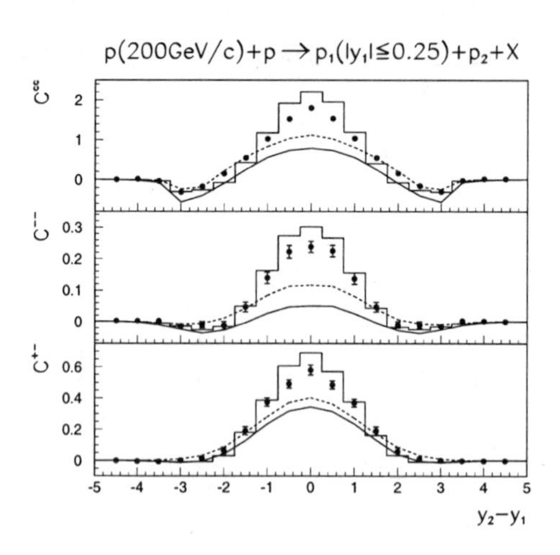

Figure 9.6. The correlation function $C_2(y_1, y_2)$ for all charged particles as well as for pairs of negative particles and positive particles produced in 200 GeV/c pp collisions [16]. The histograms correspond to the ROC model, full lines to PYTHIA and dashed lines to ROC without correlations [27].

9.1.2 Correlations in e^+e^- and μ^+p-collisions

Fig. 9.7 shows $K_2^{+-}(y_1, y_2)$ and $K_2^{--}(y_1, y_2)$ for muon-nucleon interactions at 280 GeV/c [28]. A steep peak is seen at $y_1 = y_2 = 0$ for K_2^{+-}, with two shoulders along the diagonal $y_1 = y_2$. On the other hand, K_2^{--} is below 0 for most of the distribution, but we shall see in Chapters 10, 11 that the most impressive correlation is in fact coming from $y_1 \approx y_2$, just for this case, be it only at higher resolution. As in hadron-hadron collisions, correlations are strong and depend on the two-particle charge combination.

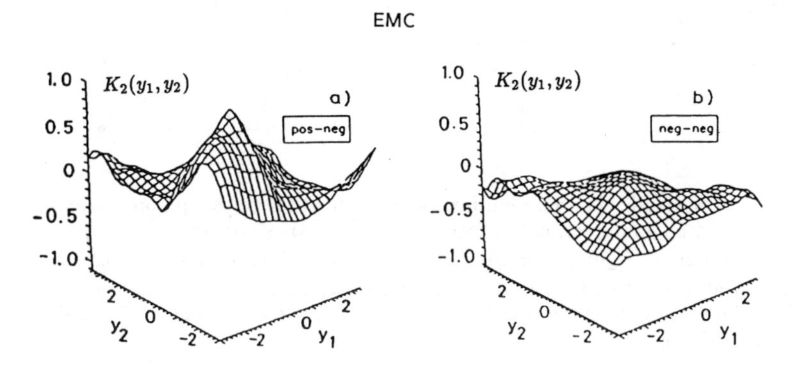

Figure 9.7. Normalized two-particle correlation function $K_2(y_1, y_2)$ for pairs of a) oppositely and b) negatively charged hadrons produced in muon-nucleon scattering at 280 GeV/c [28].

Fig. 9.8a shows $K_2(y_1, y_2)$ in μ^+p interactions at 280 GeV/c with $y_1 \in [-0.5, 0.5]$, the hadronic invariant mass W in the interval $13 < W < 20$ GeV and for $n \geq 3$ [29], together with the NA22 non-single-diffractive M$^+$p sample, $n \geq 2$ [26]. Correlations in μ^+p seem smaller than in NA22, but extrapolating from the energy dependence of $K_2(0, 0)$ published in [29], one finds quite similar values for μ^+p at 22 GeV and M$^+$p in NA22. What is seen more clearly in μ^+p than in M$^+$p correlations is a forward-backward asymmetry. The correlation is stronger in the backward hemisphere than in the forward one. It has been shown in [28] that this is not due to the presence of protons in the backward hemisphere but points to a difference in diquark and quark fragmentation.

In Fig. 9.8b, we compare the function $\tilde{K}_2(0, y)$ for the NA22 non-single-diffractive M$^+$p sample ($n \geq 2$) [26] with that for e^+e^--annihilation at the same energy ($\sqrt{s}$=22 GeV) [30]. The values of $\tilde{K}_2(0, y)$ are larger for (++) pairs than for (−−) in M$^+$p reactions; for (−−) and (+−) pairs, however, they agree with $\tilde{K}_2$ for e^+e^- annihilation in the central region.

A comparison of the correlation functions for pairs of any charge (cc) in e^+e^--annihilation and non-single-diffractive M$^+$p collisions with $y_1 \in [-1, 0]$ is shown in Fig. 9.8c. The e^+e^- data are given at $\sqrt{s} = 14$ and 44 GeV [30]. At $y_2 = y_1$, the 22

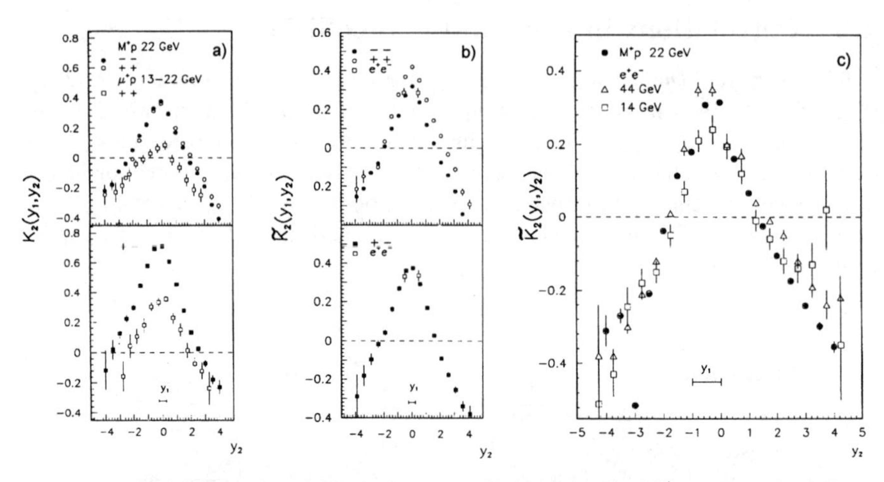

Figure 9.8. a) Normalized correlation functions $K_2(0, y)$ for the M$^+$p non-single-diffraction sample [26] and μ^+p-interactions at 280 GeV/c ($13 < W < 20$ GeV) [29]. b) Normalized correlation functions $\tilde{K}_2(0, y)$ for the M$^+$p non-single-diffraction sample [26] and e$^+$e$^-$-annihilation at $\sqrt{s}$=22 GeV [30]. c) Normalized correlation function $\tilde{K}_2^{cc}(y_1, y_2)$ at $y_1 = -1 \div 0$ in the non-single-diffraction M$^+$p sample at 22 GeV [26] and e$^+$e$^-$-annihilation at 14 and 44 GeV [30].

GeV M$^+$p correlation lies between the e$^+$e$^-$ results.

For μ^+p [28, 29] and e$^+$e$^-$ collisions [30–32], the Lund-type Monte Carlo is reported to reproduce the majority of the experimental distributions, mainly due to the inclusion of hard and soft gluon effects. However, important underestimates of $K_2(y_1, y_2)$ are still observable, in particular in the central and current fragmentation regions. For e$^+$e$^-$ [31], this is shown in Fig. 9.9, where $K_2(y_1, y_2)$ is compared to JETSET 7.2 PS as a function of $y_1 - y_2$ (dotted line), for the full sample (upper plots) and for a two-jet sample (lower plots). In all cases, the JETSET model underestimates the correlation at $y_1 - y_2 = 0$. In general, the disagreement becomes smaller when Bose-Einstein correlations are included (full lines). The main feature to note is that correlations are much weaker in the two-jet sample than in the full sample. Furthermore, correlations are larger for $y \leq 0$ (left plot), i.e. in the hemisphere opposite the most energetic jet, than for $y > 0$ (right plot). These two observations, again, point to hard gluon radiation as the main source of two-particle correlation in e$^+$e$^-$ collisions.

A systematic test of analytic QCD calculations and of QCD Monte-Carlo models for two-particle correlations has been performed by OPAL [32]. The authors study the function

$$R(\xi_1, \xi_2) = K_2(\xi_1, \xi_2) + 1 \tag{9.1}$$

with $\xi = \ln(1/x)$, $x = 2p/E_{\rm cm}$ being the Feynman variable, i.e. the particle momen-

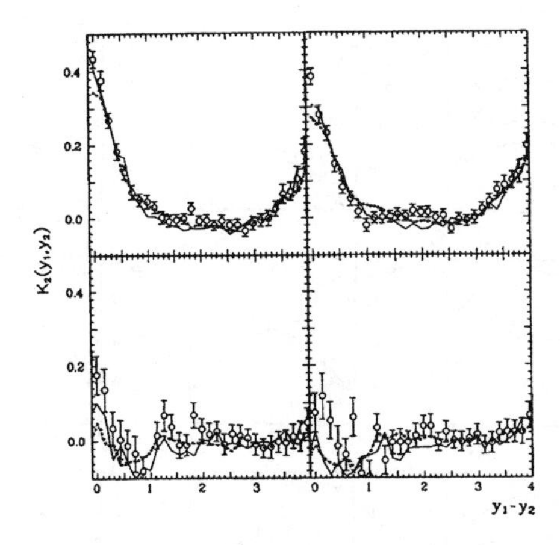

Figure 9.9. Normalized correlation functions $K_2(y_1, y_2)$ for the entire data sample (upper plots) and for selected two-jet events (lower plots) in e^+e^- annihilation at 35 GeV. The left plots show the average over the region $-1 \leq y_1 + y_2 \leq 0$ and the right plots correspond to the average over $0 \leq y_1 + y_2 \leq 1$. The data (open circles) are compared to the JETSET 7.2 PS model with (solid lines) and without (dotted lines) Bose-Einstein correlations [31].

tum p in the cms normalized to half the cms energy E_{cm}. In Fig. 9.10, R is plotted as a function of $(\xi_1 - \xi_2)$ for $(\xi_1 + \xi_2)$ centered at the values 6, 7 and 8, respectively. Fig. 9.10a proves that a next-to-leading order calculation [33] (full lines) is better than leading order (dashed), but still overestimates the overall level of the correlation for any reasonable value of Λ. Since the next-to-leading correction is large, still higher-order terms are needed.

Higher-order effects are, in an average sense, included in the existing Monte-Carlo models. In Fig. 9.10b, the same data are compared to the coherent parton shower models JETSET PS, HERWIG and ARIADNE. The latter gives an excellent fit to the data, JETSET lies slightly below (within uncertainty of parameters), but HERWIG considerably above. The agreement of JETSET could only slightly be improved by including Bose-Einstein correlations. As far as incoherent parton shower models are concerned, none of the various versions of COJETS gives a satisfactory representation of the correlation data.

All the models were tuned on the OPAL data in terms of event shapes and generally describe single-particle distributions. It is clear that correlations allow better and more discriminative tests than more integrated quantities.

String-hadronization models experience difficulties in predicting like-charged correlations in hadron-hadron collisions. It is important to verify if the otherwise successful e^+e^- models are also able to reproduce correlations between charge-separated systems such as $(+-)$ and $(\pm\pm)$ particle pairs.

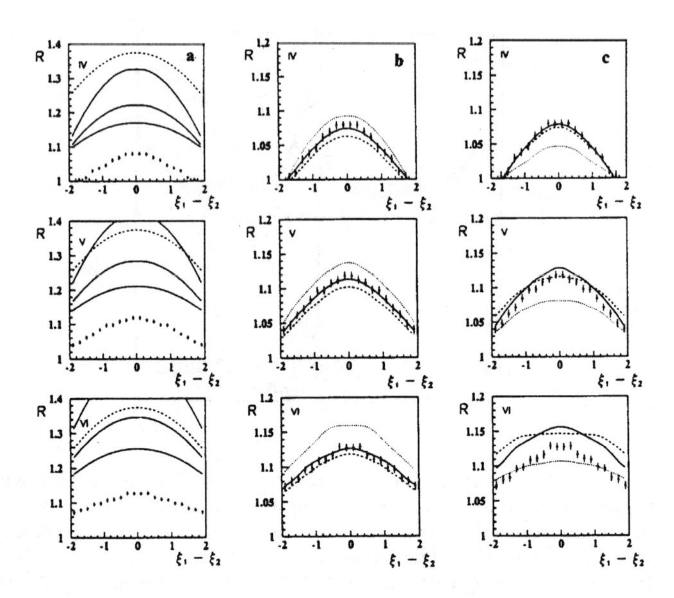

Figure 9.10. OPAL data on $R(\xi_1, \xi_2)$ as a function of $(\xi_1 - \xi_2)$, for $(\xi_1 + \xi_2)$ between 5.9 and 6.1 (IV), 6.9 and 7.1 (V), 7.9 and 8.1 (VI), respectively, compared to a) analytic QCD calculations (the dashed curves indicate leading order QCD calculations for Λ=225 MeV, the three solid curves represent next-to-leading QCD calculations with, from top to bottom, $\Lambda = 1000$, 255 and 50 MeV, respectively), and b) coherent parton-shower Monte-Carlo models ARIADNE (solid), JETSET (dashed) and HERWIG (dotted); c) incoherent parton-shower Monte-Carlo models JETSET (solid) and two versions of COJETS (dashed and dotted) [32].

9.1.3 Quantum number dependence

How C_S and $\tilde{C}_S$ depend on the charge of the pairs is shown in Fig. 9.11 for the combinations $(--), (++)$ and $(+-)$ in NA22 [26]. The short-range correlation is significantly larger for $(+-)$ than for $(--)$ and $(++)$ combinations. This is also seen in the EMC data [28]. Resonance production and local charge conservation are explanations of this difference. For like charges, a small enhancement is seen near $y_1 \approx y_2 \approx 0$ above a large negative background. This is due to Bose-Einstein interference to be studied in detail in Chapter 11.

In string-fragmentation models, direct hadrons are formed from neighbouring quark-antiquark pairs tunnelling out of the vacuum. The hadronic final states, therefore, show short-range order due to local flavor conservation. Using stable mesons only, this characteristic property is difficult to study experimentally because of the large $q\bar{q}$ combinatorial background. What is needed is a flag identifying the $q\bar{q}$ pairs created together. A suitable choice is strangeness since the number of $s\bar{s}$-pairs per event is small and the combinatorial background strongly reduced.

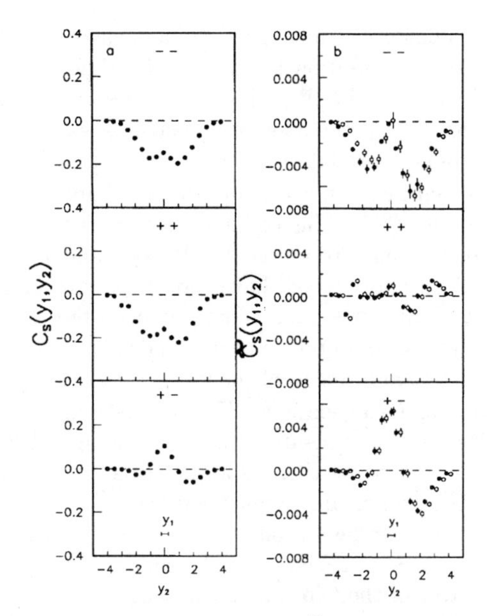

Figure 9.11. Correlation functions $C_S(0, y)$ (a) and $\tilde{C}_S(0, y)$ (b) in M^+p interactions at 250 GeV/c (open circles correspond to non-single-diffractive events and are shifted to the right to avoid overlap) [26].

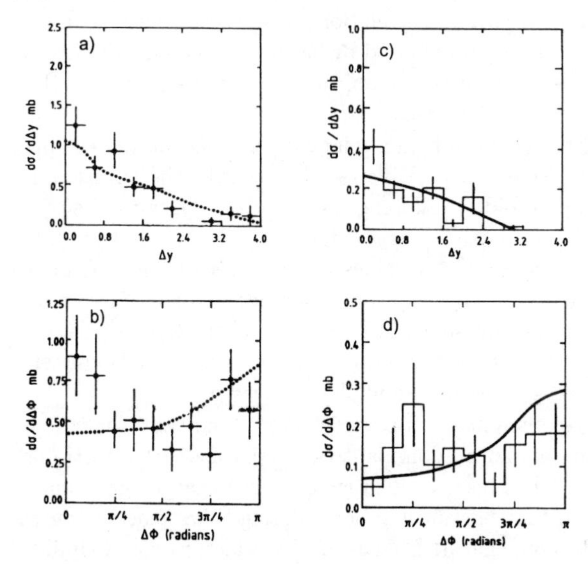

Figure 9.12. Distribution in a) the rapidity gap Δy and b) the azimuthal distance $\Delta\varphi$ for $K_s^0K_s^0$ pairs in pp collisions at 360 GeV/c. c) and d) same for $K_s^0\Lambda$ pairs [34]. The lines correspond to Lund.

In hadron-hadron collisions, strange particle pairs have been studied by the NA23 Collaboration [34]. The distribution in the rapidity difference Δy for two K^0's is given in Fig. 9.12a, for a K^0 and a Λ^0 in Fig. 9.12c. The results are compared to the single-string Lund model. As is the case for non-strange particles, the model slightly underestimates the rapidity correlation.

Good strangeness identification is available for e^+e^- annihilation e.g. in the TPC detector at $\sqrt{s} = 29$ GeV [35]. This collaboration observes significant short-range K^+K^- correlations in y, well reproduced by JETSET and by HERWIG. This and other studies of identified hadrons [36] have shown that the conservation of charge, strangeness and baryon number is predominantly local, but also long-range correlations exist between particles of opposite charge and/or strangeness in opposite jets. The latter can be understood in terms of the leading-particle effect (Sub-Sect. 4.1.2).

To study the short-range correlations, SLD [37] determined the difference of the $|\Delta y|$ distribution for six types of identified unlike charged and like charged pairs, as given in Fig. 9.13a. For all pairs, there is an excess at short range indicating local charge conservation. It is important to note that the form of the $|\Delta y|$ distribution is steeper than Gaussian but can be reproduced by JETSET (dashed lines). It has been verified that these observations are largely independent of the momentum of the two particles and of the flavor of the two primary quarks.

Similarly, to study long-range correlations, the same differences are given in Fig. 9.13b for $|\Delta y| > 1$ and particle momenta $p > 9$ GeV/c. The differences are given separately for (uds)-, c- and b-enriched samples. Light-flavor and $c\bar{c}$ events contribute about equally to correlations observed for $\pi\pi$, πp and pp pairs, while KK and Kp correlations are dominated by light-flavor events. For πK pairs, a strong anti-correlation is observed in $c\bar{c}$ events, together with a weak correlation in light-flavor ones.

Exploiting the electron beam polarization, SLD can tag the primary quark direction, which is assigned a positive rapidity (w.r.t. the thrust axis). This allows to define an ordered rapidity difference $\Delta y^{+-} = y_+ - y_-$ for pairs of unlike charged particles. Fig. 9.13c (left) shows Δy^{+-} for $\pi^+\pi^-$, K^+K^- and $p\bar{p}$ pairs. The differences between positive and negative sides of these distributions (right) indicate ordering along the event axis. The negative difference at high (Δy^{+-}) for K^+K^- pairs can be explained by leading kaons in $s\bar{s}$ events. There is a surprising positive difference for $\pi^+\pi^-$ pairs. It has been found (not shown) that this is entirely due to $c\bar{c}$ events, where the D decay gives a π^+ and the $\bar{\text{D}}$ decay a π^-.

The positive difference for $p\bar{p}$ at low $|\Delta y^{+-}|$ indicates that the baryon of a created baryon-antibaryon pair predominantly follows the quark direction rather than that of the antiquark. It has been verified (not shown) that a significant effect is present at all momenta, so that baryon number ordering takes place along the entire chain. It is important to note that JETSET is low by a factor two. A similar effect is observed for strangeness ordering in high-flavor events (not shown).

The baryon-number and strangeness ordering is confirmed by DELPHI [38], who

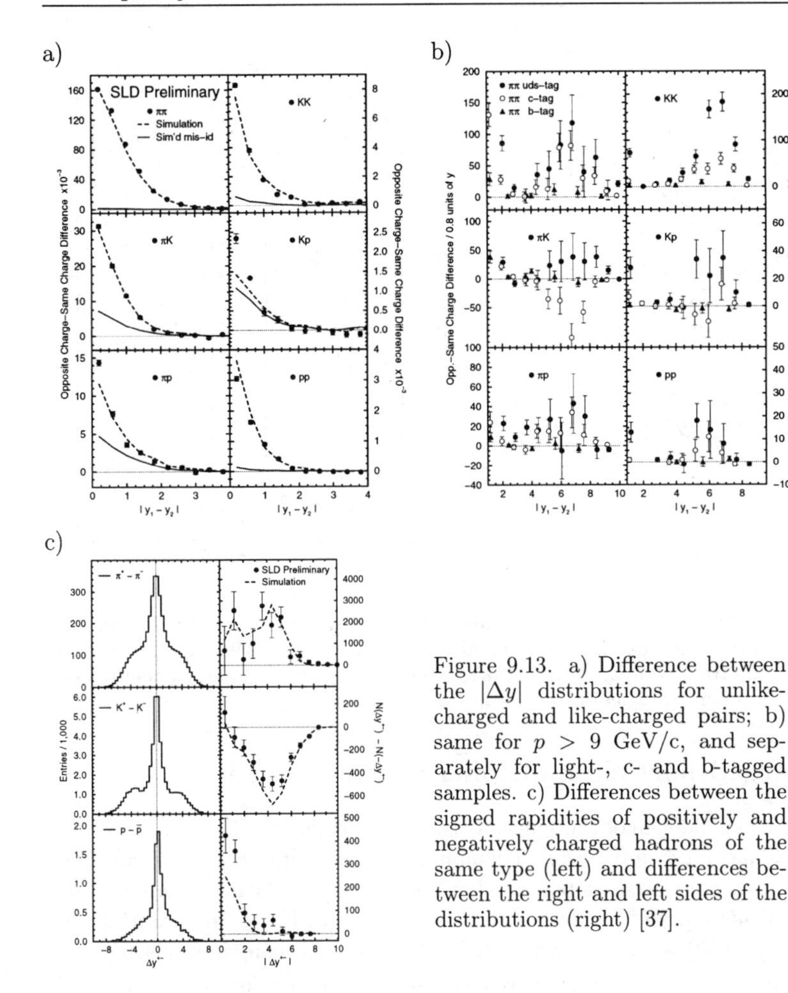

Figure 9.13. a) Difference between the $|\Delta y|$ distributions for unlike-charged and like-charged pairs; b) same for $p > 9$ GeV/c, and separately for light-, c- and b-tagged samples. c) Differences between the signed rapidities of positively and negatively charged hadrons of the same type (left) and differences between the right and left sides of the distributions (right) [37].

study the preference for either the positive or negative member of the pair to be nearer in rapidity to the fast-K^- tagged strange quark (rather than strange antiquark simultaneously tagged by a fast K^+), according to Fig. 9.14a and b.

The corresponding rapidity alignment for (additional) K^+K^- pairs and $p\bar{p}$ pairs is given as solid circles in Figs. 9.14c and d, respectively. It is seen to be positive, i.e. as expected from Figs. 9.14a and b, and to increase with increasing Δy towards a maximum possible value of 1.0. The predictions from the string model JETSET (open circles) are in fair agreement with both the K^+K^- and $p\bar{p}$ data. The cluster model HERWIG (open squares) give agreement with the K^+K^- data, however no alignment for $p\bar{p}$ pairs. This can only partially be cured by including the option of gluon splitting into diquark-antidiquark pairs (open triangles).

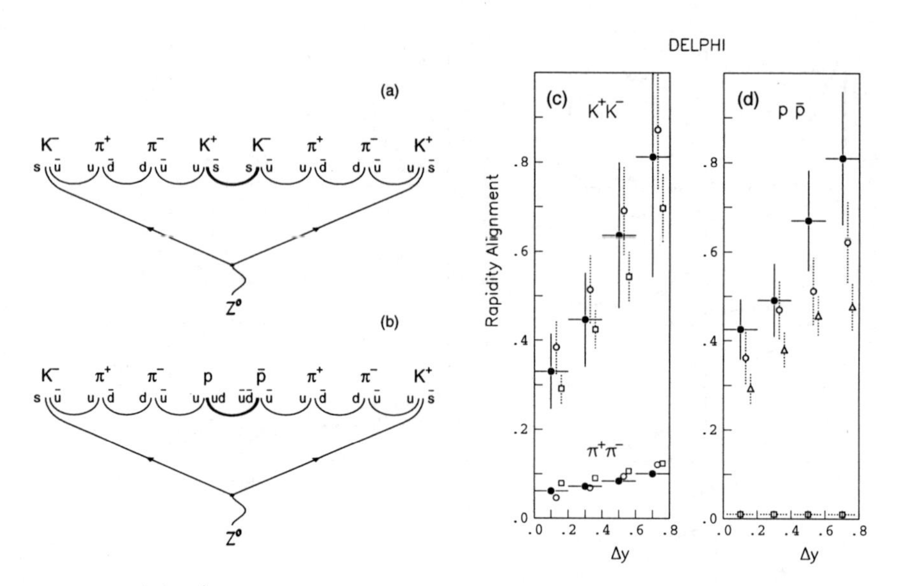

Figure 9.14. a)b) K^+K^- and $p\bar{p}$ production from an $s\bar{s}$ string. c)d) Rapidity alignment as a function of rapidity difference Δy, in integral form, i.e., all pairs with Δy greater than a given value are plotted, compared to JETSET (open circles) and HERWIG without (open squares) and with (open triangles) gluon-to-diquarks splitting [38].

9.1.4 Charged-particle multiplicity dependence

The multiplicity dependence of the semi-inclusive correlation function $\tilde{C}_2^{(n)}(0,y)$ (see Sub-Sects. 7.1.3 and 7.1.4) for the $(+-)$ combination is shown in Fig. 9.15 [26]. Near the maximum at $y = 0$ (and at bin size 0.5 units) the correlation function seems approximately Gaussian and narrows with increasing n. In Fig. 9.16a are presented the values of $\tilde{C}_2^{(n)}(0,0)$ as a function of n for three charge combinations. Within errors, $\tilde{C}_2^{(n)}(0,0)$ is independent of n, but consistently higher for $(+-)$ and $(--)$ than for $(++)$. The reason for the difference between $(--)$ and $(++)$ lies in the positive charge of both beam and target.

On the other hand, an increase of $\tilde{C}_2^{(n)}(|\eta_1 - \eta_2|)$ with $1/(n-1)$ was found [12] when averaging over a region $|\eta| < 2$ (Fig. 9.16b). Since $\tilde{C}_2$ becomes smaller when moving away from the center, and that may happen faster for higher than for lower n, this is not necessarily in contradiction with the data in Fig. 9.16a.

9.1.5 Transverse momentum dependence

The search for density fluctuations, described in the next chapter, has revealed the importance of correlations in multi-dimensional momentum space. It is, therefore,

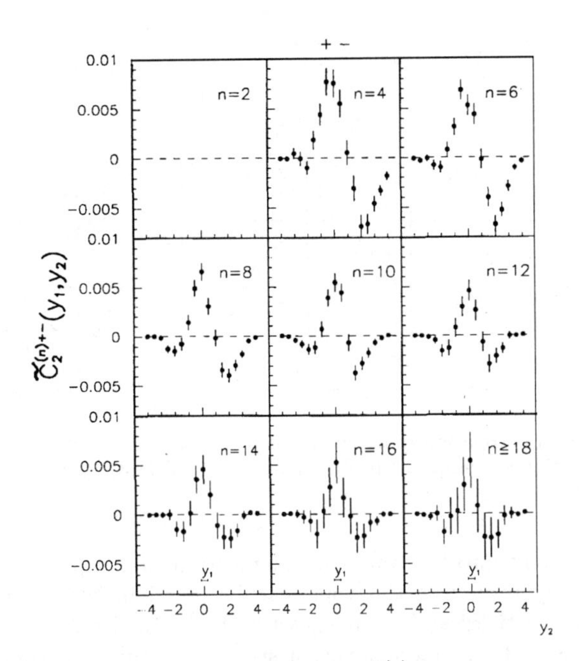

Figure 9.15. Semi-inclusive correlation functions $\tilde{C}_2^{(n)}(0,y)$ in M$^+$p reactions at 250 GeV/c for $(+-)$ pairs [26].

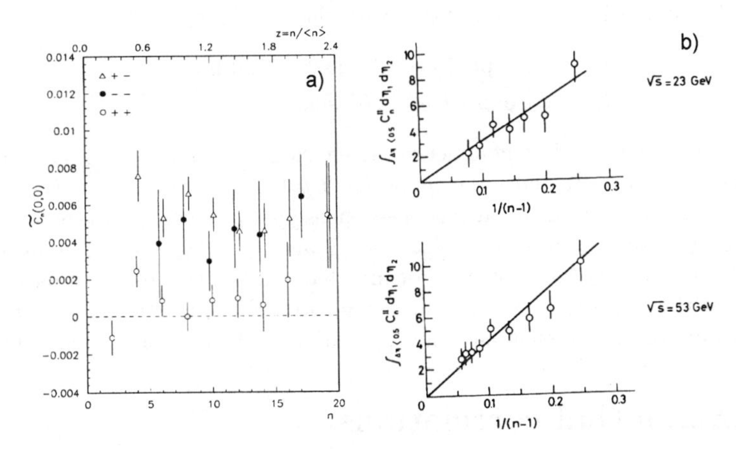

Figure 9.16. a) $\tilde{C}_2^{(n)}(0,0)$ dependence on n and $z = n/\langle n \rangle$ for M$^+$p interactions at 250 GeV/c. The last $(--)$ point corresponds to $n \geq 16$, the last $(+-)$ and $(++)$ points to $n \geq 18$ [26]. b) $\tilde{C}_2^{(n)}(|\eta_1 - \eta_2|)$ as a function of $1/(n-1)$ for pp collisions at ISR energies [12].

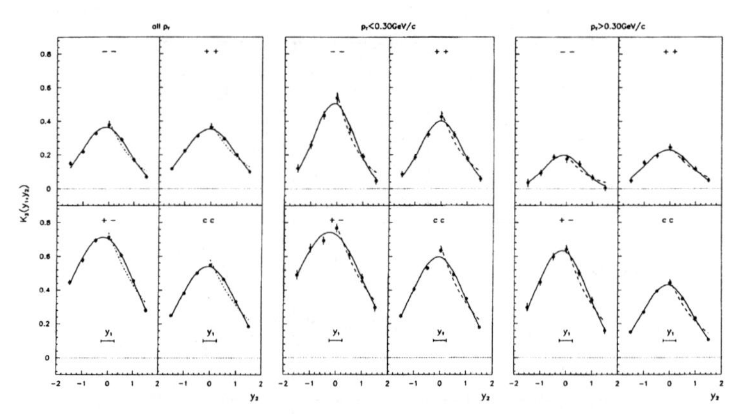

Figure 9.17. Normalized correlation functions $K_2(0, y)$ for particles with all p_T, $p_T < 0.30$ GeV/c, $p_T > 0.30$ GeV/c as compared to functions (9.2) and (9.3), solid and dot-dashed lines, respectively [39].

of interest to gain insight into the transverse momentum (p_T) dependence of rapidity correlations. Early results on this topic can be found in [40]. Later data on $K_2(0, y_2)$ [39] for all particles and for particles with p_T smaller or larger than 0.3 GeV/c, plotted in Fig. 9.17, indeed reveal a strong sensitivity to transverse momentum. The correlation function is largest, and stronger peaked near $y_2 = 0$, for $p_T < 0.3$ GeV/c, in particular for $(--)$-pairs. A similar effect was noted already in [40]. The data of Fig. 9.17 were fitted with the functions

$$f_1 = c \exp[-(y - y_0)^2/2\sigma^2] \quad \text{(full line)} , \tag{9.2}$$

$$f_2 = a \exp(-b|y|) \quad \text{(dashed)} , \tag{9.3}$$

with c, y_0, σ, a and b as free parameters. Even though for low p_T the data point at $y_2 = 0$ lies systematically above the curve, $K_2(0, y_2)$ is well fitted by the Gaussian f_1 but not by the exponential f_2, in this one-dimensional projection on rapidity.

Changing to the variables $x_1 = (y_2 + y_1)/2$ and $x_2 = (y_2 - y_1)/2$, a steepening is observed at small x_2 (not shown). For like-charge pairs, this becomes particularly sharp when the bin size is reduced to $\delta x_2 = 0.1$. For the latter, $C_2(x_1, x_2 = 0)$ increases and both a Gaussian and an exponential can fit the correlation function.

9.2 Azimuthal correlations

In interactions of unpolarised particles, no distinguished direction exists in the plane transverse to the beam and the distribution in the azimuthal angle φ is uniform. Still, a two-particle correlation exists also in φ and is visible in the distribution $W(\Delta\varphi)$ of $\Delta\varphi = |\varphi_1 - \varphi_2|$, the azimuthal angle between two particles, $\Delta\varphi \in (0, \pi)$. Due to

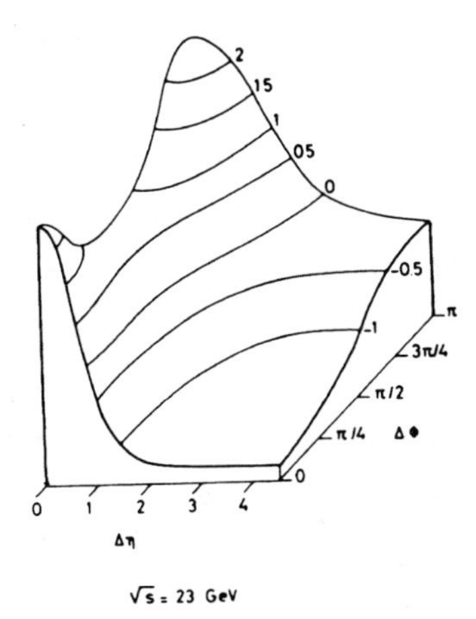

Figure 9.18. The multiplicity-averaged angular correlation function $\langle (n-1)\tilde{C}_2^{(n)}(\eta_1, \varphi_1; \eta_2, \varphi_2) \rangle$ for CC combinations in units of 10^{-3} [12].

transverse momentum conservation, $W(\Delta\varphi)$ is expected to be negative near $\Delta\varphi = 0$ and positive near $\Delta\varphi = \pi$. The actual azimuthal correlation may depend on the charge (flavor) of the particles in the pair, on the rapidity distance $\Delta y = |y_1 - y_2|$ between these particles and on their transverse momentum.

The first measurements of azimuthal correlations [41,42] as a function of rapidity y, observed an asymmetry as expected from transverse momentum conservation. The asymmetry increases with increasing p_T, but is essentially independent of y, suggesting global rather than local transverse momentum balance.

Experiments to study two-particle correlations as a function of both rapidity and azimuthal angular separation [12,43] showed that the correlation at small rapidity distance is strongest when the two particles are produced in the same or opposite directions in transverse momentum (see Fig. 9.18). The correlation length in rapidity is larger towards $\Delta\varphi = \pi$ than towards $\Delta\varphi = 0$. Furthermore, significant differences in the shape of the joint rapidity and azimuthal correlation functions have been observed for pairs of like and unlike pions [43].

In Fig. 9.19, the distribution $W(\Delta\varphi, \Delta y)$, normalized to unity, is shown as a function of $\Delta\varphi$, for all charge combinations, in the intervals $\Delta y < 1$, $1 < \Delta y < 2$ and $2 < \Delta y < 3$ [39]. A horizontal line at the average value $1/\pi$ corresponds to a flat distribution in $\Delta\varphi$. The distribution is influenced by conservation of transverse momentum, by the decay of resonances (mainly for unlike-charged particles) and by Bose-Einstein correlations (for like-charged particles). In all cases, W is larger than $1/\pi$ for $\Delta\varphi > \pi/2$ and has a maximum at $\Delta\varphi = \pi$. Except for $(--)$ pairs at $\Delta y < 1$, the W function is smaller than $1/\pi$ for $\Delta\varphi < \pi/2$. Such a global anti-correlation follows from transverse momentum conservation.

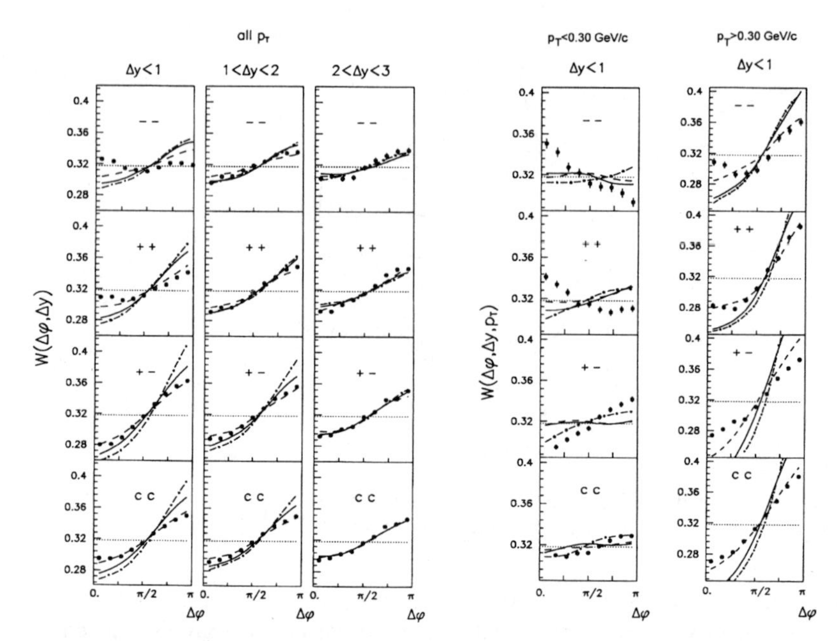

Figure 9.19. $W(\Delta\varphi, \Delta y)$ for inclusive non-single-diffractive π^+p interactions at 250 GeV/c as compared to FRITIOF 2 (dot-dashed), DPM (full) and QGSM (dashed) [39].

Figure 9.20. $W(\Delta\varphi, \Delta y, p_T)$ as compared with calculations in FRITIOF 2 (dot-dashed), DPM (full) and QGSM (dashed) for $\Delta y < 1$ and p_T cuts as indicated [39].

Model predictions are shown in Fig. 9.19 for FRITIOF 2 (dot-dashed), two-string DPM (full) and multi-string QGSM (dashed). The comparison with the data shows that it is much easier to account for azimuthal correlations at large than at small Δy. At small Δy the models differ from each other and from the experimental data. The QGSM shows somewhat better agreement with experiment than the other models. This is a consequence of the multi-string structure of QGSM, where strong azimuthal correlations in a single string are destroyed, with the result that the $\Delta\varphi$-dependence is weaker than in two-string models.

Differences between experiment and all models exist at small $\Delta\varphi$ and $\Delta y < 1$, in particular for ($--$) pairs. Bose-Einstein correlations, not included in the models, may explain this disagreement. The influence of Bose-Einstein correlation can also be observed in the (++) combination, but is smaller because of the influence of the (positive) beam particle.

Azimuthal distributions are shown in Fig. 9.20 for particles with $\Delta y < 1$, for $p_T < 0.30$ GeV/c and for $p_T > 0.30$ GeV/c, together with model calculations. A comparison of these figures reveals that azimuthal correlations have a strong p_T-

Table 9.1 Asymmetry parameter B

	Experiment	Lund	DPM
$\Lambda^0 h^+ (\Delta y < 2)$	0.18 ± 0.03	0.30 ± 0.01	0.19 ± 0.01
$h^+ h^- (\Delta y < 2)$	0.066 ± 0.003	0.126 ± 0.002	0.106 ± 0.002

dependence. Large *positive* azimuthal correlations exist at small $\Delta\varphi$ and $\Delta y < 1$ for like-charged particles with small p_T. As the transverse momentum of particles increases, the peak at $\Delta\varphi = \pi$ becomes more pronounced. This trend is reproduced (and even exaggerated) by the models and reflects momentum conservation.

Similar K^+K^- correlations are seen in the exclusive hh final state $K^-p \rightarrow pK^+K^-K^-\pi^+\pi^-$ at 32 GeV/c [44].

In the azimuthal correlation of K^0 pairs (Fig. 9.12b) and of $K^0\Lambda^0$ (Fig. 9.12d) studied by NA23 [34], the data tend to show pairs of small $\Delta\varphi$ not present in low-p_T Lund (solid line). By the same collaboration, the azimuthal correlation is studied [24] in terms of the asymmetry parameter

$$B = [N(\Delta\varphi > \pi/2) - N(\Delta\varphi < \pi/2)]/N_{\text{all}} \tag{9.4}$$

for hadron pairs with
a) opposite charge (h^+h^-)
b) equal charge ($h^+h^+ + h^-h^-$)
c) possibly opposite strangeness ($\Lambda^0 h^+, x_\Lambda < -0.2$)
d) no opposite strangeness ($\Lambda^0 h^-, x_\Lambda < -0.2$),
for $\Delta y < 2$ and for $\Delta y > 2$. No azimuthal correlation is seen for $\Delta y > 2$ in all cases and for $\Delta y < 2$ in case of no common $q\bar{q}$ pairs ($h^+h^+ + h^-h^-, \Lambda^0 h^-$). For h^+h^- and $\Lambda^0 h^+$, the parameter B is compared to low-p_T Lund and DPM predictions in Table 9.1. The parameter B is strongly overestimated in single-string low-p_T Lund and still too large in the two-string DPM. Furthermore, B increases with the sum of the transverse momenta (Fig. 9.21) but less strongly than in the models.

An azimuthal correlation has also been observed for charmed [45–49] and beauty [50, 51] pairs. Also there, the Lund model overestimates the effect and so do NLO perturbative calculations [52]. For the BEATRICE data, this is shown in Fig. 9.22a.

Agreement can, however, be obtained with a model [53] where a (Gaussian shaped) transverse component is added to the incoming parton momentum before performing the NLO perturbative QCD calculation (Fig. 9.22b).

In μp collisions, azimuthal correlations have been studied in $(+-)$ and $(++,--)$ charge combinations [28] for $|\Delta y| < 1$ and $|\Delta y| > 1$. The distribution $W(\Delta\varphi)$ is described fairly well by the Lund model including primordial k_T and gluons, except that for $|\Delta y| < 1$, where it slightly underestimates the anti-correlation for $(+-)$ and overestimates it for $(++,--)$.

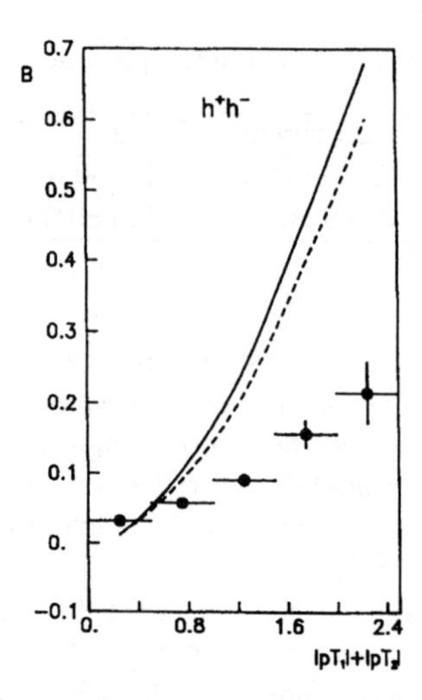

Figure 9.21. The p_T dependence of the azimuthal correlation parameter B for h^+h^- pairs in pp collisions at 360 GeV/c compared to Lund (full line) and DPM (dashed line) [24].

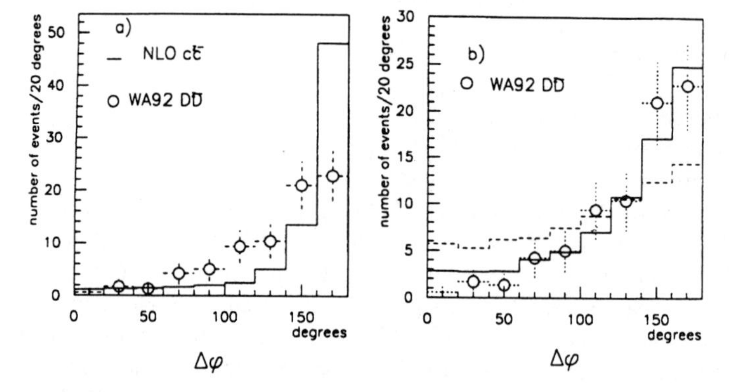

Figure 9.22. Charmed pair azimuthal correlation: BEATRICE (WA92) data are compared with predictions from (a) a NLO perturbative QCD calculation and (b) a model where a parton transverse momentum, p_T, is added to the NLO perturbative QCD predictions dotted line: $\langle p_T^2 \rangle = 1.0$ (GeV/c)^2, solid line: $\langle p_T^2 \rangle = 0.3$ (GeV/c)^2 [49].

For e^+e^- collisions, an azimuthal $\Lambda\bar{\Lambda}$ pair correlation has been observed in MARK II at 29 GeV [54].

As can be seen in Figs. 9.14a and b, local p_T compensation can best be studied in azimuthal correlations of identified $\pi^+\pi^-$, K^+K^- and $p\bar{p}$ pairs in e^+e^- collision. These are given in Fig. 9.23 (full circles), and again peaking is seen at small and large

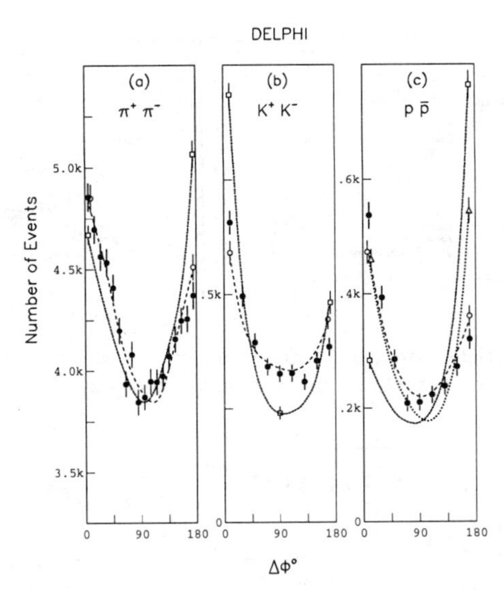

Figure 9.23. Azimuthal correlations between particles from oppositely charged particle pairs $\pi^+\pi^-$, K^+K^- and $p\bar{p}$ (full circles), compared to JETSET (dashed line), HERWIG without (dot-dashed) and with (dotted) gluon-to-diquarks process [38].

angles [38]. JETSET (dashed line) gives significantly better agreement than HERWIG (dot-dashed) because of the isotropically decaying clusters inherent to the latter. As in the Δy correlations (Fig. 9.13d), the incorporation of the gluon-to-diquarks process (dotted) reduces the discrepancy but does not cure it.

9.3 Angular correlations on the parton level

The hadronization of a quark-antiquark pair at high virtuality is currently thought to proceed via parton showering (see Chapter 6). QCD implies that this parton showering be coherent. The coherence can be incorporated into Monte-Carlo programs as angular ordering, whereby for each successive branching the gluon is emitted at a smaller angle.

Furthermore, the idea of local parton-hadron duality (LPHD) suggests that features at the parton level survive the fragmentation process. We can, therefore, expect that the coherence of the parton radiation will be reflected in angular ordering of the observed particles.

The OPAL Collaboration [55] has compared hadronic azimuthal correlations to coherent and incoherent shower models (Fig. 9.24). The coherent models JETSET PS with angular ordering, HERWIG and ARIADNE describe the azimuthal correlations in hadronic Z^0 decays, but the incoherent models JETSET PS without angular ordering

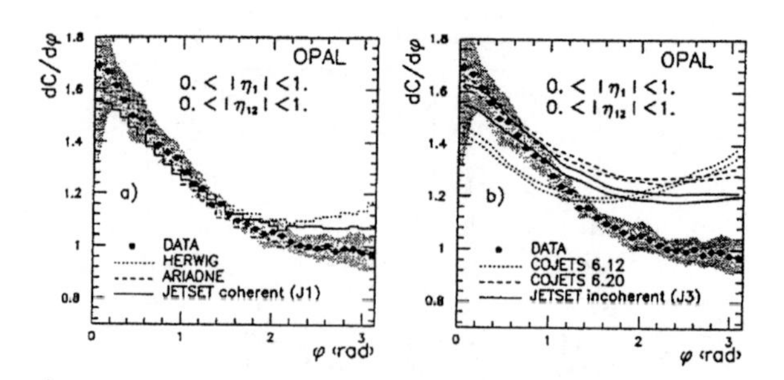

Figure 9.24. Two-particle azimuthal correlations with respect to the sphericity axis in OPAL [55] compared to coherent and incoherent MC models.

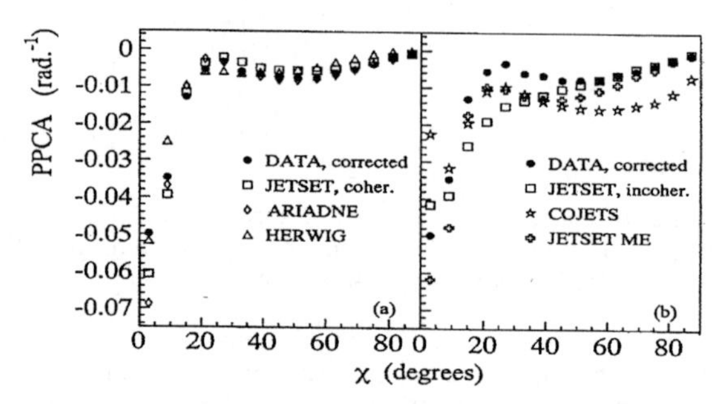

Figure 9.25. The PPCA distribution for corrected L3 data compared to a) coherent and b) incoherent Monte-Carlo models [58].

and COJETS fail for $\varphi \gtrsim \pi/2$.

As a method particularly sensitive to angular ordering [56], particle-particle correlations PPC and their asymmetry PPCA [57,58] are examined in a way analogous to the earlier study of energy-energy correlations [59],

$$\mathrm{PPC}(\chi) = \frac{1}{\Delta\chi}\langle 2\sum_{i<j}^{n}\frac{1}{n^2}\delta_{\mathrm{bin}}(\chi-\chi_{ij})\rangle \tag{9.5}$$

$$\mathrm{PPCA}(\chi) = \mathrm{PPC}(180^\circ-\chi)-\mathrm{PPC}(\chi) \ , \tag{9.6}$$

where χ_{ij} is the full spatial angle between tracks i and j, $\langle\ \rangle$ is the average over all events in the sample, n is the number of charged tracks in an event, and $\Delta\chi$ is the bin width. The function $\delta_{\mathrm{bin}}(\chi-\chi_{ij})$ is 1 if χ_{ij} and χ are in the same bin and 0 otherwise.

At $\sqrt{s} = M_{\mathrm{Z}}$, the fraction of two-jet events is still high. For two-jet events, particles in different jets will in general be separated by an angle χ greater than

90°. The PPC for $\chi > 90°$ can, therefore, serve as an indication of what the PPC *within* a jet ($\chi < 90°$) would be *in the absence* of angular ordering. By forming the asymmetry, these 'uninteresting' correlations are effectively subtracted. The effects of angular ordering should, therefore, be more directly observable in the PPCA than in the PPC. Note, however, that the sign convention following [59] leads to a *negative* sign for a *positive* correlation.

Figures 9.25a and b show the PPCA distribution of L3 data (corrected for detector effects [58]) compared to coherent and incoherent Monte-Carlo models, respectively.

In Fig. 9.25b we see that for $\chi \lesssim 60^0$ JETSET 7.3 PS without angular ordering (incoherent) disagrees strongly with the data, while being in fair agreement at larger values of χ. COJETS is seen not to reproduce the data over the entire angular range. On the other hand, in Fig. 9.25a, the coherent Monte Carlo models, JETSET with angular ordering, HERWIG, and ARIADNE all reproduce the data reasonably well over the full angular range. Note that the disagreement of the incoherent models can not be due to the Bose-Einstein effect. Turning this effect off in the non-angular ordered JETSET model does not raise but lower its PPCA points. So, the data from the L3 experiment disfavour the incoherent models.

9.4 Correlations in invariant mass

There are arguments in favor of invariant mass as a dynamical variable rather than the single-particle variables usually used in correlation studies. Resonances, the cause of most of the correlations among hadrons, and threshold effects appear at fixed values of mass; Bose-Einstein interference correlations depend on four-momentum differences; multiperipheral-type ladder diagrams are functions of two-particle invariant masses, and so on.

The idea to study correlations as a function of invariant mass was, to our knowledge, first proposed in [60, 61]. It focusses directly on the correlation functions (cumulants) rather than on the inclusive densities. Starting from the definition (7.20), one defines the correlation function

$$C_2(M_{\text{inv}}) = \rho_2(M_{\text{inv}}) - \rho_1 \otimes \rho_1(M_{\text{inv}}), \tag{9.7}$$

obtained after integration (in a suitable region of phase space) of $C_2(p_1, p_2)$ over all variables except M_{inv}. Here, $\rho_2(M_{\text{inv}})$ is the familiar normalized two-particle invariant-mass spectrum. The "background term" $\rho_1 \otimes \rho_1(M_{\text{inv}})$ is the integral of $\rho_1(p_1)\rho_1(p_2)$ with M_{inv} fixed. For the data shown below, it is obtained from "uncorrelated" ("mixed") events, built by random selection from a track pool. Higher-order correlations are obtained in a completely analogous manner. We further utilize the function $K_2(M_{\text{inv}}) = C_2(M_{\text{inv}})/\rho_1 \otimes \rho_1(M_{\text{inv}})$, the normalized factorial cumulant of order two.

The analysis in [60], based on low statistics pp data at 205 GeV/c, demonstrates that $K_2^{+-}(M_{\rm inv})$ and $K_2^{\pm\pm}(M_{\rm inv})$ follow an approximate power law, written by the authors as

$$K_2(M_{\rm inv}) = (M_{\rm inv}^2)^{\alpha_{\rm X}(0)-1}. \tag{9.8}$$

The notation reminds of an interpretation in terms of the Mueller-Regge formalism (for details see [60]). The power $\alpha_{\rm X}(0)$ is the appropriate Regge-intercept, X = R for non-exotic pairs and X = E for exotic ones. The ratio K_2^{--}/K_2^{+-} was further seen to fall as $M_{\rm inv}^{-2}$, consistent with $\alpha_{\rm R}(0) - \alpha_{\rm E}(0) = 1$. Not relying on Mueller-Regge theory, the authors argued that most of the correlations at small $M_{\rm inv}$ are due to resonance decays and to interference of amplitudes [61].

The results already obtained in [60] clarify several issues which had troubled the interpretation of correlation and intermittency (see Chapter 10) data. Among others, they demonstrate that different charge states should be treated separately since the $M_{\rm inv}$ dependence is very different.

The method of [60] has been applied by NA22 [62], UA1 [63] and DELPHI [64]. Figure 9.26 shows data on $K_2(M_{\rm inv})$ for a combined sample of non-diffractive π^+/K$^+$p collisions at 250 GeV/c in the central c.m. rapidity region $-2 < y < 2$. $K_2^{+-}(M_{\rm inv})$ has a prominent ρ^0 peak, but is quite flat near threshold. The peak in the first bin of sub-figure (a) is attributed to contamination from Dalitz decays and γ conversions. $K_2^{--}(M_{\rm inv})$ falls much faster. A fit of $K_2 \sim (M_{\rm inv}^2)^{-\beta}$ yields $\beta^{--} = 1.29 \pm 0.04$, $\beta^{++} = 1.46 \pm 0.03$, $\beta^{+-} = 0.17 \pm 0.02$, in agreement with [60] and consistent with the relation $\alpha_{\rm R}(0) - \alpha_{\rm E}(0) \approx 1$.

NA22 also finds that cuts on transverse momentum or relative azimuthal angle $\delta\varphi$ strongly affect the shape of $K_2^{+-}(M_{\rm inv})$, but have little effect on $K_2^{--}(M_{\rm inv})$ for $M_{\rm inv} < 0.5$ GeV/c^2. This means that $K_2^{--}(M_{\rm inv})$ at small $M_{\rm inv}$ is essentially a function of $M_{\rm inv}$ (or the squared four-momentum transfer $Q^2 = M_{\rm inv}^2 - 4m_\pi^2$) only, illustrating once more the advantage of $M_{\rm inv}$ compared to other variables.

Whether Bose-Einstein effects are solely responsible for the differences between $(\pi^\pm\pi^\pm)$ and $(\pi^+\pi^-)$ pairs is not evident. It suffices to consider [61] the contributions from decays of various resonances to realize that the $M_{\rm inv}$-dependence near threshold for "exotic" particle systems must be stronger than for "non-exotic" ones. In a dual Regge picture, such differences translate into very different values of the respective Regge intercepts as in (9.8). It remains, therefore, to be verified if the $M_{\rm inv}$-dependence of the data can be explained as a superposition of a "standard" Regge-type power law and a conventional Bose-Einstein enhancement.

As pointed out in [65], there is a feasible way to test this and even to give access to the relative strength of BE interference and exotic like-charge $\pi\pi$ interaction. The idea is that particle combinations exist which are either a) exotic, but not identical (e.g. K$^+\pi^+$ or K$^-\pi^-$ pairs) or b) identical, but not exotic ($I = 0$ $\pi^0\pi^0$ pairs). NA22 [66] and ALEPH [67] data indicate that very-short-range correlations are indeed absent in the exotic Kπ channel. This supports Bose-Einstein correlations rather than exotic Regge behavior, but the point deserves further investigation.

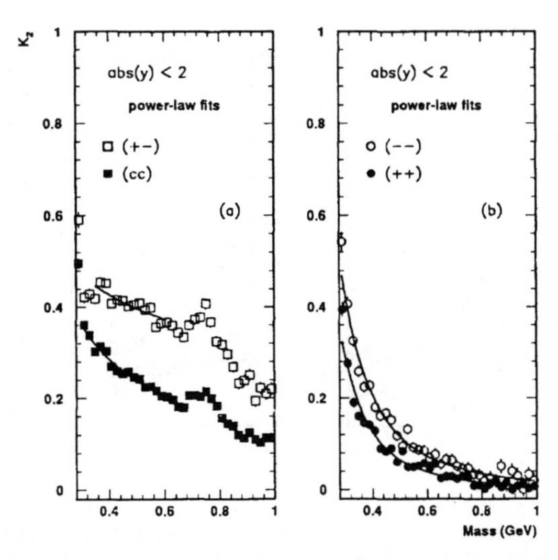

Figure 9.26. $K_2(M_{\text{inv}})$ for $(cc), (+-), (--)$ and $(++)$ pairs of tracks with c.m. rapidity $-2 < y < 2$, in $\text{K}^+/\pi^+\text{p}$ collisions at 250 GeV/c [62]. The solid line is a power-law fit (see text).

The possibility that the correlation functions depend mainly on invariant mass has interesting further consequences [68]. Taking in (9.7)

$$C_2(M_{\text{inv}}) \propto (M^2_{\text{inv}})^{\alpha-1}\,\text{BE}(M_{\text{inv}}), \tag{9.9}$$

with BE a conventional Bose-Einstein factor, exponential in Q, good agreement is obtained with the NA22 second-order correlation integral data of $(--)$-pairs. Integrating the correlation function over all variables except δy gives $F_2^{--}(\delta y)$ which also fits the data. Although C_2 does not explicitly depend on the transverse momentum of the particles, it turns out that $F_2(\delta y, p_{\text{T}1}, p_{\text{T}2})$ is larger and more steeply increasing than $F_2(\delta y)$ for small δy and small p_{T}'s. The opposite happens for large p_{T}'s. This is the "low-p_{T} intermittency effect" seen in the NA22 and UA1 data (cfr. Fig 10.23a and Subsect. 10.4.3 below). The explanation is simple: under the stated hypothesis, small p_{T} for the two particles in a pair means, on the average, smaller invariant mass than for unrestricted transverse momentum and, therefore, larger and shorter-ranged correlations in rapidity. Enhanced intermittency follows as a consequence of kinematical cuts! The influence of the Bose-Einstein factor is easily checked in this simple model. It is found to be necessary in order to reproduce the correlation integral data and F_2 for restricted p_{T} but has, as expected a priori, very little influence on the p_{T}-integrated $F_2(\delta y)$. This explains an early controversy over the role of Bose-Einstein effects in one-dimensional factorial moment analyses to be sketched in Subsect. 10.4.1 below.

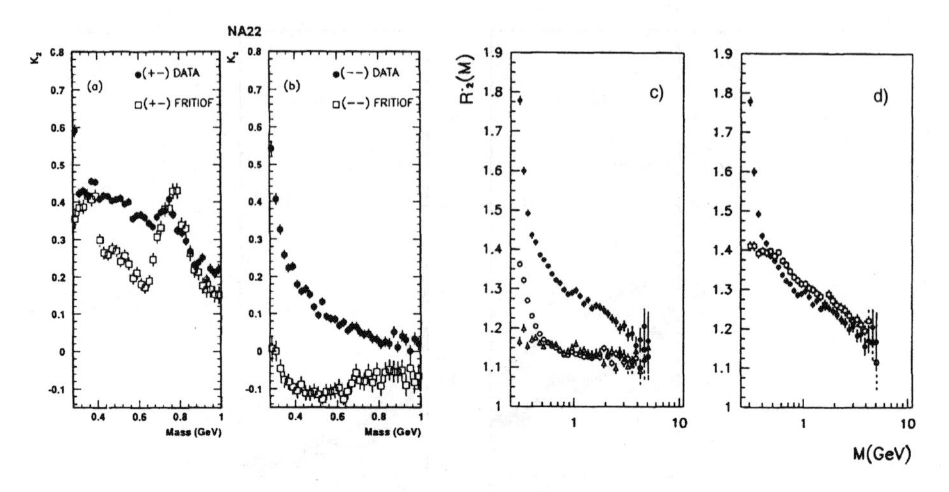

Figure 9.27. $K_2(M_{\rm inv})$ for a) $(+-)$, b) $(--)$ pairs of tracks with c.m. rapidity $-2 < y < 2$, in K^+/π^+p collisions at 250 GeV/c [62], compared to FRITIOF 2, c) $R_2(M_{\rm inv})$ for like-charge pairs for low-p_T $p\bar{p}$ collisions at 630 GeV (full) compared to PYTHIA with (circles) and without (triangles) Bose-Einstein correlations, d) to GENCL(squares) [63].

A study of the invariant-mass (or Q) dependence of the two-particle correlation function has for the first time given clear indications as to why the hadron-hadron Monte-Carlo models fare so badly when confronted with correlation data. In [28] it is demonstrated on EMC data that ρ^0 production and ω reflection are strongly overestimated in the Lund model, while low Q non-resonant pos-neg correlations are underestimated and neg-neg Bose-Einstein correlations are missing. Figures 9.27a,b show NA22 data [62] for $K_2^{--}(M_{\rm inv})$ and $K_2^{+-}(M_{\rm inv})$ compared to FRITIOF. The predicted shape of $K_2^{+-}(M_{\rm inv})$ is very different from the data, especially in the ρ^0 region. It shows an enhancement at low mass which causes the correlation function to drop much faster than seen in the experiment. In the model, this structure is traced back to reflections from η, η' and ω resonances. The model also fails to describe $K_2^{--}(M_{\rm inv})$ since correlations are very weak or even negative, except for a threshold enhancement due to η' decays.

Fig. 9.27c shows UA1 like-charge data ($0.15 \leq p_T \leq 0.3$ GeV/c) [63] of $F_2(M_{\rm inv})$ in comparison with PYTHIA, Fig. 9.27d with the UA5 cluster model GENCL.

These examples suffice to demonstrate that FRITIOF and PYTHIA (or rather JETSET) in their default version have serious shortcomings and are unable to reproduce two-particle correlations in invariant mass. For correlations in rapidity and azimuthal angle this was seen earlier, but the reasons remained obscure, mainly because of the insensitivity of these variables to dynamical correlations at small mass.

A study of the correlation function in terms of invariant mass clarifies the situation considerably. For NA22, the model was known to overestimate significantly the

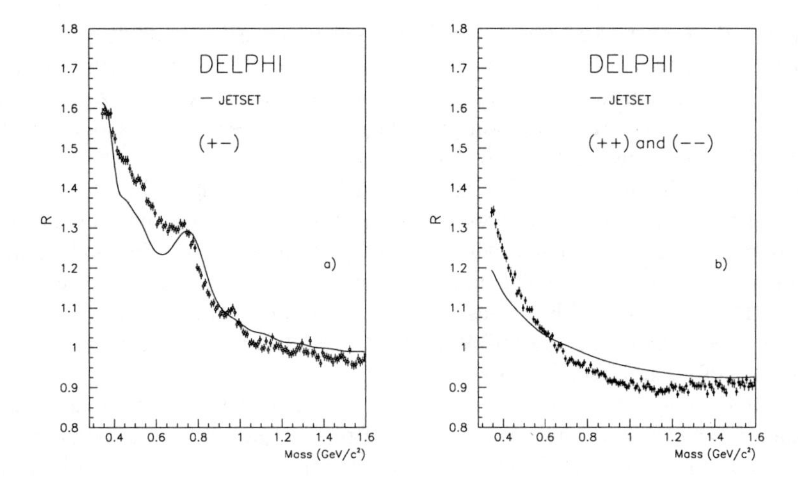

Figure 9.28. $R = K_2(M_{\text{inv}})+1$ for e^+e^- annihilation compared to the JETSET 7.3 prediction with parameters tuned to the DELPHI data for a) unlike-charged and b) like-charged combinations [64].

production rates of ρ^0 and η mesons [69] and presumably also those of η' and ω for which no direct measurements exist (see also [70] for hA collisions). This is now seen to distort heavily the M_{inv} dependence of K_2. Also Bose-Einstein low-mass enhancements, most likely responsible for the fast drop of $K_2^{--}(M_{\text{inv}})$ in the threshold region, are not included in the FRITIOF model commonly used. Finally, we note that the values of K_2 in the considered mass interval are much smaller than the data. This is related to the width of the charged-particle multiplicity distribution which is known to be too small in FRITIOF.

The like-charge distributions (Figs. 9.27b,c) look rather hopeless for FRITIOF and PYTHIA. Including Bose-Einstein correlation helps for $M_{\text{inv}} \lesssim 0.5$ GeV, but something else is needed to explain the increase with decreasing M_{inv} at large M_{inv}. Fig. 9.27d suggests that simple clustering would be a good candidate [63].

In Fig. 9.28a, the discrepancy is shown to be quite similar for $(+-)$ correlations in e^+e^- collisions [64] and JETSET. As in hh collisions, the correlation is underestimated in the mass region below the ρ^0. This discrepancy can be cured by decreasing the η' and ρ^0 production and increasing ω production in JETSET.

Fig. 9.28 gives the like-charged correlation for the data and JETSET without BE correlation. For $M_{\text{inv}} < 0.6$ GeV/c^2, the experimental data are considerably higher than JETSET. This can be attributed to Bose-Einstein interference. However, it is striking that JETSET also predicts a strong rise towards threshold even without Bose-Einstein correlations.

To summarize, the above proves that the failures of models such as FRITIOF and

PYTHIA with respect to correlation data, are not necessarily due to "novel" dynamics. They are in first instance a consequence of a variety of defects—such as incorrect resonance production rates and absence of identical particle symmetrization—which belong to "standard" hadronization phenomenology. These defects should be eliminated before "new physics" can be claimed.

For e^+e^- annihilation at LEP energies, models such as JETSET-PS are much more successful than for all other processes. Besides the evident fact that this process is much better understood theoretically, QCD effects dominate and the model parameters are much better tuned to the data. Still, serious, recently observed discrepancies of e.g. JETSET-PS with LEP measurements on particle and resonance production rates are a clear sign that hadronization in e^+e^- is in fact less well understood than commonly stated and needs improvement.

An interesting new-dynamics development in this respect is the enhancement of low-mass $\pi^+\pi^-$ correlation relative to η' and π^0 production [71] by allowing additional $\pi^+\pi^-$ pairs to be produced at string breakup due to a Goldstone-like mechanism.

9.5 Three-particle rapidity correlations

Whether dynamical correlations exist beyond the two-particle correlations discussed so far is of crucial importance for much of the present search for scaling phenomena in multiparticle processes, a subject treated in Chapter 10. With conventional techniques, this question is not easy to answer and beyond the sensitivity of many experiments.

Nevertheless, three-particle correlations in rapidity have been looked for in a number of experiments [16, 26, 72–74]. The third-order normalized factorial cumulant is defined as [cfr. (7.20) and (7.23)]:

$$K_3(y_1, y_2, y_3) = C_3(y_1, y_2, y_3) \Big/ \frac{1}{\sigma_{\text{inel}}^3} \frac{\mathrm{d}\sigma}{\mathrm{d}y_1} \frac{\mathrm{d}\sigma}{\mathrm{d}y_2} \frac{\mathrm{d}\sigma}{\mathrm{d}y_3} \ , \tag{9.10}$$

$$C_3(y_1, y_2, y_3) = \frac{1}{\sigma_{\text{inel}}} \frac{\mathrm{d}^3\sigma}{\mathrm{d}y_1 \mathrm{d}y_2 \mathrm{d}y_3} + 2 \frac{1}{\sigma_{\text{inel}}^3} \frac{\mathrm{d}\sigma}{\mathrm{d}y_1} \frac{\mathrm{d}\sigma}{\mathrm{d}y_2} \frac{\mathrm{d}\sigma}{\mathrm{d}y_3} - \tag{9.11}$$

$$- \frac{1}{\sigma_{\text{inel}}^2} \frac{\mathrm{d}^2\sigma}{\mathrm{d}y_1 \mathrm{d}y_2} \frac{\mathrm{d}\sigma}{\mathrm{d}y_3} - \frac{1}{\sigma_{\text{inel}}^2} \frac{\mathrm{d}^2\sigma}{\mathrm{d}y_2 \mathrm{d}y_3} \frac{\mathrm{d}\sigma}{\mathrm{d}y_1} - \frac{1}{\sigma_{\text{inel}}^2} \frac{\mathrm{d}^2\sigma}{\mathrm{d}y_1 \mathrm{d}y_3} \frac{\mathrm{d}\sigma}{\mathrm{d}y_2}$$

$$\text{with} \qquad \sigma_{\text{inel}} = \sum_{n \geq 8} \sigma_n \ .$$

The $\tilde{C}_{\text{S}}(y_1, y_2, y_3)$ correlation function is determined as a sum of topological correlation functions:

$$\tilde{C}_{\text{S}}(y_1, y_2, y_3) = \sum_{n \geq 8} P_n \tilde{C}_3^{(n)}(y_1, y_2, y_3) \ , \tag{9.12}$$

$$\tilde{C}_3^{(n)}(y_1, y_2, y_3) = \tilde{\rho}_3^{(n)}(y_1, y_2, y_3) - \tilde{A}_3^{(n)}(y_1, y_2, y_3) \ , \tag{9.13}$$

$$\begin{aligned}\tilde{A}_3^{(n)}(y_1, y_2, y_3) = \ & \tilde{\rho}_2^{(n)}(y_1, y_2)\tilde{\rho}_1^{(n)}(y_3) + \tilde{\rho}_2^{(n)}(y_2, y_3)\tilde{\rho}_1^{(n)}(y_1) + \tilde{\rho}_2^{(n)}(y_1, y_3)\tilde{\rho}_1^{(n)}(y_2) - \\ & -2\tilde{\rho}_1^{(n)}(y_1)\tilde{\rho}_1^{(n)}(y_2)\tilde{\rho}_1^{(n)}(y_3) \ ,\end{aligned}$$

$$\tilde{\rho}_3^{(n)}(y_1, y_2, y_3) = \frac{1}{n(1,2,3)} \frac{1}{\sigma_n} \frac{\mathrm{d}^3\sigma}{\mathrm{d}y_1 \mathrm{d}y_2 \mathrm{d}y_3} \ , \tag{9.14}$$

where $n(1,2,3)$ is the mean number of three-particle combinations in events with charged-particle multiplicity n.

The corresponding normalized function is defined as:

$$\tilde{K}_\mathrm{S}(y_1, y_2, y_3) = \tilde{C}_\mathrm{S}(y_1, y_2, y_3) / \sum_n P_n \tilde{\rho}_1^{(n)}(y_1)\tilde{\rho}_1^{(n)}(y_2)\tilde{\rho}_1^{(n)}(y_3) \ . \tag{9.15}$$

Because of small statistics, three-particle correlations were not observed in pp interactions at 200 GeV/c at FNAL [16]. In K$^-$p interactions at 32 GeV/c [72], three-particle correlations were considered using $\tilde{C}_\mathrm{S}(y_1, y_2, y_3)$ and $\tilde{K}_\mathrm{S}(y_1, y_2, y_3)$. No positive short-range correlation effect was observed. Correlations in the form of K have been observed in the central region by the ISR experiment for $n \geq 8$ [74].

Fig. 9.29 from NA22 shows $K_3(0,0,y)$ and $\tilde{K}_\mathrm{S}(0,0,y)$ for the combined M$^+$p sample at 250 GeV/c [26]. Also shown are the values of $K_3(0,0,y)$ obtained in pp-interactions at $\sqrt{s}$=31-62 GeV [74] (lines). Inclusive three-particle correlations $K_3(0,0,y)$ are indeed seen in the NA22 data. They are strongest when a third particle partially compensates the charge of a pair of identical particles. There are, however, no correlation effects visible in the function $\tilde{K}_\mathrm{S}(0,0,y)$. In FRITIOF and QGSM, three-particle rapidity correlations are absent in both $K_3(0,0,y)$ and $\tilde{K}_\mathrm{S}(0,0,y)$.

A factorization of the normalized three-particle correlation function has been proposed [75–77] under the form of a "linked-pair" structure:

$$K_3(y_1, y_2, y_3) = K_2(y_1, y_2)K_2(y_2, y_3) + K_2(y_1, y_3)K_2(y_3, y_2). \tag{9.16}$$

The comparison of the prediction of (9.16) to the data is given in Table 9.2, for $n \geq 2$, at a resolution of 0.5 rapidity units. At this resolution, the linked-pair ansatz is in agreement with the measured three-particle correlation within two standard deviations. Note, that y-correlations are much stronger for low-p_T particles and that the linked-pair ansatz continues to hold.

Table 9.2 The 3-particle correlation function compared to the prediction from the linked-pair ansatz, for non-single diffractive data ($n \geq 2$).

	all p_T		p_T <0.15 GeV/c	
	data	LPA	data	LPA
$K_3^{---}(0,0,0)$	0.23±0.10	0.30±0.03	2.3±1.7	2.0±0.4
$K_3^{+++}(0,0,0)$	0.14±0.06	0.21±0.02	1.2±0.6	1.0±0.2
$K_3^{ccc}(0,0,0)$	0.39±0.04	0.53±0.03	1.9±0.5	1.7±0.2

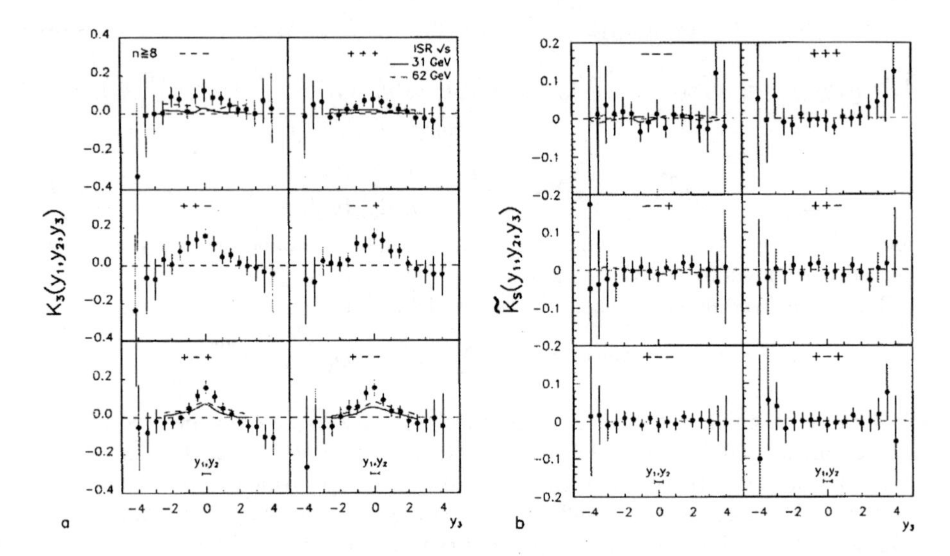

Figure 9.29. Three-particle rapidity correlations a) $K_3(0,0,y)$ [the lines correspond to the ISR results at 31 GeV (full) and 62 GeV (dashed)] and b) $\tilde{K}_3(0,0,y)$ for M$^+$p interactions at 250 GeV/c. [the FRITIOF (dot-dashed) prediction is indicated for the charge combination $(---)$, QGSM (dashed) for $(---)$ and $(--+)$] [26].

With the accuracy presently attainable for three-particle correlations, it is obvious that studies of still higher-order correlation functions require better methods. The most successful ones will be discussed in Chapter 10.

9.6 Summary and conclusions

1. The main contributions to the inclusive correlation functions C_2 and C_3 come from the mixing of events with different multiplicity and different semi-inclusive single-particle density, but some effect remains in the so-called short-range correlation part C_S.

2. $C_2(0, y_2)$ increases much faster with increasing energy than its short-range contribution C_S.

3. The short-range correlation is significantly larger for $(+-)$ than for the like-charged combinations, and is positive over a wider rapidity range in $C_2(y_1, y_2 = y_1)$.

4. Contrary to the semi-inclusive $C_2^{(n)}(0, y_2)$ correlation functions, the $\tilde{C}_2^{(n)}(0, y_2)$

are similar for different multiplicity n, except that $\tilde{C}_2^{(n)+-}$ becomes narrower with increasing n.

5. The correlation functions depend on transverse momentum and are largest for small-p_T particles. Consequently, correlations are stronger in multi-dimensional momentum space than in a lower dimensional projection, such as rapidity space. Further implications of this observation will be discussed in Sect. 10.3 below.

6. In the central c.m. region, and at comparable energy, the correlation strength observed in meson-proton (M^+p) collisions at $\sqrt{s}$=22 GeV is of similar magnitude as in e^+e^- collisions and as in μp collisions, but μp collisions show a clear forward-backward asymmetry indicating a difference between quark and diquark fragmentation.

7. Combinatorial background can be suppressed by studying the correlation of strange and heavy-flavor particles. This gives evidence for local conservation of charge, strangeness, heavy flavor and baryon number, with an alignment as expected from the quantum numbers of the leading quarks. Transverse momentum, on the other hand, is compensated more globally.

8. Gluon effects are the main source of rapidity correlations in e^+e^- (and lh) collisions. They are reproduced by ARIADNE, while JETSET is slightly low and HERWIG too high. In hh collisions, many chains as in QGSM or intermediate clusters or fireballs are needed to reproduce the correlation, even at moderately low energies.

9. Positive correlations are observed at large values of the azimuthal angle $\Delta\varphi$, as expected from transverse momentum conservation. In e^+e^- collisions, they are reproduced by ARIADNE and JETSET, but overestimated by HERWIG. In lh (and hh) collisions there is need for primordial quark transverse momentum. In hh collisions, FRITIOF and DTU overestmate the $\Delta\varphi$ dependence, while QGSM does better.

10. The correlations among like-charge particle pairs at small values of $\Delta\varphi$ and Δy, where Bose-Einstein effects should contribute, are significantly larger than predicted by FRITIOF 2, DPM and QGSM. The deviations are stronger for particles with small transverse momentum.

11. The distribution in the interparticle opening angle of e^+e^- collisions at LEP favors models with coherent parton showering.

12. The analysis of cumulants in terms of invariant mass or related variables has helped in clarifying several issues. Reasons behind failures of models are clearly revealed. They are in first instance incorrectly predicted particle and resonance

production rates and the near absence of correlations in identical-particle systems with small invariant masses. The former have led to the suggestion of a Goldstone-like mechanism for the production of additional pions in string breakup, the latter have underlined the importance of Bose-Einstein correlations.

13. Three-particle correlations are now observed in all charge combinations. They are particularly large for low-p_T particles. Within two standard deviations, they satisfy the linked-pair ansatz. No short-range contribution $\tilde{K}_S$ is observed in three-particle correlations. Other methods are needed to study higher-order correlations.

Bibliography

[1] Z. Koba, H.B. Nielsen and P. Olesen, *Nucl. Phys.* **B40** (1972) 317.

[2] K.G. Wilson, Cornell Report No. CLNS-131/1970 reprinted in: Proc. Fourteenth Scottisch Universities Summer School Conf. (1973), eds. R.L. Crawford and R. Jennings (Academic Press, New York 1974);
R.P. Feynman in: Proc. Int. Conf. Neutrino 72 (Balatonfured, 1972), eds. A. Frenkel and G. Marx (Budapest, 1973).

[3] W.D. Shephard et al., *Phys. Rev. Lett.* **28** (1972) 703.

[4] S. Stone, T. Ferbel, P. Slattery and B. Werner, Rochester preprint UR-875-439 (1971).

[5] W. Ko, *Phys. Rev. Lett.* **28** (1972) 935.

[6] E.L. Berger, B. Oh and G.A. Smith, *Phys. Rev. Lett.* **29** (1972) 675.

[7] J. Hanlon et al., *Nucl. Phys.* **B52** (1973) 96.

[8] V.V. Ammosov et al., *Sov. J. Nucl. Phys.* **23** (1976) 178.

[9] J. Erwin et al., *Phys. Rev. Lett.* **33** (1974) 1443.

[10] C. Bromberg et al., *Phys. Rev.* **D9** (1974) 1864; **D10** (1974) 3100.

[11] PSB Coll., S.R. Amendolia et al., *Phys. Lett.* **48B** (1974) 359; *Nuovo Cimento* **A31** (1976) 17.

[12] AMC Coll., K. Eggert et al., *Nucl. Phys.* **B86** (1975) 201.

[13] CHV Coll., H. Dibon et al., *Phys. Lett.* **B44** (1973) 313.

[14] CHLM Coll., M.G. Albrow et al., *Phys. Lett.* **B51** (1974) 421.

[15] R. Singer et al., *Phys. Lett.* **B49** (1974) 481.

[16] J. Whitmore, *Phys. Reports* **10** (1994) 273 and **27** (1976) 187.

[17] L. Foá, *Phys. Reports* **22** (1975) 1.

[18] E.L. Berger, *Phys. Lett.* **B49** (1974) 369; *Nucl. Phys.* **B85** (1975) 61.

[19] A. Arneodo and G. Plaut, *Nucl. Phys.* **B107** (1976) 262; **B113** (1976) 156.

[20] SFM Coll., D. Drijard et al., *Nucl. Phys.* **B155** (1979) 269; W. Hofmann, Ph.D. Thesis, Karlsruhe 1977.

[21] UA5 Coll., R.E. Ansorge et al., *Z. Phys.* **C37** (1988) 191.

[22] Ch. Fuglesang, Ph.D. Thesis, Univ. of Stockholm (1987).

[23] UA5 Coll., G.J. Alner et al., *Nucl. Phys.* **B291** (1987) 445.

[24] NA23 Coll., J.L. Bailly et al., *Z. Phys.* **C40** (1988) 13.

[25] ABCDHW Coll., A. Breakstone et al., *Phys. Lett.* **114B** (1982) 383.

[26] NA22 Coll., V.V. Aivazyan et al., *Z. Phys.* **C51** (1991) 167.

[27] H. Müller, *Eur. Phys. J.* **C18** (2001) 563.

[28] EMC Coll., I. Derado, G. Jancso and N. Schmitz, *Z. Phys.* **C56** (1992) 553; M. Arneodo et al., *Z. Phys.* **C31** (1986) 333; ibid. **C40** (1988) 347; S. Maselli, Ph.D. Thesis, TU Munich (1988).

[29] P. Malecki, Festschrift L. Van Hove, eds. A. Giovannini and W. Kittel (World Scientific, Singapore, 1990) p.159; J. Figiel, *Hadronization of Partons in Muon-Nucleon Interactions at 280 GeV/c*, Krakow preprint IFJ 1398/Ph (1988).

[30] TASSO Coll., M. Althoff et al., *Phys. Lett.* **B139** (1984) 126; *Z. Phys.* **C29** (1985) 347; J. Chwastowski, Ph.D. thesis, Hamburg (1988).

[31] CELLO Coll., O. Podobrin, Proc. Ringberg Workshop on Multiparticle Production, eds. R.C. Hwa et al. (World Scientific, Singapore, 1992) p.62; Ph.D. thesis, Hamburg (1992).

[32] OPAL Coll., P.D. Acton et al., *Phys. Lett.* **B287** (1992) 401.

[33] C.P. Fong and B.R. Webber, *Nucl. Phys.* **B355** (1991) 54; C.P. Fong, Ph.D. Thesis, Univ. of Cambridge (1991).

[34] NA23 Coll., M. Asai et al., *Z. Phys.* **C34** (1987) 429.

[35] TPC Coll., H. Aihara et al., *Phys. Rev. Lett.* **53** (1984) 2199 and **57** (1986) 3140.

[36] CHLM Coll., M.G. Albrow et al., *Phys. Lett.* **B65** (1976) 295; TASSO Coll., R. Brandelik et al., *Phys. Lett.* **B100** (1981) 357 and **B139** (1984) 126; ABCDHW Coll., A. Breakstone et al., *Z. Phys.* **C25** (1984) 21; OPAL Coll., P.D. Acton et al., *Phys. Lett.* **B305** (1993) 415; ALEPH Coll., D. Buskulic et al., *Z. Phys.* **C64** (1994) 361; DELPHI Coll., P. Abreu et al., *Phys. Lett.* **B416** (1998) 247.

[37] SLD Coll., K. Abe et al., *A Study of Correlations between Identified Charged Hadrons in Hadronic Z^0 Decays*, SLAC-PUB-9288 (July 2002), ABS971 ICHEP02.

[38] DELPHI Coll., J. Abdallah et al., *Phys. Lett.* **B533** (2002) 243.

[39] NA22 Coll., I.V. Ajinenko et al., *Z. Phys.* **C58** (1993) 357.

[40] N.N. Biswas et al., *Phys. Rev. Lett.* **37** (1976) 175.

[41] G. Neuhofer et al., *Phys. Lett.* **B37** (1971) 438.

[42] M. Pratap et al., *Phys. Rev. Lett.* **33** (1974) 797.

[43] B.Y. Oh et al., *Phys. Lett.* **B56** (1975) 400.

[44] W.G. Ma et al., *Z. Phys.* **C30** (1986) 191.

[45] NA27 Coll., M. Aguilar-Benitez et al., *Phys. Lett.* **B164** (1985) 404.

[46] WA75 Coll., S. Aoki et al., *Phys. Lett.* **B209** (1988) 113.

[47] E653 Coll., K. Kodama et al., *Phys. Lett.* **B263** (1991) 579.

[48] ACCMOR Coll., S. Barlag et al., *Phys. Lett.* **B302** (1993) 112; K. Rybicki and R. Ryłko, *Phys. Lett.* **B353** (1995) 547.

[49] BEATRICE Coll., M. Adamovich et al., *Phys. Lett.* **B348** (1995) 256.

[50] E653 Coll., K. Kodama et al., *Phys. Lett.* **B303** (1993) 359.

[51] BEATRICE Coll., Yu. Aleksandrov et al., *Phys. Lett.* **B433** (1998) 217.

[52] P. Nason, S. Dawson and R.K. Ellis, *Nucl. Phys.* **B303** (1988) 607; **B327** (1988) 49; W. Beenakker et al., *Phys. Rev.* **D40** (1989) 54; *Nucl. Phys.* **B351** (1991) 507; M. Mangano, P. Nason and G. Ridolfi, *Nucl. Phys.* **B373** (1992) 295; **B405** (1993) 507.

[53] S. Frixione, M.L. Mangano, P. Nason, G. Ridolfi, *Nucl. Phys.* **B431** (1994) 543.

[54] MARKII Coll., C. de la Vaissière et al., *Phys. Rev. Lett.* **54** (1985) 2071.

[55] OPAL Coll., P.C. Acton et al., *Z. Phys.* **C58** (1993) 207.

[56] Yu.L. Dokshitzer, V.A. Khoze, A.H. Mueller, S.I. Troyan, *Rev. Mod. Phys.* **60** (1988) 373.

[57] M.A. Chmeissani, *Study of Angular Ordering in the Hadronic Decays of the Z^0*, ALEPH internal note 93-097 (unpublished).

[58] M. Acciarri et al. (L3), *Phys. Lett.* **B353** (1995) 145.

[59] C. Louis Basham et al., *Phys. Rev. Lett.* **41** (1978) 1585.

[60] E.L. Berger, R. Singer, G.H. Thomas and T. Kafka, *Phys. Rev.* **D15** (1977) 206.

[61] G.H. Thomas, *Phys. Rev.* **D15** (1977) 2636.

[62] NA22 Coll., I.V. Ajinenko et al., *Z. Phys.* **C61** (1994) 567.

[63] UA1 Coll., B. Buschbeck, P. Lipa and F. Mandl, Proc. NATO Advanced Study Workshop on Hot Hadronic Matter: Theory and Experiment, Divonne-les-Bains, Switzerland, 1994, eds. Jean Letessier, Hans H. Gutbrod, Johann Rafelski (Plenum Press, 1995) p.251.

[64] DELPHI Coll., P. Abreu et al., *Z. Phys.* **C63** (1994) 17.

[65] A. Białas and R. Peschanski, *Phys. Rev.* **D50** (1994) 6003.

[66] NA22 Coll., M. Adamus et al., *Z. Phys.* **C37** (1988) 347.

[67] ALEPH Coll., D. Decamp et al., *Z. Phys.* **C54** (1992) 75.

[68] E.A. De Wolf, Proc. XXII Int. Symp. on Multiparticle Dynamics, Santiago de Compostela, Spain, 1992, ed. C. Pajares (World Scientific, Singapore, 1993) p.263.

[69] NA22 Coll., M.R. Atayan et al., *Z. Phys.* **C54** (1992) 247.

[70] W.D. Walker et al., *Phys. Lett.* **B255** (1991) 155.

[71] B. Andersson, G. Gustafson, J. Samuelsson, *Z. Phys.* **C64** (1994) 653.

[72] V.A. Bumazhnov et al., Serpukhov preprint IHEP 79-181 (1979);
V.A. Bumazhnov et al., *Sov. J. of Nucl. Phys.* **46** (1987) 289 and **40** (1984) 96.

[73] S.A. Azimov et al., *Doklady AN Uzbekh SSR* **9** (1980) 31.

[74] A. Breakstone et al., *Mod. Phys. Lett.* **A6** (1991) 2785.

[75] P. Carruthers and I. Sarcevic, *Phys. Rev. Lett.* **63** (1989) 1562.

[76] A. Capella, K. Fiałkowski and A. Krzywicki, *Phys. Lett.* **230B** (1989) 149.

[77] E.A. De Wolf, *Acta Phys. Pol.* **B21** (1990) 611.

Chapter 10

Multiplicity Fluctuations and Intermittency

10.1 Prelude

The study of fluctuations in particle physics already has a long history going back to early cosmic-ray observations. To our knowledge, Ludlam and Slansky [1] were the first to advocate analysis of event-to-event fluctuations in hadron-hadron collisions. Comparing rapidity distributions of single events with the sample-averaged distribution, they put in evidence strong clustering effects in longitudinal phase space, indicating "a remarkably structured phase-space density" [2]. Fluctuations in individual events were also considered in the context of Reggeon theory in the important paper establishing the AGK-cutting rules [3].

Large concentrations of the particle number in small rapidity regions for single events were reported in cosmic-ray experiments [4–6] and in pN collisions at 200 GeV beam momentum [7]. A number of particularly high-density "spikes" in rapidity space were reported in the '80ies. Fig. 10.1a shows the notorious JACEE event [8] at a pseudo-rapidity resolution (binning) of $\delta\eta = 0.1$. It has local fluctuations up to $\mathrm{d}n/\mathrm{d}\eta \approx 300$ with a signal-to-background ratio of about 1:1. The NA22 event [9] of Fig. 10.1b contains a spike at a rapidity resolution $\delta y = 0.1$ of $\mathrm{d}n/\mathrm{d}y = 100$, as much as 60 times the average density in this experiment. UA5 [10] reported spikes in $\mathrm{d}n/\mathrm{d}\eta$ up to 30 (10 times average) as early as JACEE, but found these to be in agreement with a short-range cluster Monte Carlo. Also EMU01 [11] sees events with $\mathrm{d}n/\mathrm{d}\eta = 140$, but satisfactorily explained by FRITIOF.

From an experimental point of view, there is little doubt that events with large local density fluctuations exist. The real question is whether these are of dynamical or merely statistical origin, whether the underlying probability density is continuous or intermittent.

Early attempts to answer the question of non-statistical fluctuations employing transform techniques [12] were not followed up so far. The problem resurfaced in the

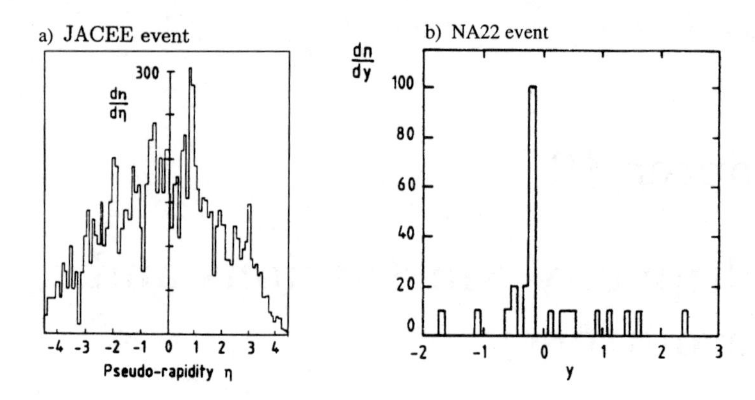

Figure 10.1. a) The JACEE event [8]; b) The NA22 event [9].

work of Białas and Peschanski [13, 14], who suggested that spikes could be a manifestation in hadron physics of "intermittency", a phenomenon well-known in fluid dynamics. The authors argued that if intermittency occurs in particle production, large density fluctuations are not only expected, but should also exhibit self-similarity with respect to the size of the phase-space volume.

Ideas on self-similarity and fractals in jet physics had earlier been formulated in [15, 16], rephrased in the language of QCD branching processes in [17] and in a simplified form in [18]. For soft hadronic processes, fractals and self-similarity were first considered in [19] and their quantitative measures in [20, 21].

In multiparticle experiments, the number of hadrons produced in a single collision is small and subject to considerable "noise". To exploit the techniques employed in complex-system theory, a method must be devised to separate fluctuations of purely statistical origin, due to finite particle numbers, from the possibly self-similar fluctuations of the underlying particle densities. The latter are the quantities of physical interest. A solution, already used in optics and suggested for multiparticle production in [13, 14], consists in measuring suitably normalized factorial moments of the multiplicity distribution in a given phase-space volume. Earlier reviews of this field are given in [22, 23].

10.2 Normalized factorial moments

10.2.1 The method

The method proposed in [13, 14] consists in measuring the dependence of the normalized factorial moments $F_q(\delta y)$ defined in (7.72-7.74) as a function of the resolution δy. For definiteness, δy is supposed to be an interval in rapidity, but the method generalizes to arbitrary phase-space dimensions.

In Sect. 7.2 we have pointed out that the scaled factorial moments enjoy the property of "noise-suppression". It is easily verified that this crucial property does not apply to ordinary moments $\langle n^q \rangle / \langle n \rangle^q$. High-order moments further act as a filter and resolve the large n_m tail of the multiplicity distribution. They are thus particularly sensitive to large density fluctuations at the various scales δy used in the analysis.

As proven in [13, 14], a *smooth* (rapidity) distribution, which does not show any fluctuations except for the statistical ones, has the property that $F_q(\delta y)$ is independent of the resolution δy in the limit $\delta y \to 0$. This follows directly from (7.104), if P_ρ is a product of δ-functions in ρ_m ($m = 1, \ldots, M$) centred around $\langle \rho_m \rangle$. On the other hand, if dynamical fluctuations exist and P_ρ is *intermittent* (i.e. regions of fluctuations exist at all scales of y), the F_q obey the power law (7.114). Eq. (7.114) is a scaling law, since the ratio of the factorial moments at resolutions L and ℓ

$$R = F_q(\ell)/F_q(L) = (L/\ell)^{\phi_q} \tag{10.1}$$

only depends on the ratio L/ℓ, but not on L and ℓ, themselves.

As mentioned in Sect. 7.4, the intermittency indices ϕ_q (slopes in a double-log plot) are related [24–26] to the anomalous (or co-) dimensions $d_q = \phi_q/(q-1)$, a measure for the deviation from an integer dimension.

We noted in Sect. 9.5 that the experimental study of correlations is difficult already for three particles. The close connection between correlations and factorial moments (Sect. 7.1) offers a possibility to measure higher-order correlations with the factorial-moment method at smaller distances than previously feasible. The method further relates possible scaling behavior of such correlations to the physics of fractal objects. Despite the advantages, it should be remembered that reliable data can only be extracted if factorial moments are averaged over a large domain of phase space. This holds the danger of obscuring important local dynamical effects.

The definition of intermittency given in (7.114) has its origin in other disciplines [27]. It rests on a loose parallel between the high non-uniformity of the distribution of energy dissipation, for example, in turbulent intermittency and the occurrence of large spikes in hadronic multiparticle final states (Sect. 10.1). In the following we use the term "intermittency" in a weaker sense, referring to the rise of factorial moments with increasing resolution but not necessarily according to a strict power law.

The suggestion that normalized factorial moments of particle distributions might show power-law behavior has spurred a vigorous experimental search for (more or less) linear dependence of $\ln F_q$ on $-\ln \delta y$. Within a surprisingly short time (one-dimensional) analyses were performed for e^+e^- [28–36], μp and μd [37], νA [38], hh [39–45], hA [46–52] and AA [46, 47, 53–61] collisions and even in AA nuclear (multi)fragmentation [62–64] and target evaporation [65]. With respect to the original objective, the early one-dimensional work allowed to accumulate valuable information and experience, but more promising insight has come from studies in two- and three-dimensional phase space. This is discussed in Sect. 10.3. Further extensions of this approach, concentrating on improved integration methods and differential studies in Lorentz-invariant variables have led to further clarification of the issues involved in intermittency. These developments are presented in Sects. 10.8-10.10, while individual events will be treated in Sect. 10.11.

10.2.2 Results on log-log plots (in one dimension)

In this and the next few sections we review experimental results and model predictions obtained from one-dimensional studies. Due to the vast amount of data available, we limit ourselves to an illustration of the major characteristics of factorial-moment behavior in various processes and at various energies.

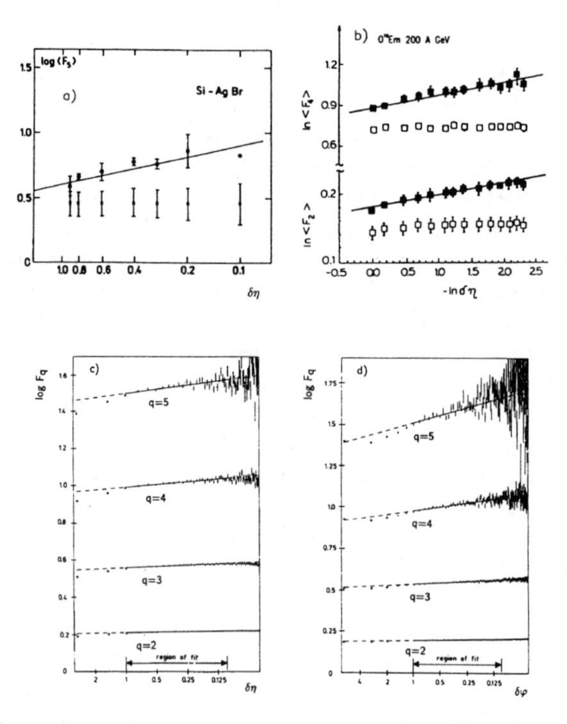

Figure 10.2. a) logF_5 as a function of $-\log\delta\eta$ for the JACEE event [13] (full circles) compared to independent emission (small crosses); b) $\ln F_2$ and $\ln F_4$ as functions of $-\ln\delta\eta$ for O^{16}Em at 200 AGeV (KLM) [46], c) logF_q on ${}_2\log\delta\eta$ and d) ${}_2\log\delta\varphi$ at 630 GeV (UA1) [40].

In Fig. 10.2a, logF_5 is plotted [13,14] as a function of -log $\delta\eta$ (η is the pseudorapidity) for the JACEE event. It is compared with an independent-emission Monte-Carlo model tuned to reproduce the average η distribution of Fig. 10.1a and the global multiplicity distribution, but has no short-range correlations. While the Monte-Carlo model indeed predicts constant F_5, the JACEE event shows a first indication for a linear increase, i.e. a possible sign of intermittency.

Further examples are given in Fig. 10.2b for KLM [46], again showing a roughly linear increase for $\delta\eta < 1$ $(-\ln\delta\eta > 0)$ instead of the flat behavior expected for independent emission, and in Figs. 10.2c and d for UA1 [40] in terms of $\delta\eta$ and $\delta\varphi$, respectively.

Anomalous dimensions d_q fitted over the range $0.1 < \delta y(\delta\eta) < 1.0$ are compiled in Fig. 10.3 [66]. They typically range from $d_q = 0.01$ to 0.1, which means that the fractal (Rényi) dimensions $D_q = 1 - d_q$ are close to one. The d_q are larger and grow faster with increasing order q in μp and e^+e^- (Fig. 10.3a) than in hh collisions (Fig. 10.3b) and are small and almost (but not fully) independent of q in heavy-ion collisions (Fig. 10.3c). For hh collisions, the q-dependence is considerably stronger for NA22 ($\sqrt{s} = 22$ GeV, no $p_{\rm T}$ cut) than for UA1 ($\sqrt{s} = 630$ GeV, $p_{\rm T} > 0.15$ GeV/c).

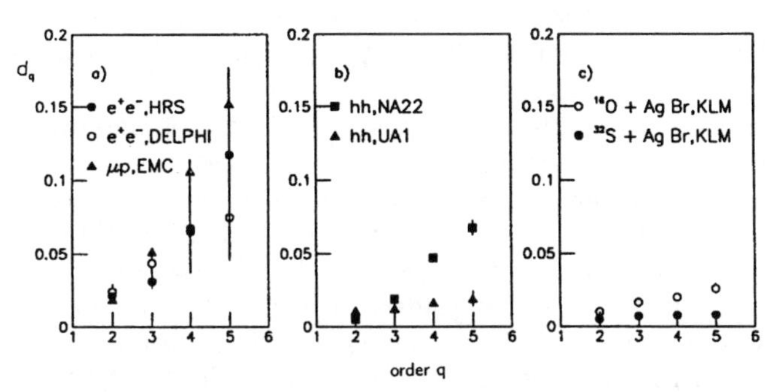

Figure 10.3. Anomalous dimension d_q as a function of the order q, for a) μp and e^+e^- collisions, b) NA22 and UA1, c) KLM [66].

10.2.3 Model predictions

10.2.3.1 Hadron-hadron collisions

A comparison to NA22 data on slopes ϕ_q (Fig. 10.4a) shows [39] that intermittency is absent at $\sqrt{s} = 22$ GeV in a two-chain DPM and underestimated by FRITIOF. In Fig. 10.4b, PYTHIA is seen to stay below the UA1 data [40], even after inclusion of Bose-Einstein interference for identical particles. The UA5 cluster Monte Carlo GENCL [67], able to reproduce conventional short-range correlations (at least in a certain range of multiplicities cf. Fig. 9.3), follows the data down to a resolution of $\delta\eta \approx 0.3$, but completely fails for smaller $\delta\eta$.

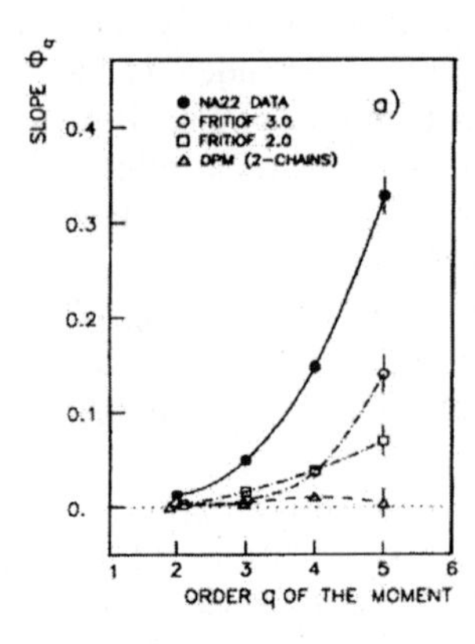

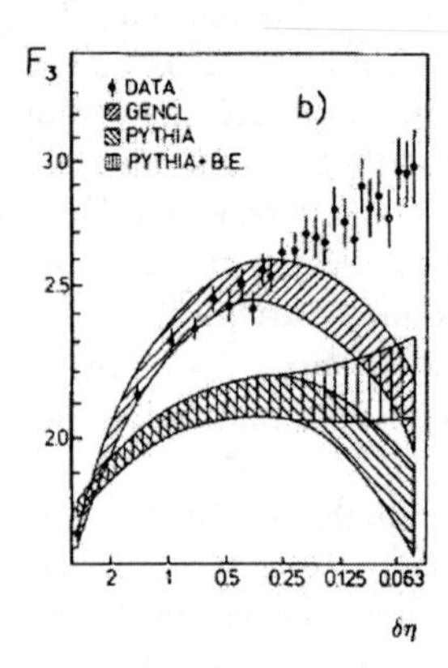

Figure 10.4. a) Slope ϕ_q as a function of order q for NA22 [39], two versions of FRITIOF and a two-chain dual parton model, b) F_3 versus $\delta\eta$ from UA1 [40] compared to GENCL, PYTHIA and PYTHIA + Bose-Einstein Monte Carlos.

Also, a multi-chain version of DPM including mini-jet production has been compared to NA22 and UA1 data [68]. The slopes are found to be too small by at least a factor of 2.

With respect to intermittency analysis, the situation is improved with the introduction of ECCO [69], an eikonal cascade model based on geometrical branching, which can account for strong fluctuations, in particular in higher dimensions (Sect. 10.3 below). However, the present version of ECCO is still less refined than the more conventional models with respect to other observables.

The above examples show that models for multiparticle production in hh collisions are unable to reproduce the magnitude and the growth of factorial moments with increasing resolution. From the discussion in Chapter 9, it is evident that model predictions for correlations in general are quite unreliable. The two-particle correlation function, measured by F_2, also determines to a large extent the higher-order factorial moments (cf. Eq. 7.76) because of the weakness of genuine high-order correlations. It is, therefore, mandatory to improve the models before evidence for "new physics" at very small (rapidity) separation can be claimed. We return to this important question in later sections.

10.2.3.2 hA and AA collisions

The intermittency indices are much smaller in hA and AA collisions than in hh collisions, but also the event samples are much smaller. Model comparisons are, therefore, less conclusive than in hh collisions.

FRITIOF is found too low in NA22 [48] for π^+/K^+ on Al and Au at 250 GeV/c, in E802 [58] for central ^{16}OAl and ^{16}OCu at 14.6 A GeV/c, in WA80 [59] for SS and

Au at 200 A GeV/c, and in NA35 [47] for pAu, OAu, SAu and SS at 200 A GeV/c. In WA80 it is shown that rough agreement can be obtained by renormalization to the leftmost point of FRITIOF on the log-log plot (essentially the shape of the overall multiplicity distribution) to the data. NA35 shows that agreement can be obtained by adding Bose-Einstein interference for like-charged particles (for a detailed analysis of the influence of BE correlations see further below).

VENUS [70] is found to agree with the (vanishing) slopes for Pbg/Br at 158 A GeV/c of KLMM [60] in η, but fails to reproduce the significant slopes in φ.

10.2.3.3 Lepton-hadron collisions

In Fig. 10.5a, EMC data [37] are compared to what is expected from an extrapolation of conventional short- and long-range correlations [71]. At small δy, the data are consistently above these expectations. As Fig. 10.5b shows, the slopes ϕ_q in the same data are considerably larger than predicted by the HERWIG and Lund models. Similarly, Fig. 10.5c shows too low $\ln F_3$ from LEPTO/JETSET, not only for νNe but also for the "simpler" νD_2 interactions [38].

We tentatively conclude that also *lepton-hadron* models as such are unable to reproduce the intermittency observed in this process.

10.2.3.4 e^+e^- annihilation

The annihilation of e^+e^- into hadrons is the best understood of all multihadron reactions. Creation of hadrons is traditionally pictured as a multistep process comprising a "hard" parton evolution phase, described by perturbative QCD - the parton shower - and a non-perturbative color-confining soft hadronization phase (Fig. 10.6). The former is a cascade process of nearly self-similar type, and is expected to show characteristics typical of a fractal object [15,16,18]. In fact, already in 1979, in a discussion of QCD jets, it was stated [16] that "the resulting picture of a jet is formally similar to that of certain mathematical objects, known as fractals, which look more and more irregular and complex as we look at them with a better and better resolution". The expectation is, therefore, that parton showers should exhibit intermittency at the parton level. However, this is not sufficient to guarantee intermittency at the hadron level. It is indeed difficult to imagine how the "re-shuffling" of the parton momenta during the hadronization phase with e.g. the formation of hadronic resonances and their subsequent decay would preserve the (supposedly singular) nature of the correlations. A local parton-hadron duality type of explanation is not satisfactory either, since "it is merely a name for a mechanism that is not at all understood" [72].

To describe the hadronization phase, all present Monte-Carlo codes rely in last instance on a large amount of e^+e^- data at different energies and are carefully tuned to these. It came, therefore, as a surprise that a first (indirect) analysis [28] of HRS results, shortly followed by TASSO data [29], revealed deviations from model predictions at small bin sizes, quite similar to those observed in lh and hh collisions

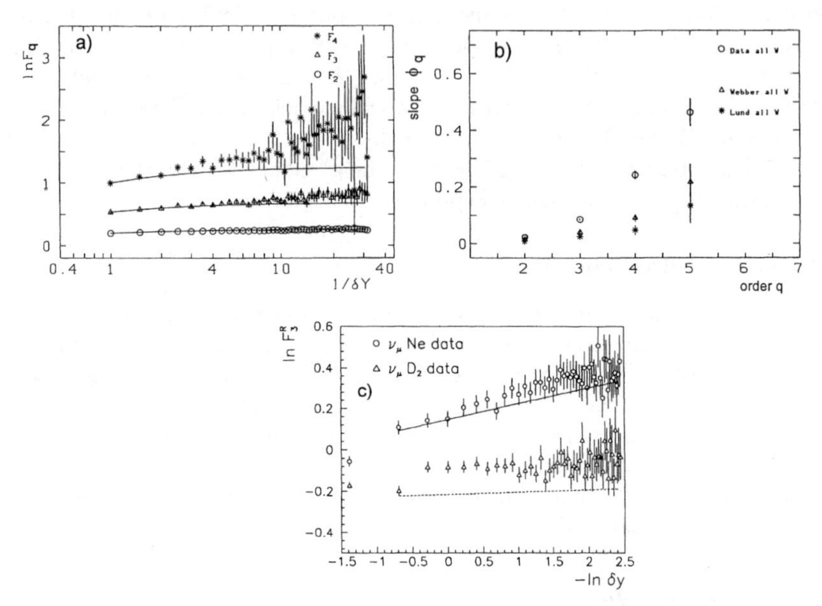

Figure 10.5. a) EMC results [37] compared to expectations from [71], b) slopes ϕ_q for EMC data as well as HERWIG and Lund models, c) νA data [38] in comparison to LEPTO/JETSET model expectations.

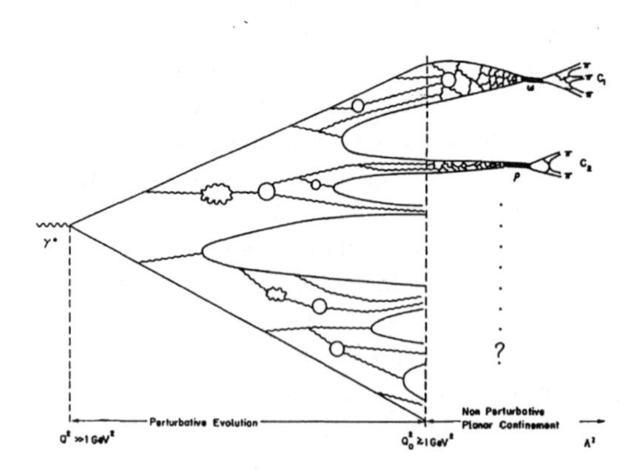

Figure 10.6. Jet evolution: the self-similarity in the parton cascade derives from the similarity of each step in the evolution [16].

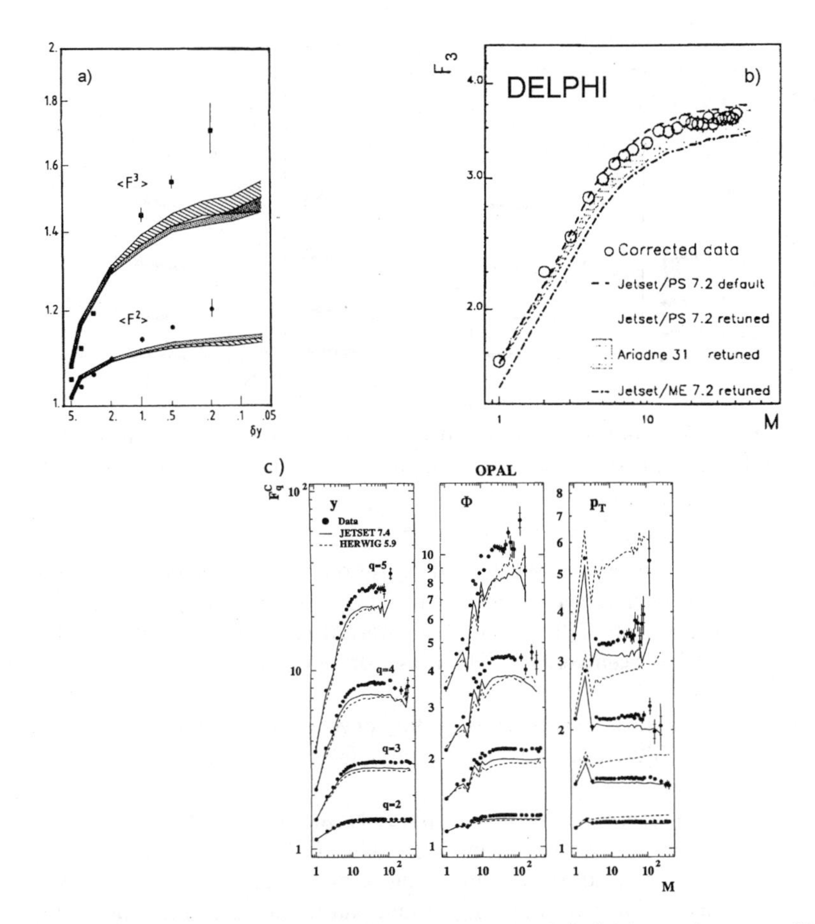

Figure 10.7. a) $\ln F_q$ as a function of $-\ln \delta y$ for HRS [28], b) as a function of log$M$ ($M = \Delta Y/\delta y$) for DELPHI [31] and c) as a function of $M(M = \Delta Y/\delta y, 2\pi/\delta\varphi, \Delta \ln p_{\rm T}/\delta \ln p_{\rm T})$ for OPAL [32] data compared to the JETSET, HERWIG and ARIADNE parton shower models.

(Fig. 10.7a). Also DELPHI [31] and in particular OPAL [32] data with an order of magnitude larger statistics show significant deviations (Figs. 10.7b,c). The models tend to fall below the data at intermediate and large M. The discrepancy rises with increasing M and order q. Note also the large difference between JETSET and HERWIG in $p_{\rm T}$ (rightmost Fig. 10.7c), where no intermittency is present in the OPAL data and in JETSET.

The origin of intermittency in the models is not quite as clear as is often stated. Indeed, comparison of the factorial moments on parton and hadron level in Figs. 10.8a,b [73], shows that in (standard) JETSET at $\sqrt{s} = 91$ GeV the increase of $\ln F_q$ at small

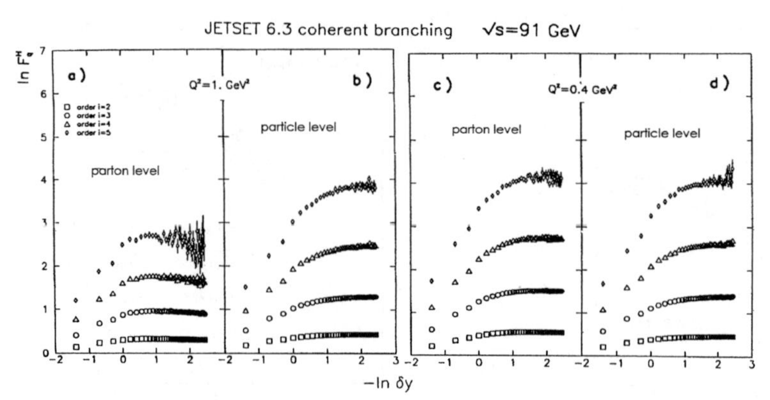

Figure 10.8. $\ln F_q^{\mathrm{H}}$ as functions of $-\ln \delta y$ for JETSET 6.3 parton shower at $\sqrt{s}$=91 GeV at the a) parton level, b) hadron level, both with cut-off Q_0^2=1 GeV^2, c) d) with cut-off Q_0^2=0.4 GeV^2 [73].

δy is not due to the parton shower, but to hadronization! Only if the parton shower is allowed to continue down to very low Q_0^2 values (Fig. 10.8c,d for Q_0^2=0.4 GeV^2), thus approaching local parton-hadron duality, is intermittency becoming visible also at the parton level. It has been verified that the sensitivity to Q_0^2 is, of course, much less important at 1 TeV.

On the contrary, intermittency seems to be fully developed on the parton level already at 91 GeV in the HERWIG model, and is then in fact smeared out by hadronization [74].

Intermittency can be increased in the soft phase by an increase of the π/ρ ratio, also required from direct measurements by NA22 [75], EMC [76] and in hA collisions [77]. The direct pions resolve the underlying parton structure better than the more massive resonances. From a tunnelling production mechanism, these pions are expected to have smaller p_{T} than other particles, a property presently neglected in the MC programs. A Goldstone-like mechanism causing additional soft direct pion production at and a p_{T}-correlation between break-up points has, therefore, been suggested by the Lund group [78] (see also Sect. 9.4).

The sensitivity to the cut-off in the perturbative QCD cascade and the role of hard and soft phases has also been discussed in terms of the dipole-radiation model [79]. This model is an alternative description of perturbative QCD and a particularly useful tool for the study of its properties, since it is incorporated in the ARIADNE Monte Carlo program (see Sect. 6.1) with full energy-momentum conservation. At high energies, the fluctuations turn out to be dominated by the perturbative phase, in particular by the first one or two gluons. Softer gluons contribute to the noise in a way rather similar to the fluctuations in the soft hadronization phase. So, the interface between the perturbative and soft phases becomes movable, thus providing a quantitative realization of local parton-hadron duality (LPHD).

10.2.4 A warning

Before going into the necessary further detail, we should mention the influence of possible experimental biases. On purpose and by its very definition, the higher factorial moments are sensitive to a small number of events in the tail of the multiplicity distribution in small phase-space bins.

Moments can be *reduced* by limited two-track resolution, by track losses from limited acceptance or bad reconstruction, or simply due to truncation of the multiplicity distribution in a finite event sample.

Moments can be *increased* due to double counting of tracks (track match failures), Dalitz decays and nearby γ-conversions or K^0/Λ decays. A dangerous increase comes from the commonly used "horizontal" averaging, where a *constant* average (pseudo-) rapidity distribution is assumed over the range ΔY. Contrary to first belief, this problem is *not* completely solved by the correction method proposed in [80] !

Further influence is to be expected from the choice of the sample (e.g. inelastic or non-diffractive), cuts on multiplicity, cuts on p_T, all events or only those with $n \geq n_0$ in ΔY, etc., the size and position of ΔY, the δy region chosen for the fit and the correlation of errors.

Many of these effects have been studied in a number of experiments and we refer to these and to [82, 83] for more details, but limit ourselves to one systematic comparison [81] based on 500.00 JETSET events. In Fig. 10.9a, a comparison is shown between results obtained for F_q in a number of phase space variables, in one and in three dimensions, with vertical (points) and horizontal (lines) averaging, Eqs. (7.72) and (7.74), respectively, using the correction factor [80], in the phase space region limited to $-2 < y < 2$, $0.1 < p_T < 2$ GeV/c. Figures 10.9b give the same F_q, but evaluated with the cumulative variables Eq. (10.2) below [84]. The F_q^V and F_q^H coincide for the cumulative variables in Fig. 10.9b, but differ for φ, p_T and in 3-D when only the correction factor is used. Furthermore, the results themselves come out very different in Figs. 10.9a and b for φ and p_T (and 3-D).

The influence of restricting phase space can be seen from a comparison of Figs. 10.9b and c, the latter corresponding to the nearly full phase space region $-5 < y < 5$, $0.1 < p_T < 3$ GeV/c. The main difference is a flattening (in y and 3-D) and even increase (in φ and φ_r) for $M \to 1$, due to momentum conservation [85]. In all subfigures, the results for F_q are seen to differ for φ (column 2) and φ_r (column 3). The first corresponds to the azimuthal angle defined with respect to the thrust-minor axis (hard gluon direction), the second with respect to an axis obtained from an essentially random rotation around the thrust-major axis.

10.3 Higher dimensions

10.3.1 The projection effect

So far, we mainly have discussed factorial moments from one-dimensional distributions in rapidity or pseudorapidity. As in Figs. 10.2d and 10.7c, the analysis can

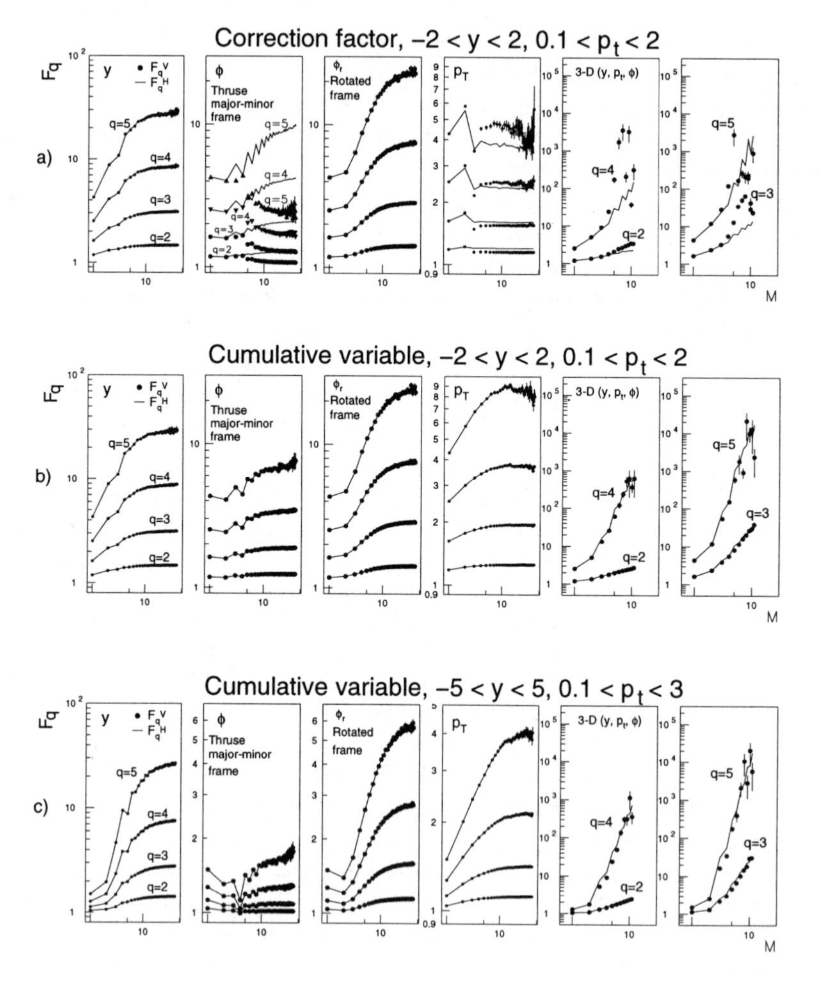

Figure 10.9. F_q^{V} (points) and F_q^{H} (lines) from JETSET7.4 in a) restricted phase space using correction factor [80], b) cumulative variables, c) nearly full phase space and cumulative variables [81].

evidently be extended to other 1D variables, such as the azimuthal angle φ in the plane perpendicular to the beam or event axis, or the particle transverse momentum p_T. It is clear from the various one-dimensional examples given (Fig. 10.2, 10.5, 10.7) that the power law holds only approximately. At closer inspection, there is a tendency of $F_q(\delta y)$ or $F_q(\delta\eta)$ to flatten at small δy or $\delta\eta$ (large M), consistent with a non-zero correlation length, after all. This is also concluded in [58, 86] on the basis of a linear extrapolation of the negative-binomial parameter

$$k(\delta\eta) = 1/K_2(\delta\eta) = \frac{1}{F_2(\delta\eta) - 1}$$

to $\delta\eta = 0$ (Fig. 8.16). Obviously, this extrapolation should give $k(0) = 0$ in case of intermittency ($F_2(0) = \infty$), but $k(0) \neq 0$ in case of saturation of $F_2(\delta\eta)$. Even though the negative binomial does not reproduce the crucial tail of multiplicity distributions (including those of [58]) and the extrapolation should not be linear, the observation of a possible saturation should be taken as a further warning.

However, real life is not one-dimensional! Given sufficient statistics, distributions can be analyzed in two- and three-dimensional phase-space domains. Common choices are $(\Delta y, \Delta\varphi)$, $(\Delta y, \Delta \ln p_T)$, $(\Delta\varphi, \Delta \ln p_T)$ and $(\Delta y, \Delta\varphi, \Delta \ln p_T)$.

Fig. 10.10a gives an example of 1D-results from UA1 [40] showing that intermittency is also present in φ. The intermittency effect is larger when two-dimensional cells $(\Delta y, \Delta\varphi)$ are studied than in 1D (Fig. 10.10b,c). This is particularly pronounced in e^+e^- annihilations (Figs. 10.11a,b), the measured slopes ϕ_q being about six times larger in 2D than in 1D. These observations are now understood to imply that intermittency "lives in 3D" [87–89]. Projection onto a lower-dimensional subspace dilutes the effect and leads to flattening of the factorial moments. This is most pleasantly demonstrated by the fact that one can enjoy a continuous (two-dimensional) shadow of a tree, in spite of the non-continuous branching of this tree in three dimensions.

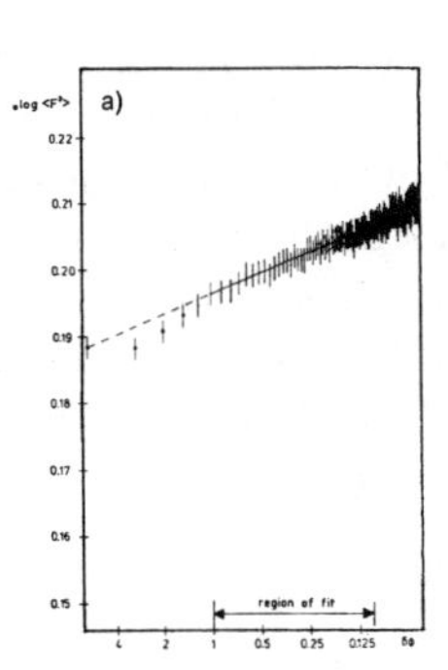

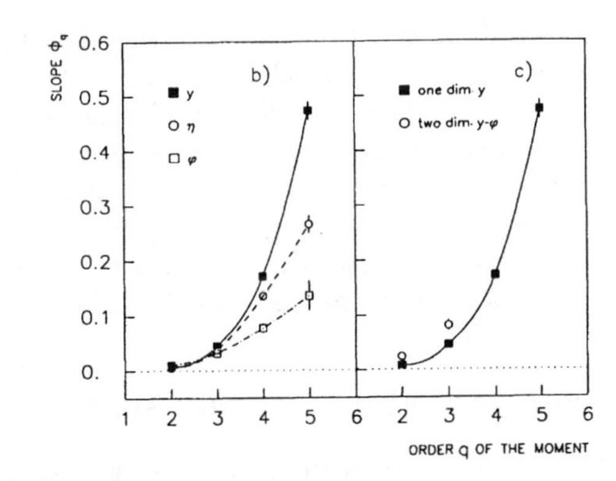

Figure 10.10. a) $\log F_2$ as a function of $-\log \delta\varphi$ for UA1 data [40], b) c) slope ϕ_q as a function of order q for y, η, φ or (y and φ) as variables for NA22 [39].

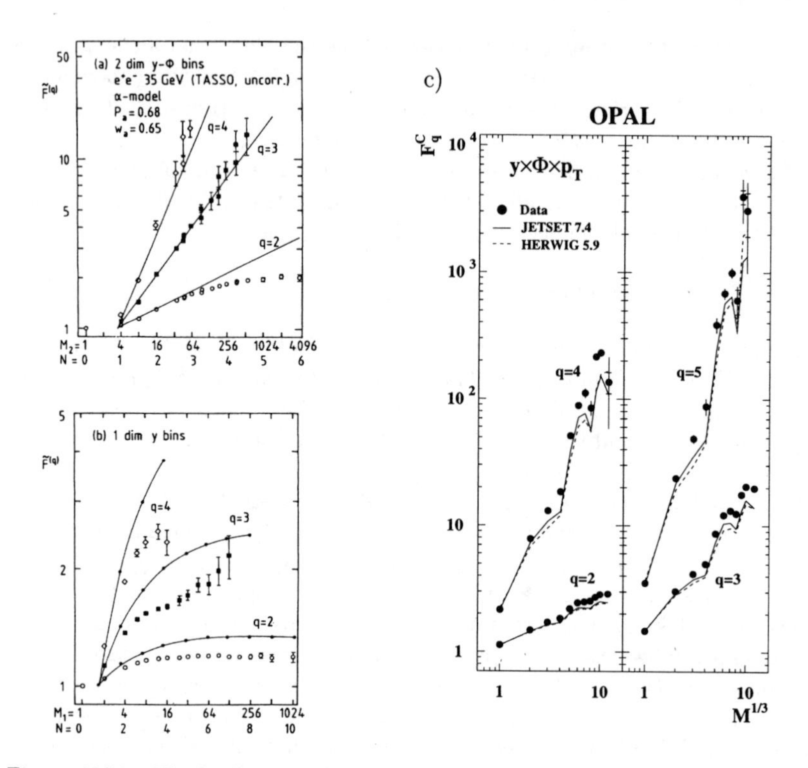

Figure 10.11. The log-log plot for a,b) TASSO data using a) two-dimensional y–φ bins, b) one-dimensional y bins, in comparison to a two-dimensional α-model [29, 87], c) OPAL data in comparison with the predictions of two Monte-Carlo models [32].

The projection effect is convincingly illustrated in Fig. 10.11a,b. The lines in Fig. 10.11a are fits by a 2D α-model; the curves in Fig. 10.11b are the projections onto rapidity space and show considerably less increase and even a flattening for $\delta y \to 0$ (note the difference in scale). Nevertheless, Fig. 10.11a still shows saturation of F_2 at large M even in 2D, an indication that an analysis in three dimensions is required. An example is given in Fig. 10.11c in the form of $\log F_q$ as a function of $M^{1/3}$ for the OPAL data. Their increase with increasing $M^{1/3}$ should be compared to the saturation observed for the three one-dimensional cases in Fig. 10.7c.

10.3.2 Transformed momentum space

To study intermittency in three-dimensional phase space, one faces the additional difficulty that the particle density is all but uniform in the usual single-particle variables y, φ and p_{T}. The distribution in p_{T} is in fact falling exponentially. Uniformity of the density is, however, an explicit assumption in the derivation of the power law (7.114). Violation of this condition renders an intermittency analysis useless.

To circumvent this problem, the authors of [84,87] have proposed to use domains in a transformed momentum space with (practically) constant density. This is accomplished by a transformation of the original variables y, φ and $\ln p_{\rm T}$ to "cumulative" variables. Thus, for a single variable, say y, one defines the new variable $X(y)$ as

$$X(y) = \frac{\int_{y_{\rm min}}^{y} \rho_1(y')dy'}{\int_{y_{\rm min}}^{y_{\rm max}} \rho_1(y')dy'} \quad . \tag{10.2}$$

For higher dimensions, it is assumed in [87] that the single-particle density factorizes as

$$\rho_1(y, \varphi, p_{\rm T}) = \rho_{\rm a}(y)\rho_{\rm b}(\varphi)\rho_{\rm c}(p_{\rm T}) \quad . \tag{10.3}$$

Under this rather strong hypothesis, one can transform each of the three variables independently. The method proposed in [84] does not assume factorization but is technically quite involved. In practice, the two techniques give satisfactorily similar results [90], as long as the correlation between the variables is small.

Data on F_2 in various dimensions are shown in Fig. 10.12 for $\rm e^+e^-$ [31] and hh collisions [39, 40]. In all cases, the data behave more power-like in 2D than in 1D, and even more so in 3D. From Fig. 10.12a, it is also evident that JETSET PS remains in good agreement with $\rm e^+e^-$ data in higher dimensions.

As mentioned in Subsect. 10.2.3 above, ECCO [69] has some success in describing the NA22 data on fluctuations in varying scales of resolution. In particular this is the case when the analysis is done in three dimensions. The basis of this model is geometrical branching for soft production at low $p_{\rm T}$. The geometrical aspect of hadrons, i.e. the fact that they are extended objects, puts the impact parameter R in a pre-eminent role. The fluctuation in R from event to event leads to fluctuations in $p_{\rm T}$ and explains the non-vanishing intermittency in $\ln p_{\rm T}$ reported by NA22. The (stronger) intermittency in rapidity can be generated only with a singular splitting function for branching in rapidity space. Since there is no branching in φ in the model, intermittency is nearly non-existent in this variable. Still, the long-range correlation due to $p_{\rm T}$ conservation leads to a decrease of F_q at large bin size, a feature also observed by NA22.

At variance with power-law behavior expected from intermittency, NA22 finds that the 3D factorial moments even show an upward bending (Fig. 10.12c). This effect persists after exclusion of Dalitz decays and γ-conversions. A rise faster than power law is also observed in 3D for μp and μd [37] and for collisions of various projectiles with Au by NA35 (Fig. 10.13a) [47]. Following a suggestion in [91], both find that the normalized factorial cumulant $K_2 = F_2 - 1$ shows much better linearity in a log-log plot than F_2 itself (Fig. 10.13b for NA35).

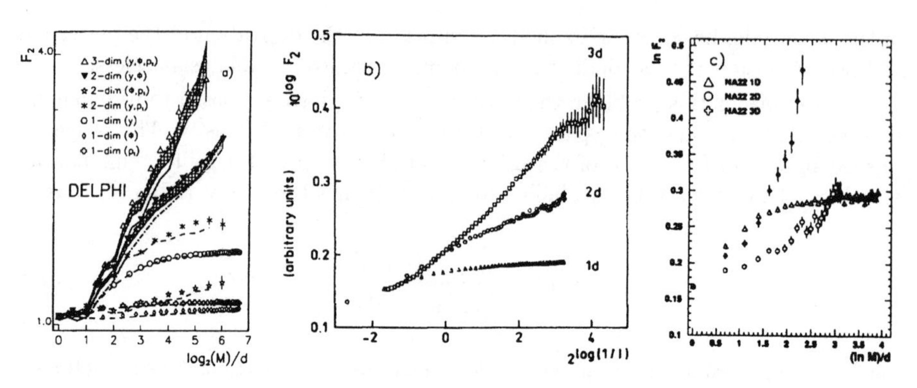

Figure 10.12. Factorial moment of order $q = 2$ for one-, two- and three-dimensional analysis for a) DELPHI [31], b) UA1 [40] and c) NA22 [39] as a function of $({}_2\log M)/d$ and $(\ln M)/d$, respectively, where M denotes the total number of cells in a d-dimensional analysis.

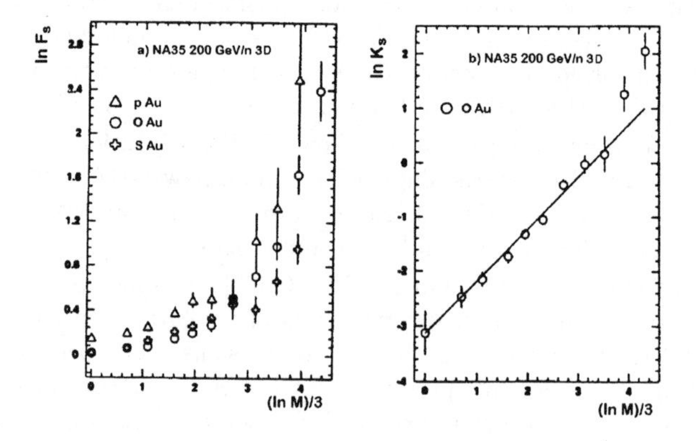

Figure 10.13. a) Factorial moment $\ln F_2$ as a function of $(\ln M)/d$ from a three-dimensional analysis of negative particles in pAu and central OAu and SAu collisions [47]; b) Factorial cumulant K_2 from the same analysis in central OAu collisions [47].

This observation, in fact, furthers considerably our understanding of the intermittency phenomenon. In [91,92] the author compared 3D data on F_2 at $\sqrt{s} \simeq 20$ GeV for μp [37], π/Kp [39], pAu, OAu and SAu [47] collisions using the parametrization

$$F_2 = 1 + c(M^3)^{\phi_2} + c' \quad , \tag{10.4}$$

where M^3 is the number of 3D phase-space cells. The constant c' accounts for long-range correlations, known to exist in hh collisions.

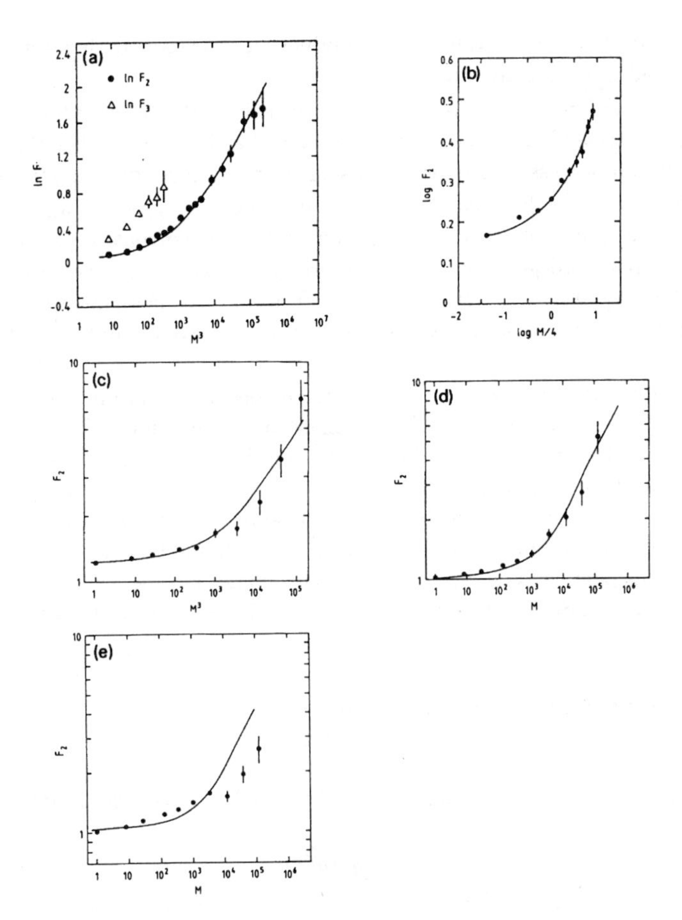

Figure 10.14. The second scaled factorial moment F_2 as a function of the number of bins M^3 in log-log scale [91] for (a) μp data [37], (b) π/Kp data [39], (c) pAu, (d) OAu and (e) SAu data [47]. Solid lines represent (10.4) with parameter values: (a) c=0.025, ϕ_2=0.45, c'=0; (b) c=0.02, ϕ_2=0.45, c'=0.16; (c) c=0.02, ϕ_2=0.45, c'=0.2; (d) and (e) c=0.01, ϕ_2=0.5, c'=0.

The comparison of (10.4) to the data is shown in Fig. 10.14; the parameters are given in the figure caption. The parameter c' is negligible for μp and heavy-ion collisions, but non-zero for meson-proton and pA collisions, in agreement with expectations. The most noteworthy result, however, concerns ϕ_2, which is seen to have a value in the range 0.4-0.5 for all processes. This is remarkable in various respects.

Firstly, if confirmed in further studies, and in particular for e^+e^- annihilation, it suggests that the resolution dependence of F_2 exhibits a high degree of "universality", is independent of specific details of the production process and thus reflects general features of hadronization dynamics.

Secondly, such universality is at variance with the hitherto accepted idea that the factorial moments and the anomalous dimensions become the smaller the more complex the collision process, due to an increasing inter-mixing of production sources [25].

Thirdly, if "universality" continues to hold in high energy e^+e^- annihilation, one must revise the commonly expressed opinion that the perturbative parton evolution, and in particular hard-jet emission, is the primary cause of the rise of factorial moments at high resolution. Needless to say, it would be most interesting to verify systematically the universality conjecture in other reactions and for three-particle correlations.

The experimental success of expression (10.4) becomes quite intriguing when one realizes that the volume $\delta \sim M^{-3}$ of a phase-space cell (for sufficiently large M) is in fact related to the invariant mass M_{inv} of the two-particle system or to Q^2, the square of their four-momentum difference. The form (10.4) implies that the two-particle correlation function behaves as a power law in M_{inv} or Q^2. The data, therefore, seem to tell that an intermittency analysis should be performed in (Lorentz-invariant) multiparticle variables, rather than single-particle variables. This was already expected in Sect. 9.4 and will be further discussed in Sects. 10.8-10.9.

10.3.3 A generalized power law

In multiplicative cascade models, the moments follow the generalized power law [93]

$$F_q \propto [g(\delta y)]^{\phi_q} , \tag{10.5}$$

where $g(\delta y)$ is a general function of δy. Expressing g in terms of F_2, one finds the linear relation

$$\ln F_q = c_q + (\phi_q/\phi_2) \ln F_2 . \tag{10.6}$$

This intriguing relation has successfully been confirmed by experiment, not only in one dimension, but up to 3D [87]. Moreover, the ratios ϕ_q/ϕ_2 are found to be largely independent of the dimension of phase space (Fig. 10.15a) and of the type of collision (Fig. 10.15b).

The ratio of the anomalous dimensions $d_q(= \phi_q/(q-1))$ and d_2 are shown in Fig. 10.16b as a function of q. The q dependence is claimed to be indicative of the mechanism causing intermittent behavior. For a (multiplicative) cascade mechanism, in the log-normal approximation (long cascades), the moments satisfy the relation [13, 14]

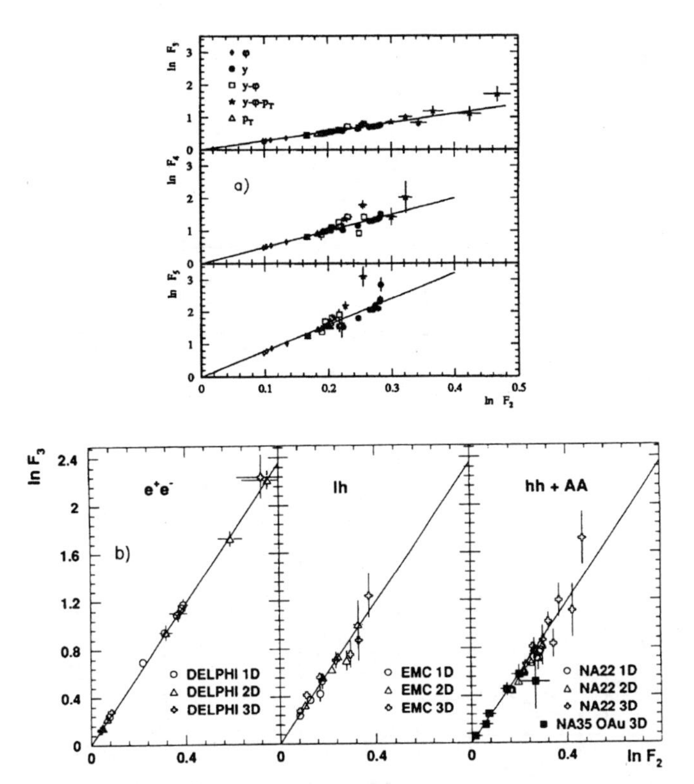

Figure 10.15. a) Illustration of the modified power-law behavior. The lines indicate an 'eye-ball' fit to the data. Only the data in bins where F_2 varies strongly are used; b) Test of the universal scaling law (10.6) for $\ln F_3$ and $\ln F_2$ in e^+e^-, lh, hh and AA collisions as indicated. The straight line is adjusted to the e^+e^- data and reproduced on the other data sets [94].

$$\frac{d_q}{d_2} = \frac{\phi_q}{\phi_2}\frac{1}{q-1} = \frac{q}{2}. \tag{10.7}$$

However, the use of the Central Limit Theorem for a multiplicative process, such as in the α-model, is a very crude approximation [95], particularly in the tails. As argued in [96], a better description might be obtained if the density probability distribution is assumed to be a long-tailed log-Lévy-stable distribution, characterized by a Lévy index μ, a continuous parameter in the range [0,2]. In that case (10.7) generalizes to

$$\frac{d_q}{d_2} = \frac{q^\mu - q}{2^\mu - 2}\frac{1}{q-1} \ . \tag{10.8}$$

For $\mu = 2$, the Gaussian case, (10.8) reduces to (10.7).

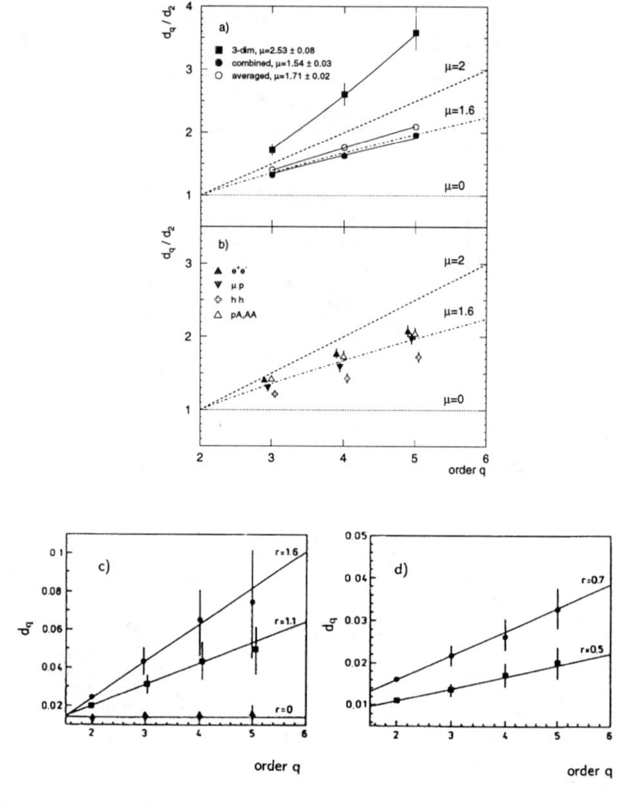

Figure 10.16. Ratio of anomalous dimensions as a function of the order q for a) NA22 results [39] and b) earlier results on all types of collisions [87]. c) and d) d_q as a function of the order q; continuous lines show the best fits using (7.138) c) • Z^0 decay, DELPHI, $r = 1.6(8)$, ■ pAg/Br, KLM, $r = 1.1(5)$, ♦ SAg/Br, KLM, $r = 0.0(2)$; d) • OAg/Br, KLM, $r = 0.7(2)$, ■ $p\bar{p}$, UA1, $r = 0.5(3)$ [97].

The multifractal behavior characterized by (10.7-10.8) reduces to a monofractal behavior [98, 99]

$$\frac{d_q}{d_2} = 1 \tag{10.9}$$

for $\mu = 0$, implying an order-independent anomalous dimension. This would happen if intermittency was due to a second-order phase transition. Consequently, monofractal behavior might be a signal for a quark-gluon plasma second-order phase transition.

The data are best fitted with the Lévy-law solution with $\mu = 1.6$ [87]. This value is inconsistent with the Gaussian approximation, and also definitely higher than expected for a second-order phase transition.

The validity of the dimension-independent generalized power behavior has been questioned in an NA22 analysis [39] shown in Fig. 10.16a. While a fit to the combined data on all variables and dimensions (full circles), as well as a weighted average over all individual fits give μ values in rough agreement with those of [87], the 3D-data have $\mu > 2$, not allowed in the sense of Lévy laws.

Even larger values of μ, ranging from 3.2 to 3.5, have been found for μp deep-inelastic scattering in [96]. According to [100,101], this is evidence that the procedure to obtain the Lévy index is used outside its domain of validity. A possible way out is self-affinity to be discussed in Sub-Sect. 10.3.5 below, but a probably more reliable way to extract the Lévy index from the fractal spectrum is given in Sub-Sect. 10.7.2.

The linear d_q/d_2 behavior in Fig. 10.16a and b gives some justification for Eq. (7.138). Fig. 10.16c and d show [97] the slope r of (7.138) for a number of experiments. All experiments, except perhaps SAg/Br, show multifractal behavior ($r > 0$).

Despite the confusion, it remains a noteworthy experimental fact that the factorial moments of different orders obey simple hierarchical relations of the type (10.6). This means that correlation functions of different orders are not completely independent but are somehow interconnected. Such situations are commonly encountered in various branches of manybody physics (see e.g. [102–104]), but a satisfactory link with particle phenomenology, let alone QCD, remains to be established. Nevertheless, on a simple example it was shown [105] that a linear relation between $\ln F_3$ and $\ln F_2$ can be obtained if the connected correlation functions are assumed to be of a factorized Mueller-Regge power-law form in two-particle invariant-masses squared s_{ij}, i.e. $C_3(1,2,3) \propto (s_{12})^{1-\alpha_1}(s_{23})^{1-\alpha_2}$ + cycl. perm.. Note that this Regge-form has the "linking" structure of Eq. (9.16) (see further in Sect. 10.5 below).

10.3.4 Thermal versus non-thermal phase transition

10.3.4.1 Second-order phase transition?

As shown e.g. in [106–108], a high-order quark-hadron phase transition can lead to strong fluctuations over a wide range of rapidity scales, i.e. to intermittency.

A simple model that can provide some hint on the nature of a second-order phase transition is the Ising model in 2D [109]. Its intermittency behavior has been studied both analytically and numerically [98,110]. The anomalous dimension is found to be $d_q = 1/8$, independent of q. Based on that finding, it has been conjectured that intermittency may be monofractal if due to a QCD second-order phase transition [99]. However, as observed in Sub-Sect. 10.3.3 above, all types of interactions, including heavy-ion collisions, show multifractal behavior.

Of course, the Ising model is very simple and the above conjecture has little basis. In [111], intermittency is, therefore, studied in the framework of the Ginzburg-Landau theory also used to describe the confinement of magnetic fields into fluxoids in a type II superconductor. In that model the anomalous dimension is not constant, but follows

$$\frac{d_q}{d_2} = (q-1)^{\nu-1} \quad , \quad \nu = 1.304 \ , \tag{10.10}$$

with ν being a universal quantity valid for all systems describable by the GL theory, independent of the underlying dimension or the parameters of the model. This is of particular importance for a QCD phase transition, since neither the transition temperature nor the other important parameters are known there.

In quantum optics, γ production at the threshold of lasing is describable as a second-order phase transition. Indeed, a photo-count experiment [112] has verified (10.10) to high precision. On the other hand, the current NA22 data on particle production in hadronic collisions give $\nu = 1.45 \pm 0.04$ [113], and similar or higher values for heavy-ion experiments [54, 61, 111].

On the other hand, it has been shown [114] that $\nu = 1.40$ can be easily obtained from a two-mechanism model [115], originally designed to describe the shoulder effect of Sub-Sect. 8.1.2, but unrelated to a phase transition.

An extension of the use of fluctuations, in the search for a second-order phase transition are the *multiplicity difference correlators* [116], designed to posses the virtues of both factorial correlators and wavelets,

$$\mathcal{F}_q = \frac{f_q}{f_1^q}, \quad f_q = \sum_{m=q}^{\infty} m(m-1)\dots(m-q+1)Q_m \ , \tag{10.11}$$

where $m = |n_1 - n_2|$, being the difference in the multiplicities n_1 and n_2 in two bins of size δ at a distance D, and $Q_m = Q_m(D, \delta)$ being the multiplicity-difference distribution for these two bins. In analogy to (10.6) and (10.10), a scaling law

$$\mathcal{F}_q \propto \mathcal{F}_2^{\beta_q}, \quad \beta_q = (q-1)^\gamma \tag{10.12}$$

is derived with $\gamma = 1.1$ for a phase transition decribable by the GL theory, while $\gamma = 1.33 \pm 0.02$ is found for the case of uncorrelated bins. Experimental study is still missing.

Phase transition or not, experimental results on ν and/or γ will be a challenge for any model of hadron production.

For a first-order phase transition, all d_q are zero and no intermittency would be observed [99]. However, it has been shown in [117] that in a generalized GL model, a first-order phase transition combined with the quantum optics analogy of lasing at threshold can lead to intermittency behavior in some regions of parameter space, with approximately the same intermittency indices as a second-order phase transition.

10.3.4.2 Non-thermal phase transition?

Of course, the phase transition does not need to be thermal, i.e., the new phase need not be characterized by a thermodynamical behavior. Such a transition could,

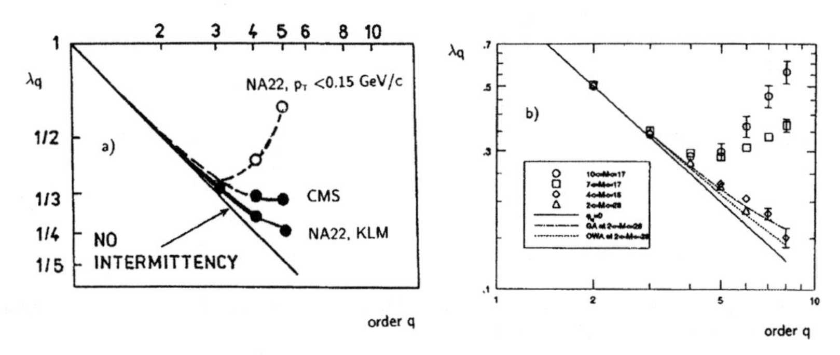

Figure 10.17. λ_q as a function of q for a) KLM, CMS and NA22 results [39, 120] and b) central CCu collisions at 4.5 A GeV/c [57].

e.g. take place during a parton-shower cascade and has been formulated in [118] for a number of "ultra-soft" phenomena, including intermittency. It leads to the co-existence of different phases, in analogy to different phases of spin-glass systems. The examples of the JACEE event (Fig. 10.1a), which contains many spikes and gaps, and that of the NA22 event (Fig. 10.1b), which consists of just one spike, indicate that such a possibility may be more than just a speculation.

The condition for the existence of such different phases of a self-similar cascade is that the function

$$\lambda_q = (\phi_q + 1)/q \tag{10.13}$$

has a minimum at some value $q = q_c$ (not necessarily an integer) [119, 120]. The regions $q < q_c$ and $q > q_c$ are dominated by numerous small fluctuations and rare large fluctuations, respectively. In the terminology of [120], the system resembles a mixture of a "liquid" of many small fluctuations and a "dust" of high density. We see either the liquid or the dust phase, depending on whether we probe the system by a moment of order $q < q_c$ or $q > q_c$, respectively.

In Fig. 10.17a, λ_q is compiled [120] from KLM, EMC and NA22 as a function of the order q. The low-p_T NA22 data [39] ($p_T < 0.15$ GeV/c) indeed show a marked minimum with q_c between 3 and 4, while the uncut data have not saturated at $q \leq 5$. Following [120], the λ_q behavior has further been studied in a number of heavy-ion experiments [50, 51, 54, 56, 57], including a two-dimensional analysis in η and φ [121]. While a saturation, but no clear minimum, is seen by experiments stopping their analysis at $q = 5$ or 6, a minimum is observed at $4 < q_c < 5$ where the analysis is carried to $q = 8$, in central C-Cu collisions at 4.5 A GeV/c [57] (Fig. 10.17b) and, be it less pronounced, in SAg/Br collisions at 200 A GeV [56]. Furthermore, minima are also observed for limited M ranges in target evaporation [65]. An indication is even seen in a limited M range in NA27 hh data [122], when $q = 7$ is reached by means of the star-integral method instead of the conventional binning (see Sect. 10.8 below).

The observation of a minimum in the λ_q-distribution may suggest a phase tran-

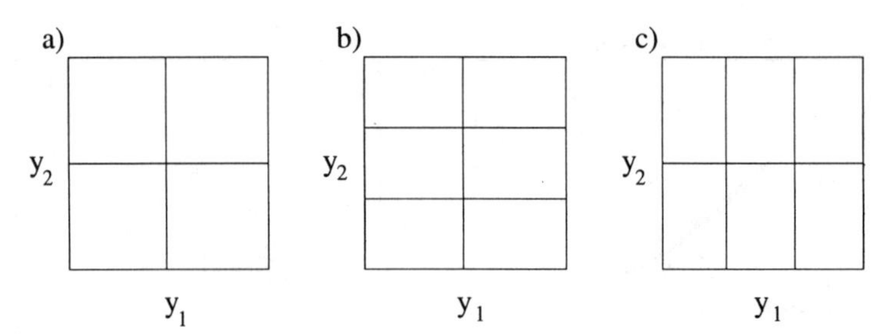

Figure 10.18. Different ways of phase-space partition: a) isotropic ($H = 1$), b) anisotropic ($H < 1$), c) anisotropic ($H > 1$) [123].

sition [119], but according to the interpretation [120] it is merely the "apparatus" changing from a sensitivity for the dominating small fluctuations at $q < q_c$ to an insensitivity for those at $q > q_c$. The two phases could coexist without a transition being necessary.

So, phase transition or not, two phases seem to coexist and it will be a challenge to find their physical interpretation in terms of the theory of strong interactions.

10.3.5 Self-affinity

Comparing log-log plots for one phase-space dimension, one notices that the $\ln F_q$ saturate, but at different F_q values for different variables y, φ or $\ln p_T$. The saturation in one dimension can be explained as projection effect of a three-dimensional phenomenon. However, also in three-dimensional analysis the power law (7.114) is not exact. In Fig. 10.12c, the 3D hh data are seen to bend upward. It has been shown in [123] (see also [124]) that this can be understood by taking the anisotropy of occupied phase space (longitudinal phase space [125]) into account. In view of this phase-space anisotropy, also its partition should be anisotropic.

If the power law (7.114) holds when space is partitioned by the same factor in different directions (Fig. 10.18a), the fractal is called *self-similar*. If, on the other hand (7.114) holds and only holds when space is partitioned by different factors in different directions (see Fig. 10.18b and c), the corresponding fractal is called *self-affine* [126].

If the phase-space structure is indeed self-affine, it can be characterized by a parameter called roughness or Hurst exponent [126], defined as

$$H_{ij} = \ln M_i / \ln M_j \quad (0 \leq H_{ij} \leq 1) \tag{10.14}$$

with M_i ($i = 1, 2, 3;\ M_1 \leq M_2 \leq M_3$) being the partition numbers in the self-affine transformations

$$\delta y_i \to \delta y_i / M_i \ , \tag{10.15}$$

of the phase-space variables y_i.

The Hurst exponents can be obtained [123] from the experimentally observed saturation curves of the one-dimensional $F_2(\delta y_i)$ distributions,

$$F_2^i(M_i) = A_i - B_i M_i^{-\gamma_i} , \tag{10.16}$$

with A_i and B_i being positive constants, as

$$H_{ij} = \frac{1+\gamma_j}{1+\gamma_i}. \tag{10.17}$$

The fluctuations are isotropic (the fractal self-similar) in the (i, j) plane if $H_{ij} = 1$ and otherwise anisotropic (the fractal self-affine). The farther H_{ij} departs from unity, the stronger the degree of anisotropy.

For hh collisions, the Hurst exponent for longitudinal-transverse combinations was indeed determined to be smaller than unity (see Table 10.1) for NA22 [127] and 0.74 ± 0.07 in the (η, φ) plane for NA27 [128], corresponding to a partitioning finer in the transverse than in the longitudinal direction, while it was found consistent with unity within the transverse plane (φ, p_T).

Table 10.1. The Hurst exponents and effective fluctuation strength for NA22 [127].

Experiments	H_{yp_T}	$H_{y\varphi}$	$H_{p_T\varphi}$	α_{eff}
NA22	0.48 ± 0.06	0.47 ± 0.06	0.99 ± 0.01	0.35 ± 0.03

The upward bending for F_q in the three-dimensional self-similar analysis is then easy to understand: Performing a self-similar analysis, phase space is not properly shrunk according to the self-affine dynamical fluctuation present in the data. So, at intermediate scales the real dynamic fluctuation cannot be fully observed and the corresponding F_q comes out smaller. At very small bins, however, this difference between self-affine and self-similar space shrinkage disappears and the F_q values obtained approach each other. As a consequence, the slope on the log-log plot has to increase to its proper value at small bin sizes and the self-similar analysis leads to an upward bending if the underlying structure is self-affine (i.e. corresponds to a power law).

Experimentally, this is verified in Fig. 10.19a,b [127,128], where the upper figures correspond to a self-similar analysis (forcing all $H_{ij} \equiv 1$) of the NA22 and NA27 data and the lower ones to a self-affine analysis. In order to minimize the influence of momentum conservation dominating at low M [85], a linear fit is applied in the self-affine analysis for $M_y \geq 2$ only. For this M_y region, the results of the self-affine analysis indeed are less upward curved and follow a steeper slope $\phi_2^{(y)}$ than those of the self-similar analysis (mind the scale).

The self-affine analysis in Fig. 10.19a,b corresponds to an extension [127–129] of the factorial-moment method to real (integer and non-integer) partition M, allowing for a larger number of data points.

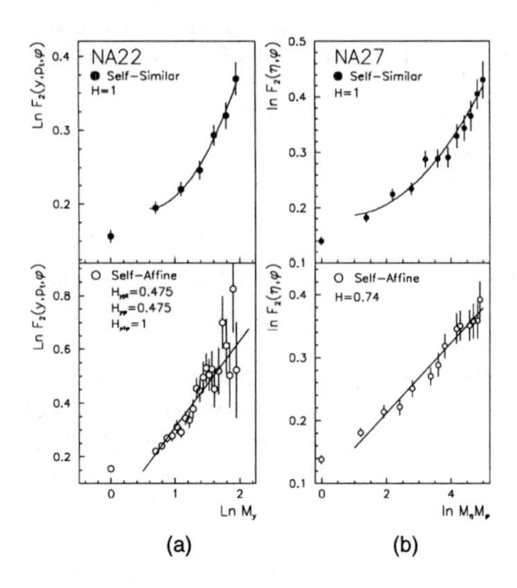

Figure 10.19. a) Factorial moment $\ln F_2(y, p_\mathrm{T}, \varphi)$ as a function of $\ln M_y$ from a three-dimensional self-similar (upper part) and self-affine (lower part) analysis [123, 127], b) $\ln F_2(\eta, \varphi)$ as a function of $\ln M_\eta M_\varphi$ from a two-dimensional analysis [123, 128].

The one-dimensional partitioning M_y is related to the three-dimensional one as

$$M_{3D} \equiv M_y M_{p_\mathrm{T}} M_\varphi = M_y^{(1+\frac{1}{H_{yp_\mathrm{T}}}+\frac{1}{H_{y\varphi}})}, \tag{10.18}$$

so that the three-dimensional intermittency index can be determined as

$$\phi_2^{3D} = \phi_2^{(y)}/(1 + \frac{1}{H_{yp_\mathrm{T}}} + \frac{1}{H_{y\varphi}}) . \tag{10.19}$$

The behavior observed in Table 10.1 means that the longitudinal direction is privileged over the transverse directions in hadron-hadron collisions, not only in particle densities but also their fluctuations. On the contrary, no upward bending is observed in the three-dimensional self-similar analysis of DELPHI and OPAL data (Figs. 10.11c, 10.12a) when excluding the first point as in Fig. 10.20 (due to the influence of energy-momentum conservation [85]). So, the self-similar analysis of e^+e^- data is compatible with a straight line and the H_{ij} are expected to be compatible with unity.

This observation is confirmed with the help of a full self-affine analysis performed with a JETSET 7.4 Monte-Carlo sample at 91.2 GeV [129] and by a full analysis of L3 data [130]. So, this method is indeed sensitive to a qualitative difference observed between the dynamics of particle production in hh and e^+e^- collisions.

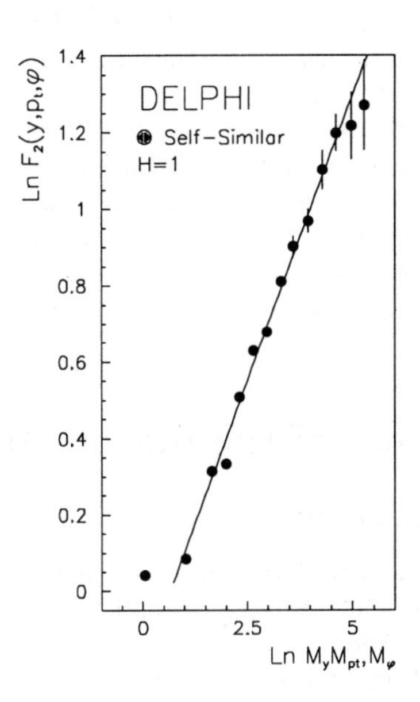

Figure 10.20. Linear fit [123] to the three-dimensional DELPHI data.

L3, therefore, also studied the behavior in 2-jet samples isolated according to $k_t = \sqrt{y_{\text{cut}} s}$ by the resolution parameter y_{cut} of the Durham jet algorithm [131]. For 2-jet events, indeed, γ_y is observed to be larger than γ_{p_t} and γ_φ, reminiscent of the situation in hadron-hadron collisions. JETSET and HERWIG give qualitatively similar results.

In principle, the intermittency index ϕ_2 or the second-order Rényi dimension $D_2 = 1 - \phi_2$ could be considered a measure for the strength of the effect. As suggested in [132], however, an effective fluctuation strength

$$\alpha_{\text{eff}} = \sqrt{2\phi_2} \ , \ 0 \leq \alpha_{\text{eff}} \leq 1 \tag{10.20}$$

can be defined as that fluctuation strength of a random cascading α-model with partition number $\lambda = 2$ which gives the same value of ϕ_2. For NA22 it is quoted in the last column of Table 10.1 and indicates that the fluctuations are about 1/3 of their maximum possible strength for hh collisions. For e^+e^- collisions it is close to 2/3, however [130].

With a self-affine Monte-Carlo branching model exactly reproducing the NA22 d_q/d_2 values of Fig. 10.16a, it is shown in [133] that it is just this upward-bending effect observed in a self-similar analysis which can cause the apparent violation of the Lévy stability $\mu \leq 2$ described in Sub-Sect. 10.3.3.

Contrary to the $H_{y\varphi} < 1$ as observed in the hh and two-jet e^+e^- events, power law fits can only be obtained for heavy-ion collisions for an effective $H_{y\varphi} \approx 2 - 3$, i.e. finer partitioning in the longitudinal than in the transverse direction [134, 135]. For smaller values of $H_{y\varphi}$, an upward bending is observed for $\ln F_q$ vs $\ln M$, the stronger the larger the nuclei. Since, in the nuclear collision, each elementary collision process is expected to resemble a hh collision, this inversion is considered as due to the superposition of independent elementary collisions scattered wider in central rapidity than in transverse momentum [134]. Such a superposition, at the same time, explains the small slopes ϕ_q observed for $H_{y\varphi} \approx 2 - 3$ in [134, 135] (see also Sub-Sect. 10.4.4 below).

10.4 Dependences of the intermittency effect

10.4.1 Charge dependence

A mechanism known to cause correlations at small distances in phase space is Bose-Einstein interference between identical particles [71, 136, 137]. For the present status of this field we refer to Chapter 11 below. From the outset it must be realized, however, that the conventional Gaussian- or exponential-type parametrizations of the Bose-Einstein effect lead to a saturation at $\delta y \to 0$ and *not* to the power law (7.114)!

In [137] it is argued that the slopes should be roughly a factor 2 larger for identical particles than for all charges combined. The experimental situation is less than clear, in particular for 1D analyses. Contrary to the prediction, TASSO [29] and DELPHI [31] see less intermittency for identical particles. EMC [37] finds an enhanced effect for positive but not much for negative particles in a one-dimensional analysis, and very similar slopes in a 3D analysis. NA22 [39] observes an enhancement for negatives, but not for positives. UA1 sees no difference, whereas NA35 sees an increase.

CELLO [30] finds Bose-Einstein interference necessary to explain the residual difference between data and JETSET 7.2, but needs an un-physically large strength-parameter λ to obtain agreement. In the DELPHI analysis, Bose-Einstein interference is insufficient to explain the difference between data and models, even with an un-physically large value of the coherence parameter λ.

Following a suggestion in [138], higher-order Bose-Einstein correlations have been studied by UA1 [139], NA22 [140] and DELPHI [141]. In this study, normalized inclusive densitites of order q,

$$R_q(Q^2_{q\pi}) = N_q(Q^2_{q\pi})/N^{\mathrm{BG}}_q(Q^2_{q\pi}), \tag{10.21}$$

are defined according to (7.22) as ratios of the distribution of like-charged q-tuplets ($q = 2, 3, \ldots, 5$) $N_q(Q^2_{q\pi})$ and a distribution of reference (background) q-tuplets

$N_q^{BG}(Q_{q\pi}^2)$ obtained from random event mixing. The variable $Q_{q\pi}^2$ is defined as a sum over all permutations

$$Q_{q\pi}^2 = Q_{12}^2 + Q_{13}^2 + \dots Q_{(q-1)q}^2 \tag{10.22}$$

of the squared four-momentum difference $Q_{ij}^2 = -(p_i - p_j)^2$ of particles i and j.

The UA1 data are shown in Fig. 10.21. A good fit is obtained if in the expansion of $R_q(Q_{q\pi}^2)$ suggested in [138], Gaussians (dashed curves) are replaced by exponentials

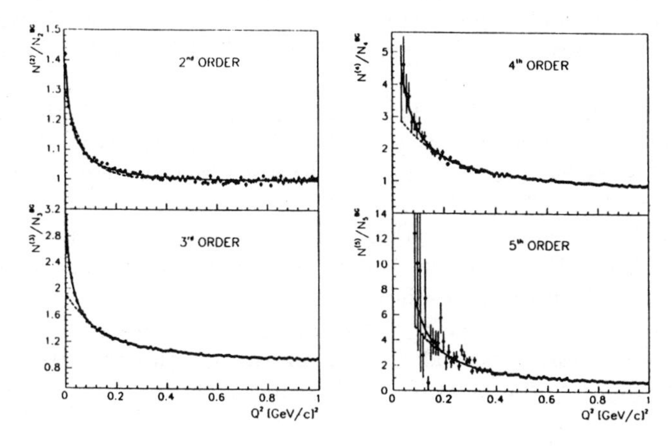

Figure 10.21. Bose-Einstein correlation of order 2 to 5, as indicated. The dashed lines represent fits by Gaussian terms, the full lines by exponential terms. All data are corrected for Coulomb interaction [139].

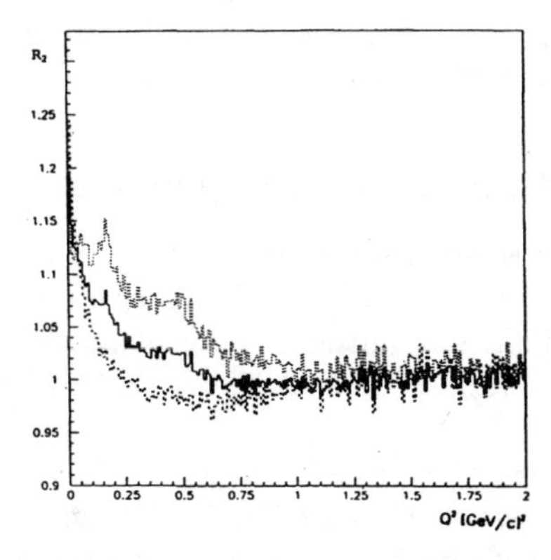

Figure 10.22. The dependence of $R_2 = N_2/N_2^{\rm BG}$ on $Q_{q\pi}^2$ for all pairs of charged particles (full line), for opposite-charge pairs (dotted) and for like-charged pairs (dashed) [40].

in $Q_{q\pi}$ (solid curves). Since low Q_{ij}^2 pairs are lost due to limited two-track resolution in the detector, the data at the smallest Q_{ij}^2 have to be regarded as a lower limit. A power law as expected from intermittency cannot be excluded.

UA1 further studied the distribution R_2 for all-charged-(cc), ($\pm\pm$)- and ($+-$)-pairs as a function of $Q^2(\equiv Q_{2\pi}^2)$ (Fig. 10.22) [40]. These results have important implications. The charge dependence of intermittency, controversial in single-particle variable analyses (see before), is now quite clear in invariant-mass variables ($Q^2 = M_{\text{inv}}^2 - 4m_\pi^2$). The data for $R_2^{\pm\pm}$ (dashed) has a much stronger Q^2-dependence than R_2^{+-} and effectively determines the small-Q^2 behavior of R_2^{cc}. This is evidence that intermittency at small Q^2 is predominantly due to like-charged particle correlations. It does not necessarily imply, however, that Bose-Einstein interference is the sole cause.

In [142] it is shown on EMC data that, especially in 3D, F_2^{--} deviates much more from Lund-model predictions than F_2^{+-}. The Lund-model version used does not include Bose-Einstein correlations. The deviation from the data is indicative for the importance of this effect.

Bose-Einstein interference must thus play a significant role at least for small Q^2. This seems in contradiction with successes in e^+e^- annihilation to tune parton shower Monte Carlos which neglect Bose-Einstein interference.

Finally, we reiterate our remark that a "conventional" Bose-Einstein effect with exponential or Gaussian Q-dependence is incompatible with intermittent power-law behavior. We return to this point in Sub-Sect. 10.8.5 below.

10.4.2 Rapidity dependence

We have already seen in Fig. 10.9 that, at least in JETSET, the factorial moments depend on the rapidity and/or transverse momentum range covered in the analysis. In [37] it is shown on μp and μd collisions, that intermittency is stronger in the central than in the fragmentation regions.

10.4.3 Transverse-momentum dependence

An interesting question is whether semi-hard effects [93], observed to play a role in the transverse-momentum behavior even at NA22 energies [143], or low-p_T effects [118, 144] are at the origin of intermittency. A first indication for the latter comes from the most prominent NA22 spike event [9], where 5 out of 10 tracks in the spike have $p_T < 0.15$ GeV/c.

In Fig. 10.23a, NA22 data [39] on $\ln F_q$ versus $-\ln \delta y$ are given for particles with transverse momentum p_T below and above 0.15 GeV/c, and with p_T below and above 0.3 GeV/c. For particles with p_T below the cut (left), the F_q exhibit a far stronger δy dependence than for particles with p_T above the cut (right). NA22 does not claim straight lines in Fig. 10.23a, but uses fits as an indicative measure of the increase of $\ln F_q$ over the region $1 > \delta y > 0.1$. In the upper half of Fig. 10.23b, the fitted anomalous dimensions d_q are compared to those obtained in the full p_{T}-range. The restriction to particles with $p_{\mathrm{T}} < 0.15$ or 0.30 GeV/c indeed leads to an *increase* of d_q; a *decrease* is observed for $p_{\mathrm{T}} > 0.15$ or 0.30 GeV/c. This observation is confirmed by IHSC [42].

FRITIOF predictions are given in the lower part of Fig. 10.23b, again for all tracks

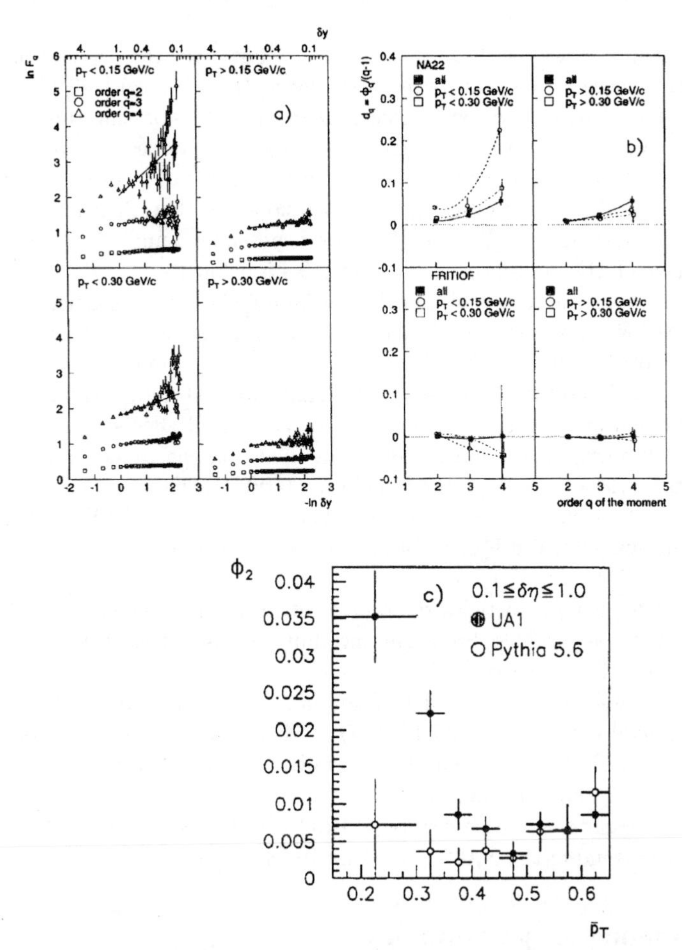

Figure 10.23. a) $\ln F_q$ as a function of $-\ln \delta y$ for various p_{T} cuts as indicated [39], b) anomalous dimensions d_q as a function of the order q, for various p_{T} cuts as indicated (lines are to guide the eye) [39], c) slope ϕ_2 as a function of the average transverse momentum $\bar{p}_{\mathrm{T}}$ in an UA1 event compared to PYTHIA 5.6 [145].

and for tracks with restricted p_T. It is known [39] that FRITIOF gives too small slopes for factorial moments integrated over p_T. Here, one notices that it also fails to reproduce their p_T-dependence.

Of course, cuts in phase space are related to the projection effects discussed in Sub-Sect. 10.3.1 above [88] and the true scaling exponents can only be extracted by means of a *finite-size* scaling analysis [146], applying the proper integration limits in the three-dimensional form of Eq. 7.73. This leads to a p_T-cut dependent shift of the curves in Fig. 10.23a and to a more universal behavior, except for the $p_T \leq 0.15$ GeV sample, however.

UA1 has a bias against tracks with $p_T < 0.15$ GeV/c, but gives the dependence of ϕ_2 on the average transverse momentum $\bar{p}_T$ of the event (Fig. 10.23c) [145]. The data show a remarkable decrease of ϕ_2 with increasing $\bar{p}_T$ and, after passing through a minimum at $\bar{p}_T \approx 0.5$ GeV/c, a slight increase at higher $\bar{p}_T$ values. Lower $\bar{p}_T$ events correspond to soft processes, while higher $\bar{p}_T$ ones correspond to events with hard jet subprocesses. Both types of events have higher slopes ϕ_2 than their mixture at intermediate $\bar{p}_T$ values. (See further in [147] for a possible connection to the multiplicity dependence to be described in Sub-Sect. 10.4.5 below.)

Fig. 10.23c also contains the results obtained from Monte-Carlo events generated with PYTHIA 5.6. At low $\bar{p}_T$ values, the PYTHIA ϕ_2 values are strongly suppressed as compared to those of the data.

We conclude that the intermittency observed in NA22 and UA1 data is enhanced at low transverse momentum and is not dominated by semi-hard effects. Hard effects dominate in high energy e^+e^- and lh collisions. Data on the p_T-dependence of factorial moments in these processes should help in clarifying the origin of intermittency. The effect of p_T-cuts on e^+e^- data has been studied by DELPHI [31]. One-dimensional data are shown in Fig. 10.24 and provide several important pieces of information:

i) The log-log plot for low-p_T particles shows less saturation (i.e. stronger intermittency) than for larger p_T particles. So, again, intermittency is strongest in the p_T region where hard gluon effects are weakest!

ii) A discrepancy between data and models (only indicative in Fig. 10.7d above) is observed in the interval $0.255 < p_T < 0.532$ GeV/c. This looks surprising at first, but we shall show in Sub-Sect. 10.4.5 that the intermittency effect can be stronger for individual mechanisms than for a mixture.

iii) The factorial moments are larger for $p_T > 0.532$ GeV/c than for $p_T < 0.255$ GeV/c, opposite to the trend of the NA22 data (Fig. 10.23a).

10.4.4 Dependence on jet topology

In their analysis, DELPHI [31] selects 2-jet and 3-jet events using the JADE/E0 invariant-mass algorithm [148], with resolution parameter values $y_{\text{cut}} = 0.04$ and 0.01, and with additional cuts to clean the 2-jet and 3-jet sample. At large bin sizes,

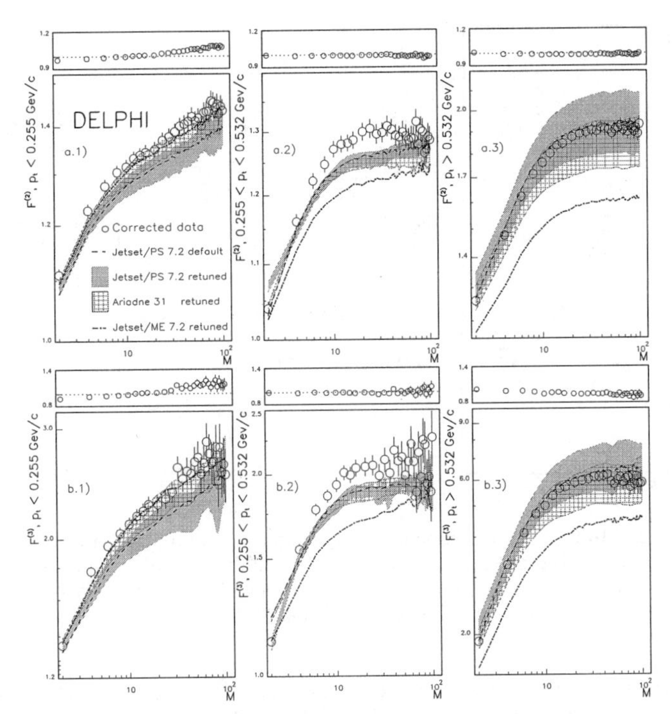

Figure 10.24. Factorial moments F_2 and F_3 as a function of resolution for three e^+e^- data sets with p_T cuts as indicated [31]. The lines correspond to the models as indicated. Correction factors are given above the corresponding sub-figures.

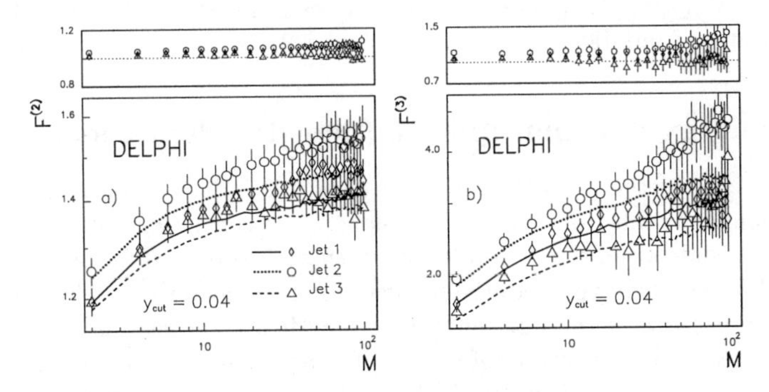

Figure 10.25. F_2 and F_3 of the first, second and third jet ordered by their energy. DELPHI corrected data (open symbols) are compared with JETSET 7.2 PS Monte-Carlo predictions with re-tuned settings [31].

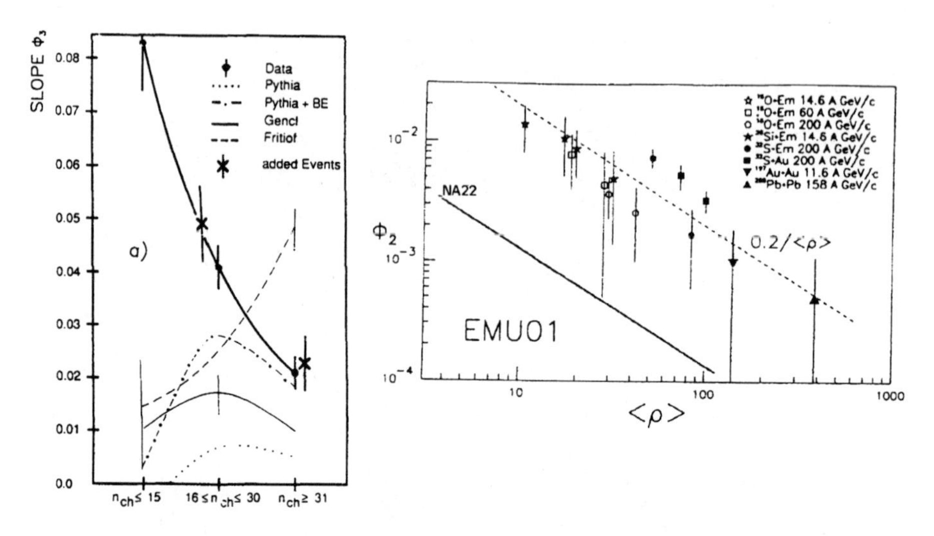

Figure 10.26. a) Multiplicity dependence of the slope ϕ_3, compared to that expected from a number of models, the crosses correspond to a combination of independent events [40], b) slope ϕ_2 extrapolated ($\propto \rho^{-1}$) as a function of particle density from NA22 (hp at 250 GeV) (solid line) and heavy-ion collisions as indicated [53].

factorial moments rise faster with decreasing bin size (and are, therefore, larger) in 3-jet than in 2-jet events. This is compatible with the (large bin size) behavior expected from hard gluons. At small bin sizes the increase is similar for 2-jet and 3-jet events.

In 3-jet events, factorial moments were calculated for tracks belonging to jet 1, jet 2 and jet 3 ordered in energy. The rapidity was defined with respect to the individual jet axis. As seen in Fig. 10.25, intermittency is weakest in jet 3 and strongest in jet 2. Also the deviation from JETSET is strongest for jet 2.

10.4.5 Energy and multiplicity (density) dependence

While an increase of the ϕ_q with increasing energy is indicated in μp data [37], a decrease is found for hh, hA and AA collisions. As seen in Fig. 10.26a, a strong multiplicity dependence of the intermittency strength is observed for hh collisions by UA1 [40]. The trend is opposite to the predictions of the models used by this collaboration. This decrease of the intermittency strength with increasing multiplicity is usually explained as a consequence of mixing of independent sources of particles [25]. The cross-over of data and FRITIOF in Fig. 10.26 at intermediate multiplicity explains the apparent success of FRITIOF in Fig. 9.3, for multiplicities close to 30, as being purely accidental.

Mixing of emission sources leads to a roughly linear decrease of the slopes ϕ_q

with increasing particle density $\langle\rho\rangle$ in rapidity [71, 149, 150]: $\phi_q \propto \langle\rho\rangle^{-1}$. This is indeed observed by UA1 [40]. Multiple emission sources are present in multichain Dual Parton models. The calculated slopes indeed depend linearly on multiplicity but are too small by a factor of two [151]. Similarly, the model studied in [152] with independent emission at fixed impact parameter finds decreasing ϕ_q with increasing multiplicity.

Also here, a study of the multiplicity dependence in e^+e^- data and JETSET allows interesting comparisons. In fact, the LEP results [31] suggest little or no n-dependence, except for the lowest multiplicities, where the slope is largest and also the difference with JETSET PS is the largest.

Fig. 10.26a helps in explaining why intermittency is so weak in heavy-ion collisions (cfr. Fig. 10.3): the density (and mixing of sources) is particularly high there. In Fig. 10.26b, EMU01 [53], therefore, compares ϕ_2 for NA22 (hp at 250 GeV) and heavy-ion collisions at similar beam momentum per nucleon, as a function of the particle density. Whereas slopes averaged over multiplicity are smaller for AA collisions than for NA22 in Fig. 10.3, at fixed $\langle\rho\rangle$ they are actually higher than expected from an extrapolation of hh collisions to high density and may even grow with increasing size of the nuclei. The trend is confirmed by KLMM [60] for intermittency in azimuthal angle φ and for slopes up to order 5. This may be evidence for re-scattering (see [38]) or another (collective) effect, but, as shown by HELIOS [55] and confirmed by EMU-01 [53], one has to be very sure about the exclusion of γ-conversions before drawing definite conclusions.

We conclude this section with an additional warning. In Sub-Sect. 10.3.2 we mentioned the Fiałkowski "universality conjecture" and noted that it is incompatible with the "mixing" hypothesis invoked to explain the multiplicity dependence of factorial moments and slopes. A different explanation of the multiplicity dependence may therefore be needed, especially since intermittency and Bose-Einstein effects are now known to be closely related.

10.5 Factorial cumulants

Normalized factorial cumulant moments, first introduced in [153] and more recently studied in [102, 103], are defined in (7.75) as integrals over the background-subtracted correlation functions. They share with factorial moments the property of noise suppression. The normalized factorial moments F_q can be expanded in terms of normalized cumulant moments K_q as given in (7.76). This expansion has been found to converge rapidly [154]. The terms in the expansion correspond to contributions from genuine $q, (q-1) \ldots, 2$-particle correlations. In [154] it is estimated that averaging over all points on the log-log plot, K_5 contributes 5% to F_5, K_4 contributes 11% to F_4 and K_3 contributes 21% to F_3.

In Fig. 10.27, the OPAL [32] results are given for the one-dimensional projections

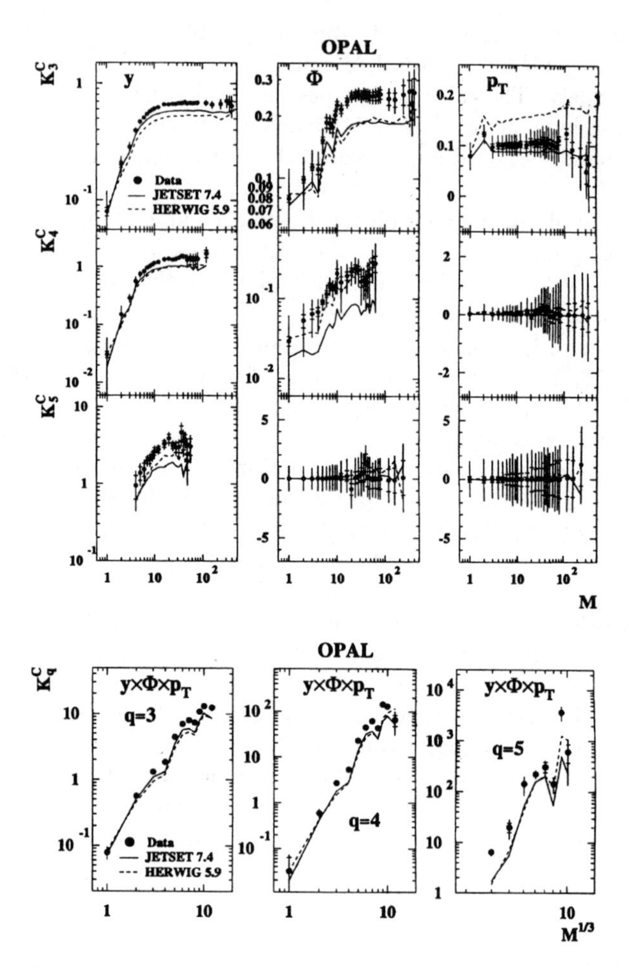

Figure 10.27. Normalized cumulants of order $q = 3$ to 5 as a function of the partition number M, in comparison with the predictions of JETSET and HERWIG [32].

and for three dimensions. Overall, the M dependence is similar to that of the factorial moments given in Figs. 10.7c and 10.11c, respectively. Particularly large positive cumulants are found in rapidity space at least up to $q = 5$, while they vanish at high orders in p_T and even φ space.

From (7.76) it is seen that the contribution $F_q^{(2)}$ to F_q from two-particle correlations alone can be expressed as

$$F_3^{(2)} = 1 + 3K_2 \tag{10.23}$$

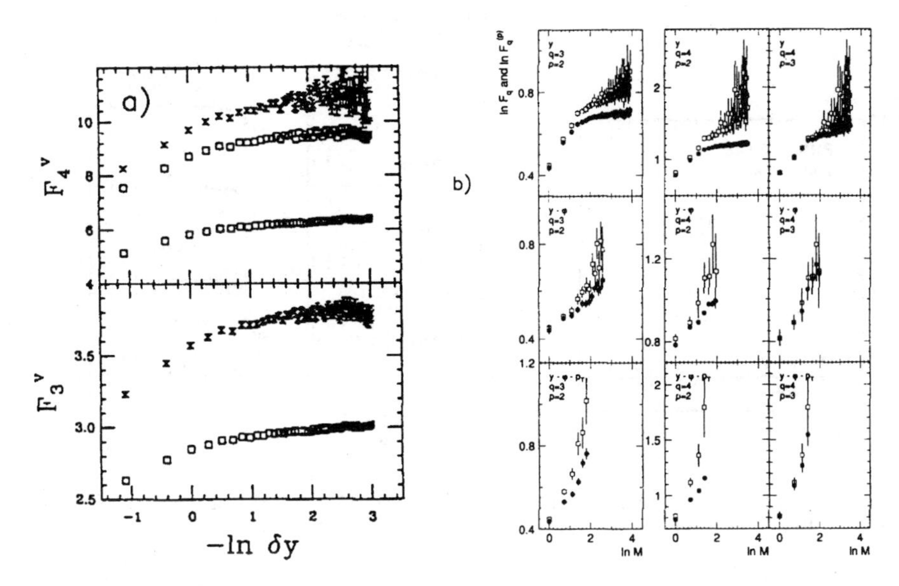

Figure 10.28. a) UA1 factorial moments (crosses) decomposed into cumulant contributions: lower squares indicate 2-particle, upper squares 2+3-particle contributions [154], b) Normalized factorial moments F_q (□) from NA22, together with contributions from $F_q^{(p)}$ (•) [39].

$$F_4^{(2)} = 1 + 6K_2 + 3\overline{K_2^2} \quad ; \tag{10.24}$$

the contribution $F_q^{(3)}$ from two- and three-particle correlations to F_4 as

$$F_4^{(3)} = 1 + 6K_2 + 3\overline{K_2^2} + 4K_3 \quad . \tag{10.25}$$

The difference $F_q - F_q^{(p)}$ is a measure for the importance of higher-order correlations.

Fig. 10.28a shows a cumulant-decomposition of F_3 and F_4 in UA1-data [154]. The differences between the curves indeed indicate large contributions from genuine higher-order correlations. Similar results are observed for NA22 [39] (Fig. 10.28b) for p =2 and 3 and q =3 and 4, in one-, two- and three-dimensional phase space (transformed $y, y - \varphi$ and $y - \varphi - \ln p_{\mathrm{T}}$) and, up to $q = 5$ for OPAL [32] (not shown).

In general, the difference increases with increasing $\ln M$ (decreasing bin size). This means that the contribution of higher-order correlations to the factorial moments increases at higher resolution. Exceptions are the variable φ in NA22, for which only two-particle correlations are found to be non-zero (not shown) and p_{T} in OPAL.[1]

[1]Absence of genuine higher-order correlations has been reported in [155], but at far too low statistics.

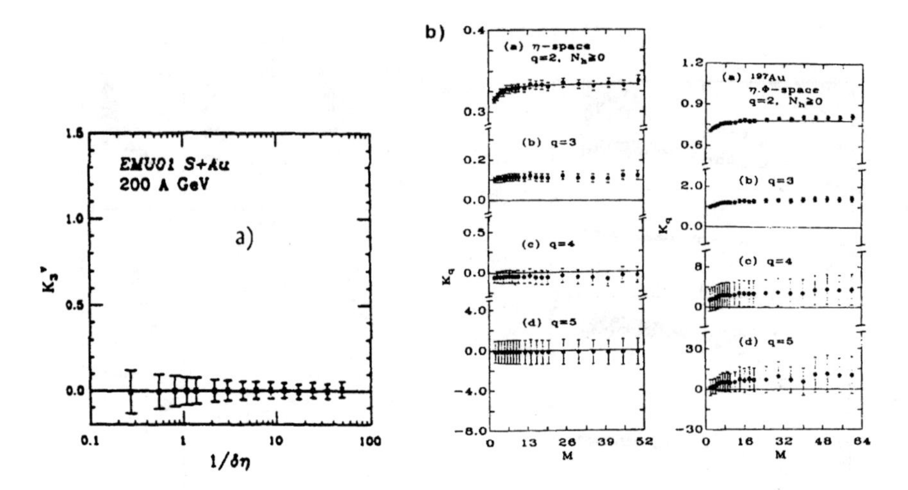

Figure 10.29. a) K_3 for SAu at 200 A GeV in η-space [156]. b) K_q for $q = 2 - 5$ for AuEm at 10.6 A GeV in η-space (left) and (η, φ)-space (right) [54].

For some time, the situation looked completely different in heavy-ion collisions where, with given accuracy, $K_q \approx 0$ for $q > 2$ (Fig. 10.29a). The factorial moments seemed completely dominated by two-particle correlations [156–158], implying that higher-order F_q contain little or no further dynamical information for this type of collisions. However, a departure of K_3 from zero is now observed in AuEm collisions at 10.6 A GeV [54]. This is shown in Fig. 10.29b for η (left) and (η, φ) space (right).

Using the linked-pair ansatz [154], higher-order cumulant functions can be expressed as products of K_2 (see also [159] for an interpretation in terms of independent superposition of sources)

$$K_q = A_q K_2^{q-1}, \tag{10.26}$$

with free constants A_q.

For a negative-binomial (NB) multiplicity distribution, $K_2 = 1/k$ and the linking parameters are fixed numbers given by $A_q^{\mathrm{NB}} = (q-1)!$ [103]. A necessary condition is stationarity, i.e. constancy of $1/k$. This works well for UA1. For NA22 [39], A_q is observed to increase with decreasing bin size. Approximately constant $A_q \approx (q-1)!$ are found only when the data are averaged over a narrow rapidity region ($-0.75 \leq y \leq 0.75$) and the most prominent spike event is excluded. The linked-pair ansatz may thus be a valid approximation for high-order correlations in small phase-space domains but not for the average over phase space. This would be consistent with the fact [160] that the negative binomial is often a good parametrization of multiplicity distributions in restricted δy-intervals.

The high-statistics data of LEP allow a detailed bin-size dependent comparison

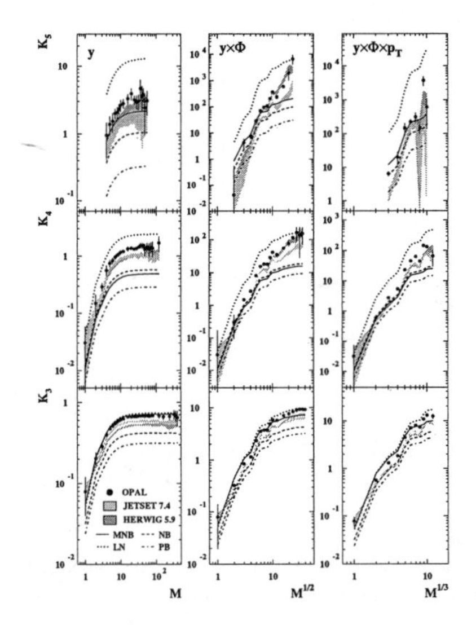

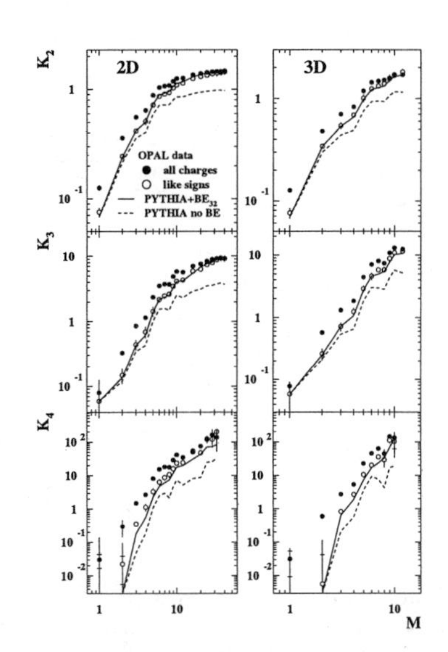

Figure 10.30. a) Cumulants of order $q = 3$ to 5 as a function of $M^{1/D}$, where M is the number of bins of the D-dimensional subspaces of the phase space of rapidity, azimuthal angle, and transverse momentum, in comparison with the predictions of various multiplicity parametrizations and two Monte Carlo models [162]. b) Cumulants of order $q = 2$ to 4 for two- and three-dimensional domains, as a function of M, for all charged hadrons (full symbols) and for multiplets of like-charged particles (open symbols) [163].

in one to three dimensions up to order $q = 5$ [161,162]. In Fig. 10.30 the normalized factorial cumulants of Fig. 10.27 [32] are compared [162] to a number of parametrizations of the multiplicity distributions, as well as to JETSET and HERWIG. One can see that the fluctuations given by the NB (dashed line) are weaker than observed in the data. At small partition number M, this agrees with the inadequacy of the NB to fit the full-phase-space multiplicity distribution (Sub-Sect. 8.1.2), but the discrepancy even increases with increasing M (decreasing bin size).

Contrary to the NB, the log-normal (LN) distribution (dotted line) overestimates the data. These observations are in agreement with those of [161]. While high multiplicities (and therefore high-multiplicity fluctuations) are underestimated by the LN. Also the cumulants of the multiplicity distribution expected by the pure birth (PB) process (Sub-Sect. 8.2.4) can be defined from the second-order cumulants K_2. They are represented by the dash-dotted line and underestimate the data even more significantly than the NB.

Among the distributions shown, the modified NB (MNBD) discussed in Sub-Sect. 8.2.4 gives the best results, even though significant underestimation is observed. Not shown in Fig. 10.30 is the generalized NB (HNBD), but good agreement is reported [162] for these. However, the LN limit ($\mu \approx 0$) is excluded in this case.

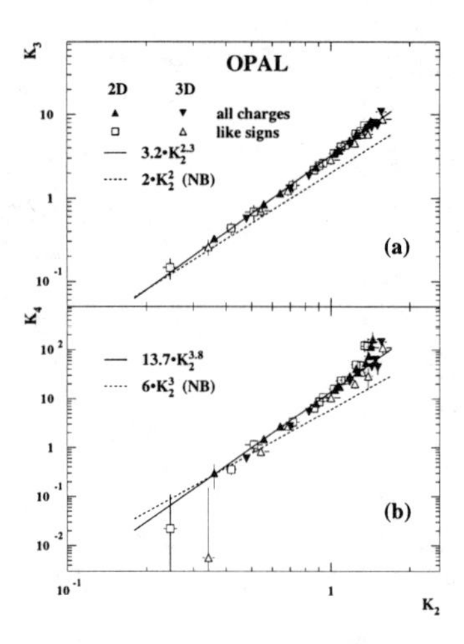

Figure 10.31. Ochs-Wošiek plot for cumulants in two- and three-dimensional domains for all charged hadrons (solid) and like-charged particles (open symbols) measured in e^+e^- annihilation at the Z^0 [163].

To study the influence of identical-boson Bose-Einstein correlations, the high-statistics OPAL data have been used to measure cumulants for multiplets of particles with the same charge [163] (Fig. 10-30b). Like-charged cumulants increase faster and approach the all-charged ones at large M. Consequently, as the cell-size decreases, the rise of all-charged correlations is increasingly driven by that of like-charged multiplets. The dashed lines in Fig. 10.30b show PYTHIA without Bose-Einstein correlations for like-charged multiplets. The model agrees with the data for small M, indicating that the overall multiplicity distribution is modelled well. For large M, however, inclusion of BEC (solid lines) is needed. Whereas the algorithm used implements pair-wise BEC only, it it noteworthy that also the higher-order correlations are reproduced. It is not clear whether this agreement is accidental and we shall come back to this in Chapter 11.

Here, we concentrate on the inter-dependence. In Fig. 10.31, K_3 and K_4 are plotted as a function of K_2 on double-logarithmic scale. Within errors, the all-charged and like-charged cumulants follow the same straight line, independently of dimension D,

$$\ln K_q = a_q + r_q \ln K_2 , \tag{10.27}$$

in analogy to the factorial moments [93] (Sub-Sect. 10.3.3). The slope r_q (exponent given in the figure) increases with increasing q. As already observed in Fig. 10.30a [162], this increase is faster than the $r_q^{\mathrm{NB}} = q - 1$ valid for the negative binomial distribution.

We shall come back to cumulants and genuine higher-order correlations in

Sect. 10.8, where they are studied by means of more sophisticated methods.

10.6 Factorial correlators

10.6.1 The method

The moments defined in (7.72-7.74) measure local density fluctuations in phase space. Additional information is contained in the correlation between these fluctuations within an event. This correlation can be studied by means of the factorial correlators defined in (7.79). Correlators are typically calculated at a given δy for each combination mm' of bins with size δy, and then averaged over all combinations separated by a given bin distance D. This is illustrated below.

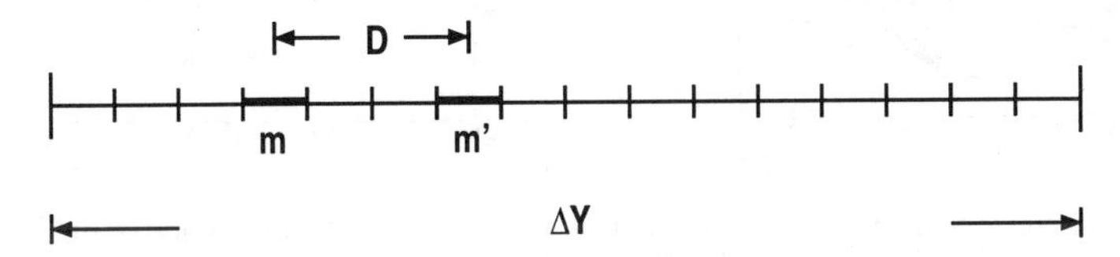

In the simple intermittency model (α-model) described in [13,14], F_{pq} depends on D but not on δy and follows the power law

$$F_{pq} \propto (\Delta Y/D)^{\phi_{pq}} \quad . \tag{10.28}$$

The powers ϕ_{pq} (slopes in a log-log plot) obey the relations [13,14]:

$$\phi_{pq} = \phi_{p+q} - \phi_p - \phi_q = pq\phi_2 \quad , \tag{10.29}$$

where the first equality sign is due to the α-model proper, the second to the log-normal approximation. According to (10.29) $\phi_{11} = \phi_2$, so that it can also be written in the form

$$\phi_{pq} = pq\phi_{11} \ . \tag{10.30}$$

10.6.2 Results

Preliminary results for pseudorapidity resolution $\delta\eta \geq 0.2$ were first reported by the HELIOS Collaboration [165]. There, however, multiplicities n_m had to be estimated from the transverse energy $E_{\mathrm{T},m}$ in bin m and the average transverse energy $\langle E_\mathrm{T} \rangle$ per particle: $n_m = E_{\mathrm{T},m}/\langle E_\mathrm{T} \rangle$ rounded to the nearest integer. The first direct measurement come from NA22 [164]. The $\ln F_{pq}$ are shown as a function of $-\ln D$ in Fig. 10.32a-d, for four values of $\delta y \geq 0.1$ (corresponding to M = 10, 20, 30 and 40). Statistical errors (estimated from the dispersion of the F_{pq} distribution) are in general smaller than the size of the symbols. F_{pq} was measured up to third order in

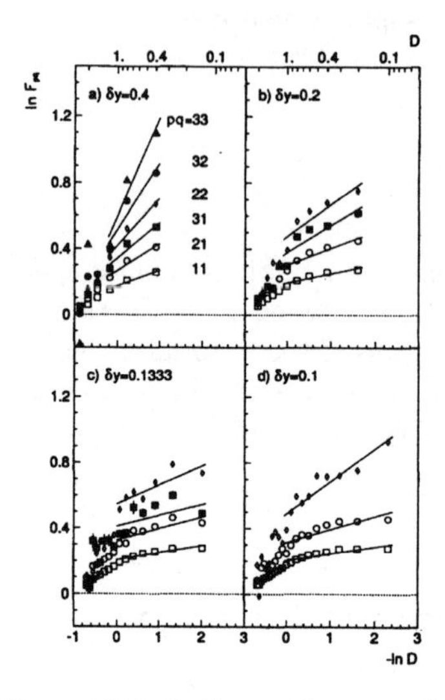

Figure 10.32. $\ln F_{pq}$ as a function of $-\ln D$ for four values of δy, as indicated [164].

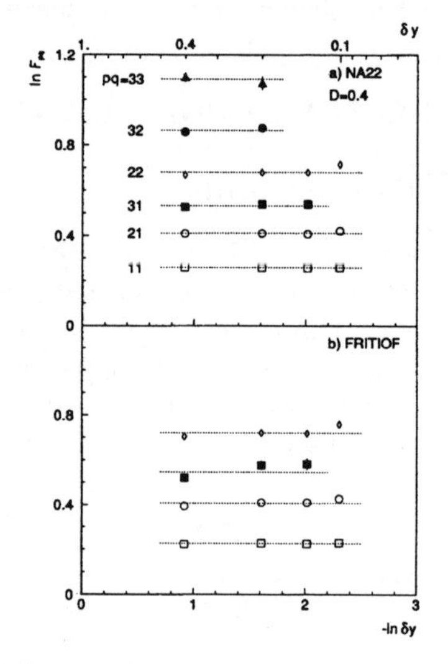

Figure 10.33. Dependence of $\ln F_{pq}$ on the bin size δy for a correlation distance $D = 0.4$, a) for NA22 data, b) for a sample of 60 000 FRITIOF Monte-Carlo events. The dashed lines correspond to horizontal-line fits through the points [164].

p and q for δy=0.4 binning (Fig. 10.32a). For δy=0.1, the analysis is possible to first and second order only (Fig. 10.32d). The smallest possible value for D being equal to the bin size δy, Fig. 10.32a extends to $D = 0.4$ and Fig. 10.32d to $D = 0.1$. In all cases, an increase of $\ln F_{pq}$ is observed with increasing $-\ln D$. Very similar results have been reported by EMC [37], EMU-01 [53], in [49,166] and [167].

In Fig. 10.33a, the $\ln F_{pq}$ are compared at fixed $D = 0.4$ for four different values of δy. The dashed lines correspond to a horizontal line fit through the data. In agreement with the α-model, the F_{pq} indeed do not depend on δy. Also this result has been confirmed on EMC data [37] and in [49,166]. The δy-independence of correlators holds exactly in the α-model, but Fig. 10.33b shows that the δy independence is also valid in FRITIOF. For the particular value of $D = 0.4$, this even happens at very similar values of $\ln F_{pq}$ as in the data. In fact, this property is far from unique to the α-model, but even holds approximately in any model with conventional short-range order [168].

For F_{11}, the δy independence is easily derived from a parametrization of the two-particle density, integrated over two regions of size δy separated by D. Using

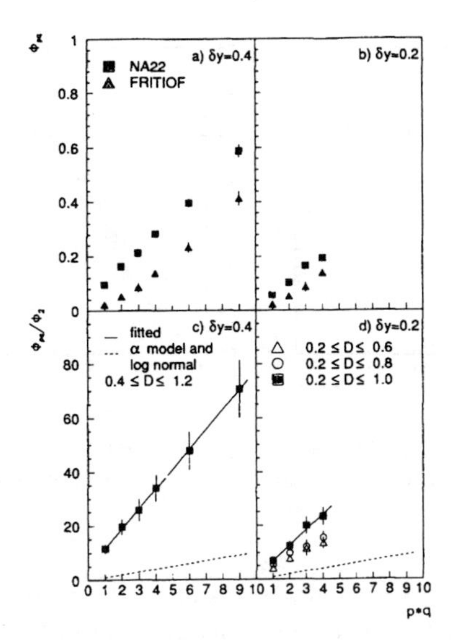

Figure 10.34. a) b) The increase of the slopes ϕ_{pq} with increasing order pq compared to the expectation from FRITIOF, for two values of δy, respectively; c) d) the increase of ϕ_{pq}/ϕ_2 with increasing order pq, compared to that expected from the α-model (dashed line), for two values of δy, respectively [164].

exponential short-range order [103], this gives

$$F_{11} - 1 \;\propto\; \frac{1}{a^2} \mathrm{e}^{-D/L} / (\mathrm{e}^a - 1)(1 - \mathrm{e}^{-a}) \ , \tag{10.31}$$

where L is a correlation length and $a = \delta y/L$. According to (10.31), F_{11} becomes independent of δy for $a \ll 1$. Since $\mathrm{e}^{-D/L} \to 1$ as $D \to 0$, this form also leads to the deviations from (10.28) observed as a bending in Fig. 10.32.

Because of the bending, fitted slopes ϕ_{pq} have no meaning, except as an indication for the increase of F_{pq} in a restricted range. The slopes for two values of δy are compared to FRITIOF predictions in Fig. 10.34a and 10.34b, respectively. As observed earlier for the case of univariate moments [39], the FRITIOF slopes are too small also for the correlators. This is not surprising since the model does not succeed in reproducing even the lowest-order (i.e. two-particle) rapidity correlation function (Chapter 9).

10.6.3 Interpretation

Factorial correlators have been analysed in [169] using a suitable parametrization of $K_2(y_1, y_2)$ and the linked-pair ansatz [102] for higher-order correlations. The relations

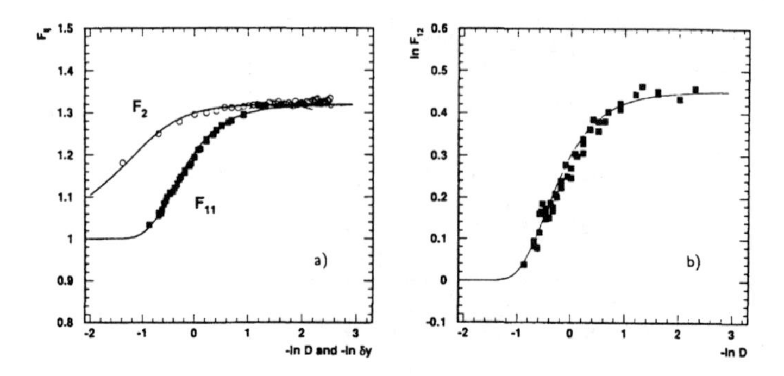

Figure 10.35. a) F_{11} versus D ($\delta y = 0.2$) and F_2 versus δy for a Gaussian-shaped two-particle correlation function [168] compared to NA22 data [164]; b) F_{12} versus D ($\delta y = 0.2$) as in a).

(7.87-7.90) then allow to express all correlators in terms of K_2 for arbitrary (p, q). Note that the expressions for F_{pq} contain many lower-order "combinatorial" terms, which effectively dominate and mask the contribution from genuine $(p+q)$-order correlations. A basically similar analysis is presented in [168], inspired by techniques used in quantum optics.

The two analyses have no difficulty to describe basic features of the NA22 data, including the (dimension independent) sum rules discussed in Sect. 7.3, which were claimed to be a unique test of random cascade models [170]. Fig. 10.35a compares NA22 data on $F_2(\delta y)$ and $F_{11}(D)$ with the calculations from [168]. $F_2(\delta y)$ is used to fix the parameters of $K_2(\delta y)$ (assuming stationarity); $F_{11}(D)$ follows after integration over the appropriate rapidity domains. With the linking ansatz of [103] all other correlators are calculated without further assumptions. An illustrative example is shown in Fig. 10.35b, which compares $F_{1q}(D)$ from NA22 to the prediction. The agreement is excellent in all cases. This observation is confirmed in [37, 49, 166].

According to (10.29), the ratio ϕ_{pq}/ϕ_2 is expected to grow with increasing orders p and q like their product pq. In Figs. 10.34c and d, this is tested for δy=0.4 and δy=0.2, respectively. In both cases, the experimental results lie far above the dashed line corresponding to the expected $\phi_{pq}/\phi_2 = pq$. Since the dependence of $\ln F_{pq}$ on $-\ln D$ is not strictly linear, this comparison depends on the range of δy and D used to determine ϕ_2 and ϕ_{pq}. In Fig. 10.34d one, therefore, compares a number of fits. Slopes are smaller when the upper limit in D is reduced, but do not reach the α-model prediction (dashed line).

It can be verified that, at least for the higher orders, the discrepancy with (10.29) is mainly due to the second equal sign, derived from a log-normal approximation to the density distribution. In [95], this approximation has been shown to be valid if the density fluctuations are weak or if the density probability distribution is log-normal. The NA22 data demonstrate that none of these conditions is fulfilled. This is supported by [167]. Approximate agreement is, however, obtained for heavy-ion collisions in [166].

We conclude that the correlators F_{pq} increase with decreasing correlation length D, but do not really follow a power law for $D \lesssim 1$. For fixed D, the values of F_{pq} do not depend on the resolution δy, a feature expected from the α-model, but also reproduced by FRITIOF and approximately true in any model with short-range order. When the increase of the correlators is roughly approximated by a straight line in a restricted interval, the powers ϕ_{pq} increase linearly with the product pq of the orders, but are considerably larger than expected from FRITIOF and, perhaps with the exception of heavy-ion collisions, from the simple α-model with log-normal approximation.

The extension of single-variate factorial moments to the multivariate case offers better insight into the complicated nature of the correlations. However, the original expectation that correlators would help in clarifying the issue of intermittency is not borne out by present data. Simple but reasonable models for higher-order correlation functions which use the experimental two-particle correlations as input, have no difficulty in reproducing the behavior of factorial correlators.

10.7 Multifractal behavior

Power-law dependence of normalized factorial moments on the resolution δ (bin size) is a signature of self-similarity in the fluctuation pattern of particle multiplicity. It suggests that the probability distribution $P(\rho, \delta)$ of the particle density ρ has fractal properties. For simple Widom-Wilson [171] type scaling, $P(\rho, \delta)$ is of the form

$$P(\rho, \delta) \sim \delta^{-\beta} P^{\star}(\rho/\delta^{\nu}) \ , \tag{10.32}$$

where β and ν are critical exponents. All qth order moments of ρ ($q = 1, 2, \ldots$) obey power laws in δ with inter-related exponents depending on q and on (β, ν). This characterizes a simple or mono-fractal. Another possibility is a multifractal behavior, in which $P(\rho, \delta)$ obeys a relation of the type (cf. [172])

$$\ln P(\rho, \delta)/\ln \delta = f(\alpha), \quad \alpha = \ln \rho/\ln \delta. \tag{10.33}$$

Multifractals, first introduced in [173] represent infinite sets of exponents—the multifractal spectrum—which describe the power-law scaling of all moments of $P(\rho, \delta)$. In principle, knowledge of the multifractal spectrum is completely equivalent to knowledge of the probability distribution.

Unlike geometrical or statistical systems, multiparticle production processes pose special problems if a multifractal analysis is to be considered. The most obvious one is the finiteness of particle multiplicity in an event at finite energy. Self-similarity, if existent, therefore cannot persist indefinitely to finer and finer scales of resolution.

$P(\rho, \delta)$ is not directly accessible. At best one can construct, for a single event of multiplicity n and for given δ, a frequency distribution which approaches $P(\rho, \delta)$ only for $n \to \infty$. For any finite (and usually small) n, the frequency distribution and its moments will be subject to statistical fluctuations.

Since the data sample contains a large number of events, it is obviously recommended to consider the event average. This averaging, however, supposes ergodicity. The applicability of the multifractality concept can, therefore, only be justified a posteriori.

10.7.1 Factorial moments of continuous order

From the definition of factorial moments (7.72), only events with $n \geq q$ can contribute to F_q. So, the F_q are well suited to study multiplicity *spikes*, but cannot reveal abnormal multiplicity *dips*. For that purpose, a method is needed to define F_q for $q < 1$. Furthermore, the study of a multifractal spectrum requires q to be continuous.

Following extensive attempts to eliminate statistical noise from the G-moments defined in Eq. (7.77) [174] (see [23] for a review of the experimental situation), an elegant extension was proposed [175] to obtain normalized factorial moments F_q of continuous order, with statistical fluctuations filtered out as in the integer-order ones. Writing the (event-averaged) factorial moment $\tilde{F}_q$ in the numerator of (7.72) as

$$\tilde{F}_q = \sum_{n=q}^{\infty} \frac{n!}{(n-q)!} P_n \qquad (q = \text{pos. integer}) \tag{10.34}$$

and

$$P_n = \int_0^\infty \mathrm{d}t \frac{t^n}{n!} e^{-t} D(t) \ , \tag{10.35}$$

with $D(t)$ representing the dynamical fluctuation and the Poisson factor describing the statistical component, one obtains, after summation over n

$$\tilde{F}_q = \int_0^\infty \mathrm{d}t \ t^q D(t) \ , \tag{10.36}$$

as the qth moment of the dynamical fluctuation $D(t)$.

For continuous order q, (10.34) is not applicable but, according to [175], P_n can be expanded into a series of negative-binomial distributions as

$$P_n = \sum_{j=0}^{N} a_j P_n^{\text{NB}}(k_j, \bar{n}_j) \ , \tag{10.37}$$

where N is the maximum observed multiplicity. Writing now

$$P_n^{\text{NB}}(k_j, \bar{n}_j) = \int_0^\infty \mathrm{d}t \frac{t^n}{n!} e^{-t} D^{\text{NB}}(t, j) \tag{10.38}$$

$$\text{with } D^{\rm NB}(t,j) = \left(\frac{k_j}{\bar{n}_j}\right)^{k_j} \frac{t^{k_j-1}}{\Gamma(k_j)} e^{-k_j t/\bar{n}_j} , \tag{10.39}$$

one obtains

$$D(t) = \sum_{j=0}^{N} a_j D^{\rm NB}(t,j) . \tag{10.40}$$

Substituting this into (10.36) and integrating over t, one obtains

$$\tilde{F}(q) = \sum_{j=0}^{N} a_j \tilde{F}^{\rm NB}(q,j) \tag{10.41}$$

$$\text{with } \tilde{F}^{\rm NB}(q,j) = \left(\frac{\bar{n}_j}{k_j}\right)^{q} \frac{\Gamma(q+k_j)}{\Gamma(k_j)} . \tag{10.42}$$

The expansion coefficients a_j can be determined, assigning $N+1$ pairs of $(\bar{n}_j, k_j)$ to calculate the $N+1$ negative binomials $P_n^{\rm NB}(k_j, \bar{n}_j)$ and solving the $N+1$ linear equations (10.37) for a_j.

The continuous-order normalized factorial moments can then be calculated from (10.41) as

$$F(q) = \tilde{F}(q)/\tilde{F}^q(1) . \tag{10.43}$$

In practice, a number J of negative binomials considerably smaller than $N+1$ is not only sufficient to describe the experimental data (see Chapter 8), but also cures [176] an instability of the original method at large and at negative values of q. The a_j $(j = 0, \dots J)$ can then be obtained from the $N+1$ experimental points P_n of the multiplicity distribution in a maximum-likelihood fit.

10.7.2 Experimental results

The improved (maximum-likelihood) method [176] was tested on NA27 data by the same authors and the intermittency index $\phi(q)$ was obtained as a function of q from a fit to the data by

$$F(q, \delta\eta) \propto \delta\eta^{-\phi(q)}. \tag{10.44}$$

Furthermore, from $\phi(q)$ the multifractal spectrum

$$f(\alpha) = q\alpha - \tau(q) \tag{10.45}$$

and the fractal dimension

$$D(q) = \tau(q)/(q-1) \tag{10.46}$$

were calculated, with

$$\tau(q) = q - 1 - \phi(q) \tag{10.47}$$

$$\alpha = \mathrm{d}\tau(q)/\mathrm{d}q . \tag{10.48}$$

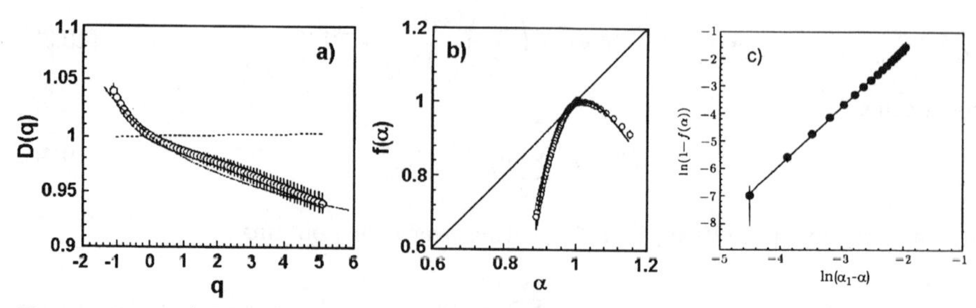

Figure 10.36. a) Multifractal dimension $D(q)$ as a function of q. The open circles are the data, the dashed line (at $D(q) = 1$) corresponds to random emission, the dotted line following the data to the α-model. b) Multifractal spectrum $f(x)$ as a function of α [176]. c) Linear fit of $\ln(1 - f(\alpha))$ versus $\ln(\alpha_1 - \alpha)$ for $q = 0.2$ to 4, for a self-affine analysis in two dimensions [177].

$D(q)$ is shown as a function of q in Fig. 10.36a. The results from a Monte-Carlo sample with random particle distribution is shown as a dashed line at $D(q) = 1$, as expected for a sample without dynamical fluctuations (Eqs. (10.46), (10.47)). For $q = 0$, according to (10.36), (10.43), (10.44), $\phi(0)$ is expected to vanish and $D(0)$ to be equal to unity. Apart from that, $D(q)$ is observed to decrease with increasing q, indicating multifractal behavior in hadron-hadron collisions.

The multifractal spectrum $f(\alpha)$ is shown in Fig. 10.36b. It is concave downward with a maximum at $q = 0$, $f(\alpha(0)) = D(0) = 1$. The straight line $f(\alpha) = \alpha$ is tangent to the $f(x)$ curve at $q = 1$. The full circle in the maximum corresponds to the Monte-Carlo sample with random particle production. The shape of the spectrum (and that of Fig. 10.36a) can be reproduced by a simple cascade model (α-model) [176].

Extension to two dimensions is straight forward, but has to take into account possible self-affine behavior. Application to NA27 data [177] in (η, φ) indeed shows that multiparticle production in 400 GeV/c hadron-hadron collisions are a self-affine multifractal process, with self-affine multifractal dimension and multifractal spectrum similar to those shown in Figs. 10.36a and b.

Extracting the Lévy index μ of (10.8) using the multifractal spectrum (10.43) [178], i.e. using factorial moments of continuous order instead of a small number of integer values of q, yields

$$1 - f(\alpha) \propto (\alpha_1 - \alpha)^{\mu/(\mu-1)} , \qquad (10.49)$$

with α_1, defined as $f(\alpha_1) = 1$. The power behavior is indeed observed [177] as a linear rise in the double-logarithmic Fig. 10.36c. A linear fit grants $\mu = 1.91 \pm 0.01$, consistent with a value for a non-thermal phase transition ($\mu > 1$).

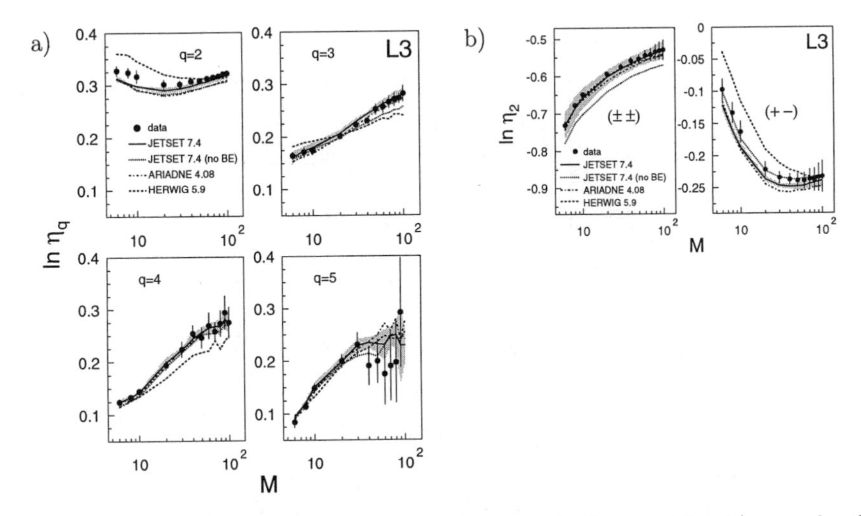

Figure 10.37. a) Bunching parameters as a function of M in rapidity, b) second-order bunching parameter for like-charged and for unlike-charged particle pairs, compared to the model predictions as indicated [179].

10.7.3 Bunching parameters

The bunching parameters described in Sect. 7.5 have been evaluated from the L3 [179] and NA27 [180] data. In both cases, the transformation to cumulative variables [84, 87] has been applied and bin averaged bunching parameters are used.

In addition to the advantages mentioned in Sect. 7.5 (locality, direct access to the fractal structure), experimentally, they are less severely affected than the NFM's by the bias from finite statistics at large n and from limited n-particle resolution.

Fig. 10.37a shows the L3 results in the rapidity variable. The second-order BP decreases with increasing M up to $M \simeq 20$. At large M, however, all BP's show a power-law increase. So, the fluctuations follow a multifractal structure, in qualitative agreement with expectations from QCD branching [181–184].

Both JETSET and ARIADNE agree well for $q \geq 3$, but not for $q = 2$. HERWIG significantly overestimates η_2 at low and intermediate M, due to its problem already with the dispersion of the multiplicity distribution (Chapter 8).

In Fig. 10.37b, η_2 is given for like-charged and unlike-charged pairs, separately. While $\eta_2^{(\pm\pm)}$ shows the expected rise, $\eta^{(+-)}$ decreases strongly at low M. This strong anti-bunching can be attributed to decay of resonances having correlation lengths of order $\delta y \sim 0.5 - 1.0$.

NA27 data have been used [180] to evaluate η_2 and η_3 both in y and φ. In both variables, both BP's show a power-like increase with incrasing M, so that a multifractal structure is indeed observed directly from the bunching parameters in

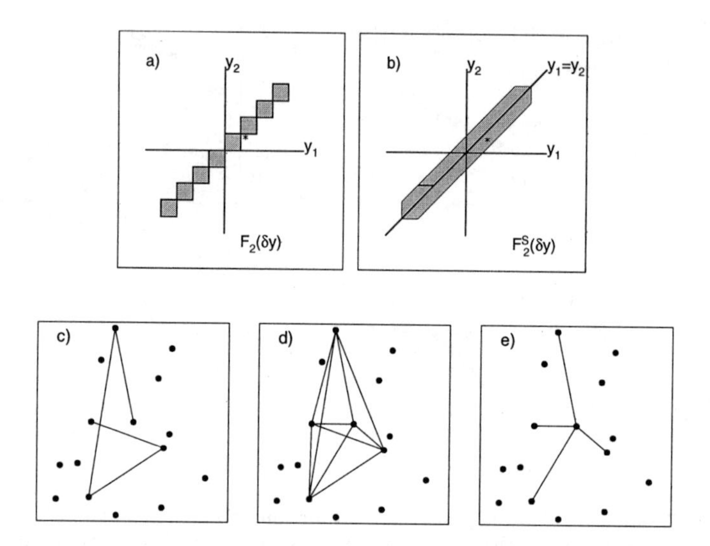

Figure 10.38. a) The integration domain $\Omega_{\rm B} = \Sigma_m \Omega_m$ of $\rho_2(y_1, y_2)$ for the bin-averaged factorial moments, b) the corresponding integration domain $\Omega_{\rm S}$ for the density integral [39], c) illustration of a q-tuple in snake topology, d) GHP topology, e) star topology [187].

hh collisions, as well.

10.8 Density and correlation strip-integrals

10.8.1 The method

A fruitful development in the study of density fluctuations is the density and correlation strip-integral method [24,185,186]. By means of integrals of the inclusive density over a strip domain, rather than a sum of box domains, one not only avoids unwanted side-effects, such as splitting up of density spikes, but also drastically increases the integration volume (and therefore the accuracy) at a given resolution.

Consider first the (vertical) factorial moments F_q defined, for an analysis in one dimension, as

$$F_q(\delta y) \equiv \frac{1}{M}\sum_{m=1}^{M}\frac{\langle n_m^{[q]}\rangle}{\langle n_m\rangle^q} = \frac{1}{M}\sum_{m=1}^{M}\frac{\displaystyle\int_{\Omega_m}\Pi_i dy_i \rho_q(y_1\ldots y_q)}{\displaystyle\int_{\Omega_m}\Pi_i dy_i \rho_1(y_1)\ldots\rho_1(y_q)}\ . \tag{10.50}$$

The integration domain $\Omega_{\rm B} = \sum_{m=1}^{M}\Omega_m$ thus consists of M q-dimensional boxes Ω_m of edge length δy. For the case $q = 2$, $\Omega_{\rm B}$ is the domain in Fig. 10.38a. A point in the mth box corresponds to a pair (y_1, y_2) of distance $|y_1 - y_2| < \delta y$ and both particles in the same bin m. Points with $|y_1 - y_2| < \delta y$ which happen *not* to lie in the same but in adjacent bins (e.g. the asterix in Fig. 10.38a) are left out. The statistics can be approximately doubled by a change of the integration volume Ω_B to the strip domain of Fig. 10.38b. For $q > 2$, the increase of integration volume (and reduction of squared statistical error) is in fact roughly proportional to the order of the correlation. The gain is even larger when working in two or three phase-space variables.

In terms of the strips (or hyper-tubes for $q > 2$), we define as (vertical) *density* integrals

$$F_q^{\mathrm{S}}(\delta y) \equiv \frac{\int_{\Omega_s} \Pi_i \mathrm{d}y_i \rho_q(y_1, \ldots, y_q)}{\int_{\Omega_s} \Pi_i \mathrm{d}y_i \rho_1(y_1) \ldots \rho_1(y_q)} \tag{10.51}$$

and, similarly, the *correlation* integrals $K_q^{\mathrm{S}}(\delta y)$ by replacing the density $\rho_q(y_1, \ldots, y_q)$ by the correlation function $C_q(y_1, \ldots, y_q)$. (Note that in the literature the term "correlation integral" is often also used for the $F_q^{\mathrm{S}}(\delta y)$.)

These integrals can be evaluated directly from the data, after selection of a proper distance measure ($|y_i - y_j|$, $[(y_i - y_j)^2 + (\phi_i - \phi_j)^2]^{1/2}$, or better the four-momentum difference $Q_{ij}^2 = -(p_i - p_j)^2$) and after definition of a proper multiparticle topology, the snake integral [102], the GHP integral [24], or the star integral [188] as shown in Figs. 10.38c-e, respectively.

10.8.2 Results

As an example, the conventional $F_4(\delta y)$ for the single NA22 spike event [9] (Fig. 10.39a) is compared to $F_4^{\mathrm{S}}(\delta y)$ (and $F_2^{\mathrm{S}}, F_3^{\mathrm{S}}$)) in Fig. 10.39b (no error bars are shown, because it is one event). Depending on whether the prominent spike lies entirely in one bin or is split across two, F_4 shows large artificial fluctuations. These are practically absent in F_4^{S}. Large improvement in one-dimensional (η) and two-dimensional ($\eta - \varphi$) analysis is also observed in [189].

How much the statistical errors are reduced can be seen on Fig. 10.40a, where the NA22 data [39] are plotted as a function of $-\ln Q^2$, with all two-particle combinations in an n-tuple having $Q_{ij}^2 < Q^2$ [24]. The following observations can be made:

i) the errors and fluctuations are indeed largely reduced, as compared e.g. to Fig. 10.23a,

ii) with the (one-dimensional) distance measure Q^2, the moments show a similarly steep rise as in the three-dimensional analysis (e.g. Fig. 10.12c),

iii) contrary to the results in rapidity, positives and negatives behave very similarly here (only negatives are shown in Fig. 10.40a), but are now much steeper than all-charged,

iv) F_2^{S} is flatter for $(+-)$ than for all-charged or like-charged combinations.

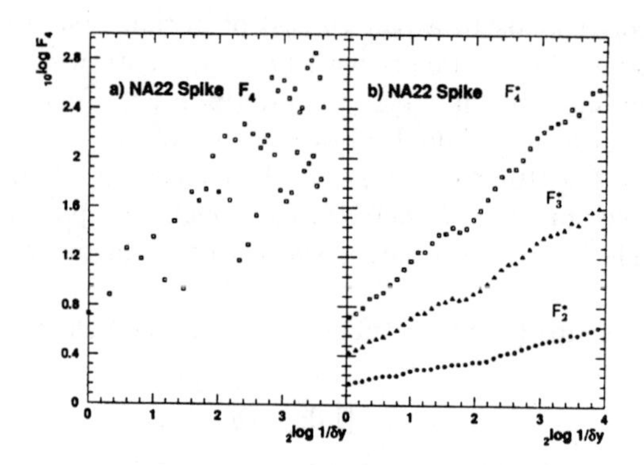

Figure 10.39. a) The fourth factorial moment of the NA22 spike event [9], b) the density strip-integrals for $q = 2 - 4$ [186].

The first two observations demonstrate the strength of the new method and the advantage of using the proper variable. The second two observations directly demonstrate the large influence of identical-particle correlations on the factorial moments. These results agree very well with results from the UA1 collaboration [40] shown in Fig. 10.40b and with lh results [142, 190].

Monte-Carlo simulations with FRITIOF 2 show the following (see Fig. 10.41 for the case of F_2^{S}). The default "plain" version is unable to describe the all-charged NA22 data, but a "biased" version (including misidentified Dalitz decay + 0.25% undetected γ conversions) comes closer to the data. However, not unexpectedly, both versions fail completely in describing the like-charged data, where the model stays way too low. On the other hand, F_2^{S} for the $(+-)$ combination is largely overestimated when γ conversions are included, but saturates without.

10.8.3 Genuine higher-order correlations

The correlation integral method turns out particularly useful for the unambiguous establishment of genuine higher-order correlations in terms of the normalized cumulants $K_q(Q^2)$, when using the star integration [188],

$$K_q^*(Q^2) = \frac{\int \Pi_i dy_i \Theta_{12} \ldots \Theta_{1q} C_q(y_1, \ldots, y_q)}{\int \Pi_i dy_i \Theta_{12} \ldots \Theta_{1q} \rho_1(y_1) \ldots \rho_1(y_q)} , \tag{10.52}$$

with $\Theta_{1j} = \Theta(Q^2 - Q_{1j}^2)$ restricting all $q-1$ distances Q_{1j}^2 to lie within a distance Q^2 of the position of particle 1. The star-integral method combines the advantage of optimal use of available statistics and minimal use of computer time. Since higher

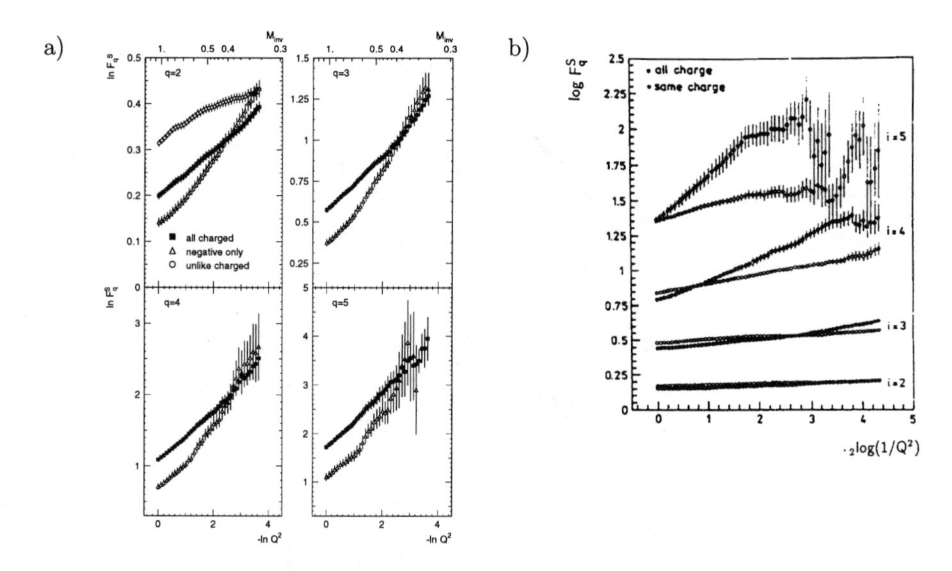

Figure 10.40. Density strip-integrals $F_q^{\rm S}$ for $q = 2 - 5$ for a) all-charged and negatives in π^+p and K$^+$p collisions at $\sqrt{s}$=22 GeV (the integral $F_2^{\rm S}$ is also given for $(+-)$ combinations) [113], b) all-charged and same-charged combinations in p$\overline{\rm p}$ collisions at 630 GeV [40].

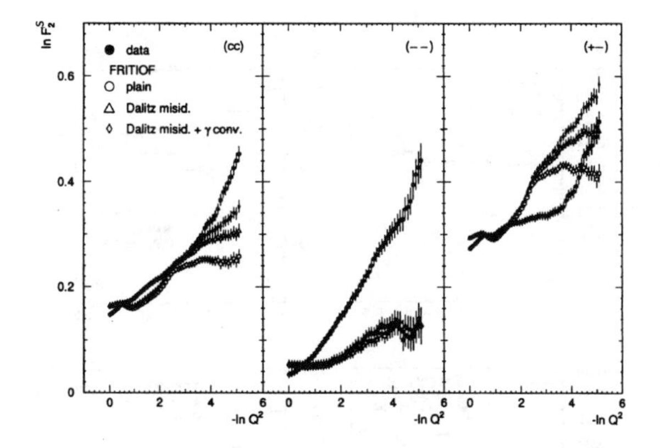

Figure 10.41. ln $F_2^{\rm S}$ in the NA22 data (full circles) compared to FRITIOF 2.0, FRITIOF 2.0 with Dalitz decay and FRITIOF 2.0 with Dalitz decay and γ-conversion (open symbols), for (cc), $(--)$ and $(+-)$ combinations, as indicated [113].

accuracy is obtainable, dynamical structures in the correlations can be studied in greater detail than with conventional methods.

Non-zero values of $K_q^*(Q^2)$ increasing according to a power law with decreasing

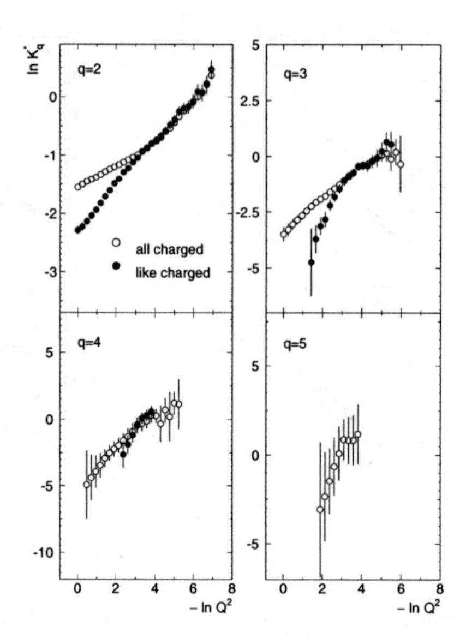

Figure 10.42. $\ln K_q^*(Q^2)$ as a function of $-\ln Q^2$ for all charged particles as well as for like-charged particles [191].

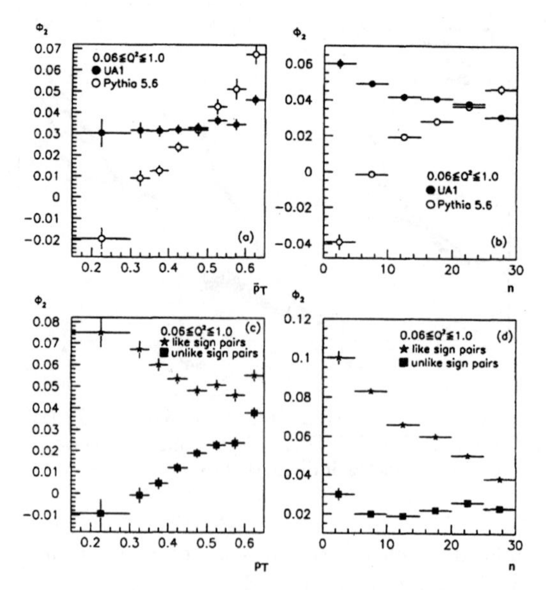

Figure 10.43. The slope ϕ_2 as a function of average transverse momentum $\bar{p}_{\rm T}$ and multiplicity n for UA1 data and PYTHIA [145].

Q^2 were first observed in NA22 up to fifth order [191] (see Fig. 10.42) and in E665 for third order [190].

10.8.4 Transverse-momentum and multiplicity dependence (revisited)

As in Sub-Sect. 10.4.2 (Fig. 10.23c), UA1 [145] has studied the ϕ_2 dependence on the average transverse momentum $\bar{p}_{\mathrm{T}}$ of the event, but now in terms of density integrals in Q^2. In contrast to the strong decrease (and subsequent slight increase) of ϕ_2 with increasing $\bar{p}_{\mathrm{T}}$ observed for the one-dimensional analysis in Fig. 10.23c, a strikingly flat behavior (and slight increase above 0.6 GeV/c) is observed for the data (full circles) in Fig. 10.43a. The discrepancy of PYTHIA (open circles) is even stronger here than in Fig. 10.23c. The PYTHIA slope ϕ_2 starts at even negative values for small $\bar{p}_{\mathrm{T}}$, but increases fast with increasing $\bar{p}_{\mathrm{T}}$ to reach values overestimating ϕ_2 at $\bar{p}_{\mathrm{T}} \gtrsim 0.5$ GeV/c.

A similar disagreement is observed for the multiplicity dependence in Fig. 10.43b. While the UA1 data (full circles) decrease with increasing n, PYTHIA predicts a strong increase. In Figs. 10.43c and d, it is shown that this violent discrepancy between PYTHIA and data is mainly due to like-charged pairs, so to the way Bose-Einstein correlations are incorporated into the model.

10.8.5 Bose-Einstein correlations versus QCD effects

Of particular interest is a comparison of hadron-hadron to e^+e^- results in terms of same and opposite charges. This is shown in Fig. 10.44 for $q = 2$ UA1 and DELPHI data in [192] (note that in this figure the derivative of (10.51) is presented in small Q^2 bins). An important difference between UA1 and DELPHI can be observed on both sub-figures: For relatively "large" $Q^2 (> 0.03$ GeV^2), where Bose-Einstein effects do not play a role, the e^+e^- data increase much faster with increasing ${}_2\log(1/Q^2)$ than the hadron-hadron results. For e^+e^-, the increase in this Q^2 region is very similar for same and for opposite sign charges. At small Q^2, however, the e^+e^- results approach the hadron-hadron results. The authors conclude that for e^+e^- at least two processes are responsible for the power-law behavior: Bose-Einstein correlations at small Q^2 following the evolution of jets at larger Q^2. In hadron-hadron collisions at present collider energies Bose-Einstein effects seem to dominate.

Since string fragmentation causes an anti-correlation between same-charged particles, it is of interest to compare e^+e^- results to JETSET in terms of strip integrals for the different charge combinations, separately. This has been done in [192] and, indeed, the Monte-Carlo results level off at small Q^2 and fall below the data for the same-charge results, while they describe the opposite-charge data well (not shown here).

DIFFERENTIAL CORRELATION INTEGRALS (Q^2)

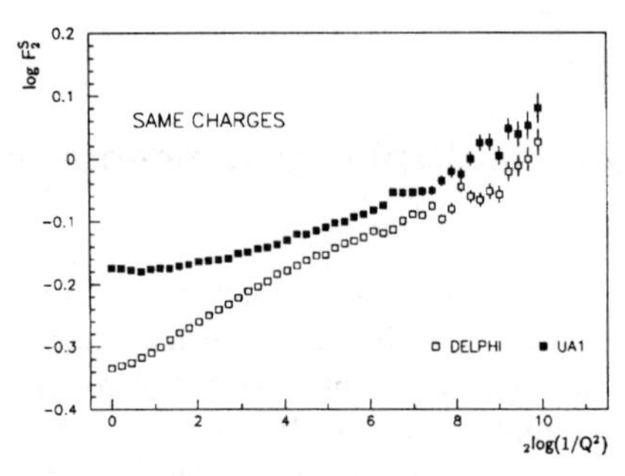

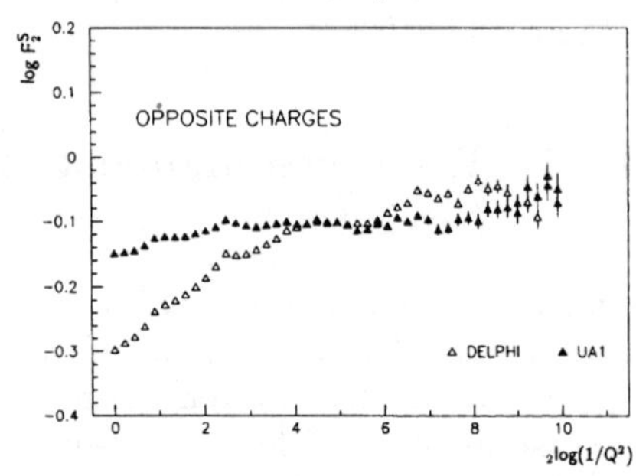

Figure 10.44. Comparison of density integrals for $q = 2$ in their differential form (in intervals $Q^2, Q^2 + \mathrm{d}Q^2$) as a function of ${}_2\log(1/Q^2)$ for $\mathrm{e}^+\mathrm{e}^-$ (DELPHI) and hadron-hadron collisions (UA1) [192].

The exact functional form of F_2^{S} is derived from the data of UA1 [40] and NA22 [39], again in its differential form,[2] in Fig. 10.45. Clearly, the data favour a power law in Q over an exponential, double-exponential or Gaussian law.

If the observed effect is real, it supports a view recently developed in [72]. There, intermittency is explained from Bose-Einstein correlations between (like-charged) pions. As such, Bose-Einstein correlations from a static source are not power behaved. A power law is obtained i) if the size of the interaction region is allowed to fluctuate, and/or ii) if the interaction region itself is assumed to be a self-similar object extending over a large volume. Condition ii) would be realized if parton avalanches were to arrange themselves into self-organized critical states [193]. Though quite speculative at this moment, it is an interesting new idea with possibly far-reaching implications. We should mention also that in such a scheme intermittency is viewed as a final-state effect and is, therefore, not troubled by hadronization effects.

The influence on the factorial moments of adding Bose-Einstein correlations in FRITIOF and RQMD is demonstrated for heavy-ion collisions in [194] and [195], respectively. Because of the large number of collision processes, other correlation effects are expected to play a much reduced role and Bose-Einstein correlations, as a collective effect, can become the dominant source of non-statistical fluctuations. Also from these results it is clear that more than one fixed interaction-volume radius is needed to reproduce the experimental results.

In perturbative QCD, on the other hand, the intermittency indices ϕ_q, are directly related to the anomalous multiplicity dimension $\gamma_0 = (6\alpha_{\mathrm{s}}/\pi)^{1/2}$ [181–184] and, therefore, to the running coupling constant α_{s}. In the same context, it has been argued [182–184] that the opening angle between particles is a suitable and sensi-

[2] In fact in this differential form $F_2^{\mathrm{S}}(Q^2)$ is identical to $R(Q^2)$ usually used in Bose-Einstein analysis. The only difference is that it is plotted on a double-logarithmic plot, here.

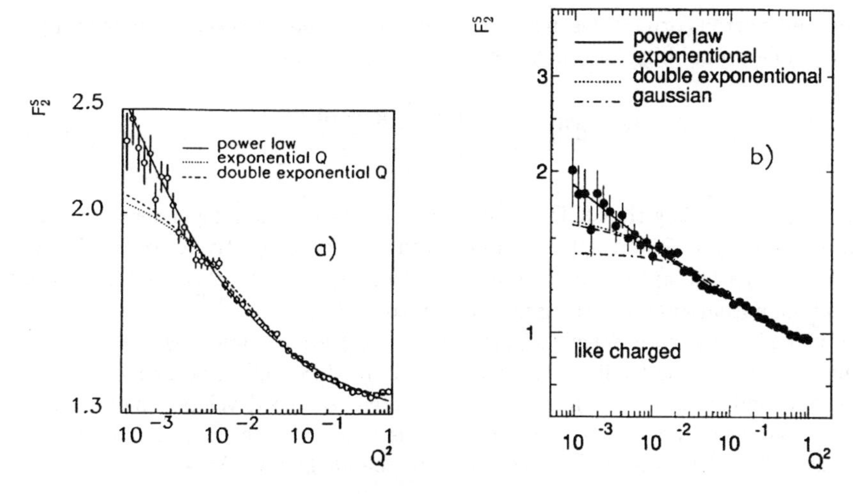

Figure 10.45. Density integrals F_2^{S} (in their differential form) as a function of Q^2 for like-charged pairs in UA1 [40] and NA22 [39], compared to power-law, exponential, double-exponential and Gaussian fits, as indicated.

tive variable to analyse and well suited for these first analytic QCD calculations of higher-order correlations. This angle is, of course, closely related to Q^2.

10.9 Analytical QCD predictions

10.9.1 The QCD framework

As we have seen above, QCD shower inspired models as the parton-shower version of JETSET reproduce the observed fluctuations quite well. These models have a perturbative-QCD shower-evolution phase and a non-perturbative hadronization phase with many parameters, however, and even though the fluctuations can be reproduced without retuning these parameters, one is inclined to try better.

A simple alternative is to extend the perturbative-QCD evolution down to a low mass scale [196] and to use the concept of Local Parton-Hadron Duality (LPHD) [197] to compare the multiparton final states to the experimentally observed multihadron final states, directly. We have seen in Chapters 4 and 8 that such an approach is quite successful in the description of the data in fully integrated variables, as the energy dependence of the average multiplicity or the single-particle inclusive distribution in ξ and its energy dependence.

Here, we shall investigate how far this success can be extended to two-particle correlations and multiparticle fluctuations.

Detailed analytical predictions based on the double-log approximation (DLLA)

[198] for multiparton correlations in angular cones have been proposed by three groups [182–184]. The probability for gluon bremsstrahlung in DLLA is

$$M(k)\mathrm{d}^3\mathbf{k} = c_a\gamma_0^2\frac{\mathrm{d}k}{k}\frac{\mathrm{d}\Theta_{\mathbf{pk}}}{\Theta_{\mathbf{pk}}}\frac{\mathrm{d}\varphi_{\mathbf{pk}}}{2\pi}\ , \tag{10.53}$$

with $\gamma_0 = \sqrt{6\alpha_s/\pi}$ being the QCD (gluon transverse-momentum dependent) anomalous dimension, $c_a = 1$ and $4/9$ for gluons and quarks, respectively, $\mathbf{p}$ and $\mathbf{k}$ the 3-momenta of parent parton and radiated gluon, respectively, $\Theta_{\mathbf{pk}}$ and $\varphi_{\mathbf{pk}}$ the polar and azimuthal angles of the gluon with respect to $\mathbf{p}$.

The inclusive n-parton densities $\rho_n(\mathbf{k}_1, \mathbf{k}_2, \ldots, \mathbf{k}_n)$ are obtained by the generating functional technique [153] developed for QCD in [198, 199]. Energy-momentum conservation and $\mathrm{g} \to \mathrm{q\bar{q}}$ production are neglected, a well developed high-energy cascade is assumed, and angular ordering [200] is taken into account. Furthermore, the lowest-order relation is used between α_s and the QCD scale Λ:

$$\alpha_\mathrm{s} = \frac{\pi\beta^2}{6}\frac{1}{\ln(Q/\Lambda)},\quad \beta^2 = 12(\frac{11}{3}n_c - \frac{2}{3}n_f)^{-1}\ , \tag{10.54}$$

where $n_c = 3$ and $n_f = 3$ are the number of colors and that of flavors effectively involved in the parton shower. The only free parameter remaining is Λ. Because of the large number of approximations necessary, this has to be considered as an effective scale, only.

10.9.2 Two-particle angular correlations

Following [182], two different two-parton relative-angular ϑ_{12} correlation functions are defined as

$$\tilde{r}(\vartheta_{12}) = \frac{\rho_2(\vartheta_{12})}{\bar{n}^2(\Theta)} \tag{10.55}$$

$$\text{and } r(\vartheta_{12}) = \frac{\rho_2(\vartheta_{12})}{\rho_1 \otimes \rho_1(\vartheta_{12})} \tag{10.56}$$

within a cone of half opening angle Θ around the jet axis, where ρ_2 and $\rho_1 \otimes \rho_1$ should be evaluated from single- and two-parton densities and $\bar{n}(\Theta)$ is the average multiplicity of partons emitted into the Θ cone. While $r(\vartheta_{12})$ corresponds to the normalized genuine two-parton correlation (plus an additive constant of unit value), $\tilde{r}(\vartheta_{12})$ is dominated by the product of single-parton distributions.

Asymptotically, in DLLA at high energies, a scaling function

$$\tilde{Y}(\epsilon) \equiv -\frac{\ln(\tilde{r}(\epsilon)/b)}{2\sqrt{\ln(P\Theta/\Lambda)}} \approx 2\beta(1 - 0.5\omega(\epsilon, 2)) \tag{10.57}$$

$$
\begin{aligned}
\text{with } \tilde{r}(\epsilon) &= \vartheta_{12}\tilde{r}(\vartheta_{12}) \ln(P\Theta/\Lambda) && (10.58)\\
b &= 2\beta\sqrt{\ln(P\Theta/\Lambda)} && (10.59)\\
\epsilon &= \ln(\Theta/\vartheta_{12})/\ln(P\Theta/\Lambda) && (10.60)\\
\omega(\epsilon, q) &= q\sqrt{1-\epsilon}(1 - \frac{1}{2q^2}\ln(1-\epsilon)) && (10.61)
\end{aligned}
$$

is expected to be independent of Θ and primary parton momentum P [182–184]. Assuming LPHD also for two-particle distributions, a similar behavior is expected on the particle level.

Figs. 10.46a,b show $\tilde{Y}(\epsilon)$ as a function of ϵ for DELPHI [201], Figs. 10.46c,d for ZEUS [202], together with the (asymptotic) analytical prediction (dashed line in Fig. 10.46a,b and full line in Fig. 10.46c,d). The particle data (as well as parton- and

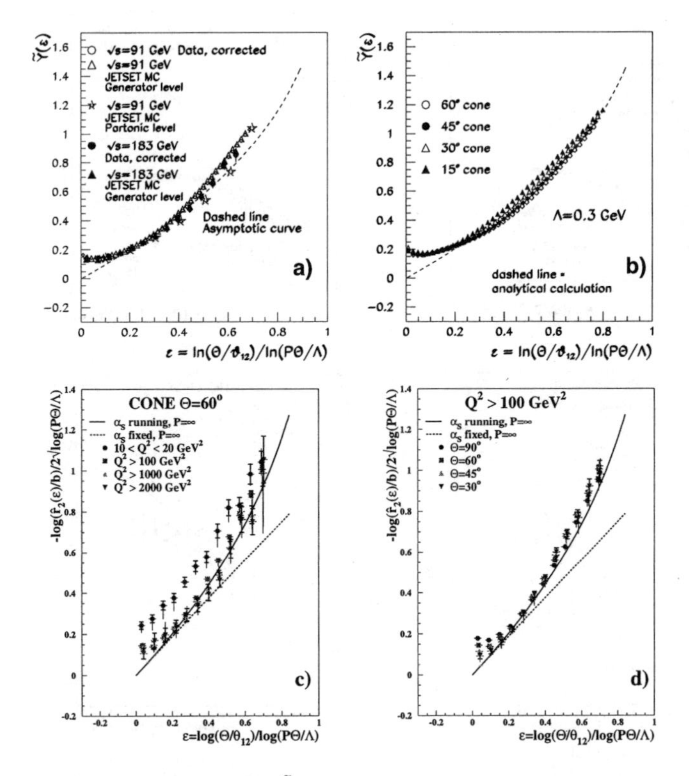

Figure 10.46. The scaling function $\tilde{Y}(\epsilon)$ as a function of ϵ for final-state particles in DELPHI (a,b) [201] and ZEUS (c,d) [202] for different values of Q or $\sqrt{s}$ (a,c) and Θ (b,d). The asymptotic analytical prediction is given as dashed line in (a) and (b) and as a full line in (c) and (d).

particle-level JETSET) scale in Q (or $\sqrt{s}$) and Θ and, at $\epsilon > 0.2$, agree surprisingly well with the parton prediction.

The situation changes, however, when the correlation-dominated function $r(\vartheta_{12})$ is used. In this case, the quantity

$$Y(\epsilon) \equiv \frac{\ln(r(\vartheta_{12}))}{\sqrt{\ln(P\Theta/\Lambda)}} \approx 2\beta(\omega(\epsilon, 2) - 2\sqrt{1-\epsilon}) \tag{10.62}$$

is expected to be asymptotically independent of Θ and P for produced partons. Comparing to the corresponding particle distribution in the data (Fig. 10.47), the Θ dependence (b,d) remains weak, but the shape differs from the prediction (in particular for DELPHI). Energy scaling is violated both in DELPHI and ZEUS.

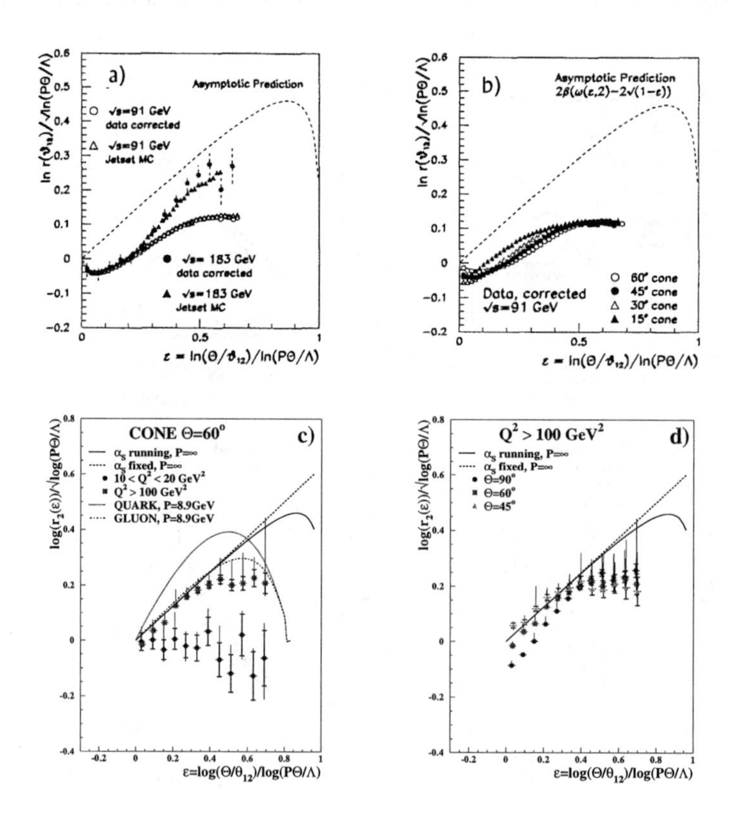

Figure 10.47. The rescaled function $Y(\epsilon) = \ln(r(\delta_{12}))/\sqrt{\ln(P\Theta/\Lambda)}$ as a function of ϵ for final-state particles in DELPHI (a,b) [201] and ZEUS (c,d) [202] for different values of Q or $\sqrt{s}$ (a,c) and Θ (b,d). The asymptotic analytical prediction is given as dashed line in (a) and (b) and as a full line in (c) and (d).

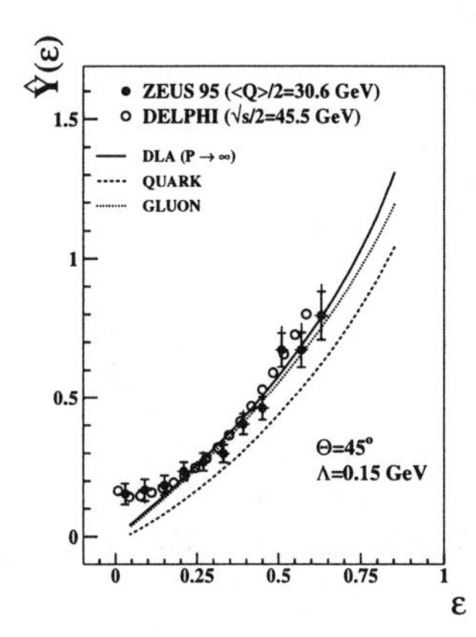

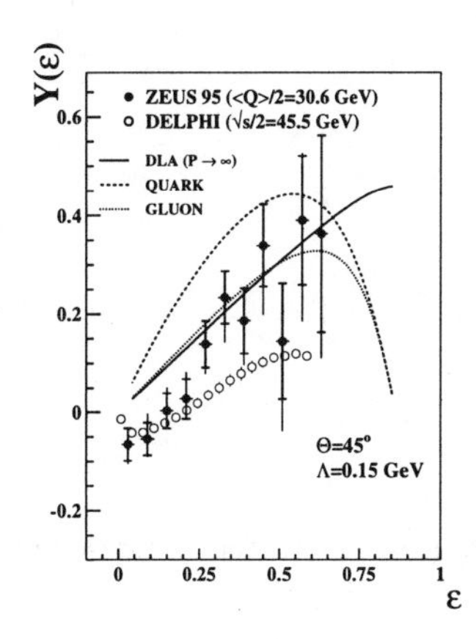

Figure 10.48. a) Normalized particle density $\tilde{Y}(\epsilon)$ and b) correlation function $Y(\epsilon)$ for ZEUS data at $P = \langle Q \rangle/2 = 30.6$ GeV compared to DELPHI results at $P = \sqrt{s}/2 = 45.5$ GeV and to analytic QCD predictions [202].

Fig. 10.48 gives a direct comparison of the ZEUS current-region data and the DELPHI data at $\Theta = 45^0$ and $\Lambda = 0.15$ GeV. No energy dependence was found for $\tilde{Y}$ by DELPHI in Fig. 10.46a and no difference is observed for $\tilde{Y}$ between ZEUS ($P = 30.6$ GeV) and DELPHI ($P = 45.5$ GeV) in Fig. 10.48a.

On the other hand, $Y(\epsilon)$ was observed to be steeper at $\sqrt{s} = 183$ GeV than at 91 GeV in Fig. 10.47a. From a quark-jet universality, one would, therefore, expect the ZEUS data to be less steep than the DELPHI data. However, the contrary is observed in Fig. 10.48b.

Also shown in Fig. 10.48 are the analytic DLLA QCD predictions [182] at infinite energy, as well as for finite energy ($P = 30.6$ GeV) quark jets and gluon jets. The fact that gluon-jet predictions describe the quark-jet results better than the quark-jet predictions, themselves, should not be taken too seriously, since they depend on the ratio 9/4, which is strongly reduced when going beyond DLLA (see Sect. 8.1).

10.9.3 Fluctuations in one- and two-dimensional angular regions

The dominant source of particle production inside a jet is gluon splitting in the QCD cascade, where the presence of a gluon enhances the probability for emission of another gluon nearby in momentum space.

QCD is inherently intermittent and QCD predictions [182–184] for normalized factorial moments for multiparton production into one- or two-dimensional angular

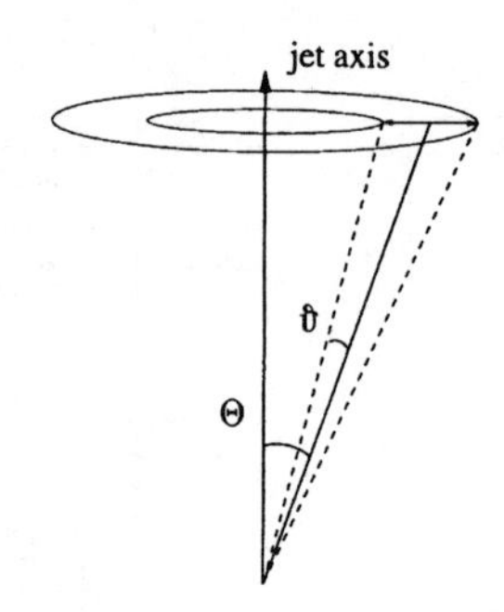

Figure 10.49. A schematic representation of the interval in the polar angle around the jet axis.

windows have the following scaling behavior:

$$F_q(\vartheta) \propto \left(\frac{\Theta}{\vartheta}\right)^{(D-D_q)(q-1)} . \tag{10.63}$$

The angular windows considered here are either rings around the jet axis (Fig. 10.49) with mean opening angle Θ and half width ϑ (topological dimension $D = 1$), or cones of half opening angle ϑ around a direction (Θ, Φ) with respect to the jet axis ($D = 2$); the D_q are the Rényi dimensions.

The analytical QCD expectations for D_q at the parton level are as follows:

1) In the fixed coupling (α_s= constant) regime, for moderately small angular bins,

$$D_q = \gamma_0 \frac{q+1}{q} . \tag{10.64}$$

2) In the running-coupling regime, for small bins, the Rényi dimensions become a function of the size of the angular window ($\alpha_s(Q)$ increases with decreasing ϑ). Three approximations derived in DLLA will be compared:

a) According to [183], the D_q have the form

$$D_q \simeq \gamma_0(Q) \frac{q+1}{q} \left(1 + \frac{q^2+1}{4q^2}\epsilon\right) . \tag{10.65}$$

with the (rescaled) scaling variable ϵ as defined in (10.60), but ϑ_{12} replaced by ϑ,

b) according to [184]:

$$D_q \simeq 2\,\gamma_0(Q) \frac{q+1}{q} \left(\frac{1-\sqrt{1-\epsilon}}{\epsilon}\right) . \tag{10.66}$$

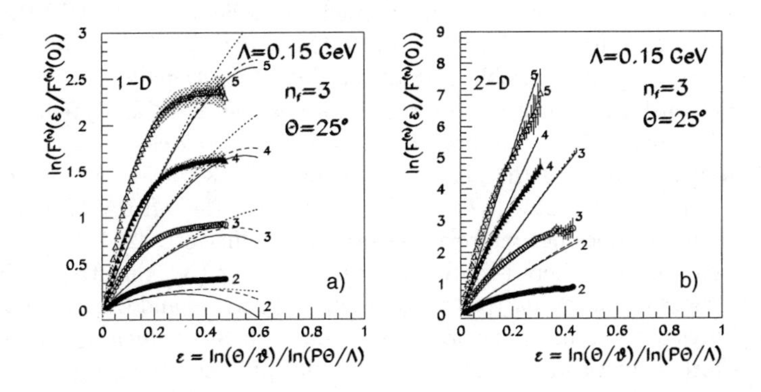

Figure 10.50. Normalized factorial moments of order $q = 2 - 5$, renormalized to their value at $\epsilon = 0$, as a function of ϵ, in a) rings ($D = 1$) and b) side cones ($D = 2$), compared with analytical calculations according to (10.67) (solid lines), (10.66) (dashed) and (10.65) (dotted) [205].

c) In [182], a result has been obtained for the cumulant moments converging to factorial moments for high energies,

$$D_q \simeq 2\,\gamma_0(Q)\frac{q - w(q,\epsilon)}{\epsilon(q-1)}, \tag{10.67}$$

with w as defined in (10.61)

In [183], an estimate for D_q has also been obtained in a Modified Leading Log Approximation (MLLA). In this case, (10.65) remains valid, but $\gamma_0(Q)$ is replaced by an effective $\gamma_0^{\text{eff}}(Q)$ depending on q.

Furthermore, analytical DLLA predictions have been presented [203] in the fixed coupling regime for factorial correlators.

Since these predictions are on the parton level, LPHD has to be assumed to be able to compare to experimental data. Note, however, that this hypothesis had been suggested to hold for sufficiently inclusive observables, and not necessarily for (higher-order) correlations.

The factorial-moment predictions have been compared to the data by L3 [204], DELPHI [205] and ZEUS [206]. Fig. 10.50 gives the DELPHI results on the normalized factorial moments for order $q = 2 - 5$ as a fuction of ϵ for $\Theta = 25^0$, together with the DLLA predictions [182–184] for $\Lambda = 0.15$ GeV. Fig. 10.50a corresponds to angular rings ($D = 1$), Fig. 10.50b to side cones ($D = 2$).

The data are not described well. In the $D = 1$ case (Fig. 10.50a), the predictions lie below the data for not too large ϵ and differ in shape. Choosing $n_f = 5$ instead of 3, increases the discrepancy, reducing Λ reduces the discrepancy at small ϵ, but overshoots at large ϵ (not shown here). Similar conclusions are drawn by L3 and

ZEUS. In the $D = 2$ case (Fig. 10.50b), the predictions lie above the DELPHI data for $q = 2$ and 3, while the behavior is similar to that for $D = 1$ for $q = 4$ and 5.

In conclusion, it is not possible to find one set of QCD parameters Λ and n_f which simultaneously minimize the discrepancies between data and predictions for the moments of order $q = 2 - 5$ in both the $D = 1$ and $D = 2$ cases.

The main shortcoming of the analytical calculations is that energy-momentum conservation is not taken into account. However, also the MLLA formula [183], partially including it, does not improve the situation sufficiently [204–206].

Still, considering that there is only one free parameter (Λ), that the predictions are asymptotic and meant for partons rather than for particles and that, nevertheless, the basic features of the predictions are in fact seen in the experimental data, further theoretical effort is justified (see also [207] on the problem of energy-momentum conservation).

10.9.4 In (transverse-)momentum cut phase space

Within DLLA, the normalized factorial moments of gluons which are restricted in (transverse) momentum, $p_t < p_t^{cut}$ or $p < p^{cut}$, are expected [208] to follow, respectively,

$$F_q(p_t^{cut}) \simeq 1 + \frac{q(q-1)}{6} \frac{\ln(p_t^{cut}/Q_0)}{\ln(P/Q_0)} \qquad (10.68)$$

$$F_q(p^{cut}) \simeq C(q) > 1 \,, \qquad (10.69)$$

where P is again the initial energy of the outgoing quark, p_t is defined relative to the direction of this quark and $C(q)$ are constants depending on q.

Again, the DLLA predictions are on the parton level and should be regarded asymptotic, i.e. valid at small p_t^{cut} and p_{cut}. Therefore, they should be considered only as qualitative predictions when compared to the data in conjungation with the LPHD hyphothesis.

Such a comparison has been made by ZEUS [206] up to order $q = 5$ (see Fig. 10.51). While DLLA (Eq. (10.68)) predicts the moments to approach unity from above as p_t^{cut} decreases, the data (corrected for Bose-Einstein correlations) show the opposite. The Monte-Carlo models follow the trend of the data, with ARIADNE giving the best overall description. A similar observation is made for the moments as a function of p^{cut}.

To check the effect of energy-momentum conservation, the moments were also determined at the parton level of ARIADNE, the physics implementation of which strongly resembles the analytic calculations [208], but which includes energy-momentum conservation explicitly. To satisfy LPHD, the cut-off parameter Q_0 was reduced to 0.27 GeV, i.e. slightly above the Λ value of 0.22 GeV, also ensuring the parton multiplicity to equal that of the hadrons. The results are given as the thin solid line in Fig. 10.51. They indeed show the behavior expected from Eq. (10.68), i.e., they

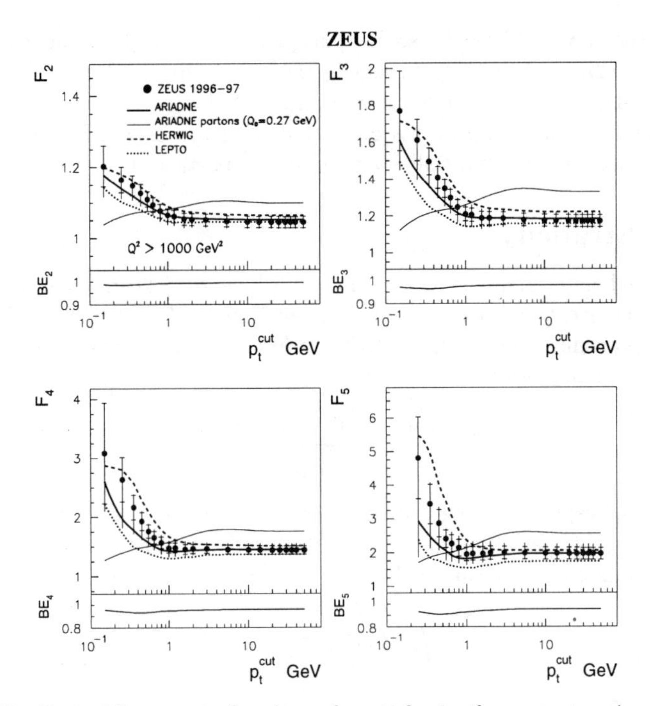

Figure 10.51. Factorial moments for charged particles in the current region of the Breit frame of e^+p collisions at HERA, as a function of p_t^{cut}, compared to Monte-Carlo models at the hadron level (thick lines) and ARIADNE with $Q_0 = 0.27$ GeV at the parton level (thin solid line). The data are corrected for Bose-Einstein correlations by the BE factor indicated [206].

still disagree with the hadronic data. Analogous differences between the hadron and parton levels of ARIADNE have been observed in e^+e^- annihilation [208]. So, one has to conclude with the authors that here the limits of LPHD are crossed.

10.10 Individual Events

10.10.1 Single-event intermittency

The original suggestion of intermittency in particle production at high energies [13] was based on a single high-multiplicity event [8]. The event-sample averaging used in this chapter so far, was necessary to cope with limited-multiplicity final states, but bears the danger of a loss of interesting effects if these are present in only part of the sample. To be able to catch these, an event-by-event analysis seems more appropriate.

No experimental results exist so far, except that obtained for the JACEE event in [13] (Fig. 10.2a) and the NA22 event in [186] (Fig. 10.39). However, simulations were made [209] on the bias and the resolution of the intermittency exponents ϕ_2 and ϕ_3 obtained from single events generated with the α-model. The results are encouraging, provided the multiplicity of produced particles is large.

10.10.2 Erraticity

The normalized factorial moments as defined in 7.72 and 7.74 contain averages over all bins within one event and over the whole event sample. Event-to-event fluctuations lost in this procedure can be used [210] by first defining a normalized horizontal factorial moment per event

$$F_q^{\rm e} = \frac{\langle n(n-1)\dots(n-q+1)\rangle_{\rm e}}{\langle n\rangle_{\rm e}^q} \tag{10.70}$$

where $\langle\dots\rangle_{\rm e}$ corresponds to the average over all M bins in the event. $F_q^{\rm H}$ of (7.74) is an average of $F_q^{\rm e}$ over all events, but $F_q^{\rm e}$ itself follows a distribution which we call $P(F_q^{\rm e})$ and assume to be normalized as

$$\int_{+0}^{\infty} P(F_q^{\rm e}){\rm d}F_q^{\rm e} = 1., \tag{10.71}$$

so that

$$F_q^{\rm H} \equiv \langle F_q^{\rm e}\rangle = \int_0^{\infty} F_q^{\rm e} P(F_q^{\rm e}){\rm d}F_q^{\rm e} \tag{10.72}$$

with $\langle\dots\rangle$ running over all events in the sample. Besides the horizontal moments, $F_q^{\rm e}$ could equally well represent a density integral or the result of a wavelet analysis.

Experimentally, as an alternative to trying to obtain the full distribution, it is expected sufficient to determine a limited number of moments of $P(F_q^{\rm e})$ as

$$C_{q,p} = \frac{\langle (F_q^{\rm e})^p\rangle}{\langle F_q^{\rm e}\rangle^p} \ , \tag{10.73}$$

in order to quantify the degree of the event-to-event fluctuation. The order p need not be integer. It may also be less than unity, but may or may not be less than 0, depending on whether there are events in the sample with $F_q^{\rm e} = 0$. For $p > 1$, $C_{q,p}$ probes the large-$F_q^{\rm e}$ behavior of $P(F_q^{\rm e})$ (spikes), for $p < 1$ the low-$F_q^{\rm e}$ behavior (dips). So, knowing $C_{q,p}$ for $0 < p < 2$ already reveals a lot about $P(F_q^{\rm e})$ not probed by conventional intermittency analysis in terms of the event averages $F_q^{\rm H}$ or $F_q^{\rm V}$.

Having generalized the F_q to the $C_{q,p}$ moments, it comes natural [210] to search for a scaling behavior similar to that of (7.114),

$$C_{q,p}(\delta y) \propto (\delta y)^{-\psi_q(p)} \ , \quad \delta y \to 0 \ . \tag{10.74}$$

To distinguish this from intermittency, it is referred to as "erraticity" of the system. If the event-to-event fluctuations scale with bin size δy, the erraticity exponents $\psi_q(p)$ represent an economic way to characterize this aspect of self-similar dynamics.

One then can extract [210] an entropy index (in event space)

$$\mu_q = \frac{\mathrm{d}}{\mathrm{d}p}\psi_q(p)\bigg|_{p=1} \tag{10.75}$$

describing the width of the fluctuation used in [210] to study chaotic behavior in QCD branching. Small μ_q corresponds to large entropy, so no chaotic behavior.

In analogy to the multifractal spectrum (10.45) an erraticity spectrum $e(\alpha)$ can be defined (for fixed q) as

$$e(\alpha) = p\alpha - \psi(p)\ , \quad \alpha = \mathrm{d}\psi(p)/\mathrm{d}p\ . \tag{10.76}$$

At $p = 1$, one has $e(\alpha) = \alpha$, since $\psi(1) = 0$.

The method has first been applied to E299 [211] and to NA27 [212] data, where the $C_{p,q}$, however, do not show a simple power-law behavior for $p \geq 1.5$. A generalized power law in analogy to (10.5) was therefore proposed in [213] and tested with the Monte-Carlo generator ECOMB [214] tuned to NA22 data [39].

In the left column of Fig. 10.52a, $\ln C_{p,q}$ is plotted as a function of $\ln M$ ($M \propto 1/\delta\eta$ being the partition number) for the NA27 data [212]. The behavior is not linear, so the right-hand side of Fig. 10.52a shows the generalized power behavior as a function of $\ln C_{2,2}(M)$ suggested in [213],

$$C_{p,q} \propto C_{2,2}^{\chi(p,q)}(M)\ . \tag{10.77}$$

The exponents $\chi(p,q)$ are shown as a function of p in Fig. 10.52b. Fitting second-order polynomials through the points for each q,

$$\chi'_q = \frac{\mathrm{d}}{\mathrm{d}p}\chi(p,q)\bigg|_{p=1}\ , \tag{10.78}$$

can be determined from the data, and the entropy index [213] from

$$\tilde{\mu}_q = \psi'_2(2)\chi'_q \tag{10.79}$$

to 0.049±0.002, 0.27±0.01 and 0.59±0.02 for $q = 2, 3$ and 4, respectively.

However, in systematic analyses of the influence of purely statistical fluctuations on the results obtained for the erraticity of a system [215, 216], it turned out that these completely dominate the erraticity results in hh collisions near $\sqrt{s} = 22$ GeV and have to be taken into account even in heavy-ion collisions.

In an analysis of SAg/Br collisions at 200 A GeV [217], the authors, therefore, compare their values of μ_q as a function of q (full circles) in Fig. 10.53a to those expected from purely statistical fluctuations (triangles). The experimental μ_q increase with increasing q much faster than the statistical ones, signifying that in heavy-ion collisions, event-to-event fluctuations are mainly dynamical and become more erratic with increasing q.

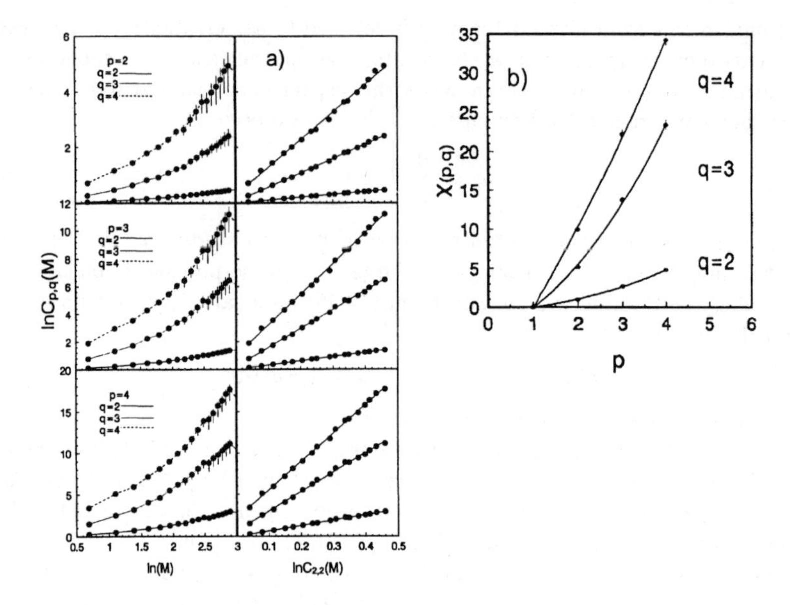

Figure 10.52. a) Log-log plots of $C_{p,q}$ versus M (left) and versus $C_{2,2}$ (right). The lines on the left are to guide the eye, those on the right are linear fits, b) slopes of these linear fits as a function of p for $q = 2, 3, 4$. The lines correspond to a second-order polynomial fit [212].

10.10.3 Void analysis

When the event multiplicity is low, the gaps between neighboring particles carry more information about an event than multiplicity spikes. In order to obtain complementary information from the low multiplicities, the method has therefore been extended to measure the rapidity gaps [218] (see also Sect. 8.4).

Consider an event with n particles $i = 1, 2, \ldots, n$, located at y_i, or better its transformed positions $X_i = X(y_i)$ of Eq. (10.2), ordered from small to large. The distances between neighboring particles are $x_i = X_{i+1} - X_i$, with $X_0 = 0$, $X_{n+1} = 1$, and $\sum_{i=0}^{n} x_i = 1$.

Then, for each event, moments

$$G_q = \frac{1}{n+1} \sum_{i=0}^{n} x_i^q \qquad (10.80)$$

are defined, which emphasize large rapidity gaps.

The moments G_q fluctuate from event to event, and the shape of the distributions $P(G_q)$ characterizes the nature of the event-to-event fluctuations. Again, this shape

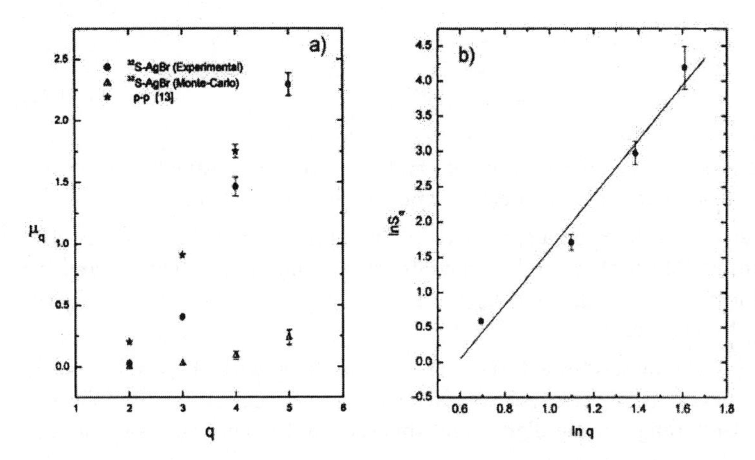

Figure 10.53. The dependence of a) μ_q on q for SAg/Br data compared to the pp results [212] and to SAg/Br Monte-Carlo events with purely statistical fluctuations [217], b) $\ln S_q$ on $\ln q$ for SAg/Br collisions [217].

is described by moments,

$$C_{q,p} = \frac{1}{N_\mathrm{e}} \sum_{\mathrm{e}=1}^{N_\mathrm{e}} (G_q^\mathrm{e})^p = \int \mathrm{d}G_q G_q^p P(G_q), \tag{10.81}$$

where e labels an event and N_e is the number of events in the sample. Although one can consider a range of order p of the moments, a large amount of information is already contained in the derivative

$$s_q = -\frac{\mathrm{d}}{\mathrm{d}p} C_{q,p}\Big|_{p=1} = -\langle G_q \ln G_q \rangle \ , \tag{10.82}$$

where $\langle \ldots \rangle$ stands for the average over all events.

Since the G_q do not filter out statistical fluctuations, one has to divide out s_q^st for a sample of same size but random distribution in X space,

$$S_q = s_q / s_q^\mathrm{st} \tag{10.83}$$

and to examine how much the S_q deviate from unity.

Alternatively, one can use [218]

$$\hat{G}_q = \frac{1}{n+1} \sum_{i=0}^{n} (1 - x_i)^{-q} \ . \tag{10.84}$$

As the G_q, the $\hat{G}_q$ emphasize large gaps, but unlike the G_q, they can become $\gg 1$. Substituting for G_q in Eqs. (10.81)-(10.82), above, one obtains

$$G_q = \langle \hat{G}_q \ln \hat{G}_q \rangle \tag{10.85}$$

and

$$\Sigma_q = \sigma_q / \sigma_q^{\rm st} \tag{10.86}$$

as a second set of rapidity-gap measures of erraticity. An important limitation for Σ_q is that n must be larger than q to ensure convergence of $\sigma_q^{\rm st}$ [218].

Both sets were tested [218] on a sample of 10^6 ECOMB Monte-Carlo events [214, 219], with the result that both S_q and Σ_q deviate from 1 and increase with increasing order q. The increase is stronger for Σ_q than for S_q and, therefore, makes Σ_q the better indicator for erraticity.

The stability of the method was further checked in [220], with the result that $S_{\rm q}$ is stable already for multiplicities and sample sizes as encountered for hh collisions in the $\sqrt{s} \approx 20$ GeV range, while higher multiplicities and larger event samples are needed for Σ_q.

Figure 10.53b gives lnS_q as a function of ln$q$ for the SAg/Br data [217] already used in Sub-Sect. 10.12.2 above. It is clear that S_q deviates significantly from unity (so from statistical fluctuations) and increases with increasing q.

For hh collisions, the erraticity measures have been studied on NA27 [221] and NA22 [222] data. Fig. 10.54 gives $\ln S_q$ and $\ln \Sigma_q$ as a function of q, for three different cuts at the low end of the multiplicity distribution. The data show an increase of the erraticity measure with increasing q, which is exponential for S_q and faster than exponential for Σ_q. FRITIOF does not quite follow this fast increase.

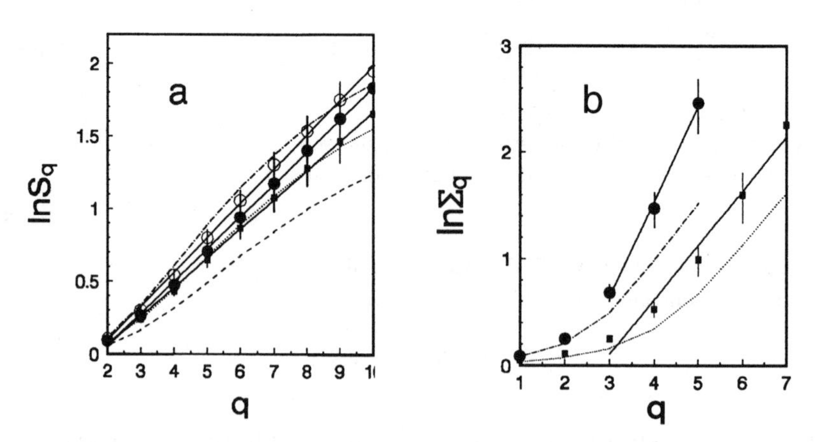

Figure 10.54. $\ln S_q$ and $\ln \Sigma_q$ as a function of q for NA27 data with $n \geq 4(\circ)$, $n \geq 6(\bullet)$ and $n \geq 8(\blacksquare)$, respectively. The solid lines are linear fits to the data, the broken lines are FRITIOF results [221].

10.10.4 Entropy

The erraticity measure based on factorial moments is sensitive to high multiplicities, that based on void sizes on low ones. A method most sensitive to the maximum of the probability distribution is a measure of entropy suggested in [223] (see also Sect. 8.3). Contrary to the first method it is not restricted to events with very large multiplicity.

Again, an originally chosen phase-space region is divided into M equal-size bins. An event is then characterized by the number of particles m_i in each bin i, i.e., by a set of integer numbers $s \equiv m_i$, $i = 1, \dots M$. These sets represent different states of the multiparticle system realized in a given experiment.

The basis of the method is the measurement of coincidence probabilities, i.e., in simply counting the number n_s of times any given set s appears in the given event sample. The total number of observed coincidences of k configurations is the kth factorial moment of the n_s distribution,

$$N_k = \sum_s n_s(n_s - 1) \dots (n_s - k + 1) \ , \quad k = 1, 2, 3 \dots \tag{10.87}$$

with only states with $n_s \geq k$ contributing. The coincidence probability of k configurations is

$$C_k = \frac{N_k}{N(N-1) \dots (N-k+1)} \ , \tag{10.88}$$

where $N = \sum_s n_s$ is the number of events in the sample, so that $C_1 = 1$.

Rényi entropies [224], defined as

$$H_k \equiv -\frac{\ln C_k}{k-1} \quad , \quad k > 1 \tag{10.89}$$

can then be used to extrapolate to $H_1 \equiv S$, the standard statistical (or Shannon) entropy, Eq. (8.38). With the proper extrapolation formula, an effective reproduction of S could be achieved for a number of typical multiplicity distributions [223].

The advantage over the standard direct calculation of S is minimization of the statistical error and increase of the stability of the result [225].

However, also the Rényi entropies themselves contain valuable information about the multiparticle system. In principle, the entropies H_k (and from their extrapolation also S) obtained depend on the method of discretization of the momentum spectrum, in particular the binning. If the bins are small enough and if the fluctuations are small (e.g. if the system is close to thermal equilibrium), one expects [223] scaling of the form

$$H_k(\ell M) = H_k(M) + \ln \ell \ , \tag{10.90}$$

with ℓ being the change in bin size. Strong fluctuations, as in cascading, on the other hand, are expected to violate this scaling property.

Furthermore, if performed independently (and simultaneously) in different phase space regions Ω_i, the entropy density distribution over phase space can be determined and the additivity property

$$H_k(\Omega) = H_k(\Omega_1) + H_k(\Omega - \Omega_1) \tag{10.91}$$

can be verified. Deviations from this propery give information about correlations between the different regions.

10.11 Lévy stable distributions

Stimulated by an original idea of Mandelbrot on the fluctuations in financial markets [226], the properties of stability, stationarity and scaling of two JACCEE events [8] have been examined [227] with the help of the quantity

$$L(\eta, \Delta\eta) = \ln \frac{\mathrm{d}n}{\mathrm{d}\eta}(\eta + \Delta\eta) - \ln \frac{\mathrm{d}n}{\mathrm{d}\eta}(\eta) \ , \tag{10.92}$$

i.e. the relative change of the particle density between successive pseudorapidity intervals of size $\Delta\eta$. Provided that the $L(\eta, \Delta\eta)$ are statistically independent, identically distributed random variables, it follows from the generalized central limit theorem (see Sect. 7.7) that the class of limit distributions should be Lévy stable distributions.

In Fig. 10.55, the cumulative absolute frequency distribution is plotted [227] for the running sample variance

$$S_m^2(\Delta\eta) = \frac{1}{m-1} \sum_{i=1}^{m} [L(\eta_i, \Delta\eta) - \bar{L}_m(\eta, \Delta\eta)]^2 \tag{10.93}$$

with respect to the running sample mean $\bar{L}_m$ and ordering number m running over the η bins of size 0.1 units of the corresponding JACEE event (full symbols). The approximate straight-line behavior suggests scale invariance (i.e. scaling) of the data. In contrast, replacing the experimental L values by Gaussian random variables with zero population mean and unit population variance results in the open squares in Fig. 10.55, which clearly approach the line $S_m^2 \to \sigma^2 = 1$, i.e. the scale set by the population variance.

In addition, stationarity is suggested from the similarity of the behavior in the two sub-figures.

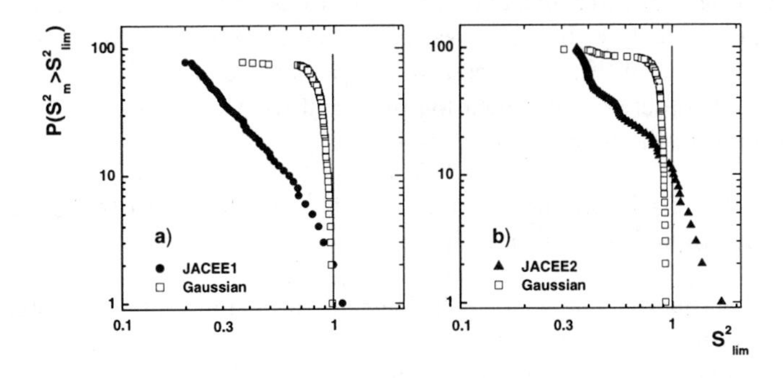

Figure 10.55. Cumulative absolute frequency distribution for the S^2_m for two JACEE events in sub-figures a) and b), respectively (full symbols), compared to that of Gaussian random samples of zero population mean and unit population variance (open symbols) [227].

10.12 Summary and conclusions

1. Intermittency, defined as an increase of normalized factorial moments with increasing resolution in phase space, is seen in all types of collisions. Intermittency is a 3D phenomenon. The anomalous dimensions are small (d_q=0.01-0.1) in a one-dimensional projection, but the factorial moments are considerably larger, and their resolution dependence more power-like, in two- or three-dimensional phase space.

2. The effect is stronger in e^+e^- collisions than in hadron-hadron collisions, and stronger in the latter than in heavy-ion collisions.

3. The generally used Monte Carlo models (DPM, FRITIOF, PYTHIA, HERWIG) strongly underestimate the effect, in lh, hh and AA collisions. Due to parton showering, the situation is better in e^+e^- collisions, but not perfect and its origin different in JETSET and HERWIG.

4. The factorial-moment method is very sensitive to biases in the data. These have to be studied in detail before final conclusions can be drawn. Because of its sensitivity, the method has in fact proven to be very helpful in detecting and tracing such biases.

5. The logarithms of factorial moments of different order satisfy a possibly dimension-independent linear inter-relation which allows to determine directly ratios of anomalous dimensions. The observed order-dependence of anomalous dimensions is multifractal and, therefore, excludes a second-order phase transition

as the origin of intermittency, but indications exist for coexistence of different phases, possibly (but not necessarily) with a non-thermal phase transition. While the full multifractal spectrum can be obtained from factorial moments of continuous order, a first indication of a multifractal structure can be obtained directly from the bunching parameters.

6. The multiparticle production dynamics is self-similar in e^+e^- collisions, but self-affine in hadron-hadron collisions.

7. In hadron-hadron and e^+e^- collisions, factorial moments and intermittency indices depend strongly on transverse momentum and are larger for low transverse momentum hadrons ($p_T < 0.3$ GeV/c) than for intermediate values ($p_T \approx 0.4 - 0.5$ GeV/c), thus indicating some soft hadron contribution. For larger p_T, the moments (and indices) show an increase again, with a saturation at large M, indicating a hard gluon contribution. The influence from pure kinematics (phase space cuts) will have to be studied more carefully.

8. The multiplicity dependence in hh collisions agrees with what is expected from mixing of independent sources. However, the observation on possible universality casts some doubt on this type of interpretation. For a given density, heavy-ion collisions show more intermittency than extrapolated from hadron-hadron collisions, possibly as a result of Bose-Einstein interference or other collective effects.

9. Factorial cumulants are direct measures of genuine higher-order fluctuations and are studied up to order 5.

10. Factorial correlators reveal bin-bin correlations. The correlators F_{pq} increase with decreasing correlation length D, but only approximately follow a power law for $D \lesssim 1$. For fixed D, the values of F_{pq} are independent of resolution δy, a property predicted in the α-model, but also shared by models with short-range order. The powers ϕ_{pq} increase linearly with the product pq of the orders, but are considerably larger than expected from the simple α-model and from FRITIOF.

11. The correlation (or density) strip-integral strongly reduces statistical errors, as well as artificial fluctuations due to splitting of spikes. Using the squared four-momentum difference Q^2_{ij} as a distance measure, an increase similar to that in three-dimensional analyses is observed. At low Q^2 this increase is caused by correlations among like-charged particles. Bose-Einstein interference must contribute significantly to the intermittency effect but is not power behaved in its conventional approach. Power-law behavior in Bose-Einstein interferometry would imply a random superposition of "emission centres" with possibly fractal

properties. Parton avalanches in a self-organized critical state are an intriguing possibility.

12. QCD is inherently intermittent and leads to fractal structure of the parton shower, with the fractal dimensions given by the QCD anomalous dimension. Saturation is however expected at small phase space distances due to the running of the coupling constant α_s (and therefore of the anomalous dimension). Analytical calculations on the *parton level* clearly show these features, but cannot reproduce the data (which are of course on the *hadron level*) in detail, in particular when the (transverse) momentum behavior of the moments is considered. Since QCD inspired models, in particular ARIADNE, reproduce the data well, one is inclined to conclude that local parton-hadron duality (LPHD) has reached (and crossed) its limits in the description of fluctuations. A great challenge, considering the high-virtuality jets to come from LHC.

13. Even though originally stimulated by individual events (JACEE event, NA22 event), the fluctuation analysis of single events was so far hampered by the relatively low multiplicity of these events. Nevertheless, a number of promising techniques have been developed and tested, as erraticity (particularly, useful for high multiplicity as in heavy ion events), void analysis (better suited for lower multiplicities), entropy (most sensitive to the maximum of the probability distribution), or Lévy stability (at first instant no t worrying about statistical fluctuations at all). A great future challenge, considering the high-multiplicity events to come from LHC!

Bibliography

[1] T. Ludlam and R. Slansky, *Phys. Rev.* **D8** (1973) 1408; J. Hanlon et al. *Phys. Lett.* **46B** (1973) 415 .

[2] R. Slansky , *Phys. Rep.* **11C** (1974) 99.

[3] V.A. Abramovskii, V.N. Gribov and O.V. Kancheli, *Yad. Fiz.* **18** (1973) 595; *Sov. J. Nucl. Phys.* **18** (1974) 308.

[4] K.I. Aleksejava et al., *J. Phys. Soc. Japan* **17** (1962) A-III, 409; D.H. Perkins and P.H. Fowler, *Proc. R. Soc.* **A273** (1964) 401; J. Iwai et al., *Nuovo Cimento* **69A** (1982) 295.

[5] N. Arata, *Nuovo Cimento* **43A** (1978) 455.

[6] A.V. Apanasenko et al., *JETP Lett.* **30** (1979) 145; I.M. Dremin and M.I. Tretyakova, paper presented at the XVIII Int. Cosmic Ray Conference, Paris (1981); S.A. Slavatinskii et al., paper presented at XVIII Int. Cosmic Ray Conference, Paris (1981); S.A. Azimov et al., paper presented at XVIII Int. Cosmic Ray Conference, Paris (1981).

[7] N.A. Marutjan et al., *Sov. J. Nucl. Phys.* **29** (1979) 804.

[8] JACEE Coll., T.H. Burnett et al., *Phys. Rev. Lett.* **50** (1983) 2062.

[9] NA22 Coll., M. Adamus et al., *Phys. Lett.* **B185** (1987) 200.

[10] UA5 Coll., P. Carlson, Proc. 4th Topical Workshop on $p\bar{p}$ Collider Physics, Bern, March 1984, CERN Yellow Report 84-09 (1984) 286 and G.J. Alner et al., *Phys. Rep.* **154** (1987) 247.

[11] EMU-01 Coll., M.I. Adamovich et al., *Phys. Lett.* **B201** (1988) 397 .

[12] F. Takagi, *Phys. Rev. Lett.* **53** (1984) 427; idem *Phys. Rev.* **C32** (1985) 1799.

[13] A. Białas and R. Peschanski, *Nucl. Phys.* **B273** (1986) 703 .

[14] A. Białas and R. Peschanski, *Nucl. Phys.* **B308** (1988) 857; A. Białas, *Multiparticle Dynamics*, Festschrift Van Hove, eds. A. Giovannini and W. Kittel (World Scientific, Singapore, 1990) p.75 .

[15] R.P. Feynman, Proc. 3rd Workshop on Current Problems in High Energy Particle Theory, Florence 1979, eds. R. Casalbuoni et al. (Johns Hopkins University Press, Baltimore, 1979).

[16] G. Veneziano, Proc. 3rd Workshop on Current Problems in High Energy Particle Theory, Florence 1979, eds. R. Casalbuoni et al. (Johns Hopkins University Press, Baltimore, 1979) p.45.

[17] K. Konishi, A. Ukawa and G. Veneziano, *Phys. Lett.* **78B** (1978) 243; *Nucl. Phys.* **B157** (1979) 45.

[18] A. Giovannini, Proc. Xth Int. Symp. on Multiparticle Dynamics, Goa 1979, eds. S.N. Ganguli, P.K. Malhotra and A. Subramanian (Tata Inst.) p. 364.

[19] P.Carruthers and Minh Duong-Van, *Evidence for a common fractal dimensionin turbulence, galaxy distributions and hadronic multiparticle production*, Los Alamos preprint LA-UR-83-2419, 1983.

[20] I.M. Dremin, *JETP Lett.* **45** (1987) 643.

[21] I.M. Dremin, Festschrift L. Van Hove, eds. A. Giovannini and W. Kittel (World Scientific, Singapore, 1990) p. 455.

[22] P. Bożek, M. Płoszajczak and R. Botet, *Phys. Reports* **252** (1995) 101.

[23] E.A. De Wolf, I.M. Dremin and W. Kittel, *Phys. Reports* **270** (1996) 1.

[24] H.G.E. Hentschel and I. Procaccia, *Physica* **8D** (1983) 435;
P. Grassberger, *Phys. Lett.* **97A** (1983) 227.

[25] P. Lipa and B. Buschbeck, *Phys. Lett.* **B223** (1989) 465.

[26] R.C. Hwa, *Phys. Rev.* **D41** (1990) 1456 .

[27] Ya.B. Zeldovich, A.A. Ruzmaikin and D.D. Sokoloff, *The Almighty Chance*, World Scientific Lecture Notes in Physics, Vol. 20 (World Scientific, Singapore, 1990).

[28] B. Buschbeck, P. Lipa and R. Peschanski, *Phys. Lett.* **B215** (1988) 788;
HRS Coll., S. Abachi et al., *Study of Intermittency in e^+e^- Annihilations at 29 GeV*, preprint ANK-HEP-CP-90-50. (unpublished).

[29] TASSO Coll., W. Braunschweig et al., *Phys. Lett.* **B231** (1989) 548.

[30] CELLO Coll., H.-J. Behrend et al., *Phys. Lett.* **B256** (1991) 97; P. Podobrin, Proc. 26th Rencontre de Moriond, Les Arcs 1991, ed. J. Tran Thanh Van (Editions Frontières 1991) p.311.

[31] DELPHI Coll., P. Abreu et al., *Phys. Lett.* **B247** (1990) 137; A. De Angelis, *Mod. Phys. Lett.* **A5** (1990) 2395; P. Abreu et al., *Nucl. Phys.* **B386** (1992) 471 .

[32] OPAL Coll., M.Z. Akrawy et al., *Phys. Lett.* **B262** (1991) 351; G. Abbiendi et al., *Eur. Phys. J.* **C11** (1999) 239.

[33] ALEPH Coll., D. Decamp et al., *Z. Phys.* **C53** (1992) 21; N. Lieske, Proc. Ringberg Workshop on Multiparticle Production, Ringberg Castle, Germany 1991, eds. R.C. Hwa, W. Ochs and N. Schmitz (World Scientific, Singapore, 1992) p. 32; V. Raab, idem p. 46; R. Barate et al., *Phys. Rep.* **294** (1997) 1.

[34] L3 Coll., B. Adeva et al., Z. Phys. **C55** (1992) 39; M. Acciarri et al., Phys. Lett. **B429** (1998) 375; S. Chekanov, Ph.D. Thesis, Univ. of Nijmegen, 1997.

[35] Mark II Coll., W.N. Murray, R.E. Frey and H.O. Ogren, *Search for Intermittency in e^+e^- Collisions at $\sqrt{s} = 29$ and 91* GeV, Oregon preprint OREXP-93-16.

[36] SLD Coll., J. Zhou, Ph.D. Thesis, SLAC-R-496 (1996).

[37] EMC Coll., I. Derado, G. Jancso, N. Schmitz and P. Stopa, *Z. Phys.* **C47** (1990) 23; I. Derado et al., *Z. Phys.* **C54** (1992) 357.

[38] WA59 and E180 Coll., L. Verluyten et al., *Phys. Lett.* **B260** (1991) 456 .

[39] NA22 Coll., I.V. Ajinenko et al., *Phys. Lett.* **B222** (1989) 306 and **B235** (1990) 373; N.M. Agababyan et al., *Phys. Lett.* **B261** (1991) 165; F. Botterweck et al., Proc. Ringberg Workshop on Multiparticle Production, Ringberg Castle, Germany 1991, eds. R.C. Hwa, W. Ochs and N. Schmitz (World Scientific, Singapore, 1992) p.125; M. Charlet, idem p.140; N. Agababyan et al., *Z. Phys.* **C59** (1993) 405.

[40] UA1 Coll., C. Albajar et al., *Nucl. Phys.* **B345** (1990) 1; B. Buschbeck, Festschrift L. Van Hove, eds. A. Giovannini and W. Kittel (World Scientific, Singapore, 1990)p. 211; P. Lipa, Ph.D. Thesis, Univ. of Vienna 1990; P. Lipa et al., Proc. Ringberg Workshop on Multiparticle Production, Ringberg Castle, Germany 1991, eds. R.C. Hwa, W. Ochs and N. Schmitz (World Scientific, Singapore, 1992) p. 111; N. Neumeister et al., *Z. Phys.* **C60** (1993) 633; *Acta Phys. Slovaca* **44** (1994) 113.

[41] NA23 Coll., J.B. Singh and J.N. Kohli, *Phys. Lett.* **B261** (1991) 160.

[42] IHSC Coll., V. Arena et al., *Nuovo Cimento* **A105** (1992) 883.

[43] Mirabelle Coll., L.V. Bravina et al., *Comparison of the Factorial Moments in* $\mathrm{p\bar{p}}$ *and* pp *Interactions at* 32 GeV/c, Moscow preprint 92-37/286 (1992).

[44] CDF Coll., F. Rimondi, *Intermittency Studies in* $\mathrm{p\bar{p}}$ *Collisions at* $\sqrt{s} = 1800$ GeV, Proc. XXI Int. Symp. on Multiparticle Dynamics, Wuhan, China, 1991, eds. Y.F. Wu and L.S. Liu (World Scientific, Singapore, 1992)p.476.

[45] S.S. Wang et al., *Phys. Rev.* **D49** (1994) 5785;
Y.X. Ye et al., *Phys. Rev.* **D55** (1997) 5641.

[46] KLM Coll., R. Hołyński et al., *Phys. Rev. Lett.* **62** (1989) 733 and *Phys. Rev.* **C40** (1989) R2449.

[47] NA35 Coll., I. Derado, Festschrift L. Van Hove, eds. A. Giovannini and W. Kittel (World Scientific, Singapore, 1990)p. 257; I. Derado, Proc. Ringberg Workshop on Multiparticle Production, Ringberg Castle, Germany 1991, eds. R.C. Hwa, W. Ochs and N. Schmitz (World Scientific, Singapore, 1992) p.184; J. Bächler et al., *Z. Phys.* **C56** (1992) 347; **C61**(1994) 551.

[48] NA22 Coll., F. Botterweck et al., *Z. Phys.* **C51** (1991) 37.

[49] D. Ghosh et al., *Phys. Rev.* **D46** (1992) 3712; ibid. **D47** (1993) 1235; ibid.**D49** (1994) 3113; ibid. **D51** (1995) 3298; *Europhys. Lett.* **29** (1995) 521.

[50] R.K. Shivpuri and V.K. Verma, *Phys. Rev.* **D47** (1993) 123; R.K. Shivpuri and Vandana Anand, *Z. Phys.* **C59** (1993) 47 and *Phys. Rev.* **D50** (1994) 287.

[51] R.K. Shivpuri and N. Parashar, *Phys. Rev.* **D49** (1994) 219.

[52] D.K. Maity et al., *Mod. Phys. Lett.* **A40** (1993) 3853; D.K. Maity, P.K. Bandyopadhyay, D.K. Bhattacharjee, *Z. Phys.* **C65** (1995) 75.

[53] EMU-01 Coll., M.I. Adamovich et al., *Phys. Rev. Lett.* **65** (1990) 412;
Phys. Lett. **B263** (1991) 539; *Z. Phys.* **C49** (1991) 395;
Nucl. Phys. **B388** (1992) 3; *Phys. Lett.* **B407** (1997) 92.

[54] EMU-08 Coll., K. Sengupta, P.L. Jain, G. Singh and S.N. Kim, *Phys. Lett.* **B236** (1990) 219; P.L. Jain and G. Singh, *Phys. Rev.* **C44** (1991) 854; *Z. Phys.* **C53** (1992) 355; P.L. Jain, G. Singh and A. Mukhopadhyay, *Phys. Rev.* **C48** (1993) R517; P.L. Jain and G. Singh, *Nucl. Phys.* **A596** (1996) 700.

[55] HELIOS Coll., T. Åkesson et al., *Phys. Lett.* **B252** (1990) 303.

[56] D. Ghosh et al., *Phys. Lett.* **B272** (1991) 5.

[57] E.K. Sarkisyan and G.G. Taran, *Phys. Lett.* **B279** (1992) 177; E.K. Sarkisyan, L.K. Gelovani, G.G. Taran and G.I. Sakharov, *Phys. Lett.* **B318** (1993) 568; E.K. Sarkisyan, L.K. Gelovani, G.L. Gogiberidze and G.G. Taran, *Phys. Lett.* **B347** (1995) 439.

[58] E-802 Coll., T. Abbott et al., *Phys. Lett.* **B337** (1994) 254; *Phys. Rev.* **C52** (1995) 2663.

[59] WA80 Coll., R. Albrecht et al., *Phys. Rev.* **C50** (1994) 1048.

[60] KLMM Coll., B. Wosiek et al., Proc. XXVth Int. Symp. on Multiparticle Dynamics, Stara Lesná 1995, eds. D. Bruncko, L. Šándor and J. Urbán (World Scientific, Singapore, 1996) p.271.

[61] R. Hasan and M.S. Ahmad, *Intermittency and Multifractality in* ^{12}CEm *Collisions*, preprint (1995).

[62] M. Płoszajczak and A. Tucholski, *Phys. Rev. Lett.* **65** (1990) 1539.

[63] P.L. Jain, G. Sing and M.S. El-Nagdy, *Phys. Rev. Lett.* **68** (1992) 1656; *Europhys. Lett.* **21** (1993) 527; *Mod. Phys. Lett.* **A7** (1992) 1113; P.L. Jain and G. Singh, *Phys. Rev.* **C46** (1992) R10; *Nucl. Phys.* **A591** (1995) 711.

[64] KLMM Coll., M.L. Cherry et al., *Phys. Rev.* **C53** (1996) 1532.

[65] D. Ghosh et al., *Z. Phys.* **C73** (1997) 269.

[66] A. Białas, *Nucl. Phys.* **A525** (1991) 345c.

[67] UA5 Coll., G.J. Alner et al., *Nucl. Phys.* **B291** (1987) 445.

[68] F.W. Bopp, A. Capella, J. Ranft and J. Tran Thanh Van, *Z. Phys.* **C51** (1991) 99.

[69] R.C. Hwa and J.C. Pan, *Phys. Rev.* **D45** (1992) 106; J.C. Pan and R.C. Hwa, *Phys. Rev.* **D46** (1992) 4890.

[70] K. Werner, *Phys. Rev.* **D39** (1989) 780.

[71] A. Capella, K. Fiałkowski and A. Krzywicki, *Phys. Lett.* **230B** (1989) 149.

[72] A. Białas, *Nucl. Phys.* **A545** (1992) 285; *Acta Phys. Pol.* **B23** (1992) 561.

[73] F. Botterweck, private communication; B. Buschbeck and P. Lipa, private communication.

[74] M. Jędrzejczak, *Phys. Lett.* **B228** (1989) 259; I. Sarcevic, private communication.

[75] NA22 Coll., N.M. Agababyan et al., *Z. Phys.* **C46** (1990) 387.

[76] EMC Coll., M. Arneodo et al., *Z. Phys.* **C33** (1986) 167.

[77] W.D. Walker et al., *Phys. Lett.* **B255** (1991) 155.

[78] B. Andersson, G. Gustafson and J. Samuelsson, *Z. Phys.* **C64** (1994) 653.

[79] P. Dahlqvist, G. Gustafson and B. Andersson, *Nucl. Phys.* **B328** (1989) 76; G. Gustafson and C. Sjögren, *Phys. Lett.* **B248** (1990) 430; B. Andersson, G. Gustafson, A. Nilsson and C. Sjögren, *Z. Phys.* **C49** (1991) 79; G. Gustafson and A. Nilsson, *Nucl. Phys.* **B355** (1991) 106 .

[80] K. Fiałkowski, B. Wošiek and J. Wošiek, *Acta Phys. Pol.* **B20** (1989) 639 .

[81] Chan Gang and Liu Lianshou, *Phys. Rev.* **D65** (2002) 094030.

[82] B. Wošiek, *Intermittency Analysis of Correlated Data*, Krakow preprint INP 1496/PH (1990); G. Ekspong, R. Peschanski and J. Wošiek, *Phys. Lett.* **B251** (1990) 455; D. Seibert, Analysis of Correlated Data, preprint (1990).

[83] E.M. Friedlander, *Mod. Phys. Lett.* **A4** (1989) 2457; P. Lipa, H.C. Eggers, F. Botterweck and M. Charlet, *Z. Phys.* **C54** (1992) 115.

[84] A. Białas and M. Gazdzicki, *Phys. Lett.* **B252** (1990) 483.

[85] Liu Lianshou, Zhang Yang, Deng Yue, *Z. Phys.* **C73** (1997) 535.

[86] M.J. Tannenbaum, *Phys. Lett.* **B347** (1995) 431.

[87] W. Ochs, *Phys. Lett.* **B247** (1990) 101; *Z. Phys.* **C50** (1991) 339.

[88] A. Białas and J. Seixas, *Phys. Lett.* **B250** (1990) 161.

[89] P. Bożek and M. Płoszajczak, *Phys. Lett.* **B251** (1990) 623.

[90] F. Botterweck, Ph.D. thesis, Univ. of Nijmegen, 1992.

[91] K. Fiałkowski, *Phys. Lett.* **B272** (1991) 139.

[92] K. Fiałkowski, *Z. Phys.* **C61** (1994) 313.

[93] W. Ochs and J. Wošiek, *Phys. Lett.* **B214** (1988) 617; **B232** (1989) 271.

[94] A. De Angelis, P. Lipa, W. Ochs, Proc. Joint Int. Lepton-Photon Symposium and Europhysics Conference on High Energy Physics, eds. S. Hegarty, K. Potter and E. Quercigh (World Scientific, Singapore, 1992) v.1, p.724.

[95] J. Alberty and A. Białas, *Z. Phys.* **C50** (1991) 315.

[96] Ph. Brax and R. Peschanski, *Phys. Lett.* **B253** (1991) 225.

[97] S.V. Chekanov and V.I. Kuvshinov, *Acta Phys. Pol.* **B25** (1994) 1189.

[98] H. Satz, *Nucl. Phys.* **B326** (1989) 613; B. Bambah, J. Fingberg and H. Satz, *Nucl. Phys.* **B332** (1990) 629.

[99] A. Białas and R. Hwa, *Phys. Lett.* **B253** (1991) 436.

[100] S.P. Ratti et al., *Z. Phys.* **C61** (1994) 229.

[101] IHSC Coll., S.P. Ratti et al., *Universal Multifractal Analysis of Multiparticle Production in h-h Collisions at* $\sqrt{s} = 16.7$ *GeV*, Proc. XXI Int. Symp. on Multiparticle Dynamics, Wuhan, China, 1991, eds. Y.F. Wu and L.S. Liu (World Scientific, Singapore, 1992)p.409; V. Arena et al., *Nuovo Cim.* **108A** (1995) 417.

[102] P. Carruthers and I. Sarcevic, *Phys. Rev. Lett.* **63** (1989) 1562.

[103] E.A. De Wolf, *Acta Phys. Pol.* **B21** (1990) 611 .

[104] P.J.F. Peebles, *The Large Scale Structure of the Universe* (Princeton, NJ, 1980).

[105] E.A. De Wolf, Proc. XXII Int. Symp. on Multiparticle Dynamics, Santiago de Compostela, Spain, 1992, ed. C. Pajares (World Scientific, Singapore, 1993) p. 263 .

[106] P. Carruthers, I. Sarcevic, *Phys. Lett.* **B189** (1987) 442.

[107] N.G. Antoniou, A.P. Contogouris, C.G. Papadopoulos, S.D.P. Vlassopulos, *Phys. Lett.* **B245** (1990) 619; *Phys. Rev.* **D45** (1992) 4034; N.G. Antoniou, E.N. Argyres, C.G. Papadopoulos, S.D.P. Vlassopulos, *Phys. Lett.* **B260** (1991) 199; N.G. Antoniou, F.K. Diakonos, I.S. Mistakidis, *Phys. Lett.* **B293** (1992) 187; N.G. Antoniou, F.K. Diakonos, I.S. Mistakidis, C.G. Papadopoulos, *Phys. Rev.* **D49** (1994) 5789; N.G. Antoniou, F.K. Diakonos, C.N. Ktorides, M. Lahanas, *Phys. Lett.* **B432** (1998) 8.

[108] I.M. Dremin, M.T. Nazirov, *Sov. J. Nucl. Phys.* **55** (1992) 197 and 2546.

[109] L.E. Reichl, *A Modern Course in Statistical Physics* (Univ. of Texas Press, Austin, 1980).

[110] J. Wošiek, *Acta Phys. Pol.* **B19** (1988) 863.

[111] R.C. Hwa and M.T. Nazirov, *Phys. Rev. Lett.* **69** (1992) 741; R.C. Hwa and L. Lesniak, *Phys. Lett.* **B295** (1992) 35; R.C. Hwa, *Phys. Rev.* **D47** (1993) 2773 and **C50** (1994) 383.

[112] M.R. Young, Y. Qu, S. Singh and R.C. Hwa, *Opt. Commun.* **105** (1994) 325.

[113] M. Charlet, private communication.

[114] I. Golyak, T. André, *Z. Phys.* **C74** (1997) 275.

[115] I. Golyak, *Z. Phys.* **C57** (1993) 421; *J. Phys.* **G20** (1994) 565; I. Golyak, S. Galayda, *Z. Phys.* **C70** (1996) 227.

[116] R.C. Hwa, *Phys. Rev.* **D57** (1998) 1831.

[117] L.F. Babichev, D.V. Klenitsky and V.I. Kuvshinov, *Phys. Lett.* **B345** (1995) 269.

[118] L. Van Hove, *Ann. Phys.* **192** (1989) 66.

[119] R. Peschanski, *Nucl. Phys.* **B327** (1989) 144; *Int. J. Mod. Phys.* **A6** (1991) 3681; Ph. Brax and R. Peschanski, *Nucl. Phys.* **B346** (1990) 65 and *Int. J. Mod. Phys.* **A7** (1993) 709.

[120] A. Białas and K. Zalewski, *Phys. Lett.* **B238** (1990) 413.

[121] D. Ghosh et al., *Phys. Rev.* **C59** (1999) 2286.

[122] Wang Shaoshun, Liu Ran, Wang Zhaomin, *Phys. Lett.* **B438** (1998) 353.

[123] Y.F. Wu and L.S. Liu, *Phys. Rev. Lett.* **70** (1993) 3197 and *Science in China* **A38** (1995) 435; Y.F. Wu, Y. Zhang and L.S. Liu, *Phys. Rev.* **D51** (1995) 6576; Liu Feng, Liu Fuming and Liu Lianshou, *Phys. Rev.* **D59** (1999) 114020.

[124] J. Wošiek, Proc. XXIV Int. Symp. on Multiparticle Dynamics, eds. A. Giovannini et al. (World Scientific, Singapore, 1995) p.99.

[125] L. Van Hove, *Phys. Lett.* **28B** (1969) 429; *Nucl. Phys.* **B9** (1969) 331.

[126] B.B. Mandelbrot, in *Dynamics of Fractal Surfaces*, eds. E. Family and T. Vicsek (World Scientific, Singapore, 1991).

[127] NA22 Coll., N.M. Agababyan et al., *Phys. Lett.* **B382** (1996) 305; **B431** (1998) 451.

[128] Wang Shaoshun, Wang Zhaomin, Wu Chong, *Phys. Lett.* **B410** (1997) 323.

[129] Gang Chen, Lian-shou Liu, Yan-min Gao, *Int. J. Mod. Phys.* **A14** (1999) 3687.

[130] L3 Coll., *Measurement of the Scaling Property of the Factorial Moments in Hadronic Z Decay*, L3 Note 2758, ABS 494, ICHEP02.

[131] Yu.L. Dokshitzer, *J. Phys.* **G17** (1991) 1537; S. Bethke, Z. Kunszt, D.E. Soper and W.J. Stirling, *Nucl. Phys.* **B370** (1992) 310.

[132] Liu Lianshou, Fu Jinghua and Wu Yuanfang, *Phys. Lett.* **B444** (1998) 563.

[133] Y. Zhang, L.S. Liu and Y.F. Wu, *Z. Phys.* **C71** (1996) 499.

[134] Liu Lianshou, Hu Yuan and Deng Yue, *Phys. Lett.* **B388** (1996) 10.

[135] EMU-01 Coll., M.I. Adamovich et al., *Z. Phys.* **C76** (1997) 659.

[136] P. Carruthers, E.M. Friedlander, C.C. Shih and R.M. Weiner,*Phys. Lett.* **B222** (1989) 487.

[137] M. Gyulassy, Festschrift L. Van Hove, eds. A. Giovannini and W. Kittel (World Scientific, Singapore, 1990)p.479.

[138] M. Biyajima et al., *Prog. Theor. Phys.* **84** (1990) 931.

[139] UA1 Coll., N. Neumeister et al., *Phys. Lett.* **B275** (1992) 186.

[140] NA22 Coll., N.M. Agababyan et al., *Z. Phys.* **C68** (1995) 229.

[141] DELPHI Coll., P. Abreu et al., *Phys. Lett.* **B355** (1995) 415.

[142] EMC Coll., I. Derado, G. Jancso and N. Schmitz, *Z. Phys.* **C56** (1992) 553; M. Arneodo et al., *Z. Phys.* **C31** (1986) 333; **40** (1988) 347.

[143] NA22 Coll., I.V. Ajinenko et al., *Phys. Lett.* **B197** (1987) 457.

[144] A. Białas et al., *Phys. Lett.* **B229** (1989) 398.

[145] UA1 Coll., Y.F. Wu et al., *Acta Phys. Slov.* **44** (1994) 141; Proc. Cracow Workshop on Multiparticle Production, eds. A. Białas et al. (World Scientific, Singapore, 1994) p.22.

[146] P. Bożek and M. Płoszajczak, *Z. Phys.* **C56** (1992) 473.

[147] Y.F. Wu and L.S. Liu, *Z. Phys.* **C53** (1992) 273.

[148] JADE Coll., W. Bartel et al., *Z. Phys.* **C33** (1986) 23; S. Bethke et al., *Phys. Lett.* **B213** (1988) 235.

[149] A. Białas, Festschrift L. Van Hove, eds. A. Giovannini and W. Kittel (World Scientific, Singapore, 1990) p. 75.

[150] D. Seibert, *Phys. Lett.* **B240** (1990) 215 .

[151] P. Aurenche et al., *Phys. Rev.* **D45** (1992) 92; F.W. Bopp et al., Proc. Ringberg Workshop on Multiparticle Production, Ringberg Castle, Germany 1991, eds. R.C. Hwa, W. Ochs and N. Schmitz (World Scientific, Singapore, 1992) p. 313.

[152] S. Barshay, *Z. Phys.* **C47** (1990) 199.

[153] A.H. Mueller, *Phys. Rev.* **D4** (1971) 150 .

[154] P. Carruthers, H.C. Eggers and I. Sarcevic, *Phys. Lett.* **B254** (1991) 258.

[155] NA27 Coll., S.S. Wang et al., *Phys. Lett.* **B321** (1994) 431.

[156] H.C. Eggers, Ph.D. Thesis, University of Arizona 1991; T. Elze and I. Sarcevic, *Phys. Rev. Lett.* **68** (1992) 1988; I. Sarcevic, Proc. Ringberg Workshop on Multiparticle Production, Ringberg Castle, Germany 1991, eds. R.C. Hwa, W. Ochs and N. Schmitz (World Scientific, Singapore, 1992) p. 206.

[157] P.L. Jain, A. Mukhopadhyay and G. Singh, *Z. Phys.* **C58** (1993) 1.

[158] EMU-01 Coll., M.I. Adamovich et al., *Phys. Rev.* **D47** (1993) 3726.

[159] J. Dias de Deus, *Phys. Lett.* **B240** (1990) 481; *Phys. Lett.* **B278** (1992) 377.

[160] A. Giovannini and L. Van Hove, *Z. Phys.* **C30** (1986) 391.

[161] K. Fiałkowski, W. Ochs and I. Sarcevic, *Z. Phys.* **C54** (1992) 621.

[162] E.K.G. Sarkisyan, *Phys. Lett.* **B477** (2000) 1.

[163] OPAL Coll., G. Abbiendi et al., *Phys. Lett.* **B523** (2001) 35.

[164] NA22 Coll., V.V. Aivazyan et al., *Phys. Lett.* **B258** (1991) 487.

[165] HELIOS, paper submitted to the EPS Conf. on High Energy Physics, Madrid 1989.

[166] D. Ghosh et al., *Phys. Rev.* **C52** (1995) 2092.

[167] G. Singh and P.L. Jain, Factorial correlators from ^{197}Au-emulsion collisions at 10.6 A GeV, SU9726 (1997).

[168] E.A. De Wolf, Proc. Ringberg Workshop on Multiparticle Production, Ringberg Castle, Germany 1991, eds. R.C. Hwa, W. Ochs and N. Schmitz (World Scientific, Singapore, 1992) p. 222 .

[169] H.C. Eggers, P. Carruthers, P. Lipa and I. Sarcevic, *Phys. Rev.* **D44** (1991) 1975 .

[170] R. Peschanski and J. Seixas, Scaling relations between fluctuations and correlations in multiparticle production, CERN-TH-5903-90; J. Seixas, Proc. XXVI Recontres de Moriond, ed. J. Tran Thanh Van (Editiones de Frontières, 1991) p.335.

[171] B. Widom, *J. Chem. Phys.* **43** (1965) 3892, 3898; K.G. Wilson, *Phys. Rev.* **179** (1969) 1499.

[172] L. Kadanoff, *Physica* **D38** (1989) 213.

[173] B. Mandelbrot, *J. Fluid Mech.* **62** (1974) 331.

[174] R.C. Hwa, *Phys. Rev.* **D41** (1990) 1456; R.C. Hwa and J. Pan, *Phys. Rev.* **D45** (1992) 1476.

[175] I.M. Dremin, *Phys. Uspekhi* **37** (1994) 715; P. Duclos and J.-P. Meunier, *Z. Phys.* **C51** (1994) 295; R.C. Hwa, *Phys. Rev.* **D51** (1995) 3323.

[176] Zhang Jie, Wang Shaoshun, *Phys. Lett.* **B370** (1996) 159; *Phys. Rev.* **D55** (1997) 1257.

[177] Wang Shaoshun, Wu Chong, *Phys. Lett.* **B473** (2000) 172; *Chin. Phys. Lett.* **18** (2001) 18.

[178] Hu Y. et al., *Chin. Phys. Lett.* **16** (1999) 553.

[179] L3 Coll., M. Acciari et al., *Phys. Lett.* **B429** (1998) 375.

[180] Wang Shaoshun, Liu Run, Wang Zhaomin, *Phys. Lett.* **B441** (1998) 473.

[181] G. Gustafson and A. Nilsson, *Z. Phys.* **C52** (1991) 533.

[182] W. Ochs and J. Wošiek, *Phys. Lett.* **B289** (1992) 159; ibid. **B305** (1993) 144; *Z. Phys.* **C68** (1995) 269.

[183] Yu.L. Dokshitzer and I.M. Dremin, *Nucl. Phys.* **B402** (1993) 139 .

[184] Ph. Brax, J.L. Meunier and R. Peschanski, *Z. Phys.* **C62** (1994) 649.

[185] I.M. Dremin, *Mod. Phys. Lett.* **A3** (1988) 1333.

[186] P. Carruthers, *Ap. J.* **380** (1991) 24; P. Lipa, P. Carruthers, H.C. Eggers and B. Buschbeck, *Phys. Lett.* **B285** (1992) 300.

[187] M. Charlet, PhD Thesis, Nijmegen, 1994.

[188] H.C. Eggers et al., *Phys. Rev.* **D48** (1993) 2040; M. Charlet, Proc. XXIIIrd Int. Symp. on Multparticle Dynamics, Aspen 1993, eds. M.M. Block and A.R. White (World Scientific, Singapore, 1994) p.302.

[189] Zhang Jie, *Phys. Lett.* **B352** (1995) 169.

[190] E665 Coll., M.R. Adams et al., *Phys. Lett.* **B335** (1994) 535.

[191] NA22 Coll., N.M. Agababyan et al., *Phys. Lett.* **B332** (1994) 458.

[192] DELPHI/UA1 Coll., F. Mandl and B. Buschbeck, *Correlation Integral Studies in Delphi and in UA1*, Proc. XXII Int. Symp. on Multiparticle Dynamics, Santiago de Compostela, Spain, 1992, ed. C. Pajares (World Scientific, Singapore, 1993), p.561.

[193] P. Bak, C. Tang and K. Wiesenfeld, *Phys. Rev. Lett.* **59** (1987) 381; P. Bak and K. Chen, *Scientific American* **264** (1991) 46.

[194] K. Kadija and P. Seyboth, *Phys. Lett.* **B287** (1992) 363.

[195] F. Kun et al., *Phys. Lett.* **B333** (1994) 233.

[196] A. Basetto, M. Ciaffaloni, G. Marchesini, *Phys. Rep.* **100** (1983) 202; Yu.L. Dokshitzer, V.A. Khoze, A.H. Mueller, S.I. Troyan, *Rev. Mod. Phys.* **60** (1988) 373.

[197] Ya.I. Azimov, Yu.I. Dokshitzer, V.A. Khoze, S.I. Troyan, *Z. Phys.* **C27** (1985) 65; *Z. Phys.* **C31** (1986) 231.

[198] Yu.L. Dokshitzer, V.S. Fadin, V.A. Khoze, *Z. Phys.* **C15** (1982) 325; ibid. **C18** (1983) 37.

[199] K. Konishi, A. Ukawa and G. Veneziano, *Nucl. Phys.* **B157** (1979) 45.

[200] B. I. Ermolaev and V. S. Fadin, *JETP Lett.* **33** (1981) 269; A. H. Mueller, *Phys. Lett.* **B104** (1981) 161; A. Bassetto, M. Ciafaloni, G. Marchesini and A. H. Mueller, *Nucl. Phys.* **B207** (1982) 189; G. Marchesini and B. R. Webber, *Nucl. Phys.* **B238** (1984) 1.

[201] DELPHI Coll., P. Abreu et al., *Phys. Lett.* **B440** (1998) 203.

[202] ZEUS Coll., J. Breitweg et al., *Eur. Phys. J.* **C12** (2000) 53; paper 802, submitted to ICHEP98, Vancouver, 1998.

[203] R. Peschanski and B. Ziaja, *Eur. Phys. J.* **C21** (2001) 649.

[204] L3 Coll., M. Acciarri et al., *Phys. Lett.* **B428** (1998) 186.

[205] DELPHI Coll., P. Abreu et al., *Phys. Lett.* **B457** (1999) 368.

[206] ZEUS Coll., S. Chekanov et al., *Phys. Lett.* **B510** (2001) 36.

[207] J.-L. Meunier and R. Peschanski, *Z. Phys.* **C72** (1996) 647; J.-L. Meunier, *Nucl. Phys.* **B** (Proc. Suppl.) **71** (1999) 238.

[208] S. Lupia, W. Ochs, J. Wosiek, *Nucl. Phys.* **B540** (1999) 405.

[209] A. Białas and B. Ziaja, *Phys. Lett.* **B378** (1996) 319.

[210] Z. Cao and R.C. Hwa, *Phys. Rev. Lett.* **75** (1995) 1268; *Phys. Rev.* **D53** (1996) 6608; ibid. **D54** (1996) 6674; R.C. Hwa, *Acta Phys. Pol.* **B27** (1996) 1789.

[211] G. Gianini, Proc. XXVI Int. Symp. on Multiparticle Dynamics, eds. J. Dias de Deus et al. (World Scientific, Singapore, 1997) p.202.

[212] Wang Shaoshun and Wang Zhaomin, *Phys. Lett.* **B416** (1998) 216; *Phys. Rev.* **D57** (1998) 3036; Wang Shaoshun, Wu Chong, Wang Zhaomin, *Phys. Lett.* **B458** (1999) 505.

[213] Z. Cao and R.C. Hwa, *Phys. Rev.* **D61** (2000) 074011.

[214] Z. Cao and R.C. Hwa, *Phys. Rev.* **D59** (1999) 114023.

[215] Liu Lianshou, Fu Jinghua, Wu Yuanfang, *Science in China* **A30** (2000) 432; Fu Jinghua, Wu Yuanfang, Liu Lianshou, *Phys. Lett.* **B472** (2000) 161; Liu Fuming et al., *Phys. Lett.* **B516** (2001) 293.

[216] NA22 Coll., M.R. Atayan et al., *Phys. Lett.* **B558** (2003) 22.

[217] D. Ghosh, A. Deb, M. Mondal, J. Ghosh, *Phys. Lett.* **B540** (2002) 52.

[218] R.C. Hwa and Qing-hui Zhang, *Phys. Rev.* **D62** (2000) 0140003.

[219] R.C. Hwa and Y. Wu, *Phys. Rev.* **D60** (1999) 097501.

[220] Liao Hongbo, Wu Yuanfang, *Int. J. Mod. Phys.* **A17** (2002) 4669.

[221] Wang Shaoshun, Wu Chong, *Phys. Lett.* **B505** (2001) 43.

[222] NA22 Coll., M.R. Atayan et al., *Phys. Lett.* **B558** (2003) 29.

[223] A. Białas, W. Czyż, J. Wosiek, *Acta Phys. Pol.* **B30** (1999) 107; A. Białas, W. Czyż, *Phys. Rev.* **D61** (2000) 074021; *Acta Phys. Pol.* **B31** (2000) 687.

[224] A. Rényi, *Acta Math. Sci. Hung.* **10** (1959) 193.

[225] K. Fiałkowski and R. Wit, *Phys. Rev.* **D62** (2000) 114016.

[226] B.B. Mandelbrot, *The Journal of Business* **36** (1963) 394 and **40** (1967) 393; *Fractals and Scaling in Finance* (Springer, 1997).

[227] Liu Qin and Meng Ta-chung, *Phys. Rev.* **D69** (2004) 054026.

Chapter 11

Bose-Einstein Correlations

As proposed by Hanbury Brown and Twiss [1] in 1954, the (angular) diameter of stars and radio sources in the universe was successfully determined by measuring the intensity (as opposed to Michelson or amplitude) correlations between separated telescopes. Likewise, in particle physics, one can in principle use Bose-Einstein intensity correlations between identical bosons to measure the space-time structure of the region from which the particles originate in a high-energy collision [2].

The first experimental evidence for Bose-Einstein correlations in particle physics dates back to 1959 when, in $p\bar{p}$ annihilation at 1.05 GeV/c, Goldhaber et al. [3] observed an enhancement at small relative angles in like-charged pion pairs not present for unlike-charged pairs. More recently, Bose-Einstein correlations have been exploited in hadron-hadron, hadron-nucleus, nucleus-nucleus, e^+e^- and lepton-hadron collisions to obtain surprisingly detailed information on the space-time development of particle production.

The recent revival of interest comes from various directions:
1. Their role in the phenomenon of intermittency discussed in Chapter 10.
2. Their application to determine the space-time development of a particle collision, to be discussed in this chapter, including the search for a long-lived source.
3. Their influence on the measurement of effective masses [4], in particular of the W mass at LEP2 to be discussed in Sect. 11.7 [5,6].
4. Their possible effect on the multiplicity distribution and on single-particle spectra [7–9], to be discussed in Sect. 11.8.

11.1 Pion interferometry

In Fig. 11.1 we illustrate the production of two identical pions with momenta $\mathbf{p}_1$ and $\mathbf{p}_2$, arising from two sources A and B with coordinates $\mathbf{x}_A$, $\mathbf{x}_B$. The pion wave functions can be written (e.g. [10]) as

$$\psi_{1A} = f_A e^{-i\mathbf{p}_1 \cdot \mathbf{x}_A} \quad , \quad \psi_{2B} = f_B e^{-i\mathbf{p}_2 \cdot \mathbf{x}_B} \tag{11.1}$$

if $\pi(\mathbf{p}_1)$ is emitted from source A and $\pi(\mathbf{p}_2)$ from source B, where f_A and f_B are arbitrary phase factors.

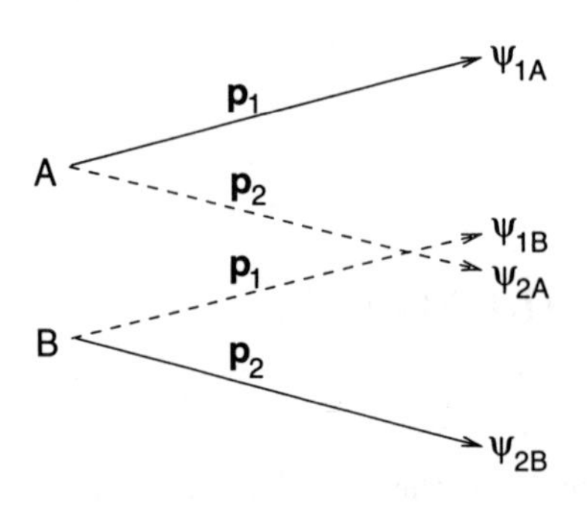

Figure 11.1. Emission of two identical bosons with momenta $\mathbf{p}_1, \mathbf{p}_2$ from two sources A, B.

If $\pi(\mathbf{p}_1)$ is emitted from source B and $\pi(\mathbf{p}_2)$ from source A, then indices A and B should be interchanged in (11.1). Since the two pions are identical bosons and the observer cannot decide from which source a particular pion was emitted, the complete amplitude for the emission of one pion with momentum $\mathbf{p}_1$ is

$$\psi_1 = f_A e^{-i\mathbf{p}_1 \cdot \mathbf{x}_A} + f_B e^{-i\mathbf{p}_1 \cdot \mathbf{x}_B} , \tag{11.2}$$

and correspondingly for the pion with momentum $\mathbf{p}_2$. The coincidence amplitude for simultaneous observation of two pions with momenta $\mathbf{p}_1$ and $\mathbf{p}_2$ then reads:

$$\psi_{12} = \psi_1\psi_2 = (f_A e^{-i\mathbf{p}_1 \cdot \mathbf{x}_A} + f_B e^{-i\mathbf{p}_1 \cdot \mathbf{x}_B})(f_A e^{-i\mathbf{p}_2 \cdot \mathbf{x}_A} + f_B e^{-i\mathbf{p}_2 \cdot \mathbf{x}_B}) . \tag{11.3}$$

The joint probability is

$$P_{12} = (f_A^2 + f_B^2 + [f_A^* f_B e^{i\mathbf{p}_1 \cdot (\mathbf{x}_A - \mathbf{x}_B)} + c.c.])(f_A^2 + f_B^2 + [f_A^* f_B e^{i\mathbf{p}_2 \cdot (\mathbf{x}_A - \mathbf{x}_B)} + c.c.]) , \tag{11.4}$$

in fact just the product of the probabilities for single pion emission. However, in a chaotic (often called incoherent) source the products $f_A^* f_B$ and *c.c.* fluctuate randomly and drop out in the expectation value, so that the density ratio becomes

$$R_2 = \frac{\langle P_{12} \rangle}{\langle P_1 \rangle \langle P_2 \rangle} = 1 + \frac{2 f_A^2 f_B^2}{(f_A^2 + f_B^2)^2} \cos(\Delta\mathbf{p} \cdot \Delta\mathbf{x}) , \tag{11.5}$$

where $\Delta\mathbf{p} = \mathbf{p}_1 - \mathbf{p}_2$ and $\Delta\mathbf{x} = \mathbf{x}_A - \mathbf{x}_B$. For two equal-strength sources, $R_2 \to 3/2$ for $\Delta\mathbf{p} \to 0$. For n sources, this is increased to $2 - 1/n$.

From (11.5) it follows that R_2 reaches a maximum value of 2 for $\Delta\mathbf{p} = 0$. Furthermore, it can be seen that the momentum difference $\Delta\mathbf{p}$ probes the source dimensions in a direction parallel to $\Delta\mathbf{p}$.

We shall, however, see later on (Sub-Sect. 11.3.5) that such a simple picture cannot be maintained due to correlation between $\mathbf{x}$ and $\mathbf{p}$ observed in the data and expected from hydrodynamical models as well as from string models.

11.1.1 The Lorentz invariant (Goldhaber) form

One step more realistic than the binary source considered in Fig. 11.1 is a source with a spherically symmetric Gaussian density distribution of emitting centres [3]

$$\rho(\mathbf{r}) \propto \exp\left[-\mathbf{r}^2/(2r_0^2)\right] , \tag{11.6}$$

which yields as Bose-Einstein ratio

$$R_2 = 1 + \exp\left[-r_0^2 \Delta \mathbf{p}^2\right] , \tag{11.7}$$

or, in its Lorentz-invariant form,

$$\begin{aligned} R_2(Q^2) &= 1 + \exp(-r_G^2 Q^2) \\ \text{with} \qquad & \\ Q^2 &= -(p_1 - p_2)^2 = M^2 - 4m_\pi^2 , \end{aligned} \tag{11.8}$$

where M is the invariant mass of the pion pair and m_π is the pion mass. This corresponds to a Gaussian shape of the source in the centre-of-mass system of the pair, where $q_0 \equiv \Delta E = 0$.

11.1.2 The Kopylov-Podgoretskiĭ parametrization

Another parametrization has been suggested by Kopylov and Podgoretskiĭ [2] in the framework of a simple model of a radiating spherical surface of radius r_K with incoherent pointlike oscillators of lifetime τ:

$$R_2(q_T, q_0) = 1 + \left[\frac{2J_1(r_K q_T)}{r_K q_T}\right]^2 \cdot \frac{1}{1 + (q_0 \tau)^2} , \tag{11.9}$$

where q_T is the transverse component of $\Delta \mathbf{p}$, i.e. $\mathbf{q}_T \perp (\mathbf{p}_1 + \mathbf{p}_2)$, and $q_0 = E_1 - E_2$, while J_1 is the first-order Bessel function. The parametrization is not Lorentz-invariant. In general, the variables are calculated in the centre of mass of the initial collision.

Due to the different assumptions on the shape of the source, the spatial dimensions r_K and r_G used in the two parametrizations have a different meaning, but are related via the exponential approximation of (11.9):

$$R_2 \approx 1 + \exp[-(r_K/2)^2 q_T^2 - \tau^2 q_0^2] , \tag{11.10}$$

so that $r_K \approx 2r_G$ can be expected from a comparison of (11.9) and (11.10).

However, jet fragmentation cannot be expected to be spherically symmetric and other shapes will be used in Sections 11.5 and 11.6.

11.1.3 Emission function and Wigner function

The picture presented in Sub-Sect. 11.1.1 corresponds to the one-dimensional treatment of a spherically symmetric static Gaussian source. However, a high-energy collision is neither spherically symmetric, nor static, nor Gaussian. The expansion of a source that is not static, but exhibits a (longitudinal) hydrodynamical scaling evolution has first been discussed by Shuryak [11] and Bjorken [12]. A formalism particularly appropriate for a fully-dimensional treatment of a dynamical emitter is the so-called Wigner-function formalism [13–15]. This is based on the emission function $S(x,p)$, a covariant Wigner transform of the source density matrix. $S(x,p)$, a function of x and p (!), can be interpreted as a quantum-mechanical analogue of the classical probability that a boson is produced at a given space-time point $x = (t, \mathbf{r})$ with a given momentum-energy $p = (E, \mathbf{p})$. In fact, it comes as close to that as allowed by the Heisenberg uncertainty relation.

In the general case, the normalized two-particle density $R_2(1,2)$ or correlation function $K_2(1,2)$ depend on the momentum components of particles 1 and 2. For the study of correlations, it is convenient to decompose the two single-particle four-vectors p_1 and p_2 into the average $k = [(E_1+E_2)/2,\ \mathbf{k} = (\mathbf{p}_1+\mathbf{p}_2)/2]$ and the relative momentum $Q = (\Delta E = E_1 - E_2, \mathbf{Q} = \mathbf{p}_1 - \mathbf{p}_2)$.

Starting from the space-time x and momentum-energy k dependent pion-emission function $S(x,k)$, the normalized density in momentum space can be written as [14,15]

$$R_2(\mathbf{Q},\mathbf{k}) \approx 1 + \frac{|\int \mathrm{d}^4x S(x,k) e^{iQ\cdot x}|^2}{|\int \mathrm{d}^4x S(x,k)|^2} = 1 + |\langle e^{iQ\cdot x}\rangle(\mathbf{k})|^2 \ . \tag{11.11}$$

In a Gaussian approximation around the mean space-time production point $\bar{x}$,

$$R_2(\mathbf{Q},\mathbf{k}) = 1 + \exp[-Q_\mu Q_\nu \langle (x-\bar{x})_\mu (x-\bar{x})_\nu \rangle(\mathbf{k})] + \delta R_2(\mathbf{Q},\mathbf{k}) \ . \tag{11.12}$$

For the case of otherwise uncorrelated emission, the variances $\langle (x-\bar{x})_\mu (x-\bar{x})_\nu \rangle$ give the size of the space-time region from which pions of similar momentum are emitted (which for Gaussian sources, coincides with the more general concept [16] of *lengths of homogeneity*, see also [17]) and δR_2 contains all non-Gaussian contributions, usually assumed to be small (however, see Sects. 11.5 and 11.6 below). It is important to keep in mind, however, that in case of strongly correlated emission (e.g. from clusters) the measured size parameters are dominated by the momentum correlation length rather than by a geometrical size in configuration space [18].

Since the four-momenta p_i of the two particles are on-shell, Q and k are in general off-shell but obey the orthogonality and mass-shell constraints

$$Q \cdot k = 0\ ,\quad k^2 - Q^2/4 = m^2\ , \tag{11.13}$$

so that only 6 linear combinations of the variances are measurable [19]. If the source is azimuthally symmetric in coordinate space, a reflection symmetry is present in momentum space with respect to the plane spanned by $\mathbf{k}$ and the event axis. As a consequence, all mixed variances linear in the direction orthogonal to this plane ("sidewards") must vanish and the correlator must be symmetric under $Q_{\text{side}} \to -Q_{\text{side}}$, so that only four linear combinations remain measurable! Note, however, that every one of them depends on $\mathbf{k}$.

11.1.4 String models

Bose-Einstein correlations have also been introduced into string models [10, 20, 21]. In these models, an ordering in space-time exists for the hadron production points within a string. Bosons close in phase space are nearby in space-time and the length scale measured by Bose-Einstein correlations is not the full length of the string, but the distance in boson-production points for which the momentum distributions still overlap.

Fig. 11.2a illustrates the production of (identical) particles 1 and 2 from a color string in x and t. The color field breaks up into quark-antiquark pairs and adjacent quarks and antiquarks recombine into mesons. The production of the same final state, but with particles 1 and 2 exchanged is shown in Fig. 11.2b.

In a color-string model, the (non-normalized) probability $\mathrm{d}\Gamma_n$ to produce an n-particle state $\{p_j\}$, $j = 1, \ldots n$ of distinguishable particles is

$$\mathrm{d}\Gamma_n = [\Pi_{j=1}^{n} N \mathrm{d}p_j \delta(p_j^2 - m_j^2)] \delta(\Sigma p_j - P) \exp(-bA_n) \ , \tag{11.14}$$

where the exponential factor can be interpreted as the square of a matrix element

$$M_n = \exp(i\xi A_n) \ , \quad \mathrm{Re}(\xi) = \kappa \ , \quad \mathrm{Im}(\xi) = b/2 \ , \tag{11.15}$$

and the remaining terms describe longitudinal phase space, with P being the total energy-momentum of the state. N is related to the mean multiplicity and b to the correlation length in rapidity. A_n corresponds to the total space-time area covered by the color field (Fig. 11.2), or to an equivalent area in energy-momentum space divided by the square of the string tension $\kappa = 1$ GeV/fm [21].

The production of two identical bosons (1,2) is governed by the symmetric matrix element

$$M = \frac{1}{\sqrt{2}}(M_{12} + M_{21}) = \frac{1}{\sqrt{2}}[\exp(i\xi A_{12}) + \exp(i\xi A_{21})] \ . \tag{11.16}$$

From Fig. 11.2 it is clear that there is an area difference and, consequently, a phase difference between M_{12} and M_{21} given by

$$\Delta A = |A_{12} - A_{21}| = \frac{1}{\kappa^2} |p_{\|1} E_2 - p_{\|2} E_1 + (p_{\|1} - p_{\|2}) E_{\mathrm{I}} - (E_1 - E_2) p_{\|\mathrm{I}}| \ , \tag{11.17}$$

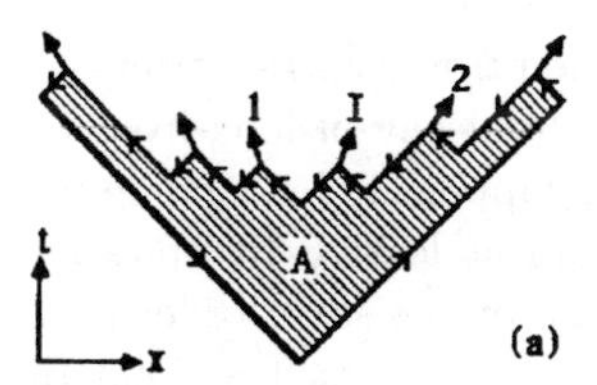

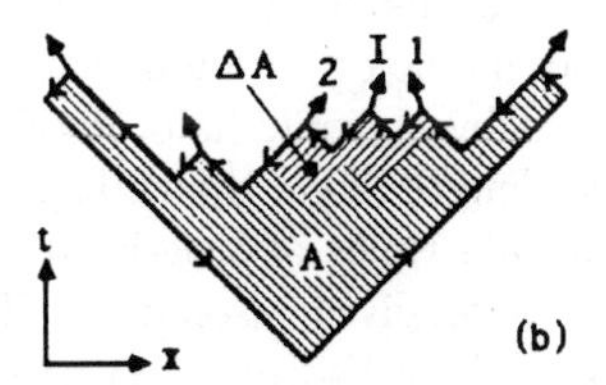

Figure 11.2. Space-time diagram for two ways to produce two identical bosons in the color-string picture [21].

where the indices 1,2 and I represent particles 1, 2 and the intermediate system I, respectively.

Using this matrix element, one obtains

$$R_{\mathrm{BE}} \approx 1 + \langle \cos(\kappa \Delta A) / \cosh(b \Delta A/2) \rangle \ , \tag{11.18}$$

where the average runs over all I. In the limit $Q^2 = -(p_1 - p_2)^2 = 0$, (11.17) gives $\Delta A = 0$ and (11.18) $R_{\mathrm{BE}} = 2$, in agreement with the results from the conventional interpretation for completely incoherent sources. However, for $Q^2 \neq 0$ follows an additional dependence on the momentum p_{I} of the system I produced between the two bosons.

Corrections to (11.18) are necessary due to non-zero mass and transverse momentum of quarks and due to the contribution of resonances to the production of particles of type 1, 2.

The model can account well for most features of the e^+e^- data [22–24]. More recently, the symmetrization has been generalized to more than 2 identical particles [25].

11.1.5 The strength parameter λ

If identical bosons are produced completely incoherently, a maximum value of $R_2 = 2$ is expected for $p_1 = p_2$, compared to $R_2 = 1$ in the case of absence of interference effects. In most experiments, the maximum effect seen is smaller than $R_2 = 2$ and a strength parameter λ [26] is introduced in front of the correlator in (11.8) or (11.9). Since a coherent source gives no enhancement, the strength of the effect is often interpreted as a measure of the incoherence of the pion emitters. On the other hand, also biases in general lead to a decrease of the effect. These may be physics induced (resonance production, local charge conservation, Coulomb repulsion) or detector induced (misidentification of the particles, wrong charge assignment, track losses). Furthermore, a large fraction of the observed pions are decay products of long-lived resonances, such as ω and η. This results in a large effective radius of the pion source (e.g. [27–31]).

The effect of resonance decays on λ is further quantified in the core-halo model of [32,33]. In this model, a central core corresponds to a direct production mechanism (e.g. hydrodynamic evolution or particle production from central strings). This core is surrounded by a pionic halo originating from the decay of long-lived resonances, with a decay length larger than 20 fm. Since there happens to be a gap in the life-time distribution, life times of long-lived resonances are at least 5-10 times longer than those of the short-lived resonances. The latter are of the same order of magnitude as the time scale of hadronic rescattering within the core, itself, so that their decay products are absorbed in the core.

Halo length scales larger than 20 fm give rise to a sharp peak of the correlation function in the $Q \leq \hbar/r \equiv 10$ MeV region, not resolved by present experiments. As a consequence, the measured value of λ is determined by the squared fraction of bosons emitted from the core [32].

11.1.6 The reference sample

Ideally, the reference sample is identical to the sample of like-pion pairs in all respects, except for the interference effect itself. A number of alternatives have been used in the literature:

i) The reference sample is formed by a so-called "mixed-event" technique, i.e. a pion from one event is combined with a pion randomly selected from a different event of the same multiplicity (see [34] for a rigorous a-posteriori justification of this method).

ii) The reference sample is formed from pairs of unlike-charged pions in the same event and resonances such as ρ^0 are excluded.

iii) The momentum components transverse to the event axis are "reshuffled" and the reference distribution is formed by combining reshuffled unlike-charged particles or reshuffled like-charged particles.

In a careful comparison [35] (where also a list of previous references is given) one finds that the application of the first two approaches leads to results consistent within statistical errors. From a Monte-Carlo study, the same two methods are found to give good estimators for the reference sample. On the other hand, method (iii) causes an artificial enhancement at low q_T, and therefore an unnaturally large λ.

As an example $R_2(Q^2)$ and $R_2(q_T)$ are given in Figs. 11.3a and b, respectively, for π^+p collisions at 22 GeV [36]. Both mixed event and the unlike-pair results are shown for both functions. The Bose-Einstein interference is clearly visible as an enhancement of R_2 at small Q^2 and q_T, respectively. The solid lines correspond to fits of the data by the forms

$$R_2(Q^2) = \gamma_G[1 + \lambda_G \exp(-r_G^2 Q^2)(1 + \delta_G Q^2) \tag{11.19}$$

$$R_2(q_T) = \gamma_K[1 + \lambda_K \left(\frac{2J_1(r_K q_T)}{r_K q_T}\right)^2 (1 + \delta_K q_T) , \tag{11.20}$$

where γ_G, γ_K are overall normalization parameters and λ_G, λ_K are the corresponding strength parameters. The last bracket is added to account for a (long-range) background correlation with δ_G and δ_K as free parameters determined by the behavior at large Q^2 and q_T, respectively.

Due to the close relation of Q^2 and squared invariant pair mass M^2, the ρ^0 signal is clearly visible as a dip near $Q^2 = 0.5$ GeV^2 for the unlike-pair method, so that this region has to be excluded from the fit. The ρ^0 signal is smeared out when q_T is used as a variable, but the bias remains. This problem is avoided in the mixed

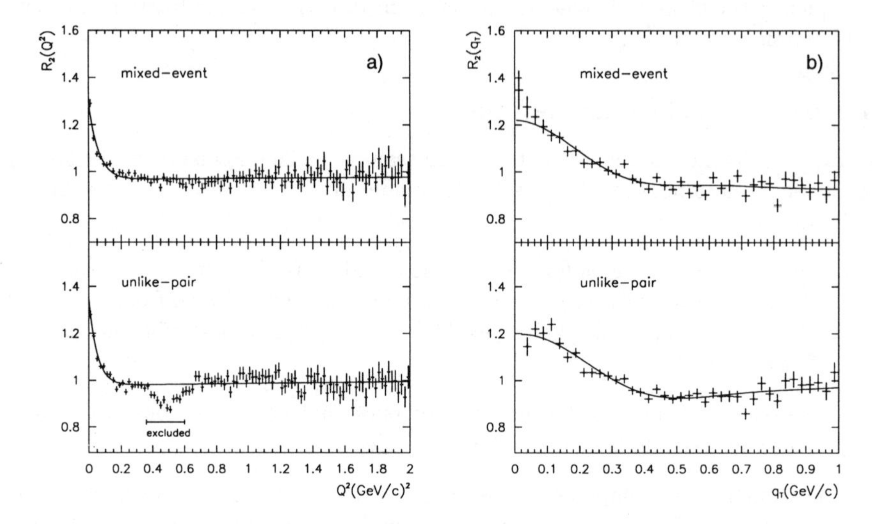

Figure 11.3. The ratio R_2 as a function of a) Q^2 and b) q_T as obtained with mixed-event technique (upper sub-figures) and the unlike-pair technique (lower sub-figures). The solid lines in a) and b) are fits by (11.19) and (11.20), respectively [36].

event technique, where no residual correlations exist, but there also those due to energy-momentum conservation are lost.

11.1.7 Coulomb correction

In order to correct for the Coulomb repulsion of identical pions, an early proposal [13] was to weight each pion pair by the Gamow factor

$$G(\eta) = \frac{2\pi\eta}{e^{2\pi\eta} - 1} \ , \qquad \eta = \pm\frac{m_\pi \alpha}{Q} \ , \tag{11.21}$$

where m_π is the pion mass and α the fine-structure constant. The sign of η is positive for like-charged pions and negative for unlike-charged ones. This (model dependent) factor is derived for an isolated pair of non-relativistic pions emitted from a small volume. In reality, Coulomb repulsion depends on the particle density in configuration and momentum space and on the size of the pion source.

The Gamow factor becomes sizable only for $Q^2 \lesssim 10^{-3}$ GeV^2, a region more important for heavy-ion collisions, where the radii are considerably larger than 1 fm, than for e^+e^- or hh colisions.

As shown for Pb+Pb and S+Ag collisions [37] by the dotted line in Fig. 11.4, G disagrees with the experimentally measured $\pi^+\pi^-$ correlations. A modification of

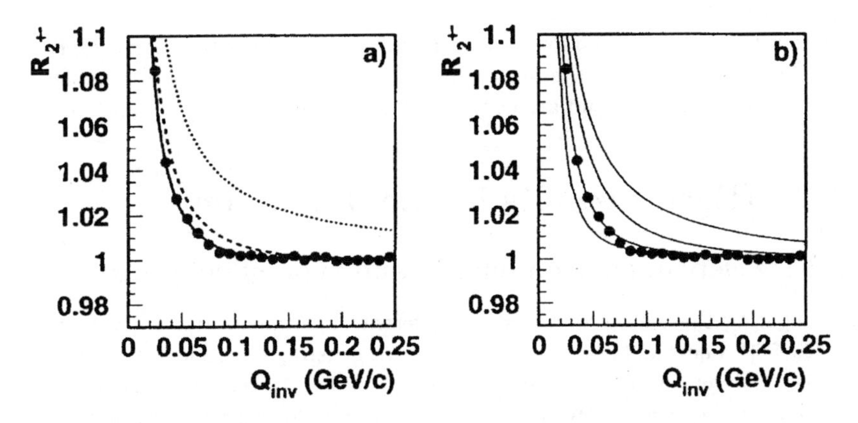

Figure 11.4. $\pi^+\pi^-$ correlation function for central Pb+Pb and S+Ag collisions (full circles), compared to a) the Gamow factor (dotted) and $F(Q)$ with $Q_{\text{eff}} = 46$ and 75 MeV (full and dashed), b) source radii r_0 of 8 fm (lowest), 5 fm, 2 fm and 0.55 fm [37].

(11.21) of the form

$$F(Q) = 1 + (G - 1)e^{-Q/Q_{\text{eff}}} \tag{11.22}$$

was suggested in [38] to fit the data by adjusting the parameter Q_{eff}. The result for central Pb+Pb and S+Ag collisions is shown in Fig. 11.4a for $Q_{\text{eff}} = 46$ MeV (full line) and 75 MeV (dashed).

The effect of the source radius under the assumption of a Gaussian source density $\rho(r) = \exp(-r^2/2r_0^2)$ can be seen in Fig. 11.4b. Best agreement with the data is obtained for $r_0 = 4.6$ fm, a radius compatible with results from interferometry in Sect. 11.5 below. The authors verified that $R_2^{--} = (R_2^{+-})^{-1}$ within a few percent, so that the measured R_2^{--} can be multiplied by the measured R_2^{+-} to obtain the Coulomb correction.

However, (11.22) is a phenomenological correction, without any theoretical basis. More precise description of Coulomb wave-function integration over the source distribution is a highly non-trivial problem, as described in [39–42] for two and more particles. The deviation from the Gamow factor Eq. (11.21) and its generalization to more particles increases with increasing source radius and increasing order of the correlation. For a 5-particle correlation the error is of the order of 10% for a radius of 1 fm, but as much as 100% for a radius of 5-10 fm. Furthermore, no Coulomb correction should be applied if one or both pions originate from long lived resonances. Corrections, in particular on λ, can be quite substantial, even for e^+e^- collisions. They, however, depend on the models used and have either not been applied at all or not been applied consistently in the various experiments. We, therefore, will quote the values as published by the experiments themselves, but refer to [39–42] for sometimes important reinterpretation.

Other final-state interactions, as strong interaction of the two pions [28,43,44] or Coulomb interaction with the nuclear remnant are expected to be negligible [45] or to effectively cancel [28] and are usually not considered.

11.2 (Early) results in one dimension

11.2.1 Dependence on energy and type of collision

In comparing the results obtained by different experiments (for early reviews see e.g. [46–50]), one should keep in mind that, in general, experimental conditions are different, different techniques are used in constructing the reference distributions and sometimes ad-hoc corrections are applied to make up for experimental shortcomings.

In hadron-hadron experiments, the Bose-Einstein enhancement at small momentum difference is often parametrized in terms of the Kopylov-Podgoretskiĭ variables q_{T} and q_0 (11.9). With this parametrization, most of the reported values [51] are in the range 1.1-1.8 fm, a reasonable average being 1.5 fm. No clear energy dependence is seen up to ISR energies [26,35,36,52–56], while UA1 [57] and E735 [58] show that no energy dependence of r or λ is observed between $\sqrt{s} = 200$ GeV and 1800 GeV as long as the particle density is kept fixed.

Also for e^+e^- annihilation such values (1.3 ± 0.1 fm) have been reported [22,23] when using parametrization (11.9).

The Lorentz invariant parametrization (11.8) commonly used in e^+e^- annihilation experiments [22–24,59–63], gives values in the range $r_{\mathrm{G}} = 0.7 - 0.9$ fm, in good agreement with a hadron-hadron result of $r_{\mathrm{G}} \approx 0.8$ fm [35]. Furthermore, similar values (0.5-0.8 fm) are reported for lepton-hadron collisions [64–68]. In particular, ZEUS [68] has used the advantage to be able to study the dependence of the radius and strength parameters on the photon virtuality $Q^2_{\gamma^*}$ with the same parametrization, reference sample and experimental conditions, from 0.1 to 8000 GeV^2. As shown in Fig. 11.5, no dependence on the virtuality is observed and no difference between target and beam fragmentation.

We can conclude that the parameter r, of course, depends on the parametrization, but, within the systematic uncertainties, neither on the total energy or virtuality, nor on the type of collision or fragmenting particle.

The strength of the effect is found to be in the range $\lambda_{\mathrm{K}} = 0.3 - 0.5$ for hadron-hadron experiments when using reference distributions based on the mixed-event or unlike-reference techniques. Significantly higher values are quoted when using a reference distribution based on the shuffling of transverse momenta. For e^+e^- annihilation it is in the range λ_{G} =0.4-0.7. It is premature, however, to associate a λ value smaller than the theoretical maximum of $\lambda = 1$ with the existence of (partially) coherent states.

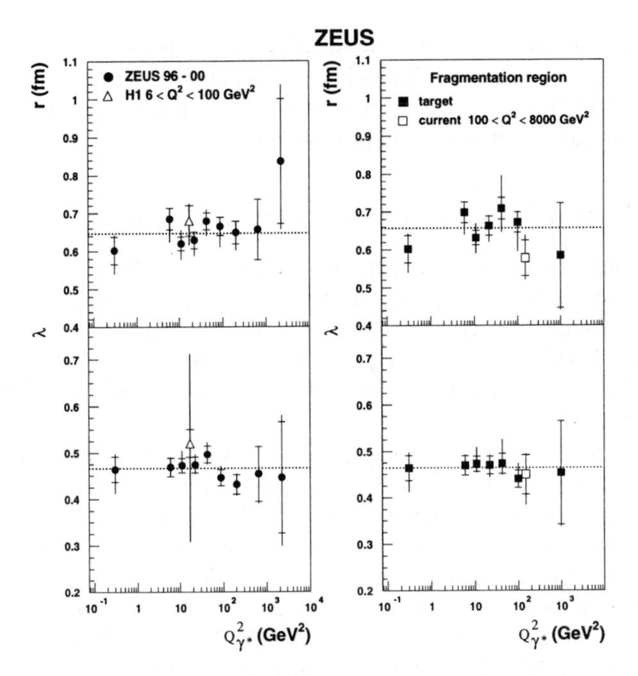

Figure 11.5. The radius and strength parameters as a function of virtuality $Q^2_{\gamma^*}$ [68].

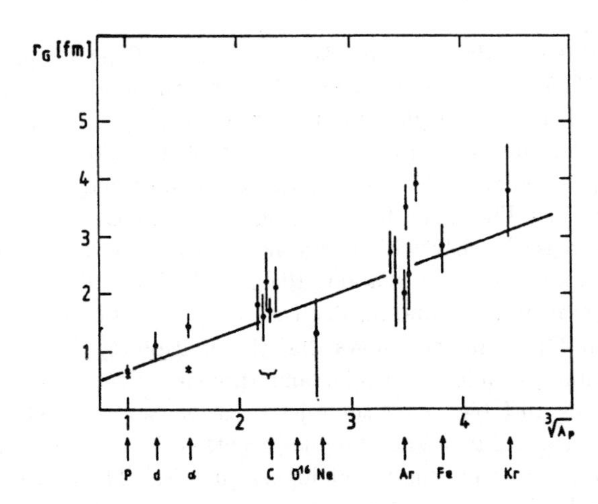

Figure 11.6. The radius parameter r_{G} as a function of the cubic root of the mass number A_{p} of the projectile. The solid line corresponds to $r_{\mathrm{G}} = 0.7\ A_{\mathrm{p}}^{1/3}$ fm [48].

The radius of the emitting region for collisions of two heavy nuclei, on the other hand, is found to be several fm [69–76] and increases linearly with $A^{1/3}$, where A is the atomic mass number (see Fig. 11.6 for r_G) [48,77]. As shown by [76], it, furthermore, increases with increasing centrality of the collision.

11.2.2 The multiplicity (or density) dependence

In nucleus-nucleus experiments [70, 73], the radius r_K was found to increase with increasing charged-particle multiplicity n. By relating r_K to the size of the overlap region of the two colliding particles, this increase can be understood in terms of the geometrical model [78]: a large overlap should imply a large multiplicity. On the other hand, no evidence for a multiplicity dependence is found in hadron-nucleus collisions at 200 GeV/c [79].

After some time of confusion, the n dependence is now clear for hadron-hadron collisions. At energies below $\sqrt{s} \approx 30$ GeV (i.e at $\sqrt{s} \approx 8$ [80], 22 [35] and 27 GeV [56]) no n-dependence is observed for r_G. At higher energies (last ref. [54] and [53]) an n-dependence starts to set in and to grow with increasing energy (see Fig. 11.7a). At the highest ISR energy ($\sqrt{s} = 62$ GeV) the increase is about 40% when the density in rapidity is doubled, but at $\sqrt{s} = 31$ GeV the increase is still very weak. The result is extended to $\sqrt{s} = 630$ GeV by UA1 [57] and to 1800 GeV by E735 [58] in Fig. 11.7b. At very large density, the increase of r_G with increasing density is shown to extrapolate well to the heavy-ion results of NA35 [73] in Fig. 11.7c. The effect is reproduced in thermodynamic and hydrodynamic models to be discussed in Sec. 11.6. The λ parameter, on the other hand, decreases with increasing n (not shown).

At the low-density side, the effect is also observed in e^+p collisions by H1 [67] (crosses in Fig. 11.7b). The results from e^+e^- experiments at lower energy [23, 69] were consistent with no multiplicity dependence as expected from the geometrical model, but also for this type of collisions a multiplicity dependence was finally established at higher energy [81], both in the Goldhaber and Kopylov-Podgoretskiĭ parametrizations (see Fig. 11.8a for the former). At 91 GeV, the radii r_G and r_K are found to increase linearly with increasing multiplicity n, showing a small but statistically significant increase of about 10% for $10 \leq n \leq 40$. As for hh-collisions, the chaoticity parameters λ_G and λ_K decrease with increasing n.

In Fig. 11.8b, OPAL further shows that the multiplicity dependence is strongly reduced in separate samples of two-jet and three-jet events, the average value of r_G, however, being 10% bigger for three-jet than for two-jet events. Folding in the multiplicity difference of two- and three-jet events, this at least partly explains the effect as due to multi-jet production at higher energies. The decrease of λ is larger in the 3-jet than in the 2-jet sample.

As shown quantitatively in [82], it is crucial to study the normalized cumulants $K_2(Q)$ rather than the normalized densities $R_2(Q)$ in a density dependent analysis

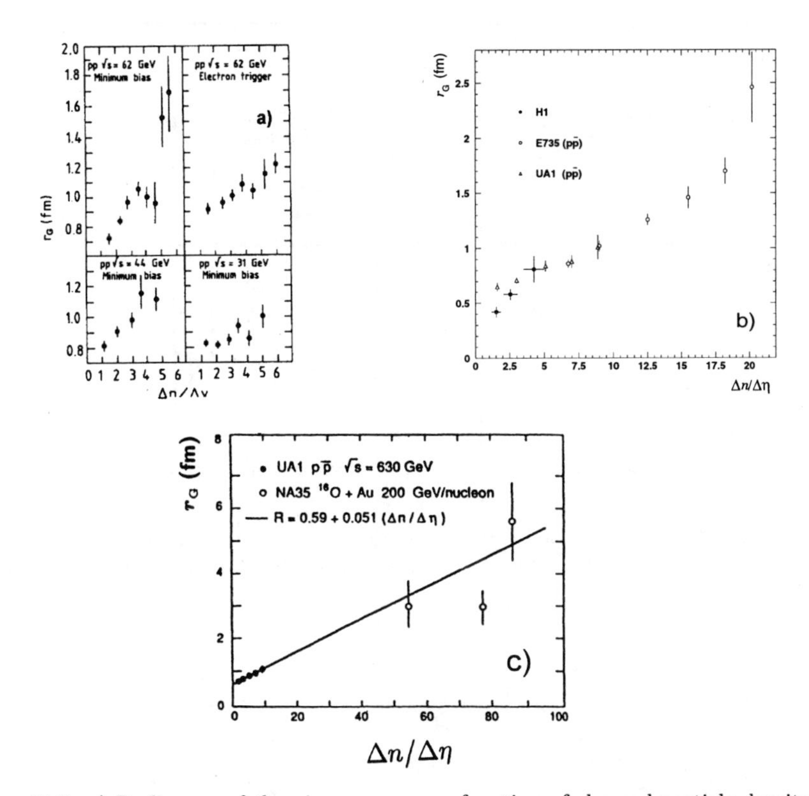

Figure 11.7. a) Radius r_G of the pion source as a function of charged-particle density for the energies indicated [54], b) same for p$\bar{\text{p}}$ collisions at 630 GeV [57] and 1800 GeV [58], as well as e$^+$p collisions at 300 GeV [67], c) Comparison of r_G as a function of charged-particle density $\Delta n/\Delta\eta$ [57] with the results of relativistic heavy-ion collisions [73].

and to correct for a well-defined multiplicity-dependent bias due to the cut in the multiplicity distribution (the point being that $K_q \neq 0$ for limited n, even in case of independent emission). In Fig. 11.9a and b [83], the bias-corrected (so-called "internal") cumulants are given for UA1 as a function of the inverse rapidity density, for small and large values of Q, respectively. The data show

i) a linear dependence (similar for like-charged and unlike-charged pairs),

ii) vanishing of the cumulant at large density for large Q,

iii) approach towards a non-zero limit for large density at small Q (where BE correlations are expected to dominate).

The large-Q behavior [points i) and ii) above] is that expected from particle emission from N fully overlapping, identical but fully independent sources (e.g. strings).

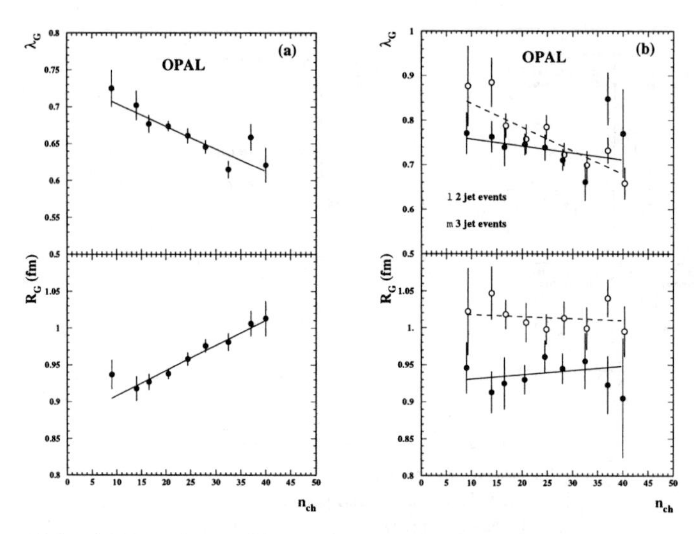

Figure 11.8. a) Dependence of λ_{G} and r_{G} on the charged-particle multiplicity n for $\mathrm{e}^{+}\mathrm{e}^{-}$ collisions at the Z mass, b) same for two-jet events (solid points) and three-jet events (open points) [81].

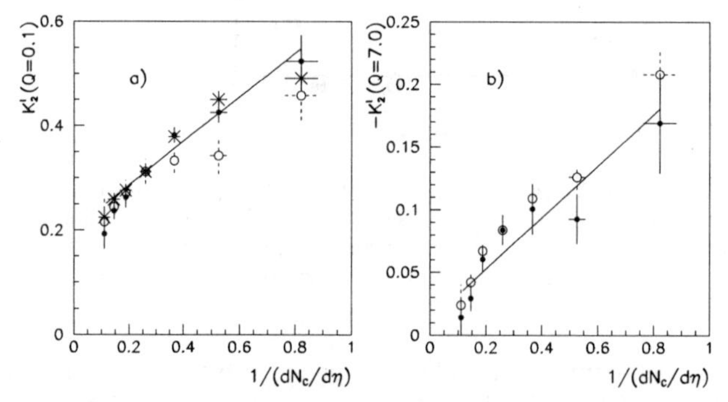

Figure 11.9. Inverse density dependence of the bias-corrected cumulant at $Q = 0.1$ GeV (Fig. 11.9a) and $Q = 7.0$ GeV (Fig. 11.9b) for like-charged pairs (full circles) and unlike-charged pairs (open circles). The crosses in Fig. 11.9a correspond to λ-values [83].

From the additivity of unnormalized cumulants (7.21) follows immediately the dilution property (7.24) (see also [84–86]). This results in a normalized cumulant inversely proportional to N or to the total density $\mathrm{d}n/\mathrm{d}y = N\mathrm{d}n^{(1)}/\mathrm{d}y$, as observed in Fig. 11.9b.

At small Q (Fig. 11.9a), however, the normalized cumulant approaches a constant different from zero at large densities. Naively, this would imply correlations between particles coming from different sources, which could be interpreted as inter-source Bose-Einstein correlations, would not a similar effect be observed for unlike-charged pairs, as well. So also resonances play an important part. One has to keep in mind, however, that (7.24) only holds for full overlap of identical sources and that at $Q \approx 0.1$ GeV the overlap is far from complete and the number of sources limited.

Also in heavy-ion collisions there is evidence that λ does not drop with increasing density forever. In collisions of ions with relatively low A, the λ values quickly decrease with increasing density at lower densities. In agreement with the expectation from overlapping but independent sources, λ drops from 0.79 to 0.32 [87] from O-C to O-Cu, O-Ag and O-Au. In high-A collisions, on the other hand, a saturation seems to set in [88–90]. For S-Pb and Pb-Pb central collisions NA44 [91] quotes $\lambda = 0.56$ and 0.59, respectively, and similar values are obtained for AuAu at RHIC [90]. Such a saturation and eventual increase of λ is indeed expected if the densely packed strings of a heavy-ion collision finally coalesce until they form a large single fire-ball (percolation of strings) [89].

11.2.3 Four types of Monte-Carlo implementation

The probabilistic structure of Monte Carlo event generators, in principle, prevents the simulation of quantum statistical effects such as the Bose-Einstein correlations. One, therefore, models the effect by properly modifying the output of these generators to reproduce the signals measured experimentally.

1. *Reshuffling*: The MC code LUBOEI [6] in JETSET treats BE correlations as a final state interaction and actually changes particle momenta according to a spherically symmetric Gaussian (or, alternatively, exponential) correlator. The advantages are that it is a fast and unit-weight (i.e. efficient) generator. The bad news are that it is imposed a-posteriori (without any physical basis), is even unphysical (since it changes the momenta), is not self-consistent (since it introduces an artificial length scale [92, 93]), is spherically symmetric, does not consider higher-order correlations, etc. The worse news is that it is generally used to correct for detector effects and that there is no real alternative, at the moment.

2. *Charge reassignment*: An alternative suggested in [94] is to use the momenta as well as the multiplicity of a given charge as generated in the original event, but change the original charge assignment, so that equal charges get bunched into elementary emitting cells according to the measured correlation function. In principle, higher-order correlations are included automatically. In practice, they are limited by the highest multiplicity reached in the elementary cells. A drawback is, however, that anticorrelations are produced for unlike charged pairs, so that strong positive correlations have to be introduced already during generation [95].

3. *Global reweighting*: Another, theoretically better justified approach is to attach to each pre-generated event a BE weight depending on its momentum configuration,

but leaving this momentum (and charge) configuration untouched. Based on the use of Wigner functions rather than amplitudes, a weight factor can be derived of the form [96]

$$W(p_1, \dots p_n) = \sum_{\{P_n\}} \prod_{i=1}^{n} K_2(Q_{iP_n(i)}) \ , \tag{11.23}$$

where n is the number of identical particles, $K_2(= R_2 - 1)$ is the two-particle correlator and $P_n(i)$ is the particle which occupies the position i in the permutation P_n of the n particles. Applications of the global weighting [97–102] are essentially all variations on this theme, with varying model assumptions on the exact form of K_2. In general, $K_2(Q)$ is still assumed to be spherical in Q. Higher-order correlations are included, but either assume [97] a quantum optical model, already shown to be wrong (see Sub-Sect. 11.4.2, below), or factorization not allowing for phases between the terms in the product above. As in [6], the weight is imposed a posteriori, so is not part of the MC model itself. Retuning is necessary, but this can, in practice, be achieved by just retuning the multiplicity distribution [102]. Problems arise from the fact that the number of permutations is $n!$, so that simplifications have to be introduced [102]. Wild fluctuations of event weights can occur, so that cuts on event weight are necessary. The weight may even change the parton distributions, while BE correlations only work on the pion level.

4. *Symmetrizing* according to Sub-Sect. 11.1.4: An ordering in space-time exists for the hadron moments within a string [20, 21]. Bosons close in energy-momentum space are nearby in space-time and the length scale measured by Bose-Einstein correlations is not the full length of the string, but the distance in boson-production points for which the momentum distributions still overlap. An event weight is calculated according to Eq. (11.18).

The model can account well for most features of the e^+e^- data, including the non-spherical shape of the BE effect. More recently, the symmetrization has been generalized to more than two identical particles. [25]

However, the extension to multigluon string fragmentation required the adaptation [103] of the new directrix scheme [104] incorporated in the ALFS (Area Law Fragmentation of Strings) Monte Carlo code (Sub-Sect. 6.1.2). Whereas it turns out unnatural to symmetrize across hard gluon corners, symmetrization across soft gluons is necessary. To calculate the BE weights, the multigluon string is treated as a simpler string of groups of gluons with relative small transverse momenta (so-called coherence chains). Symmetrization is then carried out separately for each coherence chain.

11.2.4 Conclusions so far

1. Bose-Einstein interference is visible for pairs of identical bosons in all types of collision.

2. The radius of the pion emitting region ($r_K \approx 2r_G \approx 1.5$ fm) depends on the parametrization, but not on energy, virtuality or type of collision. For heavy-ion collisions it, however, increases with increasing mass number A like $A^{1/3}$.
3. The observed strength of the effect ($\lambda \approx 0.5$) depends on the reference sample used, but not on energy or type of collision, except for heavy-ion collisions, where it depends on A.
4. At $\sqrt{s} \geq 30$ GeV an increase of r is found with increasing particle density.
5. The effect is a particular challenge for Monte Carlo model builders.

11.3 Other bosons and fermions

11.3.1 The $\pi^0\pi^0$ system

In a string model, unlike $\pi^\pm\pi^\pm$-pairs, pairs of prompt π^0's can be emitted in adjacent string break-ups. In momentum space, the correlation function is, therefore, expected to be wider for neutral pions than for charged ones. Neutral pions, furthermore, are insensitive to Coulomb effects. However, the detection of several π^0's in one event requires high efficiency of γ-detection in a wide energy range and geometrical acceptance. Furthermore, the correlation function at small Q is strongly influenced by resonance decays such as $\eta \to \pi^0\pi^0\pi^0$, $\eta' \to \pi^0\pi^0\eta$, $K^0_S, f_0 \to \pi^0\pi^0$ and other final-state interactions [105].

First evidence for Bose-Einstein correlations in $\pi^0\pi^0$ pairs was found in [106]. In a first measurement of the radius in π^-Xe interactions at 3.5 GeV [107], the size of the π^0 emission region was found compatible with that for charged pions.

The question was taken up again by L3 [108], where both r_G and λ are found to be on the low side when compared to the $\pi^\pm\pi^\pm$ results obtained under the same experimental conditions (e.g. $Q > 0.3$ GeV). The difference in λ can at least partially be explained by the contribution of resonances. The difference in size parameter is $r_{\pm\pm} - r_{00} = 0.15 \pm 0.08 \pm 0.07$ fm. A similar difference of $0.15 \pm 0.08 \pm 0.15$ is obtained by OPAL [109], be it only in a comparison of their $\pi^0\pi^0$ results to an average over the (widely scattered) LEP $\pi^\pm\pi^\pm$ results obtained under different experimental conditions.

It is interesting to note that no BEC is found in $p\bar{p} \to 4\pi^0$ at rest [110], while a small effect is seen in $p\bar{p} \to 2\pi^+2\pi^-$ [111]. Both reactions are dominated by resonances, apparently leaving no room for stochastic pion emission phases in the $4\pi^0$ final state.

11.3.2 The $K^\pm K^\pm$ system

Kaons are less affected by resonance decay than pions and provide a cleaner measure of the source. Bose-Einstein correlations among equally-charged kaons were observed in hh [53, 56], AA [91, 112–114] and e^+e^- [115, 116] collisions (see Table 11.1 for the latter).

Table 11.1. Parameters λ_G and r_G in the Gaussian parametrization in e^+e^- interactions at LEP, for different like-charged particles.

Pair	λ_G	r_G fm	Ref.	Selection	Ref. sample
$\pi^{\pm}\pi^{\pm}$	0.35 ± 0.04	0.42 ± 0.04	DELPHI [61]	2-jet	mixed
	0.40 ± 0.02	0.49 ± 0.02	ALEPH [62]	2-jet	mixed
	0.58 ± 0.01	0.79 ± 0.02	OPAL [81]	all	MC
	$0.72 \pm 0.08 \pm 0.03$	$0.74 \pm 0.06 \pm 0.02$	L3 [117]	all	mixed
$\pi^{\pm}\pi^{\pm}$	0.45 ± 0.02	0.82 ± 0.03	DELPHI [61]	all	unlike
	0.62 ± 0.04	0.81 ± 0.04	ALEPH [62]	2-jet	unlike
	$0.67 \pm 0.01 \pm 0.02$	$0.96 \pm 0.01 \pm 0.02$	OPAL [81]	all	unlike
	0.65 ± 0.02	0.91 ± 0.01	OPAL [81]	2-jet	unlike
$\pi^{\pm}\pi^{\pm}$	$1.06 \pm 0.05 \pm 0.16$	$0.49 \pm 0.01 \pm 0.05$	DELPHI [61]	prompt pions	
$K^{\pm}K^{\pm}$	$0.82 \pm 0.11 \pm 0.25$	$0.48 \pm 0.04 \pm 0.07$	DELPHI [115]	all	unlike
	$0.82 \pm 0.22^{+0.17}_{-0.12}$	$0.56 \pm 0.08^{+0.08}_{-0.06}$	OPAL [116]	2-jet	mixed
$K^0_SK^0_S$	$1.14 \pm 0.23 \pm 0.32$	$0.76 \pm 0.10 \pm 0.11$	OPAL [118]	all	MC
	$0.61 \pm 0.16 \pm 0.16$	$0.55 \pm 0.08 \pm 0.12$	DELPHI [115]	all	MC
	$0.63 \pm 0.06 \pm 0.14$	$0.57 \pm 0.04 \pm 0.14$	ALEPH [119]	all	mixed

The measured r values tend to be smaller than those of the pion emission region, in particular in AA collisions. The difference in resonance effects on $\pi\pi$ and KK correlations can explain this difference only partially, but see further in Sub-Sects. 11.3.6 and 11.6.7 below.

11.3.3 The $K^0_SK^0_S$ system

The $K^0_SK^0_S$ system is a mixture of $K^0\bar{K}^0$ and K^0K^0 ($\bar{K}^0\bar{K}^0$) pairs. At LEP1 energy, only 28% of all $K^0_SK^0_S$ pairs are estimated to come from the (identical) K^0K^0 or $\bar{K}^0\bar{K}^0$ system. What is particularly interesting is that K^0_S's can interfere even if they originate from a (non-identical) $K^0\bar{K}^0$ system [120]: An enhancement is expected in the low-Q region if one selects the $C = +1$ eigenstate of

$$|K^0\bar{K}^0\rangle_{C=\pm 1} = \frac{1}{\sqrt{2}}(|K^0(\mathbf{p})\bar{K}^0(-\mathbf{p})\rangle \pm |\bar{K}^0(\mathbf{p})K^0(-\mathbf{p})\rangle), \tag{11.24}$$

where $\mathbf{p}$ is the three-momentum of one of the kaons in their cms. In the limit $\mathbf{p} \to 0$ $(Q \to 0)$, the $C = -1$ $(K^0_SK^0_L)$ state disappears and $C = +1$ $(K^0_SK^0_S$ or $K^0_LK^0_L)$ becomes maximal.

The enhancement in $K^0_SK^0_S$ and $K^0_LK^0_L$ pairs at low Q is exactly compensated by the low Q suppression of the $K^0_SK^0_L$ state, so that no BE effect is to be expected as long as all possible final states of the $K^0\bar{K}^0$ system are considered. A full BE-like enhancement is, however, expected for the $K^0_S K^0_S$ system by itself.

Early, low statistics results come from the hh experiment [121], new results exist from DELPHI [115,122], OPAL [118] and ALEPH [119]. While the kaon-production radius is smaller for the hh experiment, it seems to agree with both those measured for charged kaons and pions in the e^+e^- experiments, within the large spread of values observed. Furthermore, the parameter λ is large in agreement with the expectation [120].

11.3.4 $\Lambda^0\Lambda^0$

An interesting generalization of the Bose-Einstein formalism used above is to consider Fermi-Dirac interference, essentially by changing the sign in front of the correlator. This leads to a destructive interference at small momentum-space distance and allows to determine the emission radius for identical fermions in a comparison of the amount of their total-spin $S = 1$ state (destructive) to that of their $S = 0$ state (constructive) as a function of Q [123]. The method does not need a further reference sample. It was applied to e^+e^- data at LEP1 in [124–126] and gives a radius of about 0.15 fm. It was verified, that the conventional method with JETSET used to generate the reference sample gives a similarly low radius.

11.3.5 pp and $\bar{p}\bar{p}$

Similarly, Fermi-Dirac interference is observed in pp and $\bar{p}\bar{p}$ pairs, be it on top of a significant background from strong interaction [127, 128]. Again, the radius comes out smaller than for pion or kaon pairs and is as low as $r_G = 0.11 \pm 0.01 \pm 0.01$ fm in e^+e^- collisions [119].

11.3.6 (Transverse) mass dependence of the radius parameter

The simultaneous comparison of the emission radii for pions, kaons, Λ's and protons suggests a decrease with increasing mass. Such a behavior has first been observed by NA44 in heavy-ion collisions [113]. In Fig. 11.10a,b [127], the longitudinal and transverse radii (for a definition see Sect. 11.6 below) from pion correlation measurements are compared to those from kaon and proton correlation measurements. The results are consistent with approximate $m^{-1/2}$ scaling, decreasing somewhat faster for the longitudinal than for the transverse radius. Such a scaling is in agreement with the expectations from a hydrodynamic model [16,19,129–132] with three-dimensional collective expansion and cylindrical symmetry [132]. This will be discussed in more detail in Sect. 11.5.

In Fig. 11.10c, the radius parameter r is shown as a function of the hadron mass m [133] for e^+e^- annihilation at the Z mass. The large error associated with $r_{\pi\pi}$ reflects the systematic uncertainty due to the choice of the reference sample. A general trend can be observed as a hierarchy

$$r_{\pi\pi} > r_{\mathrm{KK}} > r_{\Lambda\Lambda} \; . \tag{11.25}$$

Some effect is to be expected from kinematics, i.e. from the mass-dependent integration limits when transforming from $R_2(\mathbf{p}_1, \mathbf{p}_2)$ in six-dimensional momentum space to $R_2(Q)$ in one-dimensional momentum separation [134]. This effect is far too small, however.

The authors [133] show that a $1/\sqrt{m}$ behavior can be expected already from the Heisenberg principle with

$$\begin{aligned} \Delta p \Delta r &= mvr = \hbar c \\ \Delta E \Delta t &= p^2 \Delta t/m = \hbar \qquad (11.26) \\ \text{and} \quad r &= \frac{c\sqrt{\hbar \Delta t}}{\sqrt{m}} \, , \qquad (11.27) \end{aligned}$$

where m, v and p are the hadron mass, velocity and momentum and r is the distance between the two hadrons. Assuming ΔE to only depend on the kinetic energy of the produced particle and $\Delta t = 10^{-24}$ sec, independent of m, produces the thin solid line in Fig. 11.10c. The upper and lower dashed lines correspond to an increase or decrease of Δt by $0.5 \cdot 10^{-24}$ sec, respectively. (The thick solid line corresponds to a perturbative QCD cascade using the virial theorem and assuming local parton-hadron duality (LPHD).)

However, as shown in [133], a formula identical to (11.27) also holds for the radius r_z in the longitudinal direction and the average transverse mass $\bar{m}_{\mathrm{T}} = 0.5 \left(\sqrt{m^2 + p_{\mathrm{T1}}^2} + \sqrt{m^2 + p_{\mathrm{T2}}^2}\right)$. Fig. 11.10d shows DELPHI results [135] (see Sect. 11.5 for details) compared to $\Delta t = 10^{-24}$sec (dashed) and the best fitted value of $\Delta t = 2.1 \cdot 10^{-24}$sec (full line).

Alternatively, the transverse mass dependence can [136, 137] be explained by a generalized inside-outside cascade assuming (i) approximate proportionality of four-momenta and production space-time position (freeze-out point) of the emitted particles $p_\mu = a x_\mu$ and (ii) a freeze-out time distributed along the hyperbola $\tau_0^2 = t^2 - z^2$ (i.e., a generalization of the so-called Bjorken-Gottfried conditions). From the two conditions above follows directly

$$a^2 \tau_0^2 = E^2 - p_z^2 = m_{\mathrm{T}}^2 \tag{11.28}$$

and the generalized Bjorken-Gottfried condition

$$p_\mu = \frac{m_{\mathrm{T}}}{\tau_0} x_\mu \; . \tag{11.29}$$

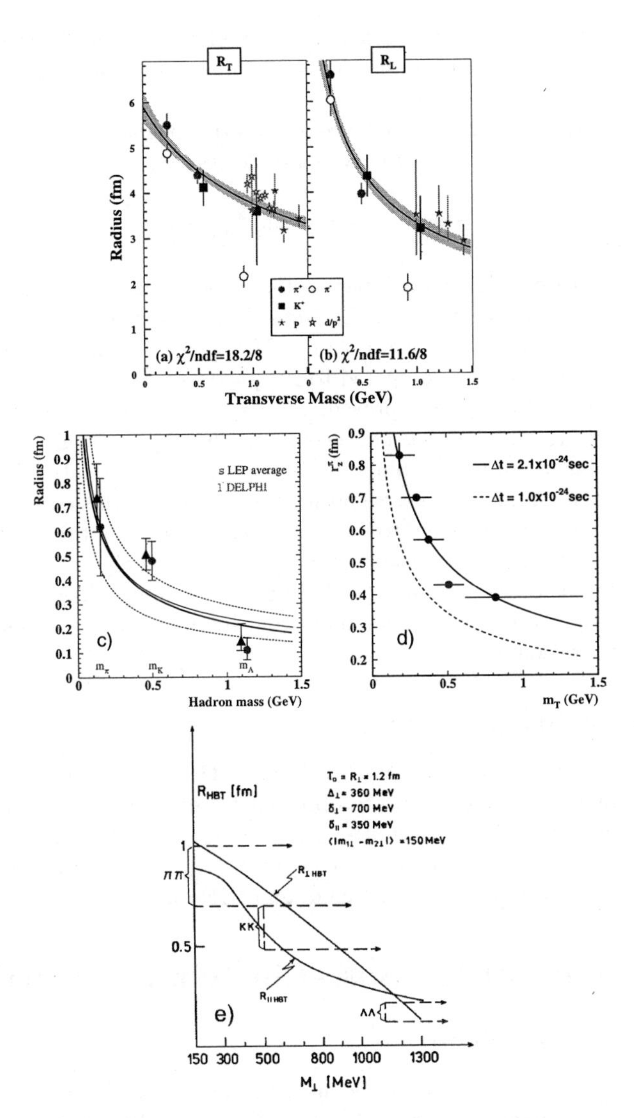

Figure 11.10. a)b) The longitudinal and transverse source-size parameters of K^+K^+ and pp, compared with those of pions as a function of m_{T}. The lines correspond to fits by $1/R^2 = p_o + p_1 m_{\mathrm{T}}$ [125]. c) The radius parameter r as a function of the hadron mass m. d) The longitudinal radius r_z as a function of m_{T} [135]. The lines are described in the text [133]. e) Longitudinal and transverse radius as a function of the transverse mass $M_\perp$ of the two-particle system, compared to the $M_\perp$-threshold values (same as data in Fig. 11.10c) [136].

Using a more rigorous formulation in terms of the Wigner representation, the authors show how this proportionality leads to an $m_{\rm T}$ dependence of the radius parameter.

Fig. 11.10e gives indeed a dependence of both the longitudinal and the transverse radius on the transverse mass of the two-particle system,

$$M_\perp^2 = \bar{m}_{\rm T}^2 + m_{\rm T1} m_{\rm T2} \sinh^2\left(\frac{y_1 - y_2}{2}\right) . \tag{11.30}$$

For a set of "reasonable" model parameters [136], the experimental results of Fig. 11.10c are reproduced rather well. Note that the experimental data are given at the threshold value of the corresponding $M_\perp$ at which transverse momenta and rapidity differences are small compared to the particle masses.

The parameters are to be improved, but $\Delta_\perp$ is closely related to the average transverse momentum and $\delta_\perp$ has to be considerably larger than $\Delta_\perp$ in the model to satisfy the uncertainty principle. Since $\delta_\perp$ corresponds to a correlation length between transverse momentum and transverse position at freeze-out, this correlation is rather weak. Nevertheless, it is sufficient to create a strong variation of the transverse radius, and suggests the existence of an important "collective flow", even in the system of particles produced in e^+e^- annihilation! Note that strong space-time momentum-energy correlations are expected not only from hydrodynamic expansion, but also from jet fragmentation.

11.3.7 Conclusions so far

So, in addition to the conclusions reached in Sect. 11.2 we conclude:
6. Bose-Einstein correlations are also observed for other bosons (and Fermi-Dirac correlations for fermions). The emitter radius scales with the inverse square root of the particle mass m (or transverse mass $m_{\rm T}$), as expected from the Heisenberg principle or from a strong position-momentum correlation.

11.4 Higher-order Bose-Einstein correlations

11.4.1 The formalism

It is convenient to use the normalized inclusive density and correlation functions already defined in Chapter 7 and used in Chapters 9 and 10,

$$R_q(1, \ldots, q) = \rho_q(1, \ldots, q)/\rho_1(1) \ldots \rho_1(q), \tag{11.31}$$

$$K_q(1, \ldots, q) = C_q(1, \ldots, q)/\rho_1(1) \ldots \rho_1(q). \tag{11.32}$$

The normalized inclusive density for two identical pions is

$$R_2(1, 2) = 1 + K_2(1, 2). \tag{11.33}$$

For a completely chaotic and static pion source, $K_2(1,2)$ reduces to the square of the Fourier transform $F(\mathbf{p}_1 - \mathbf{p}_2, E_1 - E_2)$ of the space-time distribution of the source, $K_2(1,2) = |F(1,2)|^2$, where $\mathbf{p}_i$ and E_i $(i = 1,2)$ are the three-momentum and energy of pion i, respectively.

If the Gaussian parametrization is used for $|F(Q_2^2)|^2$, then one has

$$K_2(Q_{12}^2) = |F(Q_{12}^2)|^2 = \exp(-r_\mathrm{G}^2 Q_{12}^2) \ . \tag{11.34}$$

In terms of the Q_{ij} variables and for the case of a completely chaotic source, the normalized inclusive three-pion density is [48,138]

$$\begin{aligned} R_3(1,2,3) &= 1 + |F(Q_{12}^2)|^2 + |F(Q_{23}^2)|^2 + |F(Q_{31}^2)|^2 \\ &+ 2\mathrm{Re}\{F(Q_{12}^2)F(Q_{23}^2)F(Q_{31}^2)\} \ , \end{aligned} \tag{11.35}$$

so that the genuine three-particle correlation reads

$$K_3(1,2,3) = 2\mathrm{Re}\{F(Q_{12}^2)F(Q_{23}^2)F(Q_{31}^2)\}. \tag{11.36}$$

In general, the genuine three-particle correlation $K_3(1,2,3)$ is not expressed completely in terms of the two-particle correlation function (11.34). Further information is contained in the ratio

$$\omega = \frac{K_3(1,2,3)}{2\sqrt{K_2(1,2)K_2(2,3)K_2(3,1)}} \ , \tag{11.37}$$

in which resonance halo and misidentified-particle contributions cancel. If interpreted as a phase factor, $\omega = \cos\phi$ with ϕ being a function of Q_{ij}. Assuming incoherent production, $\omega \to 1$ when all $Q_{ij} \to 0$. At $Q_{ij} \neq 0$, geometrical asymmetry in the production mechanism (emission function) due to flow or resonance decays will only lead to small (few percent) reduction of ω from unity [139]. However, Eq. (11.37) is not valid for (partially) coherent sources and more complicated expressions [138,139] for a chaotic fraction $p < 1$ lead to an intercept

$$\omega(Q_{ij} \to 0) = \sqrt{p}\frac{3-2p}{(2-p)^{3/2}} \tag{11.38}$$

smaller than unity. If ω considerably differs from unity, one can infer that partial coherence is present. An alternative explanation could be that the nominator K_3 is suppressed faster than $K_2^{1/2}$ due to dilution in the case of many independent sources (see below).

To the extent that phase factors may be neglected and the Gaussian approximation would hold, K_3 is related to K_2 via the expression

$$K_3(Q_3^2) = 2\exp(-\frac{r^2}{2}Q_3^2) = 2\sqrt{K_2(Q_3^2)} \tag{11.39}$$

with

$$Q_3^2 \equiv Q_{123}^2 = (P_1 + P_2 + P_3)^2 - 9M_\pi^2 = Q_{12}^2 + Q_{13}^2 + Q_{23}^2. \tag{11.40}$$

In a more general case, chaotic and coherent components may coexist in the pion source [13, 140–142]. Although the coherent source by itself does not cause any BE correlation, superposition of chaotic and coherent radiation changes the interference pattern and the interrelation between correlations of different order, as between (11.34) and (11.39).

Pion radiation by a partially coherent source (with the chaoticity parameter $p = \langle n_{\rm ch}\rangle/\langle n\rangle$, where $\langle n_{\rm ch}\rangle$ denotes the chaotic fraction in the pion average multiplicity) can be described in the framework of quantum statistics, applying an approach analogous to that used in quantum optics. Usually, it is assumed that the coherent source is pointlike and the chaotic source has a Gaussian form, $f(\mathbf{x}) \sim \exp(-|\mathbf{x}|^2/r^2)$, with $r(r.m.s.) = r\sqrt{3} = r_G\sqrt{3/2}$. For the simplified case of a symmetric configuration in momentum space one has

$$Q_2^2 \equiv Q_{12}^2 = Q_{13}^2 = \ldots = Q_{(q-1)q}^2 = 2Q_q^2/q(q-1) \tag{11.41}$$

(for example, $Q_2^2 = \frac{1}{3}Q_3^2 = \frac{1}{6}Q_4^2$), where Q_2^2 and Q_3^2 are defined in (11.8) and (11.40), and, in general,

$$Q_q^2 = (\sum_{i=1}^{q} P_i)^2 - (qM_\pi)^2 \ . \tag{11.42}$$

The normalized two-, three- and four-pion inclusive densities are:

$$\begin{aligned}
R_2(Q_2^2) &= 1 + 2p(1-p)\exp(-r^2Q_2^2) + p^2\exp(-2r^2Q_2^2), &(11.43)\\
R_3(Q_3^2) &= 1 + 6p(1-p)\exp(-\frac{1}{3}r^2Q_3^2) + 3p^2(3-2p)\exp(-\frac{2}{3}r^2Q_3^2)\\
&+ 2p^3\exp(-r^2Q_3^2), &(11.44)\\
R_4(Q_4^2) &= 1 + 12p(1-p)\exp(-\frac{1}{6}r^2Q_4^2) + 6p^2(7-8p+2p^2)\exp(-\frac{1}{3}r^2Q_4^2)\\
&+ 4p^3(11-9p)\exp(-\frac{1}{2}r^2Q_4^2) + 9p^4\exp(-\frac{2}{3}r^2Q_4^2). &(11.45)
\end{aligned}$$

The normalized two- and three-pion correlation functions are:

$$K_2(Q_2^2) = 2p(1-p)\exp(-r^2Q_2^2) + p^2\exp(-2r^2Q_2^2), \tag{11.46}$$

$$K_3(Q_3^2) = 6p^2(1-p)\exp(-\frac{2}{3}r^2Q_3^2) + 2p^3\exp(-r^2Q_3^2). \tag{11.47}$$

For $p \to 1$ (completely chaotic source) and symmetric configuration, (11.46) and (11.47) reduce to (11.34) and (11.39), respectively. For $p < 1$ (partially coherent source), the normalized correlation functions (11.46) and (11.47), in contrast with (11.34) and (11.39), now contain two exponential terms. The maximum values of the normalized densities (11.43) and (11.44) are smaller than, respectively, $R_2(0) = 2$ and $R_3(0) = 6$ expected for a completely chaotic source (cf. (11.43) and (11.44) at $p = 1$ with (11.33), (11.34) and (11.35)).

One should stress, that the above properties of a partially coherent source can be reproduced by a superposition of two completely chaotic sources with radii $r'_{\rm G}$ and $r''_{\rm G}$ accidentally related by $r''_{\rm G}/r'_{\rm G} \approx \sqrt{2}$ [47]. However, the predictions of quantum statistics are definite and contain, in a simplified case, only two free parameters (p and r) for the correlation of all orders. So, an experimental observation of higher-order correlations allows, in principle, to establish quantum statistics, a basic approach in various physics fields (such as quantum mechanics and field theory, condensed matter physics, nuclear physics etc.), also in multiparticle production processes.

11.4.2 Experimental results

11.4.2.1 The normalized higher-order densities

Three-particle BE correlations studied in [24, 55, 59, 143–145] were found to be consistent with (11.39). However, higher-order BE correlations measured by UA1 [145] and NA22 [146] manifest some inconsistency with the expectation from quantum statistics: while the chaoticity parameter p is practically constant, the parameter r turns out to increase with increasing order q.

The normalized q-particle densities $R_q(Q_q^2)$ are determined as

$$R_q(Q_q^2) = N_q(Q_q^2)/N_q^{\rm BG}(Q_q^2), \tag{11.48}$$

where $N_q(Q_q^2)$ is the number of q-particle combinations at given Q_q^2, $N_q^{\rm BG}(Q_q^2)$ that for a mixed-event reference sample.

The genuine three-particle correlation function $C_3(Q_3^2)$ and its normalized form $K_3(Q_3^2)$ are extracted by means of (7.23). The product $\rho_1\rho_1\rho_1$ in (7.23) is determined by combining three particles with a given Q_3^2 randomly chosen from different events with $n_{\pi^-} \geq 3$. The product $\rho_2\rho_1$ is determined by combining three particles with a given Q_3^2, two of which are chosen from the same event and the other from another event with $n_{\pi^-} \geq 3$. The density ρ_3 is determined by combining three particles with a given Q_3^2 chosen from the same event.

In Figs. 11.11a,c,e the NA22 ratios $R_q(Q_q^2) = N_q(Q_q^2)/N_q^{\rm BG}(Q_q^2)$ are shown for q=2,3,4, respectively. Figures 11.11b,d,f present the same distributions corrected for Coulomb repulsion of the like-charge pions in the final state by the factor [59] $W_q = \prod_{i<j}^q G^{-1}(Q_2)$, where $G(Q_2)$ is defined in (11.21).

11.4.2.2 The q-dependence of the radius parameter r

In Fig. 11.12, the parameters r and p of Eqs. (11.43)-(11.45) are presented as a function of q, the order of the correlation, for p$\bar{\rm p}$ collisions at $\sqrt{s}$ = 630 and 900

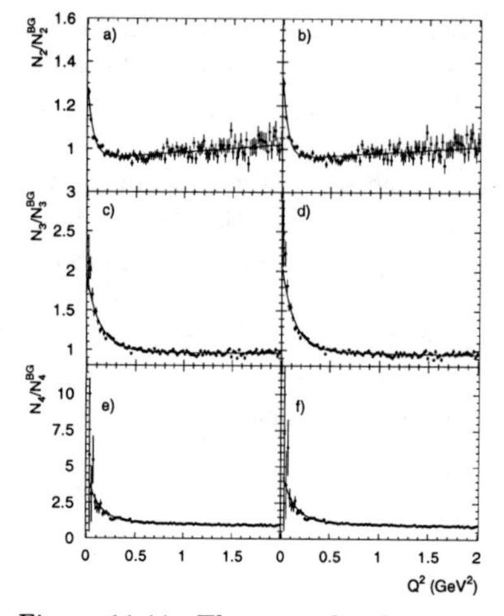

Figure 11.11. The normalized two-, three- and four-particle inclusive densities not corrected (a,c,e) and corrected (b,d,f) for Coulomb interaction in the final state, as a function of Q_q^2. Curves show the fits by expressions (11.43)-(11.45) multiplied by a background factor [146].

Figure 11.12. The extracted parameters r and p as a function of the order q of the correlation, for UA1 at 630 and 900 GeV [145] and NA22 at 22 GeV [146] compared to FRITIOF with BE at 22 GeV.

GeV [145] and π^+p collisions at 22 GeV [146]. A substantial increase of r with increasing q is observed.

Bose-Einstein correlations can be simulated in FRITIOF and JETSET (see Sub-Sub-Sect. 11.2.3.1 above). Practically, after the generation of the pion momenta the generated values of the momenta of all identical pions are modified in such a way that their momentum vector differences are reduced with a quantity determined by the chosen form of parametrization and given parameters of correlation strength λ and radius r in order to describe the like-charged *two-pion* distribution for $R_2(Q_2)$. Even though higher orders are not explicitely introduced, the FRITIOF results, also plotted in Fig. 11.12, indicate a q-dependence quite similar to the data.

In the quantum-statistical model, the parameters r and p are supposed to be the same for *all* orders, in clear contradiction with the trend of the combined data. This, however, does not necessarily invalidate the QS approach as such, in view of several simplifying assumptions underlying (11.43)-(11.45): symmetric configuration of q particles, pointlike coherent source, stationary source, additivity (as opposed to multiplicativity) of coherent and chaotic components, etc. [140, 141].

11.4.3 Genuine three-particle correlations

Non-zero genuine correlations up to order $q = 5$ were first established for all charge combinations by the NA22 collaboration [147] in terms of the cumulant moments shown in Fig. 10.42. So they must show up here, as well. The function $K_3(Q_3^2) + 1$ is given in Fig. 11.13a for like-charged triplets. A non-zero K_3 is indeed observed in the data for $Q_{3\pi}^2 < 0.2$ (GeV/c)^2 [146], but not in FRITIOF with BEC.

Both observations, the existence of genuine three-particle correlations and the underestimate in JETSET are supported by DELPHI [148]. In Fig. 11.13b, the three-particle correlation function $K_3 + 1$ is shown for like-charged triplets (upper) and unlike-charged triplets (lower), respectively, together with the prediction of JETSET with and without BE correlations. The parameters used to include the BE correlations are the same as in the two-particle correlation study of DELPHI [61]. The model is in reasonable agreement with the data for the $(++-)$ and $(+--)$ configurations but underestimates the enhancement for the $(+++)$ and $(---)$ correlations. Bose-Einstein interference in JETSET not only changes the distribution of like-charged correlations, but also the unlike-charged ones and leads to better agreement with the data. Due to particular implementation of BE correlations, particle distributions and invariant masses of jets are changed [6]. This observation is important for studies at high energies, particularly for the W mass measurement at LEP2.

Statistically better evidence for genuine three-particle BE correlations comes from OPAL [149]. This is shown in Fig. 11.13c, together with a Gaussian fit over the range $0.25 < Q_3 < 2.0$ GeV, giving $r_3 = 0.580 \pm 0.004 \pm 0.025$ fm and $\lambda_3 = 0.504 \pm 0.010 \pm 0.041$. Within two standard deviations, the value for r_3 agrees with the relation $r_3 = r_2/\sqrt{2}$ (see (11.39)) when compared to r_2 obtained in [81].

The question is, whether the observed genuine three-particle correlation can be fully expressed in terms of the simple product of two-particle correlation functions according to (11.39), or whether information can be extracted on the relative phases of (11.36). If relation (11.39) holds, the function $1 + K_3(Q_3^2)$ can be described by the parameters $r_2 = 0.85 \pm 0.01$ fm and $\lambda_2 = 0.38 \pm 0.02$ deduced from the fit of the normalized two-particle density $R_2(Q_2^2)$ in Fig. 11.11:

$$K_3(Q_3^2) + 1 = \gamma[1 + 2\lambda_2^{3/2} \exp(-\frac{1}{2} r_2^2 Q_3^2)](1 + \delta Q_3^2) \ . \qquad (11.49)$$

Within the errors of NA22, the resulting parameters r_2 and λ_2 do not contradict those of the two-particle correlations and, therefore, are consistent with an incoherent production of pions, but do not allow to reveal new information on the phase of the Fourier transform $F(Q_2^2)$. DELPHI and OPAL unfortunately did not make use of this possibility, but L3 did [117]:

Fig. 11.14 shows ω (Eq.(11.37)) as a function of Q_3 for the case that the cumulants K_2 and K_3 are parametrized in terms of a first-order Edgeworth expansion around a Gaussian (see Sect. 11.5 below). The L3 result is consistent with $\omega = 1$ for all Q_3 and therefore with full incoherence.

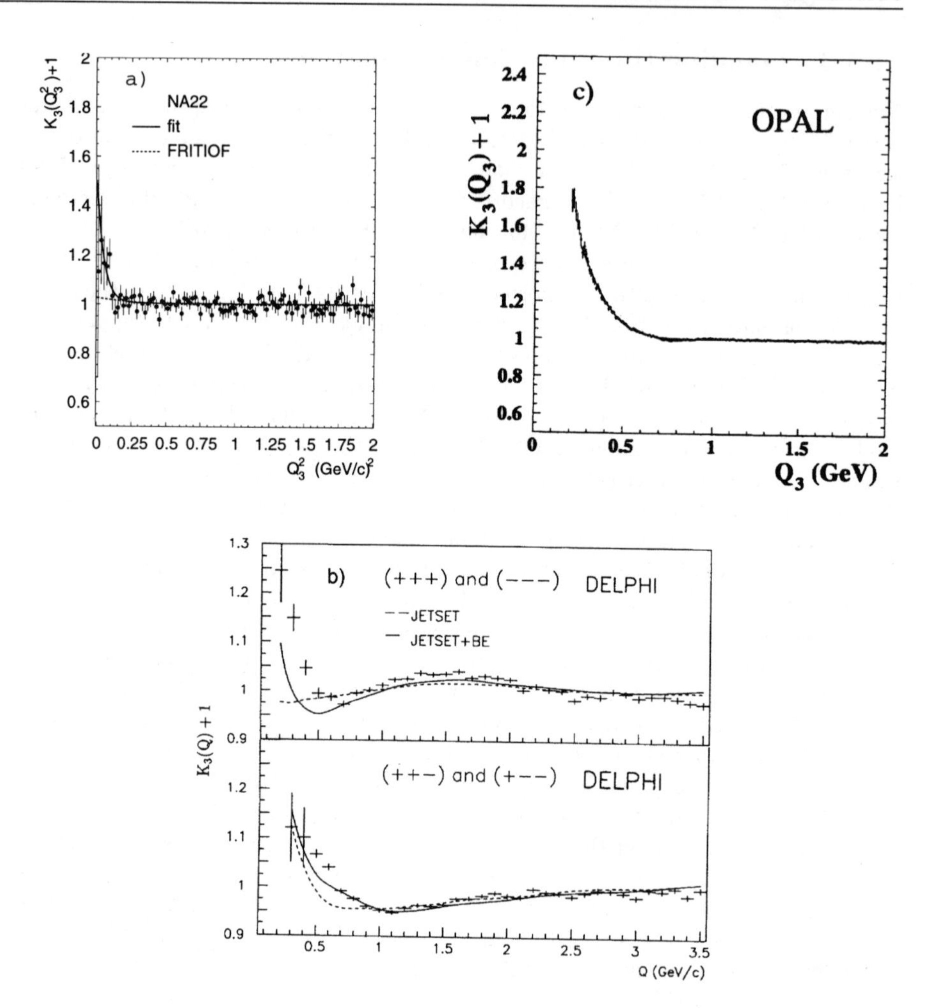

Figure 11.13. a) The normalized three-particle correlation function $K_3(Q_3^2)$ added to 1, for like-charged triplets. The full line is the result of a fit by (11.49), the dashed line corresponds to FRITIOF results with BEC [146]. b) The function $K_3(Q)+1$ for like-charged triplets and unlike-charged triplets. The predictions of JETSET without BE (dashed line) and with BE correlations (full line) are also shown [148]; c) Like-charged triplets after Coulomb correction, with a Gaussian fit (solid line) [149].

Three-pion correlations have also been studied in heavy-ion collisions [150–152]. In agreement with earlier observations from the cumulant moments in Sect. 10.5 above, $\langle\omega\rangle = 0.20 \pm 0.02 \pm 0.19$ is obtained with no clear Q_3 dependence, i.e. no genuine three-particle correlations are found outside the (large) errors for SPb by

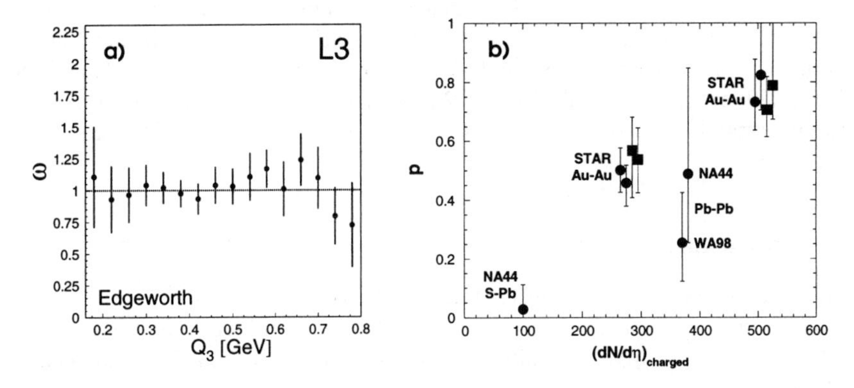

Figure 11.14. a) The ratio ω as a function of Q_3 assuming R_2 is described by the first-order Edgeworth expansion [117]. b) Chaotic fraction p calculated from Eq. (11.37) for heavy-ion data, as a function of charged-particle density [150].

NA44 [152]. The authors interpret this result as evidence for partial coherence [152]. This is confirmed in the last ref. [42].

What is particularly remarkable, however, is that the same experiment (NA44) using the same methodology, finds an average $\langle\omega\rangle = 0.85 \pm 0.02 \pm 0.21$ for PbPb collisions [152] and that this is supported by a value of $\langle\omega\rangle = 0.606 \pm 0.005 \pm 0.179$ reported earlier by WA98 [151]. At RHIC [150], ω is strongly Q_3 dependent and approaches $\omega = 0$ at $Q_3 \approx 0.1$ GeV, but the extrapolation down to $Q_3 = 0$ is consistent with unity (fully chaotic source) for central collisions, and (after proper treatment of Coulomb corrections) to only about $\omega = 0.6$ for mid-central collisions.

Fig. 11.14b shows the heavy-ion results for the chaotic fraction p calculated from Eq. (11.38) as a function of charged-particle density [150]. Within the uncertainties shown, there appears to be a systematic increase with increasing density.

So, if we trust NA44 (and we have no reason not to) we end up with a dilemma within the framework of conventional pion interferometry.

i) e^+e^- collisions are consistent with fully incoherent pion production ($\omega \sim 1$).

ii) SPb collisions are consistent with coherent pion production ($\omega \sim 0$).

iii) PbPb and AuAu are somewhere in between.

It could not be more opposite to any reasonable expectation from conventional interferometry. The hint for an alternative interpretation comes from a comparison of Eqs. (11.37) and (7.24). What conventional interferometry calls the cosine of a phase may, in fact, have nothing to do with a phase. It is simply the ratio of K_3 and twice $K_2^{1/2}$. It may be a challenge for the string model to explain why this is unity in e^+e^- annihilation. If that can be explained, the rest looks easy and very much in line with the behavior of the strength parameter λ discussed at the end of Sub-Sect. 11.2.2: For independent sources, the ratio $\omega \sim K_3/2K_2^{1/2}$ (see Eq. (11.39) decreases as $N^2/2N^{1/2} \propto N^{3/2}$, with N being the number of independent sources. As λ does, it decreases with increasing atomic mass number A up to SPb collisions. The saturation or increase of λ at and above this A value has been explained by percolation [89] of strings in Sub-Sect. 11.2.2. Exactly the same explanation can be used to understand an increase of the ratio (not phase!) ω between SPb and PbPb collisions [153].

11.4.4 Summary

In addition to observations from Sects. 11.2 and 11.3 we find:
7. Strong genuine three-particle correlations exist in hadron-hadron and e^+e^- collisions. They seem to disappear within errors in low-A ion collisions, but reappear in large-A ones.

11.5 The functional form of the correlation

More important than the parameters extracted from "forcing" the two-particle correlation function into a fit by a pre-selected parametrization, is the actual experimentally observed shape of this distribution itself.

The simple geometrical interpretation of the interference pattern based on the optical analogy as in Sect. 11.1 is invalid when emitters move relativistically with respect to each other, leading to strong correlations between the space-time and momentum-energy coordinates of emitted particles [154, 155]. Correlations of this type e.g. arise due to the nature of inside-outside cascade dynamics [12] as in color-string fragmentation [156]. In the interpretation of BEC by Andersson and Hofmann [21] (see Sub-Sect. 11.1.4 above) in the string model, the length scale measured by BEC is therefore not related to the size of the total pion emitting source, but to the space-time separation between production points for which the momentum distributions still overlap. This distance is, in turn, related to the string tension. The model predicts an approximately exponential shape of the correlation function

$$R_2(Q) = R_0(1 + \lambda \exp(-rQ)) , \tag{11.50}$$

with r independent of the total interaction energy.

In Chapter 10 we discussed the effort devoted to the study of fluctuation phenomena in multiparticle production processes. Scale-invariant dynamics is strongly connected with Bose-Einstein correlation. Scale invariance implies that multiparticle correlation functions exhibit power-law behavior over a considerable range of the relevant relative distance measure (such as Q^2) in phase space. As such, BEC from a static source do not exhibit power-law behavior. However, a power law is obtained if the size of the particle source fluctuates event-by-event, and/or, if the source itself

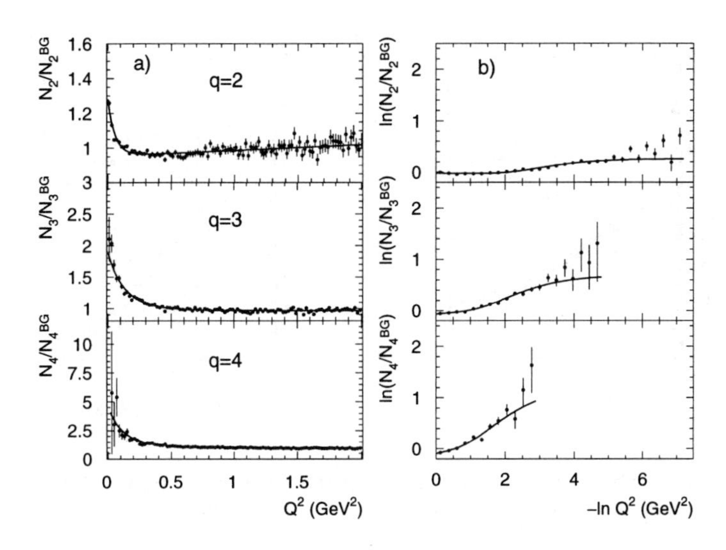

Figure 11.15. The normalized two-, three- and four-particle inclusive densities as a function of Q^2 (left) and -lnQ^2 (right). Curves show the multi-Gaussian fits according to [141].

is a self-similar (fractal-like) object extending over a large volume [157]. In these studies, R_2 is parametrized using the form

$$R_2(M) = A + B\left(\frac{1}{Q^2}\right)^{\beta} \quad . \tag{11.51}$$

In fact, is has been convincingly demonstrated [158] that already the interplay of resonance decays and some 50% directly produced pions can give such a power-law scaling at small Q, provided that the formation time of resonance and direct pions is short ($\tau_{\mathrm{f}} \approx 0.2$ fm/c).

The usually "reasonable" χ^2 values of the single-Gaussian fits hide the fact the Gaussian parametrization in general undershoots the data at low values of Q^2. For the case of two-particle correlations, this has been demonstrated convincingly by NA22 [159] and UA1 [160] (see Figs. 10.45), but deviations from a Gaussian are also observed in lepton-hadron [66,67], e^+e^- [61], and even heavy-ion [161] collisions.

Deviations from a (multi-) Gaussian form are also observed for higher-order correlations. In Fig. 11.15a the NA22 data [146] on BE correlations of order $q = 2$ to 4 are plotted as a function of Q^2, in conventional linear scale identical to Fig. 11.11. The curves are the fits by a q-fold Gaussian parametrization [141]. In Fig. 11.15b the same data and the same fits are repeated for $Q^2 < 1$ GeV^2 on ln-ln scale. Even though the statistical errors at small Q^2 are large (the very reason why small Q^2 does not contribute much to χ^2), it is obvious that the small-Q^2 points *systematically* lie

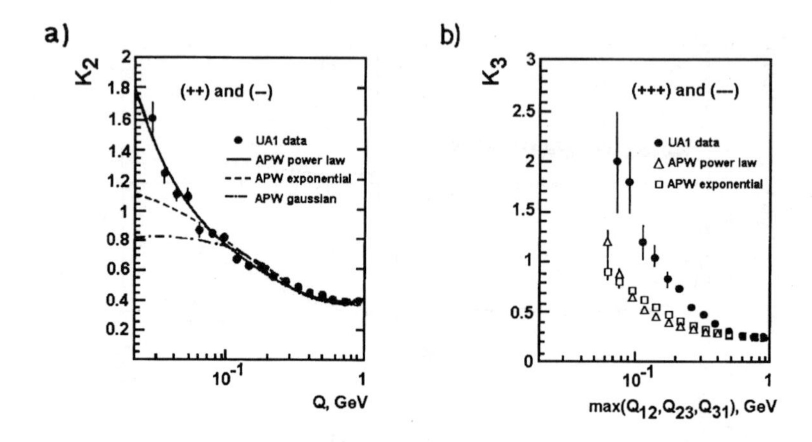

Figure 11.16. a) Second-order cumulant with fits of the forms given. b) Third-order cumulant with APW predictions based on K_2 [162].

above the multi-Gaussian fit, thus supporting a power-law behavior. This effect is even enhanced when the data are corrected for Coulomb repulsion.

Fig. 11.16a shows the second-order cumulant K_2 as a function of Q (on log-scale) compared to a general quantum statistical model by Andreev, Plümer and Weiner (APW), based on a classical source current formalism applied successfully in quantum optics [163]. It includes as special cases more specific models such as [13] and [141]. The APW normalized cumulant predictions are built from normalized correlators d_{ij}, the on-shell Fourier transforms of classical space-time current correlators. The specific parametrizations tested in Fig. 11.16 are

$$\begin{aligned} \text{Gaussian}: &\quad d_{ij} = \exp(-r^2 Q_{ij}^2) \\ \text{exponential}: &\quad d_{ij} = \exp(-r Q_{ij}) \\ \text{power law}: &\quad d_{ij} = Q_{ij}^{-2\beta}\ . \end{aligned} \tag{11.52}$$

For constant chaoticity λ and real-valued currents, the APW model predicts

$$K_2(Q_{12}) = 2\lambda(1-\lambda)d_{12} + \lambda^2 d_{12}^2\ , \tag{11.53}$$

$$K_3(Q_{12}, Q_{23}, Q_{31}) = 2\lambda^2(1-\lambda)[d_{12}d_{23} + d_{23}d_{31} + d_{31}d_{12}] + 2\lambda^3 d_{12}d_{23}d_{31}\ . \tag{11.54}$$

The data are best described by the power-law form of d_{ij} and clearly exclude a Gaussian parametrization.

Also K_3, plotted in Fig. 11.16b, shows a power-law increase which is even stronger than expected in the APW model.

So, there is ample room for improvement of the models and we believe that the recently developed methods of studying the correlations (higher-order cumulants,

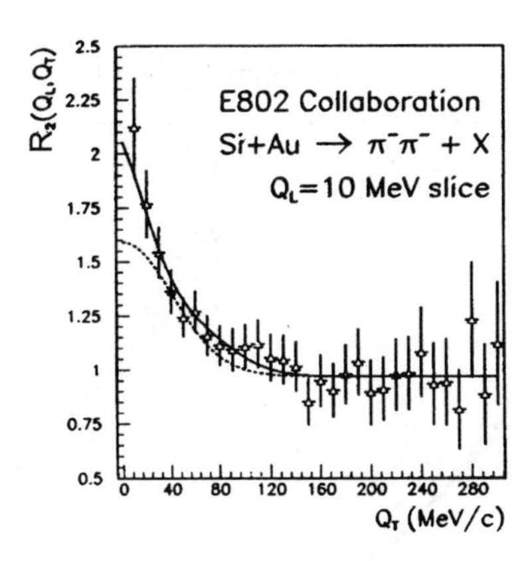

Figure 11.17. Projection onto Q_{T} of a two-dimensional Gaussian (dashed) and Edgeworth (solid) fits for a small-Q_{L} slice [165].

higher dimensionality, alternative parametrizations of the correlation function) discussed in the last three sections have opened the way for an improvement of these models.

A very interesting extension of the usual Gaussian approximation of the BE correlation function is the Edgeworth expansion [164] as suggested in [165],

$$R_2(Q) = \gamma(1 + \lambda^* \exp(-t^2/2)[1 + \frac{\kappa_3}{3!}H_3(t) + \frac{\kappa_4}{4!}H_4(t) + \ldots]) \ , \tag{11.55}$$

with $t = \sqrt{2}Q \cdot r$, H_n being the n-th order Hermite polynomial, and κ_n the n-th order cumulant moment of the correlation function. The Hermite polynomials of odd order vanish at the origin, so that

$$\lambda = \lambda^*[1 + \kappa_4/8 + \ldots] \ . \tag{11.56}$$

Similarly, the exponential approximation can be extended by a Laguerre expansion [165]. A generalization to higher dimensions is straightforward [165], except for possible correlations between the Q_i components.

The influence of the non-Gaussian shapes was studied [165] on AFS [29, 144], E802 [166] and NA44 [167] data. In Fig. 11.17, the transverse component Q_{T} of a 2D Edgeworth fit is compared to that of a 2D Gaussian fit to the E802 data. The deviation from a Gaussian (dashed) is obvious, and the Edgeworth expansion (full line) is flexible enough to describe it (with $\lambda = 1$!). Recently, results more satisfactory than those obtained with either Gaussian or exponential parametrizations were obtained in a 3D analysis of e^+e^- collisions at the Z-mass [168].

Fig. 11.18 shows that even an exponential is not steep enough to reproduce the fast increase of K_2 in hadron-hadron data. An interesting observation of [165] is,

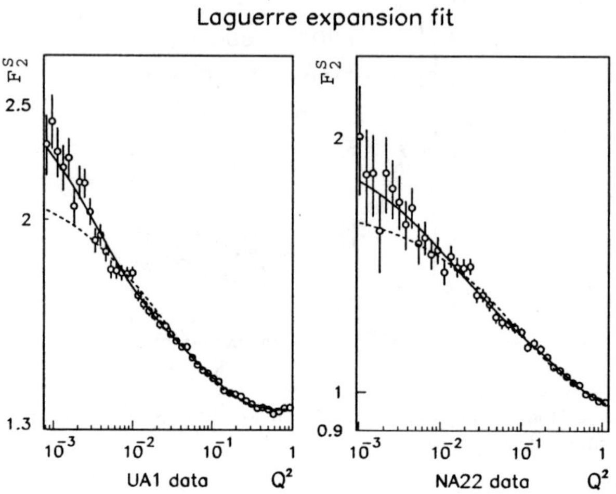

Figure 11.18. The figures show F_2^s which is proportional to the two-particle Bose-Einstein correlation function, as measured by the UA1 and the NA22 Collaborations. The dashed lines stand for the exponential fit, the solid lines for that with the Laguerre expansion [165].

that a Laguerre expansion of an exponential can reproduce these UA1 and the NA22 data. However, at low Q^2 data are still systematically above the fit and a power-law fit is reported in [165] to give similarly good χ^2/NDF with a smaller number of fit parameters. With a core-halo model [32] strength parameter of $\lambda_* = 1.14 \pm 0.10$ (UA1) and $\lambda_* = 1.11 \pm 0.17$ (NA22), i.e., at maximum possible value (unity), there are either other than BE correlations at work or all resonances are resolved at these low Q^2 values. This may imply the connection between the observed power-law behavior (intermittency) and resonance contributions of BE correlations [157].

Functions $f(r)$ which behave asymptotically as

$$f(r) \to |r|^{-1-\mu} \quad \text{for} \quad |r| \to \infty \tag{11.57}$$

with non-finite variance are naturally described by Lévy-stable distributions (see Sect. 7.7, above). A feature particularly useful for the purpose of BEC [169] is that the Fourier transform (characteristic function) of a symmetric ($\beta = 0$) stable distribution is

$$F(Q) = \exp(iQ\delta - |\gamma Q|^{\mu}), \tag{11.58}$$

from which follows that

$$R_2(Q) = 1 + \exp(-|rQ|^{\mu}) \tag{11.59}$$

with $r = 2^{1/\mu}\gamma$. Equation (11.59) is a generalization of the Gaussian ($\mu = 2$) or exponential ($\mu = 1$) distribution to $0 < \mu \leq 2$.

Fitting this form to the data of Fig. 11.18 gives Lévy indices $\mu = 0.49 \pm 0.01$ for UA1 and $\mu = 0.67 \pm 0.07$ for NA22. The quality of the fits is similar to that of the Laguerre expansion above. The advantage of the Lévy-stable fits is, however, that they relate the Lévy index μ to a power-law behavior of the configuration-space density fluctuations at large distance. Finally, we note that, according to Eq. (7.144), the ratio ω of Eq. (11.37) turns out to be sensitive to the asymmetry parameter β, since

$$\omega = \cos\left\{\frac{\beta}{2} r^{\mu} \tan\frac{\pi\mu}{2}\sum_{i,j} Q_{ij}^{\mu}\mathrm{sign}Q_{ij}\right\} . \tag{11.60}$$

11.6 Multi-dimensional parametrization and shape of the source

The final-state system we are observing in a high-energy particle collision is a dynamical system with fast expansion, possible re-scattering, shadowing, etc. Interferometric pion-source size parameters are, therefore, not simple geometrical sizes at the time hadrons cease to interact (freeze-out). They depend on kinematical variables such as the particle four-momentum components, particle species and (for heavy-ion collisions) on the centrality of the collision. A systematic measurement of Bose-Einstein interference in its fully-dimensional kinematical region is, therefore, necessary, to separate the various dynamical effects.

11.6.1 Directional dependence

Pion interferometry, indeed, not only allows to measure the average radius of a pion source segment, but also to determine its shapc [47]. The latter can be obtained from the dependence of the size on a direction e.g. given by the angle θ of the c.m.s. momentum difference $\mathbf{q} = \mathbf{p}_1 - \mathbf{p}_2$ with respect to the collision or event axis. For hadronic collisions, such shape measurements were reported in [36,53,170–172], where transverse and longitudinal radii r_{T} and r_{L} of the pion source were estimated in the framework of the surface emission model proposed by Kopylov and Podgoretskiĭ [2]. The pion emission region is found to be oblate ($r_{\mathrm{L}} < r_{\mathrm{T}}$) at the lower energies [170–172], but prolate ($r_{\mathrm{L}} > r_{\mathrm{T}}$) at higher energies [36,53]. In $\mathrm{e}^+\mathrm{e}^-$ collisions at PEP and PETRA energies [23,24,59] using a Gaussian approximation in the two-pion rest frame, it was consistent with being spherical.

A method for direct determination of the ratio $a = r_{\mathrm{T}}/r_{\mathrm{L}}$ from the angular distribution of the vector $\mathbf{q}$ itself was proposed by Podgoretskiĭ and Cheplakov [173]. As a minimal assumption on the form of the pion source, rotational symmetry is used around the interaction axis.

In general, the angular distribution of $\mathbf{q}$ for pion pairs with $\mid \mathbf{q} \mid < q_{\mathrm{cut}}$ and *very small* c.m.s. energy difference $q_0 =| E_1 - E_2 |$ is given [173] by

$$\varphi(\cos\theta) \propto \int_0^{q_{\rm cut}} | f[q^2 r_{\rm T}^2 + q^2(r_{\rm L}^2 - r_{\rm T}^2)\cos^2\theta] |^2 q^2 {\rm d}q \qquad (11.61)$$

$$= [r_{\rm T}^2 + (r_{\rm L}^2 - r_{\rm T}^2)\cos^2\theta]^{-1.5} \int_0^{x_{\rm cut}} | f(x^2) |^2 x^2 {\rm d}x \ , \qquad (11.62)$$

where $x^2 = q^2[r_{\rm T}^2 + (r_{\rm L}^2 - r_{\rm T}^2)\cos^2\theta]$ and the function $f(x^2)$ is the Fourier transform of the spatial distribution of the source, normalized to unity as $\mathbf{q} \rightarrow 0$. One can show from (11.61) that, independently of the particular form of $f(x^2)$, the function $\varphi(\cos\theta)$ becomes constant at sufficiently small $q_{\rm cut}$, $q_{\rm cut} \ll 1/r_{\rm L}$. At sufficiently large $q_{\rm cut}$ (i.e. above the correlation region) the integral in (11.61) is practically independent of $q_{\rm cut}$, and the angular distribution $\varphi(\cos\theta)$ (normalized to unity in the interval $-1 \leq \cos\theta \leq 1$) turns out to be

$$\varphi(\cos\theta) = \frac{a^2}{2[a^2 + (1 - a^2)\cos^2\theta]^{1.5}} \ . \qquad (11.63)$$

The ratio $a = r_{\rm T}/r_{\rm L}$ can be determined by fitting distribution (11.63) to the experimental angular distribution obtained after subtraction of a background (reference) distribution for which like-pion interference effects are absent.

The (large) advantages of the method described above are that it does not require a fit to any particular form of the spatial distribution of the source, is insensitive to the strength of the correlation, and smaller statistics is needed than required for separate measurement of $r_{\rm T}$ and $r_{\rm L}$. The method was successfully applied to hh-collisions at $\sqrt{s} = 53$, 63 GeV [174] and hp and hA collisions at 22 GeV [175].

The angular distribution of the correlated pion pair is presented in Fig. 11.19a. The result of a fit by (11.63) is given as solid line, the obtained parameter value $a_{\rm fit}$ is indicated in the figure.

The ratio $r_{\rm T}/r_{\rm L}$ is extracted by extrapolation of the $(q_0)_{\rm cut}$-dependence of the parameter a to $(q_0)_{\rm cut} = 0$. The result is $r_{\rm T}/r_{\rm L} = a\ ((q_0)_{\rm cut} = 0) = 0.55 \pm 0.06$. This agrees with the ratio $a = 0.56 \pm 0.08$ of the transverse and longitudinal radii, $r_{\rm T} = 1.04 \pm 0.12$ fm and $r_{\rm L} = 1.85 \pm 0.13$ fm, measured separately in the same experiment [36], but due to the method of direct measurement the error is reduced. A value of $a = 0.60 \pm 0.02$ is obtained in [174] for p($\bar{\rm p}$)p-interactions at $\sqrt{s} = 53$ and 63 GeV, while a separate measurement of $r_{\rm T}$ and $r_{\rm L}$ had given $a = 0.55 \pm 0.07$ [53].

The multiplicity dependence of the shape is studied in Fig. 11.19b (upper left), again extrapolated to $(q_0)_{\rm cut}$ =0 at $q_{\rm cut}$ = 0.5 GeV/c. An indication is obtained that the source may become more elongated for larger multiplicities. The rapidity dependence is not clear (Fig. 11.19b, upper right). The elongation increases with increasing momentum $p =| \mathbf{p}_1 + \mathbf{p}_2 |$ of the pion pair in the c.m.s. (Fig. 11.19b, lower left). For pions with small transverse momentum $p_{\rm T} < 0.2$ GeV/c the source is less elongated than for the unbiased sample (Fig. 11.19b, lower right).

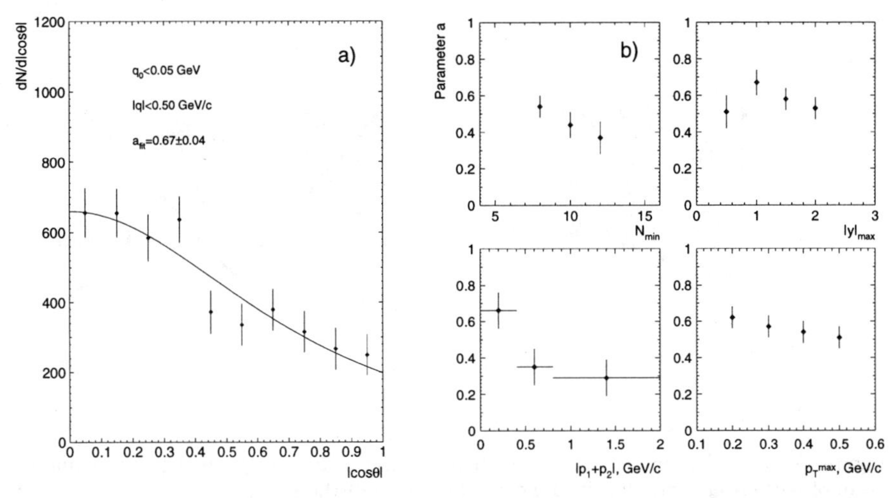

Figure 11.19. a) The angular distribution of correlated pion pairs. b) The dependence of the ratio $a = r_T/r_L$ extrapolated to $(q_0)_{cut} = 0$ at $q_{cut} = 0.5$ GeV/c on: minimum charged-particle multiplicity, maximum rapidity, pair momentum, maximal transverse momentum for pions produced in (π^+/K^+)p-interactions [175].

The ratio a extrapolated to $(q_0)_{\text{cut}}$ =0 at q_{cut} =0.5 GeV/c has a value of 0.55±0.06 for (π^+/K^+)p, 0.53±0.15 for (π^+/K^+)Al, and 0.33±0.21 for (π^+/K^+)Au interactions, i.e. the pion source is elongated along the collision axis, both for the meson-proton and meson-nucleus interactions. While no dependence of the elongation is found for the latter on the rapidity of the pair, indication for an increase is found with increasing event multiplicity, pair momentum and particle transverse momentum.

11.6.2 The Bertsch-Pratt Cartesian parametrization [176,177]

In a three-dimensional analysis, the three components of $\mathbf{Q}$ may be defined as longitudinal (Q_L), transverse sidewards (Q_{side}) and transverse outward (Q_{out}). The various transverse components are defined in Fig. 11.20a

as

$$\begin{aligned} \mathbf{Q}_T &= \mathbf{p}_{T1} - \mathbf{p}_{T2} \\ Q_{\text{side}} &= |\mathbf{Q}_T \times \mathbf{k}_T|/|\mathbf{k}_T| \\ Q_{\text{out}} &= \mathbf{Q}_T \cdot \mathbf{k}_T/|\mathbf{k}_T| \; . \end{aligned} \tag{11.64}$$

Note, however, that Q_L is not invariant to boosts along the event axis, while Q_{out} is not invariant to boosts along the direction of the transverse momentum $2\mathbf{k}_T$ of the pair.

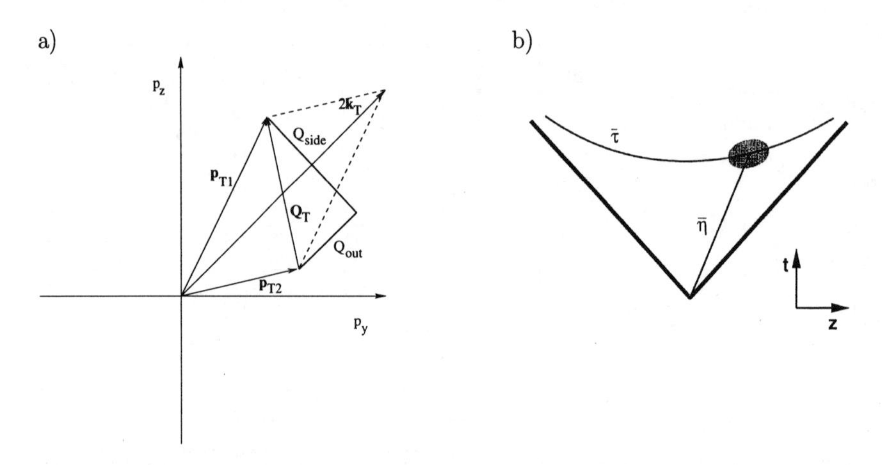

Figure 11.20. a) The momentum components in the transverse plane; b) Space-time diagram of particle emission for fixed mean momentum of the pair [178].

The normalized two-particle density Eq. (11.12) can then be parametrized in terms of the three relative-momentum components as

$$R_2(Q_{\rm L}, Q_{\rm out}, Q_{\rm side}, \mathbf{k}) = 1 + \lambda \exp[-r_{\rm L}^2(\mathbf{k})Q_{\rm L}^2 - \\ -r_{\rm out}^2(\mathbf{k})Q_{\rm out}^2 - r_{\rm side}^2(\mathbf{k})Q_{\rm side}^2 - 2r_{\rm out,L}^2(\mathbf{k})Q_{\rm out}Q_{\rm L}] \;, \qquad (11.65)$$

where $r_{\rm L}, r_{\rm out}, r_{\rm side}$ are the longitudinal, "out", and "side" effective dimensions of a source segment radiating the BE correlated pion pairs, while $r_{\rm out,L}$ represents an "out-longitudinal" cross term [179]. The other cross terms ($Q_{\rm out}Q_{\rm side}$ and $Q_{\rm side}Q_{\rm L}$) vanish for a cylindrically symmetric source.

In the Cartesian decomposition [176] and for a cylindrically symmetric source, this leads to (11.65) with

$$r_{\rm side}^2 = \langle \tilde{y}^2 \rangle \qquad (11.66)$$

$$r_{\rm out}^2 = \langle (\tilde{x} - \beta_{\rm T}\tilde{t})^2 \rangle \qquad (11.67)$$

$$r_{\rm L}^2 = \langle (\tilde{z} - \beta_{\rm L}\tilde{t})^2 \rangle \qquad (11.68)$$

$$r_{\rm out,L}^2 = \langle (\tilde{x} - \beta_{\rm T}\tilde{t})(\tilde{z} - \beta_{\rm L}\tilde{t}) \rangle \qquad (11.69)$$

$$r_{\rm out,side} = r_{\rm side,L} = 0 \qquad (11.70)$$

where $\tilde{x} = x - \langle x \rangle$ and $\boldsymbol{\beta} = \mathbf{k}/k_0$.

An advantage of the Bertsch-Pratt parametrization is that there are no kinematic constraints between $Q_{\rm side}, Q_{\rm out}$ and $Q_{\rm L}$. The Bertsch-Pratt radii contain a well-defined mixture of invariant longitudinal, temporal and transverse radii, but are

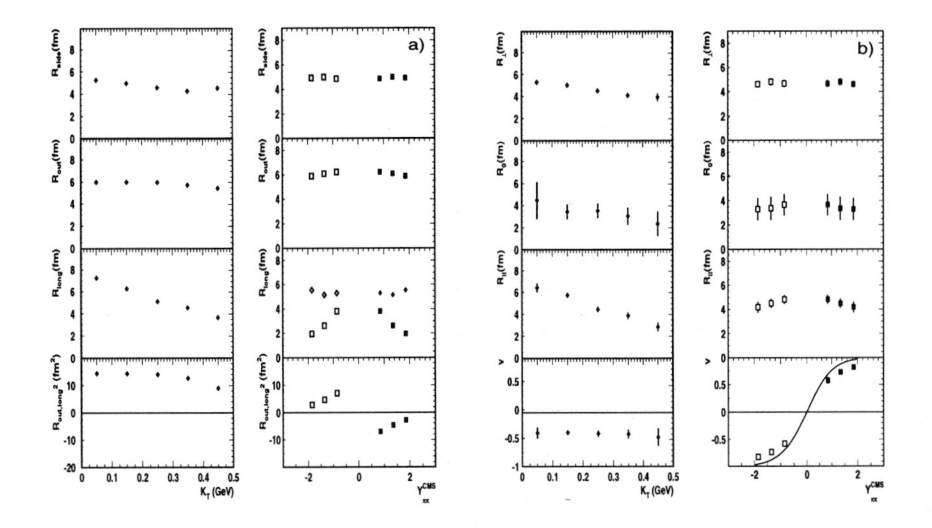

Figure 11.21. Size parameters as a function of k_T and cms $Y_{\pi\pi}$ (open squares show the result of reflection around midrapidity) [37].

not invariant themselves. They are particularly simple in the LCMS (longitudinally co-moving system or longitudinal cms) [40], in which the mean momentum of the pair has no longitudinal component ($\beta_L = 0$). In this frame, the information on the temporal scale couples only to the out direction and enters only in r^2_{out} and $r^2_{out,L}$.

The four size parameters are given for PbPb collisions in NA49 in Fig. 11.21a as a function of the average transverse momentum k_T and the rapidity $Y_{\pi\pi}$ of the pair [37]:

i) While r_{side} and r_{out} show rather little variation with k_T or $Y_{\pi\pi}$, r_L decreases both with increasing k_T and increasing $|Y_{\pi\pi}|$;

ii) r_{out} is about 20% larger than r_{side}.

iii) r_L is larger than both at small k_T, but decreases with increasing k_T and increasing $|Y_{\pi\pi}|$.

iv) $r^2_{out,L}$ is non-negligible, but decreases in absolute value with increasing $|Y_{\pi\pi}|$.

Note that $r^2_{out,L} = 0$ is expected in LCMS near midrapidity or for boost-invariant sources.

Consistent results were obtained in WA98 [161] and NA44 [180], for smaller nuclei by NA35 [181] and NA44 [113, 182] (both without considering the cross term), but the k_T (or m_T) dependence is stronger at RHIC [90] than at SPS energies.

The energy dependence of the size parameters for central PbPb or AuAu collisions at midrapidity and $p_T \approx 0.17$ GeV/c is given in Fig. 11.22 [90]. The λ parameter decreases from $\lambda \approx 1$ at low energy (see also [183]) to $\lambda \approx 0.5$ between 10 GeV and

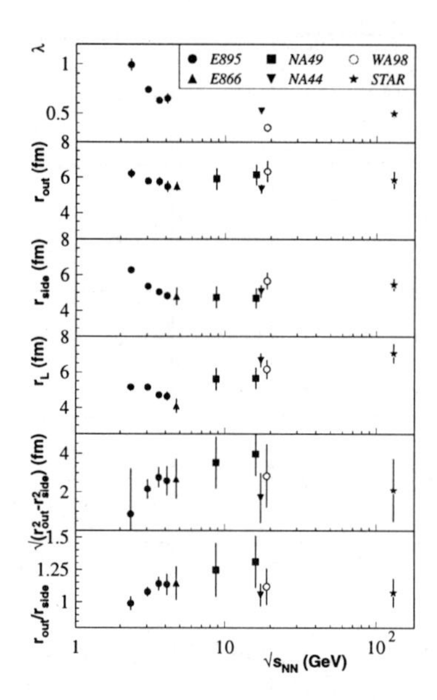

Figure 11.22. The energy dependence of the Bertsch-Pratt radius parameters for central AuAu (PbPb) collisions at midrapidity and $p_T \approx 0.17$ GeV/c [90].

RHIC. Also the three radii show some activity at low energy, but are surprisingly constant between SPS and RHIC. Above 10 GeV, an elongation $r_{\text{side}}/r_{\text{L}} < 1$ is observed.

An elongation (in the Bertch-Pratt system) is also found for meson-proton collisions by NA22 [184] (the latter with a cross term of 2.5 standard deviations), and at LEP [168, 185–187] and HERA [68]. The results are summarized in Table 11.2. In spite of the different selection criteria and reference samples, all experiments consistently demonstrate an elongated shape in the LCMS of the pion source (or rather region of homogeneity). On the other hand, $r_{\text{side}}/r_{\text{L}} = 1.08 \pm 0.03$ is found [168] for JETSET with BE. As is demonstrated by the three-dimensional analysis of NA22, L3 and OPAL, the elongation is visible in $r_{\text{side}}/r_{\text{L}}$, showing that it is not due to a Lorentz boost in the out direction.

A systematic study of the hierarchy of radii obtained from JETSET was recently performed in [188]. Starting from a spherically symmetric Gaussian correlator, the authors obtain

$$r_{\text{side}} > r_{\text{L}} > r_{\text{out}} \quad ,$$

contrary to the experimentally observed elongation, both when using momentum shifting or event weighting (see Sub-Sect. 11.2.3, above) to simulate the correlation.

Table 11.2. Elongation of the pion source ($r_{\rm T}$ corresponds to $Q_{\rm T} = \sqrt{Q^2_{\rm out} + Q^2_{\rm side}}$).

	NA22 "mixed" ref. hh	ZEUS "+−" ref. all events 4-8000 GeV^2
λ	0.25 ± 0.02	$0.44 \pm 0.01^{+0.01}_{-0.03}$
$r_{\rm L}$, fm	1.24 ± 0.14	$0.95 \pm 0.03^{+0.03}_{-0.08}$
$r_{\rm out}$, fm	0.84 ± 0.09	
$r_{\rm side}$, fm	0.54 ± 0.07	
$r_{\rm side}/r_{\rm L}$	0.43 ± 0.06	
$r_{\rm T}$, fm		$0.69 \pm 0.01^{+0.01}_{-0.06}$
$r_{\rm T}/r_{\rm L}$		$0.72 \pm 0.03^{+0.04}_{-0.03}$

	L3 "mixed" ref. all events	DELPHI "mixed" ref. 2-jet events	OPAL "+−" ref. 2-jet events	ALEPH "mixed" ref. 2-jet events
λ	$0.41 \pm 0.01^{+0.020}_{-0.019}$	$0.261 \pm 0.007 \pm 0.010$	0.443 ± 0.005	0.310 ± 0.005
$r_{\rm L}$, fm	$0.74 \pm 0.02^{+0.04}_{-0.03}$	$0.85 \pm 0.02 \pm 0.07$	$0.989 \pm 0.011^{+0.030}_{-0.015}$	0.767 ± 0.011
$r_{\rm out}$, fm	$0.53 \pm 0.02^{+0.05}_{-0.06}$		$0.647 \pm 0.011^{+0.024}_{-0.124}$	
$r_{\rm side}$, fm	$0.59 \pm 0.01^{+0.03}_{-0.13}$		$0.809 \pm 0.009^{+0.019}_{-0.032}$	
$r_{\rm side}/r_{\rm L}$	$0.80 \pm 0.02^{+0.03}_{-0.18}$		$0.818 \pm 0.018^{+0.008}_{-0.050}$	
$r_{\rm T}$, fm		$0.53 \pm 0.02 \pm 0.07$		0.470 ± 0.007
$r_{\rm T}/r_{\rm L}$		$0.62 \pm 0.02 \pm 0.05$		0.612 ± 0.010

Generalizing to asymmetric weights, the experimentally observed elongation can be reproduced, but finding a good set of input radius parameters turns out an involved procedure.

What is important to realize is that the measured longitudinal radius has nothing to do with the elongation of the $q\bar{q}$ string stretched in Z decay. We shall see in Sub-Sect. 11.6.10 that the full pion-emission function is of the order of 100 fm long, so that $r_{\rm L}$ only measures a small fraction of it. The reason for that is a strong $\mathbf{x}, \mathbf{p}$ correlation. Pions produced at a large distance on the string also have very different momenta and do not correlate. The "radii" therefore measure the effective size of the source segment radiating mesons with sufficiently small relative momentum (length of homogeneity), as shown in Fig. 11.20b.

Taking only the lowest-order non-Gaussian term into account, results in

$$r_2(Q_{\rm L}, Q_{\rm out}, Q_{\rm side}) = \\ \gamma\,(1 + \delta Q_{\rm L} + \varepsilon Q_{\rm out} + \xi Q_{\rm side})$$

$$\cdot\Big\{1+\lambda\exp\left(-r_{\rm L}^2Q_{\rm L}^2-r_{\rm out}^2Q_{\rm out}^2-r_{\rm side}^2Q_{\rm side}^2\right) \qquad (11.71)$$
$$\cdot\left[1+\frac{\kappa_{\rm L}}{3!}H_3(r_{\rm L}Q_{\rm L})\right]\left[1+\frac{\kappa_{\rm out}}{3!}H_3(r_{\rm out}Q_{\rm out})\right]\left[1+\frac{\kappa_{\rm side}}{3!}H_3(r_{\rm side}Q_{\rm side})\right]\Big\}\ ,$$

where κ_i ($i=$ L, out, side) is the third-order cumulant moment in the corresponding direction and $H_3(r_iQ_i)\equiv(\sqrt{2}r_iQ_i)^3-3\sqrt{2}r_iQ_i$ is the third-order Hermite polynomial. Note that the second-order cumulant corresponds to the radius r_i. Applying this expansion to the L3 data [168] improves the confidence level of the fit from 3% to 30%.

The non-zero values of the κ parameters indicate the deviation from a Gaussian. λ is larger than the corresponding Gaussian λ. The values of the radii, however, confirm the elongation observed from the Gaussian fit.

11.6.3 The generalized Yano-Koonin-Podgoretskiĭ scheme

Spatial and temporal properties can be decoupled [189] when describing the source as consisting of homogeneous regions parametrized by their space-time extent $(r_\parallel(\mathbf{k}), r_\perp(\mathbf{k}), r_0(\mathbf{k}))$ and their average longitudinal velocity $v(\mathbf{k})$ [190, 191]:

$$R_2(\mathbf{Q},\mathbf{k})=1+\lambda\exp[-Q_\perp^2r_\perp^2-\gamma^2(Q_0-vQ_\parallel)^2r_0^2-\gamma^2(Q_\parallel-vQ_0)^2r_\parallel^2]\ , \qquad (11.72)$$

with $\gamma=(1-v^2)^{-1/2}$, $Q_\perp$ and $Q_\parallel$ the components transverse and parallel to the event axis, and Q_0 the energy difference. Here, the size parameters do not depend on the longitudinal velocity of the reference frame.

Of course, the BP and YKP parametrizations are mathematically equivalent and the resulting size parameters obey the relations [189]

$$r_{\rm side}^2 = r_\perp^2 \qquad (11.73)$$
$$r_{\rm out}^2-r_{\rm side}^2 = \beta_{\rm T}^2\gamma^2(r_0^2+v^2r_\parallel^2) \qquad (11.74)$$
$$r_{\rm L}^2 = (1-\beta_{\rm L}^2)r_\parallel^2+\gamma^2(\beta_{\rm L}-v)^2(r_0^2+r_\parallel^2) \qquad (11.75)$$
$$r_{\rm out,L}^2 = \beta_{\rm T}[-\beta_{\rm L}r_\parallel^2+\gamma^2(\beta_{\rm L}-v)^2(r_0^2+r_\parallel^2)]\ . \qquad (11.76)$$

The results are shown in Fig. 11.21b as a function of $k_{\rm T}$ and $Y_{\pi\pi}$ [37]. Besides the $k_{\rm T}$ dependence of $r_\parallel$ indicating longitudinal expansion flow, there is very little $k_{\rm T}$ or $Y_{\pi\pi}$ dependence of the size parameters. The velocity profile is close to that expected for a boost-invariant longitudinal expansion as proposed by Bjorken [12].

A boost-invariant longitudinal velocity profile is found within the same scheme for MgMg collisions as low as 4.4 A GeV [192] as well as for SiAl and non-central SiAu collisions at 14.6 A GeV [76], but not for central SiAu collisions at 14.6 A GeV [76].

In a YKP frame where $v(\mathbf{k})=0$, the relations between the size parameters and the central moments of the emission function are (except for correction terms in $1/\beta_{\rm T}$)

$$r_\parallel^2 \approx \langle\tilde{z}^2\rangle,$$

$$
\begin{aligned}
r_\perp^2 &= \langle \tilde{y}^2 \rangle , \\
r_0^2 &\approx \langle \tilde{t}^2 \rangle .
\end{aligned} \tag{11.77}
$$

These parameters contain information[1] on e.g.
i) the expansion of the system,
ii) the boost invariance of the expansion,
iii) the expansion velocity gradient,
iv) the proper-time interval between onset of expansion and pion decoupling.

A disadvantage is that due to the constraint $0 \leq Q_0^2 \leq Q_L^2 + Q_\perp^2$, the kinematical regions in $Q_0^2 - Q_L^2$ narrow with decreasing $Q_\perp$ and the experimental analysis becomes difficult, except in the LCMS where v is small.

Theoretically [193], problems are that

a) the YKP radii contain components proportional to $1/\beta_T$, so diverge at very low k_T,

b) the YKP radii are not defined for all Gaussian sources, let alone non-Gaussian sources,

c) for expanding systems, the YKP "velocity" is not a flow velocity of a source element, but a combination of space-time variances of the source at a fixed mean momentum $\mathbf{k}$.

11.6.4 The Buda-Lund parametrization

A longitudinally boost-invariant formulation of the Bose-Einstein correlation function is given in [132, 178],

$$
R_2(Q, k) = 1 + \lambda \exp(-r_{=}^2 Q_{=}^2 - r_{\parallel}^2 Q_{\parallel}^2 - r_{..}^2 Q_{..}^2 - r_{:}^2 Q_{:}^2) , \tag{11.78}
$$

where $r_{=}$ (r-timelike) measures the width of the proper-time distribution, $r_{\parallel}$ (r-parallel) the invariant length parallel to the direction of expansion, $r_{..}$ and $r_{:}$ that in the side and out direction, respectively.

The longitudinally boost-invariant time-like, longitudinal, sideward and outward relative momenta are defined with the help of a further fit parameter, $\bar{\eta}$, the space-time rapidity of the center $\bar{x} = (\bar{t}, \bar{z})$ of particle emission (see Fig. 11.20b)

$$
\bar{\eta} = 0.5 \ln \frac{\bar{t} + \bar{z}}{\bar{t} - \bar{z}} , \tag{11.79}
$$

$$
\bar{\tau} = \sqrt{\bar{t}^2 - \bar{z}^2} \tag{11.80}
$$

transforming additively under a longitudinal Lorentz boost:

$$
Q_{=} = Q_0 \cosh \bar{\eta} - Q_L \sinh \bar{\eta} \tag{11.81}
$$

$$
Q_{\parallel} = Q_0 \sinh \bar{\eta} - Q_L \cosh \bar{\eta} . \tag{11.82}
$$

[1] Note: This interpretation is, however, unreliable if the measured $\lambda < 1$. This has been corrected for core/halo type of systems in [32].

Generalization to non-Gaussian correlations is simple in this parametrization, if one assumes the emission function to factorize into a function of τ, of η and of the transverse coordinates [178].

11.6.5 Longitudinal expansion and decoupling time

If the pion source expands longitudinally in a boost-invariant hydrodynamic way [12] and the pions decouple instantly at a proper time $\tau_{\rm f}$ with a thermal spectrum of (local effective) temperature $T_{\rm f}$, the Bertsch-Pratt parameter $r_{\rm L}$ is given [16] by

$$r_{\rm L}(y_0, Y_{\pi\pi}, \bar{m}_{\rm T}) = \sqrt{\frac{T_{\rm f}}{\bar{m}_{\rm T}}} \frac{\tau_{\rm f}}{\cosh(Y_{\pi\pi} - y_0)} , \tag{11.83}$$

when y_0 is the rapidity of the observer, $Y_{\pi\pi}$ that of the pion pair, and $\bar{m}_{\rm T}$ the average transverse mass of the two pions. As shown in [181] for a number of different nuclei and in [184] for meson-proton collisions, the $r_{\rm L}$ dependence on $Y_{\pi\pi}$ and $k_{\rm T}$ discussed for Pb+Pb in Fig. 11.21 is indeed consistent with this behavior. The $r_{\rm L}(Y_{\pi\pi})$ dependence is expected to disappear (at $\tau(Y_{\pi\pi}) = \tau_{\rm f}$ =const.) when evaluated in the LCMS [40]. This is comfirmed by the data [113, 181, 184].

Assuming a temperature of $T_{\rm f} \simeq 150$ MeV, this allows to estimate the decoupling (freeze-out) time $\tau_{\rm f}$. While $c\tau_{\rm f} = 1.3 \pm 0.2$ fm is observed for meson-proton collisions [184], it is about 4 fm for S+A collisions [181] and as much as 7 fm for Pb+Pb collisions [37].

It is important to note that $r_{\rm L}$ is the longitudinal length of homogeneity. The longitudinal dimension of the system at decoupling time can be estimated under the assumption of boost-invariant expansion all the way to the kinematic boundaries as [181]

$$r_{\rm L}^{\rm tot} \simeq 2\tau_{\rm f} \sinh(|y_o - y_{\rm max}|) \simeq 80\text{fm} , \tag{11.84}$$

i.e. 20 times larger than $r_{\rm L}$ itself.

11.6.6 Duration of pion emission

Besides the decoupling time, the duration of pion emission $\Delta\tau_{\rm f}$ [176] is an important parameter to characterize the evolution of the pion source. In the case of a boost-invariant longitudinal expansion the correlation can be analyzed in terms of the outward momentum difference $Q_{\rm out}$, which is related to the relative energy measured in the LCMS. The duration $\Delta\tau_{\rm f}$ is then given by the difference of $r_{\rm out}$ and $r_{\rm side}$ as [40]

$$\Delta\tau_{\rm f} = \frac{1}{\beta_{\rm T}} (r_{\rm out}^2 - r_{\rm side}^2)^{1/2} . \tag{11.85}$$

From Fig. 11.21 and 11.22, this difference is small and stays small even at RHIC, indicating fast pion decoupling at a proper time of $\tau_{\rm f} \sim 4 - 9$ fm/c, depending on the nuclei. This latter observation excludes a long-lived quark-gluon plasma [176,177,194] as being responsible for the hydrodynamic expansion in an average heavy-ion event. For meson-proton collisions [184], $\Delta\tau_f = 1.3 \pm 0.3$ fm/c is obtained, equal to τ_f itself. So, in meson-proton collisions pion radiation occurs during all of the hydrodynamic evolution of hadronic matter.

11.6.7 The transverse flow

How does the pion source evolve during the time τ_f in the transverse direction? Local pion emitters, such as fluid elements, resonances, strings, may also move in transverse direction. In principle, the (small) decrease of r_{side} with increasing k_T can be used to measure the transverse expansion [195]. However, as shown in [181] the measured r_{side} dependence is consistent with that expected from resonance decay. Only because of the uncertainty on the amount of resonance production, can neither transverse expansion (nor a weak transverse gradient in temperature or chemical potential [195]) be excluded.

A simple hydrodynamic picture [132] predicts the $1/\sqrt{\bar{m}_T}$ scaling of (11.83) for all r-parameters. The collective expansion leads to momentum-position correlations in both longitudinal and transverse directions. Indeed, NA44 [113] found

$$r_L \approx r_{out} \approx r_{side} \propto 1/\sqrt{\bar{m}_T} \tag{11.86}$$

for S+Pb collisions when evaluated in the LCMS, characterizing a cylindrically symmetric three-dimensionally expanding meson source. Velocity gradient, T_f and T-gradients generate a length scale in all three dimensions. If the length scales of the source are much larger than this, the three measured r-parameters become equal in the LCMS and show the $1/\sqrt{\bar{m}_T}$ scaling.

In PbPb and PbAu collisions [113,114,196–198]), however, this common scaling is violated (see Fig. 11.23a) by r_{out} and r_{side} scaling more weakly than r_L, and deviations are also observed at RHIC [90,199,200].

Fig. 11.23b gives a compilation [199] of BP radii for pion pairs as a function of transverse momentum, between $\sqrt{s}$ = 4.1 GeV and 130 GeV. The longitudinal parameter r_L (lowest plot) shows a slow increase with increasing $\sqrt{s}$ and a $1/\sqrt{\bar{m}_T}$ scaling (lines). Striking is the absence of any increase of r_{side} and r_{out} between $\sqrt{s}$ = 4.1 and 131 GeV in the upper plots and, in particular, of the ratio r_{out}/r_{side} as would have been expected from a long-lived pion source accompanying a deconfinement phase transition in some models.

Also the NA22 meson-proton data [184] are consistent with an $1/\sqrt{\bar{m}_T}$ scaling in longitudinal and transverse directions, but with $r_L > r_{out} > r_{side}$ and, of course, at smaller values. Elongation ($r_L > r_{out} \approx r_{side}$) and $1/\sqrt{\bar{m}_T}$ scaling is now also observed in e^+e^- collisions at the Z [201].

Strong evidence of a fast transverse flow in collisions of heavy nuclei comes from exponential fits performed by the NA44 [202] on the $\bar{m}_T$-distribution of pions, kaons

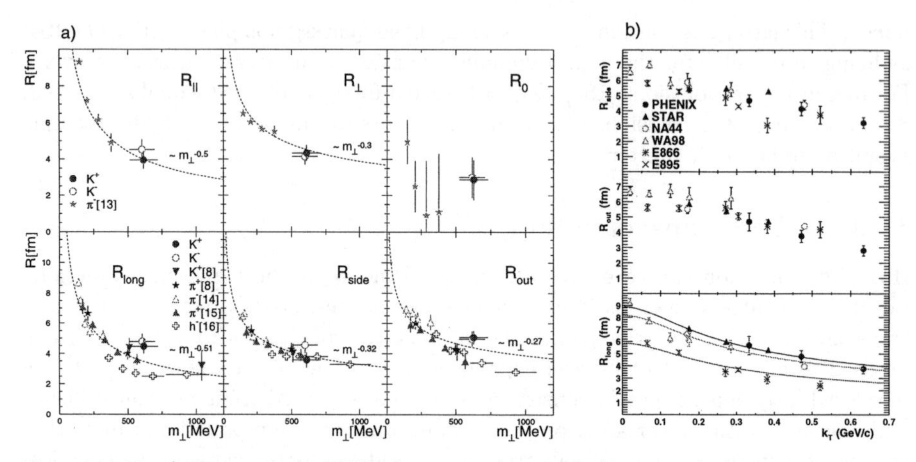

Figure 11.23. a) Comparison of YKP (upper row) and BP (lower row) radius parameters of kaons and pions as a function of m_{T}. The dashed lines indicate the fits by $r_i \propto m_{\mathrm{T}}^{-\alpha}$ leading to the α-values given in the figures [114]. b) BP radii for pion pairs in heavy-ion collisions as a function of the transverse momentum $k_{\mathrm{T}} \approx m_{\mathrm{T}}$ at mid-rapidity for $\sqrt{s_{NN}} = 4.1$ GeV (E895), 4.9 GeV (E866), 17.3 GeV (NA44, WA98) and 130 GeV (STAR, PHENIX). The lines represent $A/\sqrt{m_{\mathrm{T}}}$ fits for each energy region [199].

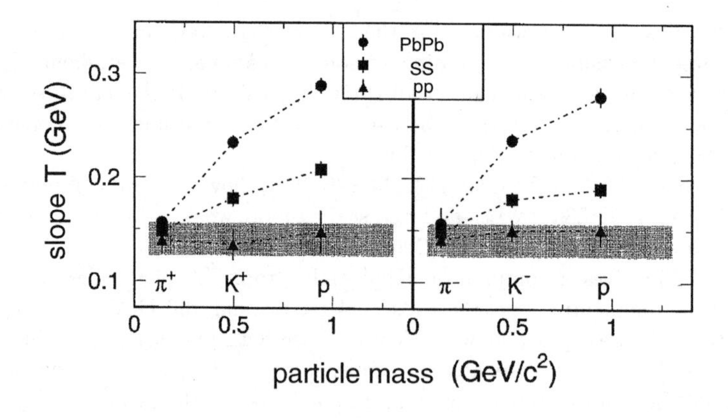

Figure 11.24. Slope parameter T as a function of particle mass m [202].

and protons produced in pp, 200 A GeV S+S and 158 A GeV Pb+Pb collisions. The resultant slopes T of these distribution are plotted in Fig. 11.24 as a function of particle mass m. While the slope is independent of particle mass for pp collisions, it increases with increasing mass for heavy-ion collisions, the more the heavier the

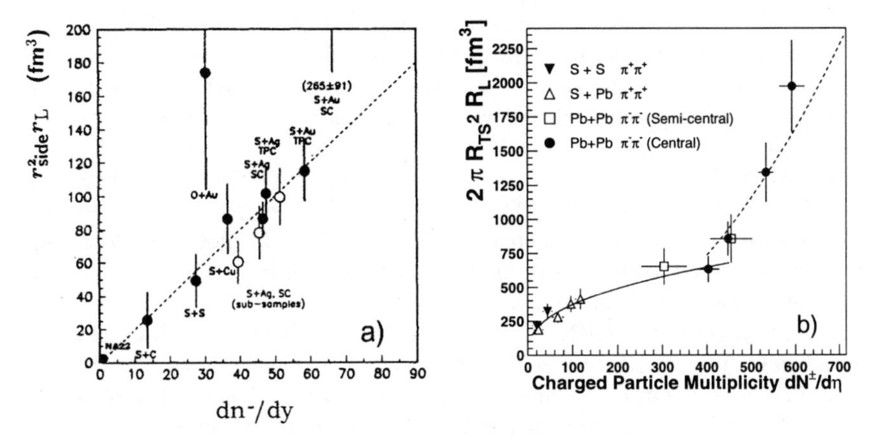

Figure 11.25. a) $r^2_{\rm side} r_{\rm L}$ (evaluated at mid-rapidity) as a function of $\mathrm{d}n^-/\mathrm{d}y$. The dashed line corresponds to $V = b \cdot \mathrm{d}\langle n^- \rangle/\mathrm{d}y$ with $b = 2.00 \pm 0.15$ fm^3 [181, 184]. b) $2\pi r^2_{\rm side} r_{\rm L}$ as a function if $\mathrm{d}n^\pm/\mathrm{d}\eta$. The lines are described in the text [180].

colliding system. The dot-dashed lines indicating the mass dependence all converge at $T \approx 140\pm15$ MeV as the particle mass approaches zero (shaded area in Fig. 11.24).

The slope parameter measures both thermal (random) and collective (e.g. rescattering) contributions. The intrinsic freeze-out temperature is determined by the thermal motion. In a hydrodynamical flow, particles of different mass all move with the same velocity, so heavier particles with larger energy. Comparing pp and heavy-ion collisions in Fig. 11.24 suggests that pp interactions are dominated by a thermal-like motion, while collective motion develops (probably due to an increased chance for rescattering) in heavy nuclei.

If the mass dependence of the slope T is parametrized in terms of the freeze-out temperature $T_{\rm f}$ and the average transverse collective-flow velocity as

$$T = T_{\rm f} + m\langle v_{\rm T}\rangle^2 , \tag{11.87}$$

one obtains $T_{\rm f}$ around 140 MeV for all types of collisions and $\langle v_{\rm T}\rangle/c$ between 0.1 ± 0.1 (pp) and 0.4 ± 0.1 (PbPb).

Within the hydrodynamic model $\langle v_{\rm T}\rangle$ can be interpreted [132] as the mean transverse expansion velocity at the freeze-out surface.

11.6.8 The decoupling volume

Assuming $r_{\rm side}$ is a good estimate for the transverse decoupling size, the decoupling (or freeze-out) volume of the system is proportional to $r^2_{\rm side} r_{\rm L}$. This quantity is shown as a function of the negative-particle density $\mathrm{d}\langle n^-\rangle/\mathrm{d}y$ in Fig. 11.25a [181, 184]. The consistency of the data with a linear rise from meson-proton (NA22) to heavy-ion collisions suggested a decoupling process at constant particle density. It is important to note, however, that the lowest density point ($b = 1.1 \pm 0.3$) [184] in fact lies 3 standard deviations below the line $b = 2$ and NA44 finds an increase of the volume smaller than linear [203], indicating a deviation from the simple model of a freeze-out at constant density.

This is shown in Fig. 11.25b [180], where the lower multiplicities are fitted by $V = C \cdot (\mathrm{d}n^{\pm}/\mathrm{d}\eta)^{\alpha}$ with $\alpha = 0.40 \pm 0.07$ (solid line). However, the high multiplicity data do not follow this trend, but increase with $\alpha = 1.97 \pm 0.14$ (dotted line). So, up to $\mathrm{d}n^{\pm}/\mathrm{d}\eta \approx 400$ a freeze-out at ever higher density is suggested, while a dilution of the freeze-out density may take over beyond that.

Taking into account the relative change of particle abundances with energy and their drastically different cross sections with pions, CERES [204] shows that thermal freeze-out in fact occurs at a critical mean free path $l_{\mathrm{f}} \approx 1$ fm, independent of energy between AGS and RHIC. This small value of l_{f} points to a considerable opaqueness of the source and may be an explanation for the weak s-dependence of the radius parameters up to RHIC and, perhaps, for the small value of r_{out}.

11.6.9 Examples of models and parametrizations

1. The SPACER (Simulation of Phase space distribution of Atomic nuclear Collisions in Energetic Reactions) event generator [14, 40, 205] is based on the string model FRITIOF assuming boost-invariant expansion. It includes resonance decay, but neglects secondary collisions. It reproduces, at least qualitatively, the k_{T} and $Y_{\pi\pi}$ dependence of the size parameters [181].

2. Also the RQMD (Relativistic Quantum Molecular Dynamics) event generator [206] (a hadron-string model which includes rescattering but no quark-gluon phase transition) qualitatively reproduces the trend of the k_{T} and $Y_{\pi\pi}$ dependence of the size parameters [113, 181]. However, the radii are too small in size and too large in temporal extent for the lower AGS energies (2-4 AGeV) [183]. This trend decreases with increasing energy, but reverses at SPS energies, where the predicted radii and the $r_{\mathrm{out}}/r_{\mathrm{side}}$ become too large [207] in the model.

Coupling the UrQMD at the hadronization hypersurface to an early-phase hydrodynamic scaling flow [208] allows to study first-order phase transition scenarios with different values for the critical temperature T_c. The radius r_{side} agrees with the RHIC data, but r_{out} and r_{L} are overpredicted [90, 199].

3. On the other hand, a simple purely hadronic rescattering model [209] gives a $r_{\mathrm{out}}/r_{\mathrm{side}}$ ratio only slightly larger than unity and also the radii themselves are roughly reproduced. Starting from a thermalized state with Gaussian rapidity distribution, hadrons are rescattering until they reach freeze-out. Unfortunately, this model needs rather strong assumptions, as a very fast hadronization and the existence of hadrons at large energy density.

4. A hydrodynamic model assuming a first-order phase transition with a large

latent heat (and, consequently, a long lifetime) [176, 177] can be excluded from the observed $\Delta\tau_f$ and τ_f [181].

5. Similarly, a hydro model [210] still assuming boost invariant longitudinal expansion (contrary to what is observed in the data, see Sub-Sect. 4.3.1.1) with Cooper-Frye freeze-out [211] fails to reproduce the radii at RHIC [150, 199].

6. On the other hand, a hydrodynamic model [33, 212] with a parametrized equation of state originating from lattice QCD gives good quantitative prediction of the $Y_{\pi\pi}$ and k_T dependence of the size parameters [181]. A moderate latent heat assumed in [33, 212] seems just large enough to prevent fast transverse expansion, but still small enough to allow for the relatively fast decoupling.

7. The MPC model [213] describes a classical gluon gas allowing for elastic collisions, but no hadronic phase. So, only pure parton dynamics is considered without worriing whether this survives in a subsequent hadronic phase with possible resonance effects. Opposite to ideal hydro models, the radii $r_{\rm out}$ and $r_{\rm L}$ come out smaller than in the data.

8. The AMPT model [214] combines initial collisions, elastic parton scattering and cascading, and comes closest to the data. In particular, $r_{\rm out}/r_{\rm side} \approx 1$, because (opposite to hydro models) positive $(x_{\rm out}, t)$ correlations give a negative contribution to $r_{\rm out}$.

9. The blast-wave model [215]: This simultaneous parametrization of $p_\perp$ spectra, elliptic flow and (azimuthally sensitive) Bose-Einstein correlations is inspired by hydrodynamic solutions and includes thermal motion superimposed on collective flow, an anisotropic transverse geometry, and a freeze-out distribution in proper time. The main assumptions are the same source for all particles, the invariance of the parameters over the freeze-out process and boost invariance.

10. The Hirano-Tsuda hydromodel [216]: This fully three-dimensional hydro model allows an early chemical freeze-out before thermal freeze-out. For $\mu_{\rm B} = 0$, this reduces $r_{\rm out}$ and $r_{\rm out}/r_{\rm side}$, but not enough to reproduce the RHIC results.

11. The Cracow single-freeze-out thermal model [217] assumes simultaneous (in the spirit of sudden hadronization) chemical and thermal freeze-out, longitudinal and transverse flow and a complete feeding from resonances. Also here, boost invariance is assumed, but it can be loosened rather easily.

12. The rather successful, not boost-invariant, Buda-Lund hydro approach will be treated in more detail below.

11.6.10 Combined analysis of two-particle correlations and single-particle spectra

As has become clear from the previous sub-sections, the correlation measurements alone do not contain the complete information on the geometrical and dynamical parameters characterizing the evolution of the hadronic matter. In particular, BEC are not measuring the full geometrical size of large and expanding systems, since that expansion results in strong correlations between space-time and momentum-energy space. More comprehensive information can be provided by a combined analysis of data on two-particle correlations and single-particle inclusive spectra [19, 132, 218–220].

11.6.10.1 The formalism

In the framework of a hydrodynamic model for three-dimensionally expanding cylindrically-symmetric finite systems (Buda-Lund hydro model) [132], the emission function corresponds to a Boltzmann (or rather Bose-Einstein) statistical approximation of the local momentum distribution. Within this model, the invariant single-particle spectrum of pions in rapidity y and transverse mass m_{T} is approximated by

$$\begin{aligned} f(y, m_{\mathrm{T}}) &= \frac{1}{N_{\mathrm{ev}}}\frac{\mathrm{d}N_{\pi}}{\mathrm{d}y\mathrm{d}m_{\mathrm{T}}^2} = \\ &= C{m_{\mathrm{T}}}^{\alpha}\cosh\eta_s \exp\left(\frac{\Delta\eta_*^2}{2}\right)\exp\left[-\frac{(y-y_0)^2}{2\Delta y^2}\right]\exp\left(-\frac{m_{\mathrm{T}}}{T_0}\right)\times \\ &\times\exp\left\{\frac{\langle u_{\mathrm{T}}\rangle^2(m_{\mathrm{T}}^2-m_{\pi}^2)}{2T_0[T_0+(\langle u_{\mathrm{T}}\rangle^2+\langle\frac{\Delta T}{T}\rangle)m_{\mathrm{T}}]}\right\}. \end{aligned} \tag{11.88}$$

with

$$\Delta y^2 = \Delta\eta^2 + \frac{T_0}{m_{\mathrm{T}}} \tag{11.89}$$

$$\frac{1}{\Delta\eta_*^2} = \frac{1}{\Delta\eta^2} + \frac{m_{\mathrm{T}}}{T_0}\cosh\eta_{\mathrm{s}}, \tag{11.90}$$

$$\eta_{\mathrm{s}} = \frac{y-y_0}{1+\Delta\eta^2\frac{m_{\mathrm{T}}}{T_0}}. \tag{11.91}$$

The width Δy of the rapidity distribution given by (11.89) is determined by the width $\Delta\eta$ of the longitudinal space-time rapidity η distribution of the pion emitters and by the thermal smearing width $\sqrt{T_0/m_{\mathrm{T}}}$, where T_0 is the freeze-out temperature (at the mean freeze-out time τ_{f}) at the axis of the hydrodynamical tube, $T_0 = T_{\mathrm{f}}(r_{\mathrm{T}} = 0)$. For the case of a slowly expanding system one expects $\Delta\eta \ll T_0/m_{\mathrm{T}}$, while for the case of a relativistic longitudinal expansion the geometrical extension $\Delta\eta$ can be much larger than the thermal smearing (provided $m_{\mathrm{T}} > T_0$).

In addition to the inhomogeneity caused by the longitudinal expansion, (11.88) also considers the inhomogeneity related to the transverse expansion (with the mean radial component $\langle u_{\mathrm{T}}\rangle$ of hydrodynamical four-velocity) and, very importantly, to the transverse temperature inhomogeneity [16], characterized by the quantity

$$\left\langle\frac{\Delta T}{T}\right\rangle = \frac{T_0}{T_{\mathrm{rms}}} - 1, \tag{11.92}$$

where $T_{rms} = T_f(r_T = r_T(\text{rms}))$ is the freeze-out temperature at the transverse rms radius $r_T(\text{rms})$ and at time τ_f.

The power α in (11.88) is related [132] to the number d of dimensions in which the expanding system is inhomogeneous. For the special case of the one-dimensional inhomogeneity ($d = 1$) caused by the longitudinal expansion, $\alpha = 1 - 0.5d = 0.5$ (provided $\Delta\eta^2 \gg T_0/m_T$). The transverse inhomogeneity of the system leads to smaller values of α. The minimum value of $\alpha = -1$ is achieved at $d = 4$ for the special case of a three-dimensionally expanding system with temporal change of local temperature during the particle emission process.

The parameter y_0 in (11.88) denotes the midrapidity in the interaction c.m.s. and can slightly differ from 0 due to different species of colliding particles. The parameter C is an overall normalization coefficient.

Note that (11.88) yields the single-particle spectra of the core (the central part of the interaction that supposedly undergoes collective expansion). However, also long-lived resonances contribute to the single-particle spectra through their decay products. Their contribution can be determined in the core-halo picture [32,212] by the momentum dependence of the strength parameter $\lambda(y, m_T)$ of the two-particle Bose-Einstein correlation function. Experimentally, the parameter is however found to be approximately independent of m_T [113,184,221]. Hence, this correction can be absorbed in the overall normalization.

The two-dimensional distribution (11.88) can be simplified for one-dimensional slices [132,221,222]:

1. At fixed m_T, the rapidity distribution reduces to the approximate parametrization

$$f(y, m_T) = C_m \exp\left[-\frac{(y - y_0)^2}{2\Delta y^2}\right], \tag{11.93}$$

where C_m is an m_T-dependent normalization coefficient and y_0 is defined above. The width parameter Δy^2 extracted for different m_T-slices is predicted to depend linearly on $1/m_T$, with slope T_0 and intercept $\Delta\eta^2$ (cf. (11.89)).

Note, that for static fire-balls or spherically expanding shells (11.93) and (11.89) are satisfied with $\Delta\eta = 0$ [222]. Hence, the experimental determination of the $1/m_T$ dependence of the Δy parameter can be utilized to distinguish between longitudinally expanding finite systems versus static fire-balls or spherically expanding shells.

2. At fixed y, the m_T^2-distribution reduces to the approximate parametrization

$$f(y, m_T) = C_y m_T^{\alpha} \exp\left(-\frac{m_T}{T_{\text{eff}}}\right), \tag{11.94}$$

where C_y is a y-dependent normalization coefficient and α is defined as above.

The y-dependent "effective temperature" $T_{\text{eff}}(y)$ can be approximated as

$$T_{\text{eff}}(y) = \frac{T_*}{1 + a(y - y_0)^2}, \tag{11.95}$$

where T_* is the maximum of $T_{\rm eff}(y)$ achieved at $y = y_0$, and

$$a = \frac{T_0 T_*}{2m_\pi^2(\Delta\eta^2 + \frac{T_0}{m_\pi})^2} \tag{11.96}$$

with T_0 and $\Delta\eta^2$ as defined above.

The approximations (11.93) and (11.94) explicitly predict a specific narrowing of the rapidity and transverse mass spectra with increasing $m_{\rm T}$ and y, respectively (cf. (11.89) and (11.95)). The character of these variations is expected [222] to be different for the various scenarios of hadron matter evolution.

11.6.10.2 The results

The Δy^2 values obtained from fits of the NA22 data [221] by (11.93) are given as a function of $1/m_{\rm T}$ in Fig. 11.26a. A fit to the widening of the rapidity distribution (i.e. increase of Δy^2) with increasing $1/m_{\rm T}$ by (11.89) gives an intercept $\Delta\eta^2 = 1.91\pm0.12$ and slope $T_0 = 159 \pm 38$ MeV. Thus, the width of the y-distribution is dominated by the spatial (longitudinal) distribution of pion emitters (inherent to longitudinally expanding systems) and not by the thermal properties of the hadron matter, as would be expected for static or radially expanding sources. Since $\Delta\eta^2$ is significantly bigger than 0, static fire-balls or spherically expanding shells, able to describe the two-particle correlation data in [184], fail to reproduce the single-particle spectra.

The $T_{\rm eff}$ values obtained from fits of the same data by (11.94) are given as a function of y in Fig. 11.26b. $T_{\rm eff}(y)$ tends to decrease with increasing $|y|$ and approximately follows (11.95) with $T_* = 160 \pm 1$ MeV, $a = 0.083 \pm 0.007$ and $y_0 = -0.065 \pm 0.039$. Note, however, an asymmetry in the $T_{\rm eff}$ distribution with respect to $y = 0$: except for the last point, $T_{\rm eff}$ is higher in the meson than in the proton hemisphere.

The values of the exponential parameter α fitted in (11.94) are near zero, corresponding to a two-dimensional inhomogeneity of the expanding system ($\alpha = 1-0.5d$). One concludes, therefore, that apart from a longitudinal inhomogeneity caused by the relativistic longitudinal flow, the hadron matter also has a transverse inhomogeneity (caused by transverse expansion or a transverse temperature gradient) or undergoes a temporal change of local temperature during the particle emission process.

11.6.10.3 The transverse direction

Further information on hadron-matter evolution in the transverse direction can be extracted from (11.88) with parameters $\langle u_{\rm T}\rangle$ and $\langle\frac{\Delta T}{T}\rangle$ characterizing the strength of the transverse expansion and temperature inhomogeneity.

A moderate value of the mean transverse four-velocity $\langle u_{\rm T}\rangle = 0.20\pm0.07$ indicates that the transverse inhomogeneity is mainly stipulated by the rather large temperature inhomogeneity $\langle\frac{\Delta T}{T}\rangle = 0.71\pm0.14$. Using (11.92), one infers that the freeze-out temperature decreases from $T_0 = 140 \pm 3$ MeV at the central axis of the hydrodynamical tube to $T_{\rm rms} = 82 \pm 7$ MeV at a radial distance equal to the transverse rms radius of the tube.

11.6.10.4 Combination with two-particle correlations

Due to the non-static nature of the source, the effective size parameters $r_{\rm L}, r_{\rm out}, r_{\rm side}$ vary with the average transverse mass $\bar{m}_{\rm T} = \frac{1}{2}(m_{\rm T1} + m_{\rm T2})$ and the average rapidity $Y = \frac{1}{2}(y_1 + y_2)$ of the pion pair. In the LCMS the effective radii can be approximated [132, 219, 222] by

$$r_{\rm L}^2 = \tau_{\rm f}^2 \Delta\eta_*^2 \tag{11.97}$$

$$r_{\rm out}^2 = r_*^2 + \beta_{\rm T}^2 \Delta\tau_*^2 \tag{11.98}$$

$$r_{\rm side}^2 = r_*^2 \tag{11.99}$$

with

$$\frac{1}{\Delta\eta_*^2} = \frac{1}{\Delta\eta^2} + \frac{\bar{m}_{\rm T}}{T_0} \tag{11.100}$$

$$r_*^2 = \frac{r_{\rm g}^2}{1 + \frac{\bar{m}_{\rm T}}{T_0}(\langle u_{\rm T}\rangle^2 + \langle\frac{\Delta T}{T}\rangle)} \quad , \tag{11.101}$$

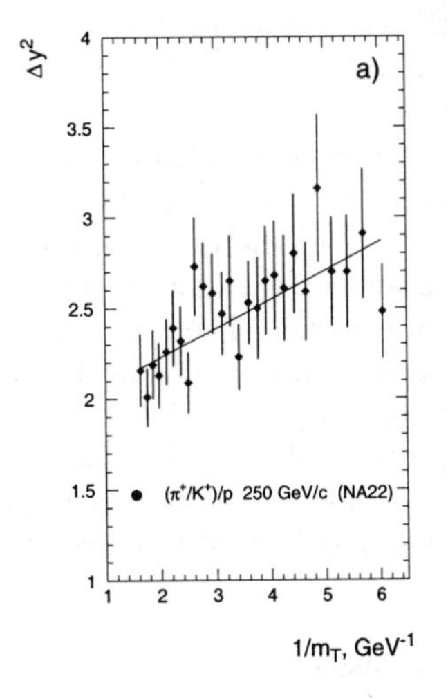

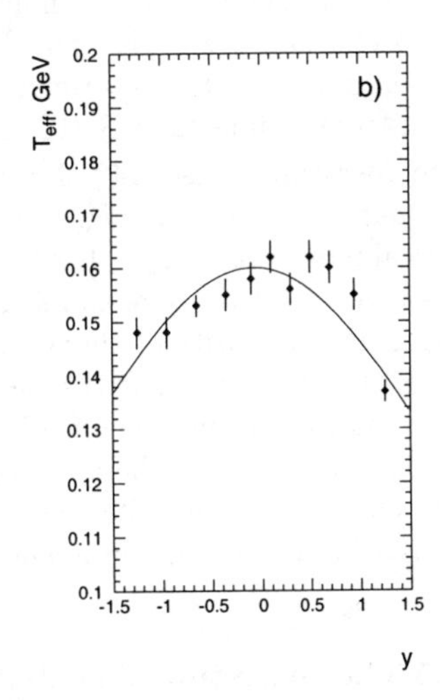

Figure 11.26. a) The $(1/m_{\rm T})$-dependence of $(\Delta y)^2$ for inclusive π^- meson rapidity distributions at $|y| < 1.5$. The straight line is the fit result according to parametrization (11.89). b) $T_{\rm eff}$ as a function of y fitted according to parametrization (11.95) [221].

where the parameters $\Delta\eta^2, T_0, \langle u_\mathrm{T}\rangle$ and $\langle\frac{\Delta T}{T}\rangle$ are defined and estimated from the invariant spectra above; r_g is related to the transverse geometrical rms radius of the source as $r_\mathrm{g}(\mathrm{rms}) = \sqrt{2}r_\mathrm{g}$; τ_f is the mean freeze-out (hadronization) time; $\Delta\tau_*$ is related to the duration $\Delta\tau_\mathrm{f}$ of pion emission and to the temporal inhomogeneity of the local temperature. If the latter has a small strength (as one can deduce from the restricted inhomogeneity dimension estimated above), an approximate relation $\Delta\tau_\mathrm{f} \geq \Delta\tau_*$ holds. The variable β_T is the transverse velocity of the pion pair.

Using (11.97) and (11.100) with $T_0 = 140 \pm 3$ MeV and $\Delta\eta^2 = 1.85 \pm 0.04$, together with r_L fitted in different $\bar{m}_\mathrm{T}$ ranges, one finds a mean freeze-out time of $\tau_\mathrm{f} = 1.4 \pm 0.1$ fm/c.

The transverse-plane radii r_out and r_side measured in [184] for the whole $\bar{m}_\mathrm{T}$ range are: $r_\mathrm{out} = 0.91 \pm 0.08$ fm and $r_\mathrm{side} = 0.54 \pm 0.07$ fm. Substituting into (11.98) and (11.99), one obtains (at $\beta_\mathrm{T} = 0.484$c [184]): $\Delta\tau_* = 1.3 \pm 0.3$ fm/c. Since the mean duration time of pion emission can be estimated as $\Delta\tau_\mathrm{f} \geq \Delta\tau_*$, the data grant $\Delta\tau_\mathrm{f} \approx \tau_\mathrm{f}$. A possible interpretation is that in meson-proton collisions the radiation process occurs during almost all the hydrodynamic evolution of the hadronic matter produced.

An estimation for the parameter r_g can be obtained from (11.99) and (11.101) using the quoted values of $r_\mathrm{side}, T_0, \langle u_\mathrm{T}\rangle$ and $\langle\frac{\Delta T}{T}\rangle$. The geometrical rms transverse radius of the hydrodynamical tube, $r_\mathrm{g}(\mathrm{rms}) = \sqrt{2}r_\mathrm{g} = 1.2 \pm 0.2$ fm, turns out to be larger than the proton rms transverse radius.

The set of parameters of the combined analysis of single-particle spectra and Bose-Einstein correlations in (π^+/K^+)p collisions [221] is compared to that obtained [223] from averaging over PbPb experiments (NA49, NA44 and WA98) of SPS [223] and pp and AuAu experiments at RHIC [224] in Table 11.3.

The temperature T_0 near 140 MeV comes out surprisingly similar for hh and PbPb collisions, but is considerably higher at RHIC, where it is even higher than the lattice QCD critical temperature $T_c = 162 \pm 2$ MeV [225]. The geometrical radius r_g and the mean freeze-out time τ_f are of course larger for heavy-ion than for hh collisions, but surprising is the similarity of the duration $\Delta\tau_\mathrm{f}$ of emission. The fact that $\Delta\tau_\mathrm{f} \approx \tau_\mathrm{f}$ in hh collisions, indicates that the radiation process occurs during all the evolution of hadronic matter in this type of collisions. On the other hand, $\Delta\tau_\mathrm{f} < \tau_\mathrm{f}$ for heavy-ion collisions suggests that there the radiation process only sets in at the end of the evolution. Other important differences are the large transverse flow velocity $\langle u_\mathrm{T}\rangle$ and small transverse temperature gradient in heavy-ion as compared to hh collisions.

11.6.10.5 The space-time distribution of π emission

Figure 11.27a gives a reconstruction of the space-time distribution of pion emission points [221], expressed as a function of the cms time variable t and the cms longitudinal coordinate z. The momentum-integrated emission function along the z-axis,

Table 11.3. Fit parameters of the Buda-Lund hydro (BL-H) model in a combined analysis of NA22 [221], NA49, NA44, WA98 [223] and RHIC [224] spectra and correlation data.

Param.	hp SPS	pp 200 GeV	PbPb SPS	AuAu 130 GeV	AuAu 200 GeV
T_0 [MeV]	140±3	289±8	139±6	214±7	200±9
T_s [MeV]	82±7	0.5 T_0	131±8	0.5 T_0	0.5 T_0
T_e [MeV]		90±42	87±24	102±11	127±13
$\langle\frac{\Delta T}{T}\rangle$	0.71±0.14		0.06±0.05		
μ_B [MeV]		8±76		77±38	61±40
r_g [fm]	1.2±0.2	1.2±0.3	7.1±0.2	28.0±5.5	13.2±1.3
r_s		1.1±0.2		8.6±0.4	11.6±1.0
$\langle u_t\rangle$	0.20±0.07		0.55±0.06		
$\langle u_t'\rangle$		0.04±0.26		1.0±0.1	1.5±0.1
τ_f [fm/c]	1.4±0.1	1.1±0.1	5.9±0.6	6.0±0.2	5.7±0.2
$\Delta\tau$ [fm/c]	1.3±0.3	0.1±0.5	1.6±1.5	0.3±1.2	1.9±0.5
$\Delta\eta$	1.36±0.02	3.0	2.1±0.4	2.4±0.1	3.1±0.1
y_0	0.082±0.06	0	0	0	0
χ^2/NDF	642/683	90/69	341/277	158/180	132/208

i.e., at $(x, y) = (0, 0)$ is given by

$$S(t,z) \propto \exp\left(-\frac{(\tau-\tau_f)^2}{2\Delta\tau_f^2}\right)\exp\left(-\frac{(\eta-y_0)^2}{2\Delta\eta^2}\right). \tag{11.102}$$

It relates the parameters fitted to the NA22 data with particle production in space-time. Note that the coordinates (t, z), can be expressed with the help of the longitudinal proper-time τ and space-time rapidity η as $(\tau\cosh\eta, \tau\sinh\eta)$.

One finds a structure resembling a boomerang, i.e., particle production takes place close to the regions of $z = t$ and $z = -t$, with gradually decreasing probability for ever larger values of space-time rapidity. Although the mean proper-time for particle production is $\tau_f = 1.4$ fm/c, and the dispersion of particle production in space-time rapidity is rather small ($\Delta\eta = 1.36$), a characteristic long tail of particle emission is observed on both sides of the light-cone, giving more than 40 fm longitudinal extension in z and 20 fm/c duration of particle production in the time variable t.

An, at first sight, similar behavior is seen in Fig. 11.27c for PbPb collisions [223] at the SPS. An important quantitative difference is, however, that particle emission

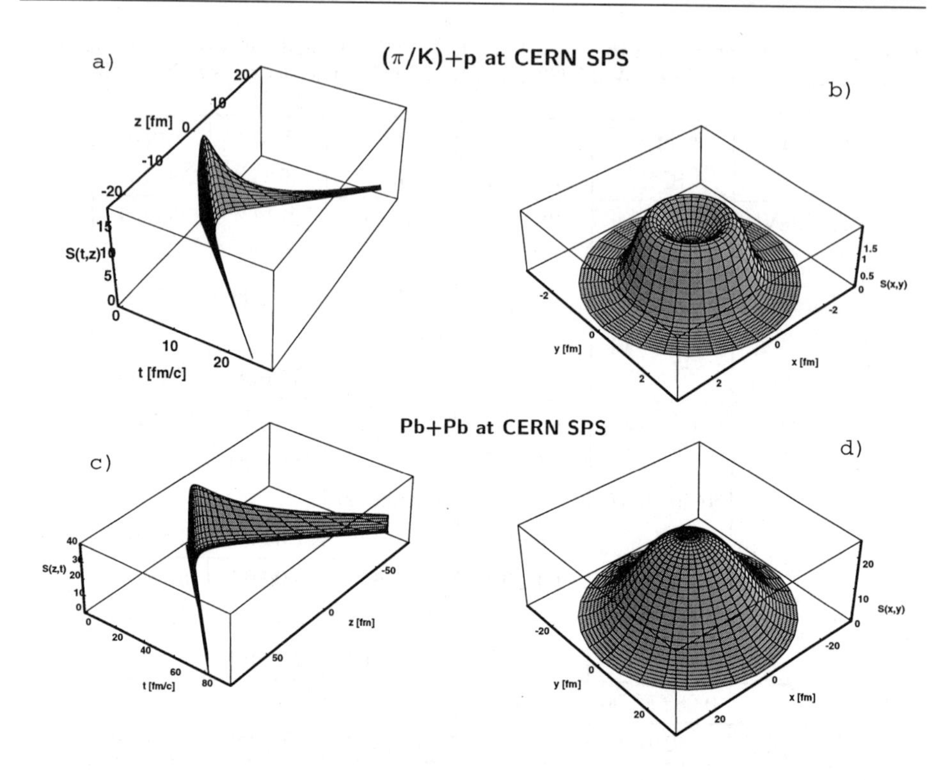

Figure 11.27. The reconstructed emission function $S(t, z)$ in arbitrary vertical units, as a function of time t and longitudinal coordinate z (left diagrams), as well as the reconstructed emission function $S(x, y)$ in arbitrary vertical units, as a function of the transverse coordinates x and y (right pictures), for hh (upper pictures) and PbPb (lower pictures) collisions, respectively [221, 223, 226].

starts immediately in hadron-hadron collision, but only after about 4-5 fm/c in PbPb collisions!

The information on $\langle u_{\rm T}\rangle$ and $\langle\frac{\Delta T}{T}\rangle$ from the analysis of the transverse momentum distribution can be used to reconstruct the details of the transverse density profile. An exact, non-relativistic hydro solution was found [227] using an ideal gas equation of state. In this hydro solution, both

$$\langle\frac{\Delta T}{T}\rangle \gtrless \frac{m\langle u_{\rm T}\rangle^2}{T_0} \tag{11.103}$$

is possible. The < sign corresponds to a self-similar expanding fire-ball, while the > sign corresponds to a self-similar expanding ring of fire (see Fig. 11.28).

Assuming the validity of this non-relativistic solution, one can reconstruct the detailed shape of the transverse density profile. The result at average freeze-out time

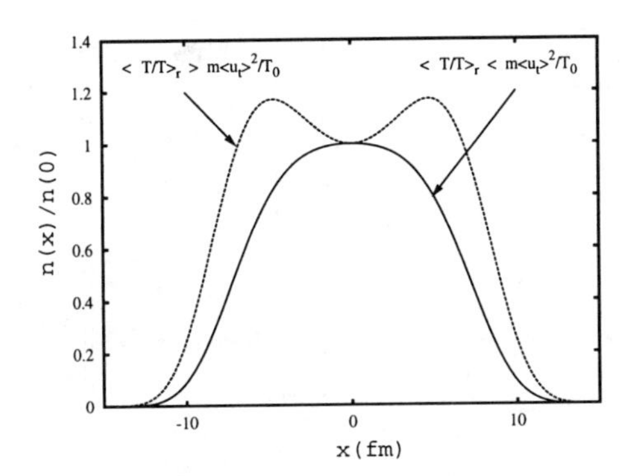

Figure 11.28. Illustration of the development of smoke-ring solutions for large temperature gradients in exact solutions of non-relativistic hydrodynamics [227].

τ_{f} and $z = 0$ looks like a ring of fire in the x, y plane in hh interactions (Fig. 11.27b), while in PbPb collisions it has a Gaussian shape (Fig. 11.27d).

The formation of a ring of fire in hh collisions is due to the rather small transverse flow and the sudden drop of the temperature in the transverse direction, which leads to large pressure gradients in the center and small pressure gradients and a density augmentation at the expanding radius of the fire-ring. This transverse distribution, together with the (approximately) scaling longitudinal expansion, creates an elongated, tube-like source in three dimensions, with the density of particle production being maximal on the surface of the tube.

The pion emission function $S(x, y)$ for PbPb collisions, on the other hand, corresponds to the radial expansion, which is a well established phenomenon in heavy-ion collisions from low-energy to high-energy reactions. This transverse distribution, together with the (approximately) scaling longitudinal expansion, creates a cylindrically symmetric, large and transversally homogeneous fire-ball, expanding three-dimensionally with a large mean radial component $\langle u_{\mathrm{T}} \rangle$ of hydrodynamical four-velocity.

Because of this large difference observed for those two types of collision, analysis of the emission function in e^+e^- collisions is of crucial importance for the understanding of the actual WW overlap and has been started.

11.6.11 Azimuthally sensitive analysis

So far, cylindrically symmetric sources have been assumed, but, in general, particle production in an individual high energy collision shows azimuthal structure. A non-central heavy-ion collision has an non-zero impact parameter and e^+e^- collisions often lead to hard gluon radiation. In both cases, this defines a reaction plane and particles

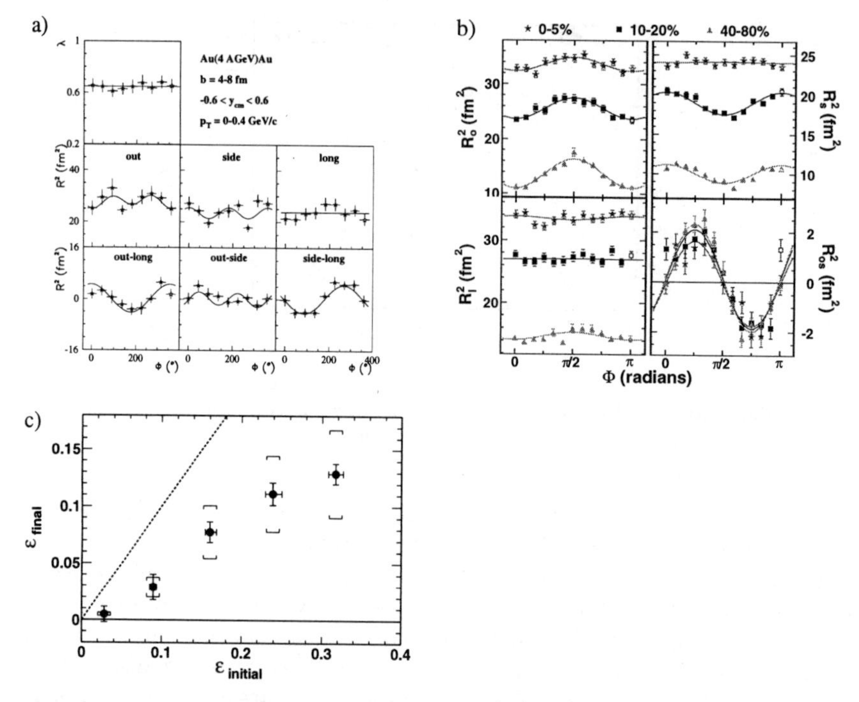

Figure 11.29. Strength parameter λ and Bertsch-Pratt radius parameters as a function of azimuthal angle Φ with respect to the impact parameter a) for 4 A GeV AuAu collisions with impact parameter 4-8 fm [236], b) for $\sqrt{s_{\rm NN}} = 200$ GeV AuAu collisions for three different impact parameter values [237]. c) Final versus initial transverse source eccentricity for the data given in Sub-figure b [237].

are emitted preferentially close to that plane.

Anisotropies are well known in momentum space as directed and elliptic flow [228–230], but theoretical studies suggest that configuration space anisotropies are equally important [231–234] for a discrimination between scenarios of system evolution.

Indeed, the radius parameters were found to depend on the azimuthal orientation Φ of the pion pair with respect to this reaction plane [235, 236] in non-central heavy-ion collisions. Fig. 11.29a shows the Bertsch-Pratt "radii" and their cross terms as a function of Φ for 4 A GeV AuAu collisions with impact parameter 4-8 fm. The transverse radii $r_{\rm out}$ and $r_{\rm side}$ show significant equal-amplitude but opposite second-order oscillation, reflecting the almond-shaped overlap region between target and projectile, with a larger spatial extent perpendicular to the impact parameter than parallel to it. The cross-term parameters show oscillations consistent with pion emission from an ellipsoidal source tilted in the reaction plane away from the beam axis [236].

In Fig. 11.29b [237], the increase of the radii with increasing centrality (decreasing impact parameter), already observed at low energies, is confirmed, but the oscillation amplitude clearly increases with increasing impact parameter. This effect is present for all transverse momenta.

As already demonstrated in [238] and further worked out in [239], information on the azimuthal structure gets lost in the Bertsch-Pratt decomposition of the correlation function. It is, however, shown to be fully retained in a reaction plane fixed coordinate system, with z along the beam, x along the impact parameter and y orthogonal to the reaction plane. In this system, the final (freeze-out) transverse eccentricity

$$\epsilon = \frac{r_y^2 - r_x^2}{r_y^2 + r_x^2} \approx 2r_{s,2}^2/r_{\text{side}}^2$$

with $r_{s,2}^2 = \langle r_{\text{side}}^2 \cos(2\Phi)\rangle$, obtained from BEC analysis in blast-wave context [215] is compared to an initial one calculated with a Glauber model. The vales of ϵ_{final} are smaller than those of $\epsilon_{\text{initial}}$ but stay positive. So, even at RHIC energies, the time scales and/or flow velocities are too low to fully quench the initial out-of-plane geometric anisotropy.

11.7 WW overlap

Among the methods to measure the W mass M_{W} from $\text{e}^+\text{e}^- \to \text{W}^+\text{W}^-$ at LEP2 (threshold cross section, Breit-Wigner shape, lepton end-point energy), the reconstruction of the Breit-Wigner resonant shape from the $\text{W}^\pm$ decay products $(\text{q}\bar{\text{q}}\text{q}\bar{\text{q}}, \text{q}\bar{\text{q}}\ell\nu)$ is the most promising under the LEP2 running conditions [5]. In a space-time picture of hadroproduction in hadronic W^+W^- decays ($\text{q}\bar{\text{q}}\text{q}\bar{\text{q}}$ channel) at $\sqrt{s} = 175$ GeV, the typical separation of the two decay vertices is of the order of 0.1 fm. The typical distance scale of hadronization is 1 fm, ten times the separation of the W^+W^- decay vertices. So, in principle, the influence of Bose-Einstein correlations of identical pions could be large [6].

What may look like a problem for the W-mass determination, could, however, even be an advantage. If indeed present, such an inter-W BE interference would provide an ideal laboratory to measure the space-time development of such an overlap.

A mathematically founded, model-independent method has been suggested in [240]. Let us first consider an uncorrelated WW decay scenario. In this, each W boson showers and fragments into final-state hadrons without any reference to what is happening to the other. In this case, the four-jet WW generating functional $\mathcal{G}^{\text{WW}}[u(p)]$ is the product of the generating functionals for the two-jet decays of the two differently charged single W's,

$$\mathcal{G}^{\text{WW}}[u(p)] = \mathcal{G}^{\text{W}^+}[u(p)]\,\mathcal{G}^{\text{W}^-}[u(p)]. \tag{11.104}$$

In terms of the generating functionals for the correlation functions, Eq. (7.18) this can be represented as

$$\ln \mathcal{G}^{\text{WW}}[u(p)] \equiv g^{\text{WW}}[u(p)] = g^{\text{W}^+}[u(p)] + g^{\text{W}^-}[u(p)]. \tag{11.105}$$

According to Eq. (7.9), the single-particle and two-particle inclusive densities for the four-jet WW hadronic decay can directly be obtained performing two successive functional differentiations of (11.104) over the probing function $u(p)$,

$$\rho_1^{\mathrm{WW}}(1) = \rho_1^{\mathrm{W}^+}(1) + \rho_1^{\mathrm{W}^-}(1), \tag{11.106}$$

$$\rho_2^{\mathrm{WW}}(1,2) = \rho_2^{\mathrm{W}^+}(1,2) + \rho_2^{\mathrm{W}^-}(1,2) + 2\rho_1^{\mathrm{W}^+}(1)\rho_1^{\mathrm{W}^-}(2). \tag{11.107}$$

Note that the latter expression differs from the sum of two-particle densities for each independent source taken separately.

Performing the same functional differentiations of (11.105), one can, according to (7.19) find the two-particle correlation function in the four-jet WW decay,

$$C_2^{\mathrm{WW}}(1,2) = C_2^{\mathrm{W}^+}(1,2) + C_2^{\mathrm{W}^-}(1,2). \tag{11.108}$$

This illustrates the fact that, in contrast to the two-particle densities, the correlation functions are additive and do not contain the contribution from lower-order inclusive densities.

Experimentally, it is advantageous to rewrite (11.107) as

$$\rho_2^{\mathrm{WW}}(1,2) = \rho_2^{\mathrm{W}^+}(1,2) + \rho_2^{\mathrm{W}^-}(1,2) + 2\rho_{\mathrm{mix}}^{\mathrm{W}^+\mathrm{W}^-}(1,2), \tag{11.109}$$

where $\rho_1^{\mathrm{W}^+}(1)\rho_1^{\mathrm{W}^-}(2)$ is replaced by the track mixing two-particle density $\rho_{\mathrm{mix}}^{\mathrm{W}^+\mathrm{W}^-}(1,2)$ obtained by pairing particles from different two-jet WW events, to ensure that particles coming from differently charged W's do not correlate. This technique leads to factorization of $\rho_{\mathrm{mix}}^{\mathrm{W}^+\mathrm{W}^-}(1,2)$ into the product of the single-particle densities.

A non-zero inter-W two-particle correlation can then be seen in the deviation from Eq. (11.109), expressed as the difference between or the ratio of the left and right-hand sides of this equation,

$$\Delta\rho(1,2) = \rho_2^{\mathrm{WW}}(1,2) - 2\rho_2^{\mathrm{W}}(1,2) - 2\rho_{\mathrm{mix}}^{\mathrm{W}^+\mathrm{W}^-}(1,2) \tag{11.110}$$

$$D(1,2) = \frac{\rho_2^{\mathrm{WW}}(1,2)}{2\rho_2^{\mathrm{W}}(1,2) + 2\rho_{\mathrm{mix}}^{\mathrm{W}^+\mathrm{W}^-}(1,2)} \tag{11.111}$$

with $\rho_2^{\mathrm{W}} \equiv \rho_2^{\mathrm{W}^+} = \rho_2^{\mathrm{W}^-}$. By definition, absence of inter-W correlation would give $\Delta\rho = 0$ and $D = 1$. Furthermore, the actual size of the genuine inter-W correlation is given by

$$\delta_{\mathrm{I}}(1,2) = \Delta\rho(1,2)/2\rho_{\mathrm{mix}}^{\mathrm{W}^+\mathrm{W}^-}(1,2) \tag{11.112}$$

One can integrate (11.106) and (11.107) to obtain the relations for average multiplicity $\langle n\rangle$ and second-order factorial moment F_2 in uncorrelated four-jet WW decays:

$$\Delta \equiv \langle n_{\mathrm{WW}}\rangle - \langle n_{\mathrm{W}^+}\rangle - \langle n_{\mathrm{W}^-}\rangle = 0, \tag{11.113}$$

$$\Delta F_2 \equiv F_2^{\mathrm{WW}} - F_2^{\mathrm{W}^+} - F_2^{\mathrm{W}^-} - 2\langle n_{\mathrm{W}^+}\rangle\langle n_{\mathrm{W}^-}\rangle = 0. \tag{11.114}$$

A deviation of $\Delta\rho$ or ΔF_2 from zero is possible only in the case of correlated WW decay. Note, however, that the opposite is not true: $\Delta\rho = 0$ is a necessary, but not a *sufficient* condition for uncorrelated WW decay.

In the case of interference effects, one can assume

$$\Delta\rho(\pm,\pm) = \Delta\rho^{\mathrm{ec}}(\pm,\pm) + \Delta\rho^{\mathrm{be}}(\pm,\pm) + \Delta\rho^{\mathrm{cr}}(\pm,\pm), \tag{11.115}$$

$$\Delta\rho(+,-) = \Delta\rho^{\mathrm{ec}}(+,-) + \Delta\rho^{\mathrm{cr}}(+,-), \tag{11.116}$$

where $\Delta\rho^{\mathrm{ec}}$ is the contribution from energy conservation and other non-interference effects, $\Delta\rho^{\mathrm{be}}$ represents the BE correlations and $\Delta\rho^{\mathrm{cr}}$-color-reconnection correlations, and directly investigate the BE interference effects by calculating the difference:

$$\delta\rho = \Delta\rho(\pm,\pm) - \Delta\rho(+,-). \tag{11.117}$$

Since the track mixing terms are very similar, one has

$$\delta\rho \simeq \rho_2^{\mathrm{WW}}(\pm,\pm) - 2\,\rho_2^{\mathrm{W}}(\pm,\pm) - \rho_2^{\mathrm{WW}}(+,-) + 2\,\rho_2^{\mathrm{W}}(+,-), \tag{11.118}$$

which no longer involves the track mixing terms since they cancel. Taking into account the fact that $\Delta\rho^{\mathrm{ec}}(\pm,\pm)$ and $\Delta\rho^{\mathrm{ec}}(+,-)$ are the same, from (11.115) and (11.116) one can see that $\delta\rho$ resolves the interference terms

$$\Delta\rho \simeq \Delta\rho^{\mathrm{be}}(\pm,\pm) + \Delta\rho^{\mathrm{cr}}(\pm,\pm) - \Delta\rho^{\mathrm{cr}}(+,-). \tag{11.119}$$

If the color-reconnection effects are charge-independent, $\Delta\rho$ is fully determined by the BE correlations, i.e., $\delta\rho \simeq \Delta\rho^{\mathrm{be}}(\pm,\pm)$.

The same arguments lead to the double ratio

$$d = D(\pm\pm)/D(+-) \tag{11.120}$$

or to

$$\Delta_{\mathrm{I}} = \delta_{\mathrm{I}}(\pm\pm) - \delta_{\mathrm{I}}(+-)\ . \tag{11.121}$$

Figure 11.30 shows the L3 results [241] for $\Delta\rho$ after subtraction of background from $\mathrm{q\bar{q}}$ events, for pairs of like charged particles (Fig. 11.30a) and pairs of unlike charged particles (Fig. 11.30b). Also shown are the predictions of KORALW [242] using JETSET with the so-called BE_{32} [6] Bose-Einstein simulation. The BEC are implemented for *all* particles (inter-W) or only for particles originating from the *same* W (no inter-W). From Fig. 11.30a, it is clear that the "no inter-W" scenario describes the data, while "inter-W" is disfavored.

For a more quantitative comparison, the integral of Fig. 11.30a

$$J \equiv \int_0^{Q_{\max}} \Delta\rho(Q)\mathrm{d}Q\ , \tag{11.122}$$

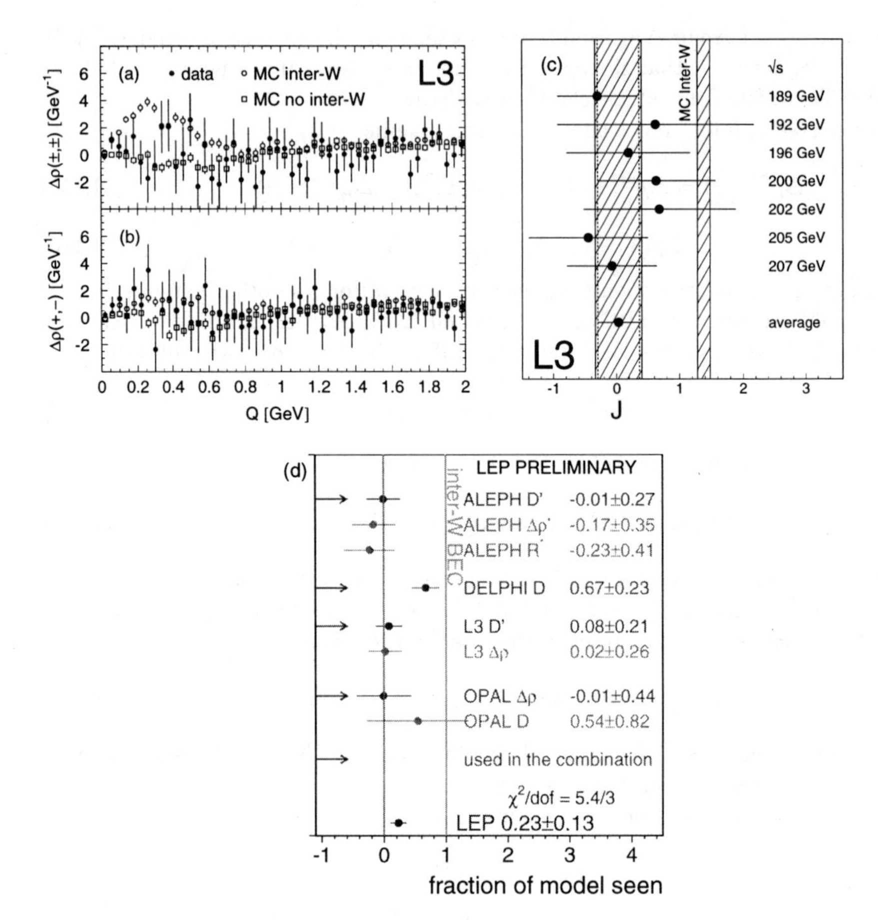

Figure 11.30. Distributions for the L3 data of (a) $\Delta\rho(\pm,\pm)$ and (b) $\Delta\rho(+,-)$. Also shown are the Monte Carlo predictions of KORALW with BEC between all particles and with only intra-W BEC. (c) Value of the integral J found at the different centre-of-mass energies. The error is statistical only, including bin-to-bin correlations. Also shown is the Monte Carlo prediction of KORALW (at the detector level) with BEC between all particles (BEA) [241]. d) Fraction of model prediction seen by the LEP experiments [243].

with $Q_{\max} = 0.68$ GeV, is computed for several LEP energies and plotted in Fig. 11.30c. Combining all energies gives $J = 0.03 \pm 0.33 \pm 0.15$, while $J = 1.38 \pm 0.10$ is expected for the "inter-W" scenario. So, an excess at small Q values of W is expected in $\Delta\rho(\pm,\pm)$ from inter-W BEC, but no evidence is seen in the L3 data, even before a subtraction as in Eq. (11.118).

These results are supported by ALEPH [244] and OPAL [245], while a 2.4σ effect is observed in $\delta_{\mathrm{I}}(Q)$ by DELPHI [246], however with a 1.4σ effect also present in $(+-)$ pairs. Fig. 11.30d gives a comparison of the results from the four LEP experiments to each other and to the JETSET/LUBEI inter-W expectation. In total, the signal observed by LEP before $\pi^+\pi^-$ subtraction so far is about 1.8σ, but as much as 5.9σ below the inter-W model expectation.

11.8 Modification of multiplicity and single-particle spectra

In the previous sections we dealt almost exclusively with the influence of BE statistics on the correlations in momentum space between identical bosons. However, the effects of quantum interference can also modify substantially the multiplicity distributions and the single-particle inclusive spectra. Bose-Einstein correlations can, under certain conditions, lead to Bose-Einstein condensation in multipion systems [7–9,247], called "pion laser" in [7] or "cold spots" in [8]. Although the effects on multiplicity distributions and on charged and neutral particle fluctuations in a single event has not yet been established in accelerator experiments, they are certainly worthy of further experimental investigation.

To appreciate the physics underlying these phenomena, it is useful to rederive the basic consequences of BE statistics, as applied to processes of *independent* particle production by employing the density-matrix formalism [9,248]. Although the assumption of independent particle production is certainly an over-simplification, it is well suited to illustrate the potentially dramatic effects of Bose-Einstein symmetrization. In this section we closely follow [8, 248].

11.8.1 Density matrix formalism

Let $\psi^{(0)}(p_1, p_2, ...p_n, \alpha) \equiv \psi^{(0)}(p, \alpha)$ be the probability amplitude for the production of n particles with momenta $[p_1, p_2, ..., p_n] \equiv p$ *calculated ignoring the identity of particles.* Here α stands for all other quantum numbers which may be relevant to the process in question. The density matrix

$$\rho^{(0)}(p, p') = \int \mathrm{d}\alpha\, \psi^{(0)}(p, \alpha)\psi^{(0)*}(p', \alpha) \tag{11.123}$$

contains all the available information about the system in question. In particular, the momentum spectrum of particles is

$$P_1^{(0)}(p) = \int \mathrm{d}\alpha \mid \psi^{(0)}(p, \alpha) \mid^2 = \rho^{(0)}(p, p) \tag{11.124}$$

with the normalization

$$\int \mathrm{d}P_1^{(0)}(p) = 1. \tag{11.125}$$

Suppose now that the particles are identical. In this case, the wave function is a sum over all permutations

$$\psi(p,\alpha) = \sum_P \psi^{(0)}(p_P,\alpha), \tag{11.126}$$

where p_P is the set of momenta $[p_1, p_2, ..., p_n]$ ordered according to the permutation P of $[1, 2, ..., n]$. Using (11.123)-(11.126), one obtains for the distribution of momenta of identical particles

$$P_1(p) = \frac{1}{n!}\int d\alpha \mid \psi(p,\alpha) \mid^2 = \frac{1}{n!}\sum_{P,P'} \rho^{(0)}(p_P, p_{P'}). \tag{11.127}$$

The factor $1/n!$ accounts for the fact that the phase space for n identical particles is $n!$ times smaller than that of non-identical ones.

Eq.(11.127) summarizes the effect of the identity of particles on the observed spectra. To evaluate its effect, it is not sufficient to know the spectrum $\Omega^{(0)}(p)$. Knowledge of the full density matrix $\rho^{(0)}(p,p')$ is needed.

It is worth noting in relation to (11.127) that, for particles emitted in a pure state, i.e. $\rho^{(0)}(p,p') = \psi^{(0)}(p)\psi^{(0)*}(p')$, and if $\psi^{(0)}(p)$ is symmetric with respect to interchange of any pair of momenta, one has

$$P_1(p) = n! P_1^{(0)}(p) \;, \tag{11.128}$$

a truely dramatic effect.

11.8.2 Independent particle emission

Consider a system of n particles emitted independently. If we ignore the identity of particles, independent emission implies that the density matrix factorizes

$$\rho^{(0)}(p,p') = \prod_{i=1}^{n} \rho^{(0)}(p_i, p'_i). \tag{11.129}$$

Introducing this into (11.127) one has

$$P_1(p) = \frac{1}{n!}\sum_{P,P'}\prod_{i=1}^{n} \rho^{(0)}((p_P)_i, (p_{P'})_i). \tag{11.130}$$

The information contained in (11.130) can be summarized in the form of the generating functional $\mathcal{G}^{\rm excl}[u]$ defined as

$$\mathcal{G}^{\rm excl}[u] = \frac{\sum_{n=0}^{\infty} P_n^{(0)} W_n[u]}{\sum_{n=0}^{\infty} P_n^{(0)} W_n[1]} \tag{11.131}$$

with

$$W_n[u] = \int \mathrm{d}p_1...\mathrm{d}p_n P_1(p_1, ..., p_n)u(p_1)...u(p_n). \tag{11.132}$$

Here $P_n^{(0)}$ is the *uncorrected* multiplicity distribution and $u(p)$ is an arbitrary real non-negative function of p. For $u(p) = const \equiv z$ the generating functional becomes the generating function of the multiplicity distribution, $G(z)$.

To find the explicit expression for $\mathcal{G}^{\mathrm{excl}}[u]$ we observe that, given Eq. (11.130) for $P_1(p)$, for each permutation P of the momenta $p_1, ..., p_n$, the integral on the right hand side of (11.132) factorizes into a product of contributions from all the cycles of P (note that each permutation can be decomposed into cycles). Let the contribution from a cycle of length k be $kc_k[u]$. One has

$$kc_k[u] = \int \mathrm{d}^3p_1...\mathrm{d}^3p_k u(p_1)\rho^{(0)}(p_1, p_2)u(p_2)\rho^{(0)}(p_2, p_3)...u(p_k)\rho^{(0)}(p_k, p_1) \,. \tag{11.133}$$

The quantities c_k (for $u = 1$) are normalized combinants [9, 249] $c_k = \tilde{c}_k/\langle n\rangle^k$ (see Sub-Sect. 7.1.6).

The rest of the calculation is combinatorics. One notes first that any two permutations which have identical partitions into cycles give equal contributions. Consider then the set of all permutations with a given partition into cycles. Denoting by n_k the number of occurrences of a cycle of length k in the set of permutations considered, the contribution from all of them can be written as

$$W'_n[u] = \prod_{k=1}^{n}(kc_k[u])^{n_k}\frac{n!}{(k!)^{n_k}}[(k-1)!]^{n_k}\frac{1}{n_k!} = n!\prod_{k=1}^{n}\frac{(c_k[u])^{n_k}}{n_k!}. \tag{11.134}$$

$W_n[u]$ is obtained by summing $W'_n[u]$ over partitions into cycles different from each other.

The sum over multiplicities can be explicitly performed if the uncorrected multiplicity distribution $P_n^{(0)}$ is Poissonian,

$$P_n^{(0)} = e^{-\nu}\frac{\nu^n}{n!}. \tag{11.135}$$

The result is an elegant formula for the generating functional (11.131) [8]:

$$\mathcal{G}^{\mathrm{excl}}[u] = \exp\left(\sum_{k=1}^{\infty}\nu^k(c_k[u] - c_k[1])\right). \tag{11.136}$$

The generating function of the multiplicity distribution is then given by

$$\tilde{G}(z) \equiv \sum_n P_n z^n = \exp\left(\sum_{k=1}^{\infty}\nu^k(z^k - 1)c_k\right). \tag{11.137}$$

(Note that $(z+1)$ in the definition 7.41 is replaced by z here).

11.8.2.1 Multiplicity distributions from independent emission

We now discuss the general properties of the multiplicity distributions obtained from Eq.(11.137). For the average multiplicity one finds

$$\langle n \rangle = \sum_{k=1}^{\infty} k\nu^k c_k, \tag{11.138}$$

and for the (un-normalized) factorial cumulants

$$f_q = \sum_{k=q}^{\infty} \frac{k!}{(k-q)!} c_k \, \nu^k. \tag{11.139}$$

Since the c_k are positive, all the cumulants are positive and the distribution is always broader than the Poisson distribution. Specific properties of the distributions defined by (11.137) depend, of course, on the value of ν and that of the cycle integrals kc_k.

In general, the resulting multiplicity distribution will bear little resemblance with the original (in this case) Poisson distribution. In particular, the observed average multiplicity may dramatically differ from the initial ν. This is a rather general phenomenon in particle systems obeying Bose-Einstein statistics.

Further discussion depends on the assumed shape of the single-particle density matrix. The evaluation of c_k is greatly simplified if one works in the basis where the density matrix $\rho^{(0)}(p, p')$ is diagonal. For the case of a discrete eigenvalue spectrum one has

$$kc_k = \sum_m \lambda_m^k, \quad c_1 = \sum_m \lambda_m = 1 \tag{11.140}$$

and the generating function of the multiplicity distribution can be represented as a product

$$\tilde{G}(z) = \prod_m \tilde{G}_m(z) \ , \tag{11.141}$$

where

$$\tilde{G}_m(z) = \frac{1 - \lambda_m \nu}{1 - z\lambda_m \nu} \tag{11.142}$$

are the generating functions of the geometric distribution.

The average multiplicity becomes

$$\langle n \rangle = \sum_m \frac{\lambda_m \nu}{1 - \lambda_m \nu} \geq \frac{\lambda_0 \nu}{1 - \lambda_0 \nu} \tag{11.143}$$

and for the factorial cumulants

$$f_q = (q-1)! \sum_m \left(\frac{\lambda_m \nu}{1 - \lambda_m \nu} \right)^q \geq (q-1)! \left(\frac{\lambda_0 \nu}{1 - \lambda_0 \nu} \right)^q , \tag{11.144}$$

where λ_0 is the largest eigenvalue of $\rho^{(0)}(p,p')$. One notes that $\nu\lambda_0 = 1$ is a critical point of the multiplicity distribution.

The case of a continuous eigenvalue spectrum can be treated along similar lines [248] leading to the conclusion that the multiplicity distribution always becomes singular when $\nu\lambda_0$, the product of the initial average multiplicity and the maximal eigenvalue of the density matrix, approaches 1.

11.8.2.2 Particle distributions

General properties of the particle distributions can be derived from the generating functional Eq.(11.136). All inclusive distributions can be expressed in terms of a single function $L(p,p')$, defined as

$$L(p,p') = \sum_{k=1}^{\infty} \nu^k [\rho^{(0)}]^k (p,p') \ , \tag{11.145}$$

where

$$[\rho^{(0)}]^k (p,p') \equiv \int \mathrm{d}^3p_2 ... \mathrm{d}^3p_k \rho^{(0)}(p,p_2)\rho^{(0)}(p_2,p_3)...\rho^{(0)}(p_k,p') \ . \tag{11.146}$$

Following the notation of Eqs. (7.13) and (7.20), the inclusive single-particle distribution is given by

$$C_1(p) = L(p,p), \tag{11.147}$$

and the two-particle correlation function is

$$C_2(p_1,p_2) = L(p_1,p_2)L(p_2,p_1). \tag{11.148}$$

The general formula for the correlation functions reads

$$\begin{aligned} C_q(p_1,...p_q) = & L(p_1,p_2)L(p_2,p_3)....L(p_q,p_1) \\ & + \text{permutations of } (p_2,.....,p_q). \end{aligned} \tag{11.149}$$

This result is quite remarkable: In an independent production model, all higher-order correlations can be derived from the two-particle correlation function.

Further discussion is greatly simplified if the matrix $\rho^{(0)}(p,p)$ is expressed in terms of its eigenvalues λ_m and its eigenfunctions $\psi_m(p)$. In case of a discrete eigenvalue spectrum, one has

$$\rho^{(0)}(p,p') = \sum_m \psi_m(p)\lambda_m\psi_m^*(p') \tag{11.150}$$

and

$$[\rho^{(0)}]^k(p,p') = \sum_m \psi_m(p)(\lambda_m)^k\psi_m^*(p'). \tag{11.151}$$

Substituting this into (11.145) and performing the sum over k, one finds

$$L(p_1, p_2) = \sum_m \psi_m(p_1)\psi_m^*(p_2)\frac{\nu\lambda_m}{1-\nu\lambda_m}. \tag{11.152}$$

It is clear from (11.152) that $L(p_1, p_2)$, and thus also all inclusive distributions, become singular when $\nu\lambda_0 \to 1$. In this limit, corresponding to Bose-Einstein condensation, $L(p, p')$ is dominated by the first term in the sum and one has

$$L(p_1, p_2) = \frac{\psi^{(0)}(p_1)\psi^{(0)*}(p_2)}{1-\nu\lambda_0} + \tilde{L}(p_1, p_2), \tag{11.153}$$

with $\tilde{L}(p_1, p_2)$ finite for $\nu\lambda_0 \to 1$. Thus, at the condensation point, all the particles, except for a negligible fraction, are in the same state described by the eigenfunction $\psi^{(0)}(p)$.

11.8.3 The case of a Gaussian density matrix

Consider now a single-particle density matrix of the Gaussian form, discussed already by Pratt [7] (see also [247])

$$\rho_0(p, p') = \rho_x(p_x, p_x')\rho_y(p_y, p_y')\rho_z(p_z, p_z') \tag{11.154}$$

with

$$\rho_x(p_x, p_x') = \left(\frac{1}{2\pi\Delta_x^2}\right)^{\frac{1}{2}} e^{-\frac{(p_x^+)^2}{2\Delta_x^2} - \frac{1}{2}R_x^2(p_x^-)^2}, \tag{11.155}$$

where

$$p^+ \equiv \frac{1}{2}(p+p'); \quad p^- \equiv p - p'. \tag{11.156}$$

Analogous formulae define ρ_y and ρ_z. As is easily seen, Δ_x^2 is the average value of the square of the x-component of the particle momentum, and R_x^2 is the average value of the square of the x-coordinate of the particle emission point. The uncertainty principle implies that, for $i = x, y, z$,

$$R_i\Delta_i \geq \frac{1}{2}. \tag{11.157}$$

11.8.3.1 The multiplicity distribution

In order to determine the multiplicity distribution, the eigenvalues of the density matrix have to be found. The eigenfunctions of Eq. (11.155) are [248]

$$g_m(p) = e^{-\frac{1}{2}\frac{R}{\Delta}p^2} H_m\left(\sqrt{\frac{R}{\Delta}}p\right), \tag{11.158}$$

where $H_m(p)$ is the Hermite polynomial of order m. Using Eq. (11.158), one obtains for the eigenvalues:

$$\lambda_m = \lambda_{rst} = \lambda_0(1-\lambda_x)^r(1-\lambda_y)^s(1-\lambda_z)^t, \quad r,s,t = 0,1,..., \tag{11.159}$$

where $\lambda_0 = \lambda_x\lambda_y\lambda_z$ and for $i = x, y, z$

$$\lambda_i = \frac{2}{(1+2\Delta_i R_i)} . \tag{11.160}$$

The generating function is given by Eqs. (11.141), (11.142) and the cumulants by (11.144) with λ_m given by (11.159).

Using Eq. (11.159), one obtains the elegant formula

$$kc_k = \sum_m \lambda_m^k = \frac{\lambda_0^k}{[1-(1-\lambda_x)^k][1-(1-\lambda_y)^k][1-(1-\lambda_z)^k]} . \tag{11.161}$$

Together with Eq. (11.139), this gives for the cumulants

$$f_q = \sum_{k=q}^{\infty} \frac{(k-1)!}{(k-q)!} \frac{(\nu\lambda_0)^k}{[1-(1-\lambda_x)^k][1-(1-\lambda_y)^k](1-\lambda_z)^k} , \tag{11.162}$$

which diverge, as expected, at $\nu\lambda_0 \to 1$.

In Fig. 11.31 the multiplicity distributions are plotted for $\langle n \rangle = 3$ and several values of λ_0. One observes radical deviations from the Poisson distribution even for fairly small λ_0. The distributions obtained are much broader and extend to large multiplicities. Therefore, as soon as $R\Delta$ becomes close to 0.75 (or smaller), one may expect strong (and thus perhaps observable) effects on multiplicity distributions.

11.8.3.2 Particle Spectra

For a Gaussian density matrix, Eqs. (11.158) and (11.159) can be used for the explicit calculation of $L(p,p')$ [c.f. (11.152)] which, in turn, determines all particle distributions. One obtains

$$L(p,p') = \sum_{k=1}^{\infty} \frac{(\nu\lambda_0)^k}{1-(1-\lambda_0)^k} \hat{\rho}_k(p,p'), \tag{11.163}$$

where

$$\hat{\rho}_k(p,p') = \left(\frac{1}{2\pi\hat{\Delta}_k^2}\right)^{\frac{1}{2}} e^{-\frac{(p^+)^2}{2\hat{\Delta}_k^2} - \frac{1}{2}\hat{R}_k^2(p^-)^2} , \tag{11.164}$$

and $\hat{\Delta}_k$ and $\hat{R}_k$ are determined from the equations

$$\frac{\hat{R}_k}{\hat{\Delta}_k} = \frac{R}{\Delta}; \quad \hat{R}_k\hat{\Delta}_k = \frac{1}{2}\frac{1+\omega_k}{1-\omega_k}; \quad \omega_k = \left(\frac{2R\Delta-1}{2R\Delta+1}\right)^k . \tag{11.165}$$

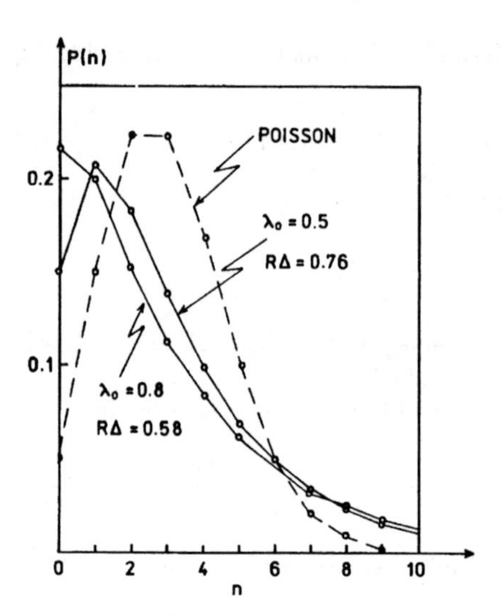

Figure 11.31. Multiplicity distributions of identical bosons for $\langle n \rangle = 3$ [8].

These results show that, even when the uncorrected distribution is described by a simple Gaussian, the resulting particle spectra are superpositions of an infinite number of Gaussians with varying width.

It follows from (11.165) that $\hat{R}_k < R$ and $\hat{\Delta}_k < \Delta$ for all $k > 1$. Consequently, the observed distributions are always *narrower* than the assumed uncorrected ones.

Another interesting quantity is the average value $\langle (p^-)^2 \rangle$ calculated from the two-particle correlation function, i.e.,

$$\langle (p^-)^2 \rangle = \frac{1}{f_2} \int \mathrm{d}p^+ C_2(p, p')(p^-)^2 \ , \tag{11.166}$$

where the cumulant f_2 is the integral of $C_2(p, p')$. In the standard analysis of the data, $2\langle (p^-)^2 \rangle$ is usually interpreted as an inverse of the average squared radius R^2_{eff} of the particle emission region.

The ratio R^2_{eff}/R^2 calculated from (11.166) is plotted in Fig. 11.32 for various values of $R\Delta$. At fixed $R\Delta$, R^2_{eff} decreases from R^2 to $R^2/2R\Delta$ when $\langle n \rangle$ varies from 0 to ∞. Thus, even fairly far from the critical point, the apparent size of the system R^2_{eff} (as determined from the two-particle correlation function) has little relation to the actual size of the system, given by R^2. One sees also that, even at the fixed particle phase-space density $\langle n \rangle / R\Delta$, the effect substantially increases with increasing $R\Delta$.

In summary: Due to symmetrization, the width of the momentum spectrum decreases whereas the width of the two-particle correlation function increases. This effect becomes stronger when the system approaches criticality. Close to the critical

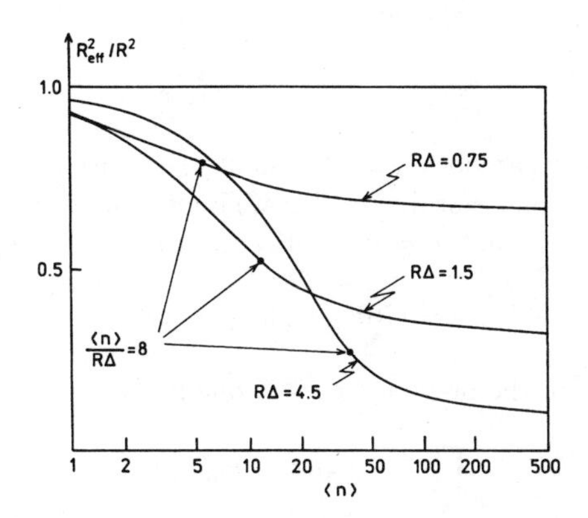

Figure 11.32. R_{eff}/R^2 plotted versus $\langle n \rangle$ for different values of $R\Delta$. The marked points correspond to a fixed value of the particle phase-space density $\langle n \rangle / R\Delta = 8$ [8].

point, the determination of the size of the emission region from the width of the correlation function is no longer possible.

For a Gaussian uncorrected density matrix, the corrected momentum distribution is Gaussian both in the low density and in the high density limit, but non-Gaussian in the intermediate region. At low density, the parameters Δ and R are determined by the widths of the momentum distribution and of the correlation function, respectively. At high density both these widths depend on the ratio R/Δ only.

11.8.4 Charge ratios

It was first pointed out by Pratt [7] that studies of the charge ratios may be an effective way to uncover the effects of HBT correlations in multiparticle systems. This issue can be readily treated using the methods developed above.

Here, we consider independent production of positive, negative and neutral pions. The main difficulty in formulating the problem is how to implement the constraint of charge conservation, as this clearly depends on the dynamics of the production process. Two cases are considered which illustrate well the main point: the charge ratios obtained may drastically differ from those expected from the uncorrected distributions.

In the first case the constraint of charge conservation is ignored altogether. This may be justified if the system of particles we consider is a small part of a very large system. The generating function of the multiplicity distribution is simply

$$\tilde{G}(z_+, z_-, z_0) = \tilde{G}(z_+)\tilde{G}(z_-)\tilde{G}(z_0), \tag{11.167}$$

where $\tilde{G}(z)$ is the generating function of the multiplicity distribution of one of the species. With $n_c = n_+ + n_-$ we have

$$\tilde{G}(z_c, z_0) = \tilde{G}^2(z_c)\tilde{G}(z_0). \tag{11.168}$$

From this equation, one can obtain the full joint distribution of charged and neutral pions by the usual methods. For illustration we quote the results for the two extreme cases $n_0 = 0$ "centauros") and $n_c = 0$ ("anticentauros"):

$$P_{n_0=0} = \tilde{G}(0), \quad P_{n_c=0} = [\tilde{G}(0)]^2. \tag{11.169}$$

Using now the results of the previous section, one finds

$$P_{n_0=0} = \prod_m (1 - \lambda_m \nu), \quad P_{n_c=0} = [P_{n_0=0}]^2. \tag{11.170}$$

One immediate consequence is that the production of centauros must be larger than that of anticentauros. For the Gaussian density matrix, according to Eq. (11.159) one has

$$P_{n_0=0} = \prod_{rst} \left[1 - \lambda_0 \nu (1 - \lambda_x)^r (1 - \lambda_y)^s (1 - \lambda_z)^t\right]. \tag{11.171}$$

The second case considered is that of charged particles produced in pairs, with the distribution of the form

$$P_{n_c,n_0} = \frac{P^2_{n_c/2} P_{n_0}}{\sum_m P^2_m}. \tag{11.172}$$

For the production of centauros the same formula is obtained as before. The generating function reads

$$\tilde{G}(z_c, z_0) = \tilde{G}_c(z_c)\tilde{G}(z_0) \tag{11.173}$$

where

$$\tilde{G}_c(z_c) = \frac{\sum_{n_c} P^2_{n_c/2} z_c^{n_c}}{\sum_m P^2_m}. \tag{11.174}$$

The probabilty of creating an anticentuaro is

$$\tilde{G}_c(z_c = 0) = \frac{[\tilde{G}(0)]^2}{\sum_m [P(m)]^2} > [\tilde{G}(0)]^2 \tag{11.175}$$

and thus the production of anticentauros is enhanced as compared to the previous case.

These results are illustrated in Fig. 11.33, where the probability of the occurence of a centauro, $P_{n_0=0}$, given by Eq. (11.171), is plotted versus the average multiplicity of the system considered for different values of the parameter $R\Delta$, which determines the maximal eigenvalue λ_0 through the relation Eq. (11.160). For $R\Delta \leq 1$ this probability remains substantial even for rather large values of the total multiplicity.

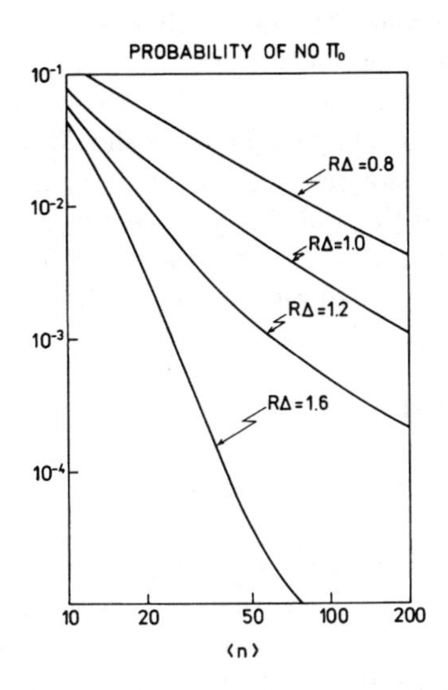

Figure 11.33. Probability of the occurence of a "Centauro" [248].

With increasing $R\Delta$, however, it drops rather fast even at moderate values of the average multiplicity.

The above results demonstrate that the effects of quantum statistics interference can modify substantially the multiplicity distributions expected naively from an uncorrelated emission of "distinguishable" particles.

The modifications become spectacular when the system approaches the critical point: the multiplicity distribution becomes very broad and does not resemble in any way the original Poisson distribution characteristic of uncorrelated emission. This means, in particular, that one expects relatively large probabilities for unusual configurations such as centauro or anticentauro events.

It is interesting to discuss the physical conditions for these phenomena to occur. Considering the example of the Gaussian density matrix, it is seen from Fig. 11.33 that the behavior of the system is mainly determined by the parameter $R\Delta$ and that spectacular effects in charge ratios occur when $R\Delta$ is of order 1 or smaller (as already noted, cfr. Eq. (11.157), $R\Delta \geq \frac{1}{2}$). The condition

$$R\Delta \sim 1 \tag{11.176}$$

implies, for large multiplicity, either a very large particle and energy density (when R is small and Δ takes its "canonical" value of $\sim 1\text{fm}^{-1}$), or a "canonical" energy density of about 1 GeV/fm^3 and a very small average momentum of the particles in the centre-of-mass of the system. For illustration, a system of 100 pions satisfying Eq. (11.176) for $\Delta = 200$ MeV would correspond to the energy density of about 35 GeV/4fm^3 $\approx$ 10 GeV/fm^3. On the other hand, for an energy density of about 1 GeV/fm^3 (and thus R correspondingly larger), Δ should not exceed 100 MeV.

Clearly, the probability of creating a centauro is enhanced if both effects cooperate. It may be concluded that the probability of creating a centauro is enhanced in an environment of high energy density and that the pions emerging from the centauros are likely to exibit abnormally small relative momenta.

11.9 Conclusions

1. Bose-Einstein interference is visible for pairs of identical bosons in all types of collision.

2. The radius of the pion emitting region ($r_K \approx 2r_G \approx 1.5$ fm) depends on the parametrization, but not on energy, virtuality or type of collision. For heavy-ion collisions it, however, increases with increasing mass number A like $A^{1/3}$.

3. The observed strength of the effect ($\lambda \approx 0.5$) depends on the reference sample used, but not on energy or type of collision, except for heavy-ion collisions, where it depends on A.

4. At $\sqrt{s} \geq 30$ GeV an increase of r is found with increasing particle density.

5. The effect is a particular challenge for Monte Carlo model builders.

6. Bose-Einstein correlations are also observed for other bosons (and Fermi-Dirac correlations for fermions). The emitter radius scales with the inverse square root of the particle mass m (or transverse mass m_T), as expected from the Heisenberg principle or from a strong position-momentum correlation. Consistently with the expectation from the string model, the radius for $\pi^0\pi^0$ correlations is found to be smaller than that for $\pi^\pm\pi^\pm$ correlations.

7. Strong genuine three-particle correlations exist in hadron-hadron and e^+e^- collisions. They seem to disappear within errors in low-A ion collisions, but reappear in large-A ones.

8. The correlation is far from Gaussian. Good results have been obtained with an Edgeworth expansion, but even power-law behavior is not excluded.

9. The correlation is far from spherically symmetric. The correlation domain (defined by the lengths of homogeneity in three space directions) is elongated along the event axis. Because of strong momentum-position correlations, this elongation is small, however, as compared to the length of the total string.

10. A $1/\sqrt{m_T}$ scaling due to momentum-position correlations was first observed in heavy-ion collisions but is now also observed in Z fragmentation.

11. Dilution of correlations in high-multiplicity hh and light-nuclei collisions as well as in WW decays suggests the lack of cross talk between neighboring strings at low density of strings, but string percolation may set in at the highest densities.

12. The full emission function in space-time can be extracted from a combination of inclusive single-particle distributions and BE correlation functions in the framework of three-dimensional hydrodynamic expansion. While a Gaussian shaped fire-*ball* is observed for AA collisions, a fire-*tube* is observed for hh collisions.

13. From the azimuthal oscillation of the radius parameters, the almond shape of the nuclear overlap is observed to be preserved in configuration space, even at RHIC energies.

14. The effects of quantum interference can substantially modify the multiplicity distribution and the single-particle spectra, when the product of initial average multiplicity and maximal eigenvalue of the density matrix approaches unity.

Bibliography

[1] R. Hanbury Brown and R.Q. Twiss, *Phil. Mag.* **45** (1954) 663; *Nature* **177** (1956) 27 and **178** (1956) 1046.

[2] G. Goldhaber, S. Goldhaber, W. Lee and A. Pais, *Phys. Rev.* **120** (1960) 300; G.I. Kopylov and M.I. Podgoretskiĭ, *Sov. J. Nucl. Phys.* **15** (1972) 219; **18** (1974) 336 and G.I. Kopylov, *Phys. Lett.* **B 50** (1974) 472;
E.V. Shuryak, *Phys. Lett.* **B 44** (1973) 387;
G. Cocconi, *Phys. Lett.* **B 49** (1974) 459;
M. Biyajima, *Phys. Lett.* **B 92** (1980) 193.

[3] G. Goldhaber, W.B. Fowler, S. Goldhaber, T.F. Hoang, *Phys. Rev. Lett.* **3** (1959) 181.

[4] OPAL Coll., P.D. Acton et al., *Z. Phys.* **C56** (1992) 521; DELPHI Coll., P. Abreu et al., *Z. Phys.* **C65** (1995) 587.

[5] A. Ballestrero et al., Determination of the mass of the W boson, Proc. Physics at LEP2, eds. G. Altarelli, T. Sjöstrand and F. Zwirner (CERN Yellow Report 96-01, 1996) V.1, p.141.

[6] L. Lönnblad and T. Sjöstrand, *Phys. Lett.* **B351** (1995) 293; *Eur. Phys. J.* **C2** (1998) 165.

[7] S. Pratt, *Phys. Lett.* **B301** (1993) 159; S. Pratt and V. Zelevinsky, *Phys. Rev. Lett.* **72** (1994) 816; S. Pratt, *Phys. Rev.* **C50** (1994) 469.

[8] A. Białas and K. Zalewski, *Phys. Lett.* **B436** (1998) 153; *Phys. Rev.* **D59** (1999) 097502.

[9] T. Csörgő and J. Zimányi, *Phys. Rev. Lett.* **80** (1998) 916; J. Zimányi and T. Csörgő, *Heavy Ion Phys.* **9** (1999) 241.

[10] M.G. Bowler, *Z. Phys.* **C29** (1985) 617.

[11] E.V. Shuryak, *Phys. Rep.* **61** (1980) 72.

[12] J.D. Bjorken, *Phys. Rev.* **D27** (1983) 140.

[13] H. Gyulassy, S. Kauffmann, L.W. Wilson, *Phys. Rev.* **C20** (1979) 2267.

[14] S. Pratt, T. Csörgő and J. Zimanyi, *Phys. Rev.* **C42** (1990) 2646.

[15] S. Chapman and U. Heinz, *Phys. Lett.* **B340** (1994) 250; S. Chapman, P. Scotto and U. Heinz, *Phys. Rev. Lett.* **74** (1995) 4400.

[16] A.N. Makhlin and Yu.M. Sinyukov, *Z. Phys.* **C39** (1988) 69; Yu.M. Sinyukov, *Nucl. Phys.* **A498** (1989) 151c; *Nucl. Phys.* **A566** (1994) 589c; in: *Hot Hadronic Matter*, eds. J. Letessier et al. (Plenum Publ. Corp., 1995) p.309; S.V. Akkelin and Yu.M. Sinyukov, *Phys. Lett.* **B356** (1995) 525.

[17] A.I. Golokhvastov, *Phys. At. Nucl.* **65** (2002) 190; M.Kh. Anikina, A.I. Golokhvastov, J. Lukstins, *Phys. At. Nucl.* **65** (2002) 573; M.Kh. Anikina et al., *Phys. At. Nucl.* **67** (2004) 406.

[18] A. Białas, K. Zalewski, *Phys. Lett.* **B591** (2004) 83.

[19] S. Chapman, J.R. Nix, U. Heinz, *Phys. Rev.* **C52** (1995) 2694.

[20] M.G. Bowler, *Z. Phys.* **C29** (1985) 617; *Phys. Lett.* **B180** (1986) 299; ibid. **185** (1987) 205; X. Artru and M. Bowler, *Z. Phys.* **C30** (1986) 355; *Z. Phys.* **C37** (1988) 293.

[21] B. Andersson and W. Hofmann, *Phys. Lett.* **B 169** (1986) 364.

[22] TPC Coll., H. Aihara et al., *Phys. Rev.* **D31** (1985) 996.

[23] CLEO Coll., P. Avery et al., *Phys. Rev.* **D32** (1985) 2294.

[24] TASSO Coll., M. Althoff et al., *Z. Phys.* **C29** (1985) 347; **C 30** (1986) 355.

[25] B. Andersson and M. Ringnér, *Phys. Lett.* **B421** (1998) 283 and *Nucl. Phys.* **B513** (1998) 627; Š. Todorova-Nová and J. Rameš, Simulation of Bose-Einstein effect using space-time aspects of Lund string fragmentation model, Strasbourg preprint IReS97-29.

[26] M. Deutschmann et al., *Nucl. Phys.* **B204** (1982) 333.

[27] P. Grassberger, *Nucl. Phys.* **B120** (1977) 231.

[28] M.G. Bowler, *Z. Phys.* **C46** (1990) 305.

[29] K. Kulka and B. Lörstad, *Z. Phys.* **C45** (1990) 581.

[30] R. Lednický and T.B. Progulova, *Z. Phys.* **C55** (1992) 295.

[31] S.S. Padula and M. Gyulassy, *Nucl. Phys.* **A544** (1992) 537c.

[32] T. Csörgő, B. Lörstad and J. Zimanyi, *Z. Phys.* **C71** (1996) 491; S. Nickerson, T. Csörgő and D. Kiang, *Phys. Rev.* **C57** (1998) 3251; T. Csörgő, B. Lörstad, J. Schmidt-Sørensen and A. Ster, *Eur. Phys. J.* **C9** (1999) 275.

[33] J. Bolz et al., *Phys. Lett.* **B300** (1993) 404; *Phys. Rev.* **D47** (1993) 3860.

[34] H.C. Eggers, P. Lipa, P. Carruthers and B. Buschbeck, *Phys. Lett.* **B301** (1993) 298; *Phys. Rev.* **D48** (1993) 2040.

[35] NA22 Coll., M. Adamus et al., *Z. Phys.* **C37** (1988) 347.

[36] NA22 Coll., N.M. Agababyan et al., *Z. Phys.* **C59** (1993) 195.

[37] P. Seyboth (NA49), Proc. 7th Int. Workshop on Multiparticle Production, Eds. R.C Hwa et al. (World Scientific, Singapore, 1997) p.13; H. Appelshäuser et al. (NA49), *Eur. Phys. J.* **C2** (1998) 661.

[38] NA35 Coll., T. Alber et al., *Z. Phys.* **C73** (1997) 443.

[39] M.G. Bowler, *Phys. Lett.* **B270** (1991) 69.

[40] T. Csörgő and S. Pratt, Proc. Workshop on Relativistic Heavy Ion Physics at Present and Future Accelerators, eds. T. Csörgő et al. (KFKO-1991-28/A, Budapest, 1991) p.75; E.O. Alt, T. Csörgő, B. Lörstad and J. Schmidt-Sørensen, *Phys. Lett.* **B458** (1999) 407; *Eur. Phys. J.* **C13** (2000) 663.

[41] Yu.M. Sinyukov et al., *Phys. Lett.* **B432** (1998) 248.

[42] M. Biyajima, T. Mizoguchi, T. Osada, G. Wilk, *Phys. Lett.* **B353** (1995) 340; **B366** (1996) 394; M. Biyajima, T. Mizoguchi, G. Wilk, *Z. Phys.* **C65** (1995) 511; T. Osada, S. Sano, M. Biyajima, *Z. Phys.* **C72** (1996) 285; T. Mizoguchi, M. Biyajima, *Phys. Lett.* **B499** (2001) 245; M. Biyajima, T. Mizoguchi, N. Suzuki, *Phys. Lett.* **B568** (2003) 237; M. Biyajima, M. Kaneyama, T. Mizoguchi, *Phys. Lett.* **B601** (2004) 41.

[43] M. Suzuki, *Phys. Rev.* **D35** (1987) 3359.

[44] M.G. Bowler, *Z. Phys.* **C39** (1988) 81.

[45] G.I. Kopylov, M.I. Podgoretsky, *Sov. J. Nucl. Phys.* **15** (1972) 219.

[46] W.A. Zajc: Bose-Einstein correlations: from statistics to dynamics. Nevis-R-1384 (1987) and Proc. NATO ASI 'Particle Production in Highly Excited Matter' Il Ciocco (Italy) 1992, eds. H. Gutbrod and J. Rafelski (Plenum Press, New York, 1993) p.435.

[47] M.J. Podgoretzkiĭ, *Sov. J. Part. Nucl.* **20** (1989) 266.

[48] B. Lörstad, *Int. J. Mod. Phys.* **A4** (1989) 2861.

[49] D.H. Boal, C.K. Gelbke, B.K. Jennings, *Rev. Mod. Phys.* **62** (1990) 553.

[50] G. Baym, *Acta Phys. Pol.* **B29** (1998) 1839.

[51] R. Carlsson et al., *Phys. Scr.* **31** (1985) 21.

[52] CCHK Coll., D. Drijard et al., *Nucl. Phys.* **B155** (1979) 269.

[53] AFS Coll., T. Åkesson et al., *Phys. Lett.* **B129** (1983) 269; **B 155** (1985) 128; **B 187** (1987) 420.

[54] ABCDHW Coll., A. Breakstone et al., *Phys. Lett.* **B162** (1985) 400 and *Z. Phys.* **C 33** (1987) 333.

[55] NA23 Coll., J.L. Bailly et al., *Z. Phys.* **C43** (1989) 341.

[56] NA27 Coll., M. Aguilar-Benitez et al., *Z. Phys.* **C54** (1992) 21.

[57] UA1 Coll., C. Albajar et al., *Phys. Lett.* **B226** (1989) 410 and **B229** (1989) 439.

[58] E735 Coll., T. Alexopoulos et al., *Phys. Rev.* **D48** (1993) 1931.

[59] MARKII Coll., I. Juricic et al., *Phys. Rev.* **D39** (1989) 1.

[60] OPAL Coll., P.D. Acton et al., *Phys. Lett.* **B267** (1991) 143; **B287** (1992) 401; *Z. Phys.* **C58** (1993) 207.

[61] DELPHI Coll., P. Abreu et al., *Phys. Lett.* **B286** (1992) 201; *Z. Phys.* **C63** (1994) 17.

[62] ALEPH Coll., D. Decamp et al., *Z. Phys.* **C54** (1992) 75.

[63] AMY Coll., S.K. Choi et al., *Phys. Lett.* **B355** (1995) 406.

[64] EMC Coll., M. Arneodo et al., *Z. Phys.* **C32** (1986) 1.

[65] BBCNC Coll., V.A. Korotkov et al., *Z. Phys.* **C60** (1993) 37.

[66] E665 Coll., M.R. Adams et al., *Phys. Lett.* **B308** (1993) 418.

[67] H1 Coll., C. Adloff et al., *Z. Phys.* **C75** (1997) 437.

[68] ZEUS Coll., S. Chekanov et al., *Phys. Lett.* **B583** (2004) 231.

[69] S. Y. Fung et al., *Phys. Rev. Lett.* **41** (1978) 1592.

[70] J.J. Lu et al., *Phys. Rev. Lett.* **46** (1981) 989.

[71] D. Beavis et al., *Phys. Rev.* **C27** (1983) 910; **C34** (1986) 757.

[72] W.A. Zajc et al., *Phys. Rev.* **C29** (1984) 2173.

[73] NA35 Coll., A. Bamberger et al., *Z. Phys.* **C38** (1988) 79.

[74] NA44 Coll., H. Bøggild et al., *Phys. Lett.* **B302** (1993) 510.

[75] E877 Coll., J. Barrette et al., *Phys. Rev. Lett.* **78** (1997) 2916.

[76] E802 Coll., L. Ahle et al., *Phys. Rev.* **C66** (2002) 054906.

[77] J. Bartke, *Phys. Lett.* **B174** (1986) 32.

[78] S. Barshay, *Phys. Lett.* **B130** (1983) 220.

[79] NA5 Coll., C. de Marzo et al., *Phys. Rev.* **D29** (1984) 363.

[80] M. Goossens et al., *Nuovo Cim.* **48A** (1978) 469.

[81] OPAL Coll., G. Alexander et al., *Z. Phys.* **C72** (1996) 389.

[82] P. Lipa, H.C. Eggers and B. Buschbeck, *Phys. Rev.* **D53** (1996) R4711.

[83] B. Buschbeck, H.C. Eggers and P. Lipa, *Phys. Lett.* **B481** (2000) 187; B. Buschbeck and H.C. Eggers, *Nucl. Phys.* **B** (Proc. Suppl.) **92** (2001) 235.

[84] P. Lipa, B. Buschbeck, *Phys. Lett.* **B223** (1989) 465.

[85] A. Capella, U. Sukhatme, C.-I. Tan, J. Tran Thanh Van, *Phys. Rep.* **236** (1994) 225.

[86] G. Alexander, E.K.G. Sarkisyan, *Phys. Lett.* **B487** (2000) 215.

[87] WA80 Coll., T.C. Awer et al., *Z. Phys.* **C96** (1995) 67.

[88] M. Biyajima, N. Suzuki, G. Wilk, Z. Włodarczyk, *Phys. Lett.* **B386** (1996) 297.

[89] M.A. Braun, F. del Moral and C. Pajares, *Eur. Phys. J.* **C21** (2001) 557.

[90] STAR Coll., C. Adler et al., *Phys. Rev. Lett.* **87** (2001) 082301; J. Adams et al., nucl-ex/0312009.

[91] NA44 Coll., H. Beker et al., *Phys. Rev. Lett.* **74** (1995) 3340; I.G. Bearden et al., *Phys. Rev.* **C58** (1998) 1656.

[92] K. Fiałkowski, R. Wit, *Z. Phys.* **C74** (1997) 145.

[93] R. Mureşan, O. Smirnova, B. Lörstad, *Eur. Phys. J.* **C6** (1999) 629.

[94] O.V. Utyuzh, G. Wilk, Z. Włodarczyk, *Phys. Lett.* **B522** (2001) 273; *Acta Phys. Pol.* **B33** (2002) 2681; O.V. Utyuzh, G. Wilk, N, Rybczyński, Z. Włodarczyk, hep-ph/0210075, hep-ph/0210328 .

[95] T. Osada, M. Maruyama, F. Takagi, *Phys. Rev.* **D59** (1999) 014024.

[96] A. Białas and A. Krzywicki, *Phys. Lett.* **B354** (1995) 134.

[97] R. Haywood, Rutherford Lab report RAL 94-074 (1995).

[98] T. Wibig, *Phys. Rev.* **D53** (1996) 3586.

[99] V. Kartvelishvili, R. Kvatadze, R. Møller, *Phys. Lett.* **B408** (1997) 331.

[100] S. Jadach and K. Zalewski, *Acta Phys. Pol.* **B28** (1997) 1363.

[101] Q.H. Zhang et al., *Phys. Lett.* **B407** (1997) 33.

[102] K. Fiałkowski, R. Wit, *Acta Phys. Pol.* B **28** (1997) 2039; *Eur. Phys. J.* **C2** (1998) 691; K. Fiałkowski, R. Wit, J. Wosiek, *Phys. Rev.* **D58** (1998) 094013.

[103] S. Mohanty, Proc. 10th Int. Workshop on Multiparticle Production, eds. N.G. Antoniou et al. (World Scientific, Singapore, 2003) p.89.

[104] B. Andersson, S. Mohanty, F. Söderberg, *Eur. Phys. J.* **C21** (2001) 631.

[105] GAMS Coll., D. Alde et al., *Phys. Lett.* **B397** (1997) 350.

[106] K. Eskreys, *Acta Phys. Pol.* **36** (1969) 237.

[107] V.G. Grishin et al., *Sov. J. Nucl. Phys.* **47** (1988) 278.

[108] L3 Coll., P. Achard et al., *Phys. Lett.* **B524** (2002) 55.

[109] OPAL Coll., G. Abbiendi et al., *Phys. Lett.* **B559** (2003) 131.

[110] O. Kortner et al., *Eur. Phys. J.* **C25** (2002) 353.

[111] CPLEAR Coll., A. Angelopoulos et al., *Eur. Phys. J.* **C1** (1998) 1.

[112] E802 Coll., Y. Akiba et al., *Phys. Rev. Lett.* **70** (1993) 1057.

[113] NA44 Coll., H. Beker et al., *Z. Phys.* **C64** (1994) 209; I.G. Bearden et al., *Phys. Rev. Lett.* **87** (2001) 112301.

[114] NA49 Coll., S.V. Afanasiev et al., *Phys. Lett.* **B557** (2003) 157.

[115] DELPHI Coll., P. Abreu et al., *Phys. Lett.* **B379** (1996) 330.

[116] OPAL Coll., G. Abbiendi et al., *Eur. Phys. J.* **C21** (2001) 23.

[117] L3 Coll., P. Achard et al., *Phys. Lett.* **B540** (2002) 185.

[118] OPAL Coll., P.D. Acton et al., *Phys. Lett.* **B298** (1993) 456; R. Akers et al., *Z. Phys.* **C67** (1995) 389.

[119] ALEPH Coll., D. Buskulic et al., *Z. Phys.* **C64** (1994) 361; S. Schael et al., *Two-particle correlations in* pp, $\bar{p}\bar{p}$ *and* $K^0_s K^0_s$ *pairs from hadronic Z decays*, CERN-PH-EP-2004-33, submitted to Phys. Lett. B .

[120] H. Lipkin, *Phys. Lett.* **B219** (1989) 474; *Phys. Rev. Lett.* **69**(1992) 3700; G. Alexander and H.J. Lipkin, *Phys. Lett.* **B456** (1999) 270.

[121] A.M. Cooper et al., *Nucl. Phys.* **B139** (1978) 45.

[122] DELPHI Coll., P. Abreu et al., *Phys. Lett.* **B323** (1994) 242.

[123] G. Alexander and H.J. Lipkin, *Phys. Lett.* **B352** (1995) 162; R. Lednický, On correlation and spin composition techniques, MPI-PhE-10, 1999.

[124] OPAL Coll., G. Alexander et al., *Phys. Lett.* **B384** (1996) 377.

[125] ALEPH Coll., R. Barate et al., *Phys. Lett.* **B475** (2000) 395.

[126] DELPHI Coll., XXIX Int. Conf. on High Energy Physics, Vancouver 1998 (paper 154).

[127] NA44 Coll., H. Bøggild et al., *Phys. Lett.* **B458** (1999) 181; I.G. Bearden et al., nucl-ex/0305014.

[128] NA49 Coll., H. Appelshäuser et al., *Phys. Lett.* **B467** (1999) 21.

[129] S. Pratt, *Phys. Rev. Lett.* **53** (1984) 1219.

[130] S.V. Akkelin and Yu. M. Sinyukov, *Phys. Lett.* **B356** (1995) 525.

[131] U.A. Wiedemann, P. Scotto and U. Heinz, *Phys. Rev.* **C53** (1996) 918.

[132] T. Csörgő and B. Lörstad, *Phys. Rev.* **C54** (1996) 1390.

[133] G. Alexander, I. Cohen, E. Levin, *Phys. Lett.* **B452** (1999) 159; G. Alexander, *Phys. Lett.* **B506** (2001) 45.

[134] M. Smith, *Phys. Lett.* **B477** (2000) 141.

[135] B. Lörstad, O.G. Smirnova, Proc. 7th Int. Workshop on Multiparticle Production, Nijmegen, eds. R.C. Hwa et al. (World Scientific, Singapore, 1997) p.42.

[136] A. Białas and K. Zalewski, Acta Phys. Pol. **B30** (1999) 359.

[137] T. Csörgő and J. Zimányi, *Nucl. Phys.* **A517** (1990) 588.

[138] R.M. Weiner, *Phys. Lett.* **B232** (1989) 278; **242** (1990) 547; V.L. Lyuboshitz, *Sov. J. Nucl. Phys.* **53** (1991) 514; H. Heiselberg and A.P. Vischer, *Phys. Rev.* **C55** (1997) 874.

[139] U. Heinz and Q. Zhang, *Phys. Rev.* **C56** (1997) 426; H. Nakamura and R. Seki, *Phys. Rev.* **C60** (1999) 064904.

[140] R. Weiner, *Phys. Lett.* **B232** (1989) 278; I.V. Andreev, R.M. Weiner, *Phys. Lett.* **B253** (1991) 416; M. Plümer, L.V. Razumov, R.M. Weiner, *Phys. Lett.* **B286** (1992) 335.

[141] M. Biyajima et al., *Progr. Theor. Phys.* **84** (1990) 931; ibid. **88** (1992) 157A.

[142] J.G. Cramer, *Phys. Rev.* **C43** (1991) 2798; J.G. Cramer and K. Kadija, *Phys. Rev.* **C53** (1996) 908.

[143] Y.M. Liu et al., *Phys. Rev.* **C34** (1986) 1667.

[144] AFS Coll., T. Åkesson et al., *Z. Phys.* **C36** (1987) 517.

[145] UA1 Coll., N. Neumeister et al., *Phys. Lett.* **B275** (1992) 186.

[146] NA22 Coll., N.M. Agababyan et al., *Z. Phys.* **C68** (1995) 229.

[147] NA22 Coll., N.M. Agababyan et al., *Phys. Lett.* **B332** (1994) 458.

[148] DELPHI Coll., P. Abreu et al., *Phys. Lett.* **B355** (1995) 415.

[149] OPAL Coll., K. Ackerstaff et al., *Eur. Phys. J.* **C5** (1998) 239.

[150] STAR Coll., J. Adams et al., *Phys. Rev. Lett.* **91** (2003) 262301.

[151] WA98 Coll., M.M. Aggarwal et al., *Phys. Rev. Lett.* **85** (2000) 2895.

[152] NA44 Coll., H. Bøggild et al., *Phys. Lett.* **B455** (1999) 77; I.G. Bearden et al., *Phys. Lett.* **B 517** (2001) 25.

[153] W. Kittel, *Acta Phys. Pol.* **B32** (2001) 3927; M.A. Braun, F. del Moral and C. Pajares, *Phys. Lett.* **B551** (2003) 291.

[154] K. Kolehmainen and M. Gyulassy, *Phys. Lett.* **B180** (1986) 203.

[155] M.G. Bowler, *Particle World* **2** (1991) 1.

[156] X. Artru and G. Mennessier, *Nucl. Phys.* **B70** (1974) 93.

[157] A. Białas, *Nucl. Phys.* **A545** (1992) 285c; *Acta Phys. Pol.* **B23** (1992) 561.

[158] J. Masarik, A. Nogová, J. Pišút, N. Pišútova, *Z. Phys.* **C75** (1997) 95.

[159] NA22 Coll., N.M. Agababyan et al., *Z. Phys.* **C59** (1993) 405.

[160] UA1 Coll., N. Neumeister et al., *Z. Phys.* **C60** (1993) 633; H.C. Eggers, B. Buschbeck and P. Lipa in Proc. XXVth Int. Symp. on Multiparticle Dynamics, eds. D. Bruncko et al. (World Scientific, Singapore, 1996), p.650.

[161] WA98 Coll., M.M. Aggarwal et al., *Eur. Phys. J.* **C16** (2000) 445.

[162] H.C. Eggers, P. Lipa and B. Buschbeck, *Phys. Rev. Lett.* **79** (1997) 197.

[163] I.V. Andreev, M. Plümer and R.M. Weiner, *Int. J. Mod. Phys.* **A8** (1993) 4577.

[164] M.G. Kendall and A. Stuart, The Advanced Theory of Statistics, Vol.1 (Ch. Griffin, London 1958).

[165] S. Hegyi and T. Csörgő, Proc. Budapest Workshop on Relativistic Heavy Ion Collisions, eds. T. Csörgő et al. (KFKI-1993-11/A, Budapest, 1991) p.47; T. Csörgő, Proc. Cracow Workshop on Multiparticle Production, eds. A. Białas et al. (World Scientific, Singapore, 1994) p.175; T. Csörgő, S. Hegyi, *Phys. Lett.* **B489** (2000) 15.

[166] E802 Coll., T. Abbott et al., *Phys. Rev. Lett.* **69** (1992) 1030.

[167] NA44 Coll., B. Lörstad, Proc. Budapest Workshop on Relativistic Heavy Ion Collisons, eds. T. Csörgő et al. (KFKI-1993-11/A, Budapest, 1993) p.36.

[168] L3 Coll., M. Acciarri et al., *Phys. Lett.* **B458** (1999) 517.

[169] T. Csörgő, S. Hegyi, W.A. Zajc, *Eur. Phys. J.* **C36** (2004) 67.

[170] M. Deutschmann et al., *Nucl. Phys.* **B103** (1976) 198;
E. Calligarich et al., *Lett. Nuovo Cim.* **16** (1976) 129;
C. Ezell et al., *Phys. Rev. Lett.* **38** (1977) 873.

[171] N. Angelov et al., *Sov. J. Nucl. Phys.* **35** (1982) 45; ibid. **37** (1983) 202.

[172] C. De Marzo et al., *Phys. Rev.* **D29** (1984) 363.

[173] M.I. Podgoretzkiĭ, A. Cheplakov, *Sov. J. Nucl. Phys.* **44** (1986) 835.

[174] R.A. Kvatadze, R. Møller, B. Lörstad, *Z. Phys.* **C38** (1988) 551.

[175] NA22 Coll., N.M. Agababyan et al., *Z. Phys.* **C66** (1995) 409.

[176] G.F. Bertsch, M. Gong and M. Tohyama, *Phys. Rev.* **C37** (1988) 1896;
G.F. Bertsch and G.E. Brown, *Phys. Rev.* **C40** (1989) 1830.

[177] S. Pratt, *Phys. Rev.* **D33** (1986) 1314; *Phys. Rev.* C42 (1990) 2646.

[178] T. Csörgő and B. Lörstad, Proc. 8th Int. Workshop on Multiparticle Production, Matrahaza (Hungary) eds. T. Csörgő et al. (World Scientific, Singapore, 1999) p.108 .

[179] S. Chapman, P. Scotto, U. Heinz, *Phys. Rev. Lett.* **74** (1995) 4400.

[180] NA44 Coll., I.C. Bearden et al., *Eur. Phys. J.* **C18** (2000) 317.

[181] NA35 Coll., D. Ferenc et al., *Nucl. Phys.* **A544** (1992) 531c; Th. Alber et al., *Z. Phys.* **C66** (1995) 77.

[182] NA44 Coll., H. Bøggild et al., *Phys. Lett.* **B349** (1995) 386.

[183] E895 Coll., M.A. Lisa et al., *Phys. Rev. Lett.* **84** (2000) 2798.

[184] NA22 Coll., N.M. Agababyan et al., *Z. Phys.* **C71** (1996) 405.

[185] DELPHI Coll., P. Abreu et al., *Phys. Lett.* **B471** (2000) 460.

[186] OPAL Coll., G. Abbiendi et al., *Eur. Phys. J.* **C16** (2000) 423.

[187] ALEPH Coll., A. Heister et al., *Eur. Phys. J.* **C36** (2004) 147.

[188] K. Fiałkowski and R. Wit, *Acta Phys. Pol.* **B32** (2001) 1233.

[189] U. Heinz et al., *Phys. Lett.* **B382** (1996) 181.

[190] F. Yano and S. Koonin, *Phys. Lett.* **B78** (1978) 556; M. Podgoretskiĭ, *Sov. J. Nucl. Phys.* **37** (1983) 272.

[191] W. Zajc, *Nucl. Phys.* **A525** (1991) 315c.

[192] GIBS Coll., M.Kh. Anikina et al., *Phys. Lett.* **B397** (1997) 30.

[193] Y.-F. Wu et al., *Eur. Phys. J.* **C1** (1998) 599;
B. Tomasik and U. Heinz, *Eur. Phys. J.* **C4** (1998) 327.

[194] D. Rischke and M. Gyulassy, *Nucl. Phys.* **A608** (1996) 479.

[195] U. Mayer, E. Schnedermann and U. Heinz, *Phys. Lett.* **B294** (1992) 69;
Yu.M. Sinyukov, *Nucl. Phys.* **A566** (1994) 589c.

[196] WA98 Coll., L. Rosselet et al., *Nucl. Phys.* **A698** (2002) 647c.

[197] CERES Coll., D. Adamová et al., *Nucl. Phys.* **A714** (2003) 124.

[198] WA97 Coll., F. Antinori et al., *J. Phys.* **G27** (2001) 2325.

[199] PHENIX Coll., K. Adcox et al., *Phys. Rev. Lett.* **88** (2002) 192302; S.S. Adler et al., nucl-ex/0401003.

[200] PHOBOS Coll., B.B. Back et al., nucl-ex/0406027.

[201] DELPHI Coll., B. Lörstad and O.G. Smirnova, Proc. 7th Int. Workshop on Multiparticle Production, eds. R. Hwa et al. (World Scientific, Singapore,1997) p.42.

[202] NA44 Coll., I.G. Bearden et al., *Phys. Rev. Lett.* **78** (1997) 2080.

[203] NA44 Coll., K. Kaimi et al., *Z. Phys.* **C75** (1997) 619.

[204] CERES Coll., D. Adamová et al, *Phys. Rev. Lett.* **90** (2003) 022301.

[205] T. Csörgő, J. Zimányi, J. Bondorf, H. Heiselberg, *Phys. Lett.* **B222** (1989) 115; *Z. Phys.* **C46** (1990) 507; T. Csörgő et al., *Phys. Lett.* **B241** (1990) 301; T. Csörgő and J. Zimányi, *Nucl. Phys.* **A525** (1991) 507c.

[206] H. Sorge, H. Stöcker, W. Greiner, *Nucl. Phys.* **A498** (1989) 567; H. Sorge, *Phys. Rev.* **C52** (1995) 3291.

[207] D. Hardtke, S.A. Voloshin, *Phys. Rev.* **C61** (2000) 024905.

[208] S. Soff, S.A. Bass, A. Dumitru, *Phys. Rev. Lett.* **86** (2001) 3981; S. Soff, S.A. Bass, D. Hardtke, S. Panitkin, *Phys. Rev. Lett.* **88** (2002) 072301; *Nucl. Phys.* **A715** (2003) 801; S. Soff, in *Proc. 30th Int. Workshop on Gross Properties of Nuclei and Nuclear Excitation*, eds. M. Buballa et al. (GSI, Darmstadt, 2002) p.222.

[209] T.J. Humanic, *Nucl. Phys.* **A715** (2003) 641.

[210] P.F. Kolb, J. Sollfrank, U. Heinz, *Phys. Rev.* **C62** (2000) 054909; U Heinz and P.F. Kolb, *Nucl. Phys.* **A702** (2002) 269c.

[211] F. Cooper and G. Frye, *Phys. Rev.* **D10** (1974) 186.

[212] B.R. Schlei, U. Ornik, M. Plümer, R.M. Weiner, *Phys. Lett.* **B293** (1992) 275.

[213] D. Molnar, M. Gyulassy, *Heavy Ion Phys.* **18** (2003) 69.

[214] Z.W. Lin, C.M. Ko, S. Pal, *Phys. Rev. Lett.* **89** (2002) 152301.

[215] F. Retière, M.A. Lisa, nucl-th/0312024; F. Retière, *J. Phys.* **G30** (2004) S335.

[216] T. Hirano, *Phys. Rev.* **C65** (2002) 011901; T. Hirano, K. Tsuda, *Phys. Rev.* **C66** (2002) 054905; *Nucl. Phys.* **A715** (2003) 821.

[217] W. Broniowski and W. Florkowski, *Phys. Rev. Lett.* **87** (2001) 272302; *AIP Conf. Proc.* **660** (2003) 177; W. Broniowski, A. Baran and W. Florkowski, *AIP Conf. Proc.* **660** (2003) 185.

[218] S.V. Akkelin, Yu.M. Sinyukov, *Z. Phys.* **C72** (1996) 501.

[219] T. Csörgő, *Phys. Lett.* **B347** (1995) 354.

[220] T. Csörgő, B. Lörstad and J. Zimányi, *Phys. Lett.* **B338** (1994) 134.

[221] NA22 Coll., N.M. Agababyan et al., *Phys. Lett.* **B422** (1998) 359.

[222] T. Csörgő, S. Nickerson, D. Kiang, Proc. 7th Int. Workshop on Multiparticle Production, Nijmegen, eds. R.C. Hwa et al. (World Scientific, Singapore, 1997) p.50 .

[223] A. Ster, T. Csörgő and B. Lörstad, *Nucl. Phys.* **A661** (1999) 419.

[224] M. Csanád, T. Csörgő, B. Lörstad, A. Ster, *Acta Phys. Pol.* **B35** (2004) 191; nucl-th/0402037; hep-ph/0406042.

[225] Z. Fodor and S.D. Katz, *JHEP* **0404** (2004) 050.

[226] T. Csörgő in *Proc. Particle Production Spanning MeV and TeV Energies*, eds. W. Kittel, P.J. Mulders and O. Scholten (NATO Science Series, 2000) p.203; *Heavy Ion Phys.* **15** (2002) 1.

[227] T. Csörgő, *Simple Analytic Solution of Fireball Hydrodynamics*, nucl-th/9809011.

[228] W. Reisdorf, H.G. Ritter, *Ann. Rev. Nucl. Part. Sci.* **47** (1997) 663.

[229] N. Herrmann, J.P. Wessels, T. Wienold, *Ann. Rev. Nucl. Part. Sci.* **49** (1999) 581.

[230] P.F. Kolb, U. Heinz, in *Quark gluon plasma*, eds. R.C. Hwa et al., p.634; nucl-th/0305084.

[231] S.A. Voloshin, W.E. Cleland, *Phys. Rev.* **C53** (1996) 896 and **C54** (1996) 3212; S.A. Voloshin, *Nucl. Phys.* **A715** (2003) 379c.

[232] D. Teaney, E.V. Shuryak, *Phys. Lett.* **83** (1999) 4951.

[233] H. Heiselberg, *Phys. Rev. Lett.* **82** (1999) 2052; H. Heiselberg, A.-M. Levy, *Phys. Rev.* **C59** (1999) 2716.

[234] J. Brachmann et al., *Phys. Rev.* **C61** (2000) 024909 and nucl-th/9912014.

[235] E877 Coll., D. Miskowiec et al., *Nucl. Phys.* **A590** (1995) 473c.

[236] E895 Coll., M.A. Lisa et al., *Phys. Lett.* **B496** (2000) 1.

[237] STAR Coll., J. Adams et al., *Phys. Rev. Lett.* **93** (2004) 012301.

[238] P. Filip, *Azimuthally sensitive HBT analysis*, hep-ex/9609001.

[239] U.A. Wiedemann, *Phys. Rev.* **C57** (1998) 266; M.A. Lisa, U. Heinz, U.A. Wiedemann, *Phys. Lett.* **B489** (2000) 287; U. Heinz, A. Hummel, M.A. Lisa, U.A. Wiedemann, *Phys. Rev.* **C66** (2002) 044903; U. Heinz, P.F. Kolb, *Phys. Lett.* **B542** (2002) 216.

[240] S.V. Chekanov, E.A. De Wolf, W. Kittel, *Eur. Phys. J.* **C6** (1999) 403; E.A. De Wolf, *Correlations in* $e^+e^- \to W^+W^-$ *hadronic decays*, hep-ph/0101243.

[241] L3 Coll., M. Acciarri, *Phys. Lett.* **B493** (2000) 233; **B547** (2002) 139.

[242] M. Skrzypek et al., *Comp. Phys. Comm.* **94** (1996) 216; M. Skrzypek et al., *Phys. Lett.* **B372** (1996) 289.

[243] S. Todorova, Proc. XXXIVth Int. Symposium on Multiparticle Dynamics, Acta Phys. Pol. (to be publ.); The LEP Electroweak Working Group et al., hep-ex/0412015 .

[244] ALEPH Coll., R. Barate et al., *Phys. Lett.* **B478** (2000) 50; J. Schael et al., CERN-PH-EP-2004-053, submitted to Phys. Lett. B .

[245] OPAL Coll., G. Abbiendi et al., *Eur. Phys. J.* **C8** (1999) 559; **C36** (2004) 297.

[246] DELPHI Coll., P. Abreu et al., *Phys. Lett.* **B401** (1997) 181; DELPHI 2004-026-CONF-701 .

[247] U.A. Wiedemann, *Phys. Rev.* **C57** (1998) 3324; **C61** (2000) 029902.

[248] A. Białas and K. Zalewski, *Eur. Phys. J.* **C6** (1999) 349.

[249] H. Gyulassy and S.K. Kauffman, *Phys. Rev. Lett.* **40** (1978) 298; S.K. Kauffman and H. Gyulassy, *J. Phys.* **A11** (1978) 1715; S. Hegyi, *Phys. Lett.* **B309** (1993) 443; D. Hegyi, *Phys. Lett.* **B318** (1993) 642.

Index

J

K

L

N

O

P

Q

R

W

Y

Figure Credits

With permission from Elsevier:

Fig. 1.3: reprinted from Physics Letters B565, E.G.S. Luna, M.J. Menon, "Extreme bounds for the soft Pomeron intercept", pp. 123-130 (2003)

Fig. 1.5: reprinted from Physics Letters B215, P.V. Chliapnikov, A.K. Likhoded, V.A. Uvarov, "Inclusive cross-sections in the central region and the supercritical Pomeron", pp. 417-420 (1988)

Fig. 1.12: reprinted from Physics Letters B389, R.J.M. Covolan, J.Montanha, K. Goulianos, "A new determination of the soft Pomeron intercept", pp. 176-180 (1996)

Fig. 1.39: reprinted from Physics Letters B388, A. De Roeck and E.A. De Wolf, "An observation on F_2 at low x", pp. 843-874 (1996)

Fig. 1.41: reprinted from Physics Letters B597, S. Sapeta, "Diffractive production at high energies in the Miettinen-Pumplin model", pp. 352-355 (2004)

Figs. 2.5 and 2.6: reprinted from Physics Letters B37, J.V. Beaupré et al., "Limiting properties in inclusive π^+p reactions at ....", pp. 432-434 (1971)

Fig. 2.18: reprinted from Physics Letters B489, M. Acciarri et al., QCD studies in e^+e^- annihilation from 30 GeV to 189 GeV", pp. 65-80 (2000)

Fig. 2.19: reprinted from Physics Letters B536, M. Acciarri et al., "Determination of α_s from hadronic event shapes in e^+e^- annihilation at $192 \leq \sqrt{s} \leq 208$ GeV", pp. 217-228 (2002)

Fig. 2.21: reprinted from Physics Letters B456, P. Abreu et al., "Energy dependence of event shapes and of α_s at LEP2", pp. 322-340 (1999)

Fig. 4.7: reprinted from Physics Letters B407, P. Abreu et al., "Observation of charge ordering in particle production in hadronic Z^0 decay", pp. 174-184 (1997)

Fig. 4.10: reprinted from Physics Letters B365, S. Lupia and W. Ochs, "Testing the running of the QCD coupling in particle energy spectra", pp. 339-346 (1996)

Figs. 4.18 and 4.34: reprinted from Physics Letters B462, P.V. Chliapnikov, "Hyperfine splitting in light-flavour hadron production at LEP", pp. 341-353 (1999)

Fig. 4.25: reprinted from Physics Letters B590, A. Białas, M. Jezabeck, "Bremsstrahlung from colour charges as a source of soft particle production in hadronic collisions", pp. 233-238 (2004)

Fig. 4.26c: reprinted from Physics Letters B530, M.C. Abreu et al., "Scaling of charged particle multiplicity in Pb-Pb collisions at SPS energies", pp. 43-55 (2002)

Fig. 4.28a: reprinted from Physics Letters B209, V.V. Aivazyan et al. (NA22), "Multiplicity dependence of the average transverse momentum in π^+p, K^+p and pp collisions at 250 GeV/c", pp. 103-106 (1988)

Figs. 4.28c and 8.26: reprinted from Physics Letters B320, R. Akers et al. (OPAL), "Multiplicity and transverse momentum correlations in multihadron final states in e^+e^- interactions at $\sqrt{s} = 91.2$ GeV", pp. 417-430 (1994)

Fig. 4.30: reprinted from Physics Letters B198, M. Adamus et al. (NA22), "A comparison of inclusive ρ^0, ρ^+ and ω production in K^+p interactions at 250 GeV/c", pp. 292-296 (1987)

Fig. 4.35c: reprinted from Physics Letters B595, C.Adler et al. (STAR), "Kaon production and kaon to pion ratio in Au+Au collisions at $\sqrt{s_{\mathrm{NN}}} = 130$ GeV", pp. 143-150 (2004)

Figs. 4.47 and 6.14: reprinted from Physics Letters B197, I.V. Ajinenko et al. (NA22), "Indication of an onset of hard-like effects in K^+p, π^+p and pp collisions at 250 GeV/c", pp. 457-462 (1987)

Fig. 5.7: reprinted from Physics Letters B115, L. Gatignon et al., "Quark distributions in the kaon: a study of hadron structure by low-p_{T} reactions", pp. 329-332 (1982)

Fig. 5.13a: reprinted from Physics Letters B82, S. Erhan et al., "Λ^0 polarization in proton-proton interactions at $\sqrt{s} = 53$ and 62 GeV", pp. 301-304 (1979)

Fig. 5.14: reprinted from Physics Letters B185, A.M. Smith et al. (R608), "Λ^0 polarization in proton-proton interactions at $\sqrt{s} = 31$ and 62 GeV", pp. 209-212 (1987)

Fig. 8.2a: reprinted from Physics Letters B444, M. Acciarri et al. (L3), "QCD results from studies of hadronic events produced in e^+e^- annihilation at $\sqrt{s} = 183$ GeV", pp. 569-582 (1998)

Fig. 8.4: reprinted from Physics Letters B550, G. Abbiendi et al. (OPAL), "Charged particle multiplicities in heavy and light quark initiated events above the Z^0 peak", pp. 33-46 (2002)

Fig. 8.7a: reprinted from Physics Letters B138, G.J. Alner et al. (UA5), "Scaling violation favouring high multiplicity events at 540 GeV cms energy", pp. 304-310 (1984)

Fig. 8.8b: reprinted from Physics Letters B435, T. Alexopoulos et al., "The role of double parton collisions in soft hadron interactions", pp. 453-457 (1998)

Figs. 8.11a-d: reprinted from Physics Letters B467, S. Hegyi, "Log-scaling: a new empirical regularity at very high energies", pp. 126-131 (1999)

Fig. 8.21b: reprinted from Physics Letters B206, V. Šimák, M. Šumbera, I. Zborovský, "Entropy in multiparticle production and ultimate multiplicity scaling", pp. 159-162 (1988)

Fig. 8.22a: reprinted from Physics Letters B274, S. Hegyi, "Analysis of the rapidity gap probability at CERN collider energies", pp. 214-220 (1992)

Fig. 9.1b: reprinted from Physics Letters B48, S.R. Amendolia et al., "Measurement of inclusive two-particle rapidity correlations at the ISR", pp. 359-366 (1974)

Fig. 9.10: reprinted from Physics Letters B287, P.D. Acton et al. (OPAL), "A study of two-particle momentum correlations in hadronic Z^0 decays", pp. 401-412 (1992)

Figs. 9.14 and 9.23: reprinted from Physics Letters B533, J. Abdallah et al. (DELPHI), "Rapidity-alignment and p_T compensation of particle pairs in hadronic Z^0 decays", pp. 243-252 (2002)

Fig. 9.22: reprinted from Physics Letters B348, M. Adamovich et al. (BEATRICE), "Study of charm correlations in $\pi^- - N$ interactions at $\sqrt{s} \approx 26$ GeV", pp. 256-262 (1995)

Fig. 9.25: reprinted from Physics Letters B353, M. Acciarri (L3), "Evidence for gluon interference in hadronic Z decays", pp. 145-154 (1995)

Fig. 10.1b: reprinted from Physics Letters B185, M. Adamus et al. (NA22), "Maximum particle densities in rapidity space of π^+p, K^+p and pp collisions at 250 GeV/c", pp. 200-204 (1987)

Fig. 10.4a: reprinted from Physics Letters B222, I.V. Ajinenko et al. (NA22), "Intermittency patterns in π^+p and K^+p collisions at 250 GeV/c", pp. 306-310 (1989)

Fig. 10.5c: reprinted from Physics Letters B260, L. Verluyten et al. (WA59 and E180), "Study of factorial moments in neutrino neon and deuteron charged current interactions", pp. 456-461 (1991)

Fig. 10.7a: reprinted from Physics Letters B215, B. Buschbeck, P. Lipa, R. Peschanski, "Signal for intermittency in e^+e^- reactions obtained from negative binomial fits", pp. 788-791 (1988)

Figs. 10.10b,c: reprinted from Physics Letters B235, I.V. Ajinenko et al. (NA22), "Intermittency effects in π^+p and K^+p collisions at 250 GeV/c", pp. 373-378 (1990)

Figs. 10.11a,b: reprinted from Physics Letters B247, W. Ochs, "The importance of phase space dimension in the intermittency analysis of multihadron production", pp. 101-106 (1990)

Fig. 10.14: reprinted from Physics Letters B272, K. Fiałkowski, "Universal intermittency in three dimensions and collective correlations in nuclear collisions", pp. 139-142 (1991)

Fig. 10.17b: reprinted from Physics Letters B347, E.K. Sarkisyan et al., "On dynamics of pseudorapidity fluctuations in central C-Cu collisions at $4.5A$ GeV/c", pp. 439-446 (1995)

Fig. 10.21: reprinted from Physics Letters B275, N. Neumeister et al. (UA1), "Higher order Bose-Einstein correlations in $\mathrm{p\bar{p}}$ collisions at $\sqrt{s} = 630$ and 900 GeV", pp. 186 (1992)

Figs. 10.23a,b: reprinted from Physics Letters B261, N.M. Agababyan et al. (NA22), "Low p_T intermittency in π^+p and $\mathrm{K^+}$p collisions at 250 GeV/c", pp. 165-168 (1991)

Fig. 10.28a: reprinted from Physics Letters B254, P. Carruthers, H.C. Eggers, I. Sarcevic, "Analysis of multiplicity moments for hadronic multiparticle data", pp. 258-266 (1991)

Fig. 10.29a: reprinted from Physics Letters B496, M.A. Lisa et al. (E895), "Azimuthal dependence of pion interferometry at the AGS", pp. 1-8 (2000)

Fig. 10.30a: reprinted from Physics Letters B477, E.K.G. Sarkisyan et al., "Description of local multiplicity fluctuations and genuine multiparticle correlations", pp. 1-12 (2000)

Figs. 10.30b and 10.31: reprinted from Physics Letters B523, G. Abbiendi et al. (OPAL), "Genuine correlations of like-sign particles in hadronic $\mathrm{Z^0}$ decays", pp. 35-52 (2001)

Figs. 10.32, 10.33 and 10.34: reprinted from Physics Letters B258, V.V. Aivazyan et al. (NA22), "Factorial correlators from π^+p and $\mathrm{K^+}$p collisions at 250 GeV/c", pp. 487-492 (1991)

Fig. 10.37: reprinted from Physics Letters B429, M. Acciarri (L3), "Local multiplicity fluctuations in hadronic Z decay", pp. 375-386 (1998)

Fig. 10.39: reprinted from Physics Letters B285, P. Lipa et al., "The correlation integral as a probe of multiparticle correlations", pp. 300-308 (1992)

Figs. 10.46a,b and 10.47a,b: reprinted from Physics Letters B440, P. Abreu et al. (DELPHI), "Two-particle angular correlations in $\mathrm{e^+e^-}$ interactions compared with QCD predictions", pp. 203-216 (1998)

Fig. 10.50: reprinted from Physics Letters B457, P. Abreu et al. (DELPHI), "Multiplicity fluctuations in one- and two-dimensional angular intervals compared with analytic QCD calculations", pp. 368-382 (1999)

Fig. 10.51: reprinted from Physics Letters B510, S. Chekanov et al. (ZEUS), "Multiplicity moments in deep inelastic scattering at HERA", pp. 36-54 (2001)

Fig. 10.52: reprinted from Physics Letters B458, Shaoshun Wang et al., "Erraticity analysis of the NA27 data", pp. 505-510 (1999)

Fig. 10.53a,b: reprinted from Physics Letters B540, Dipak Ghosh et al., "Analysis of fluctuations in ^{32}S-AgBr interactions at 200 A GeV", pp. 52-61 (2002)

Fig. 10.54: reprinted from Physics Letters B505, Wang Shaoshun, Wu Chong, "Erraticity analysis of pseudorapidity gaps of the NA27 data", pp. 43-46 (2001)

Fig. 11.2: reprinted from Physics Letters B169, B. Andersson and W. Hofmann, "Bose-Einstein correlations and color strings", pp. 364-368 (1986)

Fig. 11.6: reprinted from Physics Letters B583, S. Chekanov et al. (ZEUS), "Bose-Einstein correlations in one and two dimensions in deep inelastic scattering", pp. 231-246 (2004)

Fig. 11.7c: reprinted from Physics Letters B226, Albajar et al. (UA1), "Bose-Einstein correlations in p$\bar{\text{p}}$ interactions at $\sqrt{s} = 0.2$ to 0.9 TeV", pp. 410-416 (1989)

Fig. 11.9: reprinted from Physics Letters B481, B. Buschbeck, H.C. Eggers, P. Lipa, "Multiplicity dependece of correlation functions in $\bar{\text{p}}$p reactions at $\sqrt{s} = 630$ GeV", pp. 187-193 (2000)

Figs. 11.10c,d: reprinted from Physics Letters B506, G. Alexander, "Mass and transverse mass effects on the hadron emitter size", pp. 45-51 (2001)

Fig. 11.13c: reprinted from Physics Letters B355, P. Abreu et al. (DELPHI), "Observation of short range three-particle correlations in e^+e^- annihilations at LEP energies", pp. 415-424 (1995)

Fig. 11.14: reprinted from Physics Letters B540, P. Achard et al. (L3), "Measurement of genuine three-particle Bose-Einstein correlations in hadronic Z decay", pp. 185-198 (2002)

Fig. 11.18: reprinted from Physics Letters B489, T. Csörgő and S. Hegyi, "Model independent shape analysis of correlations in 1, 2 or 3 dimensions", pp. 15-23 (2000)

Fig. 11.23a: reprinted from Physics Letters B557, S.V. Afanasiev et al. (NA49), "Bose-Einstein Correlations of Charged Kaons in Central PbPb Collisions at $E_{beam} =$ 158 GeV", pp. 157-166 (2003)

Figs. 11.26 and 11.27a: reprinted from Physics Letters B422, N.M. Agababyan et al. (NA22), "Estimation of hydrodynamical model parameters from the invariant and the Bose-Einstein correlations of π^- mesons produced in (π^+/K^+)p interactions at 250 GeV/c", pp. 359-368 (1998)

Fig. 11.29: reprinted from Physics Letters B547, P. Achard et al. (L3), "Measurement of Bose-Einstein correlations in $e^+e^- \to W^+W^-$ events at LEP", pp. 139-150 (2002)

Fig. 11.31: reprinted from Physics Letters B436, A. Białas and K. Załewski, "Bose-Einstein condensation and independent production of pions", pp. 153-157 (1998)

Fig. 1.20: reprinted from Nuclear Physics B695, S. Chekanov et al. (ZEUS), "Exclusive electroproduction of J/ψ mesons at HERA" pp. 3-37 (2004)

Figs. 2.1 and 2.2: reprinted from Nuclear Physics B13, R. Honecker et al., "General characteristics of particle production in 16 GeV/c π^+p interactions", pp. 571-586 (1969)

Figs. 2.11, 2.14 and 2.15: reprinted from Nuclear Physics B9, L. Van Hove, "Longitudinal phase-space plots of multiparticle hadron collisions at high energy", pp. 331-348 (1969)

Figs. 2.13, 3.1 and 3.4: reprinted from Nuclear Physics B19, J. Bartsch et al., "Comparison of $\pi^\pm$p and K^-p reactions with a multiperipheral model using the Van Hove Hexagon and cuboctahedron plots", pp. 381-398 (1970)

Figs. 2.24, 2.25a,b and 4.45a,b: reprinted from Nuclear Physics B192, M. Barth et al., "Jet-like properties of multiparticle systems produced in K^+p interactions at 70 GeV/c", pp. 289-314 (1981)

Fig. 3.2: reprinted from Nuclear Physics B25, G. Rinaudo et al., "Longitudinal phase-space analysis of 5 GeV/c π^+p Reactions", pp. 351-373 (1971)

Figs. 3.9 and 3.10: reprinted from Nuclear Physics B40, K. Boesebeck et al., "Dominance of isospin zero exchange in $\pi^\pm p \to \pi(N\pi)$ at 16 GeV/c and production by it of a low mass $(N\pi)$ enhancement", pp. 39-44 (1972)

Figs. 3.11, 3.12 and 3.13: reprinted from Nuclear Physics B134, J.J. Engelen et al., "A study of non-charge-exchange $\bar{K}^0\pi^-$ production in the reactions $K^-p \to \bar{K}^0\pi^-p$ at 4.2 GeV/c", pp. 14-30 (1978)

Figs. 3.14-3.19: reprinted from Nuclear Physics B167, J.J. Engelen et al., "Multichannel analysis of the reaction $K^-p \to \bar{K}^0\pi^-p$ at 4.2 GeV/c", pp. 61-97 (1980)

Fig. 4.11: reprinted from Nuclear Physics B444, P. Abreu et al. (DELPHI), "Inclusive measurements of the $K^\pm$ and $p/\bar{p}$ production in hadronic Z^0 decays", pp. 3-26 (1995)

Figs. 4.21 and 8.3d: reprinted from Nuclear Physics B504, C. Adloff et al. (H1), "Evolution of ep fragmentation and multiplicity distributions in the Breit frame", pp. 3-23 (1997)

Fig. 4.29: reprinted from Nuclear Physics B504, G. Arnison et al. (UA1), "Analysis of the fragmentation properties of quark and gluon jets at the CERN SPS $p\bar{p}$ collider", pp. 253-271 (1986)

Fig. 4.32: reprinted from Nuclear Physics B137, P. Schmitz et al., "Production of ρ^0 and f in K$^-$p interactions at 10 and 16 GeV/c ....", pp. 13-28 (1978)

Fig. 4.33: reprinted from Nuclear Physics B124, G. Jancso et al., "Evidence for Dominant vector-meson production in inelastic proton-proton collisions at 53 GeV center-of-mass energy", pp. 1-11 (1977)

Fig. 4.38: reprinted from Nuclear Physics B118, H. Grässler et al., "Inclusive production of $\Sigma^{\pm}(1385)$ in K$^-$p interactions at 10 and 16 GeV/c", pp. 189-198 (1977)

Fig. 4.39: reprinted from Nuclear Physics B140, K. Böckmann et al., "Investigation of ρ^+, ρ^-, ρ^0 production of π^+ interactions at 16 GeV/c ....", pp. 235-248 (1978)

Fig. 4.40: reprinted from Nuclear Physics B140, J. Singh et al., "Production of high-momentum mesons at small angles at a cm energy of 45 GeV at the CERN ISR", pp. 189-219 (1978)

Figs. 4.45c,d: reprinted from Nuclear Physics B206, R. Göttgens et al., "Comparison of planar features of K$^-$p interactions with e$^+$e$^-$ annihilations and deep inelastic μp scattering", pp. 349-358 (1982)

Fig. 4.46: reprinted from Nuclear Physics B192, J.M. Lafaille et al., "Energy dependence of transverse momentum invariant distribution of pions and neutral kaons ....", pp. 18-32 (1981)

Fig. 5.8: reprinted from Nuclear Physics B204, G. Bobbink et al., "Correlations between high-momentum secondaries in pp collisions at $\sqrt{s} = 44.7$ and 62.3 GeV", pp. 173-188 (1982)

Fig. 5.17b: reprinted from Nuclear Physics B262, T.A. Armstrong et al., "Lambda polarization in the K$^-$ fragmentation region", pp. 356-364 (1985)

Figs. 5.18a,b: reprinted from Nuclear Physics B585, P. Astier et al. (NOMAD), "Measurement of the Λ polarization in ν_μ charged current interactions in the NOMAD experiment", pp. 3-36 (2000)

Figs. 9.16b and 9.18: reprinted from Nuclear Physics B86, K. Eggert et al., "Angular correlations between the charged particles produced in pp collisions at ISR energies", pp. 201-215 (1975)

Figs. 10.2c,d, 10.4b and 10.10a: reprinted from Nuclear Physics B345, C. Albajar et al. (UA1), "Intermittency studies in p$\bar{\text{p}}$ collisions in p$\bar{\text{p}}$ collisions at $\sqrt{s} = 630$ GeV", pp. 1-21 (1990)

Figs. 10.12a, 10.24 and 10.25: reprinted from Nuclear Physics B386, P. Abreu et al., "Multiplicity fluctuations in hadronic final states from the decay of the Z^0", pp. 471-492 (1992)

Fig. 10.29b: reprinted from Nuclear Physics A596, P.L. Jain, G. Singh, "Factorial, multifractal moments and short-range correlation of shower particles at relativistic energies", pp. 700-712 (1996)

Figs. 1.23, 1.24, 1.26, 1.27, 1.36 and 1.37: reprinted from Nuclear Physics (Proceedings Supplement) A99, K. Goulianos, "Diffraction in hadron-hadron interactions" pp. 9-20 (2001)

Fig. 8.11e: reprinted from Nuclear Physics (Proceedings Supplement) B92, S. Hegyi, "KNO scaling 30 years later", pp. 122-129 (2001)

Fig. 1.27: reprinted from Physics Reports 101, K. Goulianos "Diffractive interactions of hadrons at high energies", pp. 169-219 (1983)

Fig. 2.20: reprinted from Physics Reports 399, P. Achard et al. (L3), "Studies of Hadronic Event Structure in e^+e^- Annihilation from 30 to 209 GeV with the L3 Detector", pp. 71-174 (2004)

Figs. 4.35: reprinted from Physics Reports 154, G.J. Alner et al. (UA5), "UA5: A general study of proton-antiproton physics at $\sqrt{s} = 546$ GeV", pp. 247-383 (1987)

Figs. 6.6: reprinted from Physics Reports 97, B. Andersson et al., "Parton fragmentation and string dynamics", pp. 31-145 (1983)

Figs. 9.1a: reprinted from Physics Reports 27C, J. Whitmore, "Multiparticle Production in the Fermilab bubble chambers", pp. 187-273 (1976)

With permission from Springer-Verlag GmbH & Co.:

Figs. 1.6 and 1.19: reprinted from The European Physical Journal C19, J. Bartels, H. Kowalski, "Diffraction at HERA and the confinement problem", pp. 693-708 (2001)

Fig. 1.11: reprinted from The European Physical Journal C24, T. Lastovicka, "Self-similar properties of the proton structure at low x ", pp. 529-533 (2002)

Fig. 1.15: reprinted from The European Physical Journal C6, J. Breitweg, S. Chekanov, M. Derrick et al. (ZEUS), "Exclusive electroproduction of ρ^0 and J/ψ mesons at HERA", pp. 603-627 (1999)

Fig. 2.21: reprinted from The European Physical Journal C37, J. Abdallah et al. (DELPHI), "The measurement of α_s from event shapes with the DELPHI detector at the highest LEP energies", pp. 1-23 (2004)

Fig. 2.22: reprinted from The European Physical Journal C27, S. Chekanov et al. (ZEUS), "Measurement of event shapes in deep inelastic scattering at HERA", pp. 531-545 (2003)

Figs. 4.1 and 4.8: reprinted from The European Physical Journal C16, G. Abbiendi et al. (OPAL), "QCD studies with e^+e^- annihilation data at 172-189 GeV", pp. 185-210 (2000)

Fig. 4.2: reprinted from Zeitschrift für Physik C68, R. Akers et al. (OPAL), "Measurement of the longitudinal transverse and asymmetry fragmentation functions at LEP", pp. 203-213 (1995)

Fig. 4.5: reprinted from The European Physical Journal C16, G. Abbiendi et al. (OPAL), "Leading particle production in light flavour jets", pp. 407-421 (2000)

Fig. 4.9: reprinted from The European Physical Journal C27, G. Abbiendi et al. (OPAL), "Charged particle momentum spectra in e^+e^- annihilation at $\sqrt{s} = 192 - 209$ GeV", pp. 467-481 (2003)

Fig. 4.12: reprinted from The European Physical Journal C16, R. Barate et al. (ALEPH), "Inclusive production of $\pi^0, \eta, \eta'(958)$, K^0_s and Λ in two- and three-jet events from hadronic Z decays", pp. 613-634 (2000)

Figs. 4.13, 4.14 and 4.15: reprinted from The European Physical Journal C13, P. Abreu et al. (DELPHI), "Measurement of the gluon fragmentation function and a comparison of the scaling violation in gluon and quark jets", pp. 573-589 (2000)

Fig. 4.16: reprinted from The European Physical Journal C17, P. Abreu et al. (DELPHI), "Identified charged particles in quark and gluon jets", pp. 207-222 (2000)

Fig. 4.19: reprinted from Zeitschrift für Physik C76, M.R. Adams et al. (E665), "Inclusive single-particle distributions and transverse momenta of forward produced charged hadrons in μp scattering at 470 GeV", pp. 441-463 (1997)

Figs. 4.20a and 4.22: reprinted from The European Physical Journal C11, J. Breitweg et al. (ZEUS), "Measurement of multiplicity and momentum spectra in the current and target regions of the Breit frame in deep inelastic scattering at HERA", pp. 251-271 (1999)

Figs. 4.20b,c: reprinted from The European Physical Journal C2, J. Breitweg et al. (ZEUS), "Charged particles and neutral kaons in photo-produced jets at HERA", pp. 77-93 (1998)

Figs. 4.23: reprinted from The European Physical Journal C21, A. Airapetian et al. (HERMES), "Multiplicity of charged and neutral pions in deep-inelastic scattering of 27.5 GeV positrons on hydrogen", pp. 599-606 (2001)

Fig. 4.24, 4.26a,b, 4.27: reprinted from Zeitschrift für Physik C39, M.R. Adamus et al. (NA22), "Charged particle production in K^+p, π^+p and pp interactions at 250 GeV/c", pp. 311-329 (1988)

Figs. 4.31 and 5.11a: reprinted from Zeitschrift für Physik C41, N.M. Agababyan et al. (NA22), "Inclusive meson resonance production in K^+p interactions at 250 GeV/c", pp. 539-555 (1989)

Fig. 4.36: reprinted from Zeitschrift für Physik C9, D. Drijard et al., "Production of vector and tensor mesons in proton-proton collisions at $\sqrt{s} = 52.5$ GeV", pp. 293-303

(1981)

Fig. 4.37: reprinted from Zeitschrift für Physik C17, P. Malhotra and R. Orava, "Determination of strange quark suppression in hadronic vacuum", pp. 85-93 (1983)

Fig. 4.41: reprinted from Zeitschrift für Physik C3, B. Buschbeck et al., "Inclusive π^+/π^- ratio in meson-proton reactions and the Quark Recombination Model", pp. 97-100 (1979)

Fig. 4.42: reprinted from Zeitschrift für Physik C13, D. Brick et al., "Approach to scaling in inclusive π^+/π^- ratios at 147 GeV", pp. 11-17 (1982)

Fig. 4.44: reprinted from Zeitschrift für Physik C13, A. Clegg and A. Donnachie, "A description of jet structure by p_T-limited phase space", pp. 71-76 (1982)

Figs. 4.48 and 4.49a,b: reprinted from Zeitschrift für Physik C43, I.V. Ajinenko et al. (NA22), "Charge and energy flow in π^+p, K^+p and pp interactions at 250 GeV/c", pp. 37-44 (1989)

Figs. 4.49c,d: reprinted from The European Physical Journal C12, C. Adloff et al. (H1), "Measurement of transverse energy flow in deep-inelastic scattering at HERA", pp. 595-607 (2000)

Fig. 5.9a: reprinted from Zeitschrift für Physik C8, E.A. De Wolf et al., "Two-particle production in K^+p interactions at 32 GeV/c", pp. 189-197 (1981)

Fig. 5.9b: reprinted from Zeitschrift für Physik C7, M. Barth et al., "Charged pion production in 70 GeV/c K^+p interactions", pp. 187-198 (1981)

Fig. 5.10: reprinted from Zeitschrift für Physik C12, E.A. De Wolf et al., "A study of multiparticle fragmentation in K^+p interactions at 32 GeV/c", pp. 105-112 (1982)

Figs. 5.11b,c,d and 5.12: reprinted from Zeitschrift für Physik C46, N.M. Agababyan et al. (NA22), "Inclusice production of vector mesons in π^+p interactions at 250 GeV/c", pp. 387-395 (1990)

Fig. 5.17a: reprinted from Zeitschrift für Physik C1, M.L. Faccini-Turluer et al., "Λ and $\bar{\Lambda}$ polarization in $K^\pm$p interactions at 32 GeV/c", pp. 19-24 (1979)

Figs. 8.2b: reprinted from The European Physical Journal C35, A. Heister et al. (ALEPH), Studies of QCD at e^+e^- centre-of-mass energies between 91 and 209 GeV, pp. 457-486 (2004)

Figs. 8.3a,b,c and 8.15b: reprinted from Zeitschrift für Physik C72, S. Aid et al. (H1), "Charged particle multiplicities in deep inelastic scattering at HERA", pp. 573-592 (1996)

Figs. 8.5a,b: reprinted from The European Physical Journal C23, G. Abbiendi et al. (OPAL), "Particle multiplicity of unbiased gluon jets from e^+e^- three-jet events", pp. 597-613 (2002)

Figs. 8.8a, 8.17 and 8.19: reprinted from Zeitschrift für Physik C37, M. Adamus et al. (NA22), "Phase space dependence of the multiplicity distribution in π^+p and pp collisions at 250 GeV/c", pp. 215-229 (1988)

Figs. 8.9a and 8.15a: reprinted from Zeitschrift für Physik C43, R.E. Ansorge et al. (UA5), "Charged particle multiplicity distributions at 200 and 900 GeV c.m. energy", pp. 357-374 (1989)

Figs. 8.18c,d and 8.20a: reprinted from Zeitschrift für Physik C52, P. Abreu et al. (DELPHI), "Charged particle multiplicity distributions in restricted rapidity intervals in Z^0 hadronic decays", pp. 271-281 (1991)

Fig. 8.20b: reprinted from Zeitschrift für Physik C69, D. Buskulic et al. (ALEPH), "Measurements of the charged particle multiplicity distribution in restricted rapidity intervals", pp. 15-25 (1995)

Fig. 8.24b: reprinted from Zeitschrift für Physik C42, V.V. Aivazyan et al. (NA22), "Forward-backward multiplicity correlations in π^+p, K^+p and pp collisions at 250 GeV/c", pp. 533-542 (1989)

Figs. 8.27 and 10.48: reprinted from The European Physical Journal C12, J. Breitweg et al. (ZEUS), "Angular and current-target correlations in deep inelastic scattering at HERA", pp. 53-68 (2000)

Fig. 9.2: reprinted from Zeitschrift für Physik C37, R.E. Ansorge et al. (UA5), "Charged particle correlations at c.m. energies of 200, 546 and 900 GeV", pp. 191-213 (1988)

Figs. 9.4 and 9.21: reprinted from Zeitschrift für Physik C40, J.L. Bailly et al. (NA23), "Two-particle correlations in 360 GeV/c pp interactions", pp. 13-24 (1988)

Figs. 9.5, 9.8, 9.11, 9.15, 9.16a and 9.29: reprinted from Zeitschrift für Physik C51, V.V. Aivazyan et al. (NA22), "Rapidity correlations in π^+p, K^+p and pp interactions at 250 GeV/c", pp. 167-178 (1991)

Fig. 9.6: reprinted from The European Physical Journal C18, H. Müller, "Soft hadron production in pp interactions up to ISR energies", pp. 563-576 (2001)

Fig. 9.7: reprinted from Zeitschrift für Physik C56, I. Derado, G. Jancso, N. Schmitz, "Two-particle correlations and the origin of intermittency in muon-nucleon interactions", pp. 553-556 (1992)

Fig. 9.12: reprinted from Zeitschrift für Physik C34, M. Asai et al. (NA23), "Neutral strange particle correlations in 360 GeV/c pp interactions", pp. 429-435 (1987)

Figs. 9.17, 9.19 and 9.20: reprinted from Zeitschrift für Physik C58, I.V. Ajinenko et al. (NA22), "Two-particle azimuthal and rapidity correlations in intervals of transverse momentum in π^+p interactions at 250 GeV/c", pp. 357-366 (1993)

Fig. 9.24: reprinted from Zeitschrift für Physik C58, P.D. Acton et al. (OPAL),

"QCD coherence studies using two-particle azimuthal correlations", pp. 207-217 (1993)

Fig. 9.28: reprinted from Zeitschrift für Physik C63, P. Abreu et al. (DELPHI), "Invariant mass dependence of particle correlations in hadronic final states from the decay of the Z^0", pp. 17-28 (1994)

Figs. 10.5a,b: reprinted from Zeitschrift für Physik C47, I. Derado et al., "Investigation of intermittency in muon proton scattering at 280 GeV", pp. 23-29 (1990)

Figs. 10.7c, 10.11c and 10.27: reprinted from The European Physical Journal C11, G. Abbiendi et al. (OPAL), "Intermittency and correlations in hadronic Z^0 decays", pp. 239-250 (1999)

Figs. 10.16a,b, 10.28b, 10.38ab and 10.45b: reprinted from Zeitschrift für Physik C59, N.M. Agababyan et al. (NA22), "Factorial moments, cumulants and correlation integrals in π^+p and K^+p interactions at 250 GeV/c", pp. 405-426 (1993)

Fig. 11.3: reprinted from Zeitschrift für Physik C59, N.M. Agababyan et al. (NA22), "Influence of multiplicity and kinematical cuts on Bose-Einstein correlations in π^+p interactions at 250 GeV/c", pp. 195-210 (1993)

Fig. 11.7a: reprinted from Zeitschrift für Physik C33, A. Breakstone et al., "Multiplicity dependence of the average transverse momentum", pp. 333-338 (1987)

Fig. 11.8: reprinted from Zeitschrift für Physik C72, G. Alexander et al. (OPAL), "Multiplicity dependence of Bose-Einstein correlations in hadronic Z^0 decays", pp. 389-398 (1996)

Figs. 11.11 and 11.12: reprinted from Zeitschrift für Physik C68, N.M. Agababyan et al. (NA22), "Higher order Bose-Einstein correlations in π^+p and K^+p collisions at 250 GeV/c", pp. 229-237 (1995)

Fig. 11.13b: reprinted from The European Physical Journal C5, K. Ackerstaff et al. (OPAL), "Bose-Einstein correlations of three charged pions in hadronic Z^0 decays", pp. 239-248 (1998)

Fig. 11.19: reprinted from Zeitschrift für Physik C66, N.M. Agababyan et al. (NA22), "Angular dependence of Bose-Einstein correlations in interactions of π^+ and K^+ mesons .....", pp. 409-415 (1995)

Fig. 11.25b: reprinted from The European Physical Journal C18, I.G. Bearden et al. (NA44), "Space-time evolution of the hadronic source in peripheral to central Pb+Pb collisions", pp. 317-325 (2000)

Fig. 11.32: reprinted from The European Physical Journal C6, A. Białas and K. Zalewski, "HBT correlations and charge ratios in multiple production of pions", pp. 349-354 (1999)

Fig. 11.7b: reprinted from Zeitschrift für Physik C75, C. Adloff et al. (H1), "Bose-

Einstein correlations in deep inelastic ep scattering at HERA", pp. 437-451 (1997)

With permission from American Physical Society:

Fig. 1.10: reprinted from Physical Review D67, S. Chekanov et al. (ZEUS), "ZEUS next-to-leading-order QCD analysis of data on deep inelastic scattering", pp. 012007 (2003)

Fig. 1.40: reprinted from Physical Review D18, Hannu I. Miettinen, Jon Pumplin, "Diffraction scattering and the parton structure of hadrons", pp. 1696-1708 (1978)

Fig. 2.10: reprinted from Physical Review D3, C.E. DeTar, "Momentum spectrum of hadronic secondaries in the multiperipheral model", pp. 128-144 (1971)

Fig. 4.28b: reprinted from Physical Review D48, T. Alexopoulos et al. (E735), "Mass-identified particle production in proton-antiproton collisions at $\sqrt{s}$ = 300, 540, 1000 and 1800 GeV", pp. 984-997 (1993)

Fig. 5.16: reprinted from Physical Review D24, T.A. De Grand, H.I. Miettinen, "Models for polarization asymmetry in inclusive hadron production", pp. 2419-2427 (1981)

Fig. 8.14d: reprinted from Physical Review D33, Cai Xu, Chao Wei-qin, Meng Ta-chung, Huang Chao-shang, "Statistical approach to nondiffractive hadron-hadron collisions: multiplicity distributions and correlations in different rapidity intervals", pp. 1287-1299 (1986)

Figs. 10.19 and 10.20: reprinted from Physical Review D59, Liu Feng, Liu Fuming, Liu Lianshou, "Qualitative difference between the dynamics of particle production in soft and hard processes of high energy collisions", pp. 114020-1 to 6 (1999)

Fig. 1.42: reprinted from Physical Review Letters 86, A.M. Staśto, K. Golec-Biernat and J. Kwieciński, "Geometric Scaling for the Total γ^*p Cross Section in the Low x Region", pp. 596-599 (2001)

Fig. 10.24: reprinted from Physical Review Letters 78, I.G. Bearden et al. (NA44), "Collective Expansion in High Energy Heavy Ion Collisions", pp. 2080-2083 (1997)

Fig. 11.14b: reprinted from Physical Review Letters 91, J. Adams et al. (STAR), "Three-pion Hambury Brown-Twiss Correlations in Relativistic Heavy-Ion Collisions from the STAR Experiment", pp. 262301 (2003)

Fig. 11.23b: reprinted from Physical Review Letters 88, K. Adcox et al. (PHENIX), "Transverse mass dependence of two-pion correlations in Au+Au collisions at $\sqrt{s_{\rm NN}}$ = 130 GeV", pp. 192302 (2002)

Fig. 11.29b,c: reprinted from Physical Review Letters 93, J. Adams et al. (STAR), "Azimuthally sensitive Hanbury Brown-Twiss Interferometry in Au+Au collisions at

$\sqrt{s}_{\rm NN} = 200$ GeV", pp. 012301 (2004)

With permission from WSPC:

Figs. 1.8 and 1.9: reprinted from International Journal of Modern Physics A19, P. Newman, "Deep inelastic lepton-nucleon scattering at HERA", pp. 1061-1072 (2004)

Fig. 5.9d: reprinted from Proc. 12th Int. Symp. on Multiparticle Dynamics, B. Buschbeck and H. Dibon, "Fragmentation of u-quark, d-quark and the (ud)-diquark system", ISBN: 9971-950-30-8, pp. 129-136 (1982)

Figs. 10.2ab: reprinted from Multiparticle Dynamics (Festschrift L. Van Hove), A. Białas, "Intermittency and multiparticle dynamics", ISBN: 981-02-0039-0, pp. 75-114 (1990)

Fig. 10.35: reprinted from Fluctuations and Fractal Structure (Proc. Ringberg Workshop on Multiparticle Production), E.A. De Wolf, "Intermittency, correlations and correlators", ISBN: 981-02-0810-3, pp. 222-237 (1992)

Fig. 10.43: reprinted from Soft physics and fluctuations (Proc. Cracow Workshop on Multiparticle Production), Y.F. Wu et al., "The dependence of the correlation strength on average transverse momentum per event and on multiplicity", ISBN: 981-02-1588-6, pp. 22-34 (1994)

Fig. 10.44: reprinted from Proc. XXII Int. Symp. on Multiparticle Dynamics, F. Mandl and B. Buschbeck, "Correlation integral studies in Delphi and in UA1", ISBN: 981-02-1239-9, pp. 561-567 (1993)

Fig. 11.4: reprinted from Correlations and Fluctuations (Proc. 7th Int. Workshop on Multiparticle Production), P. Seyboth, "Correlations in 158 GeV Pb+Pb collisions measured by NA49", ISBN: 981-02-2887-2, pp. 13-25 (1997)

Figs. 11.6 and 11.21: reprinted from International Journal of Modern Physics A4, B. Lörstad, "Boson interferometry .....", pp. 2861-2896 (1989)

Fig. 11.17: reprinted from Soft Physics and Fluctuations (Proc. Cracow Workshop on Multiparticle Production), T. Csörgő, "Shape analysis of Bose-Einstein correlation functions", ISBN: 981-02-1588-6, pp. 175-186 (1994)

Fig. 11.20a: reprinted from Correlations and Fluctuations 1998 (Proc. 8th Int. Workshop on Multiparticle Production), T. Csörgő, B. Lörstad, "Invariant Buda-Lund particle interferometry", ISBN: 981-02-3823-1, pp. 108-127 (1999)